The Molecular Theory of Adsorption in Porous Solids

The Molecular Theory of Adsorption in Porous Solids

Yu. K. Tovbin

CISP

CRC Press
Taylor & Francis Group
Boca Raton London New York

CRC Press is an imprint of the
Taylor & Francis Group, an **informa** business

CRC Press
Taylor & Francis Group
6000 Broken Sound Parkway NW, Suite 300
Boca Raton, FL 33487-2742

First issued in paperback 2020

© 2018 by CISP
CRC Press is an imprint of Taylor & Francis Group, an Informa business

No claim to original U.S. Government works

ISBN 13: 978-0-367-57282-2 (pbk)
ISBN 13: 978-1-138-04902-4 (hbk)

Visit the Taylor & Francis Web site at
http://www.taylorandfrancis.com

and the CRC Press Web site at
http://www.crcpress.com

Contents

List of basic notation

a_f^i – local Henry constant of molecule i for the site with the number f (or a_q^i of the site of type q); for a uniform surface a_i is Henry's constant of molecule i,

a – for one substance

a_f^{i0} – pre-exponential factor of the local Henry constant of molecule i for site f.

(A similar simplification of notation in the transition from the non-uniform system to a homogeneous one and from a mixture to a single substance is retained below for all characteristics)

C_i – concentration of molecules i in the number of molecules N_i per unit volume

$C_v(f)$ – the specific heat capacity per particle in the cell f

$E(f)$ – the internal energy of the molecules in the cell f

$E_g^{iV}(\rho)$ – activation energy of hopping of molecule i from site f over distance ρ to a free site g

$E_{fgh}^{iVj}(\omega_r)$ – the energy contribution of the molecule j in the site h, located at the distance of the r-th coordination sphere from the 'central' pair of sites fg with orientation ω_r, as a function of non-ideal hopping of the molecule i from site f to site g

d_e – the ratio of the contribution of triple interactions to the contribution of the pair interaction of nearest neighbors

$d_{qp}(r)$ – the conditional probability of finding a p-type site location at a distance of the r-th coordination sphere from the type q site

$d(q\{\lambda\}R)$ – the conditional probability of finding a cluster consisting of R coordination spheres of type $q\{\lambda\}R$ around the central site of type q

$D_{i(j)}$ – the diffusion coefficient of molecules i involved in transport after the last collision with molecule j

$D_{1,2}$ – coefficient of mutual diffusion in a binary mixture

D_{ij} – diffusion coefficient of component i in a multicomponent mixture under the influence of the component j gradient

D_{fg}^* – the self-diffusion coefficient of a pure substance

f – the number of the site of the non-uniform system

F – free energy

f_q – the probability of finding the site of type q in the non-uniform system

$f_{qp}(r)$ – the probability of finding a pair of sites of the type q and p at distance r in the non-uniform system

$f(q\{\lambda\}R)$ – the probability of finding a cluster consisting of R coordination spheres of the type $q\{\lambda\}R$, around the central site of type q

F^0_i – partition function of molecule i in the gas

F^i_q – partition function of the particle i, located in a site of type q

F^{i*}_q – partition function of particle i at a site of type q, located in the transition state

$F_i(f)$ – component i of the external force created by the potential of the walls in cell f

$g(f)$ or g_e – calibration function

h – Planck constant

h_f – cluster Hamiltonian

H – width of the slit-like pore, the number of monolayers

i – particle sort, located in the cell, including the vacancy

J_i – diffusive flux of molecules of soft i

k – Boltzmann's constant

$K^{iV}_{fg}(\chi)$ – constant of the hopping velocity of molecule i from the cell f to the free cell g over distance χ

K_{ij} – equilibrium coefficient of adsorption displacement of component i by component j

l_i – the mean free path length of component i

m_i – mass of molecule i

$m(\omega_r)$ – a plurality of adjacent sites with fixed values of r and ω_r

$M(fg)$ – elasticity coefficient of bonding fg

n_i – the number density of component i per unit volume

N – number of system cells

N_{den} – number of densely packed particles per unit volume

N^i_q – the number of molecules of sort i on the sites of type q

N^{ij}_{qp} – number of pairs of molecules of sort ij on sites of type qp

P – pressure

P_i – the partial pressure of component i

$P(f)$ – the pressure in the cell f

q – the type of site (cell)

Q_N – the partition function for system states

Q^i_f – the binding energy of the molecule of sort i, located at the site f, with the adsorbent

Q^i_q – the binding energy of molecule of sort i, located in a site of type q, with the walls of the pores

$Q_{q,f}$ – the binding energy of the molecule in the site with number f of section q in a complex pore system

r – the distance between the particles (in units of λ)

R_p – radius of the pore channel, for the slit $R = H/2$

R – radius of the intermolecular interaction potential (in units of λ)

s – number of mixture components

S_m – the surface of the cell surface m, $1 < m \le z$

$S^i_{fg}(r)$ – the factor of the imperfection function $\Lambda^i_{fg}(r)$

$S(i \to k)$ – vacancy region of rotation of the molecules i from the orientation

i to orientation k

t – the number of different types of system sites

T – temperature

T_i^c – the critical temperature of component i

T_f – local temperature

$t_{fg}^{ij}(r)$ – the conditional probability of finding the molecule j on site g distance r from the molecule i on the site f

$\langle t_{f\xi}^{ij}\rangle$ – probability of finding the neighbouring particle j at site ξ on the same line with sites f and g (on the p-scale)

u – flow rate, m/s

$u(f)$ – the velocity vector in the cell f with components $u_i(f)$ in the direction $i = x, y, z$

$U(q)$ – the potential of interaction of the molecule of the layer q with the pore well

$U_{fg}^i(\chi)$ – rate of hopping of molecule i from cell f to free cell g over the distance χ

$U_{\xi fg}^{(j)iV}(\chi)$ – probability of hopping of molecule i after collision with molecule j over distance χ

$U(z)$ – the potential of interaction of the molecule with the coordinate z with the wall of the pores

V_{fg}^{iV} – the concentration component of the hopping velocity of molecule i from cell f to the free cell g

V_p – volume of the pore

v_0 – cell volume

$V_{\xi fg}^{(j)iV}(\chi)$ – the concentration component of the hopping velocity of molecule i from the cell f to the free cell g after a collision with molecule j on site ξ

W_α – the probability of occurrence of the elementary stage α

$W_i(\rho)$ – probability of hopping of molecule of sort i over distance p

w_i – average thermal velocity of the molecule i in the gas phase

$w_i(j)$ – the average relative velocity of the molecule i after collision with molecule j

w_{fg}^i – the average thermal velocity of the molecules i between the sites f and g

x_f^i – the mole fraction of component i in the site f

z – the number of nearest neighbours of the lattice structure

z_q – the number of nearest neighbours in the layer q

$z_{fg}(r)$ – the number of neighboring sites in the layer g at distance r from the considered site in the layer f

$z_{fg}^*(\chi)$ – the number of possible hops between sites f and g over the distance x

Greek symbols

α – dimensionless parameter $\varepsilon^*/\varepsilon$

α_{11} – dimensionless parameter $E_{11}^{iV}(1)/Q_1^i$, describing the height of the activation barrier pf surface hops

β – the inverse of the thermal energy, $(kT)^{-1}$

β_1 – the coefficient of sliding friction

γ_f^i – variable, describing the state of occupation of the cell (site) with number f

δ – the width of the monolayer

$\delta\varepsilon_{fh}^{ij}(r)$ – the difference of the values $(\varepsilon_{fh}^{*ij}(r) - \varepsilon_{fh}^{ij}(r))$

ε_i – the parameter of the interaction potential of the adsorbate of sort i – adsorbent

$\varepsilon_{fg}^{ij}(r)$ – parameter of interaction of molecule i in the site f with the neighbouring molecule j at the site g at the distance r

$\varepsilon_{fh}^{*ij}(r)$ – the parameter of interaction of the activated migrating complex of molecule i from the site f to the free size g with neighbouring molecule j, located in the ground state at site r at distance h

η_{fg} – local coefficient of the shear viscosity of the flow of the mixture between sites f and g

η_0 – ideal viscosity of rarefied gas in the volume at the considered temperature

η_f^q – function of correspondence of the site f and its short q

θ – the full extent of the filling the volume of the system, a dimensionless quantity, $0 \le \theta \le 1$

θ_q^i – partial degree of filling of type q sites by molecules of sort i

$\theta_{fg}^{ij}(r)$ – probability of finding particles i at the sites f and j at the site g at the distance r

$\theta_{fg}^{iV}(\chi)$ – the probability of formation of free path from the cell f to cell g with length χ

$\theta(p_1, r_1, ..., p_N, r_N)$ – complete distribution function of N molecules; p_i – momentum and r_i – the coordinate of the centre of mass of the molecule i, $1 \le i \le N$

$\kappa(f)$ – the coefficient of thermal conductivity in the cell f

$\kappa_{qp}(\omega_r|\rho)$ – the number of sites with orientation $\omega_r(\rho)$, located in the r-th coordination sphere of the central pair of particles on sites of type p and q at distance ρ

λ – the linear size of the cell

$\lambda_{qp\xi}(\omega_r|\rho)$ – the number of sites of type ξ with orientation ω_r, located in the r-th coordination sphere around the central sites of sort qp at a distance ρ

Λ_f^i – function of nonideality of the system for molecules i on the site f

$\Lambda_{fg}^i(\chi)$ – function of nonideality of the system for hopping of the molecule i from site g to the free site g over the distance χ

μ_i – chemical potential of component i

μ_{ij} – the reduced mass of the colliding molecules i and j

ν_{fg} – frequency of harmonic vibrations of molecules for bonding fg

ν_f – Single-particle contribution to the total energy of of the system of the sort i, located on the site with number f

$\xi(f)$ – the coefficient of bulk viscosity in the cell f

π – expansion pressure in the equation of state for the lattice system

$\pi_r(\rho)$ – number of different orientations of positions of neighbouring sites in the r-th coordination sphere relative to the fixed position of the dimeric molecule

Π_i – the number of contacts on the surface of the large molecule of sort i

ρ – the average distance between the molecules (in units of λ)

σ_{ij} – the distance of closest approach of the two components i and j of a mixture

$\sigma_{ik}(f)$ – the components of the viscous stress tensor of cell f

φ_i – the orientation of the molecule i

$\varphi(r) = Ar^{-n} - Br^{-m}$ – Mie pair potential (n, m), where the constants A, B, n, m are positive

χ – molecule hopping length (in units of λ)

$\omega(f)$ – enthalpy (heat function) ofmass unit, related to a given cell f

$\omega_r(\rho)$ – orientation of site h, located in the r-th coordination sphere of the 'central' pair of sites fg at a distance p; $1 \leq \omega_r(\rho) \leq \pi_r(\rho)$

Brackets

Curly: $\{\eta_{f}^{q}\}$ – indicate a complete list of variables here, for example, for η_{f}^{q}, $1 \leq f \leq N$, $1 \leq q \leq t$.

Square: $[m_{qp}]$ – fix numbers m_{qp} of sites of sort p in the coordination sphere of the central site of sort q, $1 \leq p \leq t$.

Indices

For a homogeneous system subscripts i, j, k, l – types of particles, including vacancy V.

For nonuniform systems lower indices: number of sites, if the indices f, g, h, or types of the sites of the nonunifform lattice, or the indices q, p, ξ; upper indices: i, j, k – sort of molecular grade or vacancy (V).

Abbreviations

AC – activated complex

AD – adsorption centre

BBGKY – Bogolyubov-Born-Green-Kirkwod-Yvon

CM – cluster method
c.s. – coordination sphere
KMC – kinetic Monte Carlo
QCA – quasi-chemical approximation
MD – molecular dynamics
MM – matrix method
LGM – lattice gas model
MFA – mean field approximation
TARR – theory of absolute reaction rates
TDC – thermal desorption curve
FM – fragment method

Foreword

Processes in porous bodies form the basis of many traditional and modern technologies: catalytic, sorption, membrane, electrochemical, chromatography, purification and separation of liquid and gas mixtures, capillary condensation and desorption, polyphase filtering and 'spraying', wetting, impregnating and drying of a wide range of disperse systems, both synthetic and natural, including processes for the transfer of substances in different soils, etc.

Amongst the many processes in porous bodies plays the key role is played by adsorption processes which in turn are connected to the transport process of the adsorbed molecules (the adsorbate) in a porous structure, molecular redistribution processes between different portions of the free pore volume and phase separation processes (condensation) of the molecules in the case of prevalence of the cooperative behaviour of molecules and the instability of their uniform distribution.

From the standpoint of the spatial distribution of the adsorbate by the solid we distinguish usually adsorption and absorption. It is believed that the adsorption processes occur at interfaces, while the absorption processes takes place in the absorption of molecules by the solid and/or liquid phase. In this monograph, the term adsorption is used to refer to all the situations under consideration, as the process usually begins at the exposed surface of a solid or on the walls of the pores and extends as far as the formation of multilayer films into the central part of the pores. This term can also clearly describe the unchanged state of the porous body the structure of which does not change during the adsorption process (the molecules do not penetrate into the solid outside the existing porous structure).

These processes have been actively studied since the beginning of the last century. A breakthrough in the possibility of analyzing the behaviour of adsorbed molecules has been associated with the use of thermodynamic methods. In particular, this refers to the wording of the Polanyi potential adsorption theory and to application of the Kelvin equation to estimate the saturated vapour pressure in porous materials. The well-known relationship between the change in the metastable vapour pressure with respect to the vapour pressure in the bulk phase and the size of the metastable drop was reformulated with the replacement of the droplet size by the characteristic size of the pores. Later, these approaches were partially modified, but they have remained in the thermodynamic basis.

The parallel (started by Langmuir) and especially the later development of molecular approaches allowed sequentially to construct the theory of adsorption on open uniform surfaces and then on non-uniform surfaces and then extend it to the case of cooperative processes for virtually all compositions and structures of non-uniform solid surfaces. At present, this theory for the non-porous bodies is one of the most developed areas of statistical thermodynamics of condensed phases.

Little has changed in the region of the theory of adsorption processes in porous solids. The structure of dispersed materials is extremely diverse, and it significantly affects both the equilibrium distribution of the molecules, as well as the course of phase transformations and all the dynamic processes that take place inside the material (porous materials). Here, the term 'structure' refers to the location and relationship of the constituent elements of the system in space, i.e., the concept of 'structure' refers to a set of well-defined structural solid particles and elements of the free space of the pores. Structural particles themselves may have a distinct geometric shape, or be amorphous or glassy particles, etc.

This diversity creates many options for organization of the 'volumes' of the free space – the pores.

Up to now there are no strictly formulated theoretical methods of calculating the equilibrium and dynamic characteristics of processes in porous solids. This monograph is the first step in this direction and sets out a common point of view on the molecular level for the full range of the theory of adsorption.

The prolonged absence of the theory was the impetus for the development of numerical methods for modeling using the Monte Carlo and molecular dynamics methods. These methods are inherently molecular methods, and they have been actively developed in the last thirty years. Moreover, these methods have practically superseded other theoretical approaches by the intensive development of computational capabilities. They allow a detailed study of the distribution of molecules in the pores. But here they revealed a number of factors that require independent verification of the numerical results of the study. This is due to the intrinsic properties of the numerical methods, which most people do not even realize. The central problem of stochastic methods is the question of the accuracy of the results. In the numerical experiments, the researcher cannot assess how strictly are the calculations carried out. Primarily, there is a problem of evaluation: how well studied is the coordinate space of the system and what is the statistics for each area of the pores space, which for non-uniform systems differ greatly in their energy.

Numerous examples show that the accuracy of stochastic techniques differs slightly from the theoretical calculation accuracy. Therefore, the question of the existence of the molecular theory is central to the development of molecular modelling of any processes in porous solids. For successive molecular analysis it was necessary to solve the following range of tasks.

1. Develop a method for describing the equilibrium distribution of interacting molecules in non-uniform adsorption centres of the open surface area of solids.

2. Develop a method for calculating the distribution of molecules of vapour and/or liquid in strong fields of the surface potentials of the pore walls. The surface potential can create a sharp anisotropy of the density of the molecules perpendicular to the surface of the walls. In this respect, the non-uniformity of the distribution of the molecules perpendicular to the surface may be commensurate with the difference in the non-uniformity of the fillings of different types of sites on the surface.

3. Develop a method for calculating the intermolecular interactions forming the cooperative overall behaviour of molecules which in a non-uniform field of the surface potential will delaminate in different ways depending on the properties of the porous structure and the properties of the pore walls. In particular, these properties of the non-uniform potential of the walls and the characteristic size of the channel should influence the values of the critical parameters and the form of the stratification diagrams, and also the characteristics of vapour–liquid interfaces in capillary condensation.

4. Develop a method for calculating the dynamics of vapour and/or liquid in highly heterogeneous systems where conventional hydrodynamic approaches are not applicable, in principle, due to the large velocity gradients of molecules perpendicular to the surface of

the wall. Today the description of the flow of molecules in porous bodies involves many molecular transport mechanisms, formulated for macroscopic channels, and their ranges of applicability do not overlap.

5. Ensure the self-consistency of methods for calculating the dynamics of transport processes with the methods of calculation of the equilibrium distribution of molecules in complex non-uniform external fields of the pore walls.

All of these problems have been solved in the framework of a unified approach based on the use of many discrete distribution functions. This approach is well known as the lattice-gas model (LGM) or quasi-Ising model for molecular systems, which are widely used in the problems of phase transformations. This technique is traditionally used for the bulk phase or non-uniform systems such as phase boundaries and has been generalized to arbitrary non-uniform systems. Such a generalization makes it possible to obtain a general approach to solving problems for the molecular distribution, which are implemented in porous bodies.

The method used and its application to the calculation of the equilibrium and dynamic properties of molecules in porous solids are presented in this monograph. Two new trends in the theory of statistical thermodynamics are presented for the first time: the theory of stratification in the complex micro-heterogeneous systems and microscopic hydrodynamics. The first direction generally refers to the equilibrium theory of multiphase systems, including the process of stratification on exposed surfaces. The second area relates to the theory of molecular transport in porous solids. Introduced in 1998, the term 'microscopic hydrodynamics' reflects the conventional concept – in most real-world situations the ordinary hydrodynamics does not apply, so we need to use the kinetic theory of condensed phases. This name reflects the fact that, on one hand, it has been possible to formulate such approximations which retain molecular features of the porous systems and, on the other hand, the final system of equations is similar to (but not equivalent) conventional hydrodynamic equations. Previously, the relationship between these research areas was established only through empirical adsorption isotherms binding the concentration of the molecules on the surface of the pore and in its central part. However, as indicated by the above-mentioned problems, both fields are inextricably linked. In addition, the non-equilibrium distribution of the molecules in the relaxation process should change to the equilibrium distribution, and the initial states of the dynamic processes also often correspond to the equilibrium distributions.

An important aspect is the ratio between the developed molecular theory and numerical methods of research.

1. Both approaches operate with a single source of initial information on intermolecular interactions: potential roles of the adsorbate–adsorbate interactions and adsorbate–atom of the solid wall.

2. Many comparisons have been made since the late eighties and throughout the nineties of the last century of the distributions of molecules, isotherms and phase diagrams obtained by different methods, which showed that the difference between the results of calculations obtained in the theory and numerical methods in the pores is smaller than for the bulk phase.

3. Experimental data have shown quite clearly that there are criterial situations in which the two approaches (theory and numerical methods) are not precise enough, and this required improving the theoretical calculation methods that allow to overcome this difference from the experiment for the small pore sizes.

These questions are discussed in this monograph, which aims to promote modern methods of statistical thermodynamics oriented to modeling of processes in real porous objects. For this it is necessary to be able to evaluate and compare the results of theory and numerical modeling methods, and identify the properties of molecular models which take into account all the factors of experimental systems and to achieve agreement with the available data.

The monograph is structured as follows: there are nine chapters and twelve appendices. The material is aimed at professionals who are engaged in studying the processes of equilibrium and dynamics of adsorption in porous solids, but are not experts in the theory and modeling of these processes. The book contains a large number of drawings to illustrate the results, such as phase diagrams, transfer coefficients and flow distribution with which experimenters are familiar for the bulk phase and in macroscopic pores, but for nanoscale porous systems are unknown.

All the theoretical results (formulas, discussion of meaning of certain approximations, the description of techniques of calculations) are presented in the Appendix. This simplifies the presentation of the material, reducing it to the simplest equations and drawings, and does not affect the level of theory, which is required for the transition to the modeling of real-world objects.

The introductory chapter 1 sets out the literature data on the porous structure of different materials, adsorption isotherms and models of transport in porous solids. This material gives basic information on the most popular ideas, operated by experimentalists and modellers of practical processes. Also in the first chapter there is a collection general concepts of molecular modeling based on the atom–atom potentials and the lattice models, which are widely used in the adsorption theory and the theory of liquids. It would be pointless to try to collect even a modest bibliography for various topics, so it focuses on the physical premises of the use of certain concepts. It cited the work that is 'well-established' in the literature and which may be considered fundamental. This principle emphasizes that all major issues discussed for many decades, and they are similarly heuristic as at the time of their introduction. In a sense, the choice of material in the Introduction captures those aspects of the problem of the theory that are allowed today at the molecular level. Chapter 1 refers to Appendix 1 of the statistical justification of the lattice gas model.

Chapter 2 deals with the use of LGM equations for the calculation of the equilibrium in uniform (bulk phase) and non-uniform systems. For the bulk phase it demonstrates that the use of the LGM allows to describe the equation of state in wide ranges of temperature and vapour density. Non-uniform systems are considered here as non-uniform open surfaces on which monolayer filling, the pores of the multilayer film and the pores of the slit and cylindrical geometry occur. It discusses the construction of the distribution functions for setting the non-uniform properties of surfaces and methods for describing multilayer filling on the exposed surface and inside the pores. The appendix for this chapter addresses issues of hierarchy of mathematical models, which inevitably occur in non-uniform systems due to the correlation in the occupancies of neighbouring sites that are at different distances (Appendix 2). Appendix 3 discusses ways to improve the accuracy of the calculation of the equilibrium distributions, as well as a method for reducing the dimension of the equations in the quasi-chemical approach to non-uniform systems. The second chapter and Appendix 4 discuss modifications of the LGM equations that take into account the movement of molecules within cells. As a practical application it is shown how one can use the LGM to calculate the distributions of clusters in the bulk phase and inside the pores.

The material is constructed so that the reader can consistently get into the nature of the problems without cumbersome expressions of the theory. Many sections of the book are presented on the example of equations constructed for the case of taking into account the interactions between nearest neighbouring molecules. This is not enough to allow a quantitative comparison with experimental data and with the results of calculations by the molecular dynamics and Monte Carlo methods. Expressions for the total potential of interaction are shown in the Appendices 2–4.

The same principle of successive complication is also adopted for the analysis of different pore geometries ranging from simple geometry (in the chapters 2 to 6) to more complex forms of the pore space and the pore system (Chapter 7), which should be considered for real polydisperse porous materials. Similarly, the chapters 2 to 7 consider the adsorption of a single substance, the transition to the mixtures is discussed in chapter 8.

Chapter 3 analyzes the impact of different types of inhomogeneities of the systems on the conditions of stratification and the shape of the phase diagrams. We discuss the fundamental difference between the thermodynamic and molecular perspectives on the process of stratification. At the molecular level, the surface potential of the walls refers to the 'external fields' which form the spatially inhomogeneous regions within the volume of the pore size with sharply differing energy of the molecules. The physical reason for this is the abrupt change in favour of the interaction potential of the molecules along the normal to the plane of the surface of the wall, which forms quasi-two-dimensional near-surface regions and quasi-three-dimensional central interior of the free volume of the pores.

Intermolecular interactions lead to cooperative behaviour of the ensemble of molecules, in which the molecules themselves determine the possibility of formation of a new phase. The total pore volume is divided into a number of subareas, each of which has its own critical temperature and density.

We consistently examine the properties of non-uniform systems of different nature from the open non-uniform surface to the pores of the slit and cylindrical geometry. The traditional domed shape of the curve for the bulk phase separation is structured and divided into a set of domes, which form a complete phase diagram, depending on the geometry of the pore, its characteristic size and energy of interaction of the molecule with the walls of the pore.

The calculations show a strong dependence of the phase diagrams of all the molecular properties of the porous system. Theoretical calculations were compared with the results made by the method of molecular dynamics and Monte Carlo, and showed their satisfactory agreement. Decreasing the size of the pores decreases the critical temperature, and then a situation arises for the cylindrical pores of the transition of the system to the quasi-one-dimensional state in which there are no phase transitions. Chapter 3 understands range of issues related to the analysis of the size effect for the adsorption–desorption hysteresis and discusses ways of correct description (Appendix 5). Chapter 3 also analyzes the applicability of the Kelvin equation to calculate the vapour pressure in the pores of different diameters. This question is of fundamental importance in the practical applications of adsorption porosimetry.

Chapter 4 provides a summary of the foundations of the kinetic theory of condensed phases, which allows one to build a new approach to the dynamics of the transport processes of molecules in strong locally inhomogeneous external fields. The reasoning and approach needed to build the final system of microscopic equations of hydrodynamics are described. Appendix 6 is a self-consistent description of the rates of elementary stages and the equilibrium state of the LGM. The expressions for the thermal motion of the molecules at any density are given in Appendix 7. Appendix 8 provides a brief overview of the various methods for describing the dynamics of modern methods of statistical mechanics. This survey will be useful to a wider audience, as it replies (in terms of multiscale studies of physical and chemical processes in a wide range of nanotechnology) to the question why in dynamic studies we also need, in principle, the molecular theory. Appendix 9 discusses the principles of the kinetic theory of gas and liquid to justify the need to move to a new method of describing the flows, which is applicable for arbitrary densities of the system.

Chapter 5 is devoted to a detailed exposition of the method of calculation of the transport coefficients and properties belonging to the microscopic kinetic equations of hydrodynamics. These include the coefficients of self-diffusion, shear and bulk viscosity, thermal conductivity, and the coefficient of sliding friction. All of them were first studied in detail in the framework of the new theory for the slit-like and cylindrical pores. Some situations were considered in parallel using the theory of integral equations and molecular dynamics, and they are in quite satisfactory agreement. There is also a special comparison of the description of the dynamics of the interlayer redistribution and local mobility of the molecules in the slit pore. The results show a reasonable correlation of both approaches. In the accuracy of calculating the characteristics the theory is superior to the stochastic methods, as in small systems an important role is played by the fluctuation effects that complicate the numerical estimates.

The results of the numerical analysis of the currents of molecules in the slit-shaped pores are considered in chapter 6. Circumstances are described under which it was possible to carry out molecular analyzes of: relaxation of the liquid phase, the evolution of the meniscus of a liquid layer in the vapour phase, vapor bubble dynamics of movement and dynamics of a system with two bubbles. All of them were studied after the pulsed perturbation given at the initial time. Also, the influence of the wall potential for vapour and liquid flows in narrow channels is studied. More complex examples of flows relate to processes of movement of the fluid in the pore with non-uniform walls, and the motion of the vapour–liquid meniscus in narrow slit-shaped pores. An important methodological significance has the analysis of molecular flows in narrow pores with small initial perturbations and strong non-equilibrium initial conditions during the wetting of plates with different energies due to the molecules of the liquid. Also examined are the effects of motion of the plate, the 'sink' of the film with a hydrophobic plate and contacts of the droplet with the plate.

Microscopic hydrodynamics provides information about both the local concentrations of molecules, vectors microhydrodynamic velocities and temperature, and about maintaining the complete molecular information on the local thermal velocities and energies of molecules, as well as on the distribution of clusters of molecules. The accuracy of microscopic hydrodynamics is not only comparable to the molecular dynamics method, but surpasses it, and in the speed of the calculation it exceeds by at least two orders of magnitude. The algorithm for the numerical solution of the microscopic hydrodynamics is given in Appendix 10.

Chapter 7 is devoted to the generalization of the results of the third chapter for the pores of a more complex geometry (spherocylindrical and globular) and for polydisperse systems of slit-shaped and cylindrical pores, which are characterized by a function of the pore size distribution. The LGM concept is used twice for the analysis of complex porous systems: to describe the porous supramolecular structures and to describe the distribution of molecules within each typical pore area. At the same time there are areas of the joints of pores of different types, but the overall structure of the equations and the approach are maintained. The possibility of extension of the method of adsorption porosimetry is studied, including the definition of the pore size distribution with the pore connectivity of different types taken into account. The analysis of molecular distributions and concentration dependence of the transport coefficients in porous systems of complex geometry is carried out. As before, the consideration of size effects is of principal importance, and the question of the minimum size of pores of different types, which may contain loops of adsorption–desorption hysteresis.

In Chapter 8 the molecular adsorption theory is generalized for a mixture of molecules. First, the components of a comparable size are considered, which allows one to illustrate the effects due to the difference in the interaction energies of the main component (solvent) and the solute. There are many possible relationships between the energies of interaction of the components with the walls of the pores and with each other. Another important factor is the molar fraction of components. The joint effect of these parameters allows to simulate a wide range of experimental data, particularly given the possibility of the theory to correctly reflect the non-uniformity of the solid surface and/or pore walls. In this chapter, we consistently view both the equilibrium and dynamic properties of bulk mixtures and the properties in spherocylindrical and slit pores, and also examine the specificity of calculating the characteristics of microimpurities, as their analysis by the stochastic numerical methods can not be carried out because of the large fluctuations. The last sections of Chapter 8 are devoted to the components of different sizes. The kinetic equations and the equations of the equilibrium distribution of components of different sizes are considered in Appendix 11. To calculate the mutual diffusion coefficient of the mixture we require a rigorous analysis of the definition of mutual diffusion coefficients in different phases and their method of calculation, to satisfy the conditions of self-consistency. This issue is discussed in Appendix 12.

The final chapter 9 shows the new classification of the characteristic pore size and the volume assessment of micropores and discusses the problem of the relationship microheterogeneity porous bodies with the thermodynamic approach and the theory of transport molecules. We also discuss the prospects of using the molecular theory in more complex modeling problems of equilibrium and dynamic processes in dispersed phases which require further development of the theory.

We should mention the material that is not considered in the monograph. We are talking here about the issues of the theory of adsorption at low pore filling, which are well described in the literature. This is the Henry region for the equilibrium characteristics and the Knudsen regime in dynamics, which is actually the only one that does not require revision and continues to evolve. Given this, the book focuses on dense gases and liquids. Moreover, all the concentration dependences of the transport coefficients, discussed in this monograph, are normalized to this case of extremely small fillings in the bulk phase or within the pores. This ensures a large information content of the illustrations and connection with the equilibrium and dynamic characteristics of the systems already well-known to readers. The book is not the sections devoted to the adsorption of polymer molecules.

The monograph has an interdisciplinary character. It is aimed at professionals in the field of physical chemistry, statistical thermodynamics, the physics of surface phenomena and phase transitions, the kinetic theory of condensed phases and hydrodynamics, undergraduate and graduate students in related disciplines.

The author is grateful to colleagues who have participated in joint projects. This primarily refers to the A.B. Rabinovich and R.Ya. Tugazakov and to V.N. Komarov, E.V. Votyakov, M.M. Mazo, N.K. Balabaev, E.M. Piotrovskaya and J.M.D. MacElroy.

1

Introduction

1. The structure of porous bodies

The structure of dispersed materials is extremely diverse [1–11], and it significantly affects the equilibrium distribution of the molecules, the course of phase transformations and all the dynamic processes that take place inside the material (porous materials). Here, the term 'structure' refers to the location and relationship of the constituent elements of the system in space, i.e. the concept of 'structure' refers to a set of well-defined structural elements with limited autonomy. Structural elements themselves may have a crystalline structure (micro-crystals), or be amorphous or glassy particles characteristic of such artificial materials, such as fabric and nets of regular weaving, highly ordered polymeric systems, etc. In general, the particulate materials may be composed of structural elements of both types, such as in porous carbon of various origin [2].

The concept of 'porosity' is associated with the presence in the bulk of a solid V of both its typical and relatively constant property of the free volume V_c, not filled with elementary structural particles. Then $V = V_c + V_m$, where V_m is the volume of the solid skeleton or matrix. Porosity F_p is defined as the fraction of the volume of the solid occupied by this free volume $F_p = V_c/V = 1 - F_m$, $F_m = V_m/V$ [3].

Another essential characteristic by which the solids are regarded as porous system is the discrete nature of the free volume. As the solid matrix itself, the volume may be divided into basic structural elements – the pores differing in size, shape, nature of the relationship between them and often forming the open porous structure of the solid in space. Sometimes pores are insulated from each other. The organization of the porous structure is directly related to the organization of the structure of the solid part. This structure can also be characterized by both regularity and hierarchy in the construction of structural elements.

Note that the term 'porosity' should not be used for macroscopic unconsodilated systems having the large free volume in which the hard part is made up of individual macroelements not bound in a single frame by resilient and strong ties. Examples of such systems can be primers, bulk layers of sand and gravel filters, catalysts, adsorbents, bulk fibre filters, etc.

The division of solids to porous, non-porous, low- or high-porosity applied in practice [3] is arbitrary, since it is caused by different sensitivity of controlled properties of solid substances and materials to the absolute values of both F_p and individual parameters characterizing the porous structure. If the criterion of comparative evaluation of porosity is the parameter $\xi = F_p/F_m$ [3], the low-porosity media are those in which $\xi \leq 1$ and $F_p \leq 0.5$, and highly porous ones if $\xi > 1$ and $F_p > 0.5$. The upper limit of the porosity in practice is limited by the limit of sustainable conservation of connectivity of structural elements throughout the volume of the solid above which begins its destruction. Realistically, this is determined by the values $F_p \sim 0.7\text{--}0.9$ (highest for the cellular foam materials). The lower limit can be associated with the possibility of experimental determination of individual pores as structural defects. However, this limit is evaluated more accurately by the contribution of porosity to the porous properties of the body. Porous materials are characterized by new physical qualities found only in the porous media: a large diffusion permeability, low flow resistance, filtering capacity, high adsorption properties and developed internal surface, low sound and thermal conductivity, etc. Typically, a tangible manifestation of these properties in a natural and synthetic porous materials begins with at a porosity value $F_p > 0.1$, although in the case of porous film membranes and nuclear filters this estimate is an exaggeration [1].

The difference in the porous materials in the elementary, chemical composition and structure of solid particles and in the origin of porosity determines the variety of porous structures. On the basis of genesis the porous bodies are divided into two large groups: the additive systems (corpuscular structure) and the subtraction systems (spongy structures). Formation of the former takes place by the addition of a large number of individual elements of the structure both non-porous and those with primary porosity. These systems include the vast majority of porous materials: fabrics, paper, moulded sorbent pellets, catalysts, electrodes, porous ceramics, metals, cermets, zeolite crystals, etc. (additive systems are also unconsolidated porous media). The porous structure of the additive system is formed by the spaces between the particles that make up their skeleton.

The general character of the structure of porous materials is determined by the size and geometrical factors. There are regular porous structures with regularly alternating elements in the volume of the body in the form of individual pores or cavities and their connecting channels and also interconnected ensembles of pores (clusters) of a finite number of elements and the structure of the stochastic type, in which the size of the pores or their ensembles, the relative position and the connection of the pores are random. Naturally, the largest number of the actual porous material relates to the latter type of porous bodies. The irregular stochastic structure is characterized by a set of various pore sizes differing in the size and also the shape, orientation and location in space, the type of connection (Fig. 1.1). Such concepts are important for the mechanical properties of materials. To describe the adsorption properties we need more detailed models.

Models of porous bodies [4–6]

Elucidation of the structure of porous materials usually serves two practical tasks: the development of methods of synthesis of porous systems with specified properties and the ability to manage the processes occurring in these systems. With the accumulation of evidence on the structure of porous materials the details and features of the models can be better clarified and their relevance and value to solving technological problems increased.

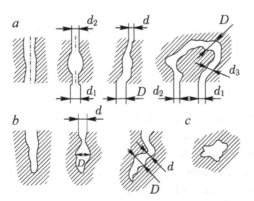

Fig. 1.1. Pore types: a – continuous open, b – open dead end, c– inner closed [10].

Table 1.1. The basic models of porous bodies [1]

Model	Examples described by the model of systems
Globular model	Black (PP)*, aerosils, aerogels, xerogels with various amorphous particles, sand, soil, brick and other systems
Model of pores between circular discs	Kaolinite, dickite, talc, pyrophyllite, mica, montmorillonite, vermiculite (PP), soot (ap)*, magnesium hydroxide (VP), modification og iron oxide, graphite, graphite oxide, various powders of lamellar crystals
Model of pores between polyhedra	Active carbon (PP), iron oxide (PP), magnesium oxide (SP), bayerite, η-Al_2O_3 (SP), porous crystals, sprayed metal films
Model of slit pores	Montmorillonite (SP), vermiculite (PP), η-Al_2O_3 (SP), the primary pores in different layered crystals, activated coals (micropores)
Model of pores between round rods	Gels of vanadium oxide (V), boehmite, γ-Al_2O_3, modification of the iron oxide (SP), gels of tungsten oxide, zirconia, sepiolite and palygorskite (SP), chrysotile asbestos (SP), endellite and halloysite (SP), paper, filters, yarn
Model of cylindrical capillaries	Endellite, halloysite (PP), palygorskite and sepiolite (PP), anodic aluminum oxide film, chrysotile asbestos (PP), the vessels of animals and plants, some porous glasses
Model ofbottle-shaped pores	Active charcoal (meso- and macropores), porous glass, reduced iron oxide, skeletal catalysts skeletal
*Letters PP and SP denote primary and secondary porosity, respectively	

It has been proposed to describe most porous materials by a set of 7 models (Table 1.1, see also Fig. 1.2) (although there is a classification that includes up to 15 types of porous systems [12]). These models represent the structure of both corpuscular porous systems (the first five models) and spongy systems, and are based on the assumption of the simplest forms of the basic elements of the structure – the particles and pores. The most practical use is made of the globular model and the hollow cylinder model (capillary model). In the first case, the porous medium is in the form of a packing of spheres of equal size, which themselves are globular elementary particles of the skeleton model, and the intervals between them

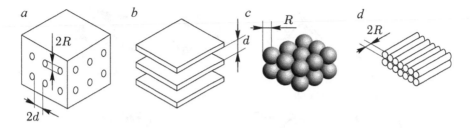

Fig. 1.2. Some models of porous systems: a – non-intersecting cylindrical capillaries, b – parallel plates, c– close-packed spheres, d – close-packed cylindrical rods [5].

simulate the porous space. All of the constructed globular models are defined by their basic parameters – the radius R_r of the globules and their packing density, characterized by the coordination number N_a.

Globular model [5]
The main parameters of the globular structure are the radius of the globule and the coordination number of the packing of the globules in a supramolecular structure (or the number of contacts with neighbouring globules). Both the regular and irregular organization of the particulates from the globules is used. The smaller the packing coordination number, the more loose the structure and higher porosity F_p.

Figure 1.3 shows the elementary pores in the regular packing of the globules. In the corpuscular bodies gaps between the particles form a complex system of intersecting pore channels with alternating contraction and expansion. In each expansion (cavity of the pores) there are a few passages (pore throats) from the neighboring expansions. If the area around the cavity is cut out so that the planes of the cuts pass through the most narrow sections of the pore throats, the porous body will be divided into elementary cells – polyhedra. The pore in such a cell is naturally called elementary. It is a cavity defined by spherical surfaces and having a plurality of throats, connecting it to the neighboring elementary pores. The shape and geometrical dimensions of the elementary pores are defined by the type of packing and particle size.

The most dense are hexagonal and face-centred cubic packing ($F_p = 0.2595$). Then there is body-centreed cubic ($F_p = 0.3198$), cubic ($F_p = 0.4764$) and tetrahedral packing ($F_p = 0.6599$). In [13, 14] other regular types of packing, consisting of densely packed layers of identical spheres, are discussed. These examples demonstrate that

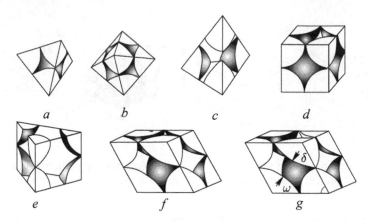

Fig. 1.3. Elementary pores in the regular packing of spheres: a, b – tetrahedral and octahedral pores in the hexagonal close packing ($N_a = 12$), c – tetrahedral pore in body-centreed cubic packing ($N_a = 8$), d – cubic pore in a simple cubic packing ($N_a = 6$), e – unit cell in the tetrahedral packing ($N_a = 4$), f – a pore in packing with $N_a = 8$ [13], g – a pore in packing with $N_a = 6$ [14].

the geometric properties of corpuscular bodies can not be determined solely by the coordination numbers of the package.

The regular packing of spheres was used as a model in the study of almost all phenomena in porous bodies, but they can not reflect the full spectrum of possible porous structures, so different ways of random packing of globules have been used widely.

Capillary models
In the capillary models the free volume is represented as a set of capillaries of different sections, length and orientation in space. In the simplest case, the porous structure of the real body is equivalent to the structure of the model with non-intersecting cylindrical capillaries. The model of independent parallel capillaries with a constant cross-section does not consider, however, the typical feature that the pores in real structures are not straight and the pores are corrugated. These properties are considered in more complex capillary models in which the basic metric characteristics – the radius of the pores, their length, and tortuosity may be random. Examples of such one-dimensional structures are shown in Fig. 1.4.

The porous structure of most materials is reflected most efficiently by the lattice models with regular and random topologies and by branching models. In a regular lattice model the porous medium is represented in the form of a regular two-dimensional or three-dimensional network whose sites are connected by the pores of

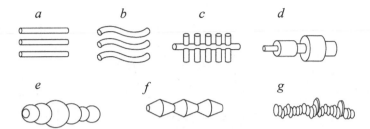

Fig. 1.4. One-dimensional capillary pore models [5]: a – straight capillaries, b – meandering capillaries, c – a capillary with dead-end pores, d – standard model of straight cavities, e – standard model of spherical cavities, f – periodic corrugated capillary; g – a pore with diffusion corrugation.

uniform cross section (Figs. 1.5 and 1.6). The symmetry of the lattice is given by a certain combination of sites and bonds. The number of bonds in a site is characterized by the site number n. For $n = 2$ the lattice model is transformed into a standard model with non-intersecting channels.

To simplify mathematical models, the cyclization of bonds, inherent to actual structures, is replaced in modelling lattices by the simplest cycles in which any two sites are connected by only one sequence of bonds (pseudolattices of the branching models). Regular lattice models allow for the presence of the distribution of the different characteristics of the main structural elements, while maintaining a certain pattern in the formation of sites and bonds; in irregular lattices their combination is already random. Thus, the properties of real systems such as the simultaneous existence of closed and connected pores, associated with the outer surface and isolated from it are also modelled.

Bidisperse model of the adsorbent [15]
To reflect the shortcomings of regular and quasi-regular structure

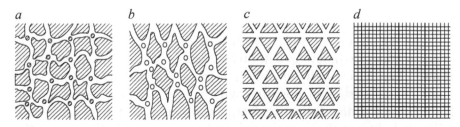

Fig. 1.5. The two-dimensional lattice models [5]: a – square ($n = 4$), b – cellular ($n = 3$), c – triple hexagonal ($n = 6$), d – bidisperse model based on a square lattice,

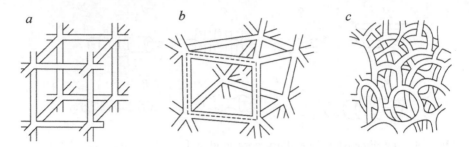

Fig. 1.6. Spatial capillary lattice models [5]: a – cubic ($n = 6$), b – random ($n = 6$) with a cycle of four pores, c – branching (pseudolattice) ($n = 3$).

models, bidisperse models have been proposed to reflect in the formal manner the presence of porous channels of different sizes. The granule of the adsorbent consists of a large number of randomly distributed microporous zones, and the gaps between them form a transport system (Fig. 1.7).

Microporous zones are not directly linked, so the kinetics of filling them depends on the change in the concentration of the sorbate in transport pores. In general, the transfer of the adsorbate in the microporous zones and the diffusion into the granules through the transport pores are developed as two parallel interdependent process. It is believed that all microporous zones have the same and regular shape and the same characteristic size. The same assumption is made with respect to the shape and size of adsorbent granules. In principle, the model can be generalized to the case of granules and microporous zones of arbitrary shape.

Another variant of the bidisperse model was proposed in [16]. This variant describes the adsorption in systems with nonuniform (two-phase) microporous zones. The difference is that in the volume of the microporous zones the adsorbate can be in two states: mobile and localized (secured with the matrix).

The classification of the pores
Due to the diversity and complexity of the structure of porous systems, their complete quantitative description has not as yet been found. The porosimetry problem is now reduced to finding such basic characteristics of porosity, which, firstly, would serve for the identification of the distinctive features of porous materials and would allow their quantitative comparison, and second, would allow to predict their structure-sensitive properties. The research results of different

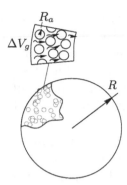

Fig. 1.7. Model of the bidisperse adsorbent and its geometrical parameters: characterizing the size of the grains R, the characteristic size of the microporous zones R_a [15].

porous materials were used to identify a set of basic parameters of the porous structure, which is necessary for the solution of practical problems. These parameters include: the total pore volume and the volume of individual species, the characteristic pore size and pore size distribution, the inner surface of the pore system.

Any classification of the pores is used to solve practical problems. Noteworthy are two basic approaches. In the first approach, proposed by Cheremski [10], the separation criterion is the principle of relativity of the pore size and the main elements of the skeleton structure of the porous body. The pores whose size considerably exceeds the size of the structural elements (more than a few micrometers) are called macropores. The micropores are commensurate with elements of the structure ($10^2 < r < 10^3$ nm). The pores substantially smaller than the structural particles are submicropores <100 nm). Among the latter there are separately allocated smaller ultrapores <1–2 nm, which may be located within the very building blocks of the porous matrix. This classification is used when considering the physical aspects of the porous state, in the study of the interaction of pores with other imperfections of structures – vacancies, dislocations, grain boundaries, etc., in the processes of their formation, growth and healing.

For highly porous systems Dubinin suggests a different classification of pores, based on the different mechanisms of adsorption phenomena occurring in the pores that differ significantly in size [1, 17–19]. In general, the porosity of the disperse systems consists of macropores, mesopores and micropores.

The size of the largest variety – macropores – is greater than 100 nm, and the magnitude of adsorption in the macropores is usually neglected. The boundaries of the linear dimensions of the mesopores are in the range of 1.6 to 100 nm, which corresponds to the limit of applicability of the Thomson–Kelvin capillary condensation equation. Mesopores have a developed surface area on which the first monomolecular and then polymolecular adsorption take place with the latter then terminated by filling the pores by the capillary condensation mechanism.

The micropore size is commensurate with the size of adsorbed molecules, so the microporous adsorbent + adsorbate system can be regarded as a single phase. For such a system the representation of the stratified filling the surface of the pores lose the physical sense, since in any interactions of the adsorbed molecules with the surface the adsorption field is created in the entire micropore volume and adsorption in the micropores proceeds according to the volume filling mechanism.

The lower limit of the radius of the micropores can be considered as the value ~0.13 nm, corresponding to the critical diameter of the He atom (0.26 nm), penetrating practically all the voids in the solid. The maximum size of the micropores in energy calculations of the dispersion interaction depends also on the critical diameter of the adsorptive molecule d_{cr} and corresponds to the equivalent radius equal to 2.5 d_{cr}. Typically, the bulk of the micropores is in the range of equivalent radii from 0.5 to 1.6 nm. To characterize the transition region between the micropores and mesopores, in which the main features of a microporous structure gradually degenerate, and the properties of the mesopores become more evident, Dubinin proposed to include a fine variety of micropores in the group of micropores (r_{eq} < 0.6–0.7 nm), and a larger variety – to supermicropores (0.6–0.7 < r_{eq} < 1.6 nm).

There is now a large number (over 60) of analytical methods for the study of the porous structure of solids. A brief but informative overview of the porosimetry methods was published by Cheremski [10] who was the first to classify these methods by basing the physical principles of measuring the parameters of the porous structure. Table 1.2 shows a few of the porosimetry methods, limited to those used in porous systems.

2. Adsorption isotherms

Interaction of molecules of gases and vapours with a solid generally leads to the absorption of molecules from the bulk phase, which

Table 1.2. Characteristics of the main methods of porosimetry [1]

Methods	The information obtained	The limits of applicability of the radii r, μm
Direct observation: visual-optical, light microscopy, electron microscopy, translucence (radiography, radioscopic, radiometric)	Identification of macropores The number, volume and pore distribution, specific surface area Identification of macropores	>10–75 0.5–100 0.01-0.5 0.1–100
Capillary: penetrant testing, capillary permeability (transpiration, capillary lift of the fluid, displacement of liquid)	The same Pore size, distribution of pore size, specific surface area	>0,1 0.01–100
Mercury porosimetry	The volume and pore distribution, specific surface	0.0015–800
Adsorption-structural	The same	0.0003–0.05
Small-angle scattering of radiation: X-ray scattering, neutron scattering	Identification of open and closed micro-irregularities pore size specific surface area	0.0005–0.7 0.002–0.1
Pycnometrically: gas pycnometer liquid pycnometry	Total porosity, the volume and size of micropores distribution of micropores	0.0002–0.001
Calorimetry: liquid immersion (wetting) thermoporometry	The specific surface area, the size of the micropores Pore distribution	0.0005–0.001 0.002–1000
The volume-weighted methods: volumetry, pore filling liquid hydrostatic interaction of liquids	The total porosity The pore volume, pore size	0.001–1000 0.001–1000
Reference porometry	Amount and distribution of the pores	0.002–1000

accumulate in the surface layer or in the bulk. In the first case we speak of *adsorption* as during concentration or condensation of molecules on the exposed surfaces. In the latter case it is *absorption* of the molecules in the crystalline solid volume (or inside the liquid). The term *sorption* has also been proposed. This term combines both variants of the concentration of the vapour molecules by the solid as compared to its volume content, including the process of capillary condensation inside pores. Currently, the term *adsorption* is probably used more often in all situations [20]. This term is used in this monograph for all situations as the description of the concentration of the molecules at the molecular level can not be carried out without the fixation of the surface potential (on the open surface of the solid or on the walls of the pores). This term reflects the steady-state condition of a porous body, the structure of which does not change during the adsorption process (the molecules do not penetrate into the depth of the solid inside the existing porous structure).

Adsorption isotherm

Adsorption is caused by the forces acting between the solid and the gas molecules. These forces are divided into two main types – physical and chemical, and they cause physical (or van der Waals) adsorption and chemisorption, respectively. The substance, adsorbed by the solid (adsorbent), is referred to as the adsorbate and the gaseous substance capable of adsorption is named the adsorptive.

The amount of the gas absorbed by the solid is proportional to the mass m of the sample and also depends on the temperature T, gas pressure P, and the nature of both the solid and the gas. If n denotes the amount of the adsorbed gas, expressed in moles per gram of solid, it is a function of P, T, the nature of the gas and the solid. The equation for the quantity of the absorbed substance of the form

$$n = f(P)_{T,\ \text{gas, solid}},\qquad(2.1)$$

is called the equation of the adsorption isotherm. If the temperature is below the critical temperature of the gas in the bulk phase T_c, another form of the adsorption equation is used:

$$n = f(P/P_0)_{T,\ \text{gas, solid}},\qquad(2.2)$$

where P_0 is the vapour pressure of the adsorptive in the bulk phase.

The large number of physical adsorption isotherms, obtained for different solids, were classified to five types (see Fig. 2.1), to which the sixth type was later added – stepped isotherms. The

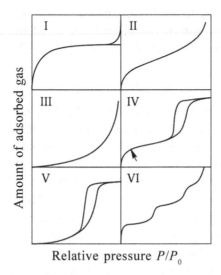

Fig. 2.1. Five types of adsorption isotherms (I–V) according to the classification, first proposed by S. Brunauer, L.S. Deming, W. Deming and E. Teller [21] and the stepped isotherm (type VI) [20].

isotherms of type IV and V have the hysteresis loop, the lower branch of which is obtained by measuring the adsorption during the sequential addition of gas to the system, and the upper arm – with its consecutive removal. Effects associated with the hysteresis are also possible for other types of isotherms. The stepped isotherms, or type VI isotherms, although relatively rare, are of particular theoretical interest and are therefore included in the classification. There are also isotherms that are difficult to attribute to any one particular type, and the number of these isotherms is not so small [20].

The experimental measurements of the adsorption isotherms have been used to make two main conclusions about the properties of porous materials: determination of the specific surface area (based on the BET isotherm [22] – see below), and the pore size distribution using the methods based on the Kelvin equation.

Langmuir equation

The first theoretical adsorption isotherm is the isotherm obtained by Langmuir on the basis of the kinetic analysis of the adsorption and desorption processes [23]. The model is based on the idea of the existence of discrete adsorption sites on which these elementary kinetic stages occur. This introduces the concept of the specific surface area, the unit of which contains a fully determined number of localization areas N_s of molecules from the gas phase. N_s corresponds to the value of the maximum volume of the monolayer V_1 of the adsorbed vapour. θ denotes the degree of surface coverage $\theta = N/N_s$, the Langmuir equation is written as

$$\theta = aP/(1 + aP), \tag{2.3}$$

where the coefficient a was expressed [23] in terms of the ratio of the constant of the adsorption rate to the constant of the desorption rate. The coefficient a is expressed as the ratio of the statistical sums functions of the molecules in the adsorbed state F to that in the vapour phase F_0; $a = \beta F \exp(\beta Q)/F_0$, if statistical thermodynamics is applied to obtain the Langmuir equation. This was first done by M.I. Temkin [24], and later repeated by Fowler [25] (see also [26]). Here $\beta = (kT)^{-1}$ and Q is the the binding energy of the molecules with the surface of the solid.

For small degrees of filling or at low pressures, when $aP \ll 1$, equation (2.3) becomes simpler and changer to the Henry equation:

$$\theta = aP, \tag{2.4}$$

and therefore the coefficient a is often called the Henry coefficient.

Multilayer adsorption
The physical adsorption of gases by non-porous solids in most cases is described by the adsorption isotherms of type II. The most successful model for the description of multilayer adsorption is considered to be the model of Brunauer, Emmett, and Teller (BET) [22]. Its expression is

$$\frac{n}{n_m} = \frac{C(P/P_0)}{(1 - P/P_0)[1 + (C-1)P/P_0]}, \tag{2.5}$$

where n_m is the monolayer capacity, expressed in moles of the adsorbate, the parameter C is expressed in terms of the kinetic coefficients, if derivation is analogous to the kinetic Langmuir conclusion or through the statistical sums of molecules in the first surface monolayer and in the second near-surface monolayer. In any case, the parameter C may be represented as $C = m \exp[\beta(Q_1 - Q_L)]$, where Q_1 is the energy of the molecule in the surface monolayer and Q_L is the energy of the molecule in the adsorbed film. Coefficient C values are usually in the range of from 0.02 to 20.

The main value of the BET equation is that the experimental adsorption isotherms of type II often have rather long straight segments (segment BC in Fig. 2.2) associated with the definition

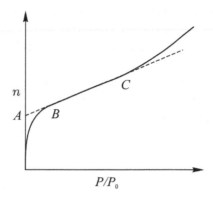

Fig. 2.2. Typical adsorption isotherm of type II with characteristic points *A* and *B* [20],

of n_m. The question of the reliability of the choice of the capacity of the monolayer n_B on the basis of the position of point B was the subject of a long debate [20]. The possibility of localization of point B depends on the shape of the isotherm curve [29]. If there is a gentle bend, the value n_B may differ significantly from n_m.

As can be seen from Fig. 2.3, the isotherms expressed by the BET equation are curves having an inflection point (if $C > 2$). The inflection point is close to the point n_m, corresponding to a monolayer filling according to BET, but not necessarily coincides with it.

Since its advent, the BET model has been criticized: 1) according to this model, all the adsorption sites on the surface are energetically identical, but homogeneous surfaces are the exception and nonuniform ones – a rule; 2) another assumption of the model is that it takes into account only the interaction forces between the adsorbate and adsorbent molecules – the so-called 'vertical' interactions – and neglects the interaction forces between the adsorbate molecules on

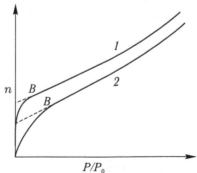

Fig. 2.3. Typical adsorption isotherms of type II with a sharp (*1*) and sloping (*2*) bends [20].

the surface in the given adsorption layer – the so-called 'horizontal' interactions; 3) the assumption that the molecules in all layers following the first can be regarded as completely identical is also questionable.

Below this is discussed in terms of the molecular theory of multilayer adsorption. The main advantage of this formula is that it allows one to introduce a measure to quantify the specific surface area of adsorbents, and this explains its widespread use. The unsuitability of the formula for higher degrees of filling was common knowledge, so its use will never go beyond the area of the filling of the second surface monolayer, that somehow justifies the values of surface areas obtained on its basis.

However, it should be noted that formula (2.5) is also incorrect for the surface coverage less than a monolayer: $n/n_m = C(P/P_0)/[1 + (C - 2) P/P_0]$, because it does not transform to the Langmuir isotherm (2.3).

Mesopores

The study of the porous structure of mesoporous solids is closely related to the interpretation of the type IV adsorption isotherms. At low pressures (*ABC* portion in Fig. 2.4) the isotherms of type II and IV are identical with each other. However, from a certain point *D* the type IV isotherm is deflected upward (*CDE*), and then at higher pressures its slope decreases (*EFG*). When approaching the saturated vapour pressure the adsorption value may vary slightly (along the curve *FGH*) or significantly increase (*GH'*). A characteristic feature of type IV isotherm is the presence of the hysteresis loop. The loop shape may be different for different adsorption systems, however, as can be seen from Fig. 2.4, the adsorption value at any given relative

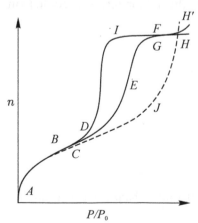

Fig. 2.4. The adsorption isotherm of type IV. The corresponding adsorption isotherm of type II is shown by the dashed curve *ABCJ* [20].

pressure for the desorption branch (*FID*) is always greater than for the adsorption branch (*DEF*). The loop is reproducible when desorption starts at a point above the point *F* which represents the upper loop point. The adsorption isotherms of type IV are frequently observed in many inorganic oxides and other porous materials [21, 30].

Zsigmondy [31] and Anderson [32] formulated a modern point of view on the interpretation of these isotherms. The Kelvin equation is used for this purpose [33]:

$$\ln(P/P_0) = -\frac{2\gamma V_L}{r_m RT},$$ (2.6)

where P/P_0 is the relative vapour pressure in equilibrium with the meniscus having a radius of curvature r_m; γ and V_L are the surface tension and the molar volume of the liquid adsorptive, respectively; R and T are used in their conventional sense. In this case, it is assumed that V_L is independent of pressure, i.e. that the fluid is incompressible.

The Kelvin equation shows that the capillary condensation of the vapour into the liquid must occur in the pores at a certain pressure determined by the value r_m of the liquid meniscus in them; this pressure must be less than the saturation vapour pressure, assuming that the meniscus is always concave (i.e. the contact angle of less than 90°).

The Zsigmondy model assumes that in the initial part of the isotherm (*ABC* in Fig. 2.4) the adsorption is limited only by the formation of a thin layer on the pore walls. Point *D* (at the base of the hysteresis loop) corresponds to the beginning of capillary condensation in the smallest pores. Therefore, in formula (2.6) the

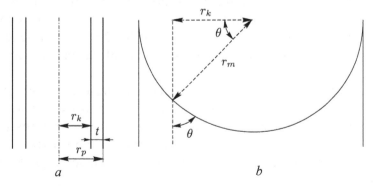

Fig. 2.5. A cylindrical pore cross-section of radius r_p, parallel to its axis; showing the inner core with radius r_k and the adsorption film thickness t [5] (*a*). The ratio between the radius r_m in the Kelvin equation and the core radius r_k for a cylindrical pore with a hemispherical meniscus, here θ is the contact angle [20] (*b*).

average radius of curvature r_m is equal to the pore radius minus the thickness of the adsorbed film on the walls (see Fig. 2.5 a). Or capillary condensation does not occur strictly in the pore, and its 'core' – the so-called 'core' (the term means the pore space bounded by the surface of the adsorbed film [32]). A simple geometric construction (Fig. 2.5 b results in a ratio of radius r_k of the core and the meniscus radius $r_m = r_k \cos(\theta)$).

If filling all the pores with the liquid adsorbate ends in the *FGH* zone the adsorbed amount corresponding to this area and expressed as the volume of the fluid (having the density of the normal fluid) should be the same for all adsorptives on this porous body (Gurvich rule) [34].

Practice has shown that this minimum radius depends on the nature of the test sample, but it is rarely less than 1 nm. The upper limit of the applicability of the Kelvin equation $r_m \sim 25$ nm is determined by the experimental difficulties of measuring very small vapour pressure drops.

The applicability of the Kelvin equation

The accuracy of determination of the dimensions of the range mesopores from 1 to 25 nm using in this range classical equations describing capillary phenomena, especially the Kelvin equation. There are the following limitations.

1. In practice, it is not possible to determine the contact angle θ. It is usually assumed that $\theta = 0$ ($\cos \theta = 1$), and this condition is the subject of constant doubts.

2. For very fine pores having a width of only a few molecular diameters, the Kelvin equation is no longer valid. We mention Guggenheim [35] who used the results of the statistical–mechanical analysis to shows that the surface tension in (2.6) should depend on the radius of curvature of the liquid surface at r smaller than about 50 nm.

3. There are practical constraints, the nature of which can be understood by considering the relative pressures of nitrogen corresponding to the size of mesopores at their upper limits (Table 1.3). These restrictions put an upper limit of application of the Kelvin equation, apart from the need that the contact angle θ should be less than 90°.

Here, the following values were used: $V_L = 34.68$ cm^2/mol; $\gamma = 8.88$ mN/m; $T = 77.35$ K; $\lg(P/P_0) = -0.416/r_m$ (where r_m is measured in nm).

Table 1.3. P/P_0 values of nitrogen at 77.35 K, corresponding to different r_m [20]

r_m		P/P_0	r_m		P/P_0
nm	μm		nm	μm	
20	0.02	0.9532	200	0.2	0.9952
50	0.05	0.9810	1000	1.0	0.9990

These relative pressure values are so close to each other that on the adsorbent having pores of such size, the adsorption isotherm will rise steeply so that it can not be measured reliably by any of the currently accepted techniques. Small variations in the temperature of the sample have a disproportionately greater impact on the computed value of r_m.

Types of hysteresis loops [20]

Commission 1.6 of the International Union of Pure and Applied Chemistry (IUPAC) published recommendations whose purpose is to draw attention to the uniformity of publications on physical adsorption and provide guidance for the evaluation and interpretation of the adsorption isotherms (Fig. 2.6). The classification of hysteresis loops, given in the IUPAC recommendations, includes four types of loops: H1–H4. In the H1 type loops adsorption and desorption branches are almost vertical and approximately parallel to each other in the estimated range of values of adsorption; in loops of the H4 type they are almost horizontal and parallel in a wide range of relative pressures. Types H2 and H3 are intermediate.

Each type of hysteresis loop is associated with a particular type of porous structure. Thus, the H1 type loop is often prepared for agglomerates or tablets of globules quite identical in size and homogeneously packed. Some corpuscular systems, e.g. certain silicagels, are characterized by the H2 type, but in this case the pore size distribution and shape is difficult to define. The types H3 and

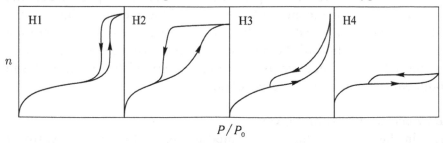

P/P_0

Fig. 2.6. IUPAC classification of hysteresis loops [20].

H4 are obtained for adsorbents having slit-like pores or (H3) consist of plane-particles. Type I isotherms are associated with the H4 type of loop and indicates the presence of microporosity.

3. Transport in porous solids

Transport processes occur in all physical and chemical processes in porous media, which are widely distributed in nature and technology [5, 36–42]. These include impregnation and drying, multiphase filtering and displacement, capillary condensation and desorption, formation of a porous structure and coatings on porous materials, etc. Particularly noteworthy are the chemical processes such as catalytic and electrochemical reactions on electrodes and porous catalysts involving liquids and gas, synthesis of carriers and catalysts, heterophase polymerization [5].

In these processes an important role is played capillary and surface forces that govern the mutual distribution of phases in the pore space and determine the conditions of the transfer, the size of interfacial surfaces, the size of phase inclusions. In addition, capillary and surface forces are responsible for such specific mechanisms for the transfer of porous media as capillary leakage, film flow, recondensation. The course of the multi-phase process is significantly affected by the geometrical features of the porous structure, which are discussed in section 1.

The general formulation of the problem can be traced to the dynamics of adsorption chromatography, which reflects the general patterns and techniques used to model the transport processes [37]. The theory of the dynamics of adsorption should take into account the following main aspects of this complex physical phenomenon: the balance of substances in the course of their movement and distribution in the adsorption medium, adsorption kinetics and statics, the hydrodynamics of the process, the relationship between the thermodynamic parameters of the state of the medium, the balance of heat and heat transfer in the adsorption process in the moving medium. The nature of the movement and distribution of substances in the adsorbing medium is also predetermined by the initial and boundary conditions of the process.

Balance equations of substances
Let in the inside of the porous medium be a flow of a mixture of adsorbed substances moving in some direction. If, in addition to

the mechanical flow there is an additional transfer of substances as a result of molecular diffusion and convection, the mass balance equation should also include additional terms that take into account these processes of transfer of substances [37]. Thus, in the theory of the dynamics of adsorption and chromatography there is a sufficiently general equation for the total balance of the transfer of substances in the form of

$$\frac{\partial n_i}{\partial t} + \frac{\partial N_i}{\partial t} + \mathrm{div}(n_i \bar{u}) = D_{m,i}(\Delta n_i + \lambda_i \Delta T), \tag{3.1}$$

where n_i is the volume concentration of the i-th component of the composition of the mobile phase per unit volume of the porous material; N_i is the volume concentration of the i-th component of the stationary phase (adsorbent) also based on the unit volume of the porous material; t is time, u i the linear velocity of flow within the porous material, it is a vector quantity; $D_{m,i}$ is the molecular diffusion coefficient, λ_i is the coefficient of thermal conductivity; Δ is the Laplace operator, T is temperature. The equations of the type (3.1) are recorded for each mixture component i ($1 \leq i \leq j$).

The following equations characterizing the dynamics of adsorption, should be the equations of adsorption kinetics, reflecting the physical and chemical nature of the adsorption process and establishing the temporal relationship between the concentrations of substances in the adsorbent and the mobile phase [38]. The kinetics of adsorption of the i-th component depends on the following basic independent factors: the concentration of all components in the composition of the mobile phase and the adsorbent (n_1, n_2,..., n_j; N_1, N_2,..., N_j); the parameters that define the so-called diffusion stage of the kinetics of adsorption (among them, we isolate isolate the impact of the mobile phase flow velocity u, temperature T and the density of the mobile phase ρ, the other parameters are symbolically denoted as K_{dif}); the parameters that define the so-called chemical stage of the kinetics of the adsorption stage (direct interaction between the adsorbent and the adsorbed particles), these parameters are symbolically denoted as K_{chem}.

The adsorption kinetics equation is formally written as

$$\frac{\partial N_i}{\partial t} = \Psi(n_1, n_2, ..., n_j; N_1, N_2, ..., N_j; \bar{u}, T, \rho, K_{dif}, K_{chem}). \tag{3.2}$$

In this equation on the left are the partial derivatives of the i-th concentration component in the adsorbent in time, as in the case of adsorption dynamics, N_i is the function of not only time, but also the coordinates (X, Y, Z). These general constructions are detailed in each specific situation.

A feature the equations of the adsorption isotherms is the lack of many parameters that affect the adsorption kinetics, such as the parameters determining diffusion. In the equations of adsorption isotherms there are only the parameters that characterize the interaction energy of the adsorbate and the adsorbent.

In general, the adsorption isotherm equation can be written in the following form of implicit functions:

$$F_i(n_1, n_2, ..., n_j; N_1, N_2, ..., N_j) = 0. \tag{3.3}$$

The hydrodynamic equation

In the adsorbates balance equation (3.1), and generally in the kinetic equation (3.2) there is the velocity of flow of the mobile phase. This means that the distribution of the adsorbed substances in the adsorption dynamics depends on the velocity distribution in a porous medium in space and time, i.e. $u(x, y, z, t)$. The function of the spatial and temporal distribution of the flow rate can be found on the basis of hydrodynamic equations [43].

The first equation of this system is the continuity equation:

$$\frac{\partial \rho}{\partial t} + \mathrm{div}(\rho \mathbf{u}) = 0. \tag{3.4}$$

This scalar equation is an expression of the law of conservation (balance) of the mass of the flowing phase (liquid or gas).

The following equation is the equation of motion of a viscous fluid:

$$\frac{\partial \mathbf{u}}{\partial t} + (\mathbf{u}\nabla)\mathbf{u} = -\frac{\nabla \rho}{\rho} + \frac{\mathbf{F}}{\rho} + \frac{\mu}{\rho}\Delta \mathbf{u} + \frac{1}{\rho}\left(\zeta + \frac{\mu}{3}\right)\nabla \mathrm{div}\,\mathbf{u}. \tag{3.5}$$

This is a vector equation – an expression of Newton's second law of dynamics: on the left is the acceleration $d\mathbf{u}/dt$, acquired by a mass unit, and on the right is the force acting on this mass. Term $\nabla \rho/\rho$ determines the hydrostatic pressure acting on the flowing phase; $\mathbf{F}$ is the external force, and the remaining terms characterize the internal friction force, μ and ζ are the coefficients of viscosity of the medium, depending on the ambient temperature T. Vector equation (3.5) projected on the coordinate axes X, Y, Z is divided into three equations.

The fifth in the system of hydrodynamic equations is the equation of state of the liquid (or gas): $\rho = f(P, T)$, which establishes the dependence of the density of the mobile phase on the pressure and temperature of the aggregate state of the environment.

The sixth equation of fluid dynamics is the heat balance and distribution equation. Velocity and temperature fields are a consequence of mechanical and thermal interactions; strictly speaking, they are interdependent. The temperature field affects the velocity field through the temperature dependence of density ρ and through the dependence of viscosity on temperature. On the other hand, the temperature field depends on the velocity field, since the transfer of the flowing phase is also accompanied by the transfer of heat. The equation of heat propagation in a moving medium, using the dissipation function of heat diss($F(\mathbf{u})$), can be written in a formal way [37]:

$$\operatorname{div}(\lambda \nabla T) + q + A\mu \operatorname{diss} F(\mathbf{u}) = c\rho \frac{dT}{dt} - A\frac{dP}{dt}, \qquad (3.6)$$

where λ is the coefficient of thermal conductivity; c is the specific heat capacity of a given thermodynamic process; q is the internal heat source (or outflow) density; A is the thermal equivalent of mechanical work.

The initial conditions are characterized by the distribution function of the concentration of adsorbed substances, density, velocity and temperature of the sorbent–mobile phase system at the initial time ($t = 0$). The boundary conditions determine these functions at the system boundaries as a whole, and also at the boundaries between the phases. Depending on the physical conditions in which the system is located, the initial and boundary conditions can be very diverse.

The structure of the dynamics equations

The considered structure of the kinetic equations is quite consistent and reasonable, but in practice it is attempted to reduce the problem to one of the two extreme cases, when the total mass flux through the pore system is described in terms of hydrodynamic equations, or diffusive transport equations.

The first case implies that the pressure difference on both sides of a porous (or membrane) is sufficient to transport molecules. The main driving force behind the transport of molecules is a direct way to transfer the momentum to the moving molecules. In addition to pressure, transfer may also be caused by external forces of gravity, electrostatic, magnetic, etc. The second case assumes that the total pressure drop on both sides of the membrane is not sufficient to transport molecules and the main driving force is the difference between the chemical potentials of the molecules on both sides of the membrane.

The term 'diffusion' is widely used in many different situations, so we must distinguish if it is a single substance, such as the flow organized in a fine-porosity in which the external momentum of the incoming material is quickly extinguished (insufficient pressure drop across the membrane). If the convective flows of the mixture are discussed, the diffusion fluxes are always there and we are talking about the relative displacements of different molecules in the general stream.

Below we confine ourselves to formal forms of both types of equations and recall the basic types of mechanisms for the transport of one-component fluids.

Diffusive transport equation [44–46]

Formally, the diffusion theory is based on Fick's law: $J_i = -D_i$ grad C_i, where D_i, the coefficient of diffusion, is the coefficient of proportionality between the flow of the substance and the concentration gradient of this substance. If these equations are used to examine the balance of diffusion fluxes in the infinitesimal volume in the transient regime of the process, we obtain the following differential equation:

$$\frac{\partial C_i}{\partial t} = \mathrm{div}\,(D_i\,\mathrm{grad}\,C_i). \qquad (3.7)$$

In the presence of interactions between the diffusing particles or with a medium there is a deviation from a linear relationship between the flow of the substance and its concentration gradient, which results in the dependence of the diffusion coefficient on the spatial coordinates or local concentration of the substance and the concentration gradient of the diffusible substance.

If the process can not be described by equations (3.7), even with a variable diffusion coefficient, then we talk about the non-Fickian diffusion. In general, the concept of the diffusion coefficient has to be clarified in accordance with the peculiarities of the specific mechanism of the process, the participating particles and the structure of the medium [47].

For the case of one-dimensional steady-state diffusion through a flat plate with a constant diffusion coefficient, the solution of (3.7) leads to a simple equation for the flow of the diffusing material:

$$J_i = -\Pi_i S \frac{P_i^{\mathrm{I}} - P_i^{\mathrm{II}}}{L},\qquad(3.8)$$

where P_i is the pressure at the surface of the membrane in the material on either side (I and II) of the membrane; S is the area of the membrane; L is its thickness, Π_i is permeability.

When considering the process of gas transport through the membrane in accordance with equation (3.8), when the flow of diffusing gas is measured in mol/s and the pressure in Pascals (Pa), the permeability is measured in (mol·m)/(s·m·Pa) or (mol·m)/(s·N). At low concentrations, the solubility of the component in the membrane material is linearly related to its concentration, i.e. $\bar{C}_i = \tilde{\sigma}_i C_i$, where $\tilde{\sigma}_i$ is the solubility coefficient, a dimensionless constant. Then, in accordance with equation (3.8), the coefficient of permeability is linearly related to the diffusion coefficient: $\Pi_i = \tilde{\sigma}_i D_i$. If the solubility of the gas in the membrane obeys Henry's law: $C_i = \sigma_i P_i$, then we have $\Pi_i = \sigma_i D_i$.

Filtering

Filtration is a fluid motion in the porous medium by the pressure gradient. The flow of an incompressible viscous fluid in a porous medium at low Reynolds number Re $= \bar{u} H\rho/\eta$ (where $\bar{u}$ is the modulus of the average flow velocity; H is the characteristic pore size; ρ is the density of the liquid; η is viscosity) is described at the microlevel in the approximation of the Stokes equations, when in the Navier–Stokes equations non-linear convective terms are neglected [48]. The system of local equations includes the continuity equation $\nabla_x u = 0$ and the linearized Navier–Stokes equations $\eta \nabla_x^2 u = \nabla_x P$. On the inner surface of the porous medium the sticking conditions $\bar{u}|_{\partial R} = 0$ must be satisfied.

The averaging procedure for the elementary physical volume of the medium, made in [49], led to the famous Darcy law establishing a linear relationship between the mean flow velocity and mean pressure gradient:

$$\langle \bar{u} \rangle = \frac{[K]}{\eta} \nabla_x \bar{P},\qquad(3.9)$$

where [K] denotes the permeability, not to be confused with the formulas (3.8). Here permeability is the dimension of the area; it does not depend on the fluid properties and is a purely geometric characteristic of the porous medium. The unit of measurement of permeability is darcy (1 darcy = $1.02 \cdot 10^{-8}$ cm²). From dimensional analysis it follows that $[K] = f (F_p,...)$, ξ^2, where $f (F_p,...)$ – the dimensional function of dimensionless parameters of the porous structure, in particular the porosity, ξ – pore size.

There are many different formulas expressing [K] through the parameters of the porous structure – both purely empirical and derived from models for the porous structure. Detailed analysis of various approaches to determine the permeability can be found in [50, 51].

Molecular transport mechanisms [5, 38, 41]

A wide range of changes in the density of the fluid in porous materials naturally raises the question about the variety of types of mechanisms of transport of molecules. As before, we restrict ourselves to only one-component fluids. The method of molecular interpretation of the mechanisms of movement of the molecules determines all the kinetic coefficients characterizing the flows in the transport equations written above.

The nature of gas movement in the forward cylindrical capillary is determined by the Knudsen parameter Kn – the ratio of the number of collisions of molecules with the walls to the number of intermolecular collisions. Up to a constant factor of the order of unity parameter Kn is equal to the ratio of the mean free path of the molecules λ to the diameter of the capillary $2r$: Kn $\sim \lambda/2r$. Depending on the number Kn, there are three characteristic regions of the gas flow: Knudsen (Kn $\rightarrow \infty$), transitional (Kn ~ 1) and molecular (Kn $\rightarrow 0$). All situations with the liquid phase also formally apply to the case Kn $\rightarrow 0$.

Faced with the wall, the gas molecule can be adsorbed, and then be reflected in an arbitrary direction. This is the so-called diffuse reflection. Diffuse reflection is the transfer of the axial component of the momentum to the capillary walls. The accommodation coefficient reflects the specificity of the interaction of the gas molecules with the walls. In diffuse reflection $\gamma_a = 1$, when mirroring $\gamma_a = 0$.

Free molecular flow (or Knudsen flow)

Studies of the free molecular or Knudsen flows were carried out on small holes in very thin plates. This ensured the absence of intermolecular collisions with gas passing through the opening and this also defines this type of flow (Kn → ∞). Molecules move completely independently of each other, and there is no significant difference between the gas flow as a whole and by diffusion as it is in the continuous medium mode.

If we have a gas with density n of molecules per cubic centimeter on the one side of the hole and vacuum on the other side, the free molecular flow expressed as a $J_K = wnv$, where v is the dimensionless probability factor and $w = (8k_B T/\pi m)^{1/2}$ is the average velocity of the molecules; J_K has the dimension molecule/(cm^2·s). Here k_B is the Boltzmann constant, T the absolute temperature of the gas and m is the mass of the molecule.

If the densities of the gas molecules are non-zero on both sides of the hole, the total flux is proportional to the difference of these densities. We write the expression for the flux of molecules in a differential form, defining thus the Knudsen diffusion coefficient D_K having dimensions cm^2/s: $J_K = -D_K \nabla n$. The minus sign indicates that the gas flow is directed against the density gradient with a positive D_K. This implies that D_K is proportional to the ratio of the average speed of molecules w. This proportionality is defined explicitly entering the parameter K_0 of the Knudsen flow or permeability,

$$K_0 = 3D_K/4\omega. \qquad (3.10)$$

Viscous flow of steam
In the case of Kn → 0 molecules constantly collide with each other and the average situation in the movement of the molecules is not very different from the movement in the bulk phase. The channel walls directly interact with only those molecules which are found in the subsurface region. The viscous flow is the flow of gas as a continuous medium, caused by the pressure gradient. Calculation of the viscous flow of gas is carried out exactly as displayed by Poiseuille's law for the fluid flow, provided that the gas is compressed. The essence of the construction of the equation is that gas moves with no acceleration and the total force exerted on any element of its volume must be equal to zero, so that the viscous damping force is fully compensated by the force due to the pressure drop at the ends of the element.

$$J_{\text{visc}} = \text{Flux}/\text{area} = -(nB_0/\eta)\nabla P, \qquad (3.11)$$

where n is the number of molecules per unit volume, B_0 is the viscous flow parameter (a constant characterizing the geometry of the channel and expressed in units of cm^2), η is the coefficient of dynamic viscosity of the gas (g/(cm·s)), P is the pressure of the gas (dynes/cm^2).

The boundary condition for viscous flow – the vanishing of the velocity of the gas at the surface of the channel walls. 'Slip' on the surface is treated as a partnership of the mechanism of free-molecular flow. Gas compressibility is taken into account by the equation of state $n = P/(k_B T)$. Thus, the coefficient at ∇P in (3.11) depends on the coordinates through pressure P.

The intermediate flow regime

At intermediate gas densities between free-molecular and viscous flows the number of collisions of molecules between the walls and between other molecules has a transition region (Kn ~ 1), in which these different types of collisions are comparable. This area lends itself to the most difficult theoretical description using primarily different empirical estimates [38, 41], which are constructed as a combination of contributions from the above trends. Capillary permeability is reduced compared to the permeability of free molecular or viscous flows. The traditional interpretation of this effect does not use any information about the properties of the channel walls, which become significant.

The importance of the existence of the flow regime was stressed by Wicca and Vollmer [53], pointing to the fact that in a very long capillary tube on one side of which there is a vacuum and on the other side there is a gas or a liquid, there must necessarily be a change in the mechanisms of steady-state steam flow.

Viscous fluid flow

This type of flow in many of its properties close to the viscous flow of vapour [38]. The difference in the mechanism of collision of molecules in the liquid and the molecules in the vapour is that the molecules of the liquid are constantly under action of each other. Description of the viscous fluid motion is in many ways similar to the description of motion of viscous vapour (3.11). It often uses the representation of the practical incompressible fluid. This mechanism was proposed by Flood for the analysis of the high speed flow of adsorbed gases [54].

Surface diffusion flow

The molecules adsorbed on solid surfaces are in continuous thermal motion, and the presence of the concentration gradient in the adsorption phase leads to a diffusion flow in the direction of the

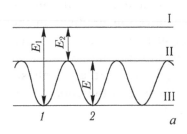 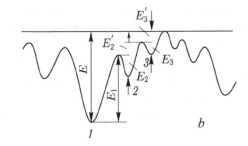

Fig. 3.1. The potential energy curve for homogeneous surface (*a*) and the potential energy curve of a nonuniform adsorbent (*b*) [38].

gradient. The authors [55] observed a greater rate of growth of thin hexagonal crystals of mercury from its vapour in the direction of the hexagonal plane. Linear growth was 1000 times more than would be expected for transfer through the gas phase, taking into account the molecular kinetic theory.

Migration of the adsorbed molecules is not free because of the presence of the energy barriers that exist even on the surface of the adsorbent with a regular crystal lattice. In such an ideal crystal the energy heterogeneity, in the range of the lattice constant, occurs because of the periodicity in the arrangement of elements of the crystal lattice.

Schematically, this change in energy is represented by the curve in Fig. 3.1 *a*. Level I corresponds to the level of vacuum, the level II – to the level of the energy peaks of migration of adsorbed particles E and the level III – to the depth of the potential well of the adsorbed molecules with energy equal to E_1. Movement of the adsorbed molecule from position 1 to 2 requires the activation energy $E = E_1 - E_2$. The value of E determines the activation energy of surface migration and the value of E_2 – the difference in the activation energies of desorption and jump at the same time, this value characterizes the dispersion of the values of the energies of the particles retained by the potential of the wall. According to Frenkel [56], the coefficient of surface diffusion D_S is expressed as

$$D_s = D_0 \exp(-E/kT), \quad D_0 = \Delta^2/4\tau_0. \tag{3.12}$$

The value D_0 can be regarded as the diffusion coefficient of the two-dimensional ideal gas on an ideally smooth homogeneous surface. The order of magnitude $\tau_0 = 10^{-13}$ s, Δ is the distance at which the molecule jumps.

The rate of mass transfer on the surface depends not only on the diffusion coefficient, but also on the concentration gradient in the adsorption phase, i.e. diffusive flux on the surface P_S can be written as $P_S = -D_S \partial a'/\partial x$, wherein a' is the adsorption per unit volume of the adsorbent. Turning to the concentration gradient in the gas phase, by performing a substitution of a' for n according to the equation of the adsorption isotherm $a' = f(n)$, we obtain

$$P_S = -D_S f'(n) \frac{\partial n}{\partial x}. \tag{3.13}$$

The surface of real adsorbents is energetically nonuniform, as shown schematically in the form of the potential energy curve in Fig. 3.1 b.

Diffusion in small pores
When the pore size is much larger than the diameter of the molecule the repulsive forces do not prevent its transport, but if the pores are so small that the fields of adsorption forces of the opposing walls overlap, the jump of the molecules from one wall to the other requires a lower activation energy than the removal to infinity as occurs in wide pores with radii are much greater than the size of the molecule.

The diffusive flux in the pores of the adsorbent in the Knudsen transport mechanism can be expressed by the equation $P = -D_a \partial a/\partial x$, where $\partial a/\partial x$ – the gradient of the concentration of the substance in the adsorption phase; D_a – the Knudsen diffusion coefficient of the adsorbed gas. It is expressed by a formula similar to equation (3.12), where E is the activation energy of molecular transition. Because of the overlap of the fields of the opposite walls there is a reduction of the activation energy of hopping to some value ΔE: $E_a = E - \Delta E$.

Diffusion in the pores of zeolites
Porous adsorbents may contain very small pores, so-called ultrapores, whose size is close to the diameters of the size of molecules and for the adsorbate with large molecular sizes are not accessible. These adsorbents include synthetic zeolites and some samples of porous glasses and coals of the sharan type. These adsorbents are not without reason called molecular sieves, because they can adsorb smaller molecules and do not adsorb molecules of larger dimensions.

If the diameter of the molecule is almost equal to the diameter of the channel, diffusion in such a channel is by its nature not longer Knudsen or surface. At very close contact of the gas molecules with the pore walls the transfer rate is controlled by the repulsive forces that impede the passage of molecules in a narrow capillary. The nature of the activation energy of the diffusion process in this case is different than in surface migration. Barrer [36] (see also [57, 58]), considering the diffusion of gases in zeolites, came to the conclusion that this type of diffusion ('zeolite diffusion') has more in common with the dissolution of gases in solids than conventional diffusion. Permeation of gas molecules through a small orifice may be promoted by the thermal vibration of the atoms in the crystal lattice of the sorbent.

Flow of polymolecular films

In wide pores more complex situations can form in which the fluid with high density and sufficiently strong adsorption may cause the formation of a multilayer film. The flow of polymolecular films is proposed to be considered by analogy with similar flows on exposed surfaces. The theory of flow of polymolecular films, proposed by B.V. Deryagin [42], is based on the concept of disjoining pressure. This theory assumes that the thin layer of a liquid on a solid substrate has the hydrodynamic properties of the fluid volume of the same configuration, with a system pressures equal to $-\Pi(h)$ applied to its surface. Disjoining pressure Π depends only on the layer thickness h. In the presence of a liquid layer of variable thickness (Fig. 3.2) under the applied pressure of the system $-\Pi(h)$ the liquid will move according to the laws of hydrodynamics [48]. Neglecting the inertial terms and assuming a sufficiently slow flow, we can write the equation for the average linear velocity in the form [42]

$$v = \frac{h^2}{3\eta} \frac{d\Pi(h)}{dz} = \frac{h^2}{3\eta} \frac{d\Pi}{dh} \frac{dh}{dz}, \quad (3.14)$$

wherein z is the coordinate along which the thickness of the film varies. Equation (3.14) has the form of Poiseuille' law (3.11), so it can

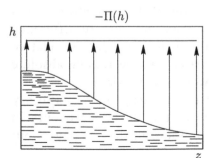

$-\Pi(h)$

Fig. 3.2. Flow of a multilayer film under the influence of the disjoining pressure $\Pi(h)$ according to B.V. Deryagin, where z is the coordinate along which changes the film thickness h [25, 42].

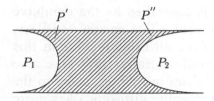

Fig. 3.3. Capillary flow of the liquid adsorbate [38].

be conventionally considered the equation of filtering of polymolecular films, and the value of $K_{film}(h) = h^2/3$ can be called the coefficient of permeability of the film thickness h.

Capillary flow
It is believed that in the capillary condensation region the adsorbed substance may move under the action of capillary forces that occur due to the different curvature of the meniscus [59, 60]. For illustration of this phenomenon Fig. 3.3 shows schematically a pore of the adsorbent at the ends of which the vapour pressure is P_1 and P_2, and inside the pore there is a capillary-condensed matter. In this case, the curvature of the meniscus in the left and right sides of the capillary is associated with relative pressure by the Kelvin equation

$$\ln\left(\frac{P_s}{P_1}\right) = \frac{\sigma}{r_{m_1}\rho}\frac{M}{RT}\cos\theta \text{ and } \ln\left(\frac{P_s}{P_2}\right) = \frac{\sigma}{r_{m_2}\rho}\frac{M}{RT}, \qquad (3.15)$$

where P_s – vapour pressure, r_m – the radius of the meniscus, σ and ρ – density and surface tension of the liquid adsorbate, θ – contact angle.

The internal fluid pressure at the ends of the capillary, P' and P'' is associated with the radius of the meniscus by the Laplace equation $P' = \sigma \cos\theta/r_{m1}$ and $P'' = \sigma \cos\theta/r_{m2}$. Hence we find that

$$\Delta P = P' - P'' = \rho RT/M \ln(P_1/P_2). \qquad (3.16)$$

Due to this pressure difference the capillary condensed substance moves in the pores of the adsorbent. Note that in equation (3.16) there is the pressure ratio P_1/P_2, rather than the pressure difference in the gas phase $\Delta P = P_1 - P_2$. From this it follows that the relative importance of this type of transfer increases with decreasing absolute pressures, i.e. substances having a low vapour tension.

Thus, if the porous body has a pressure drop in the capillary condensation region the viscous flow of the substance occurs in the pores not only due to differential pressure in the gas phase, but also due to the pressure difference in the capillary condensed matter because of different curvatures of the meniscus.

Hysteresis effects

In concluding this section, it should be noted that the hysteresis is observed in very many situations. This fact inevitably follows from the above examined adsorption hysteresis in the channels starting from a certain critical size to the micron size. The observed property of the hysteresis phenomena, for example during impregnation and drainage, is most often a contact angle hysteresis in large volume capillary flow. Direct observation of the contact angle within the porous body is impossible, so the measurement is usually carried out on flat surfaces with the properties identical with those of the inner surface. The advancing contact angle in impregnation is greater than the receding contact angle in drainage. The contact angle hysteresis is associated with the presence of soluble impurities, chemical inhomogeneities roughness on a solid surface.

4. Intermolecular interactions

The existence of dense phases – solids and liquids, as well as the interfaces between any two aggregate states is a consequence of intermolecular interactions. It is their existence that determines the whole set of phenomena in condensed phases, so the discussion of equilibrium and dynamic properties of molecules in dispersed materials is impossible without the involvement of the potential functions of the interaction of particles (atoms and molecules) with each other.

A general view of the dependence of the interaction energy $u(r)$ of two molecules of the distance between them is shown in Fig. 4.1. The interaction energy is equal to zero, if the molecules are removed at an infinite distance from each other; on approach of the molecules $(r \rightarrow 0)$, the interaction energy $u(r)$ increasing dramatically,

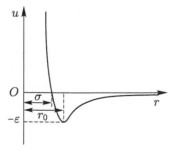

Fig. 4.1. Intermolecular interaction potential.

preventing mutual penetration of molecules with each other. At an intermediate portion of $u(r)$ there is minimum of the potential function in which both molecules have the most stable state. This distance, designated by r_0, corresponds to an energy ε. The minus sign is associated with the point of origin of the energy at an infinitely large distance r. Minimum distance σ to which two molecules can come relative to each other, characterizes their own volume, as they say, the size of the hard-sphere molecule.

This simple picture is valid for inert gases and spherically symmetric molecules. The potential energy of intermolecular interaction of polyatomic molecules depends on the positions of the molecules in space. The study of dense systems raises the question of the form of this functions for the set of a large number of particles, but the basic information about the intermolecular interactions relates to isolated pairs of particles. The potential energy of the interaction of molecules with internal rotation depends on their conformations.

Information about the potentials of intermolecular interaction is mainly derived from experiments with quantum theory [61–68]. However, the main results are related to the interaction potential of an isolated pair of molecules. The experimental information about the potential of the pair interaction is obtained mainly in the study of diluted gases (second virial coefficient and various non-equilibrium properties: viscosity, thermal conductivity, self-diffusion coefficient, molecular beam scattering, etc.). Great opportunities for the study of molecular interactions are provided by spectral methods [62, 67, 68], since these interactions affect the spectra of the molecules. Closely related are studies of intermolecular forces and crystal properties (size and shape of the unit cell, the energy of sublimation, the compressibility of the vibrational spectrum, etc.) [69, 70]. The studies of intermolecular interactions using the data on the structure and thermodynamic properties of liquids and solutions are being carried out.

The theory of chemical bonding shows that the intermolecular interactions are largely electrostatic in nature and their energy depends on the electron density distribution of the interacting molecules. Quantum-mechanical calculations of the interaction potential consist in solving the Schrödinger equation for different relative positions of the molecules and determining the potential energy for each of the considered configurations. The quantum-mechanical potentials are usually functions with a large number of parameters, which limits their use in analytical theories.

Although computer calculations by the Monte Carlo and molecular dynamics allow one to include such potentials, the time required for calculations is extremely long. Therefore, the main value of the quantum mechanical approaches is that they help finding simpler modelling empirical potentials, offering the reasonable form of functional dependences and giving some parameters.

Traditionally, there are the following major contributions to the total interaction between the molecules: 1) exchange (repulsive) interaction that occurs with a marked overlap of the filled electron shells of molecules and rapidly decreases with the increase of intermolecular distances (u_{rep}); 2) direct electrostatic interaction due to the presence in both interacting molecules of constant electric moments – dipole, quadrupole, etc., due to the difference of the charge distribution on an isolated molecule from a spherically symmetric (u_{el}) distributiun; 3) induction (polarization) interaction due to the redistribution of the electron density of the molecule in the field of another molecule having a constant electric moment (u_{ind}); 4) dispersion interaction, explained by the correlation in the instantaneous distribution of the electron density of the molecules; it occurs between any of the molecules, including non-constant electric moments (u_{disp}); 5) interaction caused by charge transfer, i.e. redistribution of the electron density between the interacting molecules ($u_{ch.tr.}$). Therefore, in general,

$$u = u_{rep} + u_{el} + u_{ind} + u_{disp} + u_{ch.tr.} \qquad (4.1)$$

Depending on the nature of the interacting molecules, some of the terms on the right-hand side of equation (4.1) can be zero or close to zero. Direct electrostatic, induction and dispersion interactions are combined in a general notion of the van der Waals, universal, non-specific interactions. In contrast, the interactions associated with charge transfer, are called specific. Specific intermolecular interactions are intermediate between the van der Waals interactions and chemical bonding.

Model potentials of central pair interactions

Molecular and statistical theories operate normally with model potentials that contain a small number of parameters. A detailed analysis of the model potentials is given in the monographs [61–64, 71–73]. We will discuss only the three simplest forms of the potential functions that have received the most widespread use.

1. Square-well potential

$$u(r) = \begin{cases} \infty, r < \sigma, \\ -\varepsilon, \sigma \le r \le \lambda\sigma, \\ 0, r \ge \lambda\sigma \end{cases}, \tag{4.2}$$

where σ is the diameter of a solid sphere; $(\lambda - 1)\sigma$ is the width of the potential well; ε is the minimum binding energy; $\lambda > 1$. Here, the first term transfers the repulsion between two molecules, and the second – their attraction.

2. The Mie power potential $(n - m)$

$$u(r) = br^{-n} - ar^{-m} = \frac{n}{n-m}\left(\frac{n}{m}\right)^{m/(n-m)} \varepsilon\left[\left(\frac{\sigma}{r}\right)^n - \left(\frac{\sigma}{r}\right)^m\right], \tag{4.3}$$

where all parameters are positive, wherein $n > m$. In the second form of writing, the following definitions are used: $u(r_0) = -\varepsilon$, and the concept of the point σ, wherein $u(\sigma) = 0$.

The values n and m usually serve as variable parameters. The most used expression (4.3) for $n = 12$ and $m = 6$ is called the Lennard-Jones (LJ) potential.

3. The Buckingham potential, which consists of three contributions: the repulsive contribution, modelled by an exponent, as well as terms due to the dispersion attraction due to the dipole-dipole interaction (r^{-6}) and the dipole–quadrupole interaction (r^{-8}):

$$u(r) = A \exp(-ar) - Br^{-6} - Cr^{-8}. \tag{4.4}$$

If we neglect the contribution of the third, then so be called a modified Buckingham potential, with three parameters, can be written as

$$u(r) = \frac{\varepsilon}{1-6/s}\left\{\frac{6}{s}\exp\left[\alpha\left(1 - \frac{r}{r_0}\right) - \left(\frac{r}{r_0}\right)^6\right]\right\}. \tag{4.5}$$

As above, the parameters ε and r_0 determine the depth of the potential well and the position of its minimum. The parameter $s = ar_0$ determines the steepness of the repulsive potential.

Atom–atom interaction scheme of the molecules

Flexible enough modelling of the interaction potentials of polyatomic molecules is provided by the atom–atom scheme, in which there are multiple interaction centres in the molecule. These centres can be identified with the atoms (atom–atom scheme) or be pseudo-particles.

The most widespread is the atom–atom potential scheme in which the energy of interaction between the molecules is represented as the sum of pair interaction potentials of the atoms included in these molecules. It is believed that the atom–atom potentials are sufficiently universal and, singling out certain classes of atoms (say, the H atoms, C atoms of alkanes or alkyl radicals), the interaction potentials within these classes are considered equal.

The development of the scheme of atom–atom potentials is largely due to the calculations of the properties of molecular crystals and conformational analysis, where these potentials are used to describe the interactions between the valence unbound atoms of the same molecule [69, 70, 75–79]. These potentials are widely used in studies of adsorption [80] and in numerical calculations of the properties of liquid systems [81, 82].

Atom–atom potentials are expressed in different forms, mostly as Lennard–Jones or Buckingham potentials. Atom–atom potentials with the inclusion of the Coulomb interaction energy of the charges on the atoms [77, 83] or otherwise disposed point charges have been proposed; the induction term is also sometimes considered. There are potentials that take into account the valence state of the atoms. The atom–atom scheme has been developed for compounds forming hydrogen bonds (see [78, 79]), in particular for water. The main features of interactions in this case are again effectively transferred by the combination of the atom–atom scheme with the model of point charges.

Non-pair interaction potential

The most famous of the three-body potential – Axilrod–Teller–Muto potential [84], which contains a term with an additional potential energy of the system of three molecules $(i - j - k)$ – compared with the sum of pair potentials (the term arises due to changes in the distribution of the electron density of the pair of molecules under the influence of a third molecule).

Three-particle and other many-particle terms have virtually no effect on the properties of diluted gases for which the probability of a simultaneous convergence of the three particles is small and where the deviations from the ideal state are almost exclusively determined by the two-particle collisions. However, many-particle terms can make a significant contribution to the properties of dense systems, in particular already in the neighborhood of critical density [85]. Introduction of non-additive corrections can reconcile the calculated properties of crystalline noble gases with the experimental data, which gives reason to draw conclusions about the determining role of the three-particle dispersion dipole interactions [72]. At the same time, the specific physical nature of the appearance of many-particle interactions remains largely unclear.

Combination rules when describing the interaction of different molecules

Interactions between dissimilar non-polar molecules with the symmetry nearly spherical are described by the potentials (4.2)–(4.5) in which the parameters of mixed interactions are substituted. In some cases, these parameters are estimated from experimental data, but most of them are calculated on the basis of the interaction parameters of the same type of molecules. The expressions relating the parameters of mixed and single-type interactions are called the combination rules. The relationships in the form of the geometric mean are approximately valid.

For the potential (12–6)

$$r_{\alpha\beta} = (r_{\alpha\alpha}r_{\beta\beta})^{1/2}, \quad \varepsilon_{\alpha\beta} = (\varepsilon_{\alpha\alpha}\varepsilon_{\beta\beta})^{1/2}. \tag{4.6}$$

The condition for the well depth for $\varepsilon_{\alpha\beta}$ is called the Berthelot rule. More often, however, the values for the minimum of the potential curve energy $r_{\alpha\beta}$ or the hard-sphere of the molecule $\sigma_{\alpha\beta}$ are found using the rule of the arithmetic mean:

$$r_{\alpha\beta} = \frac{1}{2}(r_{\alpha\alpha} + r_{\beta\beta}), \quad \sigma_{\alpha\beta} = \frac{1}{2}(\sigma_{\alpha\alpha} + \sigma_{\beta\beta}). \tag{4.7}$$

With similar values of $r_{\alpha\alpha}$ and $r_{\beta\beta}$ the relatioships (4.6) and (4.7) give approximately the same values of $r_{\alpha\beta}$. The second equation (4.7) is strictly satisfied for hard spheres, it is called the Lorentz rule. The set of the conditions for $\varepsilon_{\alpha\beta}$ (4.6) and $\sigma_{\alpha\beta}$ (4.7) is called the Lorentz–Berthelot combinational rules. Mixtures in which both rules are met are called the Lorentz–Berthelot mixtures.

Interaction of molecules with a solid [86]

The interaction of molecules with a solid body in physical adsorption is a special case of the manifestation of intermolecular forces discussed above.

Important information about the potential of the molecule–solid intermolecular interaction was obtained from the experimental values of Henry's constant K_e [87–92]. Additional information was obtained on the basis of experimental data on the scattering of atomic and molecular beams on solid surfaces [93–95]. In this way it was possible to obtain for a number of systems principally important information about the potential Φ in the potential well region: the dependence of the width averaged over the surface of the potential well of depth, the depth of this well, the values of the virial coefficient C_3, the dipole–dipole molecule–solid dispersion interaction and the change of Φ along the surface. As above, the energy of interaction of the molecule with the solid can be expressed as the sum of contributions of different nature, as in the formula (4.1) [63, 65, 96].

The atom–atom approximation (4.1) has been widely and successfully used to describe the calculation of the intermolecular interaction of structural, thermodynamic and dynamic (spectroscopic) properties of gases, liquids and molecular crystals [65, 69, 97]. Accordingly, it is used to calculate the interaction of molecules with the solid. This should take into account the general problems of application of the atom–atom approximation [65, 98–107].

Effective potentials of the interaction of a molecule with a solid
According to the atom–atom approximation (4.1), in order to calculate the potential energy Φ of the intermolecular interaction of the molecule with a solid body, it is necessary to sum up the atom–atom potentials (φ_{ij}) of the intermolecular interactions of atoms i of the molecule with the atoms of the solid j over all the atoms i and j. These calculations are performed using the lattice sums [108]. The calculation of the lattice sums were carried out in different approximations. In order to obtain accurate values of these sums, it is often necessary to carry out the numerical summation of the corresponding terms over a sufficiently large number N of atoms in a solid closest to the considered point above the surface of the solid. In the case of calculating power sums $\sum r^{-n}$ the contribution of more distant atoms of the solid is determined in several ways. Most often, the contribution of these atoms is determined by integration over the rest of the volume, assuming a continuous uniform distribution of matter in the rest of the volume [108].

Mathematically, the simplest forms of the potential Φ are obtained by replacing the summation of the potential φ_{ij} by integrating it in volume or a single outer atomic layer of the solid. In this case, φ_{ij} is expressed using the Lennard–Jones potential (12–6). In the latter case, the integration of the volume of the solid body gives the so-called potential (9–3). When integrating the same Lennard–Jones potential (12–6) over only one outer nuclear layer of the solid body, we obtain the so-called Φ potential (10–4) (see below). (For more on approximations and possible options for constructing an effective potential function of the interaction of molecules with a solid see also [80, 109]).

The energy parameters of the lattice-gas model are calculated on the basis of information on the atom–atom potentials. The contribution of the lateral interaction is described by pair interaction parameters $\varepsilon_{fg}^{ij}(r)$ of the particles i and j situated on the sites f and g at a distance of the r-th coordination sphere. It is believed that the lattice parameters $\varepsilon_{ij}(r)$, $r \le R$, do not depend on the types of sites, and they are determined by the corresponding distances r between the centres of the sites using the Lennard–Jones potential: $\varepsilon_{ij}(r) = U_{LJ}(\eta_r\lambda)$, where

$$U_{LJ} = 4\varepsilon_{ij}[(\sigma_{ij}/r_{ij})^{12} - (\sigma_{ij}/r_{ij})^6], \qquad (4.8)$$

at $r = 1$ and $\eta_r = 1$ there is a minimum of this potential equal to the potential parameters ε_{ij}, corresponding to the depth of the potential well, σ_{ij} is the distance of convergence of hard spheres of particles i and j.

If you know the distance from the site f of section q to the atoms in a solid, the potential interaction of a particle in the cell can be calculated as the sum of the Lennard–Jones potentials (4.8) reacting with all the atoms of the solid, subject to a combination of the parameters of this potential related to the different components: $\varepsilon_{As} = (\varepsilon_{AA}\varepsilon_{ss})^{1/2}$, $\sigma_{As} = (\sigma_{AA} + \sigma_{ss})/2$ (where the subscript s refers to a surface atom, while the index A – to the adsorbate). If the properties of a solid are given by the density of distribution of particles n_s in the volume of the body, the potential of interaction of the particles with the adsorbed atoms of the solid V_{sol} can be expressed as an integral over the volume of a solid ($\Omega = V_{sol}$)

$$Q_{q,f} = \int_\Omega 4\varepsilon_{As}[(\sigma_{As}/\rho)^{12} - (\sigma_{As}/\rho)^6]n_s dV, \qquad (4.9)$$

where ρ is the distance from the site f to section dV of the solid. For non-uniform surfaces and in the event of any transitional area formed by the intersection of pores of different types, the potential of the interaction of a particle at a site f of section q with the atoms of the solid is calculated numerically using the formula (4.9). Numerical integration is performed over the region Ω, formed by the intersection of a sphere of radius R centred at the site f and the volume of solid V_{sol} (assuming here that the atoms of a solid situated at a distance greater than R do not affect the value of the potential).

In the case of the simple geometry of the pore space, this integral can be taken analytically [109, 110]. Thus, the potential of the interaction of a particle located at a distance ρ over a flat solid surface is

$$Q(\rho) = (2/3)\pi n_s \sigma_{As}^3 \varepsilon_{As}[(2/15)(\sigma_{As}/\rho)^9 - (\sigma_{As}/\rho)^3]. \qquad (4.10)$$

In particular, for the basal face of graphite $n_s = 38.6$ atoms/nm², $\varepsilon_{ss}/k = 28.0$ K, $\sigma_{AA} = \sigma_{ss} = 0.34$ nm, and the distance Δ between graphite layers is 0.335 nm [109, 110].

The dependence of energy $Q(\rho)$ on the distance ρ in equation (4.10) is expressed by the terms (σ_{As}/ρ) in powers of 9 and 3. A similar analytical form of the interaction energy of the molecule with the adsorbent with the contributions of power of 10 and 4 is used to calculate the energy of the molecule in cylindrical and spherical pores with the curvature of the selected layer of the solid adsorbent taken into account of the order of λ. The summation of such deposits gives the total energy of interaction of the molecule with the adsorbent.

Slit-like pore

The slit-like geometry is formed by two infinite plates of the solid, parallel to the plane xy (in the directions x and y periodic boundary conditions are imposed on the system). The potential of interaction of the fluid with the wall was calculated as the sum $U(z) + U(h - z)$, where $U(z)$ is a model potential of the fluid–wall interaction, z is the distance to the wall, h is the width of the pore.

For the slit-like pores we considered two averaged potentials. The first variant of the potential is recorded as an average potential of the type (9–3):

$$U(z) = (3^{3/2}/2)\varepsilon_{ArC}[(\sigma_{ArC}/z)^9 - (\sigma_{ArC}/z)^3]. \qquad (4.11)$$

This potential was used for the adsorption of argon atoms in a carbon pore with the potential parameters $\sigma_{ArC} = (\sigma_{ArAr} + \sigma_{CC})/2 = 0.338$ nm, $\sigma_{CC} = 0.335$ nm, $\varepsilon_{ArC} = 9.2367\ \varepsilon_{ArAr}$.

The second variant of the potential of type (10–4–3) was used to describe the adsorption of methane in slit pores with walls with various types of

Table 1.4. Coefficients a_{jk} of function $f^{(j)}(x)$ [111]

K	a_{4j}	a_{10j}
0	4.71239	7.73126
1	−18.84855	−77.08463
2	57.64824	563.1619
3	−134.9114	−2820.991
4	253.9246	9608.343
5	−378.7086	−21794.03
6	420.1712	32000.29
7	316.4436	−28996.07
8	141.4307	14672.09
9	−27.97551	−3162.542

$$U(z) = 2\pi \rho_{As} \varepsilon_{sf} \sigma_{As}^2 \Delta (2/5(\sigma_{As}/z)^{10} - (\sigma_{As}/z)^4 - (\sigma_{As}^4/(3\Delta(0.61\Delta + z)^3))),$$
$$(4.12)$$

Energy parameter ε_{ss}, describing the effect of the adsorption field, was varied: $\varepsilon_{ss}/k = 64.51$ K (strong adsorption field) and $\varepsilon_{ss}/k = 37.21$ K (weak field). In particular, methane corresponds to the depth of the potential ε and effective diameter σ, equal to $\varepsilon/k = 148.1$ K, and $\sigma = 0.373$ nm, respectively. This potential is an average over the density within a single layer of graphite and its adjustments to the normal to the surface.

Cylindrical channel
Calculations were carried out using the 'spread' potential of interaction of the particle situated at a distance y from the cylindrical wall of the pore of radius R_p [111]:

$$U(y) = -2\pi \varepsilon_{As} n_s \sigma_{AS}^2 \times$$

$$\times \left[\sum_{j=0}^{\infty} \left(\frac{\sigma_{As}}{R_p + j\Delta - y} \right)^4 f^{(4)} \left(\frac{y}{R_p + j\Delta} \right) - \frac{2}{5} \left(\frac{\sigma_{As}}{R_p - y} \right)^{10} f^{(10)} \left(\frac{y}{R_p} \right) \right].$$
$$(4.13)$$

The coefficients of a polynomial function $f^{(j)}(x) = \sum_{k=0}^{9} a_{jk} x^k$ are defined in Table 1.4. At $R_p \to \infty$ this potential becomes the potential for the flat wall [109].

Spherical cavity

To compute the potential inside the spherical shell, formed by atoms of the solid, we use the formula

$$U_{As}^{Rp}(r_A) = -2\pi\varepsilon_{As}n_s\sigma_{As}^2 \frac{1}{x_i}\left\{\left(\frac{\sigma_{As}}{R_p}\right)^4\left[\frac{1}{(1-x_A)^4} - \frac{1}{(1+x_A)^4}\right] - \right.$$

$$\left. -\frac{2}{5}\left(\frac{\sigma_{As}}{R_p}\right)^{10}\left[\frac{1}{(1-x_A)^{10}} - \frac{1}{(1+x_A)^{10}}\right]\right\},$$

$$(4.14)$$

where $x_A = r_A/R_p$ and R_p – the radius of the spherical cavity. In the transition to a multi-layered wall of the spherical pore deposits (4.14) summed $Q(r_A) = \sum U_{As}^{Rp+\Delta n}(r_A)$ by analogy with formula (4.13), where $R_p + \Delta n$, $n = 0, 1, 2...$ is the sequence of the radii of spherical cavities, Δ is the thickness of a single shell.

The specificity of very narrow pores consist of overlapping of the interaction of the potentials between the adjacent side and/or the opposite walls of the pores. Another important factor that determines the behaviour of the molecules near the walls of the pores, is that the surface potential of the type (9–3) or (10–4) changes rapidly over distances comparable to the size of the hard-sphere of the adsorbate σ. Therefore, at the distance of the linear size of the site (the lattice constant) λ there is a great potential gradient $U_{As}(\rho)$.

Transition areas

For the transition region, formed by the intersection of spherical and cylindrical sections, the potential of interaction of a particle situated at a site f of section q with the atoms of the solid is calculated numerically, using the formula (4.8) and integrating over the domain Ω, formed by the intersection of a sphere of radius R centred at the site f and the volume of the solid V_{sol} (assuming here that the atoms of the solid, situated are at a distance greater than R, do not affect the value of the potential). In this case, we use directly formula (4.9) for the fixed structure of the free space of pores.

5. Lattice model of adsorption and the theory of fluid

Lattice models are the first and most common models in the theory of surface phenomena. The most famous Langmuir model [23] is essentially a first statement of the lattice-gas model (LGM), using the concept of free and occupied sites. This model has been used repeatedly for describing equilibrium adsorption and also for calculating the adsorption and desorption rates [112–114]. It was then naturally generalized for the case of nonuniform surfaces and the case of intermolecular interaction of adsorbed particles on a uniform surface [115–118].

Experimental data published in the thirties indicating the proximity of many liquid and crystal properties: relatively small relative changes in volume and energy at melting, similar values of the heat capacity of substances in the liquid and solid states detected by X-ray diffraction study of the short-range order in the liquid, etc. [119] were used for transferring the concept of the lattice models to describe the liquid state. In [120] the authors proposed a model of non-spherical molecules, and in a series of papers [121, 122] a theory that describes all three aggregate states (gas, liquid and solid) was developed.

Lattice models can reflect the most important part of the potential function in the neighborhood of the minimum of the potential curve, and all of the many variants of these models in one way or another are connected with the properties of the intermolecular potential in the vicinity of this minimum. This is most clearly seen in the fact that all the lattice models operate with a number of nearest neighbors z, which is directly related to the local structure of the liquid. If the number z for the adsorbed particles is determined by the potential relief of the substrate, then for the liquid medium this number is given by an average statistical characteristic that is associated with the cooperative behaviour of a large ensemble of molecules and the properties of the intermolecular interactions.

The central feature of lattice models is that the total volume of the system V is divided into some elementary volume v_0, which form a periodic lattice. The statistical problem of the spatial distribution of molecules is greatly simplified by introducing assumptions about the spatial regularity of the distribution of molecules. This replaces the continuum description of the distribution of molecules by the discrete one. Additionally introducing a number of assumptions, we can simplify the task of calculating the integral configuration of the system, taking into account the intermolecular interactions, that this will make possible its analytical solution.

For a system of many interacting bodies such opportunity is quite unique. It is for this kind of discrete problem that statements were received for a few exact solutions within the framework of the multidimensional structures [123]. One-dimensional structures always permint an exact solution [124, 125]. This was first found in Ising's study [125], and then this kind of model systems became known as the Ising models. In some situations, the one-dimensional models proved to be useful in many applications of the adsorption of molecules of faces, the base of steps stepped surfaces for linear polymers, etc.

For the first time the exact solution was obtained only for the two-dimensional Ising model in the absence of an external field. Onsager [126] studied the rectangular lattice; later solutions were found for other types of two-dimensional lattice [123]. For the two-dimensional Ising model in a non-zero field and three-dimensional model the results that can be called accurate are available only in the form of expansions with respect to density and temperature [127, 128]. Approximate statistics methods of lattices were developed in the works of Kramers and Wannier, Kikuchi and others (see, e.g., [121]).

Exact solutions for the two-dimensional case and almost exact solutions for three-dimensional systems make the Ising model (and the 'Ising' lattice gas) the most important modelling basis for the study of phase transitions and critical phenomena. Significant advances in the theory of phase transitions are associated with the consideration of the generalized Ising model in which the variable, describing the state of the site, receives more than two values and more (particularly when there is a continuous range of values).

Since their emergence the lattice theories of liquids and solutions have come a long way of development. The essence of such modifications is related to the specific properties of real fluids. In theory the Mie and the Lennard–Jones potential are used in most cases. The parameters of the potential are considered either strictly fixed, or used for the convergence of the calculated and experimental characteristics. In the latter case, there are semi-phenomenological energy parameters, the determination of which is based on experimental thermodynamic data for the system under study.

The thermodynamic properties of the melts are described by the model of the 'surrounded atom' [129, 130], which differs from the strictly regular solution method by the method for assessing the energy systems, as well as by the fact that it is possible to take into account the dependence of the vibrational statistical sum of the atom on the environment, including many-particle non-additive interactions.

The first molecular cellular theory of fluids, which solved the problem of computing energy and free space on the basis of the pair interaction potential, was the theory of Lennard–Jones and Devonshire (LJD) [121]. It is based on the assumption of a certain geometry of the lattice (usually a face-centreed lattice, in which the cell has the shape of a dodecahedron), and that the potential in the cell is created by environment molecules equally 'smeared' in the coordination sphere. The calculations were performed for the pair potential (6–12) taking into account the interaction of the central molecule with molecules of one coordination sphere and three spheres. According to the LDJ theory, the system should be governed by the law of corresponding states. Other examples of the use of harmonic and anharmonic vibrations of molecules in cell models of the liquid are given in [131].

States of the cells
Most often it is assumed that each cell contains a single particle; some variants of models take into account the possibility of finding several molecules in the cell. Attempts have been made to bring the cell models closer to real fluids, assuming that the cells in the system are of different sizes and arranged in an irregular manner [132, 133]. Models with irregular cell structures are primarily used for the calculation of the radial distribution function; good results are also obtained for the thermodynamic functions. It was assumed that the cells may contain one or more molecules and the concept of the free cells is not available.

A special situation forms when the cell is permitted either to contain a single particle or be empty – the so-called lattice-gas model, which 'came' from the theory of adsorption. It is important to clearly separate the lattice gas model from other so-called lattice models. The hole models 'automatically' allow to take into account the irregular arrangement of molecules and do not require the artificial introduction of the concept of collective entropy. These models are applicable not only to liquids but also to the gas; they are capable of providing an equation of state describing both phases and simultaneously providing the liquid–vapour transition. Note that the hole models, the lattice gas model and the model of irregular cell structures [134] allow to describe not only the liquid–vapour phase transition (this requires the presence of attraction between molecules), but also the liquid–solid phase transition, i.e., the phase transition of the melting type [135–137].

The lattice-gas model (LGM) corresponds to three situations [25]: adsorption and absorption of molecules of the gas phase, and also solid solutions. If we confine ourselves to the spherical molecules of one type, the size of the cell v_0 is directly related to the size of the molecule, which is specified by the parameter σ in the Lennard–Jones potential function (LJ). This parameter characterizes the incompressible solid sphere of the molecule, the nature of which follows from the exchange interaction of the electron shells of the two molecules as they approach, which leads to repulsion of molecules. This choice of values $v_0 = \lambda^3$ imposes a condition of single filling of the volume of the cell when a molecule is placed in it. Given the explicit form of the LJ potential, we have that $\lambda = 2^{1/2}\sigma \approx 1.12\sigma$.

The ratio $M = V/v_0$ determines the number of cells in the system: it may be physically a small element of the bulk phase, $M \gg 1$; monolayer area $V_1 = \lambda S$ of the section of the flat surface, $M = V_1/v_0 = S/\lambda^2$, where S is the area of the surface; the near surface boundary $V_h = \lambda hS$, solid – vapour or liquid with width h of the monolayers, which operates the surface potential of a solid wall and forming a variable profile of the fluid as the distance from the surface plane, $M = V_p/v_0 = hS/\lambda^2$; V_p is the volume of the pore (see section 1), filled with the adsorbate. We believe that the total number of cavities of the given geometry is large enough and can be described stastically, ignoring the possibility of a significant contribution of fluctuations. Then the number of cells attaching to the considered pore may be small, ranging from 10, for example, in the case of zeolites to $< 10^3$.

Fixation of the molecule in the centre of the cell corresponds to the state of its occupation. Mathematically, this event is described by γ_f^i, where f – cell number, $1 \leq f \leq M$, subscript i denotes the state of occupation of the cell with number f. If the site f contains an adsorbed particle A, then $\gamma_f^A = 1$. If the cell f is free, then there is a vacancy, so $\gamma_f^A = 0$, and $\gamma_f^V = 1$. For the two types of conditions of occupation $s = 2$ of any site of the lattice structure corresponding to the one-component system for which $i = $ A or V – vacancies, the random variables γ_f^i are subject to the following relations:

$$\sum_{i=1}^{s} \gamma_f^i = 1, \text{ and } \gamma_f^i \gamma_f^i = \Delta_{ij} \gamma_f^i, \tag{5.1}$$

where Δ_{ij} is the Kronecker symbol, which means that any site is unavoidably occupied by some, but only one, particle. These conditions imply the absence of multiple filling of any cell. In particular, the equality $(\gamma_f^i)^k = \gamma_f^i$ of any integer $k > 1$ is fulfilled. Averaging over the states of employment/// of the cell determines the degrees $\theta_i = \langle \gamma_f^i \rangle$ of filling them with particles of kind i.

Approximate methods
In the overwhelming number of situations where there are no exact solutions of the problem, we have to use approximate solutions, which are also quite important for understanding the behaviour of cooperative systems. The question of the nature and accuracy of approximate calculations occupies the central part of the theory of condensed phases [138].

The theory of discrete distribution functions uses different ways of approximating the probability of many-particle configurations through the probabilities of configurations of smaller dimension. For simplicity, we consider a one-component system sites of which are occupied by particles A (we assume that it is adsorbed particles A) or V (vacancies). The full range of configurations of particles A in the first coordination sphere around the central (not shown) particle

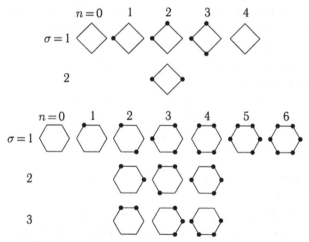

Fig. 5.1. Configurations of particles A in the first coordination sphere around the central particle of lattices $z = 4$ and 6, blackened circles – adsorbed particles A; n – the number of particles A, $\sigma(n)$ – the type of configuration for a fixed value of n. $R = 1$ [138].

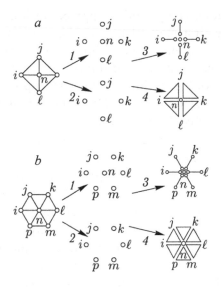

Fig. 5.2. Schemes of approximate calculations of many-particle configurations of neighbouring molecules for two-dimensional lattices $z = 4$ (*a*) and 6 (*b*): *1* – mean-field and random approximation (via the concentrations of all molecules), *2* – polynomial approximation (via the concentration of neighbouring molecules), *3* – quasichemical approximation (via the probabilities of pairs of molecules), *4* – consideration of indirect correlations in the quasichemical approximation and taking into account the triple correlations (in terms of probabilities of triples of molecules) [149].

on the lattices $z = 4$ and 6, the neighbouring particles A – blackened circles, the remaining sites of the first coordination sphere are free (or occupied by the particle V). The symbol $\theta_i(n\sigma)$ denotes the probability that the first coordination sphere of the central particle of species *i*, $i =$ A and V, contains *n* particles A with the location σ (Fig. 5.1).

Examples of such approximations are shown in Fig. 5.2 for planar lattices with $z = 4$ (*a*) and 6 (*b*). The first version of the approximation corresponds to the mean field approximation, where the probability of the many-particle configuration is expressed as the product of the probabilities of sites occupied by its particles θ_i of class *i* (or the unary distribution functions), $\theta_i(n\sigma) = \theta_i \theta_A^n \theta_V^{z-n}$, which are connected by the normalization conditions $\theta_A + \theta_V = 1$. Obviously, the method of positioning the particles, fixed by the symbol σ, does not play a role. In this approximation there is no correlation.

The second approximation variant is also a single-particle approximation – in it the many-body configuration of neighbours around the central particle is approximated by the product of unary distribution functions (polynomial approximation). The difference compared with the first method is that the correlation between the central particle and its nearest neighbour is preserved.

The third version of the approximation is that the probabilities of many-particle configurations $\theta_i(n\sigma)$ are approximated by the local probability θ_i of detection of particles i (unary distribution function), and pair distribution functions $\theta_{ij} = \langle \gamma_f^i \gamma_g^j \rangle$. This is the so-called Guggenheim quasichemical approximation [139] or the Bethe–Peierls approximation [140, 141]. Here, the correlation effect is explicitly taken into account through the pair distribution function θ_{ij}, which characterizes the probability of finding particles of type ij, ij = A, V at the adjacent sites. In this approximation we have $\theta_i(n\sigma) = \theta_i t_{iA}^n t_{iV}^{z-n}$, where $t_{ij} = \theta_{ij}/\theta_i$ is the conditional probability of finding the particle j close to the particle i; and $\theta_{iA} + \theta_{iV} = \theta_i$ or $t_{iA} + t_{iV} = 1$. Note that in the quasichemical approximation the weights of the configurations with different values of σ (with n = const) are equally probable, or symbol σ does not affect the calculation of $\theta_i(n\sigma)$.

The fourth version of the approximation is the simplest example of the correlation including the effects of three molecules. This approximation is a discrete version of the so-called Kirkwood superposition approximation [142], in which for the three particles that are separated by any distance we introduce the approximation of the probability of simultaneous realization of triplets of particles θ_{ijk} of type ijk in the form $\theta_{ijk} = \theta_{ij}\theta_{kj}\theta_{ki}/(\theta_i\theta_j\theta_k)$ through the probability of pair and unary distribution functions. As part of such an approximation the symbol σ defines the contributions from pairs of particles at different distances. When using the superposition approximation the problem is closed through the equation on the unary and pair distribution functions.

The structure of the fourth approximation allows to go beyond just the pair approximation, if we assume that for the probability of triple configurations we construct their additional equations. However, this method greatly increases the dimension of the system of equations that must be solved to find the equilibrium distribution of the particles.

Figure 5.3 shows the dependence of the pair correlation function of distribution $\theta_{AA}(T | \theta_A = 0.5)$ for the nearest neighbours on the temperature at the half-filled surface $\theta_A = 0.5$. Only for this filling we obtain an exact solution in the whole temperature range. Curve *1* corresponds to the exact solution, provided that the parameter

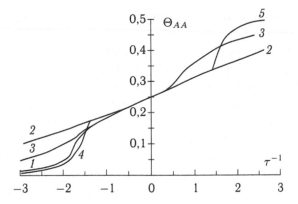

Fig. 5.3. Temperature dependence of the pair correlation function in various approximations for $\varepsilon < 0$: the exact solution (*1*), the quasichemical approximation (*2*), the superposition approximation (*3*) and the quasichemical approximation with the ordering of particles taken into account (*4*). Curve *5* corresponds to the case $\varepsilon > 0$ in the quasichemical approximation with the two-dimensional condensation of particles [142].

of interaction between two particles corresponds to their repulsion ($\varepsilon < 0$, ε is the parameter of interaction between particles).

The other curves correspond to those approximations: quasichemical (*2*), superposition (*3*), the quasichemical approximation with an additional view of the effect of ordering in the system due to the repulsion between the particles (*4*), and, finally, curve *5* corresponds to attraction between the particles ($\varepsilon > 0$) in the quasichemical approximation with an additional view of separation of molecules into two coexisting phases (sparse and dense), according to the lever rule.

The curve *4* is close enough to the curve *1*. The maximum difference between them is close to the critical point. For curves *4* and *5* taking into account the phase state of the particles strongly deflects the course of the curve for the pair distribution function $\theta_{AA}(T\,|\,\theta_A = 0.5)$ from the classical behaviour of the same function (curve *2*).

Generalizations
Currently, the Ising model and its extensions have been studied in fully and accurately over a wide temperature range (values of the given parameter ε/kT) for various two-dimensional and three-dimensional lattices. For the Ising model methods were developed to take into account for not only the pair but also many-body

interactions, methods of accounting interactions with the non-nearest neighbors [143].

The fundamental point of using discrete statistics is the possibility of reducing the problem not only to discrete cells, but also to the discrete states of their occupation. In particular, we consider discrete molecular orientation at different angles in the external field. The statistics methods of the Ising model, in particular the approximate (Bragg–Williams, Bethe–Guggenheim, Kramers, Vanier, Kikuchi, et al) have been extended to multi-component systems, systems with off-centre interactions, systems containing molecules of different sizes, etc. In some cases, we can directly use the expressions for the thermodynamic functions of the Ising model, as some of the more complex models can be reduced formally to the Ising model (for the transformations of the Ising model, see, e.g. [144–146]).

Finally, all these models were used in the theory of adsorption on homogeneous surfaces. For the theory of adsorption the situation is to a great extent more difficult because of the heterogeneity of real surfaces. Methods for the approximate calculation of the equilibrium adsorption characteristics of the system have been applied only at the end of the seventies, when both the intermolecular interactions, and the state of the nonuniform surface were taken into account for the first time (see chapter 2).

On the problem of justification of lattice models of liquids

The physical meaning of lattice models for the liquid phase remained unclear for a long time – their justification for the adsorption and absorption was not called into question, since the cells are formed by the potential relief created by the atoms of the solid.

This question was investigated using the cluster approach, which allowed to prove (Appendix 1), that the lattice models of a fluid have a strict statistical justification and are directly related by the functions of molecular distributions, which are described by the BBGKY equations. Traditional lattice models are the first step in a sequence of partitions of the space into elementary volumes, which are close to the limit of the continuum description. This process provides an arbitrarily close approach to the continuum description.

The first approximation is sufficient to obtain the thermodynamic properties of simple liquids and mixtures thereof, but for describing the structural characteristics we require subsequent approximations that can take into account not only the average density of the system, but also local density inhomogeneities. The continuum description is

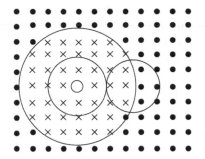

Fig. 5.4. Spherical particles that block many fine lattice sites: ○ – space occupied by the centre of the molecule; × – banned places; ● – allowed places.

approached by increasing the number of sites occupied by a molecule in the lattice, which is equivalent to a finer division into cells for a molecule of a fixed size. The molecules of the environment have access not only to the site occupied by the centre of the molecule, but also the sites of one or more coordination spheres (lattice gas of molecules with hard cores) – Fig. 5.4 [68, 132, 147, 148].

Proof of the connection of the LGM (lattice-gas model) and BBGKY (Bogolyubov–Born–Green–Kirkwood–Yvon) chain allows us to consider the lattice model as a way of renumbering the configurations, and the model approaches the continuum to the extent as the cell size decreases, i.e. increasing the number of sites held by the molecule increases. For lattice systems with molecules occupying several places, a general procedure of their statistical analysis based on the cluster approach was proposed [149]. Therefore, the partition into cells in the LGM does not impose any restrictions on the movement of the particles, the latter can be moved over the entire volume. They take into account not only the 'standard' positions of the molecules in the centre of the cell, but also the displacement of the particles from these positions that allow for the structural characteristics of the liquid.

Multicentre particles
Statistics of lattice systems with off-centre interactions is a special task not confined to the statistics of the Ising model. A particular challenge here is the statistics of systems of molecules occupying several lattice sites. This problem exists both for liquid solutions and for the adsorption of mixtures of molecules of different size. The geometric characteristics of the molecule, reflected by the repulsive potential, are transmitted by the lattice model so that the molecule is given a number of lattice sites, the location of which is

consistent with the shape of the molecule and its flexibility. A system comprising spherical particles of different size (or large spherical particles and vacancies) can be described by the lattice-gas model with excluded neighbours. The molecule–chain is a linear sequence of segments, each of which occupies one unit. A molecule consisting of r segments is called the r-mer, a molecule with one segment occupies one site is a monomer. In the case of molecules – the rigid rods all occupied sites lie on one line, if the r-mer is flexible, its segment with number $i + 1$ can take any of the sites adjacent to the i-th segment, except the sites occupied by previous segments (when $i \sim 2$ there is at least one such site). The limited flexibility of the chain can be expressed that sites, available for the next segment, are selected taking into account the angles between the latter links. Barring a site to be occupied by more than one segment, the model takes into account the repulsive potential. Counting the number of ways of placing the molecules on the lattice for their given geometrical characteristics, the evaluation of the relevant combinatorial entropy is considered a specific statistical problem. The forces of attraction can be considered the same way as it is done in the previously considered theories of systems of small spherical molecules.

For large values of the number of monomers the discussed model simulate polymer systems (melts and solutions). Discussion of the results this work is contained in the books [150–152] and a number of other books on the solutions. The approach in [153–156] should be noted: the calculations yielded good results not only for the low molecular but also for the high-molecular substances. In particular, it was possible to convey solubility diagrams of polymers with a lower critical point. Modifications were developed of the models when considering the final value of the coordination number z [157] and in the quasichemical approximation [158–160].

6. The heterogeneity of the surface of solids

In section 1 the spatial organization of the pore volume was discussed. Real porous solids are characterized by a polydispersed structure, which creates a strong heterogeneity of the properties of the solid at the level of the characteristic pore size. Another type of heterogeneity of porous materials is the heterogeneity of the surface of the pore walls at the atomic level. This kind of heterogeneity is well known in surface phenomena in the absence of pores in solids [118, 116–

118, 125, 161–164]. The presence of pores usually does not exclude the heterogeneity external and internal surfaces. The emphasis on the particular type of heterogeneity taken into account is associated with the role of each contribution to the overall cumulative effect. Moreover, with the decline of the characteristic size of the pores the value of the contributions from the near-wall region should naturally increase and the role of heterogeneity will increase.

The above discussed equations of the Langmuir adsorption isotherms (2.3) and BET (2.5) were obtained on the assumption of equivalence of all the centres. In reality, this assumption is rarely performed. The study of nonuniform surfaces is associated with the development of two parallel directions: clarification of the nature of adsorption sites and the methods of calculating the adsorption characteristics in order to describe the observed experimental adsorption data.

Below is a brief illustration of the various nature of adsorption sites on the surfaces, which together form the heterogeneity of real solids [165]. In general, the frequency of disturbance of periodicity within the bulk crystalline lattice causes a similar distortion of the surface structures. In addition to these factors, new types of adsorption centres form on the surface associated with the method of forming the surface (e.g., the angle at which the crystal is cleaved), and the availability of new types of bonds between the surface and the adsorbed molecules.

Adsorption capacity
The surface potential of any neighborhood centre of adsorption determines the local adsorption capacity of the surface. Calculations of adsorption capacities have been actively carried out with the advent of computers, and now they are the basis of all modern studies of surface processes (see section 4). It should be noted that

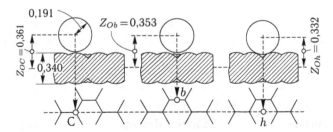

Fig. 6.1. The different positions of the adsorbed argon atom on the surface of graphite, all distances are in nanometers (from [166]).

Table 1.5. The calculated values of U_m^0 (kJ/mol) for the molecules of inert gases in three different positions on the graphite surface [166]

Gas	Position c (above C atom)	Position b (between two C atoms)	Position h (in centre of the C ring)	Average
Neon	3.22	3.51	4.64	3.81
Argon	8.16	8.66	11.05	9.29
Krypton	10.84	11.63	14.52	12.34

Table 1.6. The values of U_m^0 (kJ/mol) for the inert gas adsorbed on the (100) face of solid argon [167]

Type of calculation	Argon	Neon	Krypton
Excluding relaxation	5.67	2.48	6.60
Given relaxation	5.72	2.87	7.04

even for the simplest adsorption systems, such as the monatomic gas on the ideal graphite surface, the potential depends on the position of the adsorbate on the surface. As an example, Fig. 6.1 shows the typical position of neon, argon, krypton, in three different positions on the graphite surface [166]. Their results are shown in Table 1.5. The value of the local minimum of the potential is indicated as U_m^0. Table 1.5 illustrates two important points: 1) the growth of U_m^0 with increasing size of the adsorbed molecules, and 2) a significant dependence of U_m^0 on the position of the molecule on the surface.

In calculating [166] no account is also made of the possibility of secondary surface relaxation caused by adsorption of the inert gas atoms. Assessment of the impact of surface relaxation of solid argon in [167] showed (see Table 1.6), that although the change of U_m^0 is quite noticeable, the overall pattern for the adsorption of argon, neon and krypton is maintained. This example shows that even the

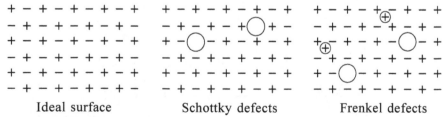

| Ideal surface | Schottky defects | Frenkel defects |

Fig. 6.2. Model of the surface of an ionic crystal 1–1 in the ideal condition, containing Schottky and Frenkel defects (top view) [165].

uniform crystal face may contain locally different positions of the adsorbed particle.

Lattice defects
Consider the surfaces of ionic solids that may have the same type of structural defects as the crystal volume. Then there must be surface analogues of Frenkel and Schottky defects, as shown in Fig. 6.2. We can assume that the Schottky defects are obtained on a perfect crystal surface with vacancies in the lattice sites, whereas Frenkel defects occur as a result of the displacement of some ions (usually cations because of their small size) into the interstices next to the lattice vacancies. If the dimensions of the anions and cations are the same, Schottky defects predominate in the crystal; if the cations are smaller than the anions the Frenkel defects form because the small ions may reside in the interstices.

Instead of vacancies, the lattice sites may contain impurity atoms or ions. The admixture of an extraneous ion in the crystal lattice site is always a cause of confusion even if the ion has the same charge as the ion of the lattice. As equally charged ions, such as various metals, differ from each other usually in the dimensions, polarizability, etc., they have different interaction energies in the lattice. Thus, the replacement of Al^{3+} on Fe^{3+} in the Al_2O_3 crystal will lead to some relaxation of the crystal; the same is true in the case of substitution of the ion on the surface. There is evidence that the concentration of these defects increases at the surface. If the valence of the foreign ion is different from the valence of the ion lattice, then the situation is very similar to non-stoichiometric crystals. Figure 6.3 illustrates the case where the impurity oxides L_2O and R_2O_3 in the oxide of the divalent metal MO (L and R- are one- and triply-charged metal ions, respectively) generate in the first case anionic and in the second

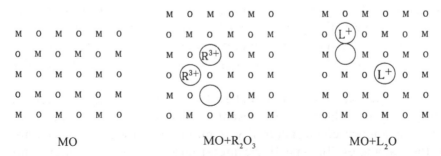

Fig. 6.3. Arrangement of impurities of L_2O monovalent metal oxide and an oxide of trivalent metal oxide R_2O_3 in the structure of the oxide of divalent metal MO [165].

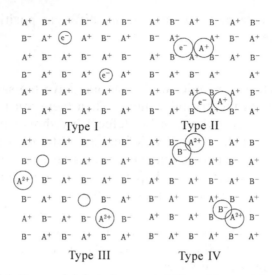

Fig. 6.4. Types of defects in non-stoichiometric crystals [165, 168].

case cationic vacancies. If the metal M may be in a higher oxidation state, for instance M^{3+}, then when adding L_2O the ionic vacancy of the oxide may be filled and the electroneutrality is maintained with the proviso that two M^{2+} ions pass into M^{3+}.

Similarly, when adding R_2O_3 a metal ion vacancy can be filled with the M^{2+} ion, if two M^{2+} ions are restored to M^+ or if two ions M^{3+} existing in the crystal recover to M^{2+}. Therefore, the substitution of low-charge metal ions leads to an increase in the number of positive holes, and highly charged metal ions – to reduce them.

An important role in the surface properties is played by the nonstoichiometry of the solid. Nonstoichiometry is a very common phenomenon, especially among metal oxides. For example, titanium oxide (IV) is often found in oxygen-depleted forms; it was found that the formula for the two forms – anatase and rutile – can be written as $TiO_{1.9-2.0}$. Schottky defects should be in the bulk of the crystal and, very likely, on the surface or very close to it in order to locally modify the surface energy. Thus, the surface properties of most samples of TiO_2, say rutile, depend on the presence of Schottky defects.

Four types of non-stoichiometric lattices are usually considered [168]. In type I, the metal is present in abundance at the expense of anion vacancies; electroneutrality is maintained by electron capture by the vacancies. In type II the metal surplus is due to interstitial metal ions; electroneutrality is maintained by the capture of electrons in the vicinity of the metal ions. This is obviously the same as Frenkel defects, but only vacancies of metal ions are absent. Type

III is characterized by the fact that the metal deficiency is caused by cationic vacancies; electroneutrality is maintained by the transition of certain metal ions to a higher oxidation state. In type IV the metal deficiency is caused by interstitial anions; electroneutrality is also preserved due to the transition of some metal ions to a higher oxidation state. All of these types are shown in Fig. 6.4.

The presence of some of these types of defects in the surface layers may cause local changes in the surface energy, pertaining to nonuniform catalysis.

Surface group

Various chemical groups formed by chemical interaction with components of the ambient gas phase can often be found on the surface. The best-studied examples are metal oxides [169], silica [170], and coal [171]. Rutile is a typical example of the behaviour of metal oxides. Changes in the spectrum of rutile with heating and evacuation have been explained in [172] in terms of bridging and terminal hydroxyl groups, formed by the dissociation of adsorbed water on the (110) face (see also [173–175]):

Silica particles can be represented as a polymer of silicic acid [176], which has on the surface either siloxane groups $\equiv Si-O-Si \equiv$ with oxygen on the surface, or one or more silanol groups $\equiv Si-OH$, formed as a result of the interaction with water. One can define the following types of groups: *isolated* – a surface Si atom is connected by a single bond with the –OH group, which has a hydrogen bond with another silanol group, *vicinal* – two single silanol groups bonded to different atoms approach each other so close as to form a hydrogen bond; *geminal* – two OH⁻ groups bonded with a single Si atom, form hydrogen bonds with each other, leaving for the Si atom only two bonds due to the bulk structure. The current state analysis of surface states for the case of study of the structure and reactivity of the Si surface is given in [177].

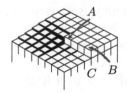

Fig. 6.5. Diagram of a screw dislocation according to Frank [180].

In general, from these studies it follows that we cannot always assume that the chemical composition and structure of the solid surface is similar to its composition and structure in the bulk.

The same statement about the differences of the composition of the surface and in the depth of the solid body is typical for all metal alloys. This led to the introduction of a special term 'surface segregation' of components in alloys.

Dislocations and growth of the crystal

Dislocations are caused by mecanical effects or in atomic and molecular processes of nucleation and crystal growth, if a solid is obtained by precipitation from a solution. There are two main types of dislocations: edge and screw. Edge dislocations appear on the surface as steps with a high local surface energy before the external pressure will restore order. Another common type of dislocation – a screw dislocation – is shown in Fig. 6.5, where each cube is a lattice site or atom [178]. Steps on the surface are formed by the dislocations are responsible for the growth of the crystal faces [178]. This process explains [179] the experimentally observed high rate of growth of close-packed faces at low supersaturation and predicts spirals on the surface of the crystal.

The dislocation density is usually expressed by the number of dislocation lines crossing the unit area in the crystal. This number ranges from 10^{12} to 1 m^2 for a 'good crystal' to 10^{16} per 1 m^2. This means that the distance between dislocations is $10-10^3$ nm, or in other words, each crystal larger than 10 nm contains on its surface dislocations, and one surface atom from a thousand atoms is situated near the dislocation.

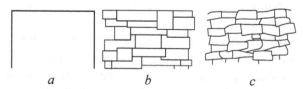

a   *b*   *c*

Fig. 6.6. Types of grain boundary (according to Lonsdale [181]): *a* – a perfect crystal; *b* – a mosaic of parallel crystallites, *c* – a mosaic of disoriented crystallites.

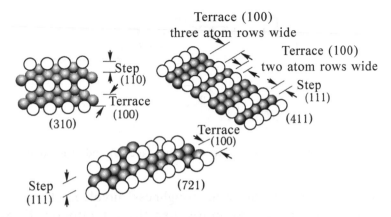

Fig. 6.7. Some of the planes of the face-centred cubic lattice formed by terraces of symmetry {111} and monoatomic steps [162, 183].

The structure around the dislocation line is in a state of stress, as a result the chemical potential of a substance is increased in this area. If the increase is large enough, a depression may appear around the exist of the dislocation to the surface [180]; the depression is also a part of the non-uniform surface.

Many real crystals are mosaics in which crystallites can be arranged with different degrees of order, as shown in Fig. 6.6 [181]. There are four main types of grain boundaries [182]: 1) low ($\varphi < 1°$) angles, consisting of parallel uncontracted edge dislocations; 2) with medium ($5 < \varphi < 200°$) symmetrical angles, consisting of parallel warped edge dislocations; 3) asymmetric ($\varphi < 200°$), including more than one family of dislocations; 4) non-coherent with large angles ($\varphi \sim 300°$), where there is no continuity along the rows of the lattice. Among these crystal mosaics the most common and stable is type 3, while the types 1 and 4 are very mobile.

Surface roughness
The exit areas dislocations are the simplest example of surface roughness (or structural heterogeneity). This term is used differently depending on the characteristic scale used in the discussion of the state of the surface structure. A wide variety of structural heterogeneities is realized in the case of crystals from different angles. As a result, the surface of the crystals contains identical adsorption sites on the terraces and a dramatic change in the local properties of atoms with dangling bonds, forming steps. Figure 6.7 shows a number of examples of planes for a face-centreed cubic lattice formed by terraces of symmetry {111} and monoatomic steps

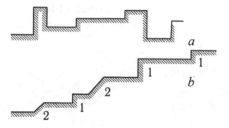

Fig. 6.8. Cleavage profiles parallel to the cubic face (*a*) and at a small angle to the cubic face (*b*) [165].

[162, 183]. In general, the term 'roughness' involves changing the geometry of the surface by more than the interatomic distance. Imagine, for example, a cubic crystal of sodium chloride, cleaved parallel to the face. The profile of the generated surface is shown in Fig. 6.8 *a*, and where planes are represented by the cube faces. However, if the same crystal is cut at a small angle to the reference edge, its surface resembles a ladder (Fig. 6.8 *b*). Although the upper face of each step is the same cubic face as in Fig. 6.8 *a*, the transition from step to step can be carried out on the same faces of the crystal (*1*) and through the other crystal planes (*2*). As the packing density in different crystal planes varies, their relaxing states on the surface will differ in surface energy.

Generally, flat surface irregularities and surface regions of the crystal may be associated with the disruption of the regularity of the distribution of solid surface atoms (structural heterogeneity) and the difference in the nature of surface atoms (chemical homogeneity). As a good example of nonuniform systems we consider real crystals with defects. Imagine that such a crystal is cut by a plane. The

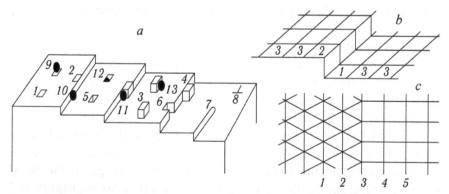

Fig. 6.9. Typical defects and adsorption sites on the crystal surface [161] (*a*). Monatomic step (*b*) and the boundary of faces (111) and (110) (*c*),

resultant surfaces will contain traces of grain boundaries, the exit point of edge and screw dislocations, vacancies and interstitials. Typical defects and adsorption centres on the surface of the crystal are shown in Fig. 6.9 [161].

Crystal molecules (A), as shown in cubes in a simple cubic lattice, can be found on the surface (1), the step (2), fracture of the step (k), in the auto-adsorbed condition (3) or adjacent to the step (4). Removal of molecules from the surface leaves the vacancy or hole in the surface (5), removal of molecules from the step leads to the formation of a vacancy in the step (6). The exit of a screw dislocation to the surface (7) is the beginning of a step, which extends to the edge of the crystal, or ends at the point of exit another, oppositely directed screw dislocation. Closing of the edge dislocation is shown by the conditional symbol (8). Adsorbate molecules (B), shown in the form of spheres, can be adsorbed on centres located on the surface (9), at steps (10), breaks of steps (11), in holes (12) and at centres alongside the auto-adsorbed molecules (13). On the right there are model schemes used below for the construction of the distribution functions: monatomic step (top) and the boundary of the (111) and (110) faces that describe the types of nonuniform systems.

The presence of the above-mentioned point heterogenities contributes to the fact that in the vicinity of each defect there is a change of the properties of the surrounding atoms, and it leads to modification of the properties of the adjacent sites of the surface structure, thereby increasing the number of types of surface sites.

An example of chemical heterogeneity of the surface can serve as a binary alloy $A_x B_{1-x}$ (for example, Pd–Ag), having unlimited solubility. The structurally homogeneous surface of the face, say (100), of this alloy is nonuniform due to the different adsorption capacity of different atoms A and B. The share of each component on the surface, depending on the temperature and concentration of particles in the volume, can vary from zero to unity. In general, the chemical and structural heterogeneities form simultaneously.

Statistical description of nonuniform surfaces
The presence of surface heterogeneities is evident in the curves of the isotherms and the heats of adsorption. The equation for the Langmuir isotherm (2.3) does not provide a quantitative description of many experimental data.

The basis for the consideration of the energy heterogeneity is a statistical approach, i.e., accounting and summation of laws for a very large number of different areas of the surface. The theory of processes on nonuniform surfaces is considering a solid surface consisting of a finite number of elementary sites (centres), the adsorption capacity of which in the general case is different [112–118]. In this surface–adsorbate system each site can be characterized by the appropriate value of the free energy of adsorption. This value is usually changed when moving from one place to another. The theory assumes that the various elementary sites of the nonuniform surface are characterized by finite values of the heat of adsorption from Q_{max} to Q_{min}. This corresponds to the change of the adsorption coefficient from a_{max} to a_{min}.

The number of elementary sites on the surface is very large (the order of 10^{15} on 1 m^2); among them are the groups of sites on which the value of a changes by no more than an infinitesimal amount. These $q–e$ sites or groups of sites are characterized by the values of the heat of adsorption from Q_q to $Q_q + dQ_q$, as well as the values of a_q: $a_q = a_q^0 \exp(\beta Q_q)$.

In the absence of any interference between the adsorbed particles, the adsorption equilibrium for the given group of sites must meet the Langmuir law. Consequently, filling θ_q of the given site or the degree of coverage of the given group of sites for the particular value of P for adsorption without dissociation is expressed by the $\theta_q = a_q P/(1 + a_q P)$.

The first consideration of adsorption equilibrium on nonuniform surfaces was published by Langmuir [115], who used the non-normalized contributions of different areas. For the 'crystalline' surface with a small number of different groups of sites and 'amorphous'surfaces with a large number of different areas, characterized by continuous change of adsorption capacity, he expressed the degree of coverage as

$$\theta = \sum_{q=1}^{t} \frac{a_q P}{1 + a_q P} \quad \text{and} \quad \theta = \int \frac{a_q P}{1 + a_q P} ds, \qquad (6.1)$$

where t is the number of different crystalline areas and s is a parameter characterizing the relative proportion of $q–x$ surface portions.

The value of s is related to the type of centre q through the distribution function f_q, which characterizes the fraction of the surface occupied by the sites of type q, i.e. $ds = f_q \, dq$, or $ds = f_q dQ_q$, if we set a rule that specifies the relationship between the binding energy of the adsorbate with the surface on the centre of type q. Then the integral equation (6.1) can be rewritten as

$$\theta = \int_{Q_{min}}^{Q_{max}} f(Q)\frac{a_Q P}{1+a_Q P}dQ. \tag{6.2}$$

Equation (6.2) is based on statistical process on nonuniform surfaces. In the particular case of small filling (6.2) it can be written in the form of the Henry isotherm $\theta = a_{av}P$, $a_{av} = \int_{Q_{min}}^{Q_{max}} f(Q)a_Q dQ$, where a_{av} is the average value of Henry's law constant for an nonuniform surface, depending on the range of values of the binding energies of the molecules with the surface and the proportion of each type of centres.

At large covering of the surface the isotherm (6.2) can be rewritten in the form $\theta = 1 - b/P$, where the parameter b is equal to $b = \int_{Q_{min}}^{Q_{max}} \frac{f(Q)}{a_Q}dQ$. The value of b also depends on the range of values of the binding energies of molecules with the surface and the proportion of each type of centre, i.e., the shape of the dependence of the adsorption equilibrium in the areas of small and larges covering of the surface in an ideal adsorbed layer on homogeneous and nonuniform surfaces is the same, and cannot be used to determine the type of surface.

Range of average surface filling
If the heterogeneity of the surface is large enough, it is always possible to select the range of equilibrium pressure P, for which the conditions $a_{max}P \gg 1$ and $a_{min}P \ll 1$ are met. These conditions also characterize the average filling area of the surface [117]. They mean that the places with low adsorption capacity remain practically empty, and places with high adsorption capacity are almost fully occupied.

Analysis of the laws of the real adsorbed layer showed [117] that in case of change of $a_Q = a_{max} \exp(-fQ)$, where $f = (Q_{max} - Q_{min})/RT = \ln(a_{max}/a_{min})$, corresponding to a linear change in the binding energy of the molecule by the surface, at a constant value a_0 in all parts of the surface, it leads to the so-called quasi-logarithmic isotherm

$$\theta = \frac{1}{f}\ln\frac{1+a_{max}P}{a+a_{min}P}. \tag{6.3}$$

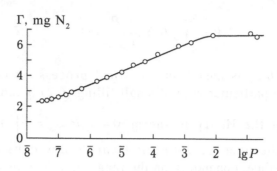

Fig. 6.10. Isotherm of nitrogen adsorption on iron at 350°C [184].

This is the simplest form of the distribution when the surface has the same number of areas of different varieties. Such a distribution is usually called uniform and the surface characterized by it the uniformly nonuniform surface [118].

Figure 6.10 shows an example of the quasi-logarithmic adsorption isotherm obtained for nitrogen adsorption on iron in [184] in the range of equilibrium pressures from $5 \cdot 10^{-5}$ to 3.4 mm Hg and in the range from $2.2 \cdot 10^{-5}$ to 49 mm Hg (where Γ is the amount of adsorbed nitrogen in mg, and the pressure is expressed in atmospheres). The same isotherms were obtained for the adsorption of oxygen [185–187] and hydrogen [188].

In the medium filling range the formula (6.3) is simplified and the well-known logarithmic adsorption isotherm is obtained:

$$\theta = \frac{1}{f} \ln(a_{max} P).\tag{6.4}$$

Figure 6.10 shows the linear range in the middle part of the curve. The angle of slope is directly related to the binding energy range for the strongest and the weakest areas of the surface.

An equation similar in form was first obtained by Frumkin in electrochemical studies [189–191]: $\theta = C_1 \lg P + C_2$, where C_1 and C_2 are empirical constants. In accordance with Faraday's law and the Nernst formula, this represents a linear relationship between the amount of hydrogen adsorbed on platinum and the logarithm of the equilibrium hydrogen pressure.

Freundlich isotherm
Experimental data show that the Langmuir equation is typically implemented with relatively small changes in the pressure range up to

two orders of magnitude. Conducting adsorption measurements over a wide range, up to 6–7 orders of magnitude, it was shown that the experimental data are described by the Freundlich empirical equation:

$$\theta = CP^{1/n}, \tag{6.5}$$

where C and n are numeric parameters, $n > 1$.

Mathematical analysis, performed for the first time by Zel'dovich [116], showed that the distribution function $f(Q)$, corresponding to equation (6.5), can be represented as $f(Q) = C' \exp(-\gamma Q)$ where C' is a constant, and $\gamma = 1/n$. This type of distribution is called exponential [118], and it leads to a logarithmic dependence of the characteristic heats of adsorption. This isotherm can be used, like logarithmic isotherm, only in the middle range of fillings.

Dubinin–Radushkevich equation

To describe the adsorption in micropores of active carbons [192], the following empirical equation was proposed:

$$W = W_0 \exp\left(-\frac{k\varepsilon^2}{\gamma^2}\right), \tag{6.6}$$

Here k – the parameter of the system, γ – the affinity factor binding the 'adsorption potential' of the given and standard adsorbate (for benzene $\gamma = 1$), $\varepsilon = RT \ln(P_s/P)$ – 'adsorption potential'– the work required to compress the vapour under isothermal conditions from the equilibrium pressure P to the saturated vapour pressure P_s. In the calculations, the value of P_s is taken from the data for the bulk phase of the experiment at a temperature T. The presupposition of this equation were the concepts of uniformity of filling the entire volume of the micropores and the Polyani thermodynamic potential theory [113]. This equation is used to determine the micropore volume W_0 [193–197].

In [195] it is shown that equation (6.6) corresponds to the distribution function (6.2) which has the form of a normalized Maxwellian:

$$f(Q) = 4k^{3/2}Q^2 \exp(-kQ^2)/\pi^{1/2}. \tag{6.7}$$

In addition to these adsorption isotherms, the experimenters often use numerous other empirical regularities, which are also functions of the distributions [198, 199].

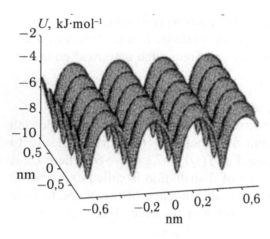

Fig. 6.11. The potential relief of the (100) face of rutile [200].

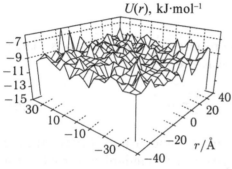

Fig. 6.12. The potential relief of the amorphous surface of rutile at $\delta = 22\%$ and $\gamma = 45\%$, $7.6 \times 7.6 \times 0.38$ nm^3 [201].

Relationship of the heterogeneity with the potential relief

Recall that the total volume of the porous body comprises a compact part of the solid (V_m) and the free pore volume (V_p), see section 1. The solid is given by the crystal lattice, i.e., a discrete way of describing, like the discussed discrete spatial distribution of the adsorbate. However, the usefulness of the overall unit of measuring the volumes V_m and V_p in terms of the cell v_0 is not obvious, except in the case of adsorption of molecules on a frozen substrate of such molecules.

At the solid–adsorbate boundary it is necessary to analyze the properties of the potential relief, calculated according to the formulas insection. 4. As illustrative examples, Figs. 6.11 and 6.12 show the potential relief of the (100) [200] face and of the [201] amorphous surface of rutile in the adsorption of argon atoms (details in section 2).

As a result of dividing the space occupied by the adsorbate into cells, each cell is characterized by its minimum energy Q_f and the characteristics of the potential relief. The total potential surface is defined by a discrete set of values η_f^Q, $1 \leq f \leq M$, which reflects the local property of the adsorbate for the site f. The entire set of values η_f^Q is predetermined and unchangeable during the adsorption process if the surface of the solid body is not reconstructed under the influence of the adsorbate. This set of values $\{\eta_f^Q\}$ is uniquely related to the chosen method of partitioning the real volume of the selected lattice constant λ (or a finer mesh λ_n).

For brevity, the value of the local minimum Q_f, the curvature of the potential energy surface in the vicinity of the minimum, and so on, will be denoted by q. All cells with the same adsorption characteristics a_q will refer to the same type q and will be denoted as η_f^q: the cell with the number of f is a cell of type q. The total number of cell types with different constants of adsorption capacity a_q are denoted as above by t.

The static weight of the cells of type q is related to their share in relation to all the cells of the surface. Formally, this can be written as $\langle \eta_f^q \rangle = M_q$, where the average is taken over all the cells of the structure sites, $1 \leq f \leq M$, then the normalized average determines the amount of cells of the type q on the surface $F_q = M_q/M$.

If the adsorbed molecule changes the state of the surface, then the question of the relationship between the values of γ_f^i and η_f^q arises. This issue is discussed in [138] in the discussion of the processes of growth and restructuring of the crystals. The LGM (lattice-gas model) allows to take into account the dynamics of changes in the state of the surface of the solid and adsorbed particles, but all the atoms of the adsorbent and the adsorbate are included in the overall system of dynamic equations, i.e., the number of components s includes, apart from the adsorbate, also the components of a solid. The general system of kinetic equations can only deal with the values of γ_f^i (its structure is discussed in section 4). The case of restructuring of the surface is often implemented in chemisorption, for which the adsorbate–adsorbent binding energy is commensurate with the energies of the atoms in the solid. For physical adsorption the energy of the adsorbate–adsorbent bond is normally not less than one order of magnitude and the introduction of value η_f^q drastically reduces the number of variables of the problem in equilibrium and dynamics of adsorbed particles, due to the fixed spatial positions of the atoms of the solid.

Information about the distribution function of different types of sites can be obtained through direct and indirect measurements. The direct methods include experiments on field-ion microscopy, electron diffraction, molecular beam scattering, especially of helium atoms,

Auger spectroscopy [164]. Based on these measurements, one can obtain information on the number of defects and auto-adsorbed atoms, structure and chemical composition of the surface, the presence of steps and the degree of roughness. From the integral characteristics, using the model considerations, one can get more rigorous evaluation of the surface layer. For example, the Auger spectroscopy signal contains information on several surface layers. Using the methods of calculating the concentration profile of the alloy [202–204], it is possible to 'distribute'the total number of its components in layers and to obtain information about the composition and structure of the surface layer. The same information can be obtained indirectly from the data for the distribution of the alloy components in the bulk and surface tension (establishment of the equilibrium state of the alloy).

Information on the nature of heterogeneity can also be based on experimental data and model representations of the mechanisms of formation of the surfaces in the process of crystal growth, deposition of films on fracture of the crystals [205–207], and mechanical and thermal processing of the samples. If the alloy during is rapidly cooled starting from a certain temperature, it is characterized by the particle size distribution corresponding to that temperature. Therefore, knowing the background of the adsorbent we can estimate the distribution function of different types of sites on the surface. The dependence of the thermal desorption spectra of CO from the Pd–Au alloy on the conditions and duration of the heat treatment of the alloy is shown in the example in [208].

References

1. Plachenov T.G., Kolosentsev S.D., Porometry. – L.: Khimiya, 1988. – 175 p.
2. Fenelonov V.B., Porous carbon. – Novosibirsk: Institute of Catalysis, 1995. – 513 p.
3. Radushkevich L.V., Main problems of the theory of physical adsorption. – Moscow: Nauka, 1970. – 270 p.
4. Simulation of porous materials. – Novosibirsk, USSR Academy of Sciences, 1976. – 190 p.
5. Kheifets L.I., Neimark A.V., Multiphase processes in porous solids. – M.: Khimiya, 1982. – 320p.
6. Neimark I.E., Sheinfain R.Yu., Silica, its preparation, properties and applications. – Kiev: Naukova Dumka, 1973. – 216 p.
7. Joffe I.I., Rechetov V.A., Dobrotvorsky A.M., Calculation methods in predicting the activity of heterogeneous catalysts. – L.: Khimiya, 1977. – 204 p.
8. Chizmadzhev Yu. A., Markin V. S., Tarasevich V. P., Chirkov Yu. G. Macrokinetics processes in porous media. – Moscow: Nauka, 1971. – 362 p.
9. Adsorbents, their preparation, properties and applications. – Leningrad: Nauka, 1985. – 158 p.

10.	Cheremskoi P.G., Methods of studying the porosity of solids. – Moscow: Energoatomizdat, 1985. – 112 p.
11.	Karnaukhov A.P., Adsorption. Texture of dispersed porous materials. – Novosibirsk: Nauka, 1999. – 469 p.
12.	de Boer J.H. // Proc. of 10th Symposium Colston Res. Soc. 1958. V. 10. P. 68.
13.	Frevel L.K., Kressey L.J. // Annal. Chem. 1963. V. 35. P. 1492.
14.	Mayer R.P., Stowe R.A. // J. Coll. Sci. 1965. V. 20. P. 893.
15.	Mamleyev B.S., Zolotarev P.P., Gladyshev P.P., Heterogeneity of sorbents. – Alma-Ata: Nauka, 1989. – 287 p.
16.	Ruckenstein E., Vaidyanathan A.S., Youngquist G.R. // Chem. Eng. Sci. 1971. V. 26. P. 1305.
17.	Dubinin M.M. // Usp. Khimii, 1955. V. 24. P. 3.
18.	Dubinin M.M. // Zh. Fiz. Khimii, 1965. V. 23. P. 1410.
19.	Dubinin M.M. // J. Colloid Interface Sci. 1980. V. 75. P. 34.
20.	Gregg, S.J. Sing, K.G.W., Adsorption, Surface Area and Porosity. – Moscow: Mir, 1984. [Academic Press, London, 1982].
21.	Brunauer S., Deming L.S., Deming W.S., Teller E. // J. Amer. Chem. Soc. 1940. V. 62. P. 1723.
22.	Brunauer S., Emmett P. H., Teller E. // J. Amer. Chem. Soc. 1938. V. 60. P. 309.
23.	Langmuir I. // J. Amer. Chem. Soc., 1916. V. 38. P. 2217.
24.	Temkin M.I. // Zh. Fiz. Khimii, 1933. V. 4. P. 573.
25.	Fowler R.H. // Proc. Camb. Phil. Soc., 1935. V. 31. P. 260.
26.	Fowler R.H., Guggenheim E.A. Statistical Thermodynamics. – Cambridge: Univer. Press, 1939.
27.	Emmett P.H., Brunauer S. // J. Amer. Chem. Soc. 1937. V. 59. P. 1553.
28.	Brunauer S., Emmett P. H. // J. Amer. Chem. Soc. 1935. V. 57. P. 1735.
29.	Young DM., Crowell A. D. Physical Adsorption of Gases. – London: Butterworths, 1962. – P. 198.
30.	Van Bemmelen J. M. // Rec. Trans. Chim. 1888. V. 7. P. 37.
31.	Zsigmondy A. // Z. Inorganic. Chem. 1911. V. 71. P. 356.
32.	Anderson J.S. // Z. Phys. Chem., 1914. V. 88. S. 191.
33.	Thomson W.T. // Phil. Mag. 1871. V. 42. P. 448.
34.	Gurvich, L. // Zh. Russkogo Fiz.-Khim. Ob-va. 1915. V. 47. P. 805.
35.	Guggenheim E.A. // Trans. Faraday Soc. 1940. V. 36. P. 407.
36.	Barrer R. M., Diffusion in Solids. – Moscow: Izd. Lit., 1948.
37.	Raczynski V.V., Introduction to the general theory of dynamics of adsorption and chromatography. – Moscow: Nauka, 1964. – 134 p.
38.	Timofeev D.P., Kinetics of adsorption. – Moscow: Publishing House of the USSR Academy of Sciences, 1962. – 252 p.
39.	Satterfield Ch. N. Mass transfer in heterogeneous catalysis. – Moscow: Khimiya, 1976. – 240 p. [MIT Press, Cambridge (Mass.), 1970].
40.	Ruthven D.M., Principles of Adsorption and Adsorption Processes. – N. Y.: J. Willey & Sons, 1984.
41.	Mason E. A., Malinauskas A. P., Gas Transport in Porous Media: The Dusty_Gas Model. – Moscow: Mir, 1986. - 200 p. [Elsevier, Amsterdam, 1983].
42.	Derjagin B.V., Churaev N.V., Muller V.M., Surface Forces. – Moscow: Nauka, 1985. – 400 p.
43.	Landau L.D., Lifshitz E.M., Theoretical Physics. V. 6. Hydrodynamics. – Moscow: Nauka, 1986. – 733 p.
44.	Nikolaev N.I., Diffusion in membranes. – M.: Khimiya, 1980. – 232 p.

72   Molecular theory of adsorption in porous solids

45.   Bird R. B., Stewart W., Lightfoot E. N., Transport Phenomena. – Moscow: Khimiya, 1974. – 687 p. [John Wiley and Sons, New York, 1965].
46.   de Groot S. R., Mazur P., Non-Equilibrium Thermodynamics. Moscow: Mir, 1964. – 456 p. [(North-Holland, Amsterdam, Amsterdam, 1962].
47.   Manning J., Kinetics of diffusion of atoms in crystals. – Academic Press, 1971. – 278 p.
48.   Slezkin N.A., Dynamics of a viscous incompressible fluid. – M.: Gostekhizdat, 1955. – 520 p.
49.   Whiteker S. // Ind. Eng. Chem. 1969. V. 61. P. 14.
50.   Scheidegger A.E., The Physics of Flow through Porous Media. – Moscow: GTTI, 1960. – 196 p. [Toronto University Press, Toronto, Canada, 1957].
51.   Leybzon L.S., Collected Works. V. 2. – Moscow: Publishing House of the USSR Academy of Sciences, 1953.
52.   Reif F., Statistical Physics. Berkeley Physics Course, Vol. 5 – Moscow: Nauka, 1977. – 352 p.
53.   Wicke E., Vollmer W. // Chem. Eng. Soc. 1952. V. 1. P. 282.
54.   Flood E.A., Huber M.E. // Canad. J. Chem. 1955. V. 33. P. 979.
55.   Volmer M., Esterman J. // Z. Physik. 1921. V. 7. P. 13.
56.   Frenkel' Ya.I., Statistical Physics. – 1950. – 760 p.
57.   Kington G. L., Laing W. // Trans. Faraday Soc. 1955. V. 51. P. 287.
58.   Rayleigh L. // Proc. Roy. Soc. A. 1936. V. 156. P. 350.
59.   Lykov A.V., Transport phenomena in capillary-porous bodies. – Moscow: GITTL, 1954.
60.   Carman P. C. Flow of gases through porous media. – London: Butterworths, 1956.
61.   Hirschfelder J. O., Curtiss Ch. F., Bird R. B., Molecular theory of gases and liquids. – Moscow: Inostr. Lit., 1961. – 929 p. [Wiley, New York, 1954].
62.   Intermolecular Forces // Adv. Chem. Phys. / Ed. by J.O. Hirschfelder – N. Y.: Interscience Publ., 1967. – V. 12. – 643 p.
63.   Kaplan I. G., Introduction to the theory of molecular interactions. – Moscow: Nauka, 1982. – 312. [Elsevier, New York, 1986].
64.   Oray S. O., Gubbins, K.E. Theory of Molecular Fluids. – Oxford: Clarendon Press, 1984. – V. 1. – 626 p.
65.   Intermolecular interactions: from diatomic molecules to biopolymers / Ed. B. Pullman. – Academic Press, 1981. – 592 p.
66.   Molecular interactions / ed. H. Ratajczak and H. Orville-Thomas. – Academic Press, 1984. – 600 p.
67.   Bakhshiev N.G., Spectroscopy of intermolecular interactions. – Leningrad: Nauka, 1972. – 264 p.
68.   Smirnova N.A., Molecular models of solutions. – L.: Khimiya, 1987. – 334 p.
69.   Kitaigorodsky A.I., Molecular crystals. – Moscow: Nauka, 1971. – 424 p.
70.   Kitaigorodsky A.I., Mirskaya K.V. // Kristallografiya. 1961. V. 6. P. 507.
71.   Reed T.M., Gubbins K.E. Applied Statistical Mechanics. – Tokyo etc.: McGraw-Hill Kogakusha, 1973. – 506 p.
72.   Barker J. A. // Rare Gas Solids / Ed. M. L. Klein, J.A. Venables – London etc.: Academic Press, 1976. – V. 1. – P. 212.
73.   Mason E., Sperling T., Virial equation of state. – Moscow: Mir, 1972. – 280 p. [The Internationsl Encyclopedia of Physical Chemistry. Topic 10. ed. J.S. Rowlinson, Vol. 2].
74.   Koide A., Kihara T. // Chem. Phys. 1974. V. 5. P. 34.
75.   Williams D. E. // Ibid. 1966. V. 45. P. 3770.

76. Williams D. E. // Ibid. 1967. V. 67. P. 4680.
77. Williams D. E. // Acta Crystallogr. A. 1974. V. 30. P. 71.
78. Dashevskii V.G., Conformations of organic molecules. – M.: Khimiya, 1974. – 432 p.
79. Timofeeva T.V., Chernikov N.Yu., Zorkii P.M.. // Usp. Khimii, 1980. V. 49. P. 966.
80. Avgul' N.N., Kiselev A.V., Poshkus D.P., Adsorption of gases and vapors on homogeneous surfaces. – M.: Khimiya, 1975. – 284 p.
81. Sarkisov O.N., Dashevsky V.O., Malenkov G.G. // Mol. Phys. 1974. V. 27. P. 1249.
82. Street W. B., Tildesley D. J. // Faraday Disc. Chem. Soc. 1978. No. 2, 66. P. 27.
83. Poltev V.I. // Kristallografiya. 1977. V. 22. P. 453.
84. Axilrod B. M., Teller E. // J. Chem. Phys. 1943. V. 11. P. 299.
85. da Silva J. D., Brandao J., Varandas A. J. C. // J. Chem. Soc. Faraday Trans. 1989. V. 85. P. 1851.
86. Kiselev A.V., Poshkus D.P., Yashin Ya.I., Molecular basis of adsorption chromatography. – M.: Khimiya, 1986. – 269 p.
87. Sams J. R., Constabaris G., Halsey G. D. // J. Phys. Chem. 1960. V. 64. P. 1689.
88. Wolfe R., Sats J. R. // J. Phys. Chem. 1965. V. 69. P. 1129.
89. Kiselev A. V., Poshkus D. P. // Adv. Coll. Interface Sci. 1978. V. 9. P. 1.
90. Steele W.A. // J. Phys. Chem. 1978. V. 82. P. 817.
91. Rybolt T.R., Pierotti R.A. // J. Chem. Phys. 1979. V. 70. P. 4413.
92. Smith B.B., Wells B.H. // J. Phys. Chem. 1983. V. 87. P. 160.
93. Hoinkes H. // Rev. Mod. Phys. 1980. V. 52. P. 933.
94. Cardillo M. // Ann. Rev. Phys. Chem. V. 32. P. 331.
95. Cole M. W., Frankl D. R., Goodstein D. L. // Rev. Mod. Phys. 1981. V. 53. P. 199.
96. Hobza P., Zahradnik R., Weak Intermolecular Interactions in Chemistry and Biology. – Prague: Academia, 1980. – 245 p.
97. Kitaigorodsky A.I. // Usp. Fiz. Nauk. 1979. V. 127. P. 391.
98. Harris S. J., Novick S. E., Klemperer W., Falconer W. // J. Chem. Phys. 1974. V. 61. P. 193.
99. Klemperer W. // Faraday Disc. Chem. Soc. 1977. No. 62. P. 179.
100. Mulder p., Huiszoon C. // Mol. Physics. 1977. V. 34. P. 1215.
101. Morris J. M. // Mol. Phys. 1974. V. 28. P. 1167.
102. Brobjer J. T., Murrell J. N. // J. Chem. Soc., Faraday Trans. 2. 1983. V. 79. P. 1455.
103. Jorgensen W. L. // J. Chem. Phys. 1979. V. 71. P. 5034.
104. Van der Avoird A., Wormer P. E. S., Mulder F., Berns R. M. Van der Waals Systems. – Berlin: Academic–Verlag, 1981. – P. 1.
105. Carlos W. E., Cole M. W. // Surface Sci. 1980. V. 91. P. 339.
106. Mattera L., Salvo C., Terreni S., Tommasini E. // Surface Sci. 1980. V. 97. P. 158.
107. Burgos E., Righini R. // Chem. Phys. Lett. 1983. V. 96. P. 584.
108. Poshkus D.P., Afreimovich A.Ya., // Zh. Fiz. Khimii. 1966. V. 40. P. 2185.
109. Steele W.A. The Interactions of Gases with Solid Surfaces. – N. Y.: Pergamon, 1974.
110. Kiselev A.V., Intermolecular interaction in adsorption chromatography. – Moscow: Vysshaya shkola, 1986. – 360 p.
111. Votyakov E.V., Tovbin Yu.K., MacElroy J.M.D., Roche A. // Langmuir. 1999. V. 15. P. 5713.
112. Trapnell B. M. W., Chemisorption. – Moscow: Publishing House of Foreign Lit., 1958. [Academic Press, New York, 1955].
113. Adamson A. W., Physical Chemistry of Surfaces. – Moscow: Mir, 1979. [3rd ed. Wiley, New York, 1976].
114. Kiperman S. Introduction to the kinetics of heterogeneous catalytic reactions. –

Moscow: Nauka, 1964. – 607 p.

115.   Langmuir I. // J. Am. Chem. Soc. 1918. V. 40. P. 1361.

116.   Zeldovich Ya.B. // Acta Phys. Chim. USSR. 1935. V. 1. P. 961.

117.   Temkin M. I. // Zh. Fiz. Khimii. 1941. V. 15. P. 296.

118.   Roginskii I.Z., Adsorption and catalysis on heterogeneous surfaces. – Leningrad: Publishing House of the USSR Academy of Sciences, 1949.

119.   Frenkel Ya.I., Kinetic Theory of Liquids. – Moscow: Publishing House of the USSR Academy of Sciences, 1945.

120.   Fowler R.H., Rushbrooke G.S. // Trans. Far. Soc. 1937. V. 33. P. 1272.

121.   Lennard-Jones J. E., Devonshire A. F. // Proc. Roy. Soc. A. 1937. V. 163. P. 53.

122.   Lennard-Jones J. E., Devonshire A. F. // Proc. Roy. Soc. A. 1939. V. 169. P. 317.

123.   Baxter R. J., Exactly solvable models in statistical mechanics. – Moscow: Mir, 1985. – 486 p. [Academic] Press, London, 1982]

124.   Kac M. Probability and related topics in physical science. – Moscow: Mir, 1965. – 407 p. [Interscience Publishers, Ltd., London, 1957].

125.   Ising E. // Zs. Phys. B. 1925. V. 31. S. 253.

126.   Onsager L. // Phys. Rev. 1944. V. 65. P. 117.

127.   Domb C. // Phase transitions and critical phenomena / Ed C. Domb, M. S. Green. – London–N. Y.: Academic Press, 1974. – V. 3. – P. 356–484.

128.   Hill T.L., Statistical Mechanics. Principles and Selected Applications. – Moscow: Izd. Inostr. lit., 1960. – 486 p. [N.Y.: McGraw–Hill Book Comp. Inc., 1956].

129.   Gerasimov Ya.I., Geyderikh V.A., Thermodynamics of solutions. – Moscow: Moscow State University Press, 1980. – 184 p.

130.   Hicter P. // J. Chim. Phys. Chim. Biol. 1967. V. 64. P. 261.

131.   Moelwyn Hughes E.A., Physical Chemistry. – Moscow: Izd. Lit., 1962. – V. 1, 2. – 1146 p. [2nd ed. Pergamon Press, London, 1961].

132.   Collins R. // Phase transitions and critical phenomena / Ed. C. Domb, M. S. Green. – London–N. Y.: Academic Press, 1972. – P. 271.

133.   Vortler H. L., Heybey J., Haberlandt R. // Physica A. 1979. V. 99. P. 217.

134.   Runnels L. K. // Phase transitions and critical phenomena / Ed. C. Domb, M. S. Green. – London–N. Y.: Academic Press, 1972. – V. 2. – P. 305.

135.   Runnels L. K. // Phase transitions and critical phenomena / Ed. C. Domb, M. S. Green. – London–N. Y.: Academic Press, 1972. – V. 2. – P. 305.

136.   O'Reilly D. E. // Phys. Rev. A. 1977. V. 15. P. 1198.

137.   Shinomoto S.-G. // Progr. Theor. Phys. 1983. V. 70. P. 687.

138.   Tovbin Yu. K., Theory of physical and chemical processes at the gas-solid interface. – Moscow: Nauka, 1990. – 288 p. [CRC, Boca Raton, Florida, 1991].

139.   Guggenheim E. A. // Proc. Roy. Soc. London. A. 1935. V. 148. P. 304.

140.   Bethe H. A. // Proc. Roy. Soc. London. A. 1935. V. 150. P. 552.

141.   Peierls R. // Proc. Camb. Phil. Soc. 1936. V. 32. P. 471.

142.   Kirkwood J. G., Monroe E. // J. Chem. Phys. 1942. V. 10. P. 395.

143.   Barber M. N. // J. Phys. A. 1979. V. 12. P. 679.

144.   Fisher M. E. // Phys. Rev. 1958. V. 113. P. 969.

145.   Wheeler J. C. // Ann. Rev. Phys. Chem. 1977. V. 28. P. 411.

146.   Andersen G. R., Wheeler J. C. // J. Chem. Phys. 1979. V. 70. P. 1326.

147.   Henderson D., Barker J. A. // Phys. Rev. 1978. V. 46. P. 587.

148.   Shulepov Yu.V., Aksenenko E.V., Lattice gas. – Kiev: Naukova Dumka, 1981. – 268 p.

149.   Tovbin Yu.K., Theoretical methods for describing the properties of the solutions // Interuniversity collection of scientific. works. – Ivanovo, 1987. – p. 44.

150. Prigogine I., Defay R. Chemical Thermodynamics. – Novosibirsk: Nauka, 1966. – 510 p. [Longmans Green, London, 1954].
151. Nesterov A.E., Lipatov Yu.S., Thermodynamics of solutions and mixtures of polymers. – Kiev: Naukova Dumka, 1984. – 300 p.
152. Guggenheim E. A. Mixtures. – Oxford: Claredon Press, 1952. – 270 p.
153. Flory P. J. Principles of Polymer chemistry. – N. Y.: Cornell Univ. Press, 1953. – 594 p.
154. Sanchez I. C., Lacombe R. H. // J. Phys. Chem. 1976. V. 80. P. 2352.
155. Lacombe R. H., Sanchez I. C. // Ibid. P. 2568.
156. Sanchez I. C., Lacombe R. H. // Macromolecules. 1978. V. 11. P. 1145.
157. Costas M., Sanctuary B. C. // Fluid Phase Equil. 1984. V. 18. P. 47–60.
158. Kut'in A.M., Zorin A.D., // Zh. Fiz. Khimii. 1984. V. 58. P. 596.
159. Panayiotou C., Vera J. H. // Polymer J. 1982. V. 14. P. 681.
160. Nies E., Kleintiens L. A., Koningsveld R., Sitha R., Jain R. K. // Fluid Phase Equil. 1983. V. 12. P. 11.
161. Dunning W., The gas – solid interface. – Moscow: Mir, 1970. – P. 230.
162. Roberts M. W., McKee C. S. Chemistry of the metal–gas interface. – Oxford: Clarendon Press, 1978.
163. Somorjai G. A. Chemistry in two-dimension surface. – L.; N. Y.; Ithaca: Cornell Univ. Press, 1981.
164. New studies of the solid surface / Ed. T. Jayadevaiah, R.M. Vanselow. – Moscow: Mir, 1977. – Issue. 1. – 316 p., Vol. 2. – 371 p. [CRC Press, Inc., Cleveland, 1974]
165. Jaycock M. Parfitt J., Chemistry of interfaces.– Moscow: Mir, 1984. – 270 p. [Wiley, New York, 1981].
166. Avgul' N., et al., // Izv. AN Ser. Khim., 1959. S. 1196.
167. Burton J. J., Jura G. // Fundamentals of Gas-Surface Interactions / Eds. H. Saltzburg, J. N. Smith, M. Rogers. – N. Y.: Academic Press, 1967. – P. 75.
168. Stone F., Solid State Chemistry / Ed. B. Garner. – Moscow: Izd. Lit., 1961. – P. 36.
169. Wisseman T. J. Characterization of Powders / Eds. G. D. Parfitt, K. S. W. Sing. – London: Academic Press, 1976. – P. 167.
170. Bardy D. // Ibid. P. 403.
171. Meddalia A. I., Rivin D. // Ibid. P. 304.
172. Jackson P., Parfitt G. D. // J. Chem. Soc. Farad. Trans. II. 1972. V. 68. P. 896.
173. Hockey J. A., Jones P. // Trans. Faraday Soc. 1971. V. 67. P. 2679.
174. Waldsax J. C. R., Jaycock M. J. // Disc. Faraday Soc. 1971. V. 52. P. 215.
175. Jaycock M. J., Waldsax J. C. R. // J. Chem. Soc. Farad. Trans. I. 1974. V. 70. P. 1501.
176. Carman P. C. // Trans. Faraday Soc. 1950. V. 36. P. 964.
177. Radzig V. A. // Physic-Chemical Phenomena in Thin Films and at Solid Surfaces / Eds.: L. I. Trakhtenberg, S. H. Lin, Olusegun J. Ilegbusi. – Amsterdam: Elsevier, 2007. – P. 233.
178. Frank F. C. // Disc. Faraday Soc. 1949. V. 5. P. 48.
179. Burton W. K., Cabrare N., Frank F. C. // Nature. 1949. V. 163. P. 398.
180. Frank F. C. // Acta Cryst. 1951. V. 4. P. 497.
181. Lonsdale K. Crystals and X-rays. – Moscow: Izd. Instr. Lit., 1952.
182. Cottrell A.D., Dislocations and Plastic Flow in Crystals. – Moscow, Metallurgy, 1958. – 215 p.
183. Nicholas J.F. An atlas of models of crystal structures. – N. Y.: Gordon and Breach, 1965.
184. Romanuškina A.E., Kiperman S.L., Temkin M.I. // Zh. Fiz. Khimii. 1953. V. 28. P. 1181.

185.   Dowden D. F. Chemisorption / Ed. W. E. Garner. – London, 1957. – P. 3.
186.   Dowden D. F. // Bull. Soc. Chim. Belg. 1958. V. 67. P. 439.
187.   Dowden D. F. // Ind. Eng. Chem. 1952. V. 44. P. 977.
188.   Ansel'm F.I. // Zh. Eksp. Teor. Fiz. 1934. V. 4. P. 678.
189.   Frumkin A.N., Shlygin A.I. // Dokl. AN SSSR. 1934. V. 2. P. 176.
190.   Frumkin A.N. et al., Kinetics of electrode processes. – Moscow: Moscow State University Press, 1950.
191.   Tverdovskiy I.P., Vert Zh.L. // Dokl. AN SSSR. 1953. V. 88. P. 305.
192.   Dubinin M.M., Radushkevich L.V.// Dokl. AN SSSR.. 1947. V. 55. P. 331.
193.   Dubinin M.M., Zaverina E.V., Radushkevich L.V. // Zh. Fiz. Khimii. 1947. V. 21. P. 1410.
194.   Dubinin M.M., Zaverina E.V. // Zh. Fiz. Khimii. 1949. V. 23. P. 1129.
195.   Radushkevich L.V. // Zh. Fiz. Khimii. 1949. V. 23. P. 1410.
196.   Dubinin M.M. // Usp. Khimii. 1955. V. 24. P. 3.
197.   Dubinin M.M. // Zh. Fiz. Khimii. 1965. V. 23. P. 1410.
198.   Jaroniec M., Madey R. Physical Adsorption on Heterogeneous Solids. – Amsterdam: Elsevier, 1988.
199.   Rudzinski W., Everett D. H. Adsorption of Gases on Heterogeneous Surfaces. – London: Academic Press, 1992.
200.   Ramalho J. P. P., Rabinovich A. B., Yeremich D. V., Tovbin Yu. K. // Applied Surface Science. 2005. V. 252. P. 529.
201.   Gvozdev V.V. Tovbin Yu.K. // Izv. AN Ser. Khim. 1997. No. 6. P. 1109. [Russ. Chem. Bull. 1997. V. 46. № 6. C. 1060].
202.   Iveronova V.I., Katznel'son A. Short-range order in solid solutions. – Moscow: Nauka, 1977.
203.   Ono S., Kondo S., Molecular theory of surface tension in liquids. – Moscow: Izd. Inostr. Lit., 1963. [Handbuch der Physik, Herausgeeben von S. Flugge. B. X. Berlin–Gottingen–Heidelberg: Springer–Verlag, 1960].
204.   Santen R.A.V., Sachtler W.M.H. // J. Catalysis. 1974. V. 33. P. 202.
205.   Laudise R. A., Parker R. L., The growth of mono-crystals. – Moscow: Mir. 1974. – 540. []Prentice-Hall, Inc., Englewood Cliffs, New Jersy, 1970; Solid State Physics, V. 25, 1970].
206.   Fransuort G. The gas–solid interface / Ed. E. Flood. – Moscow: Mir. 1970. – S. 359.
207.   Melikhov I.V., Merkulov M.S., Co-crystallization. – M.: Khimiya, 1975. – 280 p.
208.   Fukinaga Y. // Surface Sci. 1977. V. 64. P. 751.

2

Equilibrium properties of adsorption systems

Initial assumptions of the adsorption theory are completely matched with the assumptions of the LGM (lattice-gas model): the adsorbed state of the molecules corresponds to the filled cells, and the unoccupied cells are vacant, which was the basis for the development of molecular theory. The traditional interpretation of the LGM corresponds to the fixed position of the adsorbed molecules within the cell. A natural extension to non-uniform surfaces is associated with taking into account the intermolecular interactions for particles situated a different adsorption sites on the surface. This chapter discusses ways to expand the primitive model of adsorption on real surfaces due to the combined account of two main factors: the heterogeneity of the surface and the lateral interactions between adsorbed particles, and the integration of the internal motions of molecules within the cell (in the adsorption centre). This motion expands the lattice gas model, leading to the abandonment of the rigid lattice and to the modification of the molecular distributions.

7. Nonuniform systems

The main issue of the theory is the reflection of the real structure of the nonuniform surface when taking into account the lateral interactions of the adsorbate. Traditionally, the filling of the nonuniform surface is characterized by the values of θ_q^i which represent the degree of filling (mole fraction) of the centres of type q with particles of varieties i. In the absence of interaction (ideal model) this is sufficient, as each centre is 'working' independently. For interacting particles (non-ideal models) the state of occupation of any centre with the number of f depends on the type of centre and the occupation conditions of its neighbours g, which in turn depend on the types and conditions of occupation of its neighbours h, including the centre f where the counting is started, and so on (Fig. 7.1 a). As a result, probability of finding particles in specific centres depends not only on the type of these particles, but also on the type of the neighbouring points. The adsorption equilibrium models will vary depending on how accurately the influence of the neighbouring points on the distribution of particles is taken into account [1, 2]. In other words, in the case of the equilibrium distribution of the particles adsorbed on an uneven surface it is important to take into account how the sites of one type are surrounded by other sites and how the nature of environment changes at different macroscopic parts of the surface. The real character of the 'engagement' of different types of sites for each other is reflected in the construction of the corresponding distribution functions.

The second aspect of the generalization of the theory is consideration of the interaction of all neighbours within the radius of the interaction potential. Recall that in the interaction by the LJ potential of the nearest neighbours corresponds to approximately 2/3 of the total energy of interactions of the molecules in a liquid. To describe a number of effects caused by the lateral interactions, such as the formation of a number of superstructures of the orderly arrangement of adsorbed particles, observed by slow electron diffraction on homogeneous and stepped surfaces [3], it is necessary to take into account the contributions of the second and following neighbours. The need to consider further contributions also follows

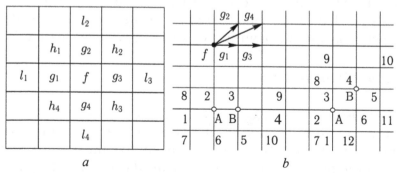

Fig. 7.1. Site f and its neighbours in the first three coordination spheres (a). Arrangement of the pairs of sites fg_k at the distance of the first four neighbours and the set of sites belonging to the first two coordination spheres of 'central' site pairs AB, which are situated at the distance of the first (left) and second (right) neighbours (b).

from the analysis of the mutual influence of adsorbed particles via the electronic subsystem on metals [4, 5] and the consideration of dispersion, dipole and other potentials of interparticle interaction [6, 7]. Therefore, one of the main problems of the theory of adsorption on nonuniform surfaces is the generalization of the theory to the case of the interaction of particles with an arbitrary radius of interaction. At the same time, if we neglect the correlations between the interacting particles, as shown in [8], the problem of interaction with any radius R is reduced to the case of considering only the nearest neighbours ($R = 1$). Therefore, in the future we will discuss only the theory in which the correlation effects between the interacting particles are preserved.

A common property of nonuniform systems is the non-equivalency of the properties of the different sites (in particular for the surface of adsorption sites) of the system. The concentration dependences of the thermodynamic characteristics of adsorption systems are defined by the combined influence of surface hetero-geneity and lateral interactions between the adsorbed particles. In nonuniform lattices the internal degrees of freedom of the particles, which determine the local Henry constants a_q and interparticle interaction potentials $\varepsilon_{qp}^{ij}(r)$ at the distance of the r-th coordination sphere depend on the types of sites q and p.

Description of nonuniform systems [1, 2, 9]

Consider a non-uniform lattice system containing t site types. Thew location of sites of different types is defined by the value η_f^q where $1 \le f \le N$, $1 \le q \le t$. The values of $\eta_f^q = 1$ if the site with the number f is a site of type q, and $\eta_f^q = 0$ if the site with the number f is not a site of type q. Values of numbers η_f^q are considered fixed and unchanging during the adsorption process. Then we deal with the problem of the distribution of the lateral interaction of the particles in the sites of the nonuniform lattice. Depending on the composition of different types of sites and their relative position (structure) the distribution of the adsorbed molecules will be different. The complete set of such numbers is $\{\eta_f^q\}$ where $1 \le f \le N$, $1 \le q \le t$ uniquely identifies the type of heterogeneity of the system.

We assume that the lateral interaction is described by a pair potential with the interaction radius R, R is any number defined in advance. The distances will be counted in the numbers of the coordination spheres (c.s.) (see Fig. 7.1 b). The number of sites in the r-th c.s. of the site with the number f of type q is denoted by $z_f(r)$

(or $z_q(r)$), $1 \leq r \leq R$. The parameter of interaction of the particles i and j, located at sites of the q and p type with indices f and g are at a distance r, is denoted by $\varepsilon_{qp}^{ij}(r)$ (or $\varepsilon_{qp}^{ij}(r)$ as the site number and its type are uniquely related by the values η_f^q).

We assume that every lattice site can be occupied by particle i, $1 \leq i \leq s$, where s is the number of states of occupation of the site, or $(s - 1)$ is the number of kinds of adsorbed particles. Appendix 2 shows the complete set of models that arises in the formulation of the problem of the equilibrium distribution of the molecules, and how each models depends on the accuracy of the description of the nature of the inhomogeneity of the lattice.

The cluster distribution functions [9]

Cluster models that reflect the local arrangement of sites of different types are used in most cases. (Under the cluster we mean a finite number of interconnected sites of the lattice). These models are obtained by mapping the original sequence of sites of the nonuniform lattice on the set of clusters M_2, consisting of a central site and its nearest neighbours z_q (or at $R > 1$ the corresponding analog for more coordination spheres). The type of cluster is determined by the type of all its sites. We denote through σ_{2q} the number of different types of clusters with a central site of type q, differing by the types of sites of its neighbours, $\sigma_{2q} \leq t^{z_q}$, as some types of clusters may be missing. This method of describing the structure takes into account the number of sites of different types, which are in the first coordination sphere

$$m_{qp}, 0 \leq m_{qp} \leq z, \sum_{q=1}^{z} m_{qp} = z_q.$$ The total number of types of clusters on which the original nonuniform lattice is displayed is $L_2 = \prod_{q=1}^{t} \sigma_{2q}$.

Here are some simple example of cluster distribution functions for point and linear surface imperfections that are more conveniently expressed by $d(q\{[m]\})$ – the conditional probability of finding in the lattice a with the coordination sphere of the type $\{[m]\}$, if its centre contains a site of type q:

$$f\left(q\{[m]\}\right) = f_q d\left(q\{[m]\}\right), \qquad \sum_{\sigma_{2q}} d\left(\{[m]\}\right) = 1. \qquad (7.1)$$

Let a point defect on the (111) face of a single crystal distorts the adsorption properties of its closest neighbours. Without specifying its nature, we assume that the position of the defect coincides with the lattice site. Then on the surface there are three types of sites: 1st –

defect, 2nd – its neighbours, 3rd – the rest of the surface sites. The number of neighbours for all types of sites is the same. Assuming that the defects are sufficiently far apart, we denote them through the concentration C so that $f_1 = C$, $f_2 = zC$, $f_3 = 1 - (z + 1) C$.

On the (111) face there are five types of clusters differing in the type of the central site and its immediate neighbours, here $\{[m]\}$ is $[m_{q1}, m_{q2}, m_{q3}]$, $1 \le q \le 3$. Sites of the 1st type are surrounded by only the 2nd type sites, i.e. the only configuration of the neighbouring sites$[0, 6, 0]$ is realized, so that $d(1[0, 6, 0]) = 1$. A site of type 2 always has one neighbour site of type 1, two adjacent sites of type 2 and three adjacent sites third type, so $d(2[1, 2, 3]) = 1$.

The sites of the third type, depending on their positions, have three different types of environment (the number of clusters is equal to three). In any case, the site type 3 has no neighbours of type $1 - m_{31} = 0$. A site of the third type can have two, one or zero neighbours of type 2; they correspond to the clusters of the type $[0, 2, 4]$, $[0, 1, 5]$, $[0, 0, 6]$. In the first two cases the type 3 sites are located alongside the sites of the 2nd type, the number of sites in both cases is equal and, therefore, $d(3[0, 2, 4]) = d(3[0, 1, 5]) = zC/f_3$. The remaining surface sites are type 3 clusters $[0, 0, 6]$. To determine $d(3[0, 0, 6])$, it is necessary on a unit area to subtract all central sites of the above-mentioned clusters: $1 - (1 + 3z)C$ and relate them to the number of type 3 sites. As a result, we obtain $d(3[0, 0, 6]) = (1 - 19C)/(1 - 7C)$.

As an example of a linear nonideality we consider the surface of the (100) face with a monatomic step (Figure 6.9), l – is the length of the terrace. We shall distinguish between sites that are adjacent to the step (type 1) and being in the front row of the terrace from the step (type 2), the remaining sites are the sites of the third type (excluding the edges along the length of the step), $t = 3$. The sites of each type occupy a whole row, so their share are characterized by the following values: $f_1 = F_2 = 1/l$, $f_3 = 1 - 2/l$.

For the (100) face the sites of the 1st and 2nd type are not the closest neighbours (particles therein are situated at the distance of the second neighbours $2^{1/2}a$). The structure of the surface is characterized by five types of clusters. Each site of type 1 has the following environment: two sites of type 1 and one site of the type 3. A similar situation holds for the sites of type 2, so $z_1 = z_2 = 3$ and $d(1[2, 0, 1]) = d(2[0, 2, 1]) = 1$. The sites of the third type have four neighbours. They differ depending on whether there are alongside the type 1 and 2 sites or not. For a number of type 3 sites, adjacent to the sites of the 1st type, the clusters have the environment [1, 0, 3], similarly for a number of sites adjacent to the sites of the 2nd type, the clusters have the environment [0, 1, 3]. Their probabilities are equal to the probabilities of accidental inclusion in one of the rows of type 3, so $d(3[1, 0, 3]) = d(3[0, 1, 3]) = 1/(l-2)$. The rest of the rows of type 3 sites do not contain in their vicinity sites of type 1 and 2, hence $d(3[0, 0, 4]) = (l-4)/(l-2)$.

Another example of a linear nonideality is the interface of the (100) and (111) faces (Fig. 6.9), which affects the properties of the sites in the adjacent layers on each side. This leads to the five types of sites: 1st – sites on the (111) face, $z_1 = 6$, 2nd – sites of the (111) face close to the interface, $z_2 = 6$, the third – the interface sites, $z_3 = 5$; 4th – sites of the (100) face close to the border, $z_4 = 4$, 5th – sites of the (100) face, $z_5 = 4$. We denote the ratios of the number of sites of the interface to the number of sites on the (111) and (100) face by $1/b$ and $1/l$, then $f_1 = (b-1)/\sigma, f_2 = f_3 = f_4 = 1/\sigma$, $f_5 = (l-1)/\sigma$, where $\sigma = b + l + 1$. Within each layer, all sites have the same environment.

The surface structure is characterized by seven types of clusters belonging to different layers: the interface itself and three layers on each side of it, and the remaining layers of the faces are equivalent to the layers adjacent to the sites of type 1 and 5 in Fig. 6.9. Repeating the arguments relating to the stepped surface, it is easy to see that

$$d(1[6, 0, 0, 0, 0]) = (b-2)/(b-1), \quad d(1[4, 2, 0, 0, 0]) = (b-1)^{-1},$$
$$d(2[2, 2, 2, 0, 0]) = d(3[0, 2, 2, 1, 0]) = d(4[0, 0, 1, 2, 1]) = 1,$$
$$d(5[0, 0, 0, 1, 3]) = (l-1)^{-1}, \quad d(5[0, 0, 0, 0, 4]) = (l-2)/(l-1).$$

Other examples of point and line imperfections are discussed in [56, 57]. A similar procedure is used to construct the distribution functions of clusters in more complex cases: accounting fracture of steps, surface roughness, the faces with high indices and so on, and also in cases of their joint presence on the surface.

The pair distribution function

Finding the distribution functions is simplified if it is known that the distribution of atoms in a solid is subject to the equilibrium distribution

$$d(q\{[m]\}) = C_{z_q}^{\{m\}} \prod_{p=1}^{t} (d_{pq})^{m_{qp}}, \quad \sum_{p=1}^{t} m_{pq} = z_q, \quad (7.2)$$

where the functions d_{qp} are the conditional probabilities of p-type sites situated next to the type q site.

These functions are similar to the functions describing the short-range order in multicomponent solutions. In the non-equilibrium conditions we can also used the functions d_{qp} to describe the surface structure. In this case, the distribution functions d_{qp} and $d(q\{[m]\})$ are related as follows

$$z_q d_{qp} = \sum_{\sigma_{2q}} m_{qp} \, d(q\{[m]\}). \tag{7.3}$$

On the right and on the left in (7.3) there are the expressions for the average number of neighbours of p-type with type q sites.

The use of (7.2) in the non-equilibrium conditions is an additional approach that simplifies the mathematical model. In some situations, using the distribution function $d(q\{[m]\})$ and their approximation (7.2) and (7.3) leads to equivalent results: in cases of random and spotty surfaces and surfaces with a regular alternation of sites of different types. In general, this is not the case, and the replacement of one function of the distribution of types of centres by another, more simple, is to be tested.

For comparison we present distribution functions d_{qp} for the above point and linear surface imperfections (functions f_q retain the same values and are not repeated).

Point defects on (111) face

Since the bonds '11' and '13' are absent, and the site of type 1 is surrounded by the sites of the 2nd type, then $d_{11} = d_{13} = d_{31} = 0$, $d_{12} = 1$. Sites of type 2 have one neighbour of type 1, two neighbours of type 2 and three neighbours of type 3, so $d_{21} = 1/6$, $d_{22} = 2/6$, $d_{23} = 3/6$. The value of d_{32} can be obtained from the obvious relation $f_q d_{qp} = f_p d_{pq}$, reflecting the equality $f_{qp} = f_{pq}$: $d_{32} = 3C/f_3$.

The same result is obtained by direct calculation. Sites of type 2 are surrounded by twelve type 3 sites. Six of them have a bond with the sites of the 2nd type and six others – two bonds. The average number of bonds between the sites of the 2nd and 3rd types for each impurity is $6 \cdot 1/6 + 6 \cdot 2/6 = 3$. Consequently, $d_{32} = 3C/f_3$. Similarly, we can calculate the average number of bonds between the type 3 sites. As a test of this calculation we use the normalization condition: $d_{33} = 1 - d_{31} - d_{32}$.

Stepped surface (100)

For sites of the 1st and 2nd type the functions d_{qp} represent the proportion of the number of neighbouring sites of each type from the total number of sites: $d_{11} = d_{22} = 2/3$, $d_{13} = d_{23} = 1/3$, $d_{12} = d_{21} = 0$. For the type 3 sites, it is necessary to take into account differences in their locations: whether they are close to the sites of the 1st and 2nd types or not. If the site of the third type belongs to the layer adjacent to the sites 1 or 2, then for this site the share of neighbouring sites of another type is equal to 1/4, and since the probability of being in such a layer is $(l-2)^{-1}$ then $d_{31} = d_{32} = [4(l-2)]^{-1}$.

Function d_{33} is determined from the following equations. At the step there are $(l-4)$ layers, where each site of the third type has four neighbours of the third type and two layers, where each site has three neighbours of the third type. The total number of bonds in the $(l-2)$ layers is $4(l-2)$, among them there are $4(l-4) + 2 \cdot 3$ bonds between the sites of the third type. Their relation gives $d_{33} = 1 - [2(l-2)]^{-1}$, and the same result is obtained from the formula $\sum_{p=1}^{t} d_{pq} = 1$.

Boundary of faces (111) and (100)

In each layer, all sites have the same environment. In the layers 2–4 as above, the functions d_{qp} represent the ratio of the number of different types of neighbouring sites to the total number of neighbouring sites $d_{21} = d_{22} = d_{23} = 1/3$, $d_{32} = d_{33} = 2/5$, $d_{34} = 1/5$, $d_{44} = 1/2$, $d_{43} = d_{45} = 1/4$. For the sites of the 1st and 5th types we should consider the possibility of the type 2 and 4 sites being in their vicinity. The sites of type 1 in layer 1 have two neighbours of type 2, that is, the probability is 1/3, and as the probability of being included in this row, remaining amongst the sites of the third type is $(b-1)^{-1}$, then $d_{12} = [3(b-1)]^{-1}$; function $d_{54} = [4(l-1)]^{-1}$ is calculated using a similar procedure. As above, it is easy to determine the number of adjacent bonds '11' and '55', it gives $d_{11} = (3b-4)/(3b-3)$, $d_{55} = (4l-5)/(4l-4)$. Other examples of the construction of the functions d_{qp} are given in [9].

The cluster approach

Static distributions of molecules on nonuniform structures have been described by the cluster approach. The essence of the cluster approach is the replacement of calculating the statistical sum of the investigated system by a solution of the system of equations relative to the cluster distribution functions which characterize the probability of realization of various local configurations of molecules $\theta(1,..., M)$ with coordinates $r_1,... r_M$. This approach, on the one hand, represents a development of the Bethe method [10] for complex nonuniform lattices. On the other hand, it is based on Bogolyubov's concept of the fundamental importance of many-particle correlations [11]. Therefore, in this approach we can consider the distribution functions of any order. At the same time it became clear that the same questions are fundamental also in the problems of chemical kinetics, both for homogeneous and nonuniform lattice systems [12, 13].

Examples of many-particle configurations in condensed phases are encountered all the time, so it is necessary to clearly allocate the dimensions of the corresponding distribution functions that characterize the probability of formation of the given configurations in the lattice system configuration data. The square lattice (Fig. 7.1, b) shows the sites g_k, where k is the number of the coordination sphere relative to the site f. Such discrete pairs $f g_k$ are the analogues of the usual continuous pair distribution functions. Here k varies from 1 to 4. The principle of partition of the set of adjacent sites to a larger number of coordination spheres is similar, and it applies to all types and dimensions of the lattice. In addition to the traditional pair distribution functions, the configuration of the molecules of any type can be considered; the number of sites included in the given configuration determines the dimension of the corresponding correlator or a local distribution function. As an example there are two sets of sites around the central sites A and B at a distance of the first and second neighbours. These sets also include neighbours at a distance of the first and second neighbours. In the first case, the number of neighbours is 10, the second – 12 (see also Appendix 2).

Sites with indices g, h, l are the sites that are closest to the site with the number f located in the first, second and third coordination sphere. The image of a fragment of the lattice as a set of clusters consisting of one central site and its closest neighbours, is shown in Fig. 7.2. For the average model in which the structure is determined by more approximately only through the pair bonds of the sites of different types, the lattice is 'split' into pairs of sites $f g_k$ – continuation of splitting.

In the cluster approach: 1) the initial lattice system is displayed on a set of clusters (Fig. 7.2) with 'memorizing' about their location on the original lattice; 2) the exact system of equations is constructed for

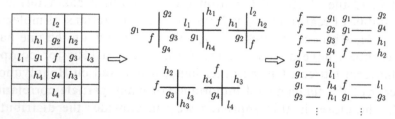

Fig. 7.2. Displaying a fragment of the lattice on the set of clusters consisting of a central site and sites of its first coordination sphere, and a pair of sites fg_k.

each cluster relative to the probabilities of different configurations of the particles that can exist on it (equations for cluster correlators); 3) the resulting systems of equations are finite, but as mentioned above, are not closed, so the same unique method of closing is introduced for all systems taking into account with the real sequence of the location of the clusters on the original lattice. The result is a system of equations for the whole lattice.

Figure 7.2 shows an example of a flat square lattice on a set of five clusters, each of which consists of five sites. Each site of the original lattice is the central site of one of the clusters and is added to the sites the coordination spheres, when the central sites of other clusters are its neighbours. The image of the same fragment of the lattice on a set of clusters with two central sites and their nearest neighbours at $R = 1$ is given in Fig. 7.3.

In general, the imaging procedure consists of isolating three lattice areas: 1 – central sites; 2 – sites surrounding the central sites; 3 – the rest of the lattice sites. The cluster comprises sites of the first and second regions. Central sites can be represented by any number of sites; we confine ourselves to discussing cases with one

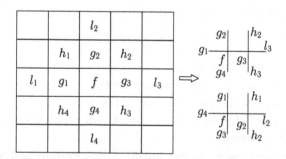

Fig. 7.3. Displaying a fragment of the lattice on a set of clusters consisting of two central sites and their immediate neighbours.

or two central sites, as it follows from the condition of simultaneous participation of the particles in the elementary process (mono- and bimolecular processes). The sites of the second area should separated the central sites from the sites of the third region so that the change in the occupation status of the third region sites does not have a direct impact on the occupation status of the central sites (and vice versa). In other words, the size of the second region must not be smaller than the radius of interaction between the particles R. The number of sites in the second region depends on the dimension of the lattice and the radius of interaction. The mapping of the square lattice (see Fig. 7.2) on a set of clusters with one central site at $R = 2$ leads to clusters consisting of nine sites (sites with indices f, g, h), and at $R = 3$ – to clusters of thirteen sites (to which sites with indices l are added).

Local isotherms

A detailed explanation of how to construct the equations that describe the equilibrium distribution of the molecules on nonuniform surfaces, including adsorption isotherms, is given in Appendix 2. Below are the simplest expressions of the adsorption isotherms in the two approximations of taking into account the lateral interaction. More accurate methods for calculating the relative distribution of molecules in nonuniform systems are considered in Appendix 3.

In the quasichemical approximation (QCA) we obtain a single closed system of equations for the entire lattice as a whole. If we only consider the direct correlations between the nearest interacting particles, the closed system of equations for θ_q – the probability of the adsorbed particle being in a site of type q – has the form [1, 2, 9]

$$a_q P \theta_q^V = \theta_q \prod_{p \in z_q} S_{qp}, \quad S_{qp} = 1 + \theta_{qp}^{AA} x_{qp}^{AA} / \theta_q^A,$$

$$x_{qp}^{AA} = \exp\{\beta(\varepsilon_{qp}^{VA} - \varepsilon_{qp}^{AA})\} - 1 = \exp(-\beta\varepsilon_{qp}^{AA}) - 1. \tag{7.4}$$

For the adsorption of particles of one type ($s = 2$), the equation for the pair functions $\theta_{fg}(r)$ – the probability of finding two particles adsorbed on neighbouring sites of type q and p, are resolved in an explicit form:

$$\theta_{qp}^{AA} = \frac{2\theta_q^A \theta_p^A}{\delta_{qp} + b_{qp}}, \quad \delta_{qp} = 1 + x_{qp}^{AA}(1 - \theta_q^A - \theta_p^A), \quad b_{pq} = \left\{[\delta_{qp}]^2 + 4x_{qp}^{AA}\theta_q^A\theta_p^A\right\} \tag{7.5}$$

$$\sum_{i=1}^{s}\theta_{qp}^{ij}=\theta_{p},\quad \sum_{j=1}^{s}\theta_{qp}^{ij}=\theta_{q},\quad \sum_{i=1}^{s}\theta_{q}^{i}=1. \qquad (7.6)$$

In the second equality for the nonideality function S_{qp}^{i} (7.4) we take explicitly into account that the interaction of the particles with free sites (these are particles with an index s_{p}) are zero. To keep the notation for the symbol s_{p} simple, we omit the site number to which it relates. It is clear in meaning: symbol p concerns the site on which the sum undertakes over a sort of prticles. In the designation of θ_{q}^{s} and θ_{sq}^{is} symbol s_{q} or s_{p} relates to a corresponding lower index q or p. The system (7.4)–(7.6) is an intermediate form between the general system of equations (A2.2a), (A2.6), (A2.7) and system (A2.13) of Appendix 2. In it instead of local sites with specific numbers f and g there are sites of different types q and p, which relate to the macroportions of the surface at $d(q\{\lambda\}R) = 1$ and $R = 1$.

Mean field approximation (MFA)

In this approximation there are no correlation effects between the interacting molecules, so it is quite rough. Equations for the adsorption isotherm in the mean field approximation were constructed in [14]. For the first time in about forty years after the construction of the equations for a homogeneous surface in this approximation [15, 16] and for isotherms on the nonuniform surfaces by Zel'dovich [17] and Temkin [18], it was possible to obtain the equations for the joint effect of both the heterogeneity and lateral interactions for all degrees fillings and for all types of non-uniform surfaces. This work has defined the direction of the subsequent work on the joint account of these factors. The principal approach was the introduction of the pair distribution function for different types of sites for arbitrary concentrations of nonuniform centres, and consideration of the dependence of the molecular parameters (Langmuir constants a_{q} and parameters of lateral interactions ε_{qp}) on the type of surface sites.

The local adsorption isotherms are written as [14]

$$a_{q}P = \frac{\theta_{q}}{1-\theta_{q}}\exp\left(-\beta z_{q}\sum_{p=1}^{t}d_{qp}\varepsilon_{qp}\theta_{q}\right),\quad 1\le q\le t, \qquad (7.7)$$

The total coverage of the surface is obtained as a function of external pressure P by weighing the contributions $\theta_{q}(P)$ from each plot using the normalized distribution function f_{q} (instead of the integral equation (6.2)):

$$\theta(P) = \sum_{q=1}^{t} f_q \theta_q(p), \quad \sum_{q=1}^{t} f_q = 1. \tag{7.8}$$

If the surface is homogeneous ($t = 1$), this equation changes to the adsorption equation [15, 16], which was empirically derived by Frumkin [18]: $ap = \theta/(1 - \theta) \exp(-2\alpha\theta)$, where α is a certain quantity that characterizes the lateral interaction between the adsorbed molecules, taking into account the lateral interactions in the mean field approximation (see also [20]). Equation (7.7) allows one to determine the meaning of the empirical coefficient $\alpha = z\beta\varepsilon/2$.

In the next section, this equation is used for the qualitative explanation of the features of condensation processes in nonuniform systems which have also been determined for the first time using this approach. However, for a quantitative description of equilibrium and, especially, kinetic experiments, this approximation should not be used because of its fundamental flaws (Appendix 6). The strictly posed system of equations (7.7) is only for $\beta\varepsilon_{qp} \to 0$, which significantly limits its potential for quantitative calculations.

In the absence of lateral interactions (ideal model) the filling the sites of type q, $1 \leq q \leq t$, where t is the number of different types of sites on the surface, is described by equation (6.1), i.e., each centre is 'working' independently.

8. Adsorption on nonuniform surfaces

Consider the typical dependences of the adsorption isotherms on nonuniform surfaces in the simplest case of two types of adsorption centres ($t = 2$), which differ in the way of mutual arrangement. The full range of locations of sites can be easily illustrated by the example of a binary alloy surface (described by the functions f_q and d_{qp}), comprising both types of sites in the same amount ($f_1 = f_2 = 1/2$). If the sites of the same type form macroscopic homogeneous sites $d_{12} = d_{21} = 0$, then it is a case of 'spotty' surface (the contribution of the boundary between the areas is neglected). The random distribution of sites is obtained with the full mixing $d_{qp} = f_q$. Finally, the regular alternation of two types of sites (checkerboard) is fully ordered surface $d_{12} = d_{21} = 1$. Between these marginal distributions of sites different types of intermediate realized how the relative position of sites. If $f_1 \neq 1/2$, so it is possible to put the same limit distributions and characterize their function d_{12} or $d_{21} = f_1 d_{12}/f_2$. If $d_{12} = 0$ we

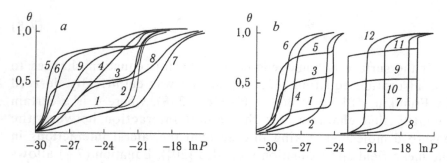

Fig. 8.1. Adsorption isotherms for attracting molecules $\beta\varepsilon = 1$ (curve *1–6*) and is the absence of interaction (curves *7–9*) (*a*): $Q_1 = 113$, $Q_2 = 80$ kJ/mol; $z = 4$, $T = 500$ K, $a_1/a_2 = \exp\beta(Q_1 - Q_2)$; $f_1 = 0.2$ (*1, 2, 7*), $f_1 = 0.5$ (*3, 4, 8*); $f_1 = 0.8$ (*5, 6, 9*). Surface – spotty (odd number) and ordered (even numbers) [21]. Adsorption isotherms of attracting molecules (*b*): $Q_2 = 88$ kJ/mol. *1–6* – $\beta\varepsilon_{11} = 1.0$, $\beta\varepsilon_{12} = 1.75$, $\beta\varepsilon_{22} = 1.5$; *7–12* – $\beta\varepsilon_{12} = 2.3$ (curves *7–12* are shifted along the horizontal axis by 7 units); $f_1 = 0.2$ (*1, 2, 7, 8*), 0.5 (*3, 4, 9, 10*), 0.8 (*5, 6, 11, 12*) [21].

analogue spotted surface, $d_{12} = f_2$ – analog chaotic surface, when $d_{12} = 1$, if $f_1 < 1/2$, then each site type 1 is surrounded by two sites of the first type and by analogy with the case of $f_1 = 1/2$ it can be considered an ordered surface. If $f_1 > 1/2$, then the ordered surface meets $d_{21} = 1$, ie, $d_{12} = f_2/f_1$.

Physical adsorption isotherms are shown in Fig. 8.1 *a* [21]; particle attraction leads to faster growth of $\theta(P)$ than in the absence of interaction. When $\theta < f_1$ attraction of particles manifested in different ways depending on the structure of the surface: the value of θ increases more rapidly with increasing pressure at the spotty surface than on the ordered surface. In the second case, the contribution of interaction plays a lesser role, as sites with higher adsorbability are surrounded by sites with a lower adsorbability, which have a lower degree of filling than the adjacent sites on the uniform portion in the first case. Therefore, the isotherms at $\varepsilon > 0$ on the ordered surfaces at $f_1 = 0.2$ and 0.5 are practically identical with the isotherms at $\varepsilon = 0$.

For $f_1 = 0.8$ the equivalent area of coincidence of the isotherms is limited $\theta < 0.07$. When $\theta \sim f_1$ the adsorption isotherms for surfaces of different structure about the same, this is due to the condition $\varepsilon_{qp} = \varepsilon$. With a further increase in pressure the isotherms for the ordered surfaces increases faster than for the 'spotty' surface because the molecules adsorbed on them have a larger number of occupied neighbouring sites. The limiting values of the pressure P^*, which correspond to the higher degrees of filling virtually unaffected changing by increasing pressure, depend on the composition of the surface, the adsorption coefficient, the values of the parameters of the interactions and the structure of the surface.

Figure 8.1 *b* shows the physical adsorption isotherms for the case $\varepsilon_{qp} \neq \varepsilon$. For homogeneous regions of the second type, wherein the attraction of the particles is stronger, the isotherms are close to critical. On the ordered surfaces the nature of the isotherms is similar to the isotherms in Fig. 8.1 *a*. The influence of the surface structure is clearly manifested in the general form of the isotherms. Conventionally, it can be assumed that on the spotted surface the filling of each plot is consistent and the ordered surface is involved completely in adsorption and, in principle, its isotherm can be approximated by the isotherm for a homogeneous surface with some effective values of the interaction parameter and the adsorption coefficient. This is most clearly seen for the curves *7–12* in Fig. 8.1 *b*. When two phase transitions are realized on the spotty surface, only one phase transition takes place on the ordered surface (at the given parameters for curve *12* the phase transition does not occur).

In the supercritical temperature range the type of isotherm can be used to make assumptions about the number of types of centres only when the difference $\beta(\varepsilon_q - \varepsilon_{q+}1) > 5$. For chemisorption, depending on the structure of the surface and the magnitude of repulsion, this difference does not always allow us to estimate the number of types of surface sites. It is therefore advisable to obtain independent information about the properties of the surface and, using the proposed theory, to check the correctness of such estimates of the distribution functions of different types of sites on the basis of the form of the calculated adsorption isotherms (and not find these distribution functions from the adsorption isotherms), or vary them in previously limited ranges. The rationale for this approach is confirmed by examination of the argon–zeolite NaX system[21].

Ordering of adsorbed particles
The interaction between the particles determines the cooperative behaviour of the system. If the magnitude of this interaction is greater than the kinetic energy of the particles, the system can show a qualitatively new distribution of the particles compared to an ideal system. Particle attraction leads to their condensation. Phase transitions of stratification are discussed in chapter 3. In the case of repulsion of particles at low temperatures there is a transition from the random distribution of the particles to their regular arrangement in the lattice sites, so that primarily occupied sites form a new

regular lattice with a period greater than the lattice constant of the surface structure. That is, the original uniform lattice is divided into a set of sublattices with different probabilities of filling with the particles. The phenomenon of ordering is realized in many chemisorbed systems [3, 22–26], as well as in the substitutional and interstitial solid solutions [27–30].

We will consider the simplest ordered system that is formed on the (100) face at average occupancies – superstructure $c(2 \times 2)$ (in the terminology of low-energy electron diffraction studies). For a description of other superstructures it is necessary to take into account the contributions from the interaction of more distant neighbours (see [26]). This superstructure is characterized by the fact that all lattice sites break up into two sublattices. The first, the sites of which are occupied, is denoted α, the second in which the sites are predominantly free, is denoted γ. Each site of the α-sublattice is surrounded by the sites of the γ-sublattice, and vice versa. This enables the ordered state of the particles to be defined by the above-introduced cluster distribution functions of different types of sites: $f_\alpha = f_\gamma = 1/2$ and $d(\alpha[0, z]) = d(\gamma[z, 0]) = 1$. The same adsorption equation are obtained by description using the pair distribution function: $d_{\alpha\gamma} = d_{\gamma\alpha} = 1$.

θ_σ ($\sigma = \alpha, \gamma$) denotes the probability of occupation type σ sites. The adsorption isotherm (7.4)–(7.6) can be written as $\theta = (\theta_\alpha + \theta_\gamma)/2$, where θ_σ is determined from the system of equations:

$y = aP = \dfrac{\theta_\alpha}{1-\theta_\alpha}(1+t_{\alpha\gamma}x)^z = \dfrac{\theta_\gamma}{1-\theta_\gamma}(1+t_{\gamma\alpha}x)^z$, where $x = \exp(-\beta\varepsilon) - 1$ and $t_{\alpha\gamma} = \theta_{\alpha\gamma}/\theta_\alpha$, $\theta_{\alpha\gamma} = 2\theta_\alpha\theta_\gamma/(\delta + b)$, $\delta = 1 + x(1 - \theta_\alpha - \theta_\gamma)$, $b = (\delta^2 + 4x\theta_\alpha\theta_\gamma)^{1/2}$.

Here, the particles are located on a homogeneous surface and their internal degrees of freedom and interaction parameters do not depend on the type of site of the sublattice. If in these equations we eliminate y, we obtain the equation linking the relative site occupations θ_α and θ_γ. Given $\theta_\gamma = 2\theta - \theta_\alpha$, this equation allows us to calculate $\theta_\alpha(\theta, \beta\varepsilon)$. It has one or three roots. In the first case, there is no ordering of the particles in the system, in the second case – the higher value of the root relates to θ_α, and the lower – to θ_γ. The temperature below which an ordered state may form is called the ordering temperature. Its value is determined by expanding the equation relating the filling of sites α and γ, for small values of the order parameter ρ, which characterizes the fraction of particles at sublattice sites α [27]: $\theta_\alpha = \theta + \rho(1 - \theta)$, $\theta_\gamma = \theta - \rho(1 - \theta)$. If all

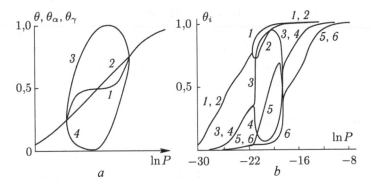

Fig. 8.2. The adsorption isotherms (*a*): the ordering of particles (*1*), disordered system (*2*) and the filling of sites α and γ of the sublattices (3, 4) [31]. The population of sites of the stepped surface of ordering particles at $\beta\varepsilon = -2$, $l = 10$, $t = 6$ (*b*): $Q_1 = Q_2 = 113$, $Q_3 = Q_4 = 97$, $Q_5 = Q_6 = 80$ kJ/mole [32].

the particles are in sublattice α, then $\rho = 1$ and $\theta_\alpha = 1$. The disordered state corresponds to $\rho = 0$ and $\theta_\alpha = \theta_\gamma = \theta$.

Taking ordering into account changes the form of the adsorption isotherms. Figure 8.2 *a* shows the filling of sites of both sublattices and the total filling of the surface, depending on the external pressure. Upon reaching the critical surface filling rapid filling of the sublattice α starts to take place, mainly due to the release of the sublattice sites γ. The value of θ_γ passes through a minimum, and then starts to increase with increasing pressure. When approaching the second critical filling, the filling of sites γ mainly occurs due to the release of the sites α. After completing the equal filling of both sublattices the system reverts to a disordered state.

The distribution pattern of repulsive particles with decreasing temperature is also illustrated by the dependence $\theta_{AA}(\beta\varepsilon)$ at half-filling of the surface (see Fig. 5.4, curve *4*). Accounting for ordering dramatically changes the curve 4 and brings it to the Onsager exact solution almost in the entire temperature range (maximum difference in the vicinity of the critical point of the order–disorder transition).

Ordering of the particles takes plane not only on the homogeneous surface. Figure 8.2 *b* shows the adsorption isotherm for a stepped surface and the filling of its sites [32]. The distribution functions f_q and d_{qp} for fixed l and t are defined by the formulas given above in section 7. For a stepped surface the number of sites of different types is doubled compared to the nonuniform surface, but without taking the ordering of particles into account, through the introduction of sites α and γ on the terrace and near to the step. Although the dependences become more complicated than for a homogeneous surface, from the analysis of local populations we can separate areas for which the heterogeneity of the sites (small and large filling) and the ordering of particles (average occupancy) are 'responsible'. For more complex situations, this procedure is more cumbersome.

9. Motion of molecules in cells

The lattice model operates with energy parameters whose values discretely change with increasing number of the coordination sphere around some site (Appendix 2). Other approaches (integral equations, numerical methods, Monte Carlo and molecular dynamics) use the continuous dependence of the potential energy of interaction of two particles on distance. In [33–35] the author proposed a method of calculating the energy parameters of the lattice model based on the information about the intermolecular interaction potentials.

We describe the same system using a lattice model and a continuum model (all coordinates, characterizing the position of the molecules, change continuously). We start from the assumption that the free energy of the system is independent of the method of description, and select in the 'pure' form the contributions made to the total free energy of the system by a single molecule and a pair of molecules in each description method. This procedure allows one to compare directly one- and two-particle contributions to the free energy and link the parameters of different description methods. As the free energy is $F = -\beta^{-1} \ln Q$, where $\beta = (k_B T)^{-1}$, Q – the statistical sum of the system, the equality of the free energies denotes the equality of the statistical sums of the system in the discrete and continuum models: $Q_{lattice} = Q_{conti}$. In Appendix 4 we consider both statistical sums and obtain expressions for the lattice parameters of the interaction between the molecules and between the molecule and the surface of the solid.

This eliminates the need to use them as parameters determined from experimental data, and calculate them on the basis of available data on the molecule–molecule and molecule–adsorbent interactions. Thus, without loss of information about the molecular nature of the system, we can use the well-known advantages of lattice models related to the simplicity of their numerical calculation: the equilibrium distribution of the molecules in the lattice model is described by algebraic systems of equations instead of integral equations and numerical experiments.

To illustrate the method of calculating the parameters of the lattice model, we limit ourselves to a mixture of polyatomic molecules having approximately the same volumes and not very different in form. (The basic idea of the calculation of the lattice parameters can be directly generalized to a mixture of molecules of different sizes.)

The obtained results allow us to offer a methodology for the use of information on the potential energy of the non-rearranged adsorbent for the calculation of lattice parameters in order to determine its equilibrium characteristics for arbitrary densities of the adsorbate. It includes the following steps (see diagram) [33].

1. Setting the coordinates of the atoms of the surface layer and the surface region of a solid body; selection of the interaction potentials of the adsorbate molecules with the surface atoms of the solid.

2. The calculation of the potential energy of interaction of the molecule of the adsorbate with the adsorbent U_p, in the region of space above the surface in which the action of the surface potential is evident.

3. Determination of the coordinates of the singular points on the maps of potential energy (minimum, maximum, saddle points, etc.) and energy values in them.

4. The calculation of the vibrational sum over states of the adsorbate in the vicinity of each local minimum n. In the harmonic approximation, $F_I(n) = \prod_{k=1}^{3}\left[1 - \exp\left(-\beta\hbar\omega_k(n)\right)\right]^{-1}$ where $\omega_k(n)$ are the natural frequencies of the vibrations on the axes $\xi_k = x, y, z, k = 1 - 3$, obtained by the matrix diagonalization of the second derivatives $\left\|\partial^2 U/\partial\xi_k^2\right\|$ at minimum n.

5. Calculation of local and macroscopic Henry constants. For each local minimum we calculate the local Henry constant (A4.22), (A4.30). The macroscopic Henry constants represent the mean values of local Henry constants, located on the unit surface of the adsorbent:

$a_I = \sum_{n=1}^{N} a_n^I / N_m = \sum_{q=1}^{t} f_q a_n^I$, where N_m is the number of local minima on the unit surface; $f_q = N_q/N_m$ (the second equality is the usual form of writing that uses the distribution function of the adsorption centres [9, 36–38] for the binding energies); N_q is the number of centres with the adsorbability, corresponding to the discrete index q; t is the number of different types of adsorption centres. If t is large, instead of the sum of the sign in the second equation we use the integral sign.

6. Using the lattice model to describe arbitrary filling of the surface. For this purpose the space of the system is broken down into sites: 1) the size of the site selected equal to the characteristic dimension (diameter) of the molecule so that each site has one molecule (the condition of the lattice model); 2) when the condition 1 is satisfied differences in the dimensions of the sites are permitted so that the volume of all sites is equal to the volume of the system. In general, the partitioning process is ambiguous, but the fulfilment of both conditions and taking into account peculiarities of the charts of the potential energy provides a sufficient accuracy of approximation. Without going into the specific details of this procedure, we note that it is based on the identification of the deepest minima (minimum coordinates correspond to the centre of the cell) and bordering of the deep minima by their closest cells. This reflects the fact that the deepest minima are filled first, and the presence of the adsorbate molecules in them affects the probability of occupation of the neighbouring cells. When partitioning the space the information about the positions of the maxima and saddle points, bordering the local minimum, is taken into account.

7. Calculation of the local Henry's law constant for each site. If a site has a local minimum, the local Henry constant for this minimum is the local Henry constant for the site. In many cases, the region of the site may include several local minima, or they may be omitted altogether. Formulas (A4.22), (P4.30) remain valid in these situations.

8. The nonuniform lattice structure is characterized by the distribution functions of sites, pairs of sites and clusters of sites. The first describe the composition of the lattice, the others – its structure. The distribution function of the sites with respect to the adsorption capacity is constructed using the data on the quantities a_f^l: if the quantities a_f^l and a_n^l are not significantly different $\left(\left|a_f^l - a_n^l\right| < \Delta a\right.$, where Δa – criterion of difference of the site types), they can be assigned to one type, otherwise they belong to different types of sites. Denoting the number of sites belonging to the same type q, for example, through N_q, we define the distribution function of the sites as f_q (as shown above). The pair distribution functions $f_{qp}(r)$ characterize the probability of finding sites of type q and p at a distance r. These functions reflect on average the structure of the system. The local structure is reflected more accurately by the cluster distribution function [1, 9]. The simplest of these are the functions that characterize the probability of formation of clusters with one or two central sites and their R coordination spheres.

9. The calculation of the lattice energy parameters using formulas (A4.18) and (A4.29) for a given division of the space to sites.

Thus, the initial information about the structure of the surface area of the adsorbent and the adsorbent–adsorbate interaction potentials allows to get all the parameters of the lattice model and use them to calculate all of the thermodynamic characteristics of the system. Similarly, the adsorption process can be studied not only at the solution–solid interface, but also in porous and non-porous adsorbents. An example of the use of this technique is given for the latter case. The process of adsorption of molecules of CH_4, CO_2 and Kr atoms in a glassy matrix base on a polycarbonate is considered. The mobility of the chains below the glass transition point is frozen so that this method can be used, since it is assumed in the method that the structure of the matrix does not change during the sorption process.

Sequence of operations	
Setting the atomic coordinates of the structure $\downarrow$	Selection of interaction potentials $\downarrow$
The calculation of the potential relief	
Analysis of the potential relief: the coordinates of $\downarrow$ extreme and other singular points $\downarrow$	
Arbitrary density	Low density
The introduction of lattice structure: $\rightarrow$ partition of the space to the sites $\downarrow$	Calculations of the vibrational sums over the states of the local minima $\downarrow$
Calculation of site local Henry constants $\downarrow$	$\leftarrow$ The calculation of the local Henry constants $\downarrow$
The construction of the distribution functions that characterize the composition and structure of the lattice structure $\downarrow$	Macroscopic Henry constant
Calculation of the effective parameters of the interparticle interaction $\downarrow$	
Calculation of the macroscopic characteristics of the system based on the lattice model	

Figure 9.1 *a* shows schematically polymer chains polycarbonatabis-phenol A forming local channels in a glassy matrix, and the cross section of these chains. The atomic coordinates were taken from [39]. The linear size $L \approx 20$ Å allowed to exclude boundary effects for the central region of the channel, and the value of $l_1 = 8$ Å is selected from the conditions of the convergence of chains such which enables the movement of the adsorbate molecules (for details see [33]).

The potential energy is calculated in the atom–atom approximation. The Lennard–Jones contributions with the simplest rules of combination were taken into account; the parameters of the potentials for atoms and groups of atoms in the polymer chain are given in [40–42]. In these systems the area of all sites contained several local minima. Figure 9.1 *b* schematically shows a site comprising three local minima (Kr-2, see Table). Also shown are the profiles of the potential energy in the vicinity of the first minimum in the coordinates x ($x =$ const) and y ($x =$ const) that illustrate the local anisotropy of the potential surface.

Calculations were carried out for two mutually arranged adjacent chains; *1* – chains parallel to each other; *2* – chains shifted relative to each other by 0.45 nm ($T = 350°C$). The calculation results are shown in Table and Fig. 9.2. Comparison of the experimental data (curve *1*) [43] and theoretical (curves *2, 3*) isotherms indicates a qualitative agreement.

Molecular calculation for small degrees of filling serves as a check of local Henry constants, presented in Table 2.1. The value of the lateral interaction corresponds to the minimum of the LJ potential. For large degrees of filling there is a deviation from the

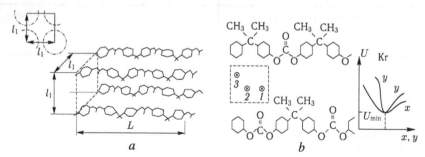

Fig. 9.1. Arrangement of the polymer chains of polycarbonate (parallel arrangement), which form channels for the migration of sorbate molecules (*a*); with offset relative to each other and the area of the site of the first type for the Kr-2 system; three of the deepest minima are identified (*b*). On the right there are energy curves for the first (largest in absolute value) minimum $U_{min} = 4.3$ kcal/mol at a site of the first type.

Table 2.1. Local Henry constant ($T = 350°C$) in glassy polycarbonate

System	t	f_q	a_q	Q_q
Kr-1	4	0.15	0.033	3.72
		0.10	0.007	2.85
		0.05	0.003	1.31
		0.70	0.0009	3.98
Kr-2	3	0.05	0.043	3.98
		0.05	0.01	3.00
		0.90	0.001	1.20
CO_2-1	3	0.05	0.31	3.92
		0.05	0.24	3.89
		.90	0.024	3.42
CO_2-2	3	0.05	0.43	4.02
		0.05	0.27	3.91
		0.90	0.026	3.46
CH_4-1	4	0.05	0.048	3.37
		0.05	0.030	3.51
		0.10	0.0063	2.54
		0.80	0.0019	2.26
CH_4-2	3	0.05	0.052	3.43
		0.05	0.034	3.46
		0.90	0.0068	2.42

Legend: t – the number of types of centres, f_q – the proportion of sites of type q, a_q – local Henry's constant, Q_q – local heat of sorption

experimental curves in Fig. 9.2, since the volume values of the lateral interaction parameters do not agree well with the mutual interaction of the molecules in a dense matrix.

This factor was not taken into account in refining the model to fit the experimental and theoretical curves. However, reasonable estimates of the local Henry constants qualitatively predict the correct behaviour of the isotherm also at large fillings. Curves 2 and 3 illustrate the effect of the molecular structure on the type of isotherm. In this case, the isotherms are sensitive to the shift of the chains, which in principle can be used to match the structural

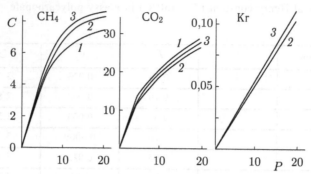

Fig. 9.2. Experimental (*1*) and calculated (*2, 3*) adsorption isotherm; *2* – chains without shift; *3* – chains shifted by 0.45 nm [33].

characteristics derived from independent experimental measurements. The calculation gives not only an assessment of the macroscopic Henry constants, but also the estimates of the number of types of sites and their local Henry constants needed to calculate the diffusion of molecules in matrices.

The lattice parameter of the continuum model [44]

Appendix 4 also shows that if we explicitly take into account the translational degrees of freedom for the centre of mass of the molecule inside the cell, the following expression is obtained for the lattice parameter of interaction between neighbouring molecules

$$\varepsilon_{fg}^{ij}(r) = \frac{1}{\theta_{fg}^{ij}(r)} \sum_{f,gr} \int_{v(f)} \int_{v(g_r)} \varepsilon_{fg_r}^{ij}(r_i\,r_j)\,\theta_{fg_r}^{ij}(r_i\,r_j)\,dr_i\,dr_j, \qquad (9.1)$$

where the sum over pairs (fg_r) is taken over all pairs of sites at distance r.

To determine the distributions of continuous functions $\theta_{fg_r}^{ij}(r_i\,r_j)$ it is necessary to solve the corresponding equations given in Appendix 4. Thus, by including information on the solid structure and the intermolecular interaction potentials, in the lattice pattern, as in other methods of the molecular level, it is possible to obtain a complete description of the thermodynamic and kinetic characteristics of adsorption.

To simplify the solution of integral equations, we can introduce a number of assumptions [45], which will reduce the solution time without the loss of accuracy. Among these simplifications are the following constructions: 1) taking into account the excluded volume by the neighbouring molecules in their displacements from the centres of their cells; 2) taking into account the softness of the lattice structure associated with the thermal motion of molecules; 3) integration of many-particle contributions to the potential functions of the intermolecular interactions.

Taking the excluded volume into account [45]

In dense phases the neighbouring molecules hinder the movement of an arbitrarily chosen central particle, resulting in part of the volume of the system not available to this particle. The value of this volume must be taken into account in the calculation of the equilibrium distribution of molecules in the solid phase. The value of the excluded volume depends on the location of the neighbouring molecules, and its definition is a difficult task (in fact, it is the problem of the liquid state of matter). The influence of neighbours on the value of the excluded volume is explained by the example of a plane lattice (Fig. 9.3). The linear cell size is of the order of λ, and the diameter of the hard sphere of the molecules is of the order σ_{AA} ($\sigma_{AA} < \lambda$). If all neighbour cells g around a central cell f, are free, the centre of mass of the molecule in this cell may be at any point of the cell. If a single cell g is occupied alongside, then, depending on its position, part of the central cell f is not available to the centre of mass of the central molecule. The shaded area shows this forbidden part. If all adjacent cells g are filled, then a small part of the cell volume f remains available for the movement of the centre of mass of the central molecule. We write $J_f^i = B_f^i V_f^i$, where from the statistical sum J_f^i in the explicit form we separate the contributions from internal degrees of freedom of the molecule B_f^i and the available cell volume V_f^i.

We construct a *geometric model* that takes into account the influence of neighbouring molecules V_f^i on the cell size of f, available to the central molecule. This value can be represented as a function of distance from the centre λ_{fg} of this site f to the centres of their neighbours g. The latter fact allows to exclude the introduction of additional (to λ_{fg}) structural parameters in the expression for the free energy. The volume V_f^i is represented as the sum of volumes of sectors V_{fg}^{ij} (available to the same particle), averaged over the states of occupation j of the neighbouring sites g, since the variety neighbouring particles determines the value of the excluded (and thus available) volume of the cell f: $V_f^i = \Sigma_{g,j} V_{fg}^{ij} \theta_{fg}^{ij} / \theta_f^i$, V_{fg}^{ij} – the volume of the sector of the cell f along the bond fg, available for particle i, when the neighbouring cell g is occupied by a particle j.

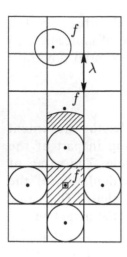

Fig. 9.3. Flat diagram illustrating an excluded volume effect. Three state of the central cell f: 1) in the cell is not occupied neighbouring sites, 2) once occupied a neighbouring site, and 3) all neighbouring sites are occupied. Shaded excluded volumes of the central cell.

For clarity, consider the case of a nonuniform lattice with $z = 6$. The volume of site f is $V_f = \lambda_f(x)\lambda_f(y)\lambda_f(z)$, where factors $\lambda_f(x) = [\lambda_{f1} + \lambda_{f3}]/2$, $\lambda_f(y) = [\lambda_{f2} + \lambda_{f4}]/2$, $\lambda_f(z) = [\lambda_{f5} + \lambda_{f6}]/2$ represent the size of a site along axes i; $i = x, y, z$. Numbers 1 and 3 are right and left of the site f along the axis x, and similarly for sites 2 and 4, along the y axis, and the sites 5 and 6 along the axis z. The volume of the sector V_{fg} along the bond fg can be expressed as $H_{fg}S_{fg}/3$, where $H_{fg} = \lambda_{fg}/2$ is the height and $S_{fg} = [\lambda_{f1} + \lambda_{f3}]$ $[\lambda_{f2} + \lambda_{f4}]/4$ is the area of the rectangular base of the 'pyramid' of this sector, the numbers of neighbouring sites h are numbered 1–4, where 1 and 3 relate to the same axis, and the numbers 2 and 4 – to another axis (both of these axes are perpendicular to the bond fg).

If the site f is free, $V_{fg}^{Vi} = V_{fg}$ – the entire volume of the site is assumed to be available for particle V for any neighbouring particle $i = A, V$. If the site f is occupied by the molecule A, then $V_{fg}^{Ai} \neq V_{fg}$, as blocking of part of the sector volume can be due to both to the neighbouring molecule A in the site g (Fig. 9.4), and due to the molecules A situated in neighbouring sites h, denoted by the above-mentioned numbers 1–4 in the expression for S_{fg}.

Figure 9.4 a shows a section along the bond fg, comprising sites $h = 1$ and 3 (sites 2 and 4 are perpendicular to the plane of the drawing). The shaded part of this sector indicates particle A in the site 1 (site 3 is free). Figure 9.4 b shows the excluded part of the same sector when all the neighbouring sites $h = 1 - 4$ are occupied by the molecules A.

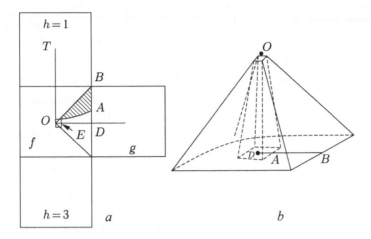

Fig. 9.4. Schematic representation of the case when the central cell f contains molecule A, and the adjacent site g is free (scale of DA and OE segments is not drawn): a – sectional view along the bond fg, comprising sites $h = 1$ and 3 (sites 2 and 4 are perpendicular to the plane of the drawing); b – the excluded part of the same sector when all neighbouring sites $h = 1$–4 are occupied by molecules A.

The excluded part of the sector along the bond fg when particle A is at the site 1 (site 3 is free) is shaded. Points O and T are the cell centres f and $h = 1$, point D is the middle of the bond fg, OE is the size of the region of thermal vibrations of molecules A in the close-packed lattice.

The OE segment characterizes the area available to the centre of mass of the molecule A in the central cell of f when the lattice is fully occupied by the molecules. This value is estimated from the average values of the thermal rms displacements $(\delta\lambda_{fg})^2$ from the molecules in the solid state in the harmonic approximation for temperatures above the Debye temperature [46]: $(\delta\lambda_{fg})^2/(\lambda_{fg})^2 = 1.5 \times 10^{-2}T/T_{\text{liq}}$, where T_{liq} is the melting point. In our evaluation of the liquid fluid we put $T = T_{\text{liq}}$. Expression $(\delta\lambda_{fg})^2 = 1.5\cdot10^{-2}(\lambda_{fg})^2$ allows us to find the linear displacement of the centre of the molecule along one of the axes: $\delta\lambda_{fg} = [(\delta\lambda_{fg})^2/3]^{1/2} = 0.071\ \lambda_{fg}$ and $OE = \delta\lambda_{fg}/2$. Note that in a nonuniform lattice the OE values for each axis are proportional to their average distance λ_{fg}. Knowing the distance OE makes it easy to find $V_{fg}^{AA} = s_{fg}h_{fg}/3$, where $h_{fg} = OE$ and $s_{fg} = (2OE)^2$.

To calculate the V_{fg}^{AV} it is necessary to consider the possible states of filling of the neighbouring sites $h = 1$–4 relative to the selected bond fg. We use the additive accounting scheme for the excluded volume, introduced by each of the four neighbours h, as shown in Fig. 9.4 b. The volume with all neighbours excluded is equal to all its neighbours $\left(V_{fg} - V_{fg}^*\right)$, where V_{fg}^* is the accessible part of the sector along the bond fg due to the central molecule in the filled sites $h = 1 - 4$ and the free site g. When calculating V_{fg}^*, we note that the length of the BA sector is maximum for the curved curve in the plane perpendicular to the bond fg, and we replace this curve and the curved portion of the curve AE by straight lines, whereas the volume of the 'interior' of the pyramid in Fig. 9.4 expressed as $V_{fg}^* = H_{fg}\left[S_{fg} - S_{fg}^*\right]/3$, where S_{fg}^* is the base of the internal pyramid limited by the segments DA^* for which from the model we can estimate $DA^* \approx 0.18\lambda_{fh}$. This value should be used to determine S_{fg}^* instead of λ_{fh} in the above formula for S_{fg}. As a result,

$$V_{fg}^{AV} = V_{fg} - \left(V_{fg} - V_{fg}^*\right)\sum_{h=1}^{4} t_{fh}/4,$$ where the conditional probabilities t_{fh} consider filling of the neighbouring sites h.

This structure expressions for V_{fg}^{AV} illustrates the many-body nature of the blocking of the central cell block by the particles surrounding it. This fact changes the form of the coefficients a_k in (7.4) and (A2.2 a) and requires modification of equations for the pair functions (A2.6), which can be rewritten as

$$\hat{\theta}_{fg}^{im}(r)\hat{\theta}_{fg}^{jk}(r) = \hat{\theta}_{fg}^{ik}(r)\hat{\theta}_{fg}^{jm}(r)\hat{V}_{fg}^{im}(r)\hat{V}_{fg}^{jk}(r)/[\hat{V}_{fg}^{ik}(r)\hat{V}_{fg}^{jm}(r)],$$ where we

can write $\hat{\theta}_{fg}^{im}(r) = \theta_{fg}^{im}(r)\exp\left[-\beta\varepsilon_{fg}^{ij}(r)\right]$ and $\hat{V}_{fg}^{im} = \sum\limits_{h\in z_f}\sum\limits_{j}V_{fh}^{ij} + V_{fg}^{im}$.

Here, the index h labels the four neighbours for the relationship fg, affecting the blocking of the site f, as indicated above. The value V_{fg}^{Vm} is defined with respect to the first index of the type of particles situated in site f. If one of the indices of the type relates vacancy, the $V_{fg}^{Vm} = v_0/z$. The geometrical model of taking into account the excluded volume in the form of pyramids can be generalized for any number z. The proposed models were used for the calculation of phase diagrams of simple fluids in the bulk phase and in cylindrical pores. The results are presented in the following chapter.

In conclusion, it should be noted that these equations are easily generalized to the mixture of components close in size. The model can significantly reduce the time of calculation of phase diagrams and other thermodynamic characteristics. An important advantage of the new model is also the possibility of accounting for the influence of intermolecular interactions on the frequency of interparticle fluctuations, since the change of the lattice constant changes both the depth of the potential well and its curvature in the vicinity of the minimum. This information is necessary for the calculation of the concentration dependences of the rates of elementary reactions in the liquid phases. All the above modifications of the lattice-gas model (LGM) can be used in the framework of the accounting adjustment correlations that improve the accuracy beyond the framework of quasichemical approximation for a wide range of applications (Appendix 3).

Softness or compressibility of the lattice structure [45]
Thermal motion of the molecules forms the average distances between the molecules, which are approximated by a lattice structure with a lattice constant λ_{fg}. Numerical values λ_{fg} should be determined by minimizing the local free energy, the general expression for which is defined in equation (A2.5). The same principle of using the minimum of free energy is saved with respect to any structural characteristics of the lattice system.

For the compressible lattice instead of the lattice parameters of the lateral interaction $\varepsilon_{fg}^{ij}(r)$ it is necessary to use the potentials of interparticle interactions that are functions of the distance between molecules. For fluid molecules we use the Lennard–Jones potential $\varphi_{AA}(r) = 4U_{AA}[(\sigma_{AA}/r)^{12} - (\sigma_{AA}/r)^{6}]$, and for the interaction of the molecules with the pore wall – the average potential of the type (9–3) or (10–4). Then the equations for local isotherms and the pair distribution functions contain, instead of fixed parameters $\varepsilon_{fg}^{ij}(r)$, the potential functions $\varphi_{AA}(r^*)$ for distances r^*, corresponding to the minimum of free energy of the system F.

Bulk phase
We assume that the lattice structure in the bulk phase retains its similarity at arbitrary fluid densities. Then all the values of $z(r)$ retain their values for a rigid lattice, and the distances between the first and subsequent neighbours are related by: $r = \lambda \eta_r$, where λ – the distance between the nearest neighbours, η_r – similarity numbers for different distances r; for example for the structure with $z = 6$ and for the four nearest neighbours $R = 4$ we have: $\eta_2 = \sqrt{2}$, $\eta_3 = \sqrt{3}$, $\eta_4 = 2$. In this case there is only one structure parameter λ, which may be determined from the condition $\partial F/\partial \lambda = 0$. The desired value of λ^* is easily determined explicitly $\lambda^* = \sigma_{AA}[2\Sigma_r z(r)\theta_{AA}(r)(\eta_r)^{-12}/\Sigma_r z(r)\theta_{AA}(r)(\eta_r)^{-6}]^{1/6}$. As a result, the lateral interaction parameters

become functions of the density of the fluid and the temperature of the system $\varepsilon_{fg}^{AA}(r) = 4U_{AA}\left[\left(\sigma_{AA}/\eta_r\lambda^*\right)^{12} - \left(\sigma_{AA}/\eta_r\lambda^*\right)^6\right]$. (Although in the particular case of $R = 1$, the value $\lambda^* = 2^{1/6}\sigma_{AA}$ not dependent on the density or temperature.)

Section 14 discusses a method for accounting for the compressibility of the lattice of slit-shaped pores.

Many-particle interactions

Influence of many-particle contributions to the potential functions of the interacting particles is considered in [9, 47]. This extension of the discussed application of the pair potential is taken into account easily in all of the expressions in this section.

10. Equation of state

The most important feature of any thermodynamic system is the equation of state which links pressure P, volume V and temperature T of a physically homogeneous system in thermodynamic equilibrium: $\Phi(P, V, T) = 0$. This equation is called the *thermal equation of state*. The thermal equation of state allows us to express pressure through volume and temperature $P = P(V, T)$. It is a necessary complement to the thermodynamic laws which permits their application to real substances. Also from the thermodynamics it follows that for the calculation of the thermal equation of state it is sufficient to know any of the thermodynamic potentials as a function of its parameters. For example, if we know the Gibbs energy $G(T, P)$, then the equation of state is derived by differentiation:

$$P = -\left(\frac{\partial F}{\partial V}\right)_T = -\left(\frac{\partial G}{\partial V}\right)_T. \tag{10.1}$$

It is important to emphasize that the equation of state can not be derived using only the laws of thermodynamics and theoretically calculated only on the basis of considerations of the structure of matter by statistical physics methods.

The equation of state $P = P(V, T)$ can be derived from the virial theorem or by differentiating the statistical function of the volume under isothermal conditions [11, 48, 49], which is reflected by the first equation (10.1). The equivalence of the two procedures for calculating the pressure was first proved by Bogolyubov [11].

The LGM traditionally uses the two-dimensional analogue of the three-dimensional pressure P — the so-called expansion pressure π [50]. We will use this symbol to distinguish it from the symbol P. In the LGM the original formulation is the formulation of the problem of the interphase exchange of particles between the system and the thermostat, so we consider a lattice system in terms of multicomponent solutions when all components have comparable sizes ($s > 2$) [51]. According to, for example, [52], the expression for the υ chemical potential is written as

$$\mu_i = u_i - T\sigma_i + P\upsilon_i, \tag{10.2}$$

where u_i, σ_i and υ_i are the internal energy, entropy and specific volume of the mixture component i; T and P are the temperature and pressure of the system; υ_i — partial molar volume per particle grade i. Gibbs potential (P, T = const) can be written as $G = \sum_{i=1}^{s} \mu_i N_i$, where the superscript s indicates the number of mixture components, N_i — the number of molecules of species i. The total number of sites in the system N coincides with the total number of particles, $N = \sum_{i=1}^{s} N_i$. In case of equal sizes of components of the mixture we have $V = N\upsilon_0$, where υ_0 is the average volume of the particle (or the cell). This construction coincides with the traditional thermodynamic determination $\upsilon_i = \partial V / \partial n_{i|T, P} = \upsilon_0$.

We pass from a multicomponent mixture to explicit accounting of the vacancies which are one of the components of the mixture, such as the component having the index s, so here we use symbol π. The total number of states of occupation of the lattice sites, including vacancies, remains equal to s, and the number of varieties of real molecules is ($s - 1$). By analogy with (10.2), we write $G = \sum_{i=1}^{s} (u_i - T\sigma_i + \pi\upsilon_0)N_i$, where we have introduced $\mu_s = u_s - T\sigma_s + \pi\upsilon_0 = -T\sigma_s + \pi\upsilon_0$, since the specificity of the vacancy as the sort of particles is seen in the fact that its internal energy u_s is identically equal to zero, and the other contributions in μ_s are preserved.

The Helmholtz energy of the system can be written as

$$F = G - \pi V = \sum_{i=1}^{s} (u_i - T\sigma_i)N_i = \sum_{i=1}^{s} \mu_i(\text{conf})N_i, \tag{10.3}$$

where $\mu_i(\text{conf}) = u_i - T\sigma_i$ is the configuration part of the chemical potential, which includes the internal energy and the entropy of the molecule i, for the vacancy $\mu_s(\text{conf}) = -T\sigma_s$, (note that in [50] the need to use the configuration part of the chemical potential $\mu_i(\text{conf})$ is not emphasized).

Using these expressions, we derive the thermodynamic expressions for the LGM. The energy change can be written as

$$dE = T\,dS + \sum_{i=1}^{s-1}\mu_i(\text{conf})\,dN_i + \mu_s(\text{conf})\,dN_s.$$

Then we obtain the Helmholtz free energy

$$dF = -S\,dT + \sum_{i=1}^{s-1}\mu_i(\text{conf})\,dN_i + \mu_s(\text{conf})\,dN_s = -S\,dT + \sum_{i=1}^{s-1}\mu_i\,dN_i - \pi\upsilon_0\,dN.$$

The difference between the Helmholtz and Gibbs potentials defines the increment

$$d(\pi\upsilon_0 N) = S\,dT + \pi\upsilon_0\,dN + \sum_{i=1}^{s-1}N_i\,d\mu_i(\text{conf}).$$

Whence the expression for the pressure of 'expansion', which plays the role of the equation of state: $Nd(\pi\upsilon_0) = S\,dT + \sum_{i=1}^{s-1}N_i\,d\mu_i(\text{conf})$ or in integral form at $T = \text{const}$:

$$\pi\upsilon_0 = \beta^{-1}\int_0^\theta \theta'(P)\,d\ln(P), \quad P = \sum_{i=1}^{s-1}P_i, \tag{10.4}$$

where P_i is the partial pressure of component i in the thermostat (in an ideal gas). Formula (10.4) connects the equation of state in the LGM and the total pressure in the thermostat through the chemical potentials of the ideal gas. For $s = 2$, the final expression (10.4) is given in [50]. When considering the two coexisting dense phases the characteristics of the thermostat act as an 'intermediate phase' of the comparison system.

Equation (10.4) reflects the above 'dualism' of the LGM: the system has a thermostat that determines the existence of cells with 'real' molecules, and there is pressure due to intermolecular interaction on the background of 'presence' of the thermostat. In this respect, we can talk about osmotic pressure of the molecules in the solvent, which is the vacuum in the LGM. The formal form of the equation (10.4) corresponds to the use of the Gibbs–Duhem equation [52–54] for the discussed situation with two particles (molecules and vacancies).

Compressibility of the lattice

Equation (10.4) shows that for a compressible lattice the calculation of the pressure P of the system requires an independent determination of $v_0(P, T)$, which requires knowledge of the concentration and temperature dependences of the cell volume. The appearance of the lattice in the LGM to simplify the calculation of configuration contributions automatically raises the question about the condition that determines the lattice parameter λ. Since the parameter λ is the internal parameter of the system, it should be determined from the internal conditions.

Traditionally, in the derivation of the equation of state for solids the lattice constant is determined by minimizing the Helmholtz free energy F (10.1) (or even just the internal energy as the potential energy contribution is decisive) [27, 30]. In general, the minimum condition F is a consequence of the 'intraphase' thermodynamic relation [52–54], which determines the pressure from equation (10.1), which must hold for any state of aggregation. However, the LGM formula (10.1) defines the functional dependence of the pressure on the selected cell size $P(\lambda)$. To eliminate the arbitrariness in the choice of this parameter, we must use an additional independent condition. This additional equation is the expansion pressure (10.4), indicating the value of the cell size λ. Thus, by analogy with single-phase systems, it is possible to calculate the pressure through joint consideration of equation (10.4) with the virial theorem (see below), or by minimizing the Helmholtz free energy F by the formula (10.1), i.e.,

$$\pi_{(4)} = P_{(\text{vir.theorem})}, \quad \pi_{(4)} = P_{(1)}, \tag{10.5}$$

where $\pi_{(4)}$, $P_{(1)}$, $P_{(\text{vir.theorem})}$ denote respectively the pressure by (10.4), (10.1) and the virial theorem (see below). Thus, any of the relations (10.5) gives the possibility to use internal links to find λ. As a result, equation (10.4) closes the system of equations for compressible (deformable) lattices and allows us to consider arbitrary pressure without introducing free (uncontrolled) parameters. (For more details see [51]).

The equation of state of a single-component system in the quasi-chemical approximation for a rigid lattice with a fixed value of the parameter λ is written as

$$\beta \pi v_0 = -\ln(1-\theta) - \frac{1}{2}\sum_{r=1}^{R} z(r)\ln\left[\theta_{vv}(r)/\theta_v^2\right], \tag{10.6}$$

where $\theta_v = 1 - \theta$ is the proportion of vacant lattice sites and $\theta_{vv}(r) = 1 - 2\theta + \theta_{AA}(r)$ is the proportion of pairs of free lattice sites, $\theta \equiv \theta_A$. The fluid compressibility factor is defined as $Z = \beta\pi v_0/\theta$. Here the fluid compressibility factor Z characterizes the change in the degree of filling rigid lattice sites when the pressure changes. For low densities equation (10.6) becomes the ideal gas equation $\beta\pi v_0 = \theta$.

In general, the expansion pressure in the LGM, which determines the equation of state, which does not depend on the type of approach between the molecules and is applicable for taking into account any internal motions of the molecules inside the cells, is given by the following expression:

$$\beta\pi = \int_0^\theta \frac{\theta}{v} d\ln P(\theta),$$ (10.7)

i.e. there is no limitation on the constancy of the parameters of lateral interactions and the lattice constant λ.

Equations (10.6) and (10.7) are derived for a homogeneous lattice of the two- or three-dimensional system, depending on the number of the nearest neighbours z. The nature of the compression process follows from the analysis of changes in the volume of the system: $dV = d(Mv_0) = M\,dv_0 + v_0\,dM$. The first term refers to a fixed number of cells, but with a variable lattice constant. The second term refers to a rigid lattice with constant parameter λ at a variable number of cells. When using the formula (10.6), the value π reflects the compressibility of the system volume V with $v_0 = $ const. Equation (10.7) considers both factors and, therefore, in the formula (10.5) it is necessary to use equation (10.7). Differences between the results of calculations by (10.6) and (10.7) depend on the specific molecular parameters and the role of the contribution of internal motions.

The equation of state for the virial theorem is expressed in the LGM in terms of the distribution function as follows [51]:

$$P = \frac{kT}{v(f)}\theta_f - \frac{1}{2dv(f)}\sum_{\chi=1}^{R}\sum_{f,g}\sum_{i,j=1}^{s-1} \int_{v(f)}\int_{v(g)} \theta_{ij}\left(r_{fg}\big|\chi\right)r_{ij}\left(\frac{\partial\varepsilon_{ij}\left(r_{fg}\big|\chi\right)}{\partial r_{ij}}\right)dr_i\,dr_j,$$ (10.8)

where $\theta_f = \sum_{i=1}^{s-1}\theta_f^i$, d is the dimension of the space. The form of equation (10.8) is a consequence of the continuum description of the positions of the molecules within discrete cells in the LGM when recording the pair distribution function in the normal equation of state with a continuous distribution of pairs of molecules. This expression can be simplified as indicated in the previous section. The partial form of equation (10.8) has been used previously in the problem of calculating vapour–liquid one-component systems ($s = 2$) in the LGM with the compressible lattice parameter as a function of density and temperature at $R = 1$ [55–57].

In homogeneous single-phase systems, there is a unique relationship between the chemical potential μ and pressure P. Phase equilibria correspond to nonuniform macroscopic systems. For them under isothermal conditions the thermodynamic equilibrium corresponds to the equality of the chemical potentials μ and pressure P in the coexisting phases. The same relationships between μ and P, which are in isolated phases, are retained.

Nonuniform systems
If we consider the adsorption of molecules on different macroscopic facets of the solid, although the system as a whole is nonuniform due to different surface potentials, each face behaves in an isolated manner and they satisfy the equilibrium equation relating the magnitude of adsorption and pressure in the surrounding medium of the vapour or liquid. For the adsorbed film (mono- or multilayer) the normal component of the pressure tensor $P_N(h) = P$ is retained, where h is the number of the monolayer in the adsorption film in the transition region between the solid surface and the ambient volume. However, due to the influence of the surface potential the mean pressure $P(h)$ does not remain equal to P. (For example, for a planar interface, we have the equality $P(h) = [P_N(h) + 2P_T(h)]/3$, where $P_T(h)$ is the tangential component of the pressure tensor.) The value of the chemical potential in equilibrium should remain strictly constant, so the relationship between the chemical potential and the pressure in the transition region will vary from layer to layer.

The same situation is observed for the curved phase boundary [58]. The pressure of the system varies from layer to layer within the transition region, while the value of the chemical potential is fixed.

The fundamental property of the molecular model in the LGM is that the molecular theory provides an expression for the local pressure in equilibrium for any nonuniform system. This is qualitatively different from all the thermodynamic constructions that imply the existence of appropriate pressure, but for it there is no way to calculate and/or experimental determination. As a result, the thermodynamic constructions with the phase interfaces do not provide a self-consistent link of the chemical potential with the pressure inside the transition area.

In general, for an arbitrary nonuniform lattice the expansion pressure π_f in the LGM is expressed by equation (A2.3) [58, 59]

$$\beta\pi_f \upsilon_f^0 = -\ln\theta_f^s - 1/2 \sum_{g \in z(f)} \ln\left(\theta_{fg}^{ss} / \left(\theta_f^s \theta_g^s\right)\right). \tag{10.9}$$

The components of the expansion pressure tensor are constructed in accordance with the symmetry of macroscopic phase boundaries.

The equation for the mean pressure according to the virial theorem (10.8) for the nonuniform lattice can be rewritten as follows:

$$P_f = \frac{kT}{\upsilon(f)}\theta_f -$$

$$- \frac{1}{2d\upsilon(f)} \sum_{\chi=1}^{R} \sum_{f,g} \sum_{i,j=1}^{s-1} \int_{\upsilon(f)} \int_{\upsilon(g)} \theta_{fg}^{ij}\left(r_{fg} | \chi\right) r_{ij} \left(\partial \varepsilon_{fg}^{ij}\left(r_{fg} | \chi\right) / \partial r_{ij}\right) dr_i dr_j. \tag{10.10}$$

The difference between (10.10) and (10.8) is only in the fixation of the numbers of cells of the nonuniform lattice f and g_r, inside which the particles i with the coordinate r_i and particles j with coordinate r_j move, whereas in the expression (10.8) the numbers are needed to fix the mutual distance between the central particle and its neighbours. Expressions for the components of the pressure tensor are written by such a rule, as specified for pressure expansion in [58, 59].

11. Bulk phase

Rarefied gas

Traditionally, the LGM is used to describe the dense phases. To show its ability to calculate the properties of the rarefied bulk phase, in [60] comparisons were made with the equation of state in the traditional form of the virial equation. Modified versions of the lattice theory, discussed in previous sections, were used. The compressibility of the lattice structure [45], pair interactions in the four coordination areas and ternary interactions of the nearest particles [9, 47] were taken into account. Calculation results are compared with similar calculations made on the basis of the 'quasi-chemical' gas-kinetic model [49, 61–64].

The calculation is performed according to equation (10.6). The magnitude of the lattice constant λ is determined from the minimum of the free energy of the system (section 9). Ternary interactions were taken into account in the form of the concentration dependence of the pair potential for the nearest neighbours [9, 47]: $\varepsilon\,(1|\theta) = \varepsilon\,(1)$ $(1 - D\theta)$, where D is the ratio of the contribution ratio of the ternary interactions to that of the pair interactions of the nearest neighbours.

The relationship of the molar fraction of dimers α_{dim} with particle concentration θ_A in the low concentration range is given by:

$$\alpha_{dim} = n_{dim}/n = \left\{\left[2\theta \exp(\beta\varepsilon)\right]^{-1} + 1\right\}^{-1}, \qquad (11.1)$$

where $n = n_{mon} + n_{dim}$, n_i is the number of atoms in the form of monomers and dimers in a unit volume, the index i denotes the monomers (i = mon) or dimers (i = dim).

In [65] the expression for the second virial coefficient was presented for a mixture of monomers and dimers in the following form: $B(T) = b_0 - K(T)$, where the equilibrium constant $K(T) = -B_b$ for gas in which the reaction of formation of the dimer $A + A + + M \leftrightarrow A_2 + M$ can take place, can be calculated using the method of statistical mechanics [50]; here $K(T) = k_f/k_{dis}$ (k_f and k_{dis} are the rate constants of formation and dissociation of the dimer), A – atoms and M – a molecule involved in a triple collision. The equilibrium constant $K(T)$ is associated with a change in the Gibbs free energy: $\Delta G_d^0 = G_d^0 - 2G_m^0 = -2\ln K(T)$. Here for the argon atoms the values of the constants of dimer Ar_2 according to [66] are: $\omega_e = 30.68$ cm^{-1}; $\omega_e \kappa_e = 2.42$ cm^{-1}; $B_e = 0.059$ cm^{-1}; $\alpha_e = 3.64 \cdot 10^{-3}$ cm^{-1}. $I_{Ar_2} = 5.86775 \times 10^{-36}$ kgm^2, $D_0 = 109.8$ K is the argon dimer dissociation energy. In further calculations we used the models of harmonic and anharmonic oscillators and a rigid rotator.

Figure 11.1 a shows the experimental and theoretical values depending on the compressibility factor for argon $Z = p/nkT$ from the numerical density $n^* = n\sigma^3$, where n is the numerical density, $\sigma = 3.405 \cdot 10^{-10}$ m at a temperature $T = 162$ K [67], and 328 K [68]. For $T = 162$ K the curves calculated in the model of the lattice gas and virial equation $Z = 1 + B(T)/V + C(T)/V^2 + D(T)/V^3 + E(T)/V^4$ are compared. The curves 1 and 2 are calculated by the LGM; calculations using virial equations [11]: 3 – consideration of the second virial coefficient, 4 – consideration of the second and third virial coefficients, and 5 – consideration from the second to fourth virial coefficients.

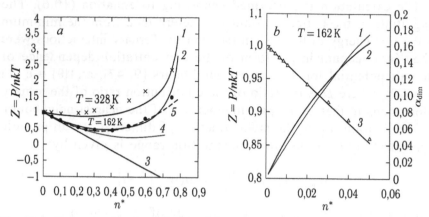

Fig. 11.1. Concentration dependence of the compressibility factor of argon (*a*): the experiment at 162 K [67] (black circles) and 328 K [68] (crosses). Compressibility factor Z at low concentrations of argon (curve *3*) and the molar fraction of argon dimers (curves *1* and *2*), triangles – experimental data [70] (*b*)

The obtained values of the lattice parameters correspond well to physical meaning: the value of ε is less than the depth of the potential well for the two argon atoms (~0.23 kcal/mol) [44] and to describe the equilibrium characteristics of inert gases it is necessary to consider the three-particle interaction (see, e.g., [69]). Apparently, in this case some overestimation of the contribution of triple interactions compensates for the neglect of indirect correlations and gives good agreement with experiment over the entire range of concentrations. It is evident that for the description of experimental data and to reach agreement with the lattice model we should use at least five members of the virial expansion. Curve *1* is calculated without changing the parameters ε and D, as their temperature dependence would lead to a closer agreement with the experimental data.

Figure 11.1 *b* shows the dependence of $Z(n^*)$ in the range $n^* < 0.05$ at $T = 162$ K. Curve *3* corresponds to the lattice-gas model. The linear dependence indicates the adequacy of use in this concentration only the second virial coefficient. Comparison of the molar fraction of bonded dimers (curves *1* and *2*) indicates a good agreement between the model and the gas-kinetic lattice model of the associating gas on the basis of the equilibrium constant $K(T)$ for the anharmonic model of the dimer Ar_2. Differences between the curves *1* and *2* in the range $n^* \sim 0.05$ and greater require taking into account the third and higher coefficients in the virial expansion.

Comparison of the proportion of the bonded dimers in a broader range of densities is given in [60] (with the coefficient of shear viscosity, see chapter 5).

A comparison of the calculations of the equilibrium characteristics of the argon atoms in a wide range of concentrations, from the dilute gas to the dense, almost liquid fluid state shows that the lattice model satisfactorily describes the experimental data and the results of calculations using the gas-kinetic model.

Other gases

A similar procedure was used to examine the compressibility coefficients of other gases over a wide range of pressures: helium, hydrogen, nitrogen and methane. Calculations of the compressibility and viscosity of helium, hydrogen, methane and nitrogen gas were carried out for these gases in the temperature range $T_c < T <$ 400 K and pressure $P_c < P <$ 100 MPa, which best represent the experimental results. This range of pressures and temperatures relates to the supercritical region of the gases.

The results of description of the experimental data for helium, hydrogen, methane and nitrogen are shown in Fig. 11.2. The ordinate gives dependence of the compressibility factor $\chi = PV/(P_s V_s)$ on pressure (index 's' denotes the standard conditions: 1 atm and 0°C). Calculations were carried out for different temperatures. The model parameters (α_0, d_e) were selected for the lowest temperature, and the curves for higher temperatures were used to determine the parameter u. The critical parameters of the considered gases (temperature T_c and pressure P_c) are shown in Table 2.2. Table 2.3 shows the parameters of the Lennard–Jones potential.

Table 2.4 summarizes the parameters corresponding to the curves shown. Values of diameters σ of the particles in Table 2.4 are chosen so as to match the theoretical and experimental values of viscosity used under standard conditions.

Figure 11.2 shows that the lattice gas model satisfactorily describes the main features of the gas compressibility factor to thousands of atmospheres for different temperatures. Figure 11.3 shows the dimensionless density of molecules θ, corresponding to the curves in Fig. 11.2, at pressures up to 1000 atm. Traditionally, this information is not used in the analysis, although it is of methodological interest. For the least compressible gas (helium), this pressure range corresponds to less than 20% volume filling. For the most compressed gas (methane) this corresponds to filling

Table 2.2. Critical values considered gases [72]

Molecule	T_c, K	P_c, MPa	V_c, cm³/g-mol	RT_c/P_cV_c
He	5.20	2.25	61.55	3.082
H₂	33.20	12.80	69.68	3.055
N₂	125.97	33.49	90.03	3.422
CO₂	304.16	72.83	94.23	3.639
CH₄	190.25	45.60	98.77	3.466

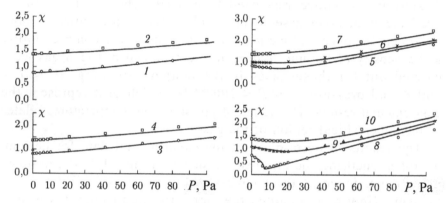

Fig. 11.2. Dependence of the coefficient of compressibility χ of some gases on pressure P (MPa). Solid curves – calculations, symbols – experiment. Helium at $T = 223$ K (*1*) and 373 K (*2*) [73, 74]; molecular hydrogen at $T = 223$ K (*3*) and 373 K (*4*) [73, 75]; molecular nitrogen at $T = 223$ K (*5*), 273 K (*6*) and 373 K (*7*) [73, 76]; methane at $T = 203$ K (*8*), 298 K (*9*) and 373 K (*10*) [73, 77].

Table 2.3. Parameters of the Lennard–Jones potential and viscosity under standard conditions ($T = 273.16$ K, $P = 0.1$ MPa) [72] (parameters are given without parentheses, they are taken as nominal in calculations) and [49] (parameters in brackets)

Molecule	Molecular weight	σ, A	ε/k, K	ε, kcal/ mol	$\eta_s \cdot 10^4$ Pa· s
He	4	2.7 (2.63)	6.03 (6.03)	0.0204	0.1887
H₂	2.016	2.968 (2.92)	33.3 (37.02)	0.0665	0.0850
N₂	28.016	3.681 (3.72)	91.46 (95.9)	0.0746	0.1674
CO₂	44	3.996 (4.57)	190.0 (185)	0.3800	0.1380
CH₄	16.031	3.822 (3.81)	136.5 (142.7)	0.2972	0.1033

Table 2.4. The model parameters corresponding to the curves in Figs. 11.2 and 11.3

Molecule	α_0	σ	d_e	u, K	α^*
He	1	2.6	20	223	0.45
H_2	0.6	2.75	0.5	288	1.9
N_2	1.25	3.7	0.5	500	0.85
CH_4	0.6	3.758	0	300	1.0

98 % of the volume, i.e. if for methane at low temperatures the system volume is almost full, then for helium the degree of filling of the volume is quite low. For molecular hydrogen we already have average degrees of filling the volume at pressures up to 100 MPa.

Particular attention should be paid to the results for helium atoms. The higher value $d_e = 20$ for helium means physically changing nature of the interaction between helium atoms at a certain value of its concentration to the opposite: the attraction of atoms at low densities is replaced by their repulsion with increasing concentration. This fact is fully consistent with the known data for liquid helium [78], in which the average distance between the atoms is 4.44 Å, which significantly exceeds the size of the cell selected at low densities of the order of the dimension σ of the Lennard–Jones potential. Therefore, in order to satisfy such a drastic change in the average distance between the atoms of helium for the rarefied and dense phases, the model parameters correspond to the effective repulsion of neighbouring atoms.

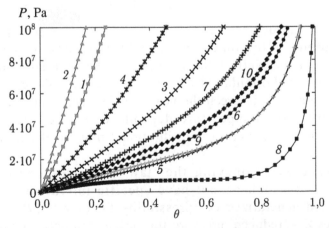

Fig. 11.3. Dependence of pressure on the degree of filling of sites of the lattice model (the numbers of curves correspond to the curves in Fig. 11.2).

For hydrogen molecules $d_e = 0.5$; these experimental data differ only slightly from similar values for helium atoms, the same parameter has the usual values (i.e. $0 < d_e < 1.0$, as for example for nitrogen molecules) and changing nature of the interaction (as determined by the sign of the energy parameters) is not happening. Thus, despite simplifications, the modified LGM provides a satisfactory agreement with the experimental data and the resultant model parameters have a strict physical sense, in particular, the specificity of the interaction between the particles of the fluid is taken into account.

Generalized compressibility factors [79]

These examples show that the agreement between the calculated curves and the experimental data can be used to employ the LGM for calculating the generalized compressibility charts, which 'are very useful for approximate calculations due to the speed and ease with which the desired results can be obtained' [80]. In other words, these charts give an idea of how this model can reflect the thermodynamic and dynamic characteristics of a large number of different molecules. This allows to extend the range of possible theoretical determination of the dissipative coefficients of dense gases from the subcritical to supercritical region.

For the analysis of the equation of state over a wide temperature range we can use a modified Lennard–Jones potential in the form

$$E_{ef} = \varepsilon_0 \left(1 - d_e \theta \delta_{1,r}\right) f(T), \quad \begin{cases} f(T) = \left(1 + u(T_r - 1)\right), & T_r < 3, \\ f(T) = \left(1 + u'T_r\right), & T_r \geq 3, \end{cases} \quad (11.2)$$

where E_{ef} is the depth of the well of the effective pair potential, which depends on the fluid density and temperature of the system, ε_0 is depth of the well of the pair potential in the soft lattice, as indicated above. Formula (11.2) takes into account the contribution of triple interactions between the nearest neighbours, and the function $f(T)$ modifies the temperature dependence of the effective pair interaction potential due to the vibrational motion of neighbouring molecules.

For dense gases obeying the 'law' of corresponding states (molecules of which do not have specific interactions) we use the so-called generalized compressibility factor $Z = \beta P v_0 / \theta$: in the coordinates Z – 'reduced' pressure for various 'reduced' temperatures all substances have some the same curves ('reduced' values are normalized to the corresponding values at the critical points).

Figure 11.4 *a* shows a generalized diagram of the compressibility factor Z as a function of the reduced pressure $P_r = P/P_c$ for various values of the reduced temperature $T_r = T/T_c$. Calculations were carried out under the following values of the energy parameters of the intermolecular interaction potential: $d_e = 0.5$, $\varepsilon_0/k = 2.75$ K, $u = -0.5$ and $u' = 0.05$. This chart agrees does not only qualitatively but also quantitatively (maximum deviations of the order of 3%) with a similar diagram by Hogen and Watson, obtained by averaging the values of Z for the following seven gases: H_2, N_2, CO_2, NH_3, CH_4, C_3H_8, C_5H_{12} (see [49, 81, 82]). The values of the energy parameters of the Lennard–Jones potential were used in the calculations as basic values. The calculation results for the compressibility show that the modified LGM, despite its simplicity, provides a satisfactory agreement with the experimental data from the values of the reduced parameters corresponding to an ideal gas, to the reduced pressure values of about 10 and the reduced temperature of about 15. A similar correspondence of the calculations and experiment for the compressibility factor χ was also obtained for neon and oxygen molecules [79].

Taking into account of the modifications of the potential function of the lateral interaction of particles improves the agreement with experiment (see Fig. 11.4 *b* for argon atoms compared with Fig. 11.1 *a*) [79].

High pressure [83]
The possibility of using the LGM throughout the density range

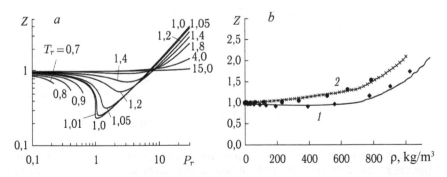

Fig. 11.4. Generalized dependence of the compressibility factor (*a*). Experimental [14] and the calculated dependences of the compressibility factor on the density of argon (*b*): $T = 273$ K (*1*) and 400 K (*2*).

allows to extrapolate the experimentally measured characteristics at relatively low pressures to at higher densities. For this purpose, it is of interest to evaluate the range of applicability of molecular parameters found experimentally in the vicinity of the critical point, at higher densities. Figure 11.5 *a* compares the results of calculation by the LGM with the experimental [84] dependence of the compressibility factor of argon on pressure in the range of up to 1000 MPa for several values of temperature: T = 308 K, 373 K and 473 K. At pressures above about 400 MPa (i.e. about $10p_c$), the calculated curves begin to deviate from the experimental ones, and the sooner the lower the temperature of the gas. To resolve this discrepancy it is necessary to consider the effect of high pressure on the deformation of the 'hard' sphere of the argon molecule.

Traditionally, the Buckingham potential (4.5) is used at high pressures. To use a single type of potential in the entire pressure range, the dependence of the effective diameter of the molecules on the density of the system was introduced using the approximation Boltzmann formula given in reference mathematical literature and electronic graphic files: $\sigma(\theta) = \dfrac{A_1 - A_2}{1 + \exp(\theta - \theta_0)/\delta} + A_2$, where for argon atoms coefficients are A_1 = 3.40411, A_2 = 2.4, θ_0 = 0.99213, δ = 0.0099.

The Buckingham and Lennard–Jones potentials in the vicinity of $\sigma \sim 1$ are shown in Fig. 11.5 *b*. Varying parameter α_B, equal to α in

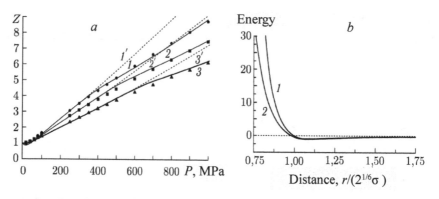

Fig. 11.5. Compressibility factor for argon atoms at different temperatures T, taking into account (solid curves) and without taking into account (dashed curves, curves with primed numbers) the concentration dependence of the particle diameter σ (*a*): T = 308 K (*1*), 373 K (*2*) 473 K (*3*); symbols – experiment [84]. Potentials (*b*) Lennard–Jones (*1*) and Buckingham (*2*) for α_B = 12.35.

(4.5), changes the phase diagram. Complete agreement of the phase diagrams is obtained by calculations with the Buckingham potential at $\alpha_B = 12.35$ and with the specified concentration dependence of the deformation of a hard sphere of the argon molecules in the Lennard–Jones potential. Accounting for this deformation allows to obtain good agreement (see Fig. 11.5 a) between the experimental data [84] and theory, with an error of less than 5% in the whole range of pressures.

Thus, the LGM satisfactorily reproduces the generalized compressibility diagram for many gases in a wide range of variation of the reduced P_r and T_r.

12. Adsorption on amorphous surfaces

Amorphous surfaces are examples of strongly nonuniform surfaces [86–88]. Analysis of the potential energy of argon atoms on the amorphous surface of rutile showed [89] that local minima are located at distances smaller than the diameter of Ar, and every argon atom can block several neighbouring local minima. The same result was obtained in [33] for the sorption of krypton atoms and molecules of CH_4 and CO_2 in amorphous glassy polymer matrices based on polycarbonate (including its different chemical modifications) (section 9).

The problem of calculating the adsorption of gases on amorphous surfaces is decomposed into two subproblems: the first – the problem of modelling the amorphous surface, the second – the methods of calculation of adsorption on the surface formed. The strong dependence of the state of the surface layer of the adsorbent heterogeneity and hence its adsorption capacity on the conditions of its formation will automatically raise the question of the need to develop methods for describing the process of the formation of a rigid body in order to reflect the final state of the surface.

For modelling of amorphous surfaces the authors of [89] used a numerical procedure: a crystalline substrate randomly coated ('deposited') with the atoms of a solid, forming the surface region. Adsorption on the surface formed was considered for small fillings. Later adsorption on the same surface was calculated in the entire monolayer using the Monte Carlo method. An additional mechanism for creating the amorphous structure of the rutile surface by removing from the crystalline surface a portion of the oxygen from the surface layer [90, 15] was also proposed. Note also [91–93, 17–19], in which molecular–dynamic studies of generation of the surfaces of amorphous SiO_2 were carried out.

In [85] the atomic relief of amorphous surfaces was studies using the approach outlined in section 9, to determine the characteristics of the amorphous surface directly by adsorption measurements (by adsorption isotherm and heat). The model of the studied system includes 1) the procedure of forming the amorphous surface of the adsorbent, 2) the atom–atom potentials used, and 3) the lattice model to calculate the adsorption characteristics.

The procedure for formation of the amorphous structure of the adsorbent

Modelling the three-dimensional structure of the amorphous adsorbent is based on the traditional geometric representation of the amorphous structure as a completely random distribution of atoms in a solid. The surface structure of the adsorbent was modelled by the completion of its basal crystal cell located in the volume at a depth of $R = 1.77$ nm, to the surface by generating a random three-dimensional structure of the atomic arrangement of TiO_2. Here R is the radius of the potential of $Ar–O^{-2}$ and $Ar–Ti^{+4}$ interactions. When $R > 1.77$ nm the interaction potential is virtually zero, and this corresponds approximately to the distance equal to six lattice constants. Completion is performed from the depth to the surface, so random displacements are continuously accumulated from layer to layer and the surface layer will be maximally amorphous. The procedure parameter is the value of δ – the maximum displacement of the atoms in % of the lattice parameter. Parameter δ allows to control a complete break of the bonds between atoms so that the short-range order exists in the amorphous substance. In contrast to [89], this approach takes into account both cationic and anionic sublattices that is fully adequate to the real amorphous material comprising cations and anions.

The second factor of surface amorphization is that after a random displacement of atoms of the structure a certain percentage γ (in %) of oxygen ions is removed in the surface layer of the sorbent, as is done in [94]. The proposed procedure reflects two basic properties of the amorphous substance: presence of the short-range order and the absence of long-range order.

Potentials of the atom–atom interaction

The potential energy of the system was calculated in the atom–atom approximation using the Lennard–Jones (12–6) potential of ions O^{2-} and Ti^{4+} and the Ar atom; the potential parameters are given in Table 2.2. In [94] three sets of parameters for O^{2-} were considered. The second set was recommended as providing the best qualitative agreement with experiment. In this paper, the potential parameters of Ar–Ti^{4+} and Ar–O^{2-} were varied (see below), calculations were performed with the parameters specified in Table 2.5. The parameters used for the Ar–Ar potential were those used in [94].

Quasi single-site model for the calculation of adsorption on amorphous surfaces

On the amorphous surface the Ar atom can block several local minima of the adsorbent–adsorbate potential energy, so to describe the monolayer filling the following LGM variant was used [95, 96]. The proposed procedure for calculating adsorption is based on breaking the surface by the lattice structure as for single-site molecules with the number of nearest neighbours z and the lattice constant equal to the molecule diameter d. The surface is broken by the set of n^2 lattice structures so that the site itself is divided into n^2 parts (each with the same lattice constant d), which are shifted relative to each other with a step d/n, where n – an integer 2, 3, ...

The centre of any small site may be taken as the centre of a 'complete' site which is then translated to the entire surface (Fig. 12.1). For each structure we obtain its own set of binding energy of the molecules with the amorphous surface of the adsorbent Q_q and local Henry constant A_q, and for them we solve systems of equations for the local fillings of sites of different type which have the same form as for the single-site particles [2, 9]. Each position of the centre of the site corresponds to a particular partition of the surface into

Table 2.5. Parameters of the Lennard–Jones potential for Ar–ion interactions

i	$\dfrac{E_{Ar-i}}{k}$, K	σ_{Ar-i}, nm	Reference
O^{2-}	226.1	0.307	94
O^{2-}	160	0.325	94
O^{2-}	130	0.336	94
Ti^{4+}	76.5	0.241	94
Ar	119.8	0.3405	94
O^{2-}	120	0.325	85
Ti^{4+}	60	0.241	85

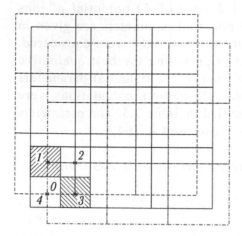

Fig. 12.1. Partitioning scheme for $n = 2$. In the initial lattice (solid lines) the site with number 0 is divided into four small pieces. The centres of the small areas are denoted 1, 2, 3 and 4. The full lattice structure with the number of nearest neighbours $z = 4$ is reproduced around each new centre. The new lattice with dotted lines corresponds to the centre 1. The lattice with the dot-and-dash lines corresponds to the centre 3.

a plurality of sites with their sets of local Langmuir constants for each site.

The usual system of equations (7.4)–(7.6) is solved in the atom – atom approximation taking into account lateral interactions $\varepsilon_{fg}^{AA} = \varepsilon$. The minimum value ε (A4.28) was estimated. To calculate the lattice parameters Q_q and A_q we use expression (A4.30)

$$Q_q = \frac{\int_{V_q} U(r)\exp\left[-\beta U(r)\right]dV_q \, / \, A_q}{\int_{V_q}\exp\left[-\beta U(r)\right]dV_q}, \quad A_q = \int_{V_q}\exp\left[-\beta U(r)\right]dV_q,$$

where V_q is the volume (size d^3) of the site with index q above the surface of the adsorbent, $U(r)$ is the adsorbate–adsorbent interaction potential obtained by summing all the contributions from the atom-atom interaction potentials of the Ar atom with the ions of the solid body inside a sphere of radius R. Since all the atoms of the adsorbent have a random distribution in three-dimensional space, the volume integrals are calculated numerically.

This calculation is repeated n^2 times ($1 \leq k \leq n^2$): each small area is the centre of the site. Then, for each particular value of pressure P it is necessary to select such a value $\theta = \theta_k(P)$, $1 \leq k \leq n^2$, which represents the minimum value of the free energy of the adsorbed molecules [59] (see (A2.5)). The adsorption isotherm corresponds to the complete dependence $\theta(P)$ constructed using this procedure. As a result, instead of solving the system of equations for many-site particles n^2 times, we solve the system of equations for the single-site particles [94, 95].

This calculation procedure is quite adequate to the physical picture. The results of calculations for $\theta \to 0$ and $\theta \to 1$ do not depend on the method of division of the surface into cells. When $\theta \to 0$ the contributions of lateral interactions are small and Henry's constant is determined only by the adsorbate–adsorbent potential. When $\theta \to 1$ adsorption occurs mainly on the weakest sites and the contribution of the lateral interactions is maximum, and the full value of contributions of the lateral interactions is virtually independent of the nature of the surface irregularities. In the intermediate range of filling the surface irregularities and lateral interactions occur simultaneously. Here there is a situation of disparities of fillings of different sites, leading to the formation of a first local and then wider (domain) close-packed arrangement of molecules in the most favourable subset of sites (on the sites with the highest binding energy) and blocking their neighbours. Thus, we have an analogue of the effect of the ordered distribution of molecules in the case of strong repulsion of the nearest neighbours (section 8).

Analysis of the potential energy of the adsorption system
Calculation of the potential relief $U(r)$ was performed with a step of 0.025 nm. The step size was chosen in light of further averaging of the resultant interaction energies and formation of the grid for the optimum time of calculations. The grid of the potential relief is positioned over the previously created amorphous structure of the adsorbent. The thickness of the grid is determined by the diameter of the Ar atoms $d = 0.38$ nm. Surface areas (fragments 10 × 10, 15 × 15 and 20 × 20 diameters of Ar), containing $N = 100$, 225 and 400 sites of the lattice structure were investigated. When calculating the potential relief of the boundary regions of the fragment the surface portion was completed on each side as a strip with the width equal to the radius of the interaction R. Local minima of the potential energy Ar/TiO_2 were determined numerically in the process of calculation: changes the signs of increments of the function while passing through the central point from the nearest neighbours in three-dimensional space were analyzed. The position of the minima was further clarified by using the three-dimensional quadratic approximation. Figure 6.2 shows the potential relief of the amorphous surface comprising 400 sites of the lattice structure. Here and in other figures we use the following parameters of the

generation procedure: $\delta = 22\%$, and $\gamma = 45\%$ (discussed below).

Analysis of potential reliefs shows that the argon atom can block up to 9 local minima. The most commonly implemented situation – blocking of 1–3 local minima of the potential energy (see data in Table 2.6). On average, each atom blocks 2–3 local minima. At the same time, almost every tenth site has no local minimum. These data show that the method [85] provides a convenient method for calculating the characteristics of adsorption: requiring a small number of partitions to account for blocking several neighbouring local minima of $U(r)$ – 2 or 3 are sufficient (systems of equations for the single-site particles is calculated very quickly).

To describe the properties of the resulting surfaces we must enter the distribution functions: unary – to describe the composition and at least binary – to describe the structure. To analyze the heterogeneity of the surface it is necessary to use the distribution $f(Q)$ (or $f(\ln [A])$), characterizing the fraction of surface sites with the binding energy in the range from Q to $Q + dQ$ (similarly for $\ln [A]$). (The calculation of the distribution functions is performed with the accuracy up to 0.5 kJ/mol.) The use of the logarithmic scale for the quantities A is due to their change by 5–6 orders.

Table 2.6. The distribution function of the number of local minima of the potential energy $f(m)$, corresponding to a single lattice site (m – number of local minima in the site, N_m – the total number of local minima, $\widehat{N}_m$ – on average lattice site)

m	$f(m)$		
	10×10	15×15	20×20
0	0.12	0.15	0.11
1	0.15	0.24	0.20
2	0.24	0.20	0.26
3	0.23	0.21	0.18
4	0.13	0.11	0.12
5	0.09	0.05	0.08
6	0.01	0.02	0.03
7	0.02	0.01	0.01
8	0.01	$< 10^{-5}$	0.0025
9	–	–	0.0025
N_m	257	492	968
$\widehat{N}_m$	2.57	2.19	2.42

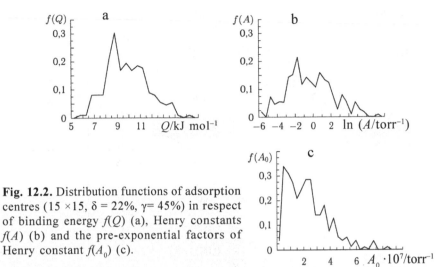

Fig. 12.2. Distribution functions of adsorption centres (15 ×15, δ = 22%, γ= 45%) in respect of binding energy $f(Q)$ (a), Henry constants $f(A)$ (b) and the pre-exponential factors of Henry constant $f(A_0)$ (c).

Comparison of the distribution functions $f(Q)$ and $f(\ln[A])$ (see Fig. 12.2 *a, b*) indicates their correlated but not identical change. Their differences are due to different values of the pre-exponential local Henry constants $A_0 = A(Q)\exp(-\beta Q_{ef})$, which may differ between by more than two orders of magnitude (distributed as shown in Fig. 12.2 *c*). The average value of A_0 for this system responds $2.35\cdot10^{-7}$ torr^{-1}. The results of calculation show that the commonly used assumption of the constant pre-exponential factor of the local Henry's constant (entropy factor) is too simplistic.

To describe the structure and the energy of the generated surface we define the two-dimensional distribution function $f(\Delta U, R)$, which characterizes the distribution of local minima of the potential energy on the basis of the difference between the values of these minima ΔU and distance between them R ($f(\Delta U, R)$ – the proportion of lows, falling at the same time and in some interval of increments $\delta\Delta U$ and in interval ΔR, in calculations it was taken into account $\delta\Delta U = 0.5$ kJ/mol and $\Delta R = 0.05$ nm). This distribution function is a generalization of the well-known binary distribution functions between atoms of the amorphous material, as it further takes into account the difference in the values of local minima. The form of such a two-dimensional distribution function for different sections for ΔU is shown in Fig. 12.3. (The vertical lines correspond to the close-packed crystal structure of Ar atoms at θ = 1.) Its analysis shows that the share of lows, belonging to the same areas for R, decreases with increasing energy difference ΔU. This conclusion is particularly important for small distances $R < d$.

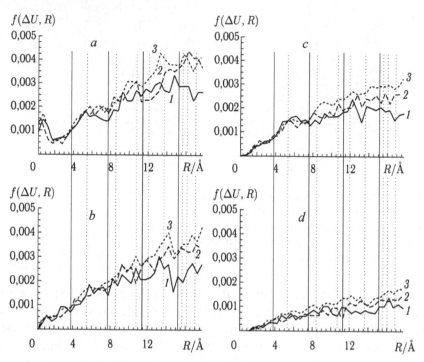

Fig. 12.3. Sections of the two-dimensional distribution function $f(dU, R)$ ($\delta = 22\%$ and $\gamma = 45\%$) at dU, $kJ \cdot mol^{-1} = 0.25$ (*a*), 1.25 (*b*), 2.25 (*c*), 4.25 (*d*) (dU – the middle of the energy range of the considered sections). Surface fragment: *1* – 10 × 10, *2* – 15 × 15, *3* – 20 × 20. Vertical lines correspond to the crystal structure of close-packed Ar atoms at $\theta = 1$.

Figure 12.4 shows the area under the curves of the radial distribution shown in Fig. 12.3 attributable to the size of the site of the lattice structure. The probability of uniting several local minima, differing greatly in the binding energies, in a single site is small. Consequently, the joining of several nearest local minima in one adsorption centre and transition to the lattice parameters averaged over the volume of the site is justified because the energy properties of the nearest local minima are relatively little discernible. This conclusion differs from the results of [89], in which the neighbouring local minima could have high values ΔU. This difference is due to rigid body models. In [89] only the anionic sublattice was studied. This model takes into account both ionic sublattices of rutile.

At large distances the sections of the function $f(\Delta U, R)$ with $\Delta U = $ const have a form similar to the radial distribution curves of amorphous solids: 'chaotic' curve shape, fluctuating about some mean value, increases with increasing R (Fig. 12.3 *a*, *b*, fragments

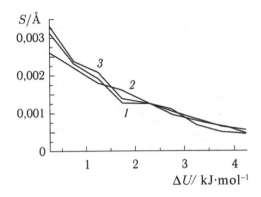

Fig. 12.4. Dependence of the area S on U for the sections shown in Fig. 2.3: 1 – a surface fragment 10×10, $2 - 15 \times 15$, $3 - 20 \times 20$.

15×15 and 20×20). However, in Fig. 12.3 c, d it is seen that the rate of increase of the average value decreases with increasing ΔU. Even more different from the default behaviour of the radial curves for amorphous solids are the curves in Fig. 12.3 for the 10×10 fragment. To understand the reason for such behaviour of the curves, we must compare the total functions $f(\Delta U, R)$.

For the 20×20 fragment the full distribution function $f(\Delta U, R)$ is shown in Fig. 12.5. At large distances ($R = $ const) with increasing ΔU the function behaves monotonically and at $\Delta U \approx 8.5$ kJ/mol passes through a maximum, which is, however, lower than the values of the function $f(\Delta U, R)$ for small ΔU. Conversely, for the 10×10 fragment at a large distance, this maximum is significantly higher than the values of the function $f(\Delta U, R)$ for small ΔU. For the 15×15 fragment the function $f(\Delta U, R)$ is similar to the function for the 20×20 fragment. These differences in the full functions $f(\Delta U, R)$ are associated only with the size of the fragments. Fragments smaller than 15×15 are apparently not sufficient to describe the properties of amorphous surfaces.

The introduced two-dimensional functions provide a sufficiently detailed description of the structure and energy of the examined sections of the amorphous surfaces and can be used for their 'certification' as they reflect differences in their final states, depending not only on the interaction potentials used, but also on the 'technological' parameters of generation of the surface. For more detail, see [85].

Analysis of the distribution function.

The calculated distribution functions, such as $f(Q)$, can be compared with the same functions obtained by solving the inverse problem of

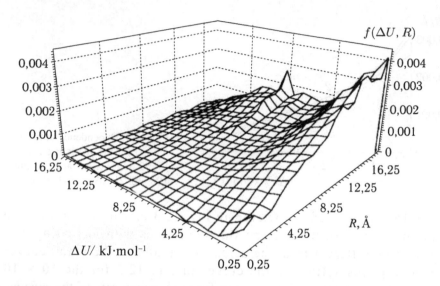

Fig. 12.5. The two-dimensional distribution function of pairs of local minima of $f(\Delta U, R)$ for the fragment of the surface ($\delta = 22\%$ and $\gamma = 45\%$, 20×20).

the description of the experimental isotherms $\theta_{exp} = \int\limits_{Q_{min}}^{Q_{max}} \theta(Q)f(Q)dQ$
or heats of adsorption. The functions $f(Q)$ for the Ar/rutile system were obtained from the data [97, 98] (see Fig. 12.6). Curves *1–3* are obtained from the heat of adsorption, curve *4* – from the conversion of data on the heat of adsorption from 85 to 0 K [97], the curve *5* was obtained from the adsorption isotherm [99]. Curves *1, 4, 5* are satisfactorily correlated. They belong to the formal description of the experimental data, which uses equation (7.4) (without additional partitioning of the sites into small pieces) with $\varepsilon = 0$, and the kernel of the integral equation $\theta_L(Q)$ is the Langmuir model, where $A(Q) = A_0 \exp(\beta Q_{ef})$, Q_{ef} is the current value of the effective heat of adsorption for this surface coverage θ. Curves *2* and *3* were obtained taking into account the lateral interactions in the quasichemical approximation. Accounting for lateral interactions changes the distribution function, allowing to single out Q_{ef} of the adsorbate-adsorbent interaction Q from the general energy contribution. These distribution functions should be taken as a basis for comparison with the atomic and molecular description.

In the atom–atom approximation it is assumed that the value of the parameter ε does not depend on the adsorbate–adsorbent interactions. Curve *2* corresponds to the value ε, obtained from the

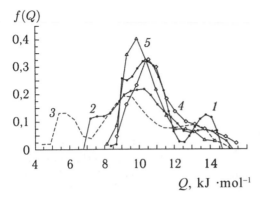

Fig. 12.6. Distribution function $f(Q)$, obtained from the inverse problem of the description of the experimental dependence Q_{is} at ε, kJ·mol^{-1}: 0 (*1*), 0.4 (*2*), 1.0 (*3*); *4* – experiment [97]; *5* – experiment [99],

formula (9.1). Calculation gives the value ε = 0.4 kJ/mol. Curve *3* corresponds to the value ε, estimated at the upper limit [100], where ε is approximately equal to the depth of the potential well: $\varepsilon \approx 1.0$ kJ/mol. Almost exactly the same estimate of ε is obtained from the critical condition in the quasichemical approximation of the lattice gas model [9]: $\beta_c \varepsilon$ = 2ln ($z/[z-2]$), where z = 6 for the volumetric gas phase, T_c – the critical temperature of the adsorbate (150.8 K for Ar [72]). Thus, in any case Q_{min} should be less than 7 kJ/mol, while not taking into account the interatomic interactions gives Q_{min} > 8 kJ/mol. All estimates for Q_{max} fit together: Q_{max} = 15.0 ± 0.5 kJ/mol. The estimate obtained for the energy range of surface irregularities of rutile ($Q_{min} \div Q_{max}$) was the basis for: 1) selecting the mode of formation of the amorphous surface of rutile, and 2) testing the potential parameters [94].

Final versions of the energy distribution functions and logarithms of the local Henry constants for different sizes of the fragments are shown in Figs. 12.2 and 12.7. These distribution functions are used to calculate the thermodynamic characteristics of adsorption of Ar/rutile.

Calculation of the heats of adsorption
Calculation of the heats of adsorption was conducted without additional adjustable parameters. The results of the comparison of experimental and theoretical curves for different sized fragments are shown in Fig. 12.8. The isosteric heat of adsorption was calculated as $Q_{is} = -(d \ln P/d\beta)_{\theta = const}$.

A small fragment of the curve does not reflect the heat of adsorption at low coverages; with medium and large monolayer coverages the theory and experiments agree qualitatively. The size of the fragment is significantly increased by the agreement between theory and experiment. Apparently, a fragment larger than 15 × 15

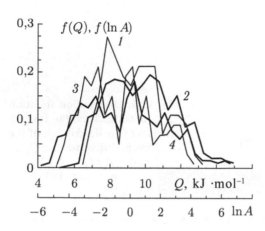

Fig. 12.7. Distribution function $f(Q)$ (1, 2) and $f(\ln A)$ (3, 4) for surface fragments ($\delta = 22\%$ and $\gamma = 45\%$): 1, $3 - 10 \times 10$ fragment, 2, $4 - 20 \times 20$ fragment.

satisfactorily reflects the basic properties of amorphous adsorbents. The average deviation between theory and experiment is about 3%. Differences at low coverages are due to differences in the distribution functions. The proximity of the two curves at monolayer coverages is due to the predominance of the contribution of lateral interactions ($z\varepsilon$) in comparison with the difference introduced by Q_{min} for the atomic and molecular distribution function and the function $f(Q)$, obtained from the experiment.

For comparison we show the curve with the value of $\varepsilon = 1$ kJ/mol — it does not correspond qualitatively to the experimental curve, as the atomic–molecular distribution function $f(Q)$ and the same function $f(Q)$, obtained from the experiment, are very different in the low-energy region. A similar conclusion follows from a comparison of the calculated and experimental adsorption isotherms (see [85]).

Thus, the introduced new two-dimensional distribution function $f(\Delta U, R)$, which is determined by the proportion of pairs of local minima of the potential energy situated at a distance R and having a difference in the values of these minima equal to ΔU, enables detailed characterization of the surface properties of amorphous adsorbents and can be used for their 'certification'. Accounting for the short-range order in the amorphous structure of TiO_2 (rutile) by modelling the structure of the anion and cation sublattices leads to the fact that the nearest local minima mainly have similar values of energy. In determining the energy distribution function $f(Q)$ (or $f(\ln [A])$) from the experimental data, it is of fundamental importance to take into account lateral interactions. And the technique, based on the LGM, allows to get new detailed information on the properties of surfaces.

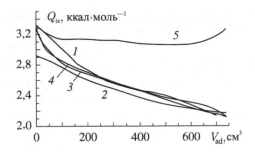

Fig. 12.8. Concentration dependences of the isosteric heat of adsorption $Q_{is}(V_{ad})$; V_{ad} – volume of adsorbate (V_{ad} = 755 cm³ – monolayer capacity) (δ = 22 %, and γ = 45 %): *1* – Experiment [97] for calculation of surface fragments of 10 × 10 (*2*), 15 × 15 (*3*), 20 × 20 (*4*), 15 × 15, with ε = 1 kJ/mol (*5*).

13. Multilayers on the exposed surface

Multilayer adsorption is described widely using the BET and FHH (Frenkel–Halsey–Hill) formulas ln (x) = $-b/\theta^s$, where x = P/P_0, P – current vapour pressure, P_0 – saturated vapour pressure at the temperature of measurement of the isotherm, b and s – parameters of the equation, $2 < s < 3$ [101]. Despite the fact that the theory of multilayer adsorption has recently been substantially developed and allows to take into account all the major real properties of multilayer adsorption systems (lateral interactions between adsorbate molecules, the heterogeneity of the adsorbent surface and the interaction of adsorbate molecules with the nonuniform surface of the adsorbent [102]), the BET and FHH formulas continue to be actively used. This is due to the simplicity of these formulas, although it is well known that the BET formula gives a poor description of the isotherms at $x > 0.30$–0.35, and the FHH formulas are not justified at coverages less than three monolayers.

Homogeneous surface
In the case of multilayer adsorption on a flat homogeneous surface the isotherm is described by equations of the local filling of sites in different monolayers discussed in section 7 at f_q = $1/t$, where t is the number of monolayers in the system $\left(\sum_q f_q = 1 \right)$ [103]:

$$\theta(P) = \sum_{q=1}^{t} f_q \theta_q(P), \quad a_q P = \frac{\theta_q}{1-\theta_q} \Lambda_q, \quad \Lambda_q = \prod_{p=q-1}^{q+1} \left(S_{qp} \right)^{z_{qp}} . \quad (13.1)$$

Here the equations take into account the energy heterogeneity of the lattice sites along the normal to the surface and the interaction between nearest neighbours ($R = 1$); Λ_q is the nonideality function, depending on the type of approximation; z_{qp} is the number of sites in a layer p near the sites in layer q. The quantity t is found by varying the upper limit of the system of equations (13.1) to get the same density profile of molecules in layers θ_q at the specified temperature T and chemical potential μ (or external pressure P).

The Henry constants reflect the character of the decrease of the attractive potential with increase of the distance from the surface plane: $a_q = a_q^0 \exp(\beta Q_q)$ (section 4). In the case of quasichemical approximation of the function $S_{qp} = 1 + xt_{qp}$, where t_{qp} is defined by (7.5) and (7.6), in the case of the mean field approximation $S_{qp} = \exp(-\beta \varepsilon \theta_p)$.

Equations (13.1) lead to the filling layer-by-layer of the surface region, but as a rule, these isotherm curves have a stepped form that differs from the experimental curves (see Fig. 13.1 [104]). This fact is well known [103–108]. (The reason for smoothing of these steps on the experimental curves for nonuniform absorbent surfaces is known [107] (see below).) However, such multilayer adsorption models may be used to test the type of potential of the molecules with the solid Q_1, as they reflect qualitatively the experiment in the first monolayer. Figure 13.1 shows nitrogen adsorption ($\varepsilon/k = 144$ K, $\sigma = 0.37 - N_2$ as a molecule having the flattened spherical shape) [109], and argon ($\varepsilon/k = 238$ K, $\sigma = 0.34$ nm) [49] on the surface of SiO_2. Lateral interactions of particles are described by the LJ-potential for a rigid lattice, $R = 1$. In accordance with the Mie potential (9–3), knowing Q_1, we can find other Q_q. In particular, parameter $Q_2 = Q_1/8$. Calculation in Fig. 13.1 was performed for $Q_{q>2} = 0$. Values of Q_1 were evaluated [104] from the experimental adsorption isotherms for nitrogen and argon on silica gel [101]. This gave $Q_1 \approx 7\varepsilon$ (nitrogen) and 4ε (argon), i.e. both nitrogen–silica gel and argon–silica gel adsorption systems represent a case of strong adsorption (see section 21 below).

The dash lines in Fig. 13.1 correspond to the so-called approximation isotherm of multilayer adsorption [103, 110] (which have been calculated with the same parameters Q_1).

Approximation isotherm

The approximation equation of the multilayer adsorption isotherm for a nonuniform surface is given by [110]

$$n(x) = n_m B(x)\gamma(x), \quad B(x) = \sum_{q=1}^{t} f_q \frac{C_q x}{1 + C_q x}, \quad \gamma(x) = \left(\frac{1 - \sigma x}{1 - x}\right)^{1/n}, \quad (13.2)$$

where $n(x)$ is the number of adsorbed molecules; n_m is the surface monolayer capacity; $B(x)$ is a function describing the surface coverage of the adsorbent; $\gamma(x)$ is a function describing the contributions of filling of the second and following layers of the gas–solid surface region. As in the BET model, $C_q = m_q \exp [\beta(Q_q - Q_L)]$, $\beta = (kT)^{-1}$, Q_q and Q_L – the binding energy per sites q in the first monolayer and on the surface of the liquid adsorbate film, respectively; $m_q = a_q^0/a_g^0$, a_q^0 and a_g^0 are the prefactors of the local Henry constants for the site of type q in the first monolayer and the liquid film of the adsorbate. The magnitude of σ depends on the structure of the surface film of the adsorbate and the approximation of taking into account the interaction between molecules. However, the interparticle interactions, defining the function $\gamma(x)$, are not considered in equation (13.2).

The approximation isotherm (13.2) is obtained based on the original equations constructed in the LGM for a film of the adsorbate. If we use the quasichemical approximation of accounting for the interaction, then $\sigma = (z_{qq} - 1)/(z - 1)$ where z_{qq} is the number of neighbouring sites in a flat monolayer, z is the total number of the nearest neighbours. If we use the mean-field approximation, then $\sigma = z_{qq}/z$. The parameter n in the lattice gas model is 2. Finally, if the equation (13.2) is used as an empirical equation, the parameters σ and n are the adjustable and their values may differ from these values. For a homogeneous surface the function $B(x)$ is simplified and

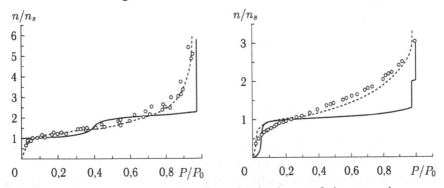

Fig. 13.1. Comparison of multilayer adsorption isotherms of nitrogen and argon on silica gel: the symbols denote the experimental data [101], the solid lines correspond to the calculation in quasichemical approximation (QCA) for a homogeneous surface of the adsorbent and the dashed lines correspond to the approximation isotherm [104].

represents a Langmuir isotherm for the monolayer surface coverage function $B(x) = Cx/(1 + Cx)$, while the dependence $\gamma(x)$ does not change.

Note that in [111] function $B(x)$, corresponding to the Langmuir equation, does not work for a homogeneous surface in the monolayer filling range in the simplified description of multilayer adsorption and this contradicts the physical sense.

The interesting feature in this type of equation is that when $x > 0.3$, they give a better description than the BET equation and they allow for a qualitative assessment of the capacity value of the monolayer by finding the inflection point of the $n(x)$ dependence [68].

The approximation isotherm was used to describe a number of experimental data [109]. Figure 13.2 shows the adsorption isotherms of nitrogen on non-porous adsorbents at $T = 77$ K, nitrogen, argon, oxygen and KCl crystals at $T = 85$ K [112] as well as the adsorption isotherms of benzene on silica [113]. Field (*a*) shows the experimental points and calculations by the BET equation and by approximation equations with $\sigma = 0$ (*2*) and 0.6 (*3*). For field (*b*) it shows the experimental points and calculations by the approximation equations with $\sigma = 0$ (dashed lines) and 0.6 (solid lines). The curve for oxygen is shifted to 0.2 on the axis x. For the field (*c*) there are the experimental points and calculations by approximation equations with the selection of parameters $n(x)$ – solid lines, the heats of adsorption and both dependences simultaneously. In all situations the approximation isotherm provides a satisfactory description of the experimental isotherms.

Nonuniform surface
Direct use of rigorous statistical models demonstrates the real complexity of the problem. Calculations were carried out by direct counting of the influence through the LJ potential of the atoms of the solid on the adsorbate located in different monolayers on the surface [102]. The surface was regarded as nonuniform, so the local energies depend on the composition and structure of the surface consisting of different types of centres. In the study of multilayer adsorption on nonuniform surfaces attention was given to examples of the surface of the adsorbent, consisting of two types of atoms ($t_s = 2$). If we measure the distance in units of the linear size (diameter) of the molecules d (then $r = \eta d$) and accept that $\sigma_{Aq} = \sigma_{AA}$ for all q, then $d = 2^{1/6}\sigma_{AA}$ and potential (12–6) can be rewritten as

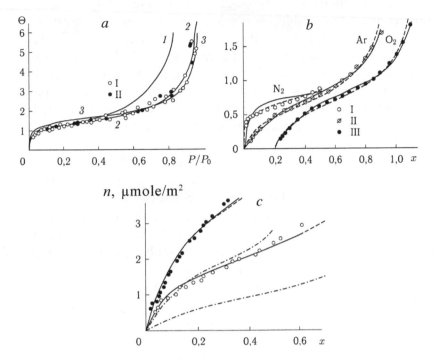

Fig. 13.2. The adsorption isotherms of nitrogen on non-porous adsorbents at $T = 77$ K [101] (*a*); nitrogen, oxygen and argon on KCl crystals at $T = 85$ K [112] (*b*); benzene on silica [113] (*c*).

$\varepsilon_{Aq}(h)/\varepsilon_{Aq}(1) = (2/\eta^6 - \eta^{-12})$, $\eta = \{r^2 + h^2\}^{1/2}$. In the calculations it was taken into account that $\varepsilon_{A1} = 3.26\varepsilon_{AA}$ and $\varepsilon_{A2} = 1.63\varepsilon_{AA}$, $\varepsilon_{AA} = 1.68$ kJ/ mol (parameters ε_{A1} and ε_{AA} were selected so as to correspond to the Kr–graphite system [103]). Details of the calculations are in [102].

The structure of the adsorbent surface was seen for three limiting cases: spotty, chaotic and regular surface. Accuracy differences in the values of binding energies, related to different types of sites, were assumed equal to 0.04 kJ/mol. This led to significant differences in the types of sites to and including the third layer (the total number of layers $t = 10$). The system temperature was set in reduced units $\tau = \beta_c/\beta$.

Figure 13.3 *a* shows the multilayer adsorption isotherms in different structures (curves *1–3*) and compositions of the nonuniform adsorbent surface. At the indicated parameters of adsorbent–adsorbate interactions on the regular–nonuniform surface in each layer we can distinguish two types of sites with similar binding energies. This leads to the fact that the form of curve *1* is close to the multilayer adsorption isotherm on a homogeneous surface [69] – each layer contains one 'step'. With the same parameters on the spotty surface in layers $h = 2$ and $h = 3$ the number of 'steps' doubles (curve *3*), as the binding energies of the two types of sites, also realized within each layers, differ significantly more than for curve *1*. On the 'chaotic' surface the number of sites in the layers $h = 2$ and $h = 3$ is equal to 10, so the number of 'steps' is further increased, but the differences between them are different and the main contribution comes from 3–4 values of $\varepsilon_k(h)$.

Curves *3–5* in Fig. 13.3 *a* illustrate the influence of surface composition for the spotty surface on the type of isotherms. Increase in the proportion of sites with some energy leads to a clearer expression of the corresponding 'step' on the integral curve of the isotherm. This fact correlates well with the known data on the influence of the adsorbent preparation technique on the type of isotherms. So, for Ar adsorption on carbon black sferon-6 [114], the change of the surface properties by changing the degree of its graphitization changes the form of the isotherm from shallow (type 2 of classification of isotherms – section 2) to stepped (6th type) ($T = 78$ K).

The model under consideration allows to consider changes in the surface layer of the adsorbent and to reflect this change in the thermodynamic characteristics of the adsorption system. Increasing the number of types of sites in the surface layers leads to a 'smearing' of the stepped isotherm characteristic of homogeneous surfaces. This was first shown in [108] for the spotted areas of the nonuniform surface, which have been weighed using the exponential

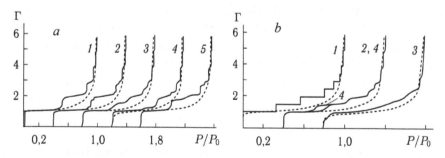

Fig. 13.3. Multilayer adsorption isotherms (*a*): on regular (*1*), chaotic (*2*), spotty (*3–5*) surfaces, dotted – approximation isotherms, $\tau = 0.6$, $f_1 = 0.5$ (*1–3*), 0.15 (*4*), 0.85 (*5*), the curves are shifted successively along the abscissa by 0.4 P/P_0. Influence of temperature on multilayer adsorption isotherms for the spotty surface with $f_1 = 0.5$ (*b*): $\tau = 0.45$ (*1*), 0.6 (*2, 4*), 0.8 (*3*). For curve *4* $\varepsilon_{Aq} = 4.5$. Dotted line – approximation isotherms.

distribution function of the sites on the basis of binding energies. The curves in Fig. 13.3 *a* show that the structure of the surface layer of the adsorbent has no lesser effect on the blurring of 'steps' than the surface composition. The discussed effect of 'blurring' also depends on the temperature of the system (Fig. 13.3 *b*). A decrease in temperature leads to a clearer identification of the steps, and their number depends on the composition and structure of the adsorbent surface. Changing the parameters of the interaction potential (curve *4*) also changes the position of the steps and the nature of their impact on blurring. At relatively high temperatures the multilayer adsorption isotherm is flatter (curve *3*).

Figure 13.3 shows by the dotted line the approximation adsorption isotherm. Calculation using formula (13.2) is much easier than according to formula (13.1). However, the approximation isotherms are quite different from the isotherms in the QCA (quasichemical approximation) in the region of filling of the second and third layers – the area of most frequent use of BET equations. Satisfactory agreement of these calculations can apparently only be achieved at high temperatures. In these examples, an even greater disparity of the isotherms is realized for the first layer (this fact is difficult to see because of the scale of the drawings). The crucial difference between the molecular calculations and approximation equations exists for the curves of isosteric heats of adsorption, which are obtained from the isotherms at variable temperatures.

Thus, the results of molecular analysis of the approximation of isotherms show that, despite the apparent agreement with the experiments for the isotherms for the area of filling $P/P_0 > 0.3$, their use can lead to significant and uncontrollable errors. Compared with the approximation estimate, the value n_m, determined by the traditional BET model gives a more accurate estimate. The essence of the apparent paradox lies in the fact that the drawback of the BET equation is well known and no one uses it at pressures greater than $P/P_0 > 0.35$. The approximation isotherm, describing the experimental data over a wider area, up to $P/P_0 \sim 0.7$–0.8, creates the illusion of satisfactory compliance with the experiment, while most of the experimental points at $P/P_0 > 0.35$ do not satisfy the correct model. Therefore, in determining the point B across the curve most of the information, related to the field of filling the second and third monolayer, provides a large statistical contribution and displaces the estimate for the position of point B.

In conclusion of this brief presentation of the correspondence of the experimental to the results of LGM calculations, we compare the argon adsorption isotherms on three crystal faces of the rutile surface with Monte Carlo calculations [115] (see Fig. 6.11). These calculations use identical interaction potentials of argon with surfaces [85, 94] and between the argon atoms with allowance for triple contributions [71] (Fig. 13.4). Both methods give good consistent results. Thus, this theory is flexible enough and, given the specific properties of the studied systems, allows us to construct adequate equation of equilibrium distribution of interacting molecules.

14. Slit-shaped pores

A more complex example of nonuniform systems are porous structure. They are characterized by multilayer filling, similar to filling on exposed surfaces, but this process is also influenced by the potential of the opposite wall.

As above, the adsorbate–adsorbent interaction energy in the sites of type q is denote Q_q, $1 \leq q \leq t$, t is the number of groups of sites in the system. The proportion of sites in the group q is denoted by f_q. In the slit-like pore having homogeneous walls, a group of sites with the same properties forms a monolayer along the entire surface [116–119], then $f_q = 1/H$, here H is the slit width in monolayers. Partition of the pore volume into the sites and their grouping are based on information about the significance of the particle–wall interaction potential with the symmetry of the internal space of the pores taken into account. This reduces the number

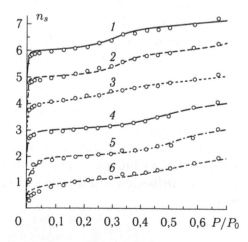

Fig. 13.4. Comparison of calculations using LGM (lines) and the Monte Carlo method (symbols) for the multilayer adsorption isotherms of argon on the three crystal faces of rutile (100) – *1*, *4*, (101) – *2*, *5*, (110) – *3*, *6*, for two potentials of argon–rutile interaction (*1–3*) [85], and (*4–6*) [94] (all curves are shifted successively up by a unit); n_s – the number of monolayers, P_0 – vapour pressure at 85 K [115].

of independent variables of the state of occupation in the case of identical pore walls: $t = H/2$ for even values of H (wherein $f_q = 2/H$, $1 \le k \le t = H/2$) and $t = (H + 1)/2$ for odd values of H (wherein $f_q = 2/H$, wherein $1 \le k \le (t - 1)$, and $f_t = 1/H$).

In the general case of unstructured and structured walls the binding energy of a particle situated in a site of type q is defined as the sum of the contributions of its atom–atom potentials with the atoms of the solid. The most important characteristic of the LGM is the binding energy of the particle A at a site f of section q with pore walls or the site f of the section q of the exposed surface $Q_{q,f}$ which is calculated by the formula (4.9) or

$$Q_{q,f} = -\beta^{-1} \ln \left[\frac{1}{V_f} \int_{r \in V_f} \exp\left(-\beta U_{As}(r)\right) dr \right],$$

where $U_{As}(r)$ is the interaction potential of particle A with pore walls. Integration is performed over the cell volume $V_q = \lambda^3$ with number f; $\lambda = 1.12\sigma$. In the case of a 'smeared' potential value Q_k is the same for all surface sites. For the structured potential the surface sites have different values of Q_q.

The local Henry constants in this case are defined as

$$a_{q,f} = \int_{\xi \in V(f)} \exp\{\beta Q_{q,f}(\xi)\} dV, \text{ integrating over all valid positions } \xi$$

of particle A inside the cell f. Note that for highly curved surfaces writing in the lattice structure can lead to the intersection by the pore wall of some of the cells of the intrapore space (for example, for cylindrical pores of radius R_p). This is accounted for in $Q_{q,f}$ by reducing the value of the cell volume available to the molecule situated in the cell q. This definition of the parameter $Q_{q,f}$ ensures agreement of the LGM results with the exact results at low density of the adsorbate.

The lattice parameters $\varepsilon(r_{ij})$, $r_{ij} \le R$, of the lateral interactions of particles i and j are determined from the corresponding distances using the Lennard–Jones potential: $U_{LJ} = 4\varepsilon_{ij} [(\sigma_{ij}/r_{ij})^{12} - (\sigma_{ij}/r_{ij})^6]$, $r_{ij} = 1$, this corresponds to a minimum of this potential (section 4).

Local isotherms

To calculate the adsorption isotherms and local filling of the sites of different groups we use a system of equations obtained in [1, 2, 9], taking into account the heterogeneity of the energy of the lattice sites and the interaction between the particles at distances $r \le R$:

$$\theta(P) = \sum_{q=1}^{t} f_q \theta_q(P), \quad a_q p = \frac{\theta_q}{1-\theta_q} \Lambda_q, \quad \Lambda_q = \prod_r \prod_{p \in z_q(r)} \left(1 + x(r) t_{qp}(r)\right),$$

$$t_{qp}(r) = 2\theta_p \Big/ \left[\delta_{qp}(r) + b_{qp}(r)\right], \quad x(r) = \exp\left(-\beta \varepsilon_{AA}(r)\right) - 1,$$

$$\delta_{qp}(r) = 1 + x(r)\left(1 - \theta_q - \theta_p\right), \quad b_{qp}(r) = \left(\delta_{qp}(r)^2 + 4x(r)\theta_q\theta_p\right)^{1/2},$$

$$(14.1)$$

where θ – the degree of filling of the pore volume at pressure P; θ_q – the molar fraction of particles in the sites of type q, $\beta = (kT)^{-1}$, $a_q = a_q^0 \exp(\beta Q_q)$ – the local Henry constant of the lattice site in layer q and the binding energy of Q_q, a_q^0 – pre-exponential factor of the local Henry constant; Λ_q – a term taking into account the lateral interactions between the particles in the neighbouring sites. Index p runs over all neighbours $z_q(r)$ of the site of type q at a distance r inside the pore.

In the case of identical layers of distributions of the adsorbate (Fig. 14.1) it is convenient to introduce the number z_{qp} – the number of the nearest neighbours of site q in the layer $p = q$, $q \pm 1$, for $z = 6$ these numbers are $z_{qq} = 4$; $z_{qq-1} = z_{qq+1} = 1$. The binding energy of a site of type q is calculated as $Q_q = Q(q) = U(q) + U(H - q + 1)$, where q – number of the layer of the pore in which the site type q is situated, $U(q)$ – the interaction potential with the pore wall: $U(q) = \varepsilon_a/q^3$, which corresponds to the potential of the type (9–3).

Due to the symmetry of the pore the system (14.1) can not be solved for the entire width of the pore H and can be solved only for the central layer, taking into account specific values t. The equilibrium distribution of particles in different types of sites was determined from the system of equations (14.1) by the standard Newton method [120].

Local distribution of the molecules

Figure 14.2 a shows the profiles of argon concentration θ_q in a slit-shaped carbon pore with width $H = 10$ monolayers (curves 1–5, q correspond to the number of the monolayer – count goes from the pore wall), and θ_1 and θ_5 for the systems with $Q_1/\varepsilon = 0$ and 16.5 (second variant ε). The inset of field (a) shows the adsorption isotherms for the three interaction potentials $Q_1/\varepsilon = 16.5$ (1), 9.24 (2), 0 (3) of argon atoms with the wall of the pore at $T = 273$ K. On the right (field b) there are three concentration profiles of the argon atoms at a total filling of the pore θ respectively equal to 0.095,

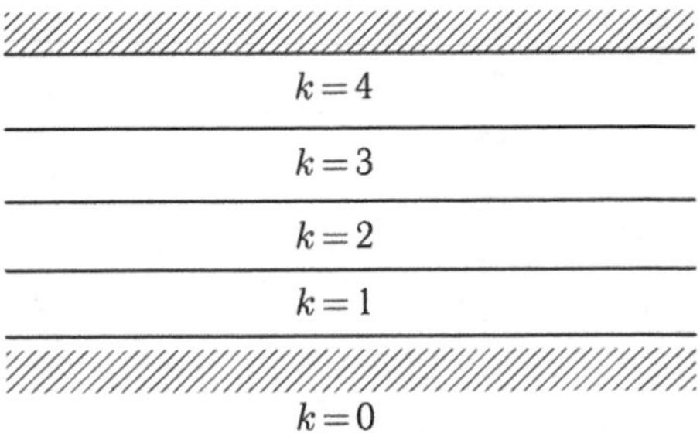

```
////////////////////////////////////////
              k = 4
_____
              k = 3
_____
              k = 2
_____
              k = 1
_____
////////////////////////////////////////
              k = 0
```

Fig. 14.1. Scheme of a slit-shaped pore of infinite length in two directions with flat walls.

0.485 and 0.905. Curves 1 and 8 indicate significant differences in the filling of the surface monolayer from the other layers at a strong attraction of the adsorbate to the walls of the pores. In the argon–carbon system the filling of the second monolayer (curve 2) is also markedly different from the first filling, and the filling of the layers 3, 4 and 5 occurs almost equally (curves 4 and 5 are the same). In the absence of wall attraction the surface layer (curve 6) has a lower degree of filling than the central layers (curve 7), although the differences between these curves are small. The anisotropy of the adsorbate concentration profiles has a significant impact on the concentration dependences of all other equilibrium and kinetic characteristics of the porous system.

Figure 14.3 a shows local isotherms in filling a carbon slit pore volume with argon atoms [122] (the numbers of curves correspond to the width of the pores, expressed in the number of monolayers) at $T = 273$ K. The field in 14.3 b shows the corresponding density profiles along the section of the pore. For narrow pores the joint action of the potential of two walls leads to more rapid filling of the pore. These differences can reach several orders of magnitude of external pressure.

In all cases, the pore walls adsorb a considerable proportion of the adsorbate and the density of the molecules near the wall is much higher than in the central portion of the pore. An exception is the case when $H = 2$, in which both layers are characterized by the same filling, coinciding with the average value θ. When $H > 2$ in the centre of the pore the average fraction of the filled volume depends on the width of the pores: the greater the width of the pores, the higher the density of the adsorbate in the centre.

Compressibility of the fluid + homogeneous walls [45]
The homogeneity of the potential of interaction of the molecule with the wall along the plane of the surface allows us to consider

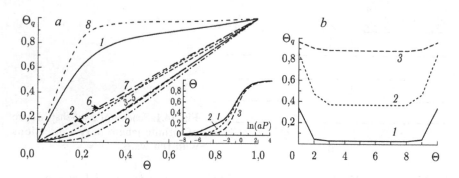

Fig. 14.2. Local layer filling θ_q (*a*): *q* = 1 (*1, 6, 8*), 2 (*2*) 3 (*3*) 4 (*4*) 5 (*5, 7, 9*), of a slit-like pore with the width of 10 monolayers of argon atoms at *T* = 273 K, depending on the degree of filling of the pore with the adsorbate θ. The concentration profiles of the argon atoms for Q_1/ε = 9.24 (*b*): θ = 0.095 (*1*) 0.485 (*2*) 0.905 (*3*) [121].

the same lattice structure parameters λ_{kk} within each layer *k*, $1 \leq k \leq t$, parallel to the surface, but the distances between adjacent layers $\lambda_{kk \pm 1}$ (layer *k* = 0 corresponds to a solid body) will be different. This is due to the nature of layer filling of the pore, and each layer has its density, and therefore, the distance between the molecules.

Equilibrium values λ_{kn} are determined from the condition $\partial F/\partial \lambda_{kn} = 0$, where $E = U_{wall} + U_{later}$, here the first term describes the interaction of molecules with the walls, and the second – the interaction between

$$U_{wall} = \sum_{k,i} \theta_k^i \varphi_{is}(k) f_k, \quad \varphi_{is}(k) = U_{is}\left[(\sigma_{is}/\rho_k)^m - (\sigma_{is}/\rho_k)^n\right];$$

molecules here $\rho_k = \sum_{n=0}^{k-1} \lambda_{n,n+1}$,

ρ_k is the distance between the molecules in the layer *k* and the surface; U_{is} and σ_{is} are parameters of the molecule–wall potential, *m* = 10 or 9 and *m* = 4 or 3, respectively; the functions f_k consider the proportions of the sites in the layers of the pore *k*, which vary depending on the local density of molecules in different layers of the pores. To determine these functions, we note that the cell size in the plane of each layer *k* is λ_{kk}^2 so we can write

$$\frac{f_k}{f_{k+1}} = \left(\frac{\lambda_{k+1k+1}}{\lambda_{kk}}\right)^2. \tag{14.2}$$

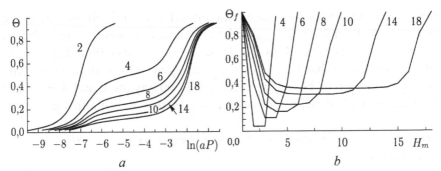

Fig. 14.3. Adsorption isotherm (here a – average value of local Langmuir constants and, product aP – dimensionless value) (a). Adsorbate concentration profiles with an average pore filling 0.5 – the abscissa gives the numbers of monolayers H_m (b) [122].

If these functions are normalized, for example, for the number of sites in the central layer of the slit indicated by t, it is easy to

obtain $f_k = \dfrac{\left(\lambda_{tt}/\lambda_{kk}\right)^2}{\sum_{k=1}^{t}\left(\lambda_{tt}/\lambda_{kk}\right)^2}$

$$U_{later} = \frac{1}{2}\sum_{r}\sum_{k,n} z_{kn}(r)\left(1+\Delta_{kn}\right)f_k \varepsilon_{kn}(r)\theta_{kn}(r),$$

$$\varepsilon_{kn}(r) = 4U_{AA}\left\{\left[\sigma_{AA}/\rho_{kn}(r)\right]^{12} - \left[\sigma_{AA}/\rho_{kn}(r)\right]^6\right\},$$

wherein $\rho_{kn}(r)$ is the distance between two molecules situated in the layers k and n at the distance of the r-th coordination sphere. Values $\rho_{kn}(r)$ are expressed in terms of average distances λ_{kn}:

$$\rho_{kk\pm1}(1)^2 = \lambda_{kk\pm1}(1)^2 + 2\Delta r_1^2;$$

$$\rho_{kk\pm1}(2)^2 = \lambda_{kk\pm1}(1)^2 + \Delta r_1^2 + \left(\Delta r_1 + \lambda_{k\pm1k\pm1}\right)^2,$$

$$\rho_{kk\pm1}(3)^2 = \lambda_{kk\pm1}(1)^2 + 2\left(\Delta r_1 + \lambda_{k\pm1k\pm1}\right)^2,$$

$$\rho_{kk\pm2}(4)^2 = \left[\lambda_{kk\pm1}(1)+\lambda_{k\pm1k\pm2}(1)\right]^2 + 2\Delta r_2^2,$$

$$\Delta r_m = \lambda_{k+mk+m}(1) - \lambda_{kk}(1), \quad m = 1 \text{ or } 2.$$

Compressibility of the lattice causes its deformation and it changes the average number of neighbouring molecules situated at different distances r. The relationship between averaged $z_{kn}(r)$ and fixed $z_{kn}^*(r)$ values of the numbers of neighbouring sites in layer n at distance r from an arbitrary site in the layer k for soft and rigid lattices is respectively defined as $z_{kn}(r) = z_{kn}^*(r)\lambda_{kk}/\lambda_{nn}$. The last ratio reflects the variation of the number of neighbours with the change of the density of molecules in adjacent layers (if layer k is denser than the layer $k + 1$, then $\lambda_{kk} < \lambda_{k+1k+1}$, and the average number of neighbours of the of site of layer k in the layer $k + 1$ is less than for a rigid lattice) and satisfies $f_r z_{kk+1}(r) = f_{k+1} z_{k+1k}(r)$, which determines the average number of bonds between adjacent layers k and $k + 1$ at a distance r at all densities of molecules in two neighbouring layers.

Compressibility of the fluid + general case of pores with nonuniform walls [45]

Description of the nonuniform porous structures using rigid lattices was given above in this chapter. It is realized by constructing the distribution functions of sites to consider the presence of structural imperfections of the surface layers of the pore walls and their chemical composition. This requires accurate definition of sites types in different layers. The number of types of sites in layer k is denoted by t_k. Function $f_k(q)$ characterizes the proportion of the sites of type q in the layer k, $\sum_{q=1}^{t_k} f_k(q) = 1$. For soft lattices, this principle is maintained. The main difference is related to taking into account the change of the number of sites of this type inherent in different layers, and the number of the nearest neighbours, depending on the local densities $\theta_k(q)$ (in the general case in equations (14.1) index q indicates both the layer number and the site type in the given layer). As in the case of homogeneous pore walls, these modifications are due to differences in the cell sizes for different types of sites in different layers of the pore. For determination of the distribution functions of different types of sites on the soft lattice within a layer k we must use the condition connecting the number $N_k(q)$ and area $s_k(q)$ of the sites of various types q with similar values for a rigid lattice (values with asterisk):

$$N_k(q)s_k(q) = N_k^*(q)s_k^*(q), \text{ and } f_k(q) = N_k(q)/\sum_{q=1}^{t_k} N_k(q).$$

As noted above, for the bulk phase, if we consider only the interactions of the nearest neighbours, the rigid and soft lattices are identical: at any density the size of the lattice constant is equal to the distance corresponding to the minimum interparticle interaction potential. Usually this lattice constant is used for all rigid lattices (i.e. the size of the rigid lattice constant is maximal and the surface is considered constant at any degree of filling). For soft lattices, these lattice parameters $N_k^*(q)$ and $s_k^*(q)$ for a rigid lattice refer to the case of extremely small amounts of the adsorbate. Compressibility of the lattice structure of the fluid is a collective effect of the interaction of molecules. To connect the number of sites in the adjacent layers, generalization of the condition (14.2) for nonuniform walls can be written in the following form:

$$\sum_{q=1}^{t_k} s_k(q) N_k(q) = \sum_{p=1}^{t_n} s_n(p) N_n(p),$$ (14.3)

where $n = k \pm 1, k \pm 2,...$ and $s_k(q) = \lambda_{kk}(q, x)\lambda_{kk}(q, y)$, $\lambda_{kk}(q, x)$ – the average size of a site of type q in the layer k along the axis x, $\lambda_{kk}(q, y)$ – similar to the axis y. x and y axes are arranged along a plane parallel to the walls of the pores, the z axis is directed along the width of the pore. $\lambda_{kk}(q, x) = [\lambda_{kk}(qq_1) + \lambda_{kk}(qq_3)]/2$, the sites q_1 and q_3 are arranged along the x axis to the left and right of the site q, and similarly $\lambda_{kk}(q, y)$ is determined for the axis y (through neighbouring sites q_2 and q_4) (see also below). It should be noted that, because the range of variation of all values $\lambda_{kn}(qp)$ is small, value $s_k(q)$ may be represented with a high degree of accuracy be $s_k(q) = \sum_p \lambda_{kk}(qp)/z_{kk}(q)$, where the sum over p extends over all $z_{kk}(qp)$ neighbouring sites p in layer k, located around the site q.

The condition for the average values of the numbers of neighbouring sites can be written as

$$N_k(q) z_{kn}(qp|r) = N_n(p) z_{nk}(qp|r),$$ (14.4)

wherein $z_{kn}(qp|r) = z_{kn}^*(qp|r)\lambda_{kk}(q)/\lambda_{nn}(p)$, $\lambda_{kk}(q) = \{s_k(q)\}^{1/2}$ is the average size of the lattice constant of site q in layer k; $z_{kn}^*(qp|r)$ is the number of neighbours of site q of layer k in layer n of type p at a distance r for a rigid lattice. The latter condition looks completely similar to the rigid lattice.

The introduction of the average values of neighbouring sites reflects the effect of disparity of the cells of different sizes in a large (infinite) ensemble of different distributions of molecules along the pore axis. In the end portions (compared to rigid lattices) the presence of additional neighbours is determined only by the possibility of introduction of the hard sphere of the molecule. Therefore, for small (finite) pore widths (along the axis z) the number of layers of pores changes discretely, and averaging over the different positions of the molecules only affects their displacement and the average distances between them. Similar remarks apply to the case of microrough surfaces: for small distances between the sites of different types the number of neighbours is preserved, as for a rigid lattice, and the possibility of introducing additional molecules at large distances between sites is taken into account by the average (non-integer) number of neighbouring sites.

15. Cylindrical pores

Cylindrical pores form the basis of the traditional description of porous systems. They, differing in geometry from infinitely long slit-shaped pores, also have the single geometrical parameter – the radius R_p (or diameter) at infinite length, which is very convenient for qualitative assessments of any characteristics. The expressions, presented in section 14, for the interaction of molecules with the wall of the pores of any geometry can be used for cylindrical pores, or averaged potential functions, described in section 4, are derived for them,

The basis of the geometric constructions of the LGM [123] is the partition of the pore volume into elementary cells (see Fig. 15.1) or separation of cylindrical monolayers [124].

In the centre of the cylindrical pores, depending on the pore radius, R_p and lattice geometry, there may be different numbers of sites. For example, for a lattice with $z = 6$ the centre of the pores may contain one $R_p = (h + 0.5)\lambda$ (a) or four ($R_p = h\lambda$) (b) sites, where h is an integer λ, $h > 1$. For a lattice with $z = 8$, in the centre of the pore there is one (d) or three (c). In the case of cylindrical monolayers the centre has one (e) or three (f) particles.

The numbering of the sites on sections of the pores reflects their geometry and the role of the surface potential. Quantity Q_1 is the criterion in partition the sites by type q, $1 \leq q \leq t$, within the specified accuracy of the description of the energy of the system.

As above, for unstructured surfaces of the pore walls the particle – wall binding is taken equal to: a) the depth of the potential minimum $U(y_{min})$ for sites adjacent to the surface of the pores (for a slit pore the set of such sites is the first monolayer), i.e. for the sites where the number of the nearest neighbours does not coincide with the bulk value z; b) the value of the potential $U(y_{min} + \lambda)$ at the point $y_{min} + \lambda$, shifted from the minimum to a distance of the site size, if the site is bordered by at least one 'surface site' (sites 6–8), c) the value of the potential $U(y)$ for all other sites. For structured walls the binding energy of a particle in a site of type k, defined as the sum of the contributions of its atom–atom potentials with the atoms of

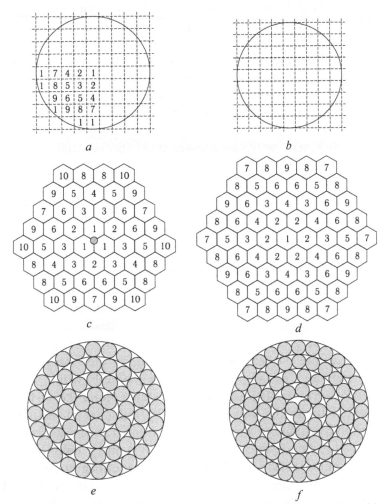

Fig. 15.1. Sections of a cylindrical pore by lattice structures: the model with $z = 6$ (*a*, *b*), the lattice $z = 8$ (*c*, *d*) [123] and the model of cylindrical layers (*e*, *f*) [124] were used.

the solid. Value Q_q can be represented as $Q_q = \varepsilon_{AA} \gamma \sum_{r=1}^{R} \left(z(r) - z_q(r) \right) \eta_r$.

Summation is over R coordination spheres (c.s.) around the site q, z (r) is the number of sites of the r-th c.s.; $z_q(r)$ is the number of sites in the r-th c.s. inside the pore around the site of type q; value $z(r) - z_q(r)$ represents the number of atoms in a solid for the site k in the r-th c.s.; γ is the parameter that specifies the ratio of the depths of the potential minima of the fluid–fluid and fluid–wall interactions, here η_r are the coefficients fixing the distance from the molecule to the atoms in a solid, forming the studied pore, $r = \eta_r \lambda$.

As usual, the pore space is divided into elementary volumes (sites) $v_0 = \lambda^3$ (λ – the lattice constant), equal to the size of the particle, and from we separate groups of sites with the same adsorption capacity, including with the same energy of the sorbate –sorbent interaction Q_q, $1 \leq q \leq t$, t is the number of groups of sites in the system. The proportion of sites in the group q is denoted by f_q.

Partition of the pore volume to the sites and their grouping shall be based on information about the value of the particle–wall interaction potential with the symmetry of the internal space of the pore taken into account. For example, Fig. 15.1 shows the types of sites for cylindrical pores with $R_p = 4.5\lambda$. The calculations are carried out most frequently using the 'smeared' potential of interaction of a particle at a distance y from the cylindrical wall of the pore with radius R_p (4.13) (section 4).

The proposed method for partitioning the pore volume into sites can also be applied to nonuniform walls of cylindrical pores (and spherical pores) because it reflects any local structural imperfections commensurate with the size of the adsorbate molecule.

The lattice parameters, defined in the previous section, are fully applicable to the given pore geometry. For highly curved cylindrical surfaces inserting the lattice structure leads to the intersection with the geometric area of the pore wall, so some cells within a cylinder of radius R_p have variable Langmuir constants along the surface. This is accounted for in the values of $Q_{q,f}$ by reducing the value of the cell volume available to the molecule situated in the given cell q.

Taking this factor into account, the system of equations (14.1) and their modifications (14.4) in the preceding section remain in force and system solutions describe the distribution of the molecules within the volume of the cylindrical pores. Transition to the cylindrical geometry increases the contribution of the surface potential, especially for small diameters of the channels. Otherwise, the layer distributions, discussed in the previous section, retain their dependence on the type of potential: an increase in the degree of filling of the case of attraction and a reduction in the case of small values of the quantity Q_1. These factors determine the phase diagrams of the adsorbate in cylindrical pores that are discussed in the following chapter.

Earlier, the lattice model to calculate the adsorption in cylindrical pores was used only in [124] taking into account the interparticle interactions in the mean-field approximation. Cylindrical layers are measured from the surface of the pores, and in the centre of the pores there may be different situations differing in the radius of the remaining part: from dense packing when the central part is a 'one-dimensional string', to a loose package containing up to five particles in the cross section. This leads to a change in the number of neighbouring sites in the centre of the pore compared to the bulk phase, which affects all of the critical parameters.

The alternative lattice model of the cylindrical pore [123] allows the use of the QCA and allows for simple generalizations of spherical pores and nonuniform walls of cylindrical and spherical pores. Using LGM provides an opportunity to consider any type of pore geometry in the same way, changing only the structural information about the geometry of the pores (f_q and $z_{qp}(r)$). Note that the structural model [124] can also be reformulated to use the QCA, taking into account the effects of direct correlations, while maintaining the number of sites in the centre as in the bulk phase (see [125]).

16. Clusters in volume and pores

Calculation of clusters in the lattice-gas model (LGM)

In recent years there has been intensive research into clusters in various bulk phases. With decreasing temperature, there is an intensive formation clusters and small particles, which play a key role in many natural and technological processes.

In the early development of numerical Monte Carlo and molecular dynamics methods attention was paid the behaviour of clusters and droplets of fluids in spherical pores with inert walls [126]. Later works studied various properties of adsorbates in different stages, but the clusters have not been studied. Interest in them appeared recently in connection with the study of solvation effects and its impact on chemical, including catalytic, reactions [127, 128]. Currently it is one of the most rapidly developing areas of research, both experimental and molecular modelling.

In section 11 it was shown that the LGM describes satisfactorily the distribution of dimers in the bulk phase. This approach allows us to explore a variety of types of clusters in the bulk phase. Knowing the equilibrium distribution of molecules in the bulk or porous systems, we can calculate the probability of formation of clusters θ_{clust} if different size and shape [129].

By analogy with the gas phase the cluster of size n will be an isolated particle consisting of n atoms (monomers) linked together by at least one bond, around which there are vacant (free) lattice sites. These vacancies ensure the existence of an isolated particle. Monomers in the cluster can be organized in different ways (to have a different structure), depending on the number of internal bonds. In the pores the clusters form of atoms situated in different layers. This leads to a dependence of the probability of cluster formation on the position of the centre of mass of the cluster along the normal to the surface and on the orientation of the cluster relative to the normal. The solution of the system of LGM equations allows us to calculate the equilibrium probabilities of the existence of clusters $\theta_{clust}(n|k)$ of size n with the centre of mass in the layer k (or simply the concentration of these clusters): $\theta_{clust}(n|k) = \theta(n|k)\Phi(n|k)$, where $\theta(n|k)$ represents the probability of filling with monomers of a certain group of linked n sites with the centre of mass in the layer k, $\Phi(n|k)$ is the probability of the existence of multiple vacant sites surrounding the said set of monomers.

As examples, we give expressions for the functions $\theta(n|k)$ and $\Phi(n|k)$, expressed in terms of θ_q and t_{qp}, for the simplest clusters. The horizontal dimer in layer k: $\theta(2|k) = \theta_k t_{kk}$, $\Phi(2|k) = \left(t_{kk-1}^{AV} t_{k+1}^{AV}\right)^2 \left(t_{kk}^{AV}\right)^6$. where $t_{km}^{AV} = 1 - t_{km}$. The vertical dimer in layers k and $k + 1$ is: $\theta(2|(k+1)/2) = \theta_k t_{kk+1}$; $\Phi(2|(k+1)/2) = t_{kk-1}^{AV} t_{k+1k+2}^{AV} \left(t_{kk}^{AV} t_{k+1k+1}^{AV}\right)^4$. The horizontal linear trimer in layer k: $\theta(3|k) = \theta_k (t_{kk})^2$ $\Phi(3|k) = \left(t_{kk-1}^{AV} t_{kk+1}^{AV}\right)^3 \left(t_{kk}^{AV}\right)^8$. The vertical linear trimer: $\theta(3|k+1) = \theta_k t_{kk+1} t_{k+1k+2}$; $\Phi(3|k+1) = t_{kk-1}^{AV} t_{k+2k+3}^{AV} t_{kk}^{AV} \left(t_{k+1k+1}^{AV} t_{k+2k+2}^{AV}\right)^4$. The corner trimer in plane k has the same expression for $\theta(3|k)$ (since only the nearest neighbours are taken into account) and function $\Phi(3|k) = \left(t_{kk-1}^{AV} t_{kk+1}^{AV}\right)^3 \left(t_{kk}^{AV}\right)^8 / \theta_k$, since one of the adjacent sites is connected with the two atoms of the cluster, and so on.

Calculation of the dimers and m-mers in the bulk phase is shown in Fig. 16.1. The range of temperatures and pressures in Fig. 16.1 a is most characteristic of prechambers in pulsed piping. The dependences are non-monotonic with a pronounced maximum, the values of which are quite close to each other and are within the range ($6.5\cdot10^{-3} < \alpha_{i,d}^{max} < 8.0\cdot10^{-3}$), and its position shifts to the right with increasing temperature. Increasing temperature also results in the expansion and rising up of the right branch of these dependences. If at $T = 273$ K the dimer concentration is reduced to virtually zero at a pressure of $P = 100$ MPa, then at $T = 2000$ K the dimer concentration is close to its maximum value at $P \geq 200$ MPa.

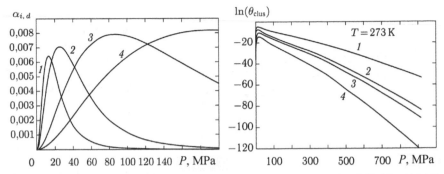

Fig. 16.1. Proportion of dimers depending on temperature (a): $T = 273$ (1), 400 (2), 10^3 (3), and $2 \cdot 10^3$ (4). Proportion of clusters of different sizes at $T = 273$ K (b).

Figure 16.1 b shows for $T = 273$ K in the logarithmic scale the concentration of some clusters (curve 1 – dimers, 2 – trimers, 3 – tetramers, 4 – semimers) depending on pressure. With increasing dimension of the cluster its concentration decreases sharply, and the position of the maximum concentration of the dimer is slightly shifted to the right.

Slit-shaped pores [129]

In [116–118] attention was paid the effect of the limited space of slit-shaped pores (section 14) on the nature of the concentration distributions of argon clusters depending on the strength of interaction of atoms with the wall. Given the complexity of cluster interaction with a solid surface, we confine ourselves to the analysis of the equilibrium state of the adsorbate when the energy exchange processes do not show themselves, and consider an infinite slit pore.

Study [129] investigated: 1) distribution of argon clusters of different sizes (from two to twenty atoms), depending on the position of the centre of mass of the cluster relative to the slit-like pore wall, 2) dependence of the distribution of clusters of different structure and orientation relative to the pore walls with their fixed size (calculations refer to the temperature range from 70 K to 300 K), and 3) the stability of the cluster within the pores.

Figure 16.2 shows the layer-by-layer distribution of the concentrations of dimers of argon atoms of different orientation depending on the degree of filling of the pore volume. Curves 1–4 correspond to argon adsorption in carbon pores (strong wetting), and for comparison there are curves 5, 6 for a second monomolecular layer of in the case of non-wettable walls. Obviously, the increase in the argon concentration increases the equilibrium concentration

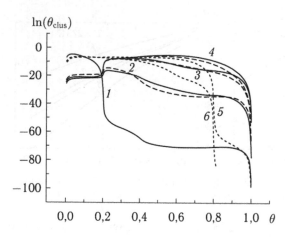

Fig. 16.2. Dependence of the dimer concentrations in layers of a slit pore with $H = 10$ on the magnitude of θ at $\tau = 0.8$ and $Q_1/\varepsilon = 9.24$. Solid curve – horizontal dimers, dashed curve – vertical dimers. Numbers: *1, 2, 3, 4* – dimer concentration of *1, 2, 3* and *5*-th monomolecular layers, respectively. The dashed curve *5* and *6* – horizontal and vertical dimers respectively in the 2nd monomolecular layer at $Q_1/\varepsilon = -8$.

of dimers. These dependences are dramatically different for different layers. The maximum number of dimers in the first layer is observed in filling of the order $\theta \leq 0.04$, for the second layer $\theta \sim 0.22$, for the third layer $\theta \sim 0.4$, for the last layer $\theta \sim 0.6$. For the first layers there is pronounced layering filling of the pore volume.

Sequential filling of each layer dramatically reduces the likelihood of the existence of isolated dimers. In the inner layers the form of the curves for vertical dimers (*1–4*) is similar to the form of the curves for horizontal dimers. For the first and last layer these differences are due to the nature of filling of adjacent layers. The second layer is thinner than the first, and the fourth layer is denser than the fifth layer, so the curves for vertical dimers are deflected in different directions from the curve for the horizontal dimers. For the non-wetting walls there is 'volume' filling of the pores, and the fraction of dimers in the second layer is sharply reduced only after filling the inner layers. In this case, the fraction of the dimers is significantly higher than in the case of highly wettable walls because the surface layer remains substantially free until θ reaches high values of $\theta \sim 0.8$.

This figure (and subsequent) has a methodological character – it shows that the method can be used to consider the most arbitrary types of clusters in the whole range of densities. It is obvious that for large fillings we can not talk about clusters, but a similar problem arises in the study of fluids in 'clusters' of vacancies in the formation of bubbles.

Figures 16.3 *a* and 16.3 *b* represent the average concentration over the cross section of the clusters of different sizes which are obtained from weighing the layer-by-layer distributions:

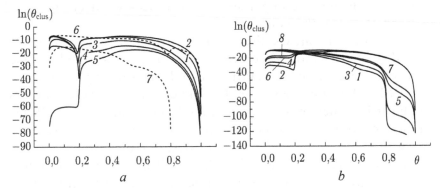

Fig. 16.3. Dependence on the average concentrations in the cross section of a slit-like pore with $H = 10$ on θ: a – different clusters at a value of $\tau = 0.8$ and $Q_1/\varepsilon =$ 9.24 (numerals indicate the concentrations of the vertical and horizontal dimers, horizontal trimers and tetramers and bulk semimers (consisting of a central site and all $z = 6$ nearest neighbours), respectively; dashed curves 6 and 7 – horizontal dimers and bulk semimers at $Q_1/\varepsilon = -8$); b – vertical trimers at $\tau= 0.65$ (1 and 2), 0.8 (3 and 4), 1.2 (5 and 6), 1.4 (7 and 8).

$$\theta_{\text{clust}}(n) = \sum_{k=1}^{H-u(k)} \theta(n|k)/[H-u(k)], \text{ where } u(k) - \text{ the number of}$$

neighbouring layers which occupies cluster whose beginning is situated in layer k. For a flat cluster in layer k $u(k) = 0$.

The layered nature of filling the wetted pore walls is reflected in these curves by the presence of minima for the horizontal clusters and vertical dimers (Fig. 16.3 a). For the semimer located in three adjacent layers (curve 5), a similar dependence has no minimum. In the low filling range the probability of formation of this cluster is small. At the same time, in the region close to the critical ($\theta \sim 0.5$–0.8), the likelihood of formation of such a semimer is not much different from the probability of formation of small clusters $n = 3$ and 4. For the non-wetting walls the probability of formation of large clusters has a maximum value at $\theta \sim 0.1$–0.3 (near θ_c) and sharply decreases with increasing density of argon. For a horizontal dimer the special feature of filling the surface (last) layer of the pore is also characterized by the presence of the minimum and maximum (as for wetted surfaces for small θ).

Note that when estimating the concentration of the clusters, we compare the extent to which the functions $\theta_{\text{clust}}(n|k)$ and $\theta_{\text{clust}}(n)$ exceed the value θ^n and how this relationship changes with θ. The fact that the values of the function $\theta_{\text{clust}}(n|k)$ and $\theta_{\text{clust}}(n)$ for areas $\theta < 0.6$–0.7 in Figs. 16.2 and 16.3 are comparable with their values for $\theta \to 0$ indicates the important role of processes of clustering of the argon atoms in the pores.

Figure 16.3 *b* shows the temperature dependences of the probability of formation of vertical trimers. For well-wetted surfaces the temperature effect is manifested only in the region of small fillings ($\theta < 0.4$). Obviously, as the temperature is lowered the first monolayer becomes most dense and the fraction of isolated clusters therein decreases. For the non-wetting walls the temperature effect is most pronounced at high bulk densities, $\theta > 0.6$.

These calculations show that the equilibrium distribution of clusters of different sizes is characterized by strong anisotropy along the normal to the surface of the pore walls and the concentration of clusters varies non-monotonically with increasing volume filling of the pores. The nature of the distribution of clusters is strongly dependent on the potential of the monomer–wall interaction of the pore and the magnitude of θ. In the critical region the probability of formation of large clusters increases.

The considered behaviour of the equilibrium ensemble of argon atoms should be compared with the behaviour of individual clusters in the field of pore walls. The attraction of the clusters to the walls of the pores causes their deformation and, depending on the position of the cluster relative to the walls, their possible degradation to smaller fragments (monomers and smaller clusters). The total potential energy U of the system consists of two components, $U = U^{(1)} + U^{(2)}$, cluster interaction with the walls $U^{(1)}$ and the cluster internal energy $U^{(2)}$. We define the mechanical stability of the cluster as a condition $U^{(2)} > U^{(1)}$. The maximum value of $U^{(2)}$ is obtained in the absence of the pore walls ($H \to \infty$). With convergence of the well-wetted pore walls the contribution of $U^{(1)}$ increases and it is possible to introduce H^*, corresponding to the equation $U^{(2)} = U^{(1)}$, which is the boundary of the stability of the cluster of a certain size, structure and orientation of the normal to the surface. The characteristic values of H^* for a number of clusters are shown in Fig. 16.4 (accuracy of determination ± 0.25 Å). To obtain H^*, the structure of the clusters in the absence of walls was initially produced in vacuum (the Davidon–Fletcher–Powell optimization procedure [120]), and then this cluster with a fixed orientation was placed in the centre of the pore and the value of $U^{(1)}$ was determined.

Analysis of local minima of the potential energy of 'non-rigid' cluster at the centre of a slit-like pore shows that the region of stability of small clusters with n from 2 to 10 (when the bond length

H^*, nm

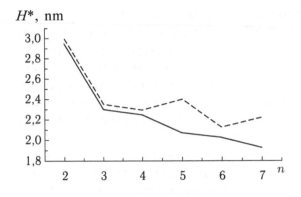

Fig. 16.4. Region of mechanical stability of horizontal and vertical (dashed line) small clusters of argon in a carbon slit-like pore.

varies by no more than 10% compared to vacuum) corresponds to $H_0 \sim 3.1$–3.2 nm. Larger clusters are large and when they are placed in a pore they are destructed (separation of part of the monomers and their transition to the wall). Thus, the calculation results show that around carbon pore walls all considered clusters are mechanically unstable in the size range of micropores ($H < 2$ nm, according to the classification by M.M. Dubinin) and undergo strong deformation in the range of pore widths $H^* < H < H_0$. When moving the positions of the centre of mass of the clusters relative to the centre of the pore the attraction of the nearest wall tends to transfer the cluster in an 'island' of a flat monolayer film.

The observed concentration dependences of clusters of different sizes and orientations in terms of the equilibrium distribution in Figs. 16.2 and 16.3 for well-wetted pores are explained by the following dynamic picture. Clusters mainly form from atoms in the interior of the pores (the question of relations of desorbed flows of argon atoms and their clusters from different monolayers requires a separate study), and as we get closer to the walls of the pores the clusters are deformed in the wall region with the size of the order of one to three monolayers on each side (for small θ) and break up on the surface, forming the first monolayer. After its formation, with increasing θ, the clusters, spreading, form a second monolayer, etc. For non-wetting of the walls all the processes of formation of clusters occur only in the internal 'volume' of the pores. These processes are similar in the first approximation to the processes in the bulk of the gas phase. The role of the walls is reduced to restricting the space containing the adsorbate, – no decay of the clusters occurs on them.

Cylindrical pores [130]

The definitions given above for the clusters hold for all types of pores. Knowing the local filling of different types of sites θ_k, we can calculate the equilibrium probability of formation of clusters $\theta_{clust}(n|k)$ of size n, with the centre of mass in the site of type k (or simply the concentration of the said clusters): $\theta_{clust}(n|k) = \theta(n|k)\Phi(n|k)$, where $\theta(n|k)$ represents the probability of filling with monomers a certain group of linked sites n with the centre of mass in the site of type k, $\Phi(n|k)$ is the probability of the existence of multiple vacant sites surrounding the specified set of monomers.

Figure 16.5 shows the concentration dependences of dimers in different pairs of central sites kn, for the case of poorly- and well-wetted walls. (the numbers of the curves indicate pairs of sites in the dimers are located, Fig. 15.1 a.) In the first case, filling of the pores drastically reduces the concentration of clusters in the volume of the pores, so the particles are not adsorbed on the walls. In the second case the near-surface region is first filled and the decrease in the concentration of clusters takes place at significantly large values of θ. The effect of the radius of the interparticle interaction on the concentrations of the cluster is shown for the dimer 44.

The analysis of distributions of clusters of different sizes (from two to seven atoms) depending on the position of the centre of mass of the clusters relative to the cylindrical wall of the pores show that the general relationships (obtained in [129] for narrow slits) remain unchanged: the equilibrium concentration of the clusters in the pores varies dramatically at various points; for the non-wetted walls the stability of the cluster inside the pore is defined by interparticle interactions, as in the bulk phase; for well-wetted walls of the narrow pores the clusters are unstable – they are deformed by the potential of the walls and flow on them. Quantitative differences between the slit and cylindrical pores are due to the increased contribution of the wall region and a significant reduction in the inner part of the pores in the second case.

17. Thermodynamic characteristics of adsorption systems

The most important thermodynamic characteristic of the adsorption systems, after the adsorption isotherm, is the isosteric heat of adsorption. (It should be noted that the largest contribution to the thermodynamics of adsorption was made by introduced T.L. Hill). The isosteric heat of adsorption is defined as [9, 37, 38, 50, 105]

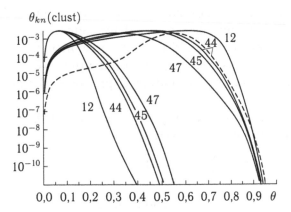

θ_{kn}(clust)

Fig. 16.5. Concentration dependences of dimers in cylindrical pores with the walls of non-wetted ($Q_{9-12}/\varepsilon_{AA} = -4.24$ – four curves in the left set) and well-wetted walls ($Q_{9-12}/\varepsilon_{AA} = 4.24$ – the other five curves), $\tau = 0.8$, $R_p = 4.5\lambda$; central site pairs kn are near the curves. Calculation of the curves was conducted at $R = 4$, the dashed curve 44 for $R = 1$ (differences due to different T_c and a more pronounced layer filling).

$$Q_{is} = -\frac{d \ln(P)}{d\beta}\bigg|_{\theta=\text{const}}. \tag{17.1}$$

Previously, this value was only used in the analysis of adsorption data on the amorphous surface in section 12. The application of the LGM for describing the isosteric heat of adsorption is discussed in many papers (see references in [2]).

We shall therefore consider other thermodynamic functions that are required in the calculation of the transport coefficients of molecules in porous materials. Their calculation requires knowledge of the equilibrium distribution function of the molecules on open surfaces and within the pores. In particular, for calculating the dynamic characteristics we must know the mean energy system, the heat capacity and frequencies of intermolecular vibrations of the molecules. As an example of using the expressions in Appendix 2 we consider the simplified Mie potential for one type of adsorbate.

Simplified Mie potential [72]
In this potential the contributions of the second and more distant neighbours are considered in an approximate way. This leads to a modification of the interaction energy of the nearest neighbours. The essence of this approach is the introduction of functions s_n and s_m, taking into account the change in the Mie potential $(n - m)$ under the influence of neighbouring particles in the medium and obtained

by averaging over the contribution of the second and following neighbours. This technique was used for defect-free crystals and a 'smoothed' model of the fluid (i.e. at constant values of s_n and s_m) [72]. In the LGM the same simplified potentials were constructed in [131]. Below are the corresponding expressions for an nonuniform lattice.

In this model, the total energy of interaction of the central particle with the surrounding can be written as

$$\varphi_f = \sum_{\zeta \in z_f} t_{f\zeta}^{AA} \left(s_{f\zeta}^n A r^{-n} - s_{f\zeta}^m B r^{-m} \right)/2, \qquad (17.2)$$

where $t_{f\zeta}^{AA}$ refers to the nearest neighbours ζ for site f, $s_{f\zeta}^{n,m} = 1 + t_{f\zeta}^{AA} \delta_{n,m}$. $\delta_{n,m} = s_{n,m} (\theta = 1)$ is the maximum contribution of the interparticle interactions to the change of the repulsive and attractive branches of the Mie pair potential at $\theta = 1$. For the Lennard–Jones potential in the case of a cubic lattice with $z = 6$, we have $\delta_{n=12} \approx 0.033$ and $\delta_{m=6} \approx 0.40$, where $a = 4D\sigma^{12}$ and $B = 4D\sigma^6$, D is the depth of the potential minimum φ at $s_{n,m} = 0$.

The expression for the coefficient of elasticity of the bond fg is

$$M(fg) = 2mn|u|_{fg} \bigg/ \left[3 \left(r_{fg}^{\min} \right)^2 \right], \qquad (17.3)$$

where $|u|_{fg}$ is the binding energy of the particles in the condensed phase; in this model

$$|u|_{fg} = \sum_{\zeta \in z_f} t_{f\zeta}^{AA} s_{fg}^n A \left(r_{fg}^{\min} \right)^{-n} (1 - n/m)/2, \qquad (17.4)$$

and the equilibrium distance of neighbouring particles

$$\left(r_{fg}^{\min} \right)^{n-m} = s_{fg}^n nA \Big/ \left(s_{fg}^m mB \right) \qquad (17.5)$$

corresponds to the minimum of the potential.

Oscillation frequencies related to $v_1 = \{mnD/\mu\}^{1/2} / 2\pi r_{fg}^{\min}$ – the frequency of harmonic oscillations of the isolated dimer, formed by fluid particles, can be written as

$$\left(\frac{v_{fg}}{v_1} \right)^2 = \frac{1}{z} \left[1 + \sum_{\zeta \in z_f - 1} t_{f\zeta}^{AA} \frac{\left[1 + t_{fg}^{AA} \delta_m \right]^{n+2/n-m}}{\left[1 + t_{fg}^{AA} \delta_n \right]^{m+2/n-m}} \right]. \qquad (17.6)$$

This ratio v_{fg}/v_1 is obtained if for any particle we 'fix' one of its neighbouring particles g and then average over all the rest $(z_f - 1)$ neighbours ζ.

Local heat capacity in slit-shaped pores

Specific heat C_v per particle located in site f, consists of two contributions $C_v(f) = C_v^{(1)}(f) + C_v^{(2)}(f)$, corresponding to the kinetic and potential energies. The contribution of $C_v^{(2)}$ from the potential energy is defined in general terms in (A2.21). For the kinetic energy we use the following approximation based on the consideration of the local state of molecules in a cluster of nearest neighbours

$$C_v^{(1)}(f) = \left(3 + \sum_{g \in z_f - \Delta_{1f}} \frac{t_{fg}^{AA}}{2} + \Delta_{1f}\right) \frac{k}{2}, \quad C_v^{(2)}(f) = \frac{dU_f}{dT}, \quad (17.7)$$

$$U_f = 0.5 \sum_{g \in z_f - \Delta_{1f}} t_{fg}^{AA} \varepsilon_{fg} + \Delta_{1f} Q_1 + \Delta_{2f} Q_2,$$

where U_f is the potential energy of the molecule located in site f. It consists of the contribution of intermolecular interactions (the first sum) and adsorbate–adsorbent interactions (two second terms for the first two surface layers – for the remaining layers the contribution of the wall potential is neglected); Δ_{ij} – Kronecker symbol: $\Delta_{ii} = 1$ and $\Delta_{ij} = 0$ for $i \neq j$; $C_v^{(2)}(f)$ – the configuration contribution to the specific heat due to intermolecular interactions in a nonuniform system.

When calculating the derivative of temperature the general density of the adsorbate in the pore remains constant. The contribution of kinetic energy differs for the translational movement of the particles in the gas volume ($C_v^{(1)} = 3k/2$) and oscillating motion in the liquid phase ($C_v^{(1)} = zkt_{AA}/2$ for the dense phase, when $t_{AA} \sim 1$, we have $C_v^{(1)} = 3kt$). If $z > 6$, then taking into account the projection of particle motion on the direction of different axes we get the same end result $C_v^{(1)} = 3kt$ for the dense phase fluid [49].

To calculate the contribution of energy transfer through collisions of particles in a dense fluid (second channel) to thermal conductivity, we consider the energy flux carried by the vibrations of neighbouring particles [82]. Every vibration along the flow direction carries a certain amount of energy by contact between the particles lying in adjacent planes f and g, at a distance of λ. Transfer occurs $2v_{fg}$ times per second for each particle θ_f on the surface (the fact that the particle and not a vacancy is near is taken into account by function t_{fg}^{AA}). It is assumed that the characteristic time of vibrational relaxation is less than the characteristic time for the local redistribution of molecules characterized by the evolution of the function t_{fg}^{AA}, discussed in chapter 4.

Calculations of v_{fg} are simplified using the Lennard–Jones potential (6–12) and the average value of the binding energy of the molecule in the central site f:

$$v_{fg} = \left\{ 48U^*(f)/\mu(f) \right\}^{1/2} / 2\pi r_{fg}^{min},$$

$$U_f^* = \left(0.5 \sum_{g \in z_f -1-\Delta_{1f}} t_{fg}^{AA} + 1 \right) \varepsilon_{fg} + \Delta_{1f}Q_1 + \Delta_{2f}Q_2,$$

(17.8)

where v_{fg} is the frequency of harmonic oscillations of the central molecule, $\mu(f)$ is the reduced mass of the molecule of the adsorbate in layer f; for a nonuniform system with variable density the value of μ is calculated for each site f according to the following relationship:

$$\mu(f)^{-1} = m^{-1} \left(\sum_{g \in z_f -\Delta_{1f}} t_{fg}^{AA} + 1 \right) + \Delta_{1f}m_s^{-1}, \quad \text{where } m \text{ is the mass of the}$$

molecule and the adsorbate m_s – mass of the atom (or atoms) solid (adsorbent), which must be taken into account if the adsorbate in the surface layer. This expression for $\mu(f)$ refers to the average of the surrounding into account all of the particles around a central molecule in site f. If the adsorbate is in the first surface layer, then one of its neighbours is necessarily an atom (or atoms) of the solid (depending on the type of bonds of the molecule with the substrate), and the remaining neighbours g contribute when averaged over all others $(z_f - 1)$ neighbours.

Note that equation (17.8) for U_f^* differs from the formula (17.7) for the average energy of the molecule U_f, to reconcile the expression (17.8) for v_{fg} with the frequency of an isolated pair of molecules (dimer) in a rarefied gas. With decreasing density U_f tends to zero, while U_f^* tends to ε. The latter does not contradict (17.7), as the number of dimers tends to zero with decreasing density of the molecules. Differences between U_f and U_f^* follow from different functional relationships for the average potential energy and the oscillation frequency of the molecules as a function of fluid density.

The concentration dependences, used in chapter 5 to calculate the local thermal conductivity (section 32), are given below.

The potential of interaction between the argon atoms found in [132] from the experimental data for the equation of state for the bulk phase was used: $\varepsilon = \varepsilon_0 (1 - d_e\theta)(1 + uT)$, where $\varepsilon_0 = 153.5$ cal·mol^{-1}, $d_e = 0.477$. The second factor takes into account the contributions of the triple interactions between the nearest neighbouring molecules that modify the pair interaction potential: d_e is the ratio of the contributions of the triple interactions to the contribution of the pair interactions of the nearest neighbours. Parameter u takes into account the temperature dependence of ε. Parameter $\varepsilon^* = \alpha\varepsilon$ was found on the basis of the dependence of the thermal conductivity coefficient on the pressure in the bulk phase ($\alpha = -0.25 + 1.05\theta$). These parameters reflect properties of the bulk phase outside the area of rarefied gas. All the curves κ_{qp} are plotted in normalized form: normalization is performed for the value of the thermal conductivity coefficient in the rarefied bulk phase (at $\theta = 0$ and $Q_1 = 0$).

Figure 17.1 a shows layered filling θ_f of a slit-shaped pore with the width of 10 monolayers with increase of its total filling θ. The inset of Fig. 17.1 a shows the adsorption isotherm. The first (surface) monolayer is the first to be filled, the central (fourth and fifth) layers are filled almost at the same time – influence of the wall is weak and can be neglected.

As the pressure of the adsorbtive increases the density of the adsorbate changes from gas to liquid. Figure 17.1 b shows specific heats $C_v(f)$ in individual layers. The main contribution to them is provided by the kinetic component $C_v^{(1)}(f)$; the component of the potential energy $C_v^{(2)}(f)$ does not exceed 10%. The curve for the first monolayer is sharply separated from the other curves that are close enough to each other (layers 3, 4 and 5 are virtually indistinguishable). In accordance with Fig. 17.1 a, increasing filling of the pore changes the heat capacity of the fluid from the heat capacity of the gas changes to a value corresponding to the liquid phase. Furthermore, an additional contribution for the first two surface layers is provided by the potential of interaction with the surface.

Figure 17.2 presents the concentration dependence of the oscillation frequencies of argon atoms in different layers of fluid (a) and the magnitude of the corresponding inverse values of the reduced mass $\mu^{-1}(f)$ (b). All curves monotonically increase with the total content of the adsorbate in the pore. Both dependences are similar to each other. The influence of the adsorption potential on the first monolayer and a weaker effect on the second monolayer are clearly shown. The dependences for the remaining monolayers are similar.

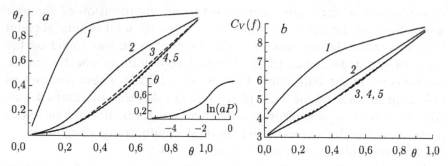

Fig. 17.1. Argon concentration profiles θ_f in the slit pore in graphite with a width $H = 10$ monolayers at $T = 273$ K (a). The numbers on the curves – the number of the monolayer (counted from the pore wall). Lower right – adsorption isotherm in dimensionless variables θ – ln (aP). Layered values of specific heat C_v (f), cal/ (mol·K) (b),

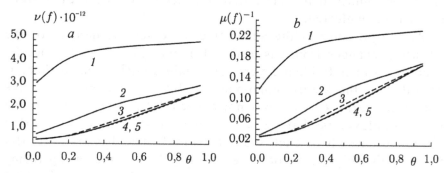

Fig. 17.2. Concentration dependences of the oscillation frequencies $v(f)$, s^{-1}, argon atoms in different layers of fluid (a) and the corresponding inverse values of the reduced masses μ^{-1} (f), g·mol^{-1}, and (b) for the argon–carbon pore system width of the pore 10 monolayers (numbers of curves correspond to the numbers of fluid layers of fluid)

References

1. Tovbin Yu.K. // Dokl. AN SSSR. 1989. V. 306, No. 4. P. 888.
2. Tovbin Yu.K. // Zh. Fiz. Khimii. 1990. V. 64. P. 865 [Russ. J. Phys. Chem. 1990. V. 64. № 4. P. 461].
3. Ohtoni N., Kao S.T., von Hove M.A., Somorjai G.A. // Prog. Surface Sci. 1987. V. 22.
4. Bol'shov L,A,, Napartovich A,P,, Naumovets A,G,, Fedorus A.G. // Usp. Fiz. Nauk. 1977. V. 122. P. 125.
5. Theory of chemisorption, Ed. J. Smith. – Moscow: Mir, 1983. – 333 p.
6. Kaplan I. G., Introduction to the theory of molecular interactions. – Moscow: Nauka, 1982. – 312. [Elsevier, New York, 1986].
7. Avgul' N.N., Kiselev A.V., Poshkus D.P., Adsorption of gases and vapours on homogeneous surfaces. – Moscow: Khimiya, 1975. – 384 p.
8. Tovbin Yu.K. // Teor. Eksp. Khimiya. 1982. V. 18, No. 4. P. 417.
9. Tovbin Yu.K., Theory of physico-chemical processes at the gas-solid interface. –

Moscow: Nauka, 1990. – 288 p. [CRC Press Boca Raton, FL, 1991.].

10. Bethe H.A. // Proc. Roy. Soc. London. A. 1935. V. 150. P. 552.

11. Bogolyubov N.N., Problems of dynamical theory in statistical physics. – Moscow: Gostekhizdat, 1946 [Wiley Interscience, New York, 1962].

12. Tovbin Yu.K. // Kinetika i kataliz. 1982. V. 23, No. 4. P. 821.

13. Tovbin Yu.K. // Kinetika i kataliz. 1980. V. 21, No. 5. P. 1165.

14. Tovbin Yu.K. // Dokl. AN SSSR. 1977. V. 235, No. 3. P. 641.

15. Fowler R.H. // Proc. Camb. Phil. Soc. 1936. V. 32. P. 144.

16. Fowler R.H., Guggenhein E. A. Statistical Thermodynamics. – Cambridge: Univ. Press, 1939.

17. Zeldovich Ya.B. // Acta Phys. Chim. USSR. 1934. V. 1. P. 916.

18. Temkin M.I. // Zh. Fiz. Khimii. 1941. V. 15. P. 296.

19. Frumkin A. N. // Z. Phys. Chem. 1925. V. 116. P. 466.

20. Damaskin B.B., Petrii O.A., Batrakov V.V., Adsorption of organic compounds on the electrodes. – Moscow: Nauka, 1968. – 334 p.

21. Tovbin Yu.K., Chelnokova O.V., Cherkasov A.V. // Zh. Fiz. Khimii. 1988. V. 62. P. 1598.

22. Somorjai G. A. Chemistry in two-dimension surface. – L.; N.Y.; Ithaca: Cornell Univ. Press, 1981.

23. Cartner D. G., Somorjai G. A. // Chem Rev. 1979. V. 79. P. 233.

24. Mozol'kov A.E., Fedyanin V.K., Diffraction of slow electrons by the surface. – Energoizdat, 1982. – 143 p.

25. Tovbin Yu.K. // Kinetika i kataliz. 1987. V. 28. P. 1148.

26. Tovbin Yu.K, Surovtsev S.Yu. // Zh. Fiz. Khimii. 1988. V. 62. P. 3354.

27. Krivoglaz A.M., Smirnov A.A. Theory of ordered alloys. – Moscow: Fizmatlit 1958.

28. Khacharutyan A.G., Theory of phase transitions and the structure of the solid solutions. – Moscow: Nauka, 1974. – 265 p.

29. Gufan M., Structural phase transitions. – Moscow: Nauka, 1982. – 304 p.

30. Smirnov A.A. Theory of interstitial alloys. – Moscow: Nauka, 1979. – 368 p.

31. Tovbin Yu.K., Chelnokova O.V. // Zh. Fiz. Khimii. 1989. V. 63. P. 2556.

32. Tovbin Yu.K, Chelnokova O.V. // Izv. AN Ser. Khim. 1989. No. 9. S. 1943.

33. Tovbin Yu.K. // Zh. Fiz. Khimii. 1995. V. 69, No. 1. P. 118 [Russ. J. Phys. Chem. 1995. V. 69. № 1. P. 105].

34. Tovbin Yu.K. // Zh. Fiz. Khimii. 1995. V. 69, No. 2. P. 214 [Russ. J. Phys. Chem. 1995. V. 69. № 2. P. 195].

35. Tovbin Yu.K. // Zh. Fiz. Khimii. 1995. V. 69, No. 2. P. 220 [Russ. J. Phys. Chem. 1995. V. 69. № 2. P. 201].

36. Roginskii I.Z., Adsorption and catalysis on heterogeneous surfaces. – Publishing House of the Academy of Sciences of USSR, 1949.

37. Jaroniec M., Madey R., Physical Adsorption on Heterogeneous Solids. – Amsterdam: Elsevier, 1988.

38. Rudzinski W., Everett D.H. Adsorption of Gases on Heterogeneous Surfaces. – London: Academic Press, 1992.

39. Perez S., Scaringe R.P. // Macromolecules. 1987. V. 20. P. 68.

40. Niketic S.R., Rasmussen K. // The Consistent Force Field: a Documentation. Lecture Notes in Chemistry. V. 3. – Heidelberg: Springer-Verlag, 1977.

41. Polozov R.V., Method of the semiempirical force field in conformational analysis of biopolymers. – Moscow: Nauka, 1981.

42. Dashevskii V.G., Conformational analysis of organic molecules. – Moscow: Khimiya, 1982. – P. 93.

43. Muruganandam N., Koros W.J., Paul D.R. // J. Pol. Sci. B. 1987. V. 25. P. 1999.
44. Tovbin Yu.K. // Zh. Fiz. Khimii. 1998. V. 72. No. 5, P. 775 [Russ. J. Phys. Chem. , 1998 V. 72, No. 5, P. 675].
45. Tovbin Yu.K., Senyavin M.M., Zhidkova L.K. // Zh. Fiz. Khimii. 1999. V. 73, No. 2. P. 304. [Russ. J. Phys. Chem. 1999. V. 73. № 2. P. 245].
46. Leibfried G., Microscopic theory of mechanical and thermal properties of crystals. – Leningrad: GIFML, 1963. – 312 p. [Handbuch der Physik, Vol. 7, Pt. 1, Ed. by S. Flugge, Springer Verlag, Berlin, 1978].
47. Tovbin Yu.K. // Zh. Fiz. Khimii. 1987. V. 61. P. 2711.
48. Green H. S. // Proc. Roy. Soc. A. 1947. V. 189. P. 311.
49. Hirschfelder J. O., Curtiss C. F., Bird R. B., Molecular Theory of Gases and Liquids. – Moscow: Izd. inostr. lit., 1961. – 929 p. [Wiley, New York, 1954].
50. Hill T.L., Statistical Mechanics. Principles and Selected Applications. – Moscow: Izd. Inostr. lit., 1960. – 486 p. [N.Y.: McGraw–Hill Book Comp. Inc., 1956].
51. Tovbin Yu.K. // Zh. Fiz. Khimii. 2006. V. 80, No. 10. P. 1753 [Russ. J. Phys. Chem. 2006. V. 80. № 10. P. 1554].
52. Prigogine I., Defay R. Chemical Thermodynamics. – Novosibirsk: Nauka, 1966. – 510 p. [Longmans Green, London, 1954].
53. Landau L.D., Lifshitz E.M., Theoretical Physics. Statistical Physics. V. 5. – Moscow: Nauka, 1964.
54. Kubo R., Thermodynamics. – Mir, Moscow, 1970. – 304 p. [North_Holland, Amsterdam, 1968].
55. Fedyanin V.K. // Surface phenomena in liquids. – Leningrad: Leningrad State University, 1975. – P. 232.
56. Batalin O., Tovbin Yu.K., Fedyanin VK // Zh. Fiz. Khimii. 1979. V. 53, No. 12. P. 3020.
57. Batalin O., Tovbin Yu.K., Fedyanin VK // Preparation and analysis of pure substances. – Gor'kii: GSU, 1979. – Issue. 4. – P. 16.
58. Tovbin Yu.K. // Zh. Fiz. Khimii. 2010. V. 84, No. 2. P. 231. [Russ. J. Phys. Chem. A, 2010. V. 84. № 2. P. 180].
59. Tovbin Yu.K. // Zh. Fiz. Khimii. 1992. V. 66. P. 1395. [Russ. J. Phys. Chem., 1992 V. 66, No. 5, P. 741].
60. Egorov B.V., Komarov V.N., Markachev Yu.E., Tovbin Yu.K. // Zh. Fiz. Khim. 2000. V. 74, No. 5. P. 882. [Russ. J. Phys. Chem. 2000. V. 74. № 5. P. 778].
61. Ferziger J. H. H., Kaper H. G., Mathematical theory of transport processes in gases. – Moscow: Mir, 1976. – 554 p. [North_Holland, Amsterdam, The Netherlands, 1972].
62. Sengers J. V. // Phys. Rev. Lett. 1965. V. 5. P. 515.
63. Mason E., Sperling T., Virial equation of state. – Moscow: Mir, 1972. – 280 p. [The Internationsl Encyclopedia of Physical Chemistry. Topic 10. ed. J.S. Rowlinson, Vol. 2].
64. The equations of state of gases and liquids, Ed. I. Novikov. – Moscow: Nauka, 1975. – 263 p.
65. Stogryn D. E., Hirschfelder J. O. // J. Chem. Phys. 1959. V. 31, No. 6. P. 1531.
66. Smirnov B.N., Yatsenko A.S., Dimers. – Novosibirsk: Nauka, 1997. – P. 13.
67. Levelt J. // Physica. 1960. V.2 ˙ 6, No. 6. P. 361.
68. Klein M., Green M. S. // J. Chem. Phys. 1963. V. 39, No. 6. P. 1967.
69. Croxton C. A., Physics of the liquid state: A Statistical Mechanical Introduction. – Moscow: Mir, 1979. – 400 p. [Cambridge University Press, Cambridge, 1974].
70. Reid R.C., Sherwood T.K. The properties of gases and liquids. (The restimation and

correlation). – N. Y.; San Francisco; Toronto, London, Sydny: McGrav-Hill Boch Comp., 1966.

71.   Tovbin Yu.K., Komarov V.N. // Zh. Fiz. Khimii. 2001. V. 75, No. 3. P. 562 [Russ. J. Phys. Chem., 2001. V. 75. No. 3. P. 490].

72.   Moelwyn Hughes E.A., Physical Chemistry. – Moscow: Izd. Lit., 1962. – V.1 and 2. – 1146 p. [2nd ed. Pergamon Press, London, 1961].

73.   Tables of physical quantities, Ed. I. Kikonin. – Moscow: Atomizdat, 1976. – 960 p.

74.   Rivkin S.L., Thermodynamic properties of gases. Handbook. – Energoatomizdat, 1987. – 312 p.

75.   Sychev V.V., Wasserman A.A., Kozlov D., Thermodynamic properties of helium. – Moscow: Publishing House of Standards, 1984. – 240 p.

76.   Sychev V.V., Wasserman A.A., Kozlov D., Thermodynamic properties of nitrogen. – Moscow: Publishing House of Standards, 1977. – 234 p.

77.   Sychev V.V., Wasserman A.A., Kozlov D., Thermodynamic properties of methane. – Moscow: Publishing House of Standards, 1984. – 197 p.

78.   Huang K., Statistical Mechanics. – Moscow: Mir, 1966. – 520 p. [Wiley, New York, 1963].

79.   Komarov V.N., Tovbin Yu.K. // Teplofiz. Vysokikh Temperatur. 2003. V. 41, No. 2. P. 217 [High Temperature. 2003. T. 41. № 2. C. 181].

80.   Thermophysical properties of technically important gases at high temperatures and pressures. Handbook. – Energoatomizdat, 1989. – 232 p.

81.   Hougen O. A., Watson K. M. Chemical Process Principles. – New York, 1947.

82.   Bird R. B., Stewart W., Lightfoot E. N., Transport Phenomena. – Moscow: Khimiya, 1974. – 687 p. [John Wiley and Sons, New York, 1965].

83.   Komarov V.N., Rabinovich A.B., Tovbin Yu.K., Teplofiz. Vysokikh Temperatur. 2007. V. 45, No. 4. P. 518. [High Temperature. 2007. T. 45. № 4. C. 463].

84.   Robertson S.L., Babb P.E. // J. Chem. Phys. M. 1969. V. 50, No. 10.

85.   Gvozdev V.V., Tovbin Yu.K. // Izv, AN, Ser. khim. 1997. No. 6. P. 1109. [Russ. Chem. Bulln. 1997. V. 46. № 6. C. 1060].

86.   Kiselev A.V., Intermolecular interactions in adsorption chromatography. – M.: Vysshaya shkola, 1986. – 360 p.

87.   Krylov O.V., Kiselev V.F., Adsorption and catalysis on transition metals and their oxides. – Moscow: Khimiya, 1981. – 288 p.

88.   Kobozev N.I. // Usp. Khimii. 1956. V. 25. P. 545.

89.   Bakaev V.A. // Surface Sci. 1988. V. 198. P. 571.

90.   Benegas E.I., Pereyra V.O., Zgrablich G. // Surface Sci. 1987. V. 187. P. L647.

91.   Feuston V.R., Gerofalini S.N. // J. Chem. Phys. 1989. V. 91. P. 564.

92.   MacElroy J.M.O., Raghavan K. // J. Chem. Phys. 1990. V. 93. P. 2068.

93.   MacElroy J. M.O, Raghavan K. // J. Chem. Soc. Faraday Trans. 1991. V. 87. P. 1971, 1993. V. 89. P. 1151.

94.   Bakaev V.A., Steele W.A. // Langmuir. 1992. V. 8. P. 1372.

95.   Tovbin Yu.K. // Izv. AN. Ser. khim. 1997. № 3. P. 458. [Russ. Chem. Bull., 1997. V. 46. № 3. P. 437]

96.   Tovbin Yu. K. // Langmuir. 1999. V. 15. P. 6107.

97.   Drain L. E., Morrison J. A. // Trans. Faraday Soc. 1952. V. 48. P. 316.

98.   Drain L. E., Morrison J. A. // Trans. Faraday Soc. 1952. V. 48. P. 840.

99.   Domnant L. M., Adamson A. W. // J. Coll. Interface Sci. 1972. V. 38. P. 285.

100. Gilyazov M.F., Kuznetsova T.A., Tovbin Yu.K. // Zh. Fiz. Khimii. 1992. V. 66. P. 305. [Russ. J. Phys. Chem., V. 66, No. 2, P. 159].

101. Gregg, S.J. Sing, K.G.W., Adsorption, Surface Area and Porosity. – Moscow: Mir,

1984. [Academic Press, London, 1982].

102.   Tovbin Yu.K., Votyakov E.V. // Zh. Fiz. Khimii. 1993. V. 67, No. 8. P. 1674. [Russ. J. Phys. Chem., 1993, V. 67, No. 8, P. 1502].

103.   Tovbin Yu.K., Votyakov E.V. // Zh. Fiz. Khimii. 1992. V. 66, No. 6. P. 1597 [Russ. J. Phys. Chem. , 1992 V. 66, No. 6, P. 848.]

104.   Tovbin Yu.K., Yeremich D.V. // Colloids and Surfaces A. 2002. V. 206. P. 363.

105.   Hill T.L. // J. Chem. Phys. 1947. V. 15. P. 767.

106.   Champion W.M., Halsey G.D. // J. Phys. Chem. 1953. V. 57. P. 646.

107.   Pace E.L. // J. Chem. Phys. 1957. V. 27. P. 1341.

108.   Champion W.M., Halsey G.D. // J. Amer. Chem. Soc. 1954. V. 76. P. 974.

109.   Tovbin Yu.K., Petrova T.V. // Zh. Fiz. Khimii. 1994. V. 68. No. 8, P. 1459 [Russ. J. Phys. Chem., 1994 V. 68, No. 8, P. 1319]

110.   Tovbin Yu.K. // Zh. Fiz. Khimii. 1992. V. 66. P. 2162 [Russ. J. Phys. Chem., 1992 V. 66, No. 8, P. 1151].

111.   Aranovich L.G. // Zh. Fiz. Khimii. 1988. V. 62. P. 3000.

112.   Keenan A. G., Holmes J. M. // J. Phys. Colloid. Chem. 1949. V. 53. P. 1309.

113.   Kiselev A. V. // Quart. Rev. Chem. Soc. 1961. V. 15. P. 116.

114.   Polley M.U., Schaeffer W.D., Smith W.R. // J. Phys. Chem. 1953. V. 57. P. 469.

115.   Ramalho J.P.P., Rabinovich A.B., Yeremich D.V., Tovbin Yu.K. // Applied Surface Science. 2005. V. 252. P. 529.

116.   Tovbin Yu.K., Votyakov E.V. // Langmuir. 1993. V. 9, No. 10. P. 2652.

117.   Tovbin Yu.K., Votyakov E.V. // Zh. Fiz. Khimii. 1993. V. 67, No. 10. P. 2126 [Russ. J. Phys. Chem., V. 67, No. 10, P. 1918].

118.   Votyakov E.V., Tovbin Yu.K. // Zh. Fiz. Khimii. 1994. V. 68, No. 2. P. 287 [Russ. J. Phys. Chem., V. 68, No. 2, P. 54].

119.   Tovbin Yu.K. // Izv. AN. Ser. khim. 2003. No. 4. P. 827. [Russ. Chem. Bull. 2003. V. 52. № 4. P. 869].

120.   Himmelblau D. Applied Nonlinear Programming. – New York: Wiley, 1975. – 534 p.

121.   Tovbin Yu.K., Gvozdeva E.E., Zhidkova L.K. // Inzh.-Fiz. Zh. 2003. V. 76, No. 3. P. 124. [J. Engin. Physics Thermophys., 2003 V. 76, No. 3, P. 619].

122.   Tovbin Yu.K., Vasyutkin N.F. // Izv. AN. Ser. Khim. 2001. No. 9. P. 1496 [Russ. Chem. Bull. 2001. V. 50. № 9. P. 1572].

123.   Tovbin Yu.K., Votyakov E.V. // Zh. Fiz. Khimii. 1998. V. 72. No. 10. P. 1885 [Russ. J. Phys. Chem. 1998. V. 72. № 10. P. 1715]

124.   Nicolson D. // JCS Faraday Trans. I. 1975. V. 71, No. 2. P. 238.

125.   Tovbin Yu.K., Petukhov A.G., Eremich D.V. // Zh. Fiz. Khimii. 2006. V. 80, No. 3. P. 488 [Russ. J. Phys. Chem. 2006. V. 80. № 3. P. 406].

126.   Nauchitel V.V., Persin A.J. // Mol. Phys. 1980. V. 40. P. 1341.

127.   Gubin S.P., Chemistry of clusters. – Moscow: Nauka, 1987. – 264 p.

128.   Smirnov B.M. // Usp. Fiz. Nauk. 1984. V. 142, No. 1. P. 31.

129.   Tovbin Yu.K., Komarov V.N., Vasyutkin N.F. // Zh. Fiz. Khimii. 1999. V. 73, No. 3. P. 500 [Russ. J. Phys. Chem., 1999 V. 73, No. 3, P. 427].

130.   Tovbin Yu.K., Votyakov E.V. // Izv. AN. Ser. Khim. 2000. No. 4. P. 605. [Russ. Chem. Bull., 2000. V. 49. № 4. P. 609].

131.   Tovbin Yu.K. // Zh. Fiz. Khimii. 1998. V. 72, No. 8. P. 1446 [Russ. J. Phys. Chem. 1998. V. 72. № 8. P. 1298].

132.   Tovbin Yu.K., Komarov V.N. // Izv. AN. Ser. Khim. 2002. No. 11. P. 1871. [Russ. Chem. Bull. 2002. V. 51. № 11. P. 2026].

Vapour–liquid phase stratifying

18. Thermodynamic conditions of phase stratifying

The process of stratifying of matter in the volume to gas and liquid is the simplest and most important example of phase transitions. About two phases we can only speak when they both exist simultaneously, *in contact* with each other, i.e. at points on the binodal curve at temperatures $T < T_c$, where T_c is critical temperature. If there is a critical point for all temperatures above the critical temperature the process of stratifying disappears and the very concept of various phases becomes conditional, since it is impossible to specify which states are single-phased and which other [1–4].

Recall that the very concept of the critical point refers to a wide range of phase transitions in multicomponent fluids and to two-phase transitions of the first kind in alloys and ferromagnetics, in lyotropic liquid crystals, polymer and micellar solutions, microemulsions, etc. [1–6]. All these phase transitions are traditionally considered in bulk phases in the absence of external fields. At the same time, as noted by Gibbs [7], the presence of external fields, changes the conditions of phase transitions and, in particular, the presence of the gravitational field changes conditions for vapour–liquid system stratifying.

The external fields can be created by the potential of different nature. An important role in surface phenomena is played by the surface potential of solids, which is created by the cumulative effects of atoms of the solid on molecules of the liquid and vapour near the boundary of the solid. Such potentials are usually relatively short-range, and the question of the conditions of vapour–liquid phase stratifying largely depends on the spatial organization (composition and structure) of the surface of solids.

The phase rule

Recall that the phase is a collection of parts of the thermodynamic system, identical in all physical and chemical properties that do not depend on the amount of the substance. Coexisting bulk phases are separated by partition surfaces, representing the layers of finite thickness, in which at least one of the parameters of the system changes in the direction from one phase to another in a finite amount (i.e. the phases are homogeneous and relatively large so that surface phenomena can be ignored).

The Gibbs phase rule is derived from the conditions of the thermodynamic equilibrium of multicomponent mixtures which in isolated system consist of the identical conditions for the existence of each of its phases (P and T values in all the phases must be equal) and the same value of the chemical potential μ for each of the components in all of the coexisting phases. The first two conditions of phase equilibrium correspond respectively to the mechanical and thermal equilibrium of the system, the last – to the chemical equilibrium (dynamic equilibrium of the exchange of molecules between the two phases).

The phase rule can be used to determine the number of independent variables f, which can be arbitrarily changed without disturbing the phase equilibrium, i.e. f is the number of thermodynamic degrees of freedom of the system. If in addition to P and T the system is also influenced by other thermodynamic forces (electric and/or magnetic field, the surface tension at the interface between phases, etc.) with the total number r, the phase rule becomes

$$n \leq k + r, \qquad (18.1)$$

where $r > 2$, then $f = k - n + r$. The phase rule is used in the study of complex systems, since it allows to calculate the possible number of phases n and the degrees of freedom f in equilibrium systems for any number of components k [1, 7].

In thermodynamics it is assumed that in the non-uniform field the inhomogeneity of the phase can be neglected, provided that its size is small and the error, associated with the introduction of this assumption, lies within the required accuracy. This reflects the situation for the gravitational field when we can assume that the elementary element contains a large number of molecules and the phase transition in this volume takes place at one value of the gravitational field. For different values of the gravitational field, which changes along the normal to the reference surface, even constant (hypothetical) values of pressure P and temperature T must result in changes of the conditions for the implementation of phase transitions. The discussed macroscopic (as regards the size) system, located in a variable gravitational field, has, according to the phase rule, a continuum of different conditions for phase transitions, depending on the distance from the ground level.

The interfaces of the coexisting phases have properties different from those of each of the adjacent phases, and are characterized by an additional parameter of state – the size of the area. On conventional bulk phase diagrams there are no areas that correspond to the surface phases. This reflects that the contribution of the interfaces to all thermodynamic functions is small. Surface contributions can be taken into account in the case of a highly developed surface of the solid material and, then, taking into account the contribution of the surface of the solid phase in the absence of external fields, the value of r in the formula (18.1) is equal to 3.

In the framework of the thermodynamic analysis of adsorption the phase rule is used to take into account the surface contributions of the properties of the system to the values of the thermodynamic properties. At the same time, the thermodynamics can not determine the number of phases in the system. This is carried out using structural, thermodynamic and kinetic information.

As regards adsorption, the thermodynamic approach can not be used to describe different phase states of the adsorbate, which are implemented in different sections of surfaces. Phase transitions were repeatedly observed in experiments [9], both on the entire surface and on its individual facets having different surface potentials. A similar situation with the presence of phase transitions of phase stratifying (capillary condensation) under different environmental conditions has been observed in the porous materials with different pore sizes. Analysis of the conditions of capillary condensation is used for practical purposes in adsorption porosimetry – in determining the function of pore size distribution [10].

The phase rule, allowing any number of phase transitions by varying the external fields at macrodistances, reduces the entire diversity of surface states to a single surface phase, although formally the concepts 'external field' and 'phase rule' are fully applicable to the surface potential and the phases stimulated by his presence.

In microheterogeneous systems, there are fundamental differences of the surface potential of the above-mentioned external electromagnetic and gravitational fields. The essence of the problem lies in the fact that the surface potential of the exposed surface and/or the walls of the porous system can not be described by a single value of 'external forces'. The point here is in the range of variation of the values of the potential of the walls and the scale of the size of the areas on which this potential changes. Areas of the nonuniform solid surface can have very wide ranges of the values of the surface potential Q_q from ~ 0.01 to 10ε, where ε is the interaction energy between the nearest molecules. Similarly, nonuniform areas on the walls of narrow pores may have the same wide range of the adsorbent–adsorbate energies and also various characteristic pore sizes (from one to one hundred nanometers). Consequently, the 'external forces' in microheterogeneous systems fluctuate much more extensively with regards to the energy of adjacent portions than in the case of external macroscopic fields. In addition, there are new questions related to different ratios between the size of inhomogeneous surface areas of solids and the minimum size of the area required for the formation of the phase.

Below the process of stratifying of the adsorbate in the micro-nonuniform systems is considered from the standpoint of molecular interpretation of the thermodynamic phase rule on the basis of the molecular–statistical theory of adsorption. The fundamental difference between the thermodynamic and molecular–statistical approaches is reflected in a drastic difference in the number of degrees of freedom used in each of them. Classical thermodynamics operates with a small number of parameters of the macroscopic state, while the statistical thermodynamics, building on the totality of the generalized momentum and generalized coordinates of the particles of the studied system, operates with any required number of parameters that reflect the molecular distributions of the system.

We confine ourselves to a single-component system ($k = 1$), for which the bulk phase may contain at the same time no more than three phases, and will only discuss the process of stratifying of a fluid into two phases: low-density (or vapour) and high-density (liquid). The phase states were studied by the LGM, developed for highly nonuniform systems of any type [11] (chapter 2).

Homogeneous systems

The homogeneous systems include bulk phases (three-dimensional system) and homogeneous surfaces (two-dimensional systems). In the homogeneous systems, the number of degrees of freedom in thermodynamic and molecular descriptions of the same, if the molecules are uniformly distributed in the volume on any spatial scale. For tasks of phase stratifying of vapour and liquid on the surface or in the bulk, using the LJ type potential, this means the presence of a single-phase state. But, if a homogeneous macroscopic system has ordered distributions of molecules (section 8), the thermodynamic approach, in principle, can not describe the molecular distribution of this kind.

The conditions of thermal equilibrium at a constant temperature leads to the inequality [1]

$$\partial P/\partial V < 0 \qquad (18.2)$$

where the pressure P is related to the volume by the equation of state or through the chemical potential with the degree of filling volume. Figure 18.1 a shows portions of the isotherms for vapour (f) and liquid (a) in the area of the two-phase state of the system. Points b and e, corresponding to the equality of the chemical potentials $\mu_1(P, T) = \mu_2(P, T)$, form a binodal curve ($bKe$) or a phase stratifying curve at a given temperature and pressure. Point K is the critical point.

Inequality (18.2) is preserved in the intersection of the area of the isotherm $P(V)$ with the binodal curve for the liquid with increasing volume of the system and for the vapour with decreasing volume of the system. This follows from the fact that each phase can exist as a metastable phase and on the other side of the transition point b or e. For each of the same functions $\mu_{1,2}(P, T)$ this point b or e is not highlighted, if we assume that outside the transition point in the metastable region, where $\mu_1(P, T) \neq \mu_2(P, T)$, chemical potentials

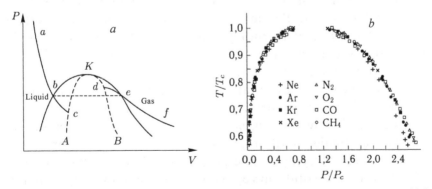

Fig. 18.1. Isotherm in the two-phase state of the substance and phase stratifying curve [1–4] (*a*). Measurements of the coexistence curve, conducted in eight liquids (by Guggenheim [13]) (*b*).

retain their meaning, introduced for strictly equilibrium states. It is obvious that the concept of metastability indicates the existence of a parameter of state by which the system is not fully equilibrium.

Condition $(\partial P/\partial V)_T = 0$ corresponds to the end portion of the isotherm with the thermodynamic inequality (18.2) – points c and d, satisfied at the points of this isotherm. This locus of points – a spinodal line AKB – corresponds to the range of parameters in which the system can not exist as a homogeneous area. The spinodal has two branches for the vapour BK and liquid AK phase. Between the spinodal and the binodal there is a metastable region. The only point on the spinodal, corresponding to the actually existing states of the homogeneous fluid, is a point of contact the spinodal curve with the binodal curve – a critical point.

Analysis of the stability of the existence of a homogeneous state of the fluid [1, 12] leads to the need to implement the following conditions at this point:

$$(\partial P/\partial V)_T = 0, \tag{18.3}$$

$$\left(\partial^2 P/\partial V^2\right)_T = 0, \tag{18.4a}$$

$$\left(\partial^3 P/\partial V^3\right)_T < 0. \tag{18.4b}$$

The two equations (18.3) and (18.4a) represent two equations with two unknowns, which can be satisfied only at a single critical point of the substance. These thermodynamic relations relate to a closed volume of the system, which can be in both the single-phase and two-phase conditions. To use the equations (18.3) and (18.4a), we must have an equation of state. Derivation of the equation of state $P(V)$, as well as determining the number of phases present in the system, is beyond the scope of thermodynamics [1, 12]. Therefore, for the nonuniform systems we must use the equations of the equilibrium distribution of molecules (section 2). The distribution pattern of the molecules depends on external fields generated by the adsorption potential of the exposed surfaces and the pore walls.

Figure 18.1 b shows the classical general curve of phase stratifying of the vapour–liquid stratifying systems, based on experimental data for eight simple substances. The projection of surface $P\rho T$ on plane ρT is shown. The solid curve, which describes the experimental data, corresponds to a cubic equation $\rho - \rho_c \sim (1 - T/T_c)^\beta$, where β is the critical exponent equal to 1/3, ρ_c is the critical density, $\rho =$

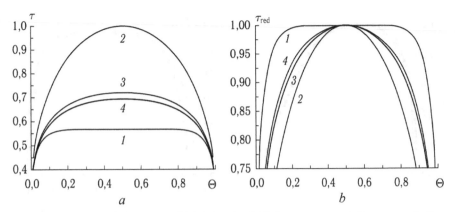

Fig. 18.2. Curves of phase stratifying in the LGM: *a* – an exact Onsager solution [17] (*1*) and the cluster method [11], the mean-field (*2*), quasichemical (*3*) and superposition (*4*) approximation, *b* – the same curves in reduced coordinates, the numbers of curves as in the field (*a*).

$\rho_L - \rho_G$ is the difference of densities of liquid ρ_L and gas ρ_G (also referred to as the order parameter) [13]. The vapour–liquid equilibria of different molecules in bulk phases has been described for many years now using the principle of corresponding states [14–16], which demonstrates the general pattern of behaviour of substances.

The LGM equation for a rigid lattice (chapter 2) lead to the well-known phase stratifying curves (Fig. 18.2), which differ in form from the curve in Fig. 18.1 *b*. The calculations were carried out for the plane square lattice $z = 4$, when the nearest-neighbour interactions are taken into account ($R = 1$). They have the symmetric form with respect to the degree of filling $\theta = 1/2$, and the values of the critical parameters T_c depend on the approach used when taking into account the intermolecular interactions.

The field in Fig. 18.2 compares Onsager's exact solution for a given lattice (curve *1*) [17] and calculations [11] in the mean-field (curve *2*), quasichemical (*3*) and superposition (*4*) approximations. With increasing accuracy of taking at the interaction into account by increasing the role of correlation effects the approximate curves converge to the exact solution. Relative or reduced coordinates are often used in practice in which the values $\tau_{red} = T/T_c$ are plotted on the ordinate, where T_c is calculated in the given approximation. This allows us to consider the values of the critical exponents β in the calculation method used.

Accounting for LGM modifications (section 9 and Appendix 4) enables to consider the length of the LJ potential ($R = 4$), the softness of the lattice and the internal motion of the molecules within

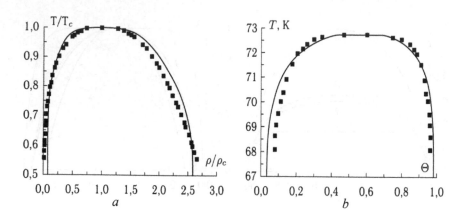

Fig. 18.3. Argon stratifying curve (points – experiment, calculations – calibrated QCA): *a* – in the bulk phase [18], *b* – two-dimensional phase stratifying in the second monolayer on the surface of $CdCl_2$ [19].

the cells. In addition, to improve the accuracy of calculations near the critical point it is necessary to introduce calibration functions, reflecting the impact of large fluctuations. The essence of the introduction of calibration functions is set out in Appendix 5 (see section 24). Figure 18.3 shows the result of modifications of the LGM to calculate the curves of phase stratifying of argon in the bulk phase [18] and for the two-dimensional system in argon adsorption at the surface of $CdCl_2$ [19].

In the analysis of the properties of nonuniform systems below we used the version of the theory, taking into account the lateral interaction of only the nearest neighbours to illustrate the impact of major energy factors of complex nonuniform systems. Examples in Fig. 18.3 show how taking into account the LGM modifications improves the accuracy of the calculation of the thermodynamic properties.

To conclude this section, we present experimental data on the phase diagrams obtained for porous systems. Figure 18.4 show experimental diagrams of Xe atoms and molecules of CO_2 and SF_6 [20, 21].

A comparison of the phase diagrams for the pores and in the volume shows that the curves differ dramatically due to the substantial influence of the structure and the surface potential. The phase stratifying curves for the pores have shifted to higher densities. However, in a first approximation, the law of corresponding states, as follows from a comparison of the curves for Xe atoms and molecules

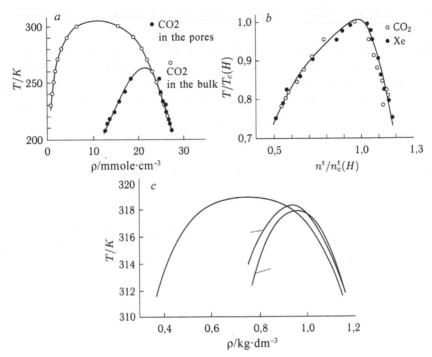

Fig. 18.4. Phase diagrams: a – for CO_2 in the pores and in the bulk phase [20]; b – CO_2 and for Xe (in reduced units) [20] c – SF_6 coexistence curves in the bulk phase and CPG pores for medium pore diameters 31 and 24 nm [21].

of CO_2, is also preserved for porous bodies. Here we have a complete analogy with the imposition of curves on each other for different substances in Fig. 18.1 b. The increase in the pore size shifts the curves for the pores in the direction of the phase stratifying curve for the bulk phase.

Later in this section attention is paid to the influence of heterogeneity properties of the adsorption systems on the conditions for the occurrence of phase stratifying of fluids, which determines the number and parameters of the critical points on the vapour–liquid phase stratifying curves. The nonuniform systems are considered to be nonuniform surfaces and porous solids where the internal volumes of the pores are bounded by walls. During adsorption of molecules in these systems the molecular density can vary in a wide range and conditions for their phase stratifying may form. The external fields are produced by the adsorption potential of open surfaces and the walls of the pores.

19. Critical conditions for nonuniform surfaces

The equations of the local isotherms for different types of sites, given the direct correlation of the interacting particles in the quasichemical approximation, are presentec in section 2. Knowing the local filling of sites of different types, we can find the average degree of filling of the general porous system $\theta(P)$, weighing the relevant contributions by the introduced distribution functions (7.8).

The mutual attraction of the molecules of the adsorbate with a reduction of the temperatures of the adsorption system leads to their two-dimensional condensation. On a homogeneous surface the critical values of filling of the surface $\theta_c = 0.5$, and temperatures depend on the approximation of taking their interaction into account. For the quasichemical approximation (QCA) $\beta_c \varepsilon = 2 \ln (z/(z - 2))$, and for the mean-field approximation (or self-consistent) (MFA) $\beta_c \varepsilon = 4/z$.

The two-dimensional pressure π of molecules is expressed in the LGM by the total adsorption isotherm $\theta(P)$ in the form (section 10)

$$\beta v_0 \pi = \int_0^P \theta(P) d \ln P, \qquad (19.1)$$

where v_0 is the cell volume.

The critical temperature is determined by the inflection point of the dependence $\pi(\theta, T)$, which leads to a system of equations with respect to θ and T:

$$\left(\frac{\partial \pi}{\partial \theta} \right)_T = \left(\frac{\partial^2 \pi}{\partial \theta^2} \right)_T = 0. \qquad (19.2)$$

Given the explicit form of π (19.1), this system can be rewritten as

$$\left(\frac{\partial \ln P}{\partial \theta} \right)_T = \left(\frac{\partial^2 \ln P}{\partial \theta^2} \right)_T = 0, \qquad (19.3)$$

which is equivalent to the systems (18.3) and (18.4a).

The relationship between pressure and filling the surface is given by the implicit relation (7.8). Differentiating the system (7.8) with respect to θ, we obtain the following expression:

$$\left(\frac{\partial \ln P}{\partial \theta} \right)_T = \frac{\Delta(t)}{\sum\limits_{q=1}^t f_q \Delta_q (1)}. \qquad (19.4)$$

Here, $\Delta(t)$ is the determinant of matrix M with the dimension t, the elements of which depend on the type of model, and $\Delta_q(1)$ is the determinant of the matrix M, whose q-th column is replaced by units. For example, for a QCA model (A2.14) the matrix elements are [22]

$$M_{qp} = \frac{\Delta_{qp}}{\theta_q(1-\theta_q)} + \sum_{r=1}^{R} \frac{z_q(r)}{D_q(r)} \sum_{p=1}^{t} D_{qp}(r),$$

wherein

$$D_{qp}(r) = d_{qp}(r)x_{qp}^{AA}(r)t_{qp}(r)\{\Delta_{qp}x_{qp}^{AA}(r)(1-t_{qp}(r)/b_{qp}(r) + \left[1 + x_{qp}^{AA}(r)\right.$$

$$\times \theta_q\left(1-t_{qp}(r)\right)/b_{qp}(r)\right]/\theta_p\},$$

and Δ_{qp} is the Kronecker delta, and $b_{qp}(r)$ and $t_{qp}(r) = \theta_{qp}^{AA}(r)/\theta_q^A$ are defined by (7.5), in which the indices indicate the type of sites instead of site numbers.

If we use the MFA, then $M_{qp} = \Delta_{qp}/\left[\theta_q(1-\theta_q)\right] - \beta z \varepsilon_{qp} d_{qp}$ [23].

Taking the derivative with respect to θ of the expression (19.4), we find that the system (19.3) can be rewritten as

$$\Delta(t) = 0, \quad \Delta'_\theta(t) = 0, \tag{19.5}$$

where $\Delta'_\theta(t)$ is the derivative of θ from the determinant of the matrix M, where only on the diagonal elements containing θ_q depend on θ. This allows the second equation of the (19.5) system in the form

$$\sum_{q=1}^{t} \frac{(1-2\theta_q)}{\left[\theta_q(1-\theta_q)\right]^2} \Delta_q(1)\Delta^q(t-1) = 0. \tag{19.6}$$

where $\Delta^q(t-1)$ is the dimensional determinants $(t-1)$ obtained from $\Delta(t)$ deleting from the q-th column and row.

Thus, the system of non-linear equations of the $(t+1)$ relative to θ_{qc} and β_c, allows us to analyze the critical temperature and surface coverage, depending on the properties of the nonuniform surface defined by the function f_q and $f_{qp} = f_q d_{qp}$, on the relationship between the adsorption coefficients a_q and the values of the parameters of interactions between adatoms ε_{qp}. The closed system consists of two equations (19.6) and $(t-1)$ (7.4) or (7.7) interconnecting θ_1 and θ_q, $2 \le q \le t$, with the equilibrium pressure P excluded from them.

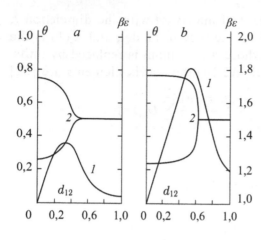

Fig. 19.1. Critical values of filling the nonuniform surface and temperature at $a_1 = 5a_2$ (a) and $a_1 = 35a_2$ (b). Curves *1* relate to values $\beta_c \varepsilon$, curves *2* – to θ_c [23].

Two types of centres

The combined effect of heterogeneity of the surface and the lateral interaction of the particles is shown most clearly on the simplest example of a surface with two types of sites ($t = 2$) in the MFA [23].

In this case, the system (19.5) for a surface containing two types of sites can be written as (here we use the notation $x_i = \beta z \varepsilon_{ii} \theta_i (1 - \theta_i)$)

$$(1 - x_1 \varepsilon_{11} d_{11})(1 - x_2 \varepsilon_{22} d_{22}) - x_1 \varepsilon_{12} d_{12} x_2 \varepsilon_{21} d_{21} = 0,$$

$$(1 - 2\theta_1)\left[1 - x_2(\varepsilon_{22} d_{22} - \varepsilon_{12} d_{12})\right](1 - x_2 \varepsilon_{22} d_{22}) +$$

$$+ (1 - 2\theta_2)\left[1 - x_1(\varepsilon_{11} d_{11} - \varepsilon_{21} d_{21})\right](1 - x_1 \varepsilon_{11} d_{11}) = 0. \quad (19.7)$$

The full range of mutual arrangements of sites of different types for different compositions and surface structures is specified in section 8.

Figure 19.1 shows the influence of the surface structure in the case of $f_1 = f_2 = 0.5$ for the critical values of θ and $\beta \varepsilon$ (calculation for all interaction parameters being equal $\varepsilon_{qp} = \varepsilon$) for two ratios $a_1 = 5a_2$ and $a_1 = 35a_2$. In the range of complete 'stratifying' of different types of sites ($d_{12} = 0$) there are two phase transitions, in the range of total ordering of the surface ($d_{12} = 1$) – one phase transition. The symmetric behaviour of θ in the region of the two phase transitions is due to the equality of compositions $f_1 = f_2$. At full 'phase stratifying' each area of the surface behaves independently and the critical conditions in the mean field approximation of a homogeneous surface are satisfied for each area, so here $\beta \varepsilon = 1$ (since $z = 4$).

With mixing of different types of sites the critical value of $\beta \varepsilon$ passes through a maximum the value and position of which depend on the ratio a_1/a_2. This behaviour is explained by the necessity of cooperation of a large number of molecules which are preferably adsorbed on strong centres randomly distributed over its surface in different spatial regions. Around them there form preferentially occupied closest and, with increasing filling, more distant neighbours. Cooperation of the molecules in phase must occur in such a random distribution of small 'clusters'. Excluding discussing the effects of the metastability of the state of clusters, the temperature must be greatly reduced to obtain a gain in the free energy of the system during the formation of a common phase. That is why the maximum of the dependence of $\beta_c \varepsilon(d_{12})$ on the surface structure is located in the vicinity of the change of the number of solutions of critical fillings from two to one. An important role is played by the binding energy of the molecules with the surface (the ratio a_1/a_2). Obviously, if the bond with the surface is much greater than the lateral interactions, their role in the possibility of the formation of the phase is small.

Complete ordering adcentres corresponds to the unique critical filling of the surface, which coincides with the value of the critical filling in the homogeneous regions. However, condensation occurs at temperatures lower than the critical temperature for homogeneous areas ($\beta_c \varepsilon (d_{12} = 1) > \beta_c \varepsilon (d_{12} = 0)$). The differences in the binding energies of the molecules with centres of various types complicates the process of forming a common phase, and to obtain a gain in the free energy of the system it is required to decrease the temperature.

Number of types of centres of more than two

For a surface consisting of separate macroscopically homogeneous regions, the system (19.5) has the form

$$\prod_{q=1}^{t}\left[1-\beta z\varepsilon_{qq}\theta_q(1-\theta_q)\right]=0, \quad \sum_{q=1}^{t}\frac{1-2\theta_q}{\left[1-\beta z\varepsilon_{qq}\theta_q(1-\theta_q)\right]}=0. \quad (19.8)$$

Here the bond between θ_q is defined by a system of equations for the local isotherms (7.7) at $d_{qp} = \Delta_{qp}$. Since $d_{q\neq p} = 0$, the process of adsorption on each section takes place in an independent manner, and the critical conditions of the condensation for each q, $1 \leq q \leq t$, remain the same as for the homogeneous isolated area. This fact is evident by its physical picture. It follows from Langmuir's assumptions imposed on independent plots of adsorption sites [24], which are usually associated with the existence of different faces of microcrystals. However, little attention had been paid to this matter prior to discussing the condensation conditions on surfaces of this type, 'homotactic' as defined by Ross and Olivier [25]. In [26, 27] the existence of stepped isotherms, with more than one 'jump', was confirmed experimentally. It is the adsorption of the noble gases, ethane, nitrogen oxide and other gases on carbon, boron nitride, and pure cadmium and its halides, nickel and sodium chlorides, and other adsorbents.

Ordered structures
The influence of the role of the binding energy of the molecules in different types of centres on the critical condensation conditions for ordered surface structures was compared in [28]. Considered were the following four surfaces for a flat square lattice with $z = 4$, containing two types of centres, but with different topography. The first topography corresponds to the case discussed above: the first (strong) and the second (weak) centres are distributed in a staggered manner (structure $c(2 \times 2)$) with $f_1 = f_2 = 0.5$, $f_{11} = 0$, the second structure, denoted as $p(2 \times 1)$ has $f_1 = f_2 = 0.5$ and $f_{12} = 0.25$; third and fourth structure $p(2 \times 2)$, denoted as $p(2 \times 2)_s$ and $p(2 \times 2)_w$, respectively, with a small fraction of the strong or weak points, where $p(2 \times 2)_s$ corresponds to $f_1 = 0.25$, $f_2 = 0.75$, $f_{12} = 0.25$ and $p(2 \times 2)_w - f_1 = 0.75$, $f_2 = 0.25$, $f_{12} = 0.25$. Note the difference in terms of ordered structures: here we discuss the case of nonuniform adsorption sites located in a regular way. Their ordering is the result of the formation of the surface, which does not depend on the degree of filling with the adsorbate. Above in section 8 we discussed the case of the ordered arrangement of the adsorbate, which was formed as a result of the cooperative behaviour of molecules adsorbed on the homogeneous surface. As a rule, the ordered distribution of the molecules exists in a range of densities and is not found at low and high degrees of filling.

Calculations in Fig. 19.2 were carried out by QCA at $\varepsilon_{qp} = \varepsilon$. The difference between the binding energies of the molecules on strong and weak centres is designated by $\Delta Q = Q_1 - Q_2$. If the ratio $\Delta Q/(z\varepsilon)$ is small, such a surface can be considered homogeneous, while if the ratio is greater than some adopted criterion, such a surface is highly nonuniform. Such criteria are generally functions of the structure.

The critical temperatures of condensation on ordered surfaces as a function of the ratio $\Delta Q/(z\varepsilon)$ are shown in Fig. 19.2. It can be seen that the increase of the degree of heterogeneity of the surface decreases the critical temperature for the structures $c(2 \times 2)$ and

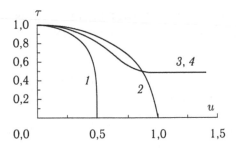

Fig. 19.2. Critical temperature $\tau = T_c/T_c$ (hom), where T_c (hom) is the critical temperature for a homogeneous surface, depending on the relationship $u = \Delta Q/z\varepsilon$ for structures: $p(2 \times 1)$ (*1*), $c(2 \times 2)$ (*2*) $p(2 \times 2)_W$ (*3*) and $p(2 \times 2)_S$ (*4*) [28].

$p(2 \times 1)$. Temperature becomes zero at some finite value of the ratio $\Delta Q^*/(z\varepsilon)$, while for the structures $p(2 \times 2)_S$ and $p(2 \times 2)_W$ reduction of the critical temperature with increasing degree of heterogeneity stops and it goes to a constant value common to both structures. The shape of the curves for critical densities is different [28]: θ_c is equal to 1/2 as for a homogeneous surface for the structures $c(2 \times 2)$ and $p(2 \times 1)$ and greatly differs from it for the structures $p(2 \times 2)_S$, and $p(2 \times 2)_W$. In both cases the strong centres are filled first.

From an analysis of the critical conditions on the ordered surfaces it follows, in contrast to the spotted surfaces having the maximum number of phase transitions and critical points, that the ordered surface 'takes part' in cooperative condensation processes as a whole and has a single critical point. Its parameters depend on the type of repeated fragment (from the local structure) and on the difference in binding energies on different sites of the fragment. Obviously, the size of the fragment must be smaller than the characteristic size of the region of formation of the phase.

If there are several macroregions (e.g., crystal faces) with different types of ordering, then, as for the set of sites macroheterogeneous surface areas, there is an additive pattern of phase transitions at different pressures of the adsorptive. Each critical point will have the same critical temperature and local filling as for a single ordered region and its gross density will depend on the weight of different crystal facets [28].

Random surfaces
The random arrangement of the centres removes the correlation in the distributions of different types of sites. If we also neglect the

correlations between the interacting molecules and assume that $\varepsilon_{qp} = \varepsilon$, then the system of equations (19.3) in the MFA can be written as

$$\beta z \varepsilon \sum_{q=1}^{t} \frac{f_q a_q^* P}{(1 + f_q a_q^* P)^2} = 1, \quad \sum_{q=1}^{t} \frac{f_q a_q^* P(1 - a_q^* P)}{(1 + f_q a_q^* P)^3} = 0, \quad (19.9)$$

where the notation $a_q^* = a_q \exp(\beta z \varepsilon \theta)$ is used, and in the index there is the total degree of filling the surface.

Under the above assumptions, the adsorption isotherm equation has the form $\theta = \sum_{q=1}^{t} \frac{f_q a_q^* P}{1 + f_q a_q^* P}$, that allows to define the required critical parameters from the system of three equations with three unknowns β_c, θ_c and P_c.

In the equations (19.9) the number of types of centres t can be anything. If we transfer to the limit of infinitely many types of centres, assuming their share to be the same (equidistant-nonuniform surface with $f(x) = (x_2 - x_1)^{-1}$, $x_2 > x_1$), we can replace the summation by integration (then for the Henry constants we have $a(x) = a_0 \exp(\beta x)$, $a_0 = $ const, x is the binding energy of the adsorbed atom). As a result, we obtain

$$\theta_c = 1/2, \quad \beta_c \varepsilon z = (x_2 - x_1) \left[1 + \frac{2}{\exp[(x_2 - x_1)/2] - 1} \right], \quad (19.10)$$

which coincides with Hill's results [29] (however, the result of [29] for the QCA is not true).

Thus, on a uniformly nonuniform surface a single phase transition is realized with a constant value of the critical filling for all values of x_2 and x_1. The effect of heterogeneity influences only the magnitude of the critical temperature. With increasing difference $(x_2 - x_1)$ the critical temperature at a constant interaction decreases. However, the correct use of MFA for the description of critical phenomena is questionable. It is well known that the critical parameters in such approaches differ from the experimental ones, which was the reason for the development of fluctuation approaches [1–6, 16, 30].

The numerical values of the critical temperature reduction in nonuniform systems depend on the approximation of the interaction used in this case. The general procedure for the calculation of phase diagrams on nonuniform surfaces is described in [31].

The greatest differences are observed precisely for the surfaces with a random distribution of sites of different types. The problem

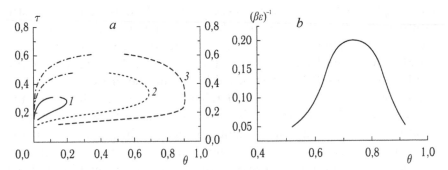

Fig. 19.3. Phase diagrams: a – for the random surface at $\Delta Q/z\varepsilon = 1.0$, the proportion of strong adsorption centres $f_1 = 0.85$ (*1*), 0.90 (*2*), 0.95 (*3*) [28]; b – for argon, calculated in QCA on the partially random surface of rutile, Q – the Ar–rutile binding energy (accuracy of the calculation of this function $\Delta Q = 0.25$ kJ/mol).

is that when considering different configurations of the adsorption centres of fragments it is necessary to consider large fragments. Figure 19.3 a shows the phase diagrams, obtained in the QCA, for two types of centres by varying the proportion of strong centres f_1 [28]. In the range $f_1 < f_1^*$ where $f_1^* = 0.83$ – the threshold proportion of strong centres, and strong surface heterogeneity ΔQ, phase transitions are not observed. However, if $f_1 > f_1^*$ the area of coexistence of two phases begins to grow sharply (there is a growth in the left corner of the phase diagram) and the phase diagram acquires a quite unexpected appearance. There are two branches of the phase stratifying curve: the upper (with a maximum) and lower (monotonous), which come together at a certain density. The upper branch has its usual meaning of phase stratifying of vapour and liquid – increase in temperature increases the entropy in the free energy of the system. The lower branch nevertheless remains monotonous and the second (lower) critical point does not appear on it. When $f_1 \to 1$, the lower branch falls on the x-axis and the coexistence curve changes to a conventional diagram of for a homogeneous surface. Evidently, the nature of the lower branch is related to the prevalence at low temperatures of the impact of the surface heterogeneity in comparison with the contribution of the lateral interactions.

A more complex example is the question of the presence of a critical point of argon atoms on the surface of rutile (Fig. 19.3 b). The constructed surface of rutile, studied in [32], contains a large number of different types of centres, whose distribution function has the form shown in Fig. 12.7 (fragment 20 × 20). This distribution function has been derived from experimental data on the heats of adsorption of argon atoms (see section 12). Its characteristic feature is the fact that despite the random generation procedure of the given surface, the structure of the surface turns out to be partially correlated (i.e. not strictly random). As a result of the calculation in the QCA it was possible to find the critical point (Fig. 19.3 b) in the range of fillings θ from 0.52 to 0.92. However, the resultant critical temperature is more than three times lower than the critical temperature on a homogeneous surface. Given that the crystallization temperature in a homogeneous system is about 0.55 T_c [33], then assuming the preservation of the relationship, at least as a first approximation, also for nonuniform systems, the observed large decrease of the critical point should be considered a sign of going beyond the range of permissible values. Therefore, this result should be viewed as indicating the absence of the phase transition and the critical point even on the partially random surface (energy heterogeneity of the surface prevails over the lateral contributions).

20. Critical conditions for porous systems

Porous systems are an important example of nonuniform systems. In [34, 35], the molecular adsorption theory was extended to the porous system in order to build self-consistent equations for the calculation of equilibrium and dynamic features. Any pore is limited by walls which can be either homogeneous or nonuniform. The surface potential of the walls is directed along the normal to the surface of the wall, which creates a heterogeneity along the normal. The potential decreases rapidly, and a large part of the pore volume is outside the area of the direct effect of the surface potential. Nevertheless, the potential of the walls influences the condensation conditions as the condensation is a cooperative process implemented at distances greater than the size of the molecules. For the first time this has been pointed out in [36] for the slit-like pores, when lowering the critical temperature $T_c(H)$ with an increase in the width of the pore H was approximated as the scaling relationship: $\Delta\tau\,(H) = (T_c - T_c\,(H))/T_c = \mathrm{const}/(1 + H)^2$, where T_c is the critical temperature in the bulk phase. The calculations [36] were carried out in the MFA for the strong and weak bonding of the molecules with the walls of the pores (see Fig. 20.1).

A similar dependence of the critical temperature of the fluid, depending on the diameter of the cylindrical pores was constructed in [37]: $\Delta\tau(D) = (T_c - T_c\,(D))/T_c = \mathrm{const}/D$, where $T_c(D)$ is the critical

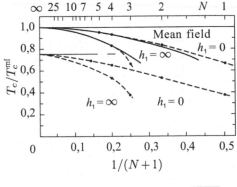

Fig. 20.1. Critical temperature as a function of the number of layers N of slit-like pores for weak ($h_1 = 0$) and strong ($h_1 = \infty$) bonding of the molecules with the walls. The upper family of curves – MFA, bottom – scaling [36].

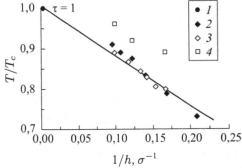

Fig. 20.2. Calculation of the critical points of methane in carbon pores with width $h = 11\lambda$ (2), 9λ (3), 7λ (4) [38]: 1 – critical point, 2 – Monte Carlo (strong adsorption), 3 – lattice-gas model, 4 – Monte Carlo (weak adsorption).

temperature of the cylinder with diameter D, on the basis of density functional calculations.

Figure 20.2 shows the dependence of the critical points of methane in a graphite pore of different widths $h = 11\lambda$ (2), 9λ (3), 7λ (4) calculated by the LGM and Monte Carlo simulations [38]. The value of h is the width of the pores measured in diameters of the hard sphere of the molecules σ. The results are qualitatively consistent with previous assumptions of the linear or near-linear decrease in the critical temperature with decreasing pore size. Both methods give close results and satisfy the above scaling dependences. Figure 20.2 shows that for the dependence $Tc/T_c^{\infty}(1/h)$ in the case of a weak adsorption field the number of the received points is not sufficient to draw definite conclusions on the dependence of T_c on the width of the pore.

In the case of a strong adsorption field in the range $h = 6{-}10.5\sigma$, according to both methods, this dependence is close to linear (correlation coefficient of 0.996 and 0.973 for the results of the Monte Carlo and the LGM methods, respectively). The data obtained suggest the possibility that in sufficiently wide pores $h \approx 8{-}10$ nm,

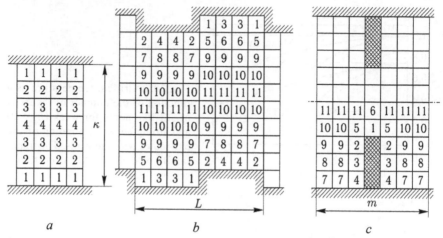

Fig. 20.3. Schemes of pores with flat homogeneous (*a*), stepped (*b*) and columnar (*c*) walls [39].

when the state of methane in the central part of the pore is close to its state in the equilibrium bulk phase, its critical temperature is still substantially (by several degrees) lower than the bulk temperature, at least in the pores with a strong adsorption field. This pattern was also observed experimentally for SF_6 in porous glasses [21].

Much more complex is the behaviour of the critical parameters in the case of nonuniform pore walls [39–41]. This issue is discussed in detail in [39]. Examples of the chemical heterogeneity of planar walls, consisting of two kinds of atoms, and examples of structural heterogeneity (walls containing monoatomic steps and 'columns') are examined.

A slit-like pore with width κ (in units of monolayers) had mirror symmetrical walls with a given composition and structure of the surface. Then, the gas molecule being in layer k, $1 \leq k \leq \kappa$, interacts with two walls of the pores. We took into account the type of centre on the surface of the walls and the distance from the surface of the given monolayer (details in [39]), which reflects the potential of type (10–4) [42]. In addition to the attraction from the side of the walls, the molecule is also subjected to the interaction with other molecules inside the pores. The lateral interaction between the molecules was calculated in the QCA.

Figure 20.3 *a* shows the cross section of a slit-like pore with flat uniform walls. Each cell is a site, and the number inside the cell – the type of the site. By varying types of adsorption centres in the surface layer of the walls we can change the number of types of

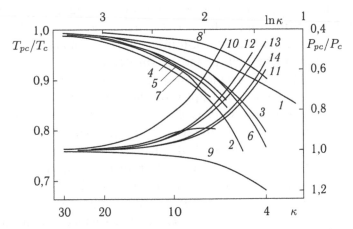

Fig. 20.4. Critical temperature (*1–8*) and pressure (*9–15*) in the slit pores with different types of walls [39]. See the text for explanation.

centres in the first surface layer of the fluid – their number depends on the number of types of adsorption centres. In the calculation the composition and structure of the surface of the wall for $t = 2$ were varied. By assigning appropriate types to the sites it is easy to identify structural inhomogeneities on the surface.

Figure 20.3 *b* shows the cross-section of a pore containing a monatomic step with width *L*, directed perpendicular to the plane of the figure. In this example, the first type has been assigned to the sites located at the bottom of the step, the second type – to the sites at the edge of the terrace, the third type – to the sites arranged between the sites of the first type, the fourth type – the sites positioned between the second type sites, etc. The scheme for identifying the sites for walls containing a single 'column' on the site $m \times m$, is shown in Fig. 20.3 *c*: the first and second types relate to the of sites at the top of 'column', a third type – the sites in the central part of the 'column', the fourth – the sites at the bottom of the 'column', etc.

Figure 20.4 illustrates the shift of the critical temperature (T_{pc}/T_c) (curves *1–8*) and pressure (P_{pc}/P_c) (curves *9–15*) depending on the width of the pore. Curves *1–3* and *9–11* relate to uniform flat walls. They demonstrate the effect of the surface potential on the critical values of temperature and pressure for the molecular parameters: $Q_1 = 0$ (*1, 9*) – no attraction to the wall and 16.8 (*2, 10*) – strong attraction to the wall; $n = 3$ (*1, 2, 9, 10*) for $Q_k = Q_1/k^n$, *k* is the number of the monolayer. For curves *3, 11* it is assumed $Q_1 = 16.8$

and for others $Q_{k>1} = 0$. Here and below, all the energies Q_q are given in kJ/mol; missing values of Q_q are equal to zero. The larger the area of the wall–molecule potential (curves 2 and 3), the greater the deviation from the bulk phase. A similar effect is exerted by the energy of attraction of the molecules to the wall (curves 2 and 1).

The influence of the topography of chemically nonuniform planar surfaces is illustrated by curves 4, 5, 12, 13 at $f_1 = 0.5$: for spotty (4, 12) and regular (5, 13) locations of the adsorption centres of two types. Regardless of the flat surface and the equality of the adsorption centres of both types, different dispositions of the two types of centre give different dependences of T_{pc}/T_c on the width of the pores. The difference in the critical parameters for regular (lines 5, 13) and spotted (lines 4, 12) of the surfaces of the walls is maintained to $\kappa = 10$ or more.

Curves 6, 14 relate to the stepped walls with molecular parameters equal to $Q_1 = 21$, $Q_2 = 4.2$, $Q_3 = Q_4 = (Q_1 + Q_2)/2$, $L = 10$. Curve 6 is close to curve 3, although the structural heterogeneity of the stepped surface has a much greater effect on the energy of the surface region than the sudden change of the surface potential for a flat homogeneous surface.

Curves 7, 15 relate to the columnar walls with molecular parameters equal to $Q_1 = Q_2 = 21$, $Q_3 = Q_4 = 8.4$, which corresponds to stronger attraction of the columns than the attraction of the flat portions (here $m = 7$).

Curve 8 is for comparison. It corresponds to the calculation in [36] for flat, homogeneous and non-attracting walls in the mean field approximation; the bulk critical parameters for this curve are given above.

All curves T_{pc}/T_c asymptotically approach unity with increasing pore width κ, but the rate of approach is affected by the nature of the walls and the fluid–wall interaction potential. Differences from the bulk properties for the same κ increases with increase of the degree of attraction of the walls (curves 1, 2 and 9, 10) and increasing heterogeneity.

It should be noted that in all variants of the molecular parameters the critical temperature in the pore always decreases with decreasing pore size. For the critical pressure the ratio P_{pc}/P_c is also reduced in the case of attraction between the molecules and the wall, but it increases in the absence of the wall–molecule attraction (curve 9).

The analysis of curves in Fig. 20.4 shows that the non-uniform properties of the pore walls can significantly affect the critical properties of fluids. In most cases these properties are more sensitive than the adsorption isotherm behaviour [40]. The direct effect of inhomogeneities on the walls on the adsorption isotherms is noticeable in the first two or three surface monolayers, while the influence of heterogeneity of the walls on the critical characteristics can be continue to κ more than 10. This was demonstrated by the regular and spotty surfaces (lines *4*, *5*), and also by an example of stepped walls (curve *6*). Therefore, in the analysis of critical features, as well as the analysis of the sorption isotherms, we must consider the nonuniform properties of the walls of micro- and mesopores.

Equations (19.3) imply the possibility of the multiplicity of critical points for nonuniform systems instead of a single point for the bulk phase. In general, the situation for porous systems is more complicated than clarified by the critical analysis of the terms of this section so that complete phase stratifying curves must be analyzed.

21. The curves of the vapour–liquid phase stratifying in slit-like pores

The phase stratifying curves give much more information about the phase behaviour of the adsorbate than the critical parameters. Hence, they are the main subject of research in the molecular theory, like the phase diagrams in the bulk phase. The phase stratifying curves for different gases in dependence on the size of the slit pores with homogenous walls were studied in [43, 44].

Figure 21.1 shows the density jumps, plotted on the basis of the Maxwell's rule, on the adsorption isotherm of argon atoms in a carbon pore (*a*) at $T = 100$ K (where ln (aP) is the dimensionless pressure related to $a = a_{k=t}$ – the Henry constant for the central layer of the pore) and the phase diagram (*b*) for the slit pore $H/\lambda = 9$. The calculations were performed at $z = 6$, which corresponds to the value $\beta\varepsilon = 1.19$ for argon atoms; in a graphite pore we have $Q_1 = \varepsilon_a = 9.24\varepsilon$ and $Q_2 = Q_1/8$. Similarly, the phase stratifying curves for a pore with width $H = 11$ (*c*) and 15 (*d*) are shown. Here are the full phase diagrams containing small (sub-surface) domes associated with local filling and phase stratifying processes within individual monolayers and a large ('central') dome relating to the filling of the central part of the pore, in which the influence of the surface potential is practically negligible. The main attention is usually paid to the properties of the central dome.

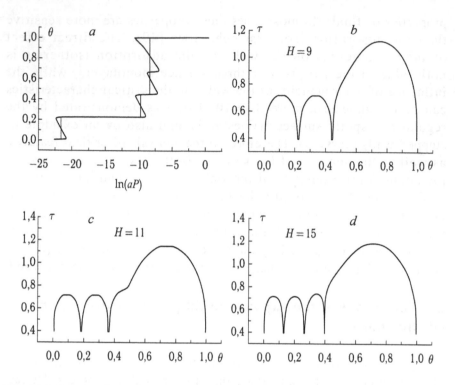

Fig. 21.1. Density 'jumps' constructed by Maxwell's rule on the adsorption isotherm of argon atoms in a carbon pore (*a*) and phase diagrams for different slit pore width $H/\lambda = 9$ (*b*), 11 (*c*) 15 (*d*), with $Q_1 = 9.24\varepsilon$, $R = 1$, $z = 6$, $T = 100$ K.

Figure 21.2 shows the phase stratifying curves in a slit pore with the width of 10 monolayers for different interaction potentials of argon atoms with the walls of the pore. The curves are presented in the reduced coordinates $\tau - \theta$, where $\tau = T/T_c$, T_c – the critical temperature in the bulk phase. Varying the adsorbate–wall potential in the first place changes the value of the binding energy of the molecule with the pore wall Q_1. The strong binding of the adsorbate with the adsorbent corresponds to the carbon wall, for which $Q_1 = 9.24\varepsilon_{AA}$. As an example of weakly attracting walls attention was given to the argon–Teflon polymer matrix system (curve *3*), for which $\varepsilon_a = 0$. Curve *2* corresponds to the wall of the adsorbate atoms. The lateral interaction parameter was taken to be equal to the depth of the potential well U_{LJ}: $\varepsilon = 0.987$ kJ/mol. Curve *4* refers to the effective repulsion of the argon atoms, located in the surface monolayer, from the polymer material. This case is conditional, taking into account the presence of groups around the polymer chain which prevent the

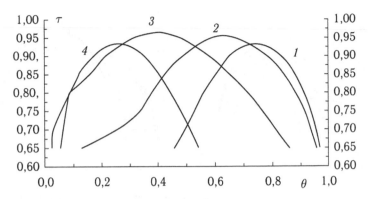

Fig. 21.2. The phase diagram in the coordinates (τ, θ) for a slit pore at $H = 10$ at values $Q_1/\varepsilon = 9.24$ (*1*), 1.00 (*2*), 0.00 (*3*), –8.00 (*4*) [43].

adsorption and accumulation of the adsorbate near the chain, so the adsorbate is condensed in the central part of the pore.

Weak and strong adsorption

The change of the nature of the adsorbate allows us to consider the effect of the depth of the potential well on the shape of the phase diagrams. The main parameter for comparing different systems is the dimensionless parameter $\gamma = \varepsilon_{sA}/\varepsilon_{AA}$ (where ε_{ij} is the parameter of potential functions), or the ratio Q_1/ε_{AA}, which gives an indication of how interactions between the adsorbate and the wall and the interaction between the adsorbed molecules differ from as regards energy. The higher the value of this parameter, the stronger the effect of the wall potential on the condensation processes of the adsorbate in the pores. Lowering the value of γ evidences the decisive role of intermolecular interactions, where the effect of the pore walls occurs only in the fact that the volume, available for the molecules in condensation, decreases.

The range of values of the parameter $\gamma = \varepsilon_{As}/\varepsilon_{AA}$ can be divided into areas for the strong and weak physical adsorption: when $\gamma < 0.2$ ('weak' adsorption and phase diagrams have the form shown in Fig. 21.3 *b*) and $\gamma > 0.4$ ('strong' adsorption and phase diagrams have the form shown in Fig. 21.3 *a*). In the intermediate range $0.2 < \gamma < 0.4$ it is not possible to predict reliably in advance the form of the phase diagram on the basis of the parameters of the potential curves and specific numerical analysis of the total energy of the adsorbate interactions with the walls and with each other must be carried out. The upper limit of the weak adsorption range overlaps

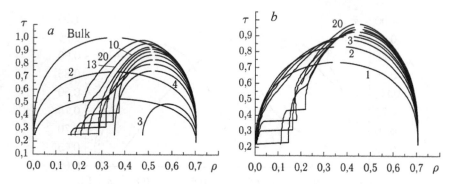

Fig. 21.3. Phase diagrams: a – methane in graphite slit pores of different width, b – a model system with weak attraction for the slit pores $Q_1 = 0.5$ (the numbers on the phase diagrams indicate the width of the pores) [44].

with the region describing the case of 'strong' adsorption $\gamma > 0.29$. This fact is explained as follows. For example, if we consider that the atomic radius of neon is about half the radius of a molecule of CCl_4, then a specific calculation of the total energy of the lateral interaction of the molecules and their interaction energy with the walls of the pores explains the differences in overlapping of the ranges of the values of parameter γ, relating to 'weak' and 'strong' adsorption.

Influence of pore width
Calculations of the phase diagrams for the pore width H varying from 1 to 20 monolayers are shown in Fig. 21.3 (the near-surface of the domes are omitted) in the reduced coordinates $\tau - \rho$, where $\rho = \theta(\sigma/\lambda)^3$ is the numerical density of the fluid when the volume of the system is measured in units of the volume of the hard-sphere particles (units of ρ are commonly used in numerical methods). For a hard (incompressible) lattice $\rho = \theta/1.41$, since $\lambda = \sigma\,(2)^{1/6}$. The use of the reduced coordinates is associated with carrying out the calculations in the quasichemical approximation (QCA). It is known that the use of QCA in the critical area gives only qualitative results. Thus, taking into account the nearest neighbours ($R = 1$) we have $T_c^\infty \approx 1.2\varepsilon_{AA}/k_B$, whereas when $R = 4 - T_c^\infty \approx 2.26\varepsilon_{AA}/k_B$.

All adsorption systems with strong bond have qualitatively similar curves of the phase diagrams. As an example, Fig. 21.3 a shows the calculated phase diagrams of methane in slit-shaped pores of the graphite. Under the influence of the adsorption capacity of the walls the phase stratifying curve of the adsorbate is shifted down and to the right relative to curve the vapour–liquid stratifying in the bulk. A strong field of of the adsorbent results in a significant increase in the density of the gas branch. As a result, the density of both coexisting phases, in particular the vapour phase, in the pores greater than the density of the coexisting phases in the bulk.

For smaller pore sizes when $H = 1$ and 2 the phase stratifying curves are symmetrical as the curve for the bulk. This is due to the fact that all the lattice sites are equivalent to each other and at different H differ only in the number of nearest neighbours z ($z = 4$ for the layer $H = 1$, $z = 5$ for the pore with $H = 2$ and $z = 6$ for the bulk phase for which it can be assumed that $H = \infty$). For the H range from 3 to 20 increase of the size of the pores monotonically increases the critical temperature. At the same time, critical density depends on the pore size in a more complicated manner. When we increase the width of the pore H from 3 to 8 there is initially a minimum (reduction of the critical density), then a maximum (an increase of the critical density), after which the critical density monotonically tends to its value characteristic of the bulk phase. This 'zig-zag' is observed for all four systems [44]: such diagrams are obtained for all other inert gases (Ar, Kr, Xe), and also molecules such as nitrogen, hydrogen, oxygen, carbon monoxide, carbon dioxide and others. This means that the non-monotonic variation of the critical density with the change of the pore width from micro- to mesopores is universal. At the same time, decreasing the parameter γ results in a slight broadening of the two-phase region, which is slightly noticeable for $H = 3$ and best seen at large H. The phase diagrams for $\tau < 0.5$ show steps, which are caused by gradual filling of the pore walls.

Figure 21.3 b is a phase diagram for a system interacting with methane molecules in which the potential of the interactions with the wall is reduced about 3.6 times. The curves in Fig. 21.3 b differ significantly from the curves in Fig. 21.3 a for pores with a width of $H = 3$ and $H = 8$. In this case, there is a monotonic increase of the critical temperature from $H = 1$ and the critical density has a single maximum at $H = 8 \div 10$. Moreover, all the curves are shifted less markedly to high densities than the curves in Fig. 21.3 a and have considerably larger two-phase regions. The type of phase diagrams for other systems that have weak adsorption is similar. In particular, this type of diagram includes many real systems, starting with systems containing inert gas and methane in weakly adsorbing solid porous bodies (including polymeric matrices) to substances such as Hg ($\varepsilon_{AA}/k_B = 851$ K) and SnCl$_4$ ($\varepsilon_{AA}/k_B = 1550$ K) [14] interacting non-specifically with the majority of adsorbents. The calculated phase diagrams explain the phase state of the mercury, which is actively used in the porosimetry mesoporous systems in narrow pores of almost all adsorbents.

The qualitative difference of adsorption systems for values of the parameter γ can be clearly seen in Fig. 21.3 to micropores (H value of from 1 to 3). For $H = 1$ and $H = 2$ curves are symmetric and are similar to the phase stratifying curve for the bulk phase, but the phase stratifying effect operates at a lower critical temperature due to the smaller number of neighbouring molecules. For pores with $H = 3$ the form of the phase stratifying curves is qualitatively different in the case of weak (b) and strong (a) adsorption. For the weak adsorption the critical temperature increases monotonically from $H = 2$ to $H = 3$, and the critical density is slightly shifted to higher values. In the case of strong adsorption there is a sharp change of both the critical temperature and the critical density in the transition from $H = 2$ to $H = 3$. The critical temperature for $H = 3$ is close to the critical temperature for pores with $H = 1$. Thus, with filling of the pores there is initially stratified filling of the first two layers and the total density is shifted to higher values of ρ, then the condensation of the adsorbate takes place in the last layer. In other words, the pre-adsorbed molecules reduce the effective width of the pore at $H = 3$. A similar situation occurs in the case of weak and strong adsorption in filling the pore with $H = 4$.

However, this does not mean that the condensation takes place in the pores regardless of the state of the previously adsorbed molecules and the surface layer is not involved in the condensation process and does not respond to the phase state of the rest of the pore. Figure 21.4 shows how the degree of filling of the surface and intermediate layers and the total degree of filling of the pore with $H = 3$ change for weak cases (b) and strong (a) adsorption. For the weak adsorption the behaviour of all three curves is similar to one another: as the temperature increases the loop disappears. In the case of strong adsorption the three curves are very different from each other. At the lowest temperature (curve 1) the phase transition is accompanied not only by the density change in the second (intermediate) layer, but also an additional 'loop' forms for the surface layer (and hence a density jump) at the same pressure at which condensation occurs in the intermediate layer. This means that all the molecules of the adsorbate of this system collectively participate in the phase transition of phase stratifying. Otherwise, if the surface layer is not

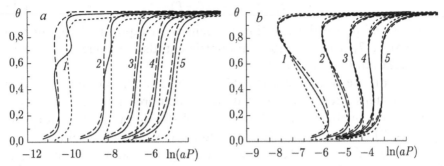

Fig. 21.4. Isotherms (solid line), filling of the surface (dashed curves) and intermediate (dotted lines) of layers in a slit-like pore $H = 3$, $\tau = 0.4$ (*1*) 0.5 (*2*) 0.6 (*3*) 0.7 (*4*), and 0.8 (*5*).

involved in the phase transition of the intermediate layer molecules, its density, after reaching a certain value, would remain constant.

Comparison with calculations by the Monte Carlo method
Studies [38, 45] compared the phase diagrams for $H < 10$ monolayers, and the dependence of the critical temperature on the width of slit-like pores for methane using the Monte Carlo (MC) method in the Gibbs ensemble [46, 47], and for argon in the lattice gas model [11, 39–41]. In the methods of molecular modelling these adsorbates can usually be considered as spherical Lennard–Jones particles [14]. Determination of the conditions of condensation of methane molecules is of great practical importance for physical and chemical analysis of the processes in coal beds. The method of calculation of the potential functions is specified in section 4.

The MC method in the Gibbs ensemble is the direct simulation of two coexisting phases between which there is thermal, mechanical and chemical equilibrium [46, 47]. To estimate the critical temperature with the help of the MC method, the methodology applied repeatedly for macroscopic systems was used. The MC method in the Gibbs ensemble was used to calculate the equilibrium density of the liquid and vapour phases at subcritical temperature $T = 0.75$ ε/k. The temperature is then gradually increased until the disappearance of the two-phase equilibrium (conclusion on the existence of equilibrium was made on the basis of the statistical distribution of the states with respect to density in both cells of the Gibbs ensemble). Next, the critical temperature T_c and density ρ_c are evaluated by the method of least squares as parameters in correlation

$$(\rho_l + \rho_v)/2 = \rho_c + A(T_c - T), \tag{21.1}$$

where ρ_l and ρ_v are the equilibrium density of liquid and vapour at a temperature T, A is an empirical coefficient also evaluated by means of correlation (21.1) [20].

It should be noted that the MC method in the Gibbs ensemble includes steps such as an instantaneous compression and expansion of the fluid in the plane xy, which is difficult in small adsorption systems. Therefore, for each of the studied systems the values ρ_l and ρ_v were tested by the MC method in the grand canonical ensemble. Furthermore, it is shown that the introduction of periodic boundary conditions in the directions x and y can significantly distort the estimate of T_c [48]; in this connection, when approaching the critical point the total volume of the system and the number of the molecules in it were regulated in such a manner that the number of molecules in both cells was about equal, and the size of each cell in the x and y directions exceeded 9σ. In the calculations it was assumed that $R = 5.0\sigma$, and the interactions at large distances were taken into account by analytical summation (its contribution is less than 0.3 %).

Phase diagrams of two-phase liquid–vapour equilibrium for pores of different widths, calculated by the LGM (lattice-gas model), are shown in Fig. 21.5. For comparison, the curves obtained by the MC method with $h = 10\sigma$ for the adsorption fields of different force as well as for the macroscopic Lennard–Jones system are also shown. The potential of the walls shifts the phase stratifying curve downwards compared to the bulk phase, so the density of both coexisting phases, in particular the vapour phase, in the pores is higher compared to the bulk values. A strong field of the adsorbent results in a significant increase in the density of the gas branch. Curves *3* and *4*, obtained by various methods, differ quantitatively, but qualitatively they show the same behavior in the near critical region. For low temperatures the phase diagram shows a step associated with layer-by-layer filling of the surface layer of the adsorbent (for the curves *1* and *2* the steps are at lower temperatures).

The concentration profiles, obtained by the MC method in both coexisting phases inside the pores, are strongly non-uniform [45], which is in line with layer filling the pores in the LGM (Fig. 21.6). Numerical studies actually confirm the validity of the lattice model. The critical properties of methane in the pores, produced in this work, are presented in Fig. 20.2. Obviously, in all the above conditions, the critical temperature of the adsorbate in the pores is

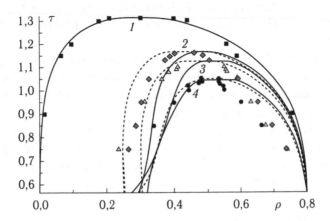

Fig. 21.5. Density–temperature phase diagrams for the Lennard–Jones fluid in slit-shaped pores, calculated using the Monte Carlo method and the LGM with calibration. Curve *1* – bulk phase, *H* = 7 (*4*), 9 (*3*) and 11 (*2*).

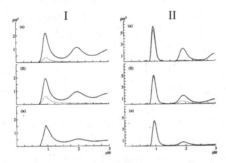

Fig. 21.6. Density profiles of methane in liquid (solid line) and vapour (dotted line) phases in a pore with the width *h* = 6.0 σ: with weak (I) and strong (II) adsorption field. Curves I: a – *T* = 0.85 ε/*k*; b – *T* = 1.0 ε/*k*; c – *T* = 1.17 ε/*k*; curves II: a – *T* = 0.75 ε/*k*; b – *T* = 0.95 ε/*k*; c – *T* = 1.05 ε/*k*.

smaller than the bulk temperature and steadily approaches the latter when increasing the width of the pores, and the critical density in the adsorption pores is higher than the bulk density and also approaches it with increasing width. A comparison of the results obtained by the MC and the LGM under similar conditions shows good agreement for the reduced critical temperature T_c/T_c^{∞} in the range *h* = 6–10.5σ. In [49] the method of the free energy differences for the Lennard–Jones fluid in a graphite pore *h* = 5.8 σ gave T_c = 1.0 ε/*k*, which is also in good agreement with the data of [45].

The differences at *h* ≈ 5σ due to insufficient accuracy of the calculations in the Gibbs ensemble using the Monte Carlo method in the grand canonical ensemble due to the high density of the liquid phase and the presence in both phases of very dense adsorbed layers. This greatly complicates obtaining equilibrium in the Gibbs ensemble.

In the case of a weak adsorption field the minimum of the adsorption potential approaches that which would be generated by a wall formed from methane at a density close to the critical bulk density. The critical temperature of methane in pores with a weak adsorption field is significantly higher than in the pores of a strong field, but substantially less than the bulk temperature, indicating the significant influence of the inhomogeneity structure of the adsorbate, produced by the walls, on the critical properties.

Saturated vapour pressure

To assess the physical content of (6.6), the saturation pressure of the pores $P(H)$ in different adsorption systems was studied. Figure 21.7 shows the curves of the values $\eta(\tau) = \ln{(P(H))}/(\beta\varepsilon_{AA})$ as a function of the reduced temperature τ for different values of the width of the pore H [44]. The limiting value for all families of the temperature dependence $\eta(\tau)$ for different gases with increasing H is the value of $\eta_s(1) = \ln{(P_s)}/(\beta\varepsilon_{AA}) = -4.79$. Three families $\eta(\tau)$ refer to adsorption of helium atoms and molecules of methane and carbon tetrachloride in graphite pores.

For a pore width $H = 20$ the values of $\eta(\tau)$ tend to limits $\eta_s(1)$. For $H = 6$–7, corresponding to the size of about 2 nm, $p(H)$ differs from p_s by up to three times or more. For small H, these differences

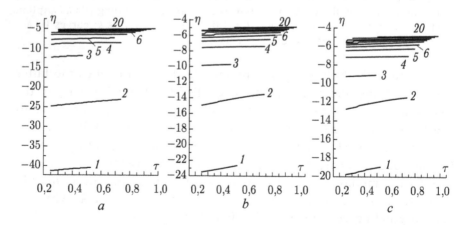

Fig. 21.7. Temperature dependence of the condensation pressure of the adsorbate: helium (*a*), methane (*b*), CCl$_4$ (*c*) in the slit-shaped pores of width $H = 1$–10, 13, 20 monolayers [44].

increase many time. They are highest for $H = 1$, i.e., we can say that the volume filling of pores of the monomolecular width occurs at very low pressures.

All the curves show rapid filling of the pores with decreasing temperature (relative to $\eta_s(1)$). The curves are 'torn' on the left (low temperatures) – for their continuing we need to know the position of the curve of the liquid–solid phase transition and for this it is therefore necessary to calculate the melting point of the substance in the respective pores (it should be noted that the latter problem has no solution at present and the exact position of the triple points A^H is not known for any of the systems). The end point of the curves $\eta(\tau)$ on the right are determined by the critical temperatures that meet the relevant phase diagrams. The graph shows how much influence the potential of the walls exerts on the condensation pressure of the adsorbate at different pore width H.

The calculated dependence $\eta(\tau)$ can be considered linear with good accuracy. A slight deviation from linearity of $\eta(\tau)$ is only observed at $H = 2$ for methane and carbon tetrachloride. Figure 21.7 shows that the size of the region of the maximum decrease of the critical pressure at $H = 1$ is determined by the features of the system and the temperature range. For systems with cylindrical pores the calculated pressure change ranges are about 50% smaller. The resulting dependences $\eta(\tau)$ reflect all the above-mentioned features of the effects of interactions of the adsorbate molecules with the walls. These data allow us to assess the conditions of the volume filling of porous systems, especially to test the applicability of the empirical Dubinin–Radushkevich equation and its generalizations.

Estimates of condensation pressure

The molecular theory allows to obtain [50] a qualitative assessment of pressures $p_1(H)$ and $p_2(H)$, corresponding to two situations: capillary condensation inside a pore width H and the condition of filling the pores without capillary condensation. In the first situation, the formula

$$p_1(H) = p_s \exp(-b_1), \ b_1 = \beta\left[z_{12}\varepsilon_{AA} + 2Q_1\varphi(H)\right]/(H-2). \quad (21.2)$$

The value b_1 decreases with increasing width of the slit H and the pressure at which the filling of the pore takes place tends exponentially to its bulk value p_s.

In the second situation, from the condition required for the last stage of gradual filling of the pore when the preceding layers are filled, it is possible to obtain two estimates (upper and lower) for the pressure corresponding to the filling of the last layer. For the upper estimate, we have

$$p_2^+(H)/p_s = \exp\{-b_2^+(H)\}, \quad b_2^+(H) = \beta[z_{12}\varepsilon_{AA}C(H)+Q_1D(H)],$$
(21.3)

where $C(H)$ equals 1 for odd H and 1/2 for even H, and $D(H)$ is equal to $[1/(H/2)^3 + 1/(1 + H/2)^3]$ for even H and $2/((H + 1)/2)^3$ for odd H.

The minimum estimate $p_2^-(H)$ is obtained in the form

$$p_2^-(H)p_s = \exp\{-b_2^-(H)\}, \quad b_2^-(H) = \beta[z\varepsilon_{AA}/2+Q_1D(H)]. \quad (21.4)$$

The resulting estimates (21.2)–(21.4) do not depend on the type of approximation used when considering the intermolecular interaction. In all cases, the filling of micropores occurs at pressures less than the saturated vapour pressure at a given temperature. Estimates of the coefficients $b_{1,2}(H)$ for spherically symmetrical particles (inert gases and molecules such as CH_4, N_2, O_2) with no specific interactions show [50] that b_1 changes from 4.9 at $H = 3$ to 1.7 at $H = 6$, and similarly b_2^+ varies from 4.9 to 1.2 and b_2^- varies from 7.7 to 2.8.

In determining the volume of micropores by adsorption isotherms the coordinates of the experimental data are of no importance The graphics of the isotherms of microporous systems in the coordinates $\theta - p/p_s$, with a long plateau [27], are well known. These curves are no less useful for determining the volume of micropores than the equation (6.6). For example, the plateau for systems CCl_4, benzene, cyclohexane (all at 25°C) and isopentane (0°C) on ammonium phosphomolybdate, benzene in the anthracite coal, propane on zeolite 5A ($T = 273$, 323 and 398 K), argon on chabazite (138 to 195 K), starts from 0.1 to 0.3 p/p_s. These data correspond to the values of b from 2.3 to 1.1.

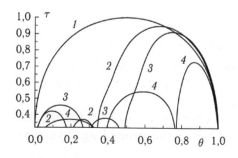

Fig. 22.1. Phase diagrams. Curve 1 – bulk phase. Isolated cylinders of different sizes: $D/\lambda = 20$ (2), 14 (3), 8 (4).

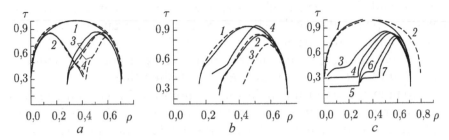

Fig. 22.2. Phase diagrams: a – in the coordinates (τ, ρ) for the bulk phase (1) and cylindrical pores with R_p = 4.5 λ; b – for the pores of the type of slit (1), cylinder (2.4) and sphere (3); R_p = 4.5λ (1–3) and 8.5λ (4), R = 4; unstructured walls ε_{AA}/k_B = 750 K (solid) and structured walls with γ = 1.5 (dashed lines), c – for the bulk phase (1, 2) and for cylindrical pores with R_p = 4.5λ, R = 4 (3–7); 1 – rigid lattice, 2 – soft lattice; ε_{AA}/k_B = 1250 (3), 1000 (4), 750 (5), 250 (6) and 125 K (7).

22. Curves of vapour–liquid phase stratifying in cylindrical pores

The phase diagrams in cylindrical pores were calculated using the model formulated in section 15 [51]. Figure 22.1 show the full phase diagrams for cylindrical pores of different diameters (the argon-silica gel system) [52]. Curve 1 refers to the bulk phase outside the pores for the cubic lattice structure. Changing the diameter of the cylinder increases the width and height of the dome corresponding to the filling of the central part of the pore at high degrees of filling the pore and to a reduction of the height and width of the dome, corresponding to the filling the surface monolayer at low degrees of filling the pores. The presence of the intermediate dome is due to the filling of the second layer of the pore wall, where the surface potential is still strong enough.

The effect of interaction potentials of particles with the wall of the cylindrical pore and with each other (the depth of the potential wells was caried), the radii of interaction potentials and the effect of structure and the type of cross sections of micropores on the phase stratifying curves was studied.

Figure 22.2 a shows the phase stratifying curves, calculated for different depths of the particle–wall potential wells on the example of the pores with R_p = 4.5λ. For comparison, similar curves for the bulk phase (curve 1) are also shown. The value of θ = 1 corresponds to $\rho \approx 0.7$. Variation of the properties of the particle–wall potential primarily illustrates the change $U(y_{min})$. As above, as an example of strongly wetted walls attention was paid to the argon particle -carbon system, and the example of poorly wetted walls (curve 2) was the particle (spherical)–perfluorinated Nafion membrane system [43]. The parameters the 'truncated' potential $Q_{12-6}/\varepsilon_{AA} = -3.28$ (2), 4.24 (3), the remaining $Q_k = 0$, and the 'full' averaged potential with $\varepsilon_{AA}/K_B = 750$ (4); $R = 4$ (solid) and $R = 1$ (dashed lines).

Strong attraction of the molecules to the wall forms a surface film (mono- or bilayer, depending on the properties of the particle-wall and particle–particle potentials) and the condensation process occurs in the remainder of the pore space (with radius $R_p - 1$ or $R_p - 2$, respectively). The dashed curves refer to the case of the interaction between the closest particles only. Reduction of R is equivalent to reducing the depth of the potential well. For non-wetting walls the phase diagrams change relatively slightly (the main role played by the interaction between the particles), whereas for strongly wetted walls otheir changes are more significant (see also Fig. 22.2 b).

The influence of the relationship of the potentials of the particle-wall lateral interaction kind on the form of the phase stratifying curve is shown in more detail in Fig. 22.2 c. With the increase in the contribution of the interparticle interaction the effect of the wall is weakened and the process of condensation changes from layer to 'volume'; in this case, the critical density decreases and the critical temperature is increased, approaching their values in the bulk phase. When $\tau \sim 0.3$ at small ε_{AA} (curve 7) the second layer is filled, at medium ε_{AA} (curve 4) the first layer is filled, and for the curve 3 the first layer is filled at a high temperature with simultaneous filling of the pore volume. For curve 5 wall attraction is strong enough to form the first monolayer at lower temperatures, but weak to form a second monolayer, and after filling the first layer bulk (capillary) condensation takes place.

Figure 22.2 b shows the phase stratifying curves for pores of different cross sections at a fixed value of $R_p = 4.5\lambda$ (for the slit 9 monolayers). The critical density increases in the slit–cylinder–sphere direction, and the critical temperature decreases in the reverse order. For a sphere, the surface layer contributes most to the total number of sites in the system, and for the slit – the least. Accordingly, the phase stratifying curves are shifted to higher densities curves at $\tau \sim 0.4$. When $\tau < 0.3$ 'step' form from surface monolayer (they are absent in the figure). The smaller the contribution of the surface area, the nearer the phase stratifying curve is to the same bulk curve. The pair of the curves 1 and 2 illustrates the effect of the structurization of the walls of the pores – the simplest case of their inhomogeneity.

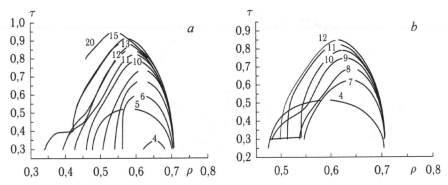

Fig. 22.3. Phase diagrams for helium (*a*) and methane (*b*) in the cylindrical pores of graphite. For field *b* the curves for pores with diameters equal to 5 and 6 lattice constants are not shown.

In general, an increase in the microroughness of the walls changes the energy of interaction with the wall for a range of its positions and this leads to the successive filling of the sites of the surface monolayer with different binding energy. When filling the second and subsequent layers the difference for the structured and unstructured walls are relatively small: the presence of 'sagging' depends on the ratio of the lateral interactions and interactions of the particle with the wall for the second monolayer. Increasing the diameter of the cylinder (curve *4*) increases the critical temperature and the critical density decreases.

These effects should be substantially dependent on the geometry of the pore cross-section, as this determines the ratio of the surface or the inside of the pores [44]. We consider the cylindrical pores, differing in the diameter. The minimum value of the diameter due to the specifics of the lattice model for a cylinder containing kinks instead of a circle, is equal to four [51]. Figure 22.3 shows the phase stratifying diagrams of helium and methane in the pores of graphite. As for the slit-like pores, we may note: the presence of 'zigzag' with increasing diameter, steps at low temperatures and even greater shift to the density of the liquid phase for the critical density. Thus, the non-monotonicity for critical densities observed previously for slit pores and characteristic of strong adsorption, is preserved for cylindrical pores. Similarly, the previously observed patterns for the slit-like pores are also preserved for weak adsorption.

Accounting for the softness of the lattice
The curves for the phase stratifying of small diameter cylindrical

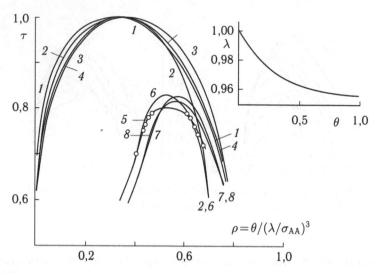

Fig. 22.4. Comparison of the phase diagrams. Curves 1–4 are for the bulk phase: *1* – 'exact' calculation by the molecular dynamics (MD) method for the LJ fluid in the bulk phase [54], *2* – rigid lattice system, *3* – soft lattice system, *4* – soft lattice system with additional consideration of the excluded volume due to the filling of the neighbouring sites. Curves 5–8 refer to the porous system: *5* – 'exact' calculation by the MD method for the LJ fluid at $R_p = 5\sigma$ (critical region is built on the basis of round points when using the scaling approximation of numerical data), *6* – rigid lattice system ($R_p = 4\lambda$); *7* – soft lattice system, *8* – soft lattice, taking into account the excluded volume.

pores were compared with calculations using the molecular dynamics method [53]. The calculations were conducted using the modifications of the LGM (section 9). The results are presented in Fig. 22.4 in the normalised coordinates $\tau - \rho$, where $\tau = T/T_c$, T_c is the critical temperature of the bulk phase, as calculated in the corresponding model; $\rho = \theta(\sigma/\lambda)^3$ is the density fluid, measured in solid spheres, for a rigid (incompressible) lattice $\rho = \theta/1.41$, since $\lambda = 2^{1/6}\sigma$. All calculations were performed with the same cutoff radius of the LJ-potential equal to 2.5σ, which corresponds to the interactions in the first four coordinate spheres in the lattice system.

Curves *1–4* refer to the bulk phase: *1* – 'exact' calculation by the MD method for the LJ fluid in the bulk phase [54], *2* – rigid lattice system, *3* – soft lattice system, *4* – soft lattice system with additional consideration of the excluded volume due to the filling of the neighbouring sites. The inset in Fig. 22.4 shows the variation of the lattice constant for the bulk fluid phase, depending on the degree of filling of the lattice structure θ. Curve *2* for the bulk phase in the LGM is symmetrical relative to the position of the critical temperature (the critical point – the maximum of the dependence $\tau(\rho)$ (or $\tau(\theta)$). Curves *3* and *4* are calculated taking into account the excluded volume and the softness of the lattice, do not exhibit symmetry. An important property of taking into account the softness lattice is the change of the position of the liquid-phase branch (relative to the same branch for a rigid lattice), which at low temperatures tends to $\rho = 0.8$, which practically coincides completely with the results of numerical simulation by the MD method.

Phase stratifying curves *5–8* in the pore were calculated at $\varepsilon_{AA}/k = 750$ K, the average potential of the walls is indicated in section 4. In all variants the position of the critical points shifts to higher densities compared with the bulk phase and the critical temperature decreases. The gas branch in all cases is almost identical. Curve *5* is characterized by greater flattening of the maximum, compared with the other curves, obtained in the lattice model. It is well known that in the quasichemical approximation (QCA) the critical temperatures are overestimated in comparison with exact solutions. Curves *6–8* fit this pattern. Calculations show that even the most primitive version of the rigid lattice (curve *6*, $R_p = 4\lambda$ corresponds to four diameters of argon atoms inside the pores) gives a fairly close quantitative agreement with the exact MD curves ($R_p = 5\sigma$ corresponds to the distance from the centre of the pore to the centres of carbon atoms). Accounting for changes in the lattice constant and the additional consideration of the excluded volume improve this agreement – the liquid-phase branch is more accurately reproduced. For the rigid lattice, as for the bulk phase, at low temperatures curve *6* tends to a density of 0.7, whereas the curves *7* and *8* (for the soft lattice) converge to the value 0.8. Accounting for the excluded volume shifts the curve *8* in the direction of the curve *5* compared to the curve *7*.

It should be emphasized that the differences in critical temperatures between the exact and approximate solutions in the reduced coordinates for the narrow pores are of the order of 7%, which is significantly better than in the bulk phase, where a similar discrepancy is of the order of 20–25 %. This is due to an increase in the contribution of the surface compared with the cooperative contributions of interacting particles. In order to be able to achieve the same result in ordinary (non-normalized) coordinates $T - \rho$ (or θ), we should additionally use the calibration functions [55], allowing to change the curvature of the phase stratifying curve in the vicinity of the critical point (see below).

However, as shown below, these calculations are exemplary in nature, showing the effect of molecular parameters, as capillary condensation does not occur in such narrow channels of (see below).

23. Hysteresis loops at capillary condensation

The phenomenon of capillary condensation is reflected on adsorption and desorption isotherms in the form of an reversible hysteresis loop. This phenomenon is widely used in practice to find the pore size distribution from the experimental data on the adsorption and/or desorption branches of the adsorption isotherms [10, 27]. Previously, all the methods of theoretical description of the pore size distribution were based on the relationship of the equilibrium value of the vapour pressure in the isolated pore with its size according to the Kelvin equation.

Recently, however, the adsorption–desorption hysteresis has been described using the concept that considers the capillary condensation process as a first-order phase transition [56]. The physical reason for the existence of the adsorption branch of the spinodal is related to the fluctuation instability of the adsorption film with an increase in its thickness during the condensation of the adsorbate. The desorption branch of the spinodal is due to the cooperative behaviour of the fluid in the pore with a decrease in the external pressure. The stability region of this branch is determined by the terms of the beginning of the desorption process at the end of the pores, i.e. the temperature, width of the pore and the vapour pressure at which desorption begins. The magnitude of this pressure depends on the ratio of the interactions of molecules with the surface of the fluid and between them. These calculations involved Monte Carlo methods, integral equations, functional density and the LGM.

Analysis of the phase diagrams shows that in the case of narrow pore adsorbents the full phase diagrams, i.e., diagrams describing the entire range of the degrees of filling of the pore system, consist of a number of 'domes'. This is due to the fact that the spatial regions in which the phase stratifying the adsorbate to low- and high-density phases occurs at a given pressure in complex adsorption fields is determined by the nature of the potential adsorbate–adsorbent interaction. Accordingly, we should expect the existence of a 'multi-dome' diagram of the spinodal state [57].

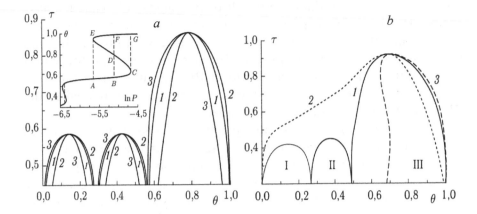

Fig. 23.1. Phase diagrams of argon: a – a slit pore in carbon with a width of $H/\lambda = 7$ (*1*) and analogues for the vapour (*2*) and liquid (*3*) branches of the spinodal ($z = 6$); the insert shows spinodal branches for vapour and liquid with $\tau = T/T_c = 0.625$; b – in a cylindrical pore $D = 15\delta$, $z = 12$ (Ar–SiO$_2$ system) [57].

The principle of building the spinodal curve for the bulk phase is shown in Fig. 18.1 a. Figure 23.1 a shows a phase diagram (curve *1*) and spinodal diagrams corresponding to adsorption (curve *2*) and desorption (curve *3*) branches of the adsorption isotherms in the whole temperature range [57]. Calculation is made for argon present in an infinite slit pore with a width of 7 monolayers ($z = 6$). In narrow pores the critical temperature $T_c(H)$ for the slits or $T_c(D)$ for the cylinders depends on their characteristic dimensions. With a decrease in the characteristic size H, D the critical temperature decreases [36–40].

The principle of construction of spinodal diagrams shown in the inset of Fig. 23.1 a. The BDF density jump in the isotherm in the condensation of the adsorbate is shown by a solid line and straightens the non-equilibrium (or metastable) isotherm plot BCDEF, which has a traditional look of the van der Waals loop [11, 17]. The areas DEF and DBC, constructed in accordance with Maxwell's rule, are equal, and the densities at the points B and F correspond to the equilibrium densities of the coexisting gas and liquid phases. The dashed line AE corresponds to the density jump at the transition from the liquid phase spinodal EF to the vapour phase in the desorption branch and the dashed line CG – to the density jump at the transition from the gas spinodal BC to the liquid phase on the adsorption branch. The pressure difference at the points F and E determines the width of the metastable liquid phase, and the pressure difference between the points B and C determines the width of the metastable vapour phase region. However, for relatively long isolated pores it is assumed that the pressure at the desorption branch coincides with the equilibrium pressure, so the hysteresis loop width is represented by the width of the metastable vapour phase region BC. In addition to the loop for the central dome on the left there is a similar loop related to the filling of the second surface monolayer.

Figure 23.1 b shows the same equilibrium phase diagrams (curves 1) and spinodal diagrams (curves 2 and 3) for the adsorbate in a cylindrical pore with a diameter of 15 monolayers ($z = 12$) located at temperatures below critical. The first two domes (designated by Roman numerals) relate to the phase stratifying of the adsorbate close to the surface in the first and second cylindrical monolayers. The critical temperatures for these two domes are substantially lower than for the third dome. The difference between the type of the spinodal diagram (only the branches for the central dome are shown) in a cylindrical pore and the analogous diagram for the slit pores is due to the dominant contribution of the surface areas, the type of lattice z structure z and lower temperatures. Thus, the form of the spinodal diagrams is affected by all molecular properties of the adsorbate–adsorbent system.

Since the process of capillary condensation is controlled by the filling of the central regions of the pores, the properties of hysteresis loops will be discussed only for the central dome.

Influence of the molecular properties of the system on the hysteresis loop [57]

Figure 23.2 a shows the influence of the nature of the surface of the pore walls, which is modelled by the energy of the adsorbent–adsorbate interaction, on the normalized width of the hysteresis loop in the slit ($H = 10\delta$) and cylindrical ($D = 15\delta$) pores. Here, $\Delta P = P_a(H, D) - P_0$, where $P_a(H, D)$ is the pressure of the gas branch of the spinodal in a slit-pore with width H or a cylindrical pore with diameter D, P_0 is the saturated vapour pressure. Both values, $P_a(H, D)$ and the vapour pressure P_0, are functions of temperature. In the cylindrical pores calculations were carried out made with the wall potential of length 2 (3) and 3 (4) monolayers ($z = 12$). The greater the length of the potential wall, the smaller the central region of the pore with $Q_f = 0$ in which capillary condensation of the adsorbate takes place, and the smaller the width of the hysteresis loop.

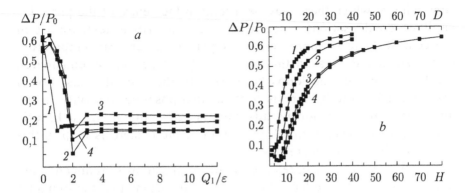

Fig. 23.2. The influence of the molecular parameters for the slit-pores with width $H = 10\delta$, $z = 6$ (*1*) and 12 (*2*) and the cylindrical pores with a diameter $D = 15\delta$, $z = 12$ (*3, 4*) on the normalized width of the hysteresis loop: *a* – the dependence on the adsorbent–adsorbate interaction energy; *b* – on the characteristic pore size [57].

In the case of a strong absorbate–adsorbate interaction the nature of the surface of the pore walls has only a small effect on the width of the hysteresis loop for the slit-shaped and cylindrical pores, or more precisely at the dimensionless ratio $\alpha = Q_1/\varepsilon > 3$ ($z = 12$) and $Q_1/\varepsilon > 2$ ($z = 6$). When $Q_1/\varepsilon < 3$ for $z = 12$ and $Q_1/\varepsilon < 2$ for $z = 6$ the saturated vapour pressure (according to [57]) approaches the P_0 value or is greater than P_0. This energy region is called as corresponding to 'weak adsorption'. Equality $P_0(H, D) = P_0$ holds at about $Q_1/\varepsilon \sim$ 0.6–1.5 for the slit and cylindrical pores. The ratio $P_0(H, D) > P_0$ at $Q_1/\varepsilon < 0.6 \div 1.5$ responds poorly wettable walls (or non-wetted when $Q_1/\varepsilon \sim 0$) and the inverse value $P_0(H, D) < P_0$ at $Q_1/\varepsilon > 0.6 \div 1.5$ refers to the wetted walls of the pores. For weak absorption the width of the loop increases sharply with decreasing energy of the adsorbent–adsorbate bond. (For both types of pores at $Q_1/\varepsilon > 3.0$ all materials can conventionally be considered equally well wetted.)

It should be emphasized that the reduction or increase in the saturated vapour pressure is determined by the molecular parameters of the adsorbate–adsorbent system and is not *explicitly* associated with the curvature of the meniscus at the vapour–liquid interface inside the pore (as is generally assumed when using the Kelvin equation).

The influence of the size factor of the pores is illustrated in Fig. 23.2 *b* by the dependence of the width of the hysteresis loop characteristic pore size H or D (Ar–SiO$_2$ system) for the same molecular parameters as in Fig. 23.2 *a*. The value of $H = 6\delta$ is the first value at which there may be a loop in the centre of the pore (i.e. capillary condensation), since for $H = 5\delta$ the critical temperature of the central dome is approximately equal to the critical temperature of the surface domes. In this case $\Delta P/P_0 \geq 0.55$ can be considered the lower limit of the loops belonging to the central areas. For H less than 6δ H the hysteresis loop is also possible, but only in the adsorption in microporous adsorbents [27], and it has a kinetic nature. For H/δ from 6 to 10, the width of the hysteresis loop in this region of H sharply decreases.

For the cylindrical pores the width of the loop is reduced more quickly. At $D = 6\delta$ we have $\Delta P/P_0 \approx 0$. This means that the diameter range of 2 nm there is almost no hysteresis and the hysteresis has a distinct form only when $\Delta P/P_0 > 0.1$, which corresponds to cylinders with a diameter greater than 3 nm. In order to carry out a quantitative comparison of the calculated data with the experimental data for the adsorption of nitrogen and argon in mesoporous materials such as MCM-41 [58], it is necessary to include the calibration function.

The influence of the inhomogeneity of the pore walls on the capillary condensation conditions is shown in Fig. 23.3 [57]. The degree of inhomogeneity of the walls of cylindrical pores with two diameters MCM-41 was varied [59–62] (group II curves are shifted

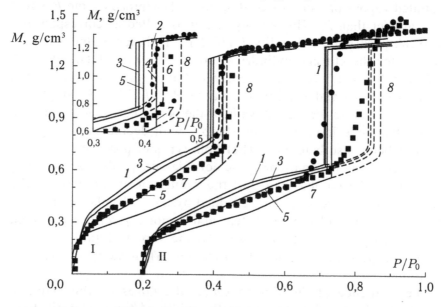

Fig. 23.3. The dependence of adsorption and desorption isotherms of argon M [g/cm³] of the degree of inhomogeneity of the walls of cylindrical pores in MCM-41 with a diameter of 4.15 nm (I) [61] and 5.54 nm (II) [60].

to the right on the abscissa by $0.2P/P_0$). Presented are four options for calculating the adsorption isotherms with the adsorption and desorption spinal branches (odd-numbered curves correspond to the desorption branch, even numbers – adsorption). The calculated curves at 87.3 K ($\tau = 0.58$) differ in the degree of amorphization of the surface material. Homogenous pore walls (without amorphization) correspond to the curves *1* and *2*. The energy parameters of the inhomogeneous walls *5* and *6* were obtained in [63] from the experimental data for the adsorption of argon at a flat surface of silica gel ([27, p. 84]), with the effect of the curvature of the walls in different cylindrical channels taken into account. The parameters of the curves *3* and *4* correspond to halving of the degree of amorphization, and the parameters of curves *7* and *8* were obtained when the degree of amorphization was doubled compared with the parameters of curves *5* and *6*. These curves show that, although the capillary condensation occurs in the central part of the pores, the change of the properties of the pore walls influences the position and width of the loop of the adsorption–desorption hysteresis. For these variations in the properties of the walls the pores the $\Delta P/P_0$ values vary from 0.031 to 0.046 for the pores of 4.4 nm and from 0.127 to 0.138 for the pores of 5.4 nm.

These calculations show that the theory qualitatively represents the experimental data both in the field of existence and the width of the hysteresis loop. A quantitative description requires the inclusion of the polydispersity of the MCM-41 material: the experimental hysteresis loops change monotonically with increasing pressure, whereas the theoretical calculations for monodispersed systems indicate the presence of density 'jump' on the hysteresis branches (see Chapter 7).

Thus, the width of the loop of the adsorption–desorption hysteresis is dependent on the molecular properties of the adsorbate–adsorbent system and demonstrates the complex nature of the process of capillary condensation in the narrow pores. The width of the hysteresis loop is also affected by the temperature at which measurements are taken.

Figure 23.4 *a* shows the temperature dependence of the saturated vapour pressure P_0 for a wide macropore (*1* and *2*) and the width of the hysteresis loop (ΔP) for a wide (*3*) and a cylindrical pore $D = 14$ monolayers (*4*). The dependence of the width of the hysteresis loop in a cylindrical pore of 14 monolayers for various lattice structures

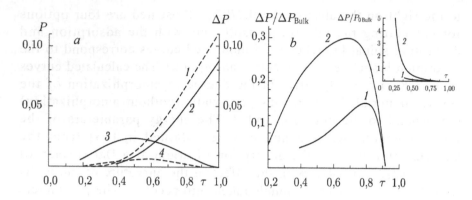

Fig. 23.4. Temperature dependence: a – the pressure of saturated vapour for a wide pore $z = 12$ (*1*) and 6 (*2*) and the width of the hysteresis loop for a wide (*3*) and a cylindrical pore with $D = 14$ monolayers (*4*) $z = 12$, b – the width of the hysteresis loop in the pore $D = 14$ monolayers for various lattice structures with $z = 6$ (*1*) and 12 (*2*). The inset shows normalized ratios $\Delta P(T)/P_0(T)$.

$z = 6$ (*1*) and 12 (*2*), normalized for the width of the loop in a very wide pore ΔP_{bulk} (*T*), is given in Fig. 23.4 *b*.

All of the temperature dependences of the loop width $\Delta P(T)$ have a maximum. This is due to the fact both at $T \to 0$ and at $T \to T_c$ (*H*, *D*) the hysteresis loop disappears when $\Delta P(T) = 0$. The peak position depends on the molecular properties of the adsorbate–adsorbent system. The inset shows normalized ratios $\Delta P(T)/P_0(T)$. Increase of the ratio $\Delta P(T)/P_0(T)$ with decreasing temperature in the inset in Fig. 23.4 *b* shows that the saturated vapour pressure decreases with temperature much faster than the width of the loop on the descending branch. For Fig. 23.4 *b* the width of the loop is normalized to the corresponding value of the width of the hysteresis loop in the wide pore (the bulk phase). Such normalization can be carried out due to the fact that the critical temperature of the pore is lower than in the volume.

Model calculations of the temperature dependences of the width of the hysteresis loop in Fig. 23.4, passing through a maximum, agree qualitatively with experimental results for capillary condensation phenomena in a wide range of temperature and pressure [64].

In general, when solving the problem of increasing the accuracy of determining the function of the pore size distribution from the data on the experimental adsorption isotherms with a hysteresis, it is necessary to take into account the specifics of the studied mesoporous adsorbent–adsorbate system and also the experiment conditions.

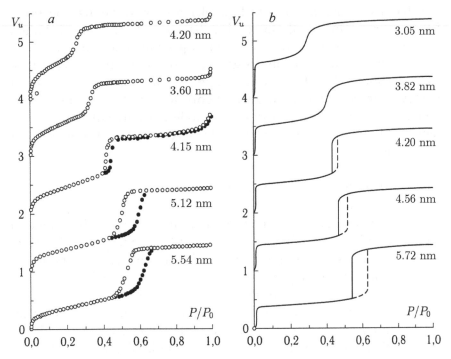

Fig. 24.1. Isotherms of argon atoms in the cylindrical pores of different diameters. The adsorption isotherms and the hysteresis loops in the size range of channels MCM-41 3–6 nm $T = 87.3$ K: a – the experiment [55] and b – theoretical isotherms at $g_e = 0.3$.

24. Size effects

The last four sections demonstrate the strong influence of the characteristic pore size H or D on the phase stratifying curves, including the critical parameters. The dependence on the size is shown through consistent linking of individual equations for local isotherms (7.4) or (14.1) describing the molecular distribution. Near-surface monolayers, reflecting the influence of the surface potential of the walls, affect the nature of the filling of the internal monolayers and, consequently, the distributions of the molecules differ from their isotropic distribution in the bulk phase. The size range to which the influence of the surface potential is discussed in section 25. This section discusses the accuracy of the description of the molecular distribution within the narrow pores. This problem has arisen on the basis of experimental data obtained from measurements of the adsorption hysteresis for mesoporous materials MCM-41 [59–62] (see Fig. 24.1 a).

Isotherms for argon atoms in Fig. 24.1 *a* showed the disappearance of the hysteresis loop with decreasing diameter of the channel. Values V_u are the number of adsorbed particles per unit pore volume.

Experiment show that the hysteresis loops exist to characteristic sizes of about $D^* \sim 4$ nm and disappear with further decrease of D. However, all of the modern theoretical methods (molecular dynamics, Monte Carlo, lattice-gas model, the density functional theory) indicate the presence of the hysteresis loops at diameters of $D^* \geq 2$ nm.

Concept of quasi-one-dimensional fluid in narrow cylindrical pores [65]

To explain this large discrepancy between the values of D^* attention should be paid to the fact that the specified size range D refers to small systems for which the existing mathematical theorems (including the theorem proposed by S.B. Frobenius) indicate [66] that no phase transitions can take place in them.

This theorem corresponds to a textbook example of the exact result for a one-dimensional chain [16, 17] $D = 1\lambda$ (where $\lambda = 2^{1/6}\sigma$ – the size of the monolayer, σ – the size of the hard-sphere molecules in the Lennard–Jones potential). Increasing the diameter to two to three cylindrical layers maintains the quasi-one-dimensional nature of the system – this can be proved by direct exact calculations by the fragment method (see Appendix 3) [67, 68]. However, with a further increase in D the fluid has to transform into the three-dimensional state in which phase transitions are realized. Therefore, there is a certain critical size $D_1 = D/\lambda$, where the system still retains the property of the quasi-one-dimensional form, while at $D_2 = D_1 + 1$, the system loses its quasi-one-dimensional properties and there should be a phase transition with its critical temperature $T(D_2) > T$, which can be verified by the presence of the hysteresis loop in the isotherm.

The transition from the 'one-dimensional' properties to the 'three-dimensional' ones in the cylindrical channels is formally implemented at $D \rightarrow \infty$. However, phase transitions also occur for the intermediate two-dimensional systems [16, 17]. The main thing – the criterion should reflect the loss of quasi-one-dimensional properties. This criterion can be formulated by using the knowledge of the phase diagrams in porous solids (sections 22 and 23). An important property of these diagrams is the presence of several domes, each of which corresponds to the sequence of filling the cylindrical layers. Naturally, the criterion shold be such that the fraction of the inside of the pore space, which is characterized by one common dome, was not less than the fraction attributable to the near-wall monolayers of the pore. If the fraction of the surface areas of the pore is greater than the central area, such a pore retains the quasi-one-dimensional properties and phase transition in it is not possible.

This criterion is easily expressed in a continuum model as $K = ((D_1 - 2n_w)/D_1)^2 \geq 1/2$, where n_w is the number of near-wall surface monolayers. Their number coincides with the number of the corresponding domes of the complete phase diagram: $n_w = 1$–3, depending on the adsorbate–wall potential (formally $n_w = 0$ corresponds to the hydrophobic walls when the filling of the narrow pores depends on the presence of impurities in them, and they are not considered in the continuum model). This criterion allows us to estimate the size range of the cylindrical pores retaining the property of one-dimensionality. For these values n_w we get that the condition $K \approx 0.5$ corresponds respectively to $D_1 = 8$, 14 and 20λ. (Analysis of the different choices of conditions for the quasi-one dimensionality criterion is discussed in [67].)

The determined pore diameters can be compared with the experimental data for the adsorption of argon and nitrogen molecules in the mesoporous materials such as MCM-41. For both gases $n_w = 2$, and therefore, given the size $\sigma(Ar) = 0.3405$ and $\sigma(N_2) = 0.37$ nm, we get for the close-packed structure of the fluid with the number of nearest neighbors 12, respectively, $D^* = 3.79$ and 4.12 nm. These numbers are in good agreement with the disappearance of the hysteresis loops: in the region between 3.6 and 4.14 nm for Ar and the beginning of formation of the hysteresis loop at ~ 4.4 nm for N_2.

Thus, the concept of quasi-one dimensional fluid located in narrow cylindrical pores reflects the size effects of the system and helps to explain the gap in the estimates of D^* between theory and experiment (so in theory we need to account more accurately for size effects). Matching estimates of D^* in theory and experiment paves the way for finding the function of the pore size distribution from the data on the hysteresis loops, and the central question of the theory is the issue of taking into account the size factor in the calculation of the phase state of the fluid in the narrow channels. For this it is necessary to use more accurate description methods of the molecular distribution than the approximate cluster approaches used above.

The calibration function [69]
The calibration function has been introduced to increase the accuracy

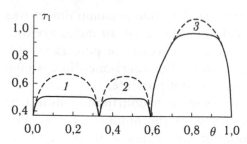

Fig. 24.2. Phase diagrams in the cylindrical pore $D = 11\lambda$.

of calculations in the quasichemical approximation near the critical areas. Detailed description of the principle of introduction of the calibration function is given in Appendix 5.

In this variant of the theory the equations [70, 71] for the local isotherms (7.4) and (14.1) can be rewritten as

$$a_f P = (\frac{\theta_f}{1-\theta_f})^{1+g(f)} \Lambda_f, \qquad (24.5)$$

where $g(f)$ is the calibration function for the sites of type f; the kind of the function of non-ideality of the system Λ_f, describing the local correlation effects (in this case in the QCA) does not change. Given the influence of the calibration function we can obtain almost the exact distributions of molecules for the bulk phase and in adsorption on the surface (Fig. 14.2) and for methane molecules in the slit-shaped pores of graphite (Fig. 21.5).

Similar use of the calibration functions for cylindrical pores gives the following correction of the phase stratifying curves (Fig. 24.2): the dashed lines are the phase stratifying curves without the use of calibration (here $\tau_1 = \varepsilon_{ArAr}/kT$). Domes 1 and 2 are near-surface domes, and they are calibrated to two-dimensional exact solutions, and the dome 3 – central or volume in the pore, so its calibration is carried out on the bulk phase.

This description essentially changes the shape of the curves of phase stratifying, especially the critical temperature for each of the domes. However, the construction of calibration functions is performed for infinite-sized two- and three-dimensional systems. When reducing the diameter of cylindrical pores such calibration leads to the disappearance of the hysteresis loop at $D < 3$ nm, which is less than the value found in the experiments. Therefore, to be consistent with experimental data, the calibration function, which depends on the size of the pores, was constructed.

Dimensional calibration

To account for the influence of the pore size, in equation (24.1) we use the additive relationship $g_e = g_{e1} + g_{e2}$, where g_{e1} is the contribution for the open system of dimension $d = 2$ or 3 (which are discussed in Appendix 5), $g_{e2} = g_{e2}(D)$ is the dimensional component of the calibration function, which depends on the diameter of the channel D. To construct a calibration function, we must have exact solutions. The absence of such solutions across a range of sizes from small diameters to macrosizes (for which experimental data can be used) raises the question about the selection of diameters used for constructing the calibration function.

In the one-dimensional system $D = 1$ QCA corresponds to the exact solution, so it can be used to build the calibration function. For the mean-field approximation a first order phase transition takes place in the one-dimensional chain, so based on this principle it is impossible to build calibration functions. The same applies to any approximation in which there are no correlation effects, and in particular the density functional method.

To transfer to higher values of diameters we use the fragment method (see Appendix 3). It was used for accurate direct calculations of molecular distributions for cylindrical pores with a width from 2 to 6 lattice constants (for $z = 6$, 8 and 12) [67] Figure 24.3 shows the isotherms calculated by the fragment method and QCA (dotted line). It is seen that the slope of the isotherms obtained using the fragment metho is positive in the entire temperature range, which excludes the presence of phase transitions. Similar curves, calculated by the quasichemical approximation (QCA), are shown by dotted lines – they have a density 'jump', indicating a phase transition. That is the

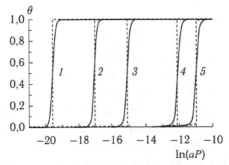

Fig. 24.3. Argon adsorption isotherms in cylindrical channels of the MCM-41 type with diameter 4λ ($z = 6$) with a binding energy at the flat wall $Q_1 = 3.9\ \varepsilon_{ArAr}$ at temperatures $T = 54$ (*1*), 62 (*2*), 70 (*3*), 87.3 (*4*), 96 (*5*) K.

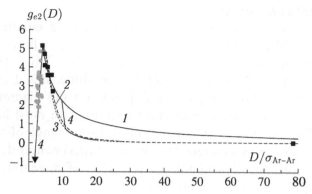

Fig. 24.4. Interpolation curves $g_{e2}(D)$ for structure $z = 12$.

usual accuracy of the QCA is not sufficient for a correct description of the molecular distribution in a narrow channel.

The direct calculations carried out in the range of up to $D = 6\lambda$ show exactly the same difference between the fragment method and the QCA. This result clearly indicates that all of the theoretical methods (molecular dynamics, Monte Carlo, lattice-gas model, the density functional method), indicating the possibility of phase transitions (by the presence of hysteresis loops) for values of diameter $D^* \sim 2$ nm, are equally flawed. The exact calculation indicates the absence of the hysteresis loop to the values of the cylinder diameter of at least 2.3 nm.

On the other hand, the calculations by the fragment method yielded accurate results [67], needed to construct the calibration function at small diameters. This enabled, using as the other limiting size of the bulk phase the value for g_{e1}, i.e. $g_{e2} \to 0$ at $D \to \infty$, to perform the scaling approximation of the behavior $g_{e2}(D)$ in the entire pore diameter range. Its appearance is shown in Fig. 24.4 (a procedure of its construction is described in [68]). Dependence $g_{e2}(D)$ is non-monotonic. For $D = 1$, it 'compensate' the contribution g_{e1}. Increase of $g_{e2}(D)$ and the existence of a maximum of g_{e2} emphasizes the impossibility of such approximations in the approximations, neglecting the effects of correlations.

To increase the accuracy, approximation of the fragment method to 8λ was carried out. Round (on the ascending branch) and square (on the decreasing branch) symbols are the data for the fragment method, a triangular point – the one-dimensional system. Curves *1–4* refer to different scaling dependences used for the interpolation

$g_{e2}(D)$ to the properties of the macrosystem that have been attributed to $D = 80\lambda$ (details in [68]).

The uncertainty in the form of the scaling relationships of the type 2–4 in the area from 6 to 10λ changes the value of the size D^*, from which a hysteresis loop should appear from 4.03 to 4.2 nm at $T = 87.3$ K, which is in good agreement with experiments in the loop at $D^* = 4.4$ nm and $T = 100$ K.

The constructed calibration curves allowed us to calculate the adsorption isotherms in Fig. 24.1 b, which properly reflect the range of existence of the hysteresis loops, as in the experiment. The calculations were performed for a monodisperse system of cylindrical pores of a fixed size, so the 'jumps' of the hysteresis loops are vertical. In real materials, there is a scatter of pore sizes, so the experimental curves are smoothed. (This is discussed in section 7).

Dimensional contributions affect the conditions of condensation of molecules in narrow channels. The constructed calibration functions with the dimensional contributions taken into allow us to calculate the critical temperatures in the entire range of sizes of cylindrical pores [72]. Figure 24.5 shows the dependence of the critical temperature of the capillary condensation of the adsorbate in the cylindrical pores of the MCM-41 type on the pore diameter. The numbers refer to the compared methods of calculation: 1 – normal QCA; 2 – QCA calibrated to an infinite volume; 3 – QCA calibrated taking into account the contribution of the size effect. The first two

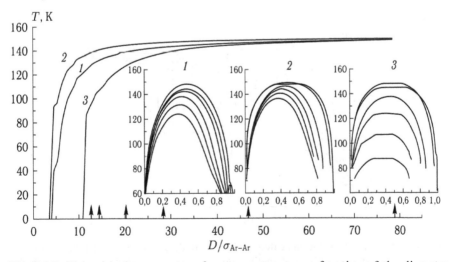

Fig. 24.5. The critical temperature for argon atoms as a function of the diameter of the cylindrical channel, calculated in QCA: no calibration (1), calibrated for the bulk phase (2), and calibrated with the size effect of pores taken into account (3).

curves show no hysteresis loop at $D = 4$–5 monolayers, whereas for the curve *3* the critical size of the pore diameter corresponds to 13 monolayers.

The behavior of the curves of the vapour–liquid coexistence is sensitive to the used method of calculation. In the inset is a family of curves of vapour–liquid coexistence of the central domes for six diameters marked by arrows on the *x*-axis (measured in the hard-sphere diameters of the argon atoms), for each of the above variants of the calculation method.

Accounting for calibration functions with the size changes the form of the contribution of each of the domes and the dependence of the coexistence curve on the value of the diameter of the pores. The calculation results are fully consistent with the experimental data [59–63].

With the decrease of the pore size and reaching the critical size there is a qualitative change in the distribution of molecules. In Fig. 24.6 the comparison results indicate that in this range of diameters of the channel there is a large difference in the adsorption isotherms [73], The curves are different in the size of the density jump and the vapour pressure corresponding to such a jump. In the area of 13 monolayers the system loses its capacity to form a cooperative ensemble, which can be treated as a separate phase. The phase transitions disappear from the system and the hysteresis loop disappears on the curves of the adsorption isotherms. In fact,

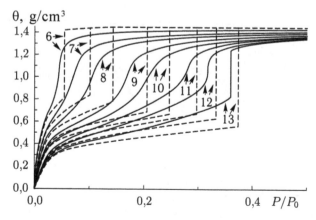

Fig. 24.6. Adsorption isotherms calculated for cylindrical pores of different diameters D/δ (diameter size is indicated in the figure) from 6 to 13 monolayers with the full calibration function taken (solid line) and not taken (dotted line) into account.

Table 3.1. The critical sizes of the infinite cylindrical and square channels, z – number of nearest neighbours of the lattice structure, n_w – number of domes for the pores of the given geometry [74]

z	n_w	Cylindrical channel				Square channel			
		I		II		I		II	
		A	B	A	B	A	B	A	B
	1	5*	5*	6	7	8*	8*	8	8
6	2	12	13	12	14	13	14	13	14
	3	18	19	18	21	19	21	19	21
	1	5*	6*	6	7	8*	8*	8	8
12	2	12	13	12	14	13	14	13	14
	3	21	21	18	21	19	21	19	21

at diameters of less than 12 monolayers the slope of the isotherm is markedly different from the vertical jump (Fig. 24.6).

The results lead to the conclusion that there is the minimum size of the pores at which the phase stratifying into a rarefied and a dense phase takes place. A similar restriction should apply to the pores of uniform length but difference cross-sectional shapes. Analysis of the conditions of the hysteresis loops for square or rectangular channels is also found to be related to the type of phase diagrams [74]. Table 3.1 shows the results of evaluation of the critical dimensions of the channels in the case of continuum (II) and discrete (I) models, depending on the value of the structural units z. Values of the critical dimensions of the pores of different geometry (in units of δ, rounded to the nearest whole number); column A refers to the criteria for $K = \varphi^* = 0.444$, B – a criterion $K = \varphi^* = 0.5$; dimension values for $n_w = 1$, marked with an asterisk, were obtained while maintaining local binding energies to $Q_q = 0.1\varepsilon$ [74]. The procedure and details of estimates are discussed in section 7.

This table shows the commensurability of the critical values of the pore size of these two geometries.

The differences in saturated vapour pressures, obtained with calculation methods of different accuracy, for the same pore size affect the accuracy of describing the function of the pore size distribution in the interpretation of the hysteresis loops.

25. Wide pores

Wide pores and Kelvin equation

Up to the present day, most methods for determining the distribution functions of the pore size use of the Kelvin equation (KE) (2.6) [27, 75]. The question of the validity of using this equation, despite numerous discussions, has not been fully resolved until recently. In [76], this question was investigated in the framework of the theory in question to determine the upper limit of the use of the Kelvin equation. The solution to this problem has a practical value associated with adsorption porosimetry for pore sizes greater than 10 nm, and for improving the accuracy of this method by using different adsorbates, including for measurements at elevated temperatures, up to room temperature.

Analytically it has been shown (section 21) that the vapour pressure in the limit of large pores can be expressed as

$$P(H,D) = P_0 \exp(-\beta \varepsilon b), \quad b = \gamma_s/H \quad \text{or} \quad b = \gamma_c/R, \qquad (25.1)$$

where the constants for the slit-shaped γ_s and cylindrical γ_c pores have the form $\gamma_s = 2z_{12}/(H-2)$ and $\gamma_c = z_{12}/(R-1)$. Value b decreases with increasing width H or R, and the pressure at which the filling of the pores takes place, tends exponentially to its bulk value P_0. For micropores coefficient b in (25.1), obtained for spherically symmetric particles without specific interactions, but with values Q_{pore} taken into account [50], is in good agreement with the experimental data [27]. In all cases the pore filling occurs at pressures less than the vapour pressure at a given temperature. This is consistent with the formula (2.6). Another result of the expression (25.1) is that its exponential structure does not depend on the geometry of the slit-like or cylindrical pores and constant γ_s and γ_c are the same as $R = H/2$.

Attention was also given to the dependences of the saturated vapour pressure (the appearance of capillary condensation) on the diameter of cylindrical pores in the adsorption of argon atoms and molecules of nitrogen and carbon tetrachloride, calculated by the LGM and the Kelvin equation. The pore diameters were measured in monolayer numbers δ, where $\delta = \sigma_{hs} \cdot 1.12 \cdot (2/3)^{1/2}$ is the width of the cylindrical monolayer for the lattice structure with the number of the nearest neighbours 12. This number corresponds to the dense packing of the spherical adsorbates. Parameter σ_{hs} is equal to the diameter of a hard sphere of the adsorbate whose value is equal to

Table 3.2. The parameters of the adsorbate molecules [27]

Gas	V, cm²/mol	T, K [3]	ε_0, cal/mol	σ_{tvs}, nm	δ, nm	σ, mN/m
Ar	4.30	87.5	238	0.34	0.3125	13.2
N_2	3.95	78	192	0.37	0.339	8.9
CCl_4	8.75	293	654	0.539	0.541	26.7

the corresponding parameter of the Lennard–Jones potential. The molecular parameters of the considered adsorbate molecules are shown in Table 3.2. The surface of the pores was considered as consisting of groups of SiO_2, and its corresponding parameters are $Q_1 = 5\varepsilon_{AA}$, $Q_2 = Q_1/8$ and $Q_{q>2} = 0$ [73, 74].

Figure 25.1 shows the curves of the dependence of the saturated vapour pressure of three gases, calculated by the LGM and the KE for the same molecular parameters. For the argon atoms and nitrogen molecules with larger dimensions the KE calculations (solid lines) and LGM calculations (solid lines and symbols) are in good agreement, but with a decrease in the diameter this agreement deteriorates. For CCl_4 molecules the agreement is about the same over the entire range of sizes. It should be noted that for larger CCl_4 sizes the ratio P/P_0 is significantly shifted to lower values. Therefore, their use for adsorption porosimetry is preferred for wide pores.

To determine D^*, after which it can be assumed that both methods give the same results, we introduce and analyze the function of the difference between the results of the LGM and the KE. For a

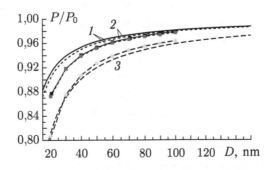

Fig. 25.1. Dependence of the ratio P/P_0 on the diameter of cylindrical pores (D) in monolayers for Ar at $T = 87.5$ K (1), for N_2 at $T = 78$ K (2), for CCl_4 at $T = 293$ K, (3), calculated using the Kelvin equation (lines) and LGM (lines and symbols).

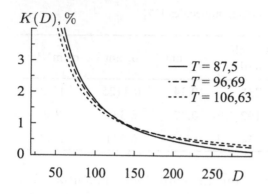

Fig. 25.2. Dependence of the function $K(D)$ on the diameter of the pores at three temperatures, $T = 87.5$, 96.69 and 106.63 K. The four level values of $K(D)$, equal to 0.5, 1, 2 and 3% are considered.

comparative analysis of the LGM and the KE we use the function

$$K(D) = \frac{(P/P_0)^{KE} - (P/P_0)^{LGM}}{(P/P_0)^{KR}} \cdot 100\% \Bigg|_D \,,$$

constructed depending on the pore diameter D.

The temperature dependence of the function $K(D)$ for the argon atoms is shown in Fig. 25.2. The following inequality between the ratios of the saturated vapour pressure in the LGM and the KE is implemented: $(P/P_0)^{LGM} < (P/P_0)^{KE}$ for a fixed diameter D, that is, the curve for the KE is higher. This is due to the different temperature dependence of both methods. At lower temperatures the agreement is better for large diameters, and at higher temperatures – at smaller diameters. Based on the results we can determine sizes from which the difference between the methods would not exceed a certain value of K^*. The Figure allocated four levels of K^* values from 0.5 to 3%. In all cases with the increase the pore size the function $K(D)$ decreases. The corresponding values of the pore size D^*, corresponding to these values of $K(D)$, are presented in Table 3.3 (the first three columns). The curves for Ar and N_2 behave in the same manner. They pass into each other at the appropriate scale transition, but belong to different temperatures. Comparison of the values of D^* for argon and nitrogen for the four levels of the function $K(D)$ is shown in Table 3.3 (columns 4 and 5). Comparison of the vapour pressure obtained by the two methods of calculation makes it possible determine the pore size for which both methods give nearly the same results, depending on the accepted level of accuracy determined by the procedure and equipment used for adsorption measurements.

Kelvin equation (2.6) was derived assuming equilibrium between the metastable vapour phase and a liquid having a curved surface in the absence of the potential of the pore walls. By analogy with a droplet in the bulk phase for porous solids we use the same concept

Table 3.3. The diameters of cylindrical pores D^* (dimensions are given in numbers of monolayers δ), corresponding to the level of accuracy of K^* of matching the LGM and the EK for the atoms argon at temperatures $T = 87.5$, 96.7 and 106.6 K (columns 1–3), and argon and nitrogen near triple, points in nm (columns 4 and 5)

K^*, %			T, K		
	87.5	96.7	106.6	87.5	78
				(D^* to Ar)	(D^* to N_2)
3	72	68	64	21.4	22.4
2	92	89	83	28	29.4
1	137	139	135	45	48
0.5	194	213	225	67	74

of the existence of an equilibrium between the liquid adsorbate and the vapour outside the porous system, and accordingly, the condition of metastability of the liquid–vapour system in the bulk phase is applied to them. This means that fixing temperature T in the isothermal experiment conditions and the pressure P/P_0 by the Kelvin equation inside the pore does not make it possible to determine the density of the vapour inside the porous systems since there is no equation of state for them. In molecular models the equation of state is constructed explicitly and it allows to determine the actual density of the vapour inside the pores.

In porous solids the vapour–liquid boundary has a curved meniscus the curvature of which depends on the contact angle φ at the solid–vapour–liquid three-phase boundary. Its value is determined by the Young rule $\sigma_{sv} = \sigma \cos \varphi + \sigma_{sf}$, where σ_{sv} and σ_{sf} are the interfacial free energies of a solid body covered with vapour and liquid, respectively. For concave menisci, which form in porous bodies with a strong molecular attraction to the wall of the pores, the saturated vapour pressure must be less than above the vapour-liquid flat interface. Moreover, the presence in the pore of the adsorbed film with a strong molecular attraction to the pore wall reduces the effective pore width. This leads to the following form of equation (2.6): $P/P_0 = \exp(-2K/R_k)$, $K = \sigma \beta V_L$, where $r_k = r_m \cos \theta$, where r_k – the effective width of the pores, r_m – radius of a semicircular meniscus. The traditional use of the assumption $\theta \sim 0$, which is justified for good wetting of the pore walls by the adsorbate gives $r_k = r_m$. This means neglecting the thickness of the adsorbed film, i.e., (2.6) is justified at $r_k = r_m \gg 1$.

Equation (2.6) gives some correlation between the size of the pores r_k and the value P/P_0, but not a strict relationship, based on the equation of state, as in the LGM. This is shown particularly clearly by the calculations of the form of the transition region between the vapour and the liquid in the slit-like pores (section 26). By increasing the width of the pores H the meniscus of the boundary in the centre of the pore is flattened instead of having a spherical shape, assumed when deriving equation (2.6). The conclusions about the flattening of the meniscus in the central part of the pore with increasing pore diameter D hold for cylindrical pores at approximately $D > 2H$. As a result of the flattening of the meniscus in the slit and cylindrical pores the law of the variation of P/P_0 is the same at larger sizes, as shown by the formula (25.1). This fact is the main factor of the divergence between molecular theory and the thermodynamic approach in the Kelvin equation.

The calculations show that the influence of the walls on the values of the saturated vapour pressure extends to sufficiently large values of the pore diameter. Use of molecules having a larger diameter than the conventionally used nitrogen and argon gases is preferred because the accuracy of adsorption measurements away from values $P/P_0 \sim 1$ increases.

As a generalization of previous results, Fig. 25.3 shows the values of saturated vapour pressure over a wide range of mesopores of the

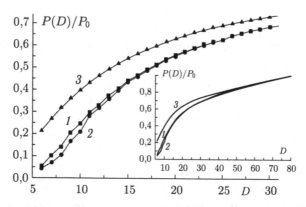

Fig. 25.3. Dependence of pressure on the diameter corresponding to the complete filling of the pores, calculated without taking into account (*1*) and taking into account (*2*) the calibration function, as well as the Kelvin equation (*3*) at $T = 87.3$ K. The inset shows a wide range of diameters of cylindrical pores,

MCM-41 type, calculated in the QCA (*1*), calibrated QCA (*2*) and the Kelvin equation (*3*) [73].

The section of the curve *2* for $D < 13$ is constructed on the basis of the position of the maximum change in pressure as the density jump itself is missing (see Fig. 24.4). For channel diameters up to 15 monolayers the differences between the exact calculations and without the calibration function taken into account equal up to 10 %. This is due to the fact that the value of the saturated vapour pressure is a sufficiently rough characteristic which is not very sensitive to the molecular properties of the system. This allows the use of such calculations to estimate the volume filling of the pores. The value of the pore volume in a first approximation obeys the Gurvich law due to the low compressibility of the liquid phase. All other features of the filling process of the pores are reflected much more noticeably on the adsorption isotherms so their description requires the inclusion of calibration functions.

The curve *3*, calculated from the Kelvin equation, is very different from the molecular calculations in the range of pore diameters most important for porosimetry tasks.

In the tasks of adsorption porosimetry the absence of the equation of state is not crucial as we are interested in the filled pore volume, which is fixed by the number of absorbed substance (e.g., weight change) for a given external pressure. As shown in section 24, in the conditions of complete filling of the pores the mechanism of filling (presence or absence of condensation) plays a secondary role. The Kelvin equation in the construction of the distribution function of the pore size in the range of large size leads to overestimated values of the pore diameter for argon and nitrogen compared with the pore sizes determined by the calculation of adsorption isotherms using strict theoretical models that take into account explicitly the influence of the pore walls. With increasing temperature, the agreement between the two methods of assessing the size improves. At the same time, the difference between the methods is high as 1%, even for pores up to 50 nm wide, which exceeds the accuracy of the experimental measurements, so in adsorption porosimetry tasks (section 7) it is inappropriate to use the Kelvin equation even for such wide pores..

26. The vapour–liquid interface in the pores

The transition region of the liquid–vapour interface in a slit-like pore is a complex nonuniform system, in which the density of matter θ_q^m changes both along the width of the pores as well as along the transition area between the vapour and the liquid, consisting of L sections of pores [77, 78], i.e., the molecular distribution has an extended form: $1 \leq q \leq H$, $1 \leq m \leq L$. It is believed that each section of the pore has a thickness equal to one monolayer.

The procedure for calculating the distribution of the molecules at the vapour–liquid interface will be demonstrate on an example of the bulk phase. The length of the transition region L in this case represents the maximum number of monolayers at which the local density values change monotonically from $\theta_m^{(V)}$ in the vapour phase to $\theta_m^{(L)}$ corresponding to the liquid phase, for all m, $1 \leq m \leq L$.

Figure 26.1 shows the results of calculation of the transition region for the bulk phase and the influence of the number of monolayers L on the free energy F. Figure 26.1 a shows the density θ change in the transition from the vapour to the liquid phase. The abscissa gives the number of the layer (in the volume it is simply the number of the site) of the transition region. The vertical axis gives the density θ_m at site m, $1 \leq m \leq L$, $L = 12$, with $T = 120$ K. The points on the lines reflect the discrete nature of the transition region consisting of L sites. Figure 26.1 b shows the values of the free energy of the transition region F, calculated according to the formulas in [79] (Appendix 2) for values of L, obtained during the iterative determination of the length of the transition region. Figure 26.1 c shows the behavior of the increment $\Delta F = F(L) - F(L - 1)$

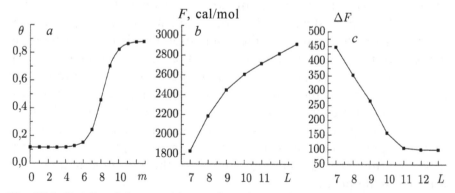

Fig. 26.1. Details of the transition region of argon in the volume at $T = 120$ K: a – density profile θ_m of the vapour–liquid interface in the volume for $L = 12$, the abscissa gives m – the number of the layer in the transition region, b – the free energy $F(L)$ as a function of the length of the transition region L; c – increment ΔF of the free energy as a function of the length of the transition region L.

with the increase of L. Because of discrete changes of L have $\Delta L = 1$ and increment ΔF is an analogue of the derivative dF/dL.

The curve in Fig. 26.1 c shows that for the values $L \geq 11$ the derivative of the free energy along the length of the transition region is essentially constant. Further increase of the length of the region leads to a certain limiting value of the specific free energy of the system and the concentration profile is no longer dependent on the length of the transition region, which determines the size of the transition region at a predetermined temperature.

The initial part of θ_m for $m < 5$ on the profile of Fig. 26.1 a remains almost constant. The real density difference between the liquid and the vapour is realized over the length κ, equal to only 5–6 monolayers, i.e. $\kappa \sim L/2$. Further increase of L (in volume, and even more so in the pore) can lead to non-monotonic profiles of local densities in the transition region. This is due to the possibility of formation of the structures of the fluid having a more developed surface between the vapour and the liquid than the monotonic structure of the transition region at the vapour–liquid interface.

The same principle of calculating the width of the transition region is also used within the pores of a given size. The following are estimates of the transition region with argon adsorption in the pores with a strong ($Q_1 = 5.0\varepsilon$) and weak ($Q_1 = 0.50\varepsilon$) attraction of the walls corresponding silica gel and non-wetted polymers, respectively. Recall that the pore assumed to be symmetric, i.e., the attraction of the walls is the same.

Figure 26.2 shows the profiles of the density of matter for two temperatures in a slit pore with the width of 20 monolayers in the case of strong adsorption. These profiles define the marginal distribution of molecules in isolated vapour and liquid phases, i.e., in single-phase states of matter $\theta_m^{(V)}$ in the vapour phase and $\theta_m^{(L)}$ in the liquid phase. Between the two profiles (1 and 3) and (2 and 4)

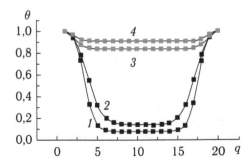

Fig. 26.2. Dependence of the local density of argon θ_q for stratified vapour (curves 1 and 2) and liquid phases (3, 4) when $Q_1 = 5.0\varepsilon$ for temperature $T = 112$ K (1, 4) and 126 K (2, 3) on the number of the monolayer in a slit pore q, $H = 20$.

there are transition regions θ_q^m forming two-dimensional non-uniform distributions of the molecules.

Figure 26.3 shows the concentration distributions of the molecules in the pores of the variable width $H = 10$ (a), 30 (b) and 60 (c) in

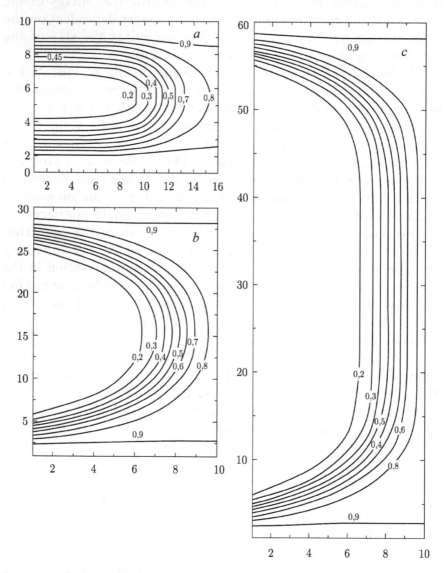

Fig. 26.3. Form of the meniscus of the liquid–vapour in slit pores of varying width $H = 10$ (a), 30 (b) and 60 (c) in the case of strong adsorption $Q_1 = 5.0\varepsilon$. Along the abscissa – the number of the layer of the transition region m, the vertical axis – number of the monolayer in the transition region of the liquid–vapour interface q.

the case of strong adsorption. These distributions are in the form of isoline curves of constant density, which allows to visualize the shape of the transition region. The axis of ordinates gives the numbers of monolayers in a slit pore, measured from one of the walls, and the abscissa – the number of the sections of the liquid–vapour transition region. The values of the local densities of the molecules in the transition region on the density isolines are marked with numbers.

Increasing the width of the pore increases its volume located outside the area of the action of the surface potential. In the case of strong adsorption (Q_1 = 5.0ε) the meniscus shape in Figs. 26.3 a and c and is very different from the shape of the semicircle. By increasing the width of the pore the meniscus shape in the central portion of the pore becomes flatter. The meniscus at the transition region also becomes flatter. This conclusion is in complete contradiction with the traditional form of the meniscus in mesoporous systems [9–19]. Of interest is also the dependence of the length L of the transition region on the width of the pores at small H: L (H = 10) = 16 and L ($H \geq 30$) = 11. Strong attraction of the walls stabilizes the system and increases the length of the transition region. The distance in the central portion of the meniscus between the isolines with the maximum and minimum density for wide pores tends to the length of the interface in the bulk phase. We have κ = 6–7 (a), 5–6 (b) and 5 (c).

Similar curves for weak adsorption (Q_1 = 0.5ε) are shown in Fig. 26.4. In this case, the isoline curves correspond to substantially smaller local density values than in the case of strong adsorption. There is no increase of the concentration of the adsorbate on the pore walls. The maximum density of the adsorbate is typical for the central part of the pores. The isolines corresponding to density values intermediate between the coexisting concentrations of the vapour and the liquid pass from one pore wall to the other. The course of the isolines is much more flattened in comparison with the same isolines in Fig. 26.3. This behaviour is qualitatively different from the traditional image of the meniscus for both the wetted and non-wetted surfaces at the macrolevel. The length L of the transition region is relatively weakly dependent on the width of the pores for small H: L (H = 10) = 13, and L ($H \geq 30$) = 11. With larger pores the meniscus becomes flat and characteristic of the bulk phase, regardless of the nature of the pore walls, and the value of κ, as in the volume, is close to 6 (a) and 5–6 (b, c).

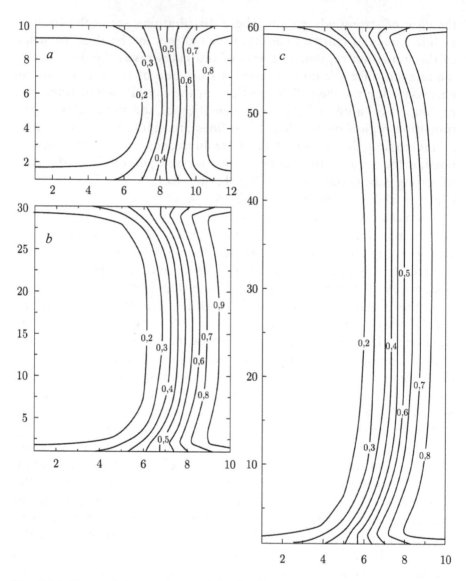

Fig. 26.4. Form of the meniscus of the liquid–vapour interface in slit pores of varying width $H = 10$ (a), 30 (b) and 60 (c) in the case of weak adsorption $Q_1 = 0.5\varepsilon$. The axes as shown in Fig. 26.3

Thus, the analysis of the shape and curvature of the meniscus, made on the basis of information about the molecular distribution of the adsorbate in the multilayer transition (distributed) region between the dense and diluted phases, excludes the shape of the meniscus made in thermodynamic constructions. It was found that with increasing pore size the length of the transition region become close to the length of the vapour–liquid interface in the bulk. This trend does not depend on the type of wall surface potential, the central portion of the meniscus is flattened and the maximum curvature is shifted from the centre of the narrow pores to both near-surface regions of the wide pores. The main difference density between the coexisting phases is the critical temperature region is localized within a fairly narrow field of $\sim \kappa/2$. With increasing temperature the length of the transition region is growing rapidly.

The nature of the molecular distribution near the walls of the pores depends on the potential of the adsorbate–adsorbent interaction. The attraction of the wall forms a meniscus which at certain correlations between the length of the transition region L and slit width H can be qualitatively approximated by a semicircle. This relationship between L and H depends on the temperature. The density isolines begin and end in the vapour phase. Weak attraction that corresponds to poorly wettable walls forms two types of density isolines. Some of them begin and end in the vapour phase others begin and end in the region of the liquid phase, and between them there are isolines connecting the wall pores. The adsorbate density near the walls is lower than the density in the centre of the liquid phase.

For the wide pores the relationship between a plane interface in the centre of the pore and the area of curvature of the meniscus is determined by the length of the adsorbate–adsorbent interaction potential R_h. In these figures we analyzed the case of a short-range potential of the Lennard–Jones type, so the results do not allow us to estimate the extent of this influence in detail. In the case of strong adsorption the increase of R_h reduces the magnitude of the size of the transition region L, and with weak absorption – increases for the relatively narrow pores.

The calculation results qualitatively demonstrate that the maximum curvature of the meniscus for the large pores is associated with the domain of wall wetting. Increasing the width of the pores does not influence the maximum curvature of the meniscus when the central area of the pore is at a distance much greater than the radius of the wall potential. On the contrary, with the decrease of the width of the pore for the sizes of the order of twice the radius of the potential of the wall, the curvature of the meniscus is maximized. With the decrease of the width of the pore for the sizes less than twice the radius of the potential of the wall the curvature decreases again, but in the case of such small sizes it is difficult to speak of the meniscus as such.

The change of the sphere-like (cylindrical geometry) forms of the meniscus at the vapour–liquid interface contradicts completely the traditional views, according to which the meniscus in mesoporous systems is a semi-circle in the cross-section for the slit-shaped pores [27, 75, 80–87]. These intuitions were laid the foundation of all estimates of the so-called disjoining pressure [83], using the Kelvin equation. The fundamental error in the construction of the shape of the meniscus is that maximum curvature is postulated at the centre of the pore. Furthermore, these constructions were repeated [88] and were also used in the calculation of molecular distributions by the density functional method [89–91].

The form of the meniscus is important also for models describing the molecular transport of molecules in the pores, as the equilibrium distribution of this section are often used as initial conditions in the molecular theory of transport. Further details on the calculation of the molecular distribution at the vapour–liquid interface in the slit-shaped and cylindrical pores, including the problem of the definition of the surface tension in porous systems, were considered in [92, 93].

References

1.   Landau L.D., Lifshitz E.M., Theoretical Physics. V. 5. Statistical Physics. – Moscow: Nauka, 1964. – P. 316.
2.   Krichevsky I.P. Phase equilibria in solutions at high pressures. – Leningrad: Goskhimizdat, 1952.
3.   Wilson K., Kogut G., The renormalization group and the ε-decomposition. – Moscow: Mir, 1975. – 256 p. [North-Holland Publ. Comp., Phys. Reports 12 C No. 2. pp. 75 – 199, 1974].
4.   Patashinskii A.Z., Pokrovsky V.L., Fluctuation theory of phase transitions. – Moscow: Nauka, 1982.
5.   De Gennes, Scaling Concepts in Polymer Physics. – Moscow: Mir, 1982. – 368 p. [Cornell Univer. Press, Ithaca – London, 1979].
6.   Anisimov M.A., Critical Phenomena in Liquids and Liquid Crystals. – Moscow: Nauka, 1987. – 271 p.
7.   Gibbs G.V. Thermodynamics. Statistical mechanics. – Moscow: Nauka, 1982.
8.   Tovbin Yu.K. // Zh. Fiz. Khimii. 2008. V. 82. P. 1805. [Russ. J. Phys. Chem. A. 2008. V. 82. № 10. P. 1611]
9.   Matecki M., Thomy A., Duval X. // Surface Sci. 1978. V. 75. P. 142.
10.   Plachenov T.G., Kolosentsev S.D., Porometry. – L.: Khimiya, 1988.
11.   Tovbin Yu.K., Theory of physical and chemical processes on the gas – solid interface. – Moscow: Nauka, 1990. – 288 p. [CRC Press Boca Raton, FL, 1991.]
12.   Bazarov I.P., Thermodynamics. – Moscow: Moscow State University Press, 1991. – 376 p.
13.   Guggenheim E. A. // J. Chem. Phys. 1945. V. 13. P. 253.
14.   Hirschfelder J. O., Curtiss C. F., Bird R. B., Molecular Theory of Gases and Liquids.

– Moscow: Izd. inostr. lit., 1961. – 929 p. [Wiley, New York, 1954].

15. Prigogine I.R., Molecular theory of solutions. – Moscow: Metallurgiya, 1990. – 360 p. [Interscience, New York, 1957].

16. Stanley H. E., Introduction to Phase transitions and critical phenomena. – Moscow: Mir, 1973. – P. 42. [Clarendon, Oxford, 1971].

17. Hill T.L., Statistical Mechanics. Principles and Selected Applications. – N. Y.: McGraw-Hill Book Comp. Inc., 1956.

18. Rivkin S.L., Thermodynamic properties of gases. Handbook. – Moscow: Energoatomizdat, 1987. – 312 p.

19. Laher K. // Surface Sci. 1978. V. 154. P. 218.

20. Burgess C.G.V., Everett D.H., Nuttall S. // Pure Appl. Chem. 1989. V. 61. P. 1845.

21. Thommes M., Findenegg G.H. // Langmuir. 1994. V. 10. P. 4270.

22. Tovbin Yu.K. // Zh. Fiz. Khimii. 1990. V. 64, No. 4. P. 865. [Russ. J. Phys. Chem. 1990. V. 64. № 4. P.461].

23. Tovbin Yu.K. // Dokl. AN SSSR. 1981. V. 260, No. 3. P. 679.

24. Langmuir I. // J. Amer. Chem. Soc. 1918. V. 40. P. 1361.

25. Ross S., Olivier J.P., On Physical Adsorption. – N. Y.: Wiley Interscience, 1964.

26. Avgul' N.N., Kiselev A.V., Poshkus D.P., Adsorption of gases and vapours on homogeneous surfaces. – Moscow: Khimiya, 1975.

27. Gregg, S.J. Sing, K.G.W., Adsorption, Surface Area and Porosity. – Moscow: Mir, 1984. [Academic Press, London, 1982].

28. Tovbin Yu.K., Votyakov E.V. // Physics of Low-Dimensional Structures. 1995. V. 10/11. P. 105.

29. Hill T.L. // J. Chem. Phys. 1949. V. 17. P. 762.

30. Ma Shang-keng, Modern theory of critical phenomena. – Moscow: Mir, 1980. – 298 p. [Benjamin, Inc., London, 1976].

31. Votyakov E.V., Tovbin Yu.K. // Zh. Fiz. Khimii. 1993. V. 67. P. 391. [Russ. J. Phys. Chem., 1993 V. 67, No. 1, P. 351].

32. Gvozdev V.V., Tovbin Yu.K. // Izv. AN. Ser. khim. 1997. No. 6. P. 1109. [Russ. Chem. Bul.. 1997. V. 46. № 6. P. 1060].

33. Ubbelohde A.R., The molten state of matter. – Moscow, Metallurgiya, 1982. – 513 p. [Wiley, New York 1978].

34. Tovbin Yu.K. // Izv. AN. Ser. khim. 2003. No. 4. P. 827. [Russ. Chem. Bull. 2003. V. 52. № 4. P. 869].

35. Tovbin Yu.K. // Zh. Fiz. Khimii. 2003. V. 77, No. 10. P. 1839. [Russ. J. Phys. Chem. 2003. V. 77. № 10. P. 1839].

36. Nakanishi H., Fisher M.E. // J. Chem. Phys. 1983. V 78, No. 6. P. 3279.

37. Tarasona P., Marconi U.M.B., Evans R. // Mol. Phys. 1987. V. 60. P. 573.

38. Vishnyakov A., Piotrovskaya E.M., Brodskaya E.N., Votyakov E.V., Tovbin Yu.K. // Langmuir. 2001. V. 17. P. 4451.

39. Votyakov E,V,, Tovbin Yu.K. // Zh. Fiz. Khimii. 1994. V. 68, No. 2. P. 287 [Russ. J. Phys. Chem., V. 68, No. 2, P. 254].

40. Tovbin Yu.K., Votyakov E.V. // Langmuir. 1993. V. 9, No. 10. P. 2652.

41. Tovbin Yu.K., Votyakov E.V. // Zh. Fiz. Khimii. 1993. V. 67, No. 10. P. 2126 [Russ. J. Phys. Chem., 1993, V. 67, No. 10, P. 1918].

42. Steele W.A., The Interactions of Gases with Solid Surfaces. – N. Y.: Pergamon, 1974.

43. Tovbin Yu.K., Komarov V.N., Vasyutkin N.F. // Zh. Fiz. Khimii. 1999. V. 73, No. 3. P. 500. [Russ. J. Phys. Chem., 1999 V. 73, No. 3, P. 427].

44. Tovbin Yu.K., Votyakov E.V. // Izv. RAN. Ser. khim. 2001. No. 1. P. 48 [Russ.

Chem. Bull. 2001. V. 50. № 1. P. 50].
45. Vishnyaakov A.M., Piotrovskaya E.M., Brody E.N., Votyakov E.V., Tovbin Yu.K. // Zh. Fiz. Khimii. 2000. V. 74, No. 2. P. 221. [Russ. J. Phys. Chem., 2000. T. 74. № 2. C. 162].
46. Panagiotopoulos A.Z. // Mol. Phys. 1987. V. 62. P. 701.
47. Panagiotopulos A.Z. // Mol. Phys. 1987. V. 61. P. 813.
48. Valleau J.P. // J. Chem. Phys. 1998. V. 108. P. 2962.
49. Forsman S., Woodward C.E. // Mol. Phys. 1997. V. 90. P. 637.
50. Tovbin Yu.K. // Izv. AN. Ser. khim. 1998. No. 4. P. 659. [Russ. Chem. Bull. 1999. V. 48. № 8. P. 1450].
51. Tovbin Yu.K., Votyakov E.V. // Zh. Fiz. Khimii. 1998. V.72, No. 10. P. 1885 [Russ. J. Phys. Chem. 1998. V. 72. № 10. P. 1715].
52. Tovbin Yu.K., Eremich D.V. // Izv. AN. Ser. khim. 2003. No. 11. P. 2208. [Russ. Chem. Bull. 2003. V. 52. № 11. P. 2334].
53. Tovbin Yu.K., Senyavin M.M., Zhidkova L.K. // Zh. Fiz. Khimii. 1999. V. 73. No. 2. P. 304. [Russ. J. Phys. Chem. 1999. V. 73. № 2. P. 245].
54. Votyakov E.V., Tovbin Yu.K., MacElroy J.M.D., Roche A. // Langmuir. 1999. V. 15. P. 5713.
55. Tovbin Yu.K., Rabinovich A.B., Votyakov E.B. // Izv. AN. Ser. Khim. 2002. No. 9. P. 1531 [Russ. Chem. Bull. 2002. V. 51. № 9. P. 1667].
56. Evans R. // J. Phys.: Condens. Matter. 1990. V. 46. P. 8989.
57. Tovbin Yu.K., Petukhov A.G., Eremich D.V. // Izv. AN. Ser. Khim. 2007. No. 5. P. 813 [Russ. Chem. Bull. 2007. V. 56. № 5. P. 845].
58. Beck J. S., et al. // J. Am. Chem. Soc. 1992. V. 114. P. 10834.
59. Morishige K., Fujii M., Uga M., Kinukama D. // Langmuir. 1997. V. 13. P. 3494.
60. Morishide K., Shikimi M. // J. Chem. Phys. 1998. V. 108. P. 7821.
61. Kruk M., Jaroniec M. // Chem. Mater. 2000. V. 12. P. 222.
62. Neimark A.V., Ravilovich P. I., Vishnyakov A. // Phys. Rev. E. 2000. V. 62A. P. R1493.
63. Vishnyakov A., Neimark A.V. // J. Phys. Chem. B. 2001. V. 105. P. 7009.
64. Nguyen-Thi Minh Hien, Serpinskiy VV // Izv. AN. Ser. Khim. 1987. No. 11. P. 2421.
65. Tovbin Yu.K. // Izv. AN. Ser. Khim., 2004. No. 12. C. 2763 [Russ. Chem. Bull. 2004. V. 53. № 12. P. 2884].
66. Montroll E., Stability and phase transitions. – Moscow: Mir, 1973. – P. 92. [Statistical, Phys. Phase Transitions and Superfluidity, ed. M. Chertiln et al., Gordon and Dreach Sci. Publishers, Vol. 1 , 2, New York, 1968].
67. Tovbin Yu.K., Petukhov A.G., Eremich D.V., // Zh. Fiz. Khimii. 2006. V. 80. No. 3. P. 488. [Russ. J. Phys. Chem. 2006. V. 80. № 3. P. 406].
68. Tovbin Yu.K., Petukhov A.G., Eremich D.V. // Zh. Fiz. Khimii. 2006. V. 80. No. 12. P. 2250. [Russ. J. Phys. Chem. 2006. V. 80. № 12. P. 2250].
69. Tovbin Yu.K. // Zh. Fiz. Khimii. 1998. V. 72. No. 5. P. 775. [Russ. J. Phys. Chem. , 1998 V. 72, No. 5, P. 675].
70. Tovbin Yu.K., Rabinovich A.B. // Langmuir., 2004. V. 20. No. 12. P. 6041.
71. Tovbin Yu.K., Rabinovich A.B., Votyakov E.B. // Izv. AN. Ser. Khim. 2002. No. 9. P. 1531. [Russ. Chem. Bull. 2002. V. 51. № 9. P. 1667].
72. Tovbin Yu.K. // Zh. Fiz. Khimii. 2008. V. 82. No. 10. P. 1805. [Russ. J. Phys. Chem. A. 2008. T. 82. № 10. C. 1611].
73. Tovbin Yu.K. Petukhov A.G. // Zh. Fiz. Khimii. 2007. V. 81. No. 8. P. 1527. [Russ. J. Phys. Chem. A. 2007. T. 81. № 8. C. 1349].
74. Tovbin Yu.K., Petukhov A.G. // Izv. AN. Ser. Khim. 2008. No. 1. P. 18. [Russ. Chem.

Bull. 2008. V. 57. № 1. P. 18].

75. Plachenov T.G., Kolosentsev S.D., Porometry. – Leningrad: Khimiya, 1988. – 175 p.

76. Tovbin Yu.K. // Fizikokhimiya poverkhnosti i zashchita materialov. 2010. V. 46. No. 2. P. 165. [Protection of Metals. 2010. T. 46. № 2. C. 197].

77. Tovbin Yu.K., Rabinovich A.B. // Dokl. AN. Ser. Fiz. Khim. 2008. V. 42, No. 1. P. 59. [Doklady Physical Chemistry. 2008. T. 422. № 1. C. 234].

78. Tovbin Yu.K., Rabinovich A.B. // Izv. AN. Ser. Khim. 2008. No. 6. P. 1118. [Russ. Chem. Bull. 2008. V. 57. № 6. P. 1138].

79. Tovbin Yu.K. // Zh. Fiz. Khimii. 1992. V. 66, No. 5. P. 1395. [Russ. J. Phys. Chem., 1992 V. 66, No. 5, P. 1309].

80. Everett D.H. // The Solid-Gas Interface, Ed. by E. A. Hood. – N. Y.: Dekker, 1967. – 1055 p.

81. Karnaukhov A.P., Adsorption. Texture of dispersed porous materials. – Novosibirdk: Nauka. IK SO RAN, 1999. – 469 p.

82. Experimental methods in molecular adsorption and chromatography / Ed. A.V. Kiselev, V.P. Dreving. – Moscow: Moscow State University Press, 1973. – 447 p.

83. Deryagin B.V. // Zh. Fiz. Khimii. 1940. V. 14. P. 157.

84. Timofeev D.P., Kinetics of adsorption. – Moscow: Publishing House of the USSR Academy of Sciences, 1962. – 252 p.

85. Carman P.C. Flow of gases through porous media. – London, 1952.

86. Ruthven D.M. Principles of Adsorption and Adsorption Processes. – N. Y.: John Wiley, 1984.

87. Lykov A.V., Heat and Mass Transfer. – Moscow: Energiya, 1978. – 480 p.

88. Deryagin B.V., Churaev N.V., Muller V.M., Surface Forces. – Moscow: Nauka, 1985. – 400 p.

89. Tarasona P., Marconi U.M.B., Evans R. // Mol. Phys. 1987. V. 60. P. 573.

90. Vruno E., Marconi U.M.B., Evans R. // Physica A. 1987. V. 141. P. 187.

91. Evans R., Marconi U.M.B., Tarasona P. // J. Chem. Phys. 1986. V. 84. P. 2376.

92. Tovbin Yu.K., Petrova T.V. // Zh. Fiz. Khimii. 1995. V. 69, No. 1. P. 127. [Russ. J. Phys. Chem. 1995. V. 69. № 1. P. 114].

93. Tovbin Yu.K., Eremich D.V., Komarov V.N., Gvozdeva E.E. // Khim. Fizika. 2007. V. 26, No. 9. P. 98 [Russ. J. Phys. Chem. B, 2007, V. 1 No. 4. P. 88].

4

Transport equations

27. Kinetic equations in the condensed phase

Transport of molecules through porous materials is carried out by a variety of flow mechanisms (section 3). Except for the Knudsen vapour flow, dense vapors and/or liquid phase take part in the discussed mechanisms. In dense phases the major role is played by particle interactions, and the contribution of the translational motion is small. In these conditions, the kinetic processes of migration of particles in space are described by the master equation for the total distribution function (the so-called Glauber-type equation [1, 2]). Movement of molecules in space takes place through elementary particle jumps to the neighboring free sites (vacancies), – the stage of migration in multi-step description of physical and chemical processes. The thermal velocity of migration of the molecules is calculated using the transition state model, which treats displacements as the activation process of overcoming a barrier created by the neighbouring particles. This model was proposed for gases by Eyring [3, 4], and later transferred to the condensed phases in [5–9].

Currently, the kinetic theory in atomic and molecular level in the lattice-gas models (LGM) [8–10] can be used in almost the entire time range, from the characteristic times of atomic vibrations to macroscopic, including equilibration times. The theory considers the full set of elementary processes of movements of molecules and their chemical reactions occurring in the system on the set of lattice sites. To construct the general structure of the kinetic equations of the lattice model, we assume that the lattice sites are not equivalent to each other. The nature of inhomogeneity of the lattice sites is considered known and constant over time. From a physical standpoint the nonuniform distribution of particles is due to both the interaction between the particles of the system and to the possible additional influence of external fields or interactions (e.g. substrate potential field) (section 7 and Appendix 2). The formulation of the problem under consideration allows to cover from a unified point of view a wide range of issues related to the nonuniform distribution of particles at the gas–solid interface: the spatial distribution of the particles on a uniform surface, the presence of ordering in them, change of the distribution of atoms of the solid adsorbed particles along the normal to the interface, the distribution of particles laterally interacting on the nonuniform surfaces and, correspondingly, in any three-dimensional volume of the porous body.

We denote by $\left\{\gamma_f^i\right\} = \gamma_1^i, \gamma_2^j, ..., \gamma_N^n$ the complete set (or full list) of values γ_f^i of all lattice sites, which uniquely determine the complete configuration of the locations of the particles on the lattice at the time τ, and by $P\!\left(\left\{\gamma_f^i\right\}, \tau\right)$ – the probability of finding the system at this time in a state $\{\gamma_f^i\}$. For the sake of brevity, this state is denoted as $\{I\} \equiv \{\gamma_f^i\}$. Let the common studied process consists of many stages and through α we denote the number of elementary stages in the process. The master equation for the evolution of the full distribution function of the system in a state $\{I\}$, due to the implementation of the elementary processes α in condensed phases has the form [1, 2, 7, 8]

$$\frac{d}{d\tau}P(\{I\}, \tau) = \sum_{\alpha, \{II\}} \left[W_\alpha\left(\{II\} \to \{I\}\right) P\left(\{II\}, \tau\right) - W_\alpha\left(\{I\} \to \{II\}\right) P\left(\{I\}, \tau\right) \right],$$

$$(27.1)$$

where $W_\alpha\left(\{I\} \to \{II\}\right)$ is the probability of the elementary process α (the probability of transition via channel α), which resulted at time τ in the transfer of the system from the initial state $\{I\}$ to the final state $\{II\}$. In equation (27.1) the sum is taken over the different types of direct processes (index α) and all reversed processes $\{II\}$, in the state of occupation of each site in the system changes.

If the elementary process occurs at one site, the lists of states of occupation of the sites of the system $\{I\}$ and $\{II\}$ differ only for this site. Single-site processes are processes associated with changes in the internal degrees of freedom of the particle, the adsorption and desorption of non-dissociating molecules, and with the reaction by the impact mechanism. If the elementary process occurs in two neighboring lattice sites, then lists of the states $\{I\}$ and $\{II\}$ differ in the conditions of occupation of these two sites. Two-site processes are exchange reactions, adsorption and desorption of the dissociating molecules, migration processes by the vacancy and exchange mechanisms, etc. The sum of the states {II} corresponds to the change of states of occupation for all lattice sites. The interconnection of the states {I} and {II} depends on the mechanism of the process that defines a set of elementary stages α.

Equation (27.1) is written in the Markov approximation for which it is assumed that the relaxation processes of the internal degrees of freedom of all particles are faster than the process of changes of the state of occupation of different sites of the lattice system.

The transition probabilities W_α subject to the condition of detailed balance

$$W_\alpha\left(\{I\} \to \{II\}\right)\exp\left(-\beta H\left(\{I\}\right)\right) = W_\alpha\left(\{II\} \to \{I\}\right)\exp\left(-\beta H\left(\{II\}\right)\right), \quad (27.2)$$

where $H(\{I\})$ – the total energy of the lattice system in the state {I}, as described in Appendix 2. In equilibrium, $P\left(\{\gamma_f^i\}, \tau \to \infty\right) = \exp\left(-\beta H\left(\{\gamma_f^i\}\right)\right)\big/Q$, where Q – the statistical sum of the system (Appendices 1–5).

Expressions for $W_\alpha(\{I\} \to \{II\})$ are constructed with all the molecular features of the system taken into account: 1) each site f is characterized by a certain set of the number of particles s_f, which can be in it; 2) internal degrees of freedom F_j^i particles of some sort i depend on the number of the lattice site; 3) the interaction parameters $\varepsilon_{fg}^{ij}(r)$ of the particles i and j, that are in the sites with the numbers f and g at a distance r from each other, depend on the numbers of the sites.

The large dimension of the system (1) does not allow to use it to study the dynamics of macroscopic systems by direct integration, so the kinetic equations are based on the functions of distributions of a lower order through which the distribution functions of high order are closed. To this end, instead of the full distribution function $P(\{\gamma_f^i\},\tau)$, the evolution of the system is described using a shortened way of defining it by time distribution function (correlators) determined by

$$\theta_{f_1\ldots f_m}^{i_1\ldots i_m}(\tau) = \left\langle \gamma_{f_1}^{i_1}\ldots\gamma_{f_m}^{i_m}\right\rangle = \sum_{i_1=1}^{s}\cdots\sum_{i_N=1}^{s}\prod_{n=1}^{m}\gamma_{f_n}^{i_n}P\left(\{\gamma_f^i\},\tau\right), \quad (27.3)$$

where the angle brackets denote the averaging procedure over all lattice sites from $f = 1$ to $f = N$ and for each site f the sum over the occupation states i_f is taken as all sorts of particles from 1 to s.

Correlators (27.3) characterize the probability of a cluster consisting of m particles in which at the time τ the particle of type i_n is the site with the number f_n; $1 \leq n \leq m$; m – the dimension of the correlator. These correlation functions with arbitrary values of m reflect on the atomic and molecular level any configuration of particles differing in the conditions of occupation of the sites of the lattice structure.

The entered local time functions (27.3) imply that at every moment of time averaging is performed on the *full ensemble of copies* of the considered non-uniform lattice system in all its implemented states. This definition is similar to the definition of the means of the Gibbs equilibrium statistical theory and the non-equilibrium theory of gases and liquids. The difference in comparison with the non- equilibrium theory of gases and liquids is that the contemplated local inhomogeneities are implemented in the microscopic atomic scale (instead of small elementary volumes containing macroscopic quantities of matter in the theory of gases and liquids).

Correlators (27.3) of different dimensions are linked by relations

$$\theta^{i_1 \cdots i_m}_{f_1 \cdots f_m}(\tau) = \sum_{i_{m+1}=1}^{s} \theta^{i_1 \cdots i_m i_{m+1}}_{f_1 \cdots f_m f_{m+1}}(\tau), \quad \sum_{i=1}^{s} \theta^{i}_{f}(\tau) = 1, \tag{27.4}$$

Kinetic equations for the correlators is derived by multiplying expression (27.1) by $\prod_{n=1}^{m} \gamma^{i_n}_{f_n}$ and by averaging over all states of the system. Types of particles in the sites f_n, leading to a non-zero contribution to the change of the correlators (3.27) and depend on the direction of the elementary process and are identified by the initial {I} or {II} final states. This leads to kinetic equations of the form

$$\frac{d}{d\tau}\theta^{i_1 \cdots i_m}_{f_1 \cdots f_m}(\tau) = \sum_{\alpha,\{\text{II}\}} \left[\left\langle \prod_{n=1}^{m} \gamma^{i_n^*}_{f_n} W_\alpha(\{\text{II}\} \to \{\text{I}\}) \right\rangle - \left\langle \prod_{n-1}^{m} \gamma^{i_n}_{f_n} W_\alpha(\{\text{I}\} \to \{\text{II}\}) \right\rangle \right], \tag{27.5}$$

where the particle of species i_n^* in the first term on the right-hand side correspond to states of occupation of the lattice sites in the state {II}. Equations (27.5) are initial equations for the kinetic equations of processes in condensed phases.

Figure 5.1 shows the configuration of neighbouring molecules in the lattices with $z = 4$ and 6, which influence through the lateral contributions the rates of adsorption–desorption stages and reactions on a 'central' site. Similar configurations around the two neighbouring molecules on a square lattice are shown in Fig. 27.1 for

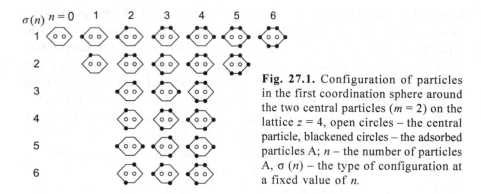

Fig. 27.1. Configuration of particles in the first coordination sphere around the two central particles ($m = 2$) on the lattice $z = 4$, open circles – the central particle, blackened circles – the adsorbed particles A; n – the number of particles A, $\sigma(n)$ – the type of configuration at a fixed value of n.

the elementary processes of adsorption–desorption with dissociation and migration of molecules in a neighboring site, which are realized on the two 'central' sites. The diagrams 5.1 and 27.1 reflect the central point of the theory of elementary processes [5–9]: all the surrounding neighbouring molecules at the same time affect the rates of these processes at the central sites ($m = 1$ or 2) through the probability of formation of the activated complex (AC) of the stage.

The construction of the kinetic equation for the local concentration of molecules (correlators (27.3) of the first order, $m = 1$) results in a large increase in the order of the unknown correlation functions on the right side of the equation. This dimension is equal to the number of all neighbouring molecules affecting the rate of this stage. Recall that in the kinetic chain of BBGKY (Bogoliubov–Born–Green–Kirkwood–Yvon) the dimension of the correlators in the right part of the kinetic equations sequentially increases on unit. By this structure the kinetic equation of the LGM is different from the traditional form of the kinetic chains BBGKY, and, respectively, in the kinetic theory there is a problem of calculating the correlators of high dimension. In the discrete version of the theory on the basis of the LGM the system of kinetic equations has a high dimension, so in practice, it is necessary to limit the minimal dimension preserving the effects of correlations between the molecules.

The need to consider the higher order correlation functions is due to the fact that the interaction of the AC with the surrounding particles not only quantitatively changes the dynamic characteristics of the transients, but can qualitatively change the evolution of the system, such as increasing the number of stationary states of the system [12]. In addition, it is shown [13] that the form of the transition probabilities affects the calculated values of the critical dynamic index for the correlation length, which is contrary to the dynamic universality hypothesis [2, 14]. Sequential extension of the theory of absolute reaction to the theory of kinetic equations for the arbitrary nature of the mobility of the particles can be found in [7–11] in considering the processes on smooth and rough surfaces.

The closed system of equations for the first $\left(\theta_f^i = \langle \gamma_f^i \rangle\right)$ and second $\left(\theta_{fg}^{ij}(r) = \langle \gamma_f^i \gamma_g^j \rangle\right)$ correlation functions in the general form can be written as

$$\frac{d}{dt}\theta_f^i = I_f^i = \sum_\alpha \left[U_f^b(\alpha) - U_f^i(\alpha)\right] + \sum_r \sum_h \sum_j \sum_\alpha \left[U_{fh}^{bd}(r|\alpha) - U_{fh}^{ij}(r|\alpha)\right],$$

(27.6)

$$\frac{d}{dt}\theta_{fg}^{ij}(r) = I_{fg}^{ij}(r) = \sum_\alpha \left[U_{fg}^{bd}(r|\alpha) - U_{fg}^{ij}(r|\alpha)\right] + P_{fg}^{ij}(r) + P_{gf}^{ji}(r),$$

(27.7)

$$P_{fg}^{ij}(r) = \sum_\alpha \left[U_{fg}^{(b)j}(r|\alpha) - U_{fg}^{(i)j}(r|\alpha)\right] + \sum_h \sum_m \sum_\alpha \left[U_{hfg}^{(cb)j}(r|\alpha) - U_{hfg}^{(mi)j}(r|\alpha)\right],$$

where $U_f^i(\alpha)$ – the speed of the elementary single site processes $i \leftrightarrow b$ (here $h \in z_f$), $U_{fg}^{ij}(r|\alpha)$ – the speed of the elementary two-site processes $i + j_\alpha \leftrightarrow b + d_\alpha$ ($h \in z(r)$) on the sites at a distance r; the second term in $P_{fg}^{ij}(r)$ describes the stage $i + m \leftrightarrow b + c$ on neighbouring sites f and h at a distance r. All the rates of the elementary stages $U_f^i(\alpha)$ and $U_{fg}^{ij}(r|\alpha)$ are calculated in the framework of the theory of absolute reaction rates for non-ideal reaction systems written in the quasi chemical approximation of the interparticle interaction.

For the equations (27.6), (27.7) we have the normalizing ratios (27.4), which are executed at any time.

The important question of the division of the dynamical variables, describing the state of the molecules, into fast and slow is solved depending on the characteristic time scale defined by the values of W_α ($\{I\} \rightarrow \{II\}$). The slow variables are described by the kinetic equations, and the fast variables are described by algebraic equations, so they refer to particles having an equilibrium distribution. In general, the fast particles (described by the fast variables) form an adapted subsystem that has an impact on the energy of slow elementary processes in the kinetic equations. In turn, the slow particles (described by the slow variables) determine the distribution of the fast particles. They form a spatial region in which the fast processes take place (excluding from the total area the spaces occupied by the slow particles), and have a potential impact on the distribution of the fast particles.

Usually, the slow variables include the spatial coordinates of molecules or quantum numbers related to their electronic terms, and the fast variable – the quantum numbers related to the rotational and vibrational states of molecules [15]. The latter does not rule out the possibility of considering as slow variables the population of the vibration states of the molecules in the gas phase. On the other hand, the spatial movement of the molecules can be fast. For example, at a small activation energy compared with the desorption energy the surface migration is a fast rapid stage compared to the desorption stage.

In this chapter, we confine ourselves to the process of the transfer of simple molecules of the spherical shape, the index α refers only to the stage of the migration of molecules. In this approach, the elementary processes of adsorption, desorption and surface migration will automatically be included in the general dynamic model. The desorption stage is regarded as 'migration' of the molecule from the surface layer on the wall of the pore inside the pore, or rather in the second monolayer, with breaking of the adsorption bond without association. The adsorption stage is a reverse migration process of the molecule toward the centre of the pore wall without dissociation. Surface migration is the migration (displacement) of the molecules along the surface of the pore wall with overcoming the corresponding activation barrier. Finally, the redistribution of the molecules within the pore is their migration between the different layers (normal to the wall), or along the layers (parallel to the wall of the pores).

In equilibrium, the distribution of particles in the quasichemical approximation is described by algebraic equations, described in detail in chapter 2 and Appendix 2. The fundamental question is self-consistent description of the equilibrium distributions of the molecules and the rates of elementary reactions in the forward and reverse directions in the quasichemical approximation. This issue is discussed in Appendix 6.

In discussing the general properties of the kinetic equations in the LGM we should point out the correlation between γ_f^i and η_f^q, entered in section 6. The quantities η_f^q characterize the energetics of the region of localization of adsorbed molecules (probability of filling them is characterized by the magnitude of γ_f^i). Formation of the given adsorption centre connected with formation of the surface and this the process can also be described by a system of kinetic equations (27.6) and (27.7), if the components of the system are the atoms or molecules from which the surface is constructed. Here we are concerned with the reactive system with a different (larger) number of components than in the task of transferring molecules in a given porous body. The summary approach in the LGM reflects both the non-equilibrium distribution of the particles on the surface and rearrangement of the surface itself, if this happens during adsorption.

These models describe a change of the state of the surface region of adsorbents and catalysts, in which the influence of the adsorption of the molecules of the gas phase changes the distribution of the atoms in a solid due to their migration by the vacancy mechanism. The case of solid–gas phase planar boundary is considered in [16]. Study [17] gives a generalization to the case of a rough surface, and work [7] describes a generalization of the theory of processes with limited mobility of particles for a pair potential with an arbitrary radius of interaction. On the basis of these studies it is possible get a hierarchy of kinetic of equations, which differ in the scale of the spatio–temporal averages. It allows to simulate the time evolution of the functions f_q, f_{qp} and $f(q\{p\})$, describing the surface composition and the structure of the non-uniform surface, which are discussed in Appendix 2. It is important that all levels of the hierarchical structure of the equations are built on the same principle, and consistently go over into each other on the relevant scales of time and have the a single set of interaction parameters and the rate constants of elementary processes.

28. Rates of elementary stages of adsorption and diffusion

Fundamentally important for the construction of probabilities $W_\alpha(\{I\} \rightarrow \{II\})$ is the use of models that describe the elementary process in the framework of the theory of absolute reaction rates (TARR) or the transitional state theory. TARR was used for the first time for functions W_α in equation (27.1) for non-ideal systems [5, 18, 19]. The rates of elementary movements will be described in the framework of this theory.

Eyring theory for elementary stages
Recall the basic idea of the TARR, consisting in the fact that there is a relationship between the concentration θ^* (number of activated

complexes per unit volume) and the rate of the elementary process $U = \theta^* v$, where v is the frequency of crossing the activation barrier (s^{-1}). For translational motion [25] $v = u_t / b$, where u_t is the average speed of the AC of mass m^*, which is equal to $u_t = (kT/2\pi m^*)^{1/2}$, b is the length of the activation barrier.

The concentration of AC on top of the barrier can be written as follows: $\theta^* = \theta F_t$, where θ is the concentration of AC, after replacing one vibrational degree of freedom by translational movement, $F_t = (2\pi m^* kT/)^{1/2} b/h$; h is Planck's constant. This implies that the rate of translational motion is $U = \theta kT/h$. Or by entering the usual specific rate of the elementary process $K_i = U/\theta_i$ for the movement of the activated complex, formed from the particle i, we have that $K_i = \theta kT/(h\theta_i)$ [25].

The ratio $\theta/\theta_i = F^*/F_i$ is expressed [11, 12, 24, 25] by the sum of the states of AC (F^*) and the molecule in the ground state (F_i), which leads to the standard form for the rate constants of the elementary movement $K_i = kTF^*/(F_i h)$ for the gas phase.

As a result, the movement type is specified only in the expressions for the sums of the states of the AC and the molecule in the ground state. For the rarefied phase $F^*/F_i = F_t$. For non-ideal reaction systems this ratio changes and the effect of neighbouring molecules must also be taken into account.

The speed of the reaction in the transitional state theory is calculated as [3, 4]

$$U_i = \kappa w \theta_i^* / b, \qquad (28.1)$$

where κ is the transmission factor, w is the average rate of passing through the barrier with length b, $w = (2\pi M^* \beta)^{-1/2}$, M^* is the reduced mass of the complex, U_i is the speed related to one site of the structure. The problem is reduced to calculating the concentration of AC θ_i^* for the monomolecular process and θ_{ij}^* for the bimolecular process.

For ideal reaction systems (in the absence of the influence of lateral interactions) the speed of the elementary stages on the mono- and bimolecular reactions are described in the framework of the law of acting masses:

$$U_i = K_i \theta_i, \quad U_{ij} = K_{ij}\theta_i\theta_j. \qquad (28.2)$$

where K_i and K_{ij} are the rate constants of the elementary processes (stages) that characterize the specific rates of elementary processes:

$$K_i = K_i^0 \exp\left(-\frac{E_i}{k_B T}\right) = \kappa \frac{kT}{h} \frac{F_i^*}{F_i} \exp\left(-\frac{E_i}{kT}\right),$$

$$K_{ij} = K_{ij}^0 \exp\left(-\frac{E_{ij}}{k_B T}\right) = \kappa \frac{kT}{h} \frac{F_{ij*}}{F_i F_j} \exp\left(-\frac{E_{ij}}{kT}\right), \qquad (28.3)$$

where K_i^0 and K_{ij}^0 are pre-exponential factors of the rate constants, E_i and E_{ij} are the activation energies of the reaction $i \rightarrow$ *product* and $i + j \rightarrow$ *products*; the transmission coefficient κ in most cases can be taken to be 1; F_i and F_j are the statistical sums (sums over the internal states) of the parent molecules, F_i^* and F_{ij}^* are the statistical sums of the AC, calculated over all degrees of freedom but 'the reaction path'.

Here it is assumed that the surface area does not change during the reaction and the concentration of particles may be characterized as 'the degree of filling' the surface θ_i. Equations (28.1)–(28.3) suggest that the equilibrium distribution of the molecules forms in the reaction system, and that this limits the stage of chemical transformation. It is also assumed: 1) the absence of diffusion transport at the macroscopic level (uniform distribution in the macrovolume), 2) the absence of the effect of external fields, 3) the absence of a diffusion controlled reaction regime at the molecular level, 4) the lack of influence of intermolecular interactions, and 5) the proportion of particles, reacting in unit time, is so small that it does not distort the equilibrium distribution of molecules on the surface.

According to TARR [4], the intermolecular interactions in the non-ideal reaction systems are taken into account by thermodynamic relations, known from the theory of non-ideal solutions [20]. This approach involves retaining in the expression for the reaction rate the concentration factor used in the form of acting masses (through the product of the concentrations of the reactants) and in the formula (28.2), and with the change of the rate constants in the form

$$K_i(ef) = K_i^0 \frac{\alpha_i}{\alpha_i^*} \exp(-E_i/k_B T) = K_i \alpha_i / \alpha_i^*,$$

$$K_{ij}(ef) = K_{ij}^0 \frac{\alpha_i \alpha_j}{\alpha_{ij}^*} \exp(-E_{ij}/k_B T) = K_{ij} \alpha_i \alpha_j / \alpha_{ij}^*, \qquad (28.4)$$

where α_i is the activity coefficient of molecules of type i, α_i^*

and α_{ij}^* are the activity coefficients of the activated complex. The activity coefficient is defined as the ratio of $\alpha_i = a_i/n_i$, here a_i is the activity of component i in the solution, depending on the concentrations of all components of the solution (subject to determination of the total concentration as $n = \sum n_i$) and all i their molecular properties, including the intermolecular interaction energies. As in [4,20], the chemical potential of an ideal system is written as follows: $\mu_i = \mu_i^0 + k_B T \ln(n_i)$, where μ_i^0 is the chemical potential of the standard state. For non-ideal systems $\mu_i = \mu_i^0 + k_B T \ln(a_i) = \mu_i^0 + k_B T \ln(\alpha_i n_i) = \mu_i^{id} + k_B T \ln(\alpha_i)$.

However, the use of thermodynamic approaches in the problems of the kinetics for non-ideal systems leads to contradictions. This question is below when compared with the TARR assumptions with the results of the molecular theory based on the lattice gas model (LGM) [8, 9], which is applicable to all aggregate states of matter and the interfaces.

Molecular theory of non-ideal reaction systems
In the molecular theory it is essential to sequentially examine the entire spectrum of the configurations of the neighbouring molecules which may affect the reaction in the considered sites (as in Fig.27.1) and weigh the probability of implementation of the elementary stage for each of the configurations of the neighbors on the surface. Lateral contributions affect the probability of formation of the AC through the change of the activation energy, so the number of neighbours and their arrangement are important for the rate of the process (see [8, 9]).

As a result of this averaging, the following expressions can be written for the mono- and bimolecular elementary processes:

$$U_i = K_i \theta_i S_i^z = K_i \theta_i \Lambda_i^*, \quad S_i = \sum_{k=1}^{s} t_{ik} \exp(\beta \delta \varepsilon_{ik}), \qquad (28.5)$$

$$U_{ij} = K_{ij} \exp(-\beta \varepsilon_{ij}) \theta_{ij} (S_i S_j)^{z-1}, \qquad (28.6)$$

where $\delta \varepsilon_{ik} = \varepsilon_{ik}^* - \varepsilon_{ik}$, , and ε_{ik}^* is the interaction parameter between the activated complex (particles i located at a site f in the transition state) of the elementary process with the neighbouring particle k on the next site g; where $t_{ij} = \theta_{ij}/\theta_i$ is the conditional probability of finding the particle j close to the particle i, calculated in the quasi–chemical approximation. In the derivation of expressions (28.5) and

(28.6) it was assumed that the rate constants K_i and K_{ij} are weakly dependent on the density.

In function S_i summation is carried out over all species of the neighbouring particles. The functions S_i are the cofactors of the non-ideality function of the reaction system $\Lambda_i^* = S_i^z$ which reflect the influence of each neighbouring particle on the height of the activation barrier of the reaction. In the absence of intermolecular interaction the cofactors $S_i = 1$ and the formula (28.5) change to the rate of the reaction for an ideal reaction system (28.2). Equations (28.5) indicate that the mutual influence of the reactants in the case of non-ideal systems complicates their cooperative behaviour (unlike traditional chemical kinetics), and this behaviour changes the 'sensitivity' of the change of the rate on the concentration and temperature. Both of these factors simultaneously influence each other, so the concentration dependences become as decisive as the temperature ones.

A similar situation occurs in the case of bimolecular reaction (28.6). The structure of the functions S_i is not related to the type of potential functions of the lateral interaction and the radius of the potential, it is determined by using the quasi–chemical approximation for taking the interactions into account. The number of functions S_i, included in the rate of the reaction in the form of the cofactors, determined by the size of the coordination sphere z. For bimolecular reactions the number of neighbours around the two reactants, equal to $(z - 1)$, is practically doubled. The difference between the expression (28.6) and the formula (28.5) is that the effect of intermolecular interactions is manifested not only through the non-ideality function, but also a change in the probability of meeting of the reagents described by the function θ_{ij}. If the length of the interparticle potential exceeds the nearest coordination sphere, the contributions of the following coordination spheres are included in the expression for U_i and U_{ij} in the form of additional cofactors of the functions S_i, belonging to different distances [8, 9]. Moreover, the larger the radius of the intermolecular interaction potential, the longer it takes to reach the equilibrium distribution of the molecules within such a local region around the reactants.

In deriving the expressions for the non-ideality functions it is important to consider the effects of correlation between the interacting molecules. If we consider at least correlations between the nearest neighbours, this ensures the self-consistency of the description of the reaction rate and the equilibrium distribution of

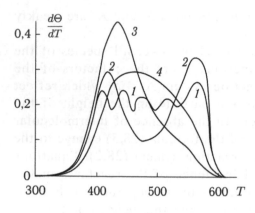

Fig. 28.1. Influence of taking into account the effect of correlation on the form of TDC calculated in polynomial (*1*), the quasi-chemical (*2*), random (*3*) and the mean-field (*4*) approximations on a uniform surface [21].

the components of the reaction mixture (see [6, 8, 9] and appendix 6). If there are no correlation effects, such self-consistency does not form and the model parameters,, determined from the equilibrium and kinetic measurements, do not coincide with each other. The presence of correlation effects also leads to significant differences in the concentration dependences of the rates. As an illustrative example, Fig. 28.1 shows the so-called thermodesorption curves (TDC), which are obtained by linear heating of the surface by the law $T = T_0 + bt$, where b – heating rate (deg/s), T_0 – initial temperature, for which it is assumed that at the initial time the initial degree of surface coverage θ_0 is given. Here we solve the differential equation $d\theta/dt = -U_A$, where U_A is the desorption rate, at the given initial population $\theta(t = 0) = \theta_0$.

Heating the surface accelerates the desorption process (see equation (28.3)), and as the substance is exhausted the desorption rate decreases. The calculations are performed at the same values of all model parameters. The differences relate only to the method of calculating the non-ideality function.

The vertical axis gives the amount of substance that is desorbed at any given time. Methods of decoupling of the correlation functions of the high order through the lower-order correlators are shown in Fig. 5.2. The random and mean-field approximations do not take into account the correlation, while the quasi-chemical and polynomial approximation retain the correlation effects. If this is not so, both curves have one peak, while taking the correlations into account leads to splitting of the curve to several peaks.

Incorrect use of the activity coefficients in the kinetics [22, 23]
The molecular theory of reaction rates for non-ideal systems is used

to evaluate the degree of correctness of the use of activity coefficients in the kinetics. For this purpose, we compare monomolecular desorption rates for two variants of considering the relaxation of the medium in the non-ideal reaction system. In the first case, the calculation is made using the formula (28.5), i.e. in the absence of an equilibrium adjustment of the medium for the desorption process. In the second case, the calculation is carried out using the same formula with the full equilibrium distribution of the neighbouring molecules around the desorbed particles (which is necessary for the use of the concept of the chemical potential for the AC of the reagent [4]). In this formulation of the problem the process of reorganization of the neighbouring molecules is reduced to a change in local concentrations due to the thermal mobility of molecules. According to expressions [22, 23], in the second case there will be a different non-ideality function S_i, which is expressed as

$$S_i^* = \exp \sum_{j=1}^{s} \left\{ t_{ij} \ln \left[t_{ij} \exp\left(-\beta\varepsilon_{ij}\right) \right] - t_{ij}^* \ln \left[t_{ij}^* \exp\left(-\beta\varepsilon_{ij}^*\right) \right] \right\},$$

$$t_{ij}^* = t_{ij} \exp(\beta\delta\varepsilon_{ij}) / \sum_{k=1}^{s} t_{ik} \exp(\beta\delta\varepsilon_{ik}),$$

(28.7)

where the energy difference $\delta\varepsilon_{ij}$ was defined above, i.e. both models have the same set of energy parameters and the desorption rate constant.

The effect of the relaxation of the medium on the difference in the rates of monomolecular processes is indicated by calculations of thermal desorption curves carried out by the formulas (28.5) and (28.7). Figures 28.2 a–c show families of thermal desorption curves (TDC) obtained at values θ_0 = 0.99, 0.7, 0.5 and 0.3. All calculations were performed at one and the same value of the pre-exponential factor $K_A^0 = 10^{12}\,\text{s}^{-1}$ and the activation energy E_A = 23 kcal/mol, which qualitatively corresponds to the desorption process in the CO–Pt system [24]. The parameter of the lateral interaction ε_{AA} is –2 kcal/mol, and $\varepsilon_{AA}^* = \varepsilon_{AA}/2$. The minus sign corresponds to the repulsion of the nearest neighbours, which is also corresponds to the system [24]. Calculation of the fields in 28.2 a and 28.2 c were carried out using equation (28.5) and on the field 28.2 b – according to equation (28.7). The molecular parameters for the curves on the field in Fig. 28.2 c corresponds qualitatively (so as not to change the relationship between the selected molecular parameters) to the desorption process in the Hg–W system [24]: between the adsorbed atoms there is the attraction of the nearest neighbours, ε_{AA} = 2 kcal/mol (the plus sign)

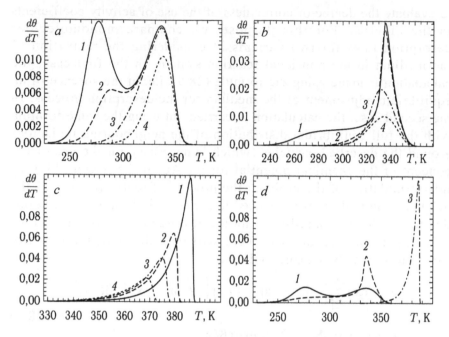

Fig. 28.2. Thermal desorption curves for fast (*a*, *c*) and slow (*b*) monomolecular desorption at initial filling $\theta_0 = 0.99$ (*1*), 0.7 (*2*) 0.5 (*3*) and 0.3 (*4*). The calculation parameters are specified in the text. In the field *d* maximum peaks of three curves with initial filling of $\theta_0 = 0.99$ and carried from the field *a* (*1*), *b* (*2*) and *c* (*3*) are compared. The abscissa – the temperature in Kelvin on the ordinate $d\theta/dT$.

and also $\varepsilon_{AA}^{*} = \varepsilon_{AA}/2$. The calculation was performed for the desorption process on a square lattice ($z = 4$).

In repulsion of the chemisorbed particles TDCs are split into two peaks when the degree of filling the surface is close to the monolayer and there is a second peak when $\theta_0 > 0.7$ (28.2, *a*). For smaller degrees of filling ($\theta_0 < 0.7$) the repulsion of chemisorbed particles does not cause splitting. For the same system the curves in the field 28.2, *b* show the absence of splitting for any initial surface filling θ_0. Instead, there is a single peak, the height of which is significantly superior to both peaks on the field 28.2, *a* (comparison of three TDCs at $\theta_0 = 0.99$ given on Fig. 28.2, *d*), i.e. the existence of a significant repulsion between the adsorbed particles does not lead to necessary kinetic effects, which contradicts the experimental data [25]. For comparison, the field 28.2, *c* shows TDCs for a system with strong attraction between the adsorbed particles. It also has a single peak, which is slightly shifted to higher temperatures with increasing θ_0.

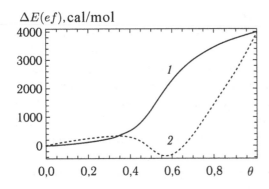

$\Delta E(ef)$, cal/mol

Fig. 28.3. Concentration dependence of E_A (*ef*) for monomolecular desorption at $T = 300$ K, calculated in the case of the fast (*1*) and slow (*2*) elementary stage.

Figure 28.3 illustrates the discussed qualitative difference the concentration dependences of reaction rates for different environment relaxation under isothermal conditions (curves correspond to the system of type CO-Pt). The effective activation energy of desorption E_A(*ef*), defined by the ratio $U_A/\theta_A K_A^0 = \exp(-\beta E_A$ (*ef*)), monotonically increases with the degree of filling, if the calculation is carried out using equation (28.5) – there is no relaxation of the medium, and the velocity changes monotonically, if the calculation is conducted using (28.7) – equilibrium relaxation of the medium takes place. The non-monotonic behaviour of the desorption rate contradicts the physics of the process, especially given the fact that in the near distance the chemisorbed particles repel each other, and, in general, in non-ideal systems, the non-monotonic behaviour of the reaction rate is only possible in the presence of an attractive potential [8–11].

The comparison shows the strength of the influence of the effect of the relaxation of the medium on the rate of monomolecular process. Therefore, the validity of the use of a procedure of averaging the contribution of neighbours plays a fundamental role in the dynamics of the elementary stages. Choosing a monomolecular reaction as an example for detailed consideration does not limit the generality of the findings. This conclusion is fully transferred to a) the reaction mixture with any number of components; b) any type of interaction potentials between molecules, and c) the rate of bimolecular reactions (for more details see [22]). The procedure for statistical averaging over the neighbouring molecules does not depend on the specific form of the potential. It is determined only by the ratio between the characteristic times of occurrence of the elementary reaction act and the local rearrangement of the molecular environment.

Thermal velocity of the molecules

If i is the molecule A and j – vacancy V, then the expression (28.6) describes the probability of molecule A jumping from one site to the next one. In the LGM instead of the average velocity of the particles w we use the concept of particle hopping probability per unit time $U(\chi)$ between the selected site occupied by a particle and a free site at a distance χ (or the rate of hopping of particles per unit time). From the dimension matching condition it follows that $w = \chi U(\chi)/\theta_A$ and the value of w is the thermal velocity of the particles in the equilibrium state of the fluid. (The expression for the thermal velocity of the molecules in the general case is discussed in Appendix 7). We present here, for clarity, the simplest version of this expression for a rigid lattice, taking into account the influence of the lateral interaction of nearest neighbours in the quasichemical approximation on the probability of moving particles:

$$U(\chi) = K_\chi V_\chi, \quad K_\chi = F_\chi^* \exp(-\beta E)/(\beta h F), \quad V_\chi = \theta_{AV}(\chi) T_\chi,$$

$$\theta_{AV}(\chi) = \theta_{AV}(1)\left[t_{VV}(1)\right]^{\chi-1}, \quad T_\chi = \left(S_A\right)^{z-1}\left(S_V\right)^b,$$

$$S_A = 1 + tx^*, \quad S_V = 1 + ty^*, \tag{28.8}$$

where K_χ is the rate constant of hopping particles to a free site, $V\chi$ is the concentration dependence of the hopping velocity, it is described by the cofactor $\theta_{AV}(\chi)$ – the probability of the existence of the free path with length χ; T_χ function takes into account the influence of particle–particle interactions on the hopping probability on the path with length χ, cofactor S_A refers to $(z - 1)$ nearest neighbours of the particle in the initial state, and S_V relates to the neighbors of free sites of the path, including the final free site $b = (\chi - 1)(z - 2) + (z - 1)$. For single-site migrating particles the vacancy region consists of at least one neighbouring vacancy to which the molecule moves (Appendix 7). Movement over longer distances $\chi > 1$ requires a vacancy region for the trajectory of hopping of the molecule consisting of a linked sequence χ of vacancies (as described cofactor $[t_{VV}(1)]^{\chi-1}$ and multiplier $(S_V)^{(\chi-1)(z-2)}$ reflects the influence of neighbours around a given trajectory). In the formula (28.8) we use the notation $x^* = \exp[\beta(\epsilon^* - \epsilon)] - 1$, $y^* = \exp(\beta\epsilon^*) - 1$, where the energy parameter ϵ^* describes the interaction of the particle in the transition state with adjacent particles in the ground state.

The rate constant K_χ is expressed in terms of the theory of absolute reaction rates, F^*, and F are the statistical sums of the particles in the transition and ground states, E is the activation energy of hopping (for sites far away from the pore wall $E = 0$; close to the walls where their adsorption potential operates E is non-zero), h is Planck's constant. Away from the walls of the pores the ratio F/F^* is the translational degree of freedom in the direction of motion of the particle and is equal to $(2\pi m \beta^{-1})^{1/2}\chi/h$, then $K_\chi = \chi(2\pi m \beta)^{-1}$ or $K_p = \chi w/4$, where $w = (8/\pi m \beta)^{1/2}$ [26, 27]. In the absence of lateral interactions, formula (28.8) can be written in the form of $U(\chi) = K_\chi \theta (1 - \theta)^x$.

Diffusion

The speed of thermal motion is used to calculate different diffusion coefficients. The most commonly considered are the self-diffusion coefficients $D^*(\theta)$ (describing the flow of labelled particles in the equilibrium state of the system) and the diffusion mass transport $D(\theta)$ in non-equilibrium conditions. For a non-uniform lattice, which is discussed in Appendix 2, the expressions for these coefficients can be written as [28]

$$D^*(\theta) = \lambda^2 \sum_{q=1}^{t} \tilde{z}_q f_q \sum_{p=1}^{t} d_{qp} V_{qp} / \theta_q, \qquad (28.9)$$

where $\tilde{z}_q$ is the number of sites in the direction of the flow of the label, V_{qp} is the concentration component of the hopping rate of the molecule between the sites of type q and p.

$$D(\theta) = \lambda^2 \sum_{q=1}^{t} \tilde{z}_q f_q \sum_{p=1}^{t} d_{qp} V_{qp} T_{qp}, \quad T_{qp} = \frac{d}{d\theta}\left[\frac{a_q \theta_q^V p}{\left(1 - t_{pq}^{AA}\right) S_{qp}} \right], \qquad (28.10)$$

which is obtained from the differentiation of the system of equations describing the equilibrium distribution of the molecules on the non-uniform surface area, the average density of θ.

We illustrate the expression (28.9) by calculating the dependence $D(\theta)$ in Fig. 28.4 ($t = 2$, one type of particle) on the crystal (100) face, taking into account the interaction of the first and second neighbours ($R = 2$) in the case of formation of an ordered structure such as $c(2 \times 2)$ (in terms of the theory of surface diffraction of slow electrons). The inhomogeneity of the local distribution of the particles is due to the cooperative properties of their interaction.

The ordering process leads to an abrupt change of the diffusion coefficient in the areas of the order–disorder transition and at

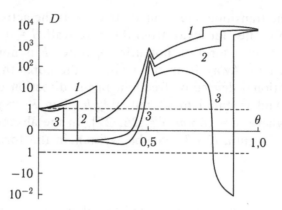

Fig. 28.4. The concentration dependences of the diffusion coefficient of interacting particles for $z_1 = z_2 = 4$ ($\beta\varepsilon_{AA}(1) = -3.0$, $\beta\varepsilon_{AA}^*(1) = 0.1$, $\beta\varepsilon_{AA}^*(2) = \beta\varepsilon_{AA}^*(2)/2$, $\beta\varepsilon_{AA}^*(2) = 0$ (*1*), 0.6 (*2*) 1.2 (*3*). On the ordinate at $|D| < 1$ the scale is linear, at $|D| > 1$ the scale is in decimal logarithms.

$\theta \sim 0.5$. At $R = 1$ and $\varepsilon_{AA}^*(1) = 0$, the curves coincide with the curves of [29]. At $R = 1$ outside the ordering range the curves are described by the formula [30], if $\varepsilon_{AA}^*(1) = 0$, and by formula [31] for arbitrary $\varepsilon_{AA}^*(1)$. Contributions of the second neighbours strongly change the form of $D(\theta)$. The attraction of the particles leads to negative values of D. The sign change is due to the fact that the flow direction is determined by the gradient of the chemical potential and not by the concentration gradient. The flow directed to the side with the lower free energy. For non-interacting particles the main contribution to the flow comes from the entropy factor, which leads to a spreading of the ensemble of particles, hence $D > 0$. In strong interaction the energy contribution can be controlling. When $\theta < 0.5$ the negative values of D are associated with the formation of a more stable ordered phase by increasing the concentration of particles and the flow is directed toward higher values of θ. The attraction of the particles can also cause their condensation [31]. In curve *3* in Fig. 28.4 this area corresponds to $\theta \sim 0.8$. More detailed information on the distribution of the particles can be obtained if the process is not described by the total value of the local density of θ, and the local distribution of low- and high-density phases are considered.

The expressions obtained for the local characteristics of particle transport in nonuniform systems describe the macroscopic flow in the homogeneous model: there are no macroscopic inhomogeneities in the system. If there are macroinhomogeneities, to calculate the flows it is necessary to separate homogeneous regions and complement the diffusion equation by appropriate boundary conditions.

As a second example we consider the diffusion of argon atoms on the amorphous surface of rutile [32]. The equilibrium properties of this system have been considered in Section. 12. It was shown above that the required size of the surface fragment corresponds to the value of the order of 15×15 diameters of argon. As a result, it was possible to construct a full potential relief of the fragment of an amorphous surface and on its basis to calculate all the values of A_q and Q_q for each quasi-single-site lattice model. Similarly, for the calculation of the diffusion coefficients of Ar all saddle points were determined from the potential relief of the amorphous surface of rutile, and analysis of paths of the atomic transition between adjacent sites was carried out to determined the activation energy of migration E_{qp}. The resulting concentration dependence of self-diffusion and mass transfer coefficients are shown in Fig. 28.5 (curves 3 and 7).

The curves shown in Fig. 28.5 for $\varepsilon = 0$ (curves 2 and 6) correspond qualitatively to similar dependences obtained for the random surface with two types of centres. The presence of the large number of types of centres on an amorphous surface makes these curves flatter. The concentration dependence of the self-diffusion coefficient $D^*(\theta)$ decreases with increasing surface filling θ due to the reduction of vacant sites. The diffusion coefficient increases with increasing filling of the amorphous surface, since the strong centres are first to be filled, and the hopping rate of particles to

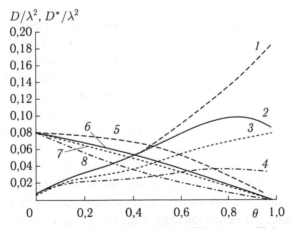

Fig. 28.5. The concentration dependence of the diffusion coefficients $D(\theta)$ (curves 1–4) and self-diffusion coefficients $D^*(\theta)$ (curves 5–8) of argon on the amorphous surface of rutile at $\beta\varepsilon = 0.4$ kJ/ mol ($\varepsilon^* = \alpha\varepsilon$), where $\alpha = 0.75$ (1, 5), 0.5 (3, 7) 0.25 (4, 8), and curves 2 and 6 are calculated for $\varepsilon = 0$.

free adjacent centres increases for weak centres. However, when the fraction of vacancies becomes very small ($\theta \sim 0.85$) and, accordingly, the probability of hopping decreases, the curve 2 shows a decrease of $D(\theta)$ when $\theta > 0.85$. Attraction of atoms prevents hopping of the particle to a free neighboring site (so curves 3, 4, 7, 8 are below the curves for $\varepsilon = 0$).

The resultant value $\varepsilon = 0.4$ kJ/mol [32] corresponds to the case of weak lateral interaction (this values is an order of magnitude smaller than the minimum binding energy of Ar with the substrate), so it is natural to expect that the real dependences $D(\theta)$ and $D^*(\theta)$ will not differ greatly from the curves for $\varepsilon = 0$. This condition corresponds to $\alpha = 0.5$. (It should be emphasized that this value α (or close thereto) describes most frequently the diffusion and reaction rates of many other non-ideal systems [9].) Variation of parameter α within a wide range from 0.25 to 0.75, provides a measure of the uncertainty of the predicted curves. For the concentration dependence of the self-diffusion coefficient we obtain a fairly narrow range of possible values. For the diffusion coefficient, this interval is much wider. Nevertheless, in the range of values for the filling of the surface θ, not exceeding 0.6–0.7, the calculated curve can serve as a relatively reasonable estimate. As a result of this theory it is possible to predict the concentration dependence of the diffusion coefficients for such complex systems as the amorphous solid surfaces, for which direct experimental evaluation of the characteristics of surface migration has been difficult up to date.

Commensurability of adsorption and migration rates
The kinetic equations of the LGM can be used to consider any relation between the characteristic times of migration atoms in a solid, adsorbed particles and chemical reactions [8]. In this case, the kinetic equations are used for all components that have comparable mobility. This applies not only to the unary distribution functions, but also to the pair distribution function, which reflect the evolution of changes in the state of neighbours.

As a simple example Fig. 28.6 shows the effect of particle mobility on the form of TDC for a nonuniform surface in the absence of lateral interactions (*a*) and for a homogeneous surface, taking into account the lateral interactions (*b*). Figure 28.6 *a* compares the curves in the case of stationary (curve *1*) and the equilibrium distribution adsorbed particles (curve *2*) [33] on the surface, consisting of two types of adsorption centres with a random distribution. Both

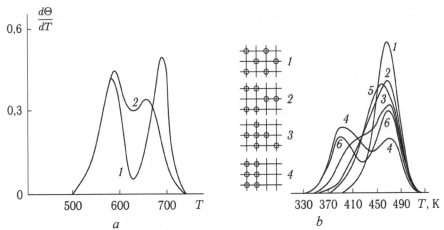

Fig. 28.6. TDC on an nonuniform surface (*a*) in the case of immobile (*1*) and equilibrium distributed (*2*) adsorbed particles. The calculation is performed for the following parameters: $f_1 = f_2 = 0.5$; $k_1^{D_0} = k_1^{D_0} = 10^{12}$ s^{-1}, $E_1 = 168$ kJ/mol, $E_2 = 126$ kJ/mol, the heating rate $b = 50$ deg/s [33]. TDC in the case of limited surface mobility of the particles (adsorption without dissociation) on the homogeneous surface (b), $k_D = 112$ s^{-1}, $E_D = 63$ kJ/mole, $\varepsilon = -8.4$, $\varepsilon^* = -4.2$ kJ/mol, $\theta_0 = 0.5$, $b = 50$ deg/s. Initial distribution: *1* – isolated particles $\theta_{AA}(0) = 0$, *2* – isolated pairs of particles $\theta_{AA}(0) = \theta_0/z$, *3* – random distribution $\theta_{AA}(0) = \theta_0^2$, *4* – macrodomains $\theta_{AA}(0) = \theta_0$ [34].

curves are very different from each other. In the absence of surface migration each type of adsorption site 'works' regardless, so there are sharply separated peaks. At high surface mobility of the particles they flow from the strongly bonded AC to weakly bonded AC. As a result of this constant replenishment of the weakly bonded AC the main flow from the surface goes through it, so the high-temperature peak shifts to lower temperatures and decreases in height. Under certain power relations and the properties of the inhomogenous surface (composition and structure) the form of the TDC can greatly vary [33].

Figure 28.6 *b* shows the TDC calculated in quasichemical approximation for different initial values of the pair functions $\theta_{AA}(0)$ [34]. The analysis used the equations of the function θ, describing the change in the overall extent of surface filling, and function θ_{AA}, which reflects the change in the number of pairs of adjacent particles in the desorption process. The slow mobility of the particles leads to new effects: TDC is split even at low initial filling, if the particles form domains. This fact describes the 'background' of the formation of the initial filling θ_0. Both functions θ_0 and $\theta_{AA}(0)$ have a significant impact on the dynamics of the process. Under similar conditions, the use of mean-field approximation (excluding the effects of correlation) leads to the degeneration of all four curves.

For comparison, curves *5* and *6*, corresponding to the initial equilibrium distribution and the case of macrodomains with the increase of the rate of surface mobility by 20 times as compared with the curve *4* (i.w. this increase in the mobility does not correspond to the equilibrium distribution).

These examples illustrate the real properties of nonuniform systems. In some systems, the surface mobility of the particles is comparable to the rate of chemical reactions, in particular their desorption. We consider the experimental data for the systems CO/Pd (111) [35] and O_2/Pt [36], wherein depending on the formation conditions of a fixed value of the initial filling the TDC has a different shape. In this case, it is necessary to move to models that take into account commensurability of surface migration rates (limited mobility) of the adsorbed particles and surface reconstruction. The oxidation of CO on platinum group metals has also become a textbook for the strong influence of surface reconstruction [37–39]. The effects of limited mobility of the reagents on the rate of bimolecular surface reactions were also considered in [40–42].

The existing methods for the study of the dynamics of atomic and molecular processes in the various phases are discussed in Appendix 8.

29. Microscopic hydrodynamics

Above we discussed the impact of channel width on the equilibrium distribution of the molecules. In particular, the scope of so-called narrow channels was determined, for which the dependence of the critical parameters of the characteristic size of the pores was determined. In these pores the potential of the walls influences the physical state of the fluid and, accordingly, the mechanism of its transport – all transport characteristics of the adsorbate differ from their values in the volume of vapour and liquid phases.

The importance of the role of the pore width for dynamic processes is indicated by the following assessment. From the kinetic theory of gases it follows [26, 27] that in the pores up to 10 nm wide, the frequency of collisions of the Ar atoms with the walls is approximately two orders of magnitude higher than the frequency of collisions between them in gas phase at $T = 300$ K and 1 atm:

v_{Ar} (pore) $\approx 10^2 v_{Ar}$ (gas). This value is close enough to the frequency of collisions in the dense fluid v_{Ar} (liquid) $\approx 10^3 v_{Ar}$ (gas) (referring to clashes of hard spheres, because the liquid molecules themselves are always in the field of action of the neighbours), that is, in narrow pores, even for the rarefied gas the collision frequencies are close to the frequencies for dense phases. Accordingly, in the narrow pores the equilibrium is established faster as a result of various collisions of the molecules both with the pore wall and with each other. From this standpoint, the rarefied gas in narrow pores is sufficiently close to the condensed phase. This fact is used in the construction of unified transport equations for a fluid of arbitrary density at the molecular level within the framework of the lattice gas model (LGM) [43–46].

To describe the flow in real porous materials we need an unified set of equations for the vapour and the liquid in the pores with nonuniform walls which can not be provided by the kinetic equations for gas and liquid phases (Appendix 9; There are also formulated other problems that do not allow the use of these theories in micro-nonuniform porous solids). The kinetic equations for gas and liquid are based on such mutually exclusive concepts as a function of the velocity distribution of the gas and the distribution function of the molecules in the coordinate space for the liquid. To resolve this contradiction we require special matching of the kinetic theories of gas and liquid. We are talking about the construction of the kinetic theory which will work throughout the range of densities. For this it is necessary to use the LGM which provides a description of both gas and liquid. As an alternative, we propose a new approach to describe the transport of molecules in the pores, which allows one to bypass these fundamental difficulties [43–46].

It is also useful to discuss the theory of the microscopic level, when the minimum considered volume of the system is equal to the volume of the molecule. It is necessary to describe the development of balance flows of mass, momentum and energy at the microlevel with the same precision of detailing the pore space as for the equilibrium distributions that provide high accuracy of calculations of thermodynamic characteristics. Only in this case can we consider the impact of all the microscopic irregularities of the exposed surfaces and pore walls, and the phase boundaries, on the flow characteristics.

It is important for the new approach to retain the number of dynamic variables, i.e. the final equations must be the equations of the hydrodynamic type. These requirements also define the concept of 'microscopic' hydrodynamics.

As a basis for the development of the theory we used the concept of the thermal velocity of the molecules, which is used in all states of aggregation, ranging from a solid in which there are vacancies (low concentration), and for the gas in which the vacancies constitute the bulk of the volume. For the intermediate liquid phase we introduce the concept of vacancy areas through which molecules can move with the thermal velocity. A physical prerequisite of the concept of vacancies in the fluid is the reduced density in comparison with the solid. This rarefaction is of the fluctuation nature in the average volume of the system, and its value can be compared with the volume of the molecule which is sufficient to shift the molecule. This view is suitable for any density and molecules of any size, as well as their rotational movement.

The concept of the thermal velocity of the molecules is a traditional 'gas' point of view in the kinetic theory. For dense phases this concept is retained when reducing the length of the displacement of the molecule, suggesting the specificity of each trajectory (its length and direction). In the calculation of the thermal velocity we use the probabilistic description of what is natural from the kinetic theory of liquids. The thermal velocity is introduced through the hopping probability $W_a = W_{fg}^{AV}(\chi)$ over the considered distance χ. The meaning of the hopping probability $W_{fg}^{AV}(\chi)$ is the average of the continuous process of movement of the molecule A (located at the initial time τ_0 in the note f) to the free cell g, which the particle reaches at time τ. The average is taken over all its paths, taking into account the lateral interaction with all neighbors and collisions with them. The length of the jump by the vacancy mechanism depends on the density: $\chi \gg 1$ is typical for the vapour, $\chi \sim 1$ – for the liquid. The calculation of the hopping probability W_a is traditional in the LGM model, which is discussed in section 28 and Appendix 7.

Note the analogy between the hopping probabilities W_a and the probabilities of changes of molecular states in the theory of fluids $W_{\Delta p}$ (Appendix 9). Both probabilities use the concept according to which the local thermalization of the system is achieved quickly enough and in the description of the spatial distribution of molecules we can use the Maxwell distribution function of their velocities. In both approaches, we use the following representation for the many-particle distribution functions:

$$\theta_{(s)}\left(\left\{i, g, \mathbf{r}_g^i, \mathbf{v}_g^i\right\}, t\right) = \theta_{(s)}\left(\left\{i, g, r_g^i\right\}, t\right) \prod_{g \in (s)} f_g^0(v), \qquad (29.1)$$

where $f_g^0(v)$ is the Maxwell velocity distribution function of the molecules in the cell g, the set of sites g corresponds to the order s of the considered distribution function; $\theta_{(s)}(\{i, g, \mathbf{r}_g^i\}, t)$ – the distribution function of the order (s) averaged over the velocities of all the molecules. In the LGM equations all the many-particle correlators of any order (s) are closed through the paired correlation functions $\theta_{(2)}$.

The difference between W_a and $W_{\Delta p}$ in the LGM and Smoluchowski equations is associated with a different way of defining the movement of the molecules and the difference in the physical sense. Probability $W_{\Delta p}$ refers to the process of exchange of momentum between the molecule and its environment. The averaged description of this process enables to introduce the friction coefficient characterizing the force acting on the moving particle. In the LGM the transition probability W_a describes the thermal velocity during movement of the molecule. This difference leads to the fact that the probability W_a may be expressed explicitly in terms of the local characteristics of the material and potential of the inter-molecular interactions, and the friction coefficient is not directly related with this potential.

The exponential form of the expressions for W_a (27.2) is a consequence of the thermalization of the molecule with the thermostat. The thermostat for any molecule is represented by the atoms of the solid of the walls of the pores and the neighbouring molecule. The thermostat is stationary in the absence of the flow of molecules and moves if the system contains a convective flow of molecules (the walls of the pores are always considered fixed). In the initial assumptions of the LGM it is always assumed that the thermostat is fixed and there are only processes of the redistribution of the molecules inside the thermostat.

On the example of the surface monolayer we can explain the differences between these two concepts if we consider the lattice structure of the surface adsorption sites with variable height of the activation barrier for hopping from one centre to another. If the barrier is high, then the jump of each particle is determined by the temperature and the probability that the molecule attains sufficient energy to overcome the barrier. This energy comes from the substrate which represents the thermostat. All the other molecules of the adsorbed monolayer are in a similar situation. Their mutual influence is manifested through the lateral interactions that change the height of the barrier, but the individual molecules move all the time, staying in different centers. After the jump of the molecule its excess energy is given back to the thermostat (in the case of the diffusion mechanism of any flow).

We decrease the height of the barrier for hopping. In the limit, it can be reduced so that it will formally determine only the position of the molecule. In this case, the state of the molecule is mainly determined by its lateral interactions with the neighbours, and their relative position plays a major role in the state of the adsorbed film. The thermostat for a randomly selected molecule is a substrate (substrate potential was common to all the molecules) and the surrounding molecules. The energy to move is received by the molecule from both the substrate and from the neighboring molecules, exactly as the molecule transfers excess to both the substrate and the neighbours. In this case, the any organization of the flow will be accompanied by a total redistribution of molecules in which the isolated molecule and its neighbors are equally involved in the collective motion (so the thermostat is considered to be moving).

Averaging the thermal velocities of the molecules over the cell volume (i.e. the forward and backward directions along the axes $\alpha = x, y, z$), which gives the resultant flow, allows to introduce the new feature at the microlevel – microhydrodynamic velocity $\mathbf{u}_f$, which is similar to the macroscopic flow velocity. The average value of the microhydrodynamic velocity is determined as

$$\mathbf{u}(f) \equiv \mathbf{u}_A(f) = \int \mathbf{v}_f^A \theta_f^A \left(\mathbf{v}_f^A \right) d\mathbf{v}_f^A / \theta_f^A. \qquad (29.2)$$

Components $u_{f\alpha}$ describes the average velocity of the molecules in the cell f in the direction α at a characteristic length scale, equal to the diameter of the molecule. It is thus possible to take into account the smallest flows (average displacement of molecules) in the convective flow of a substance that appears in the continuity equation. Accordingly, to describe it is necessary to enter all the traditional transfer coefficients.

If we formally considered in a fixed thermostat forward and backward jumps of the molecules in the form of an algebraic sum

$$I_f = \lambda \sum_{g \in z(f)} \left[U_{fg}^{VA} - U_{fg}^{AV} \right], \qquad (29.3)$$

they form diffusion [47–49] and not convective flows relative to the thermostat. Therefore, defining a pressure gradient (or molecular density) in the fixed thermostat, analysis of the non-equilibrium flow can be carried out to determine the diffusion coefficient.

The kinetic theory determined for a single component substance the only type of flow of the moving molecules – convective flow, described by the continuity equation. To study the convective flow, we should match the movement of the thermostat with the thermal motion of the molecules. This is achieved by using the equation for the distribution of pair functions which describes their relative position in the coordinate space (as in the theory of liquids) and depends on the state of the thermostat. The presence of the convective flow at the microscopic level changes the process of redistribution of the molecules in comparison with the fixed thermostat, however, there are two micro-channel transport molecules: microconvective (described by the microhydrodynamic velocity $\mathbf{u}_f$) and thermal.

To determine the ratio of these flows, we should considered the local stationary state in the evolution of the pair function. Then we can find the correction to the equilibrium pair distribution function, through which we can re-calculate all of the real distributions of the molecules. Considering the shifts of neighbouring regions of matter, from the analysis of the non-equilibrium momentum flux we can find the shear viscosity coefficient (similar to the determination of the bulk viscosity during deformation/compression of the volume under consideration). This allows one to calculate the speed of the convective flow, using the Navier–Stokes type equations, in which the viscosity coefficients are calculated within the framework of the molecular theory on the basis of the LGM, but with the movement of the thermostat taken into account. The same similar principle is used to derive the microscopic equations for the process of energy transfer.

The expressions for the transport coefficients are constructed as part of the dynamics of the locally nonuniform fluid using thermal velocities through the modified Eyring theory (see section 28 and Appendix 7). The approach keeps the self-consistent description of equilibrium and dynamics, which is discussed in Appendix 6. It should be noted that the concept of self-consistent description includes the condition of the absence of the convective flow. If the system has a small convective flow, the system does not have complete equilibrium. To achieve self-consistency it should be $u_i = 0$, only then $P(\theta, T)$ is the equilibrium equation of state, which is discussed in section 10. Otherwise, the pressure differs from the equilibrium due to non-equilibrium amendments.

As a natural measurement time unit small systems use the passage time of molecules of its diameter. If the average velocity of motion of molecules in the gas phase is about $4 \cdot 10^4$ m/s, then the molecules pass through its size of about 4 Å at $T = 300$ K for a time equal to 10^{-12} s = 1 ps.

In general, this approach leads to a system of ordinary differential equations for the conventional dynamic variables in hydrodynamics (density, three components of the vector $\mathbf{u}_f$ and temperature) as defined in the sites of the lattice structure. Because of the complexity of the problem with the strongly nonuniform distribution of molecules in narrow pores the use of the system involves numerical studies. Given the possibility of detailed description of the molecular distribution in a wide range of time from picoseconds to microseconds without the loss of all molecular information and the calculation of microhydrodynamic information for the dynamic variables and transfer coefficients, this method is an alternative to the molecular dynamics method.

30. Local transport equation

Use in the microhydrodynamic approach for any density of the concept of the thermal velocity of the molecules can formally keep in the dense phase the conservation equations, introduced in gas dynamics. The conservation equations in the conventional form is used for the gas where the strict reversible trajectories should be considered, as in the Boltzmann equation [50–54]. With their help the laws of conservation of the property S are: mass, momentum, energy, etc., allowing to enter gradually the following equations of the hydrodynamic type at the microlevel.

The equations of conservation of the property S in each cell, obtained by averaging the kinetic equation (A9.2) over the velocity of the molecules, can be written as

$$
\frac{\partial \langle \theta_f S \rangle}{\partial t} + \sum_{j=1}^{3} \left\{ \Delta \langle \theta_f v_{fj} S \rangle / \Delta q_{fj} - \theta_f \langle v_{fj} \Delta S / \Delta q_{fj} \rangle - \right.
$$

$$
- \theta_f \left\langle F(f)_j \frac{\partial S}{\partial v_{fj}} \right\rangle / m - \theta_f \left\langle S \frac{\partial F(f)_j}{\partial v_{fj}} \right\rangle / m -
$$

$$
\left. - \theta_f \left\langle F_h(f)_j \frac{\partial S}{\partial v_{fj}} \right\rangle / m - \theta_f \left\langle S \frac{\partial F_h(f)_j}{\partial v_{fj}} \right\rangle / m \right\} = 0,
$$

(30.1)

where the index $j = 1$–3 relates to vectors directed along the axes x, y, z of the cell f, and the symbol $F_h(f)$ indicates the Vlasov terms (for vacancies $\mathbf{v}_f^v = 0$, and for simplicity it is taken $\mathbf{v}_f^A = \mathbf{v}_f$ and and arguments of the velocities in unary functions are omitted). All the functions of (30.1) are defined in Appendix 9.

Consider a non-equilibrium state of the system when the axial pressure drop leading to fluid flow is created. In the lattice gas model the element of the volume is the cell volume v_0 and each cell has z faces. We describe the flows in each cell of the lattice structure. The amount of the fluid flowing through the surface S_m of the face m, $1 \le m \le z$, of the f cell per unit time is equal $\theta_f \mathbf{u}(f) S_m$, where $\mathbf{u}(f)$ is the local velocity vector of the average molecular flow in the cell f with the components $u_i(f)$ in the direction $i = x, y, z$, where the averaging is performed over all directions and the absolute values of the velocity of the molecule A in the cell (29.2). It is usually assumed that the value of the flow is positive if the fluid flows out of the cell, and negative if the fluid flows into it. The total number of the fluid flowing from the cell f per unit time is $\sum\limits_{m=1}^{z} \theta_f \mathbf{u}(f) S_m$. On the other hand, a decrease of the fluid in the cell can be written as $-\partial\theta/\partial\tau$. Then $\partial\theta_f/\partial\tau = -\sum\limits_{m=1}^{z} \theta_f u(f) S_m$. If we enter the density of distribution θ_f^* of the molecules in the volume $v_0(f)$ (which is constant in the cell f and equal to $\theta_f^* = \theta_f/v_0$), so that $\theta_f = \int \theta_f^* dv_0(f)$, then the sum $\sum\limits_{m=1}^{z} \theta_f \mathbf{u}(f) S_m$ can be represented as $\oint \theta_f^* u(f) dS = \int \text{div}\left(\theta_f^* u(f)\right) dv_0(f)$, wherein the symbol div (A) is the difference derivative div $(A) = \Delta(A)/\Delta x + \Delta(A)/\Delta y + \Delta(A)/\Delta z$, here $\Delta x, \Delta y, \Delta z \sim \lambda$ (equal to the linear size of the cell).

As a result, if S corresponds to the mass of the molecule, we obtain the following equation of continuity for the cell f ($i = x, y, z$):

$$\partial\theta_f / \partial t = -\sum_i \left[u_i(f) \frac{\Delta\theta}{\Delta i} + \theta_f \frac{\Delta u_i(f)}{\Delta i} \right], \qquad (30.2)$$

for the spatial variables i in the neighboring cells we consider the difference derivatives $\Delta/\Delta i$ because of the discreteness of describing the positions of the particles on a lattice. The introduced microhydronamic speed is a new concept, since it is determined on a scale of about one diameter of the molecule, which contradicts conventional hydrodynamic definitions using the condition of the derivative increment $i \gg \lambda$, for which the increments are described by differential derivatives $\partial/\partial i$.

Equation (30.2) by its form is identical to the continuity equation in continuum mechanics [55, 56]. It refers to the lower limit of the volumes for the transport equations applicable to randomly selected small macrovolumes, so it contains the difference derivatives of the local values of the density and the average speeds of the molecules for the distance defined on the sites of the lattice structure. The relationship of $\mathbf{u}$ with the continuity equation defines the type of flow, as in gas dynamics.

Similarly, for S, corresponding to the momentum and energy, on the basis of (30.1) we can write the other two difference equations for the local transport of momentum and energy.

The equation of viscous motion (for component i):

$$m\theta_f\left(\frac{\partial u_i(f)}{\partial t}+\sum_k u_k\frac{\Delta u_i(f)}{\Delta k}\right)=-\frac{\Delta P(f)}{\Delta i}+\theta_f F_i(f)+\sum_k\frac{\Delta\sigma_{ik}(f)}{\Delta k},$$

(30.3)

where $F_i(f)$ is the component of the external force in the cell f, produced by the potentials differing in nature from the adsorbent–adsorbate and adsorbate–adsorbate interactions taken into account in the transfer coefficients (in narrow pores the gravity field may be neglected), $P(f)$ – the pressure in the cell f, $\sigma_{ik}(f)$ – the components of the viscous stress tensor (here $\eta(f)$ – the shear viscosity coefficient, $\xi(f)$ – the bulk viscosity coefficient in the cell f):

$$\sigma_{ik}(f)=\eta_{ik}(f)\Delta u_i(f)/\Delta k, \quad \sigma_{ii}(f)=\xi_{ii}(f)\Delta u_i(f)/\Delta i. \quad (30.4)$$

This form uses the non-symmetric version of the definition of the stress tensor [55, 56]. This is due to the need to consider local changes in the density and velocity of flow at the microlevel. Components of the stress tensor involved in the symmetrization correspond to different local shear viscosity coefficients (see sections 35 and 36). (Recall that the symmetrized form of writing the stress tensor corresponds to the components of tensor $\mathbf{\eta}'$, recalculated as $\mathbf{\eta}' = \xi - 2/3\eta$ [55, 56]).

The energy transfer equation can be written as

$$m\frac{\partial}{\partial t}\left(\frac{\theta_f u(f)^2}{2}+\theta_f E_f\right)=\theta_f\sum_i u_i(f)F_i(f)-$$

$$-\sum_i\frac{\Delta}{\Delta i}\left[m\theta_f u_i(f)\left(\frac{u(f)^2}{2}+\omega_f\right)-\sum_k\left(u_k(f)\sigma_{ik}(f)\right)-\kappa_i(f)\frac{\Delta T(f)}{\Delta i}\right],$$

(30.5)

where $\kappa(f)$ is the thermal conductivity vector, $E(f)$ is the internal energy and $\omega(f) = E(f) + P(f)v_0$ is the enthalpy (heat function) per unit mass, related to the cell f.

The transport equations (30.2)–(30.5) contain the coefficients of viscosity and thermal conductivity (which are discussed below) and the characteristics of the local state of the fluid: local pressure, determined from the equation of state and internal energy. The last two characteristics are determined by the equilibrium relations obtained in the lattice model for the case of the arbitrary non-uniform distribution of molecules (Chapter 2).

31. The equations for the pair distribution functions

To close the equations (30.2)–(30.5), it is necessary to construct the kinetic equations for the pair distribution functions, which are used to express the transfer coefficients. In equilibrium, the pair function correspond strictly functionally to the unary function because of the relationship between them. If there was a deviation of the specified pair function the indicated equilibrium correspondence, a relaxation process takes place by which this correspondence is restored. If the unary function changes over time, the pair function should also change in accordance with this change ('synchronously'). However, these two processes can be separated in time only if the characteristic time scales of the relaxation process for the pair function and the process of the evolution of the unary function are incommensurate (this is also the meaning of time hierarchy), and the first scale is much smaller than the second one. Accordingly, the change in space is also incommensurate. In the non-equilibrium states the formula (7.5) does not apply – under the influence of the molecular flow this functional relationship is changing.

To analyze the situation, it is necessary to link the resulting equations with the well-known hierarchy of the time in the evolution of unary and the pair functions of the Bogolyubov distribution for macroscopic bulk phases [47, 50]. Under the principle of hierarchy of time fo the pair wer should distinguish between the 'fast' relaxation process and the 'slow' synchronous process, and with respect to the considered problem we can writte $\theta_{fg} = \theta_{fg}(t, \theta_f(t), \theta_g(t))$. The explicit dependence on the time t and the coordinates of the sites reflects the relaxation process, and the synchronous process is associated with an implicit dependence on t due to the functional dependence on θ_f and θ_g, and the latter – on t.

If we consider the macroscopic flow of matter, such as the diffusion equation for the coarse-grained coarsening of space and time scales in which the details of the evolution over time intervals characteristic of relaxation processes become evident, then in this equation, we need to take into account the dependence of θ_{fg} on time t only of the $\theta_{fg} = \theta_{fg}(\theta_f(t), \theta_g(t))$. However, at the microlevel for dense phases the Bogolyubov principle of hierarchy, especially in the conditions of strong local inhomogeneity, ceases to be valid in a wide range of time as the pair distribution functions appear along with unary functions as equal dynamical variables (see Fig. 28.6, b). At the starting time for any violation of the initial equilibrium state of matter (in which there is a clear connection between paired and unary functions) there is a certain range of time (the chaotization stage according to Bogolyubov's terminology), during which we must explicitly take into account the time dependence of the pair distribution function. It is this stage that is actively used in the LGM in consideration of complex, nonuniform processes with a very wide range of characteristic times of various atomic–molecular processes. For dense phases there are also violations of the principle of total attenuation of spatial correlations, because of the strong interaction between the molecules, which was formulated only for justifying the Boltzmann equation for rarefied gases. Accordingly, there are differences in the construction of the equation of continuity for the unary and pair distribution functions. This leads to the situation in which equation (30.1) does not contain the contribution (29.3) of the exchange of molecules with adjacent cells due to their thermal motion.

By analogy with (30.1), we construct the equations of conservation of the property S, relating to two cells f and g. To do this, we average equation (A9.3) with respect to the velocity of the molecules in these cells. This leads to the following equation (here the velocity arguments are omitted when the pair functions and $\theta_{fg} \equiv \theta_{fg}^{AA}$ are used):

$$\partial\langle\theta_{fg}S\rangle/\partial t+\sum_{j=1}^{3}\left\{\Delta\langle\theta_{fg}v_{fj}S\rangle/\Delta q_{fj}-\theta_{fg}\langle v_{fj}\Delta S/\Delta q_{qfj}\rangle-\right.$$

$$-\theta_{fg}\langle F(f)_{j}\,\partial S/\partial v_{fj}\rangle/m-\theta_{fg}\langle S\partial F(f)_{j}/\partial v_{fj}\rangle/m-$$

$$-\theta_{fg}\langle F_{h}(f)_{j}\,\partial S/\partial v_{fj}\rangle/m-\theta_{fg}\langle S\partial F_{h}(f)_{j}/\partial v_{fj}\rangle/m+$$

$$+\theta_{fg}\langle\Delta U_{fg}/\Delta q_{fj}\,\partial S/\partial v_{fj}\rangle/m+\qquad(31.1)$$

$$\left.+\theta_{fg}\langle S\partial\left[\Delta U_{fg}/\Delta q_{fj}\right]/\partial v_{fj}\rangle/m\right\}+\{\}_{g}=\langle SI_{fg}\rangle,$$

where the expression I_{fg} represents the right side of equation (27.7), which takes into account only the stages of displacement of the molecules; the sum over j is constructed relative to the contributions of the variables of the cell f, $\{\}_{g}$ denotes the same sum, constructed for the cell g (it is obtained by replacing the indices f and g). In the derivation of (31.1) it was assumed: 1) the contribution of simultaneous triple collisions of molecules can be ignored, 2) the property S is preserved under pairwise collisions of molecules, which are implemented in the potential field created by a third molecule, located next (i.e. in the collisions there are no changes in the internal states molecules), and 3) we take into account the exchange of molecules with neighbouring cells due to the thermal motion of them.

If S corresponds to the mass of the two molecules, equation (31.1) yields an analogue of the continuity equation for the flow of pair molecules. In this approach, we must explicitly take into account that the evolution of pairs in adjacent cells is realized in two ways. There is a fast relaxation mechanism of the changes of the pair distribution function due to the thermal motion of the molecules, and a slow mechanism by molecular flows (movement of the thermostat). The fast mechanism is associated with the contributions I_{fg} – these contributions are described by the equations developed in the LGM (sections 27 and 28). The slow mechanism is associated with the contributions J_{fg}, following from the left-hand side of (31.1). They have the same structure as the continuity equation for the unary functions. Due to different local velocities in the neighbouring cells, local microflows may affect the distribution of adjacent pairs of molecules (separate and form these pairs). Note that in the conventional hydrodynamic equations the local volumes and flows relate to the scale of distances greater than the radius of the correlations between the molecules, so within the local volume it is always assumed that the condition of local equilibrium is satisfied there. As a consequence, these equations do not change the paired function and carry them as 'whole'.

In general, formula (31.1) yields the following expressions for the contributions J_{fg}, containing the correlations between the average local velocities $\mathbf{u}(f)$ and $\mathbf{u}(g)$ in the neighbouring cells. To describe them should be considered to preserve the properties of S velocities and energies of the two molecules – this gives the missing equations to account for the correlations of velocity and energy, but dramatically increases the dimension of the system of equations to be solved.

We confine ourselves to the simplest case – we neglect correlations between microlocal velocities in adjacent cells and then [43–45]

$$\frac{\partial \theta_{fg}}{\partial t} + J_{fg} = I_{fg},$$

$$I_{fg} = \sum_{h \in z_f^*} \left(U_{hfg}^{(AV)A} - U_{hfg}^{(VA)A} \right) + \sum_{h \in z_g^*} \left(U_{fgh}^{A(VA)} - U_{fgh}^{A(AV)} \right), \tag{31.2}$$

$$J_{fg} = \sum_i \left[(u_i(f) + u_i(g)) \frac{\Delta \theta_{fg}}{\Delta i} + \theta_{fg} \frac{\Delta}{\Delta i} (u_i(f) + u_i(g)) \right]. \tag{31.3}$$

Here function $U_{hfg}^{(VA)A}$ and others have the form $U_{hfg}^{(ij)A} = U_{hf}^{ij} \Psi_{fg}^{jA}$, $\Psi_{fg}^{jA} = t_{hg}^{jA} \exp\left(\beta \delta \varepsilon_{fg}^{jA} \right) / S_{fg}^j$, i.e. all summands I_{fg} are the themal velocities of the molecules (function t_{hg}^{ij}, $\delta \varepsilon_{fg}^{ij}$ and S_{fg}^i are determined in section 28 for homogeneous systems). The expression for term I_{fg} (31.2) corresponds to the right side of equation (27.7) for two closest particle AA, if it does not contain all processes, with the exception of migration of the the molecules of one sort.

The convective flow of molecules disturbs their equilibrium distribution. To find out how the pair distribution functions change under the influence of the molecular flow, we consider the quasi-stationary state of the system at imposing a certain density gradient of molecules. This allows to determine the non-equilibrium corrections to the pair distribution function due to the influence of molecular microflows.

Imagine θ_{fg}^{AA} as $\theta_{fg}^{AA} = \theta_{fg}^{*AA}(1 + \delta_{fg}^{AA})$, where the asterisk denotes the pair function belonging to the equilibrium distribution, and its absence – to the non-equilibrium one, δ_{fg}^{AA} is the sought correction to the equilibrium distribution. Substituting θ_{fg}^{AA} in I_{fg} leads to an expression that depends on a variety of different local densities θ_h^A

of the sites h, surrounding a given pair AA between which jumps take place. Given that in the local equilibrium state all direct and inverse hopping of the molecules between the sites fh and gh lead to zero contributions, the overall structure of the expression for I_{fg} can be represented as $I_{fg} = \psi \delta_{fg}^{AA}$, where the function ψ is dependent on all values of θ_h^A of the nonuniform system and other unknown pair corrections δ_{hf}^{AA} and δ_{gh}^{AA}. The system of (linear) equations for paired amendments θ_{fg}^{AA} is very large because we have obtained the linked system of equations. To reduce it, we should restrict ourselves to only the nearest neighbours (i.e., we take into account the sites h between which the exchange of molecules takes place at the given time – this is a one-step approximation in time) and the differences in the values δ_{hf}^{AA} at this local site. This allows to obtain the following solution for this pair fg:

$$\delta_{fg}^{AA} = -J_{fg} / \left(bU_{fg} \right), \qquad (31.4)$$

where $U_{fg} = \sum_{h \in z_f} U_{hfg}^{(VA)A} + \sum_{h \in z_f} U_{hfg}^{A(AV)}$, and b is a coefficient approximating the contribution of concentration of functions generated by the ratio of functions Ψ_{fg}^{VA} and Ψ_{fg}^{AA} in this local area. Coefficient b depends on the lattice structure z. Near the pore wall the hopping rate constants reflect the influence of the surface potential, and the relevant links are not included in the expression U_{fg} in the summation over h.

Formula (31.4) is the ratio of local microscopic velocities of the fluid flow and thermal velocity of the molecules in the cell f. Typically, in the gas phase these values differ by 3–5 orders of magnitude, so the first order of the expansion in the small parameter is well substantiated. However, the ratio (31.4) increases with the local density as a result of the decrease of the thermal velocity of the molecules, and for the high fluid densities the correction value can be substantial, especially for the initial time with pulse perturbations.

The simple form of formula (31.4) is due to the use of the formulated approximations. Upon closer examination (by analogy with [47]) it is necessary to solve a linear system of equations for the total set of all the unknowns amendments δ_{fg}^{ij}, but this structure of the ratios of the microdrodynamic and thermal velocity is maintained. Below, the formulas for local transport coefficients use everywhere the non-equilibrium function θ_{fg}^{ij} (instead of the equilibrium function θ_{fg}^{*ij}). In the absence of the convective flow the non-equilibrium nature of the pair function disappears and all transport coefficients are transformed to the expressions obtained in the framework of the elementary kinetic theory of the micro-nonuniform systems.

The need to use as a function of S the momentum and energy for a pair of molecules is associated with the assumption of the existence of correlated velocities and excited local states. When using equation (27.1) it is assumed that the characteristic time of local vibrational relaxation is shorter than the characteristic time of the local redistribution of molecules which determines the evolution of the pair distribution functions. Otherwise, we must use a shorter characteristic time and explicitly consider the vibrational motion with the cooperative nature. To describe the combined dynamic behaviour of the neighbouring molecules we must take into account the correlation in their movements. It is necessary to go beyond the Einstein approximation for vibrational motion used below and in [57, 58] to calculate the thermal conductivity. In non-isothermal conditions when calculating θ_{fg}^{ij} to a first approximation we can confine ourselves to a linear relationship $T_{fg} = [T_f + T_g]/2$ (rigorous analysis of the formula (31.1) leads to cumbersome expressions)

As a result of these simplificantions we obtain a closed system of equations whose dimension coincides with that of a similar differential system of hydrodynamic equations, but the given system retains the character of the inhomogeneity of the distribution of the mixture of molecules in narrow pores and their spatial correlation. In this situation, we use the Bogolyubov principle of the hierarchy of time: any convective flows are small perturbations compared with high-intensity exchange of the polulation of the sites due to thermal motion of the molecules. Otherwise, the dynamic system of equations must contain equations for the paired functions that are written in section 27 and used in the calculation of Fig. 28.6.

32. Boundary and initial conditions

Differential transport equation (30.2)–(30.5) after closing, which is done by the transport coefficients that depend through the thermal velocities of the molecules on the current local density and pair functions (see Sec. 5), are solved appropriately by the given initial conditions for temperature, density and microhydrodynamic velocitiesin the entire calculation field and by boundary conditions that distinguish the boundary area of the system from the inside area.

Specificity of the boundary region

The transport equations are written for each cell of the region under consideration, and in their form they coincide with the difference form of the differential Navier–Stokes equations. All cells inside the considered region or at its boundary are completely equal – the structure of the transport equation for the boundary cells has the same form as their structure within the system.

The behaviour of molecules in the cells at the border with the pore wall differs from the similar behaviour of molecules in the cells of the inner region by the presence of the potential of interaction of molecules with the walls of the pores (section 4). The boundary cells, located on the border of the region under consideration have a different number of neighbouring cells as some of them are replaced by the pore wall (other values of z_{qp}). The boundary restricts the region, i.e., in the condition of impermeable walls, and specifies of the flows of energy and momentum through *the outer edge* of the cells close to the walls. All other differences in the properties of the fluid near the walls are caused only by the difference of potential functions between the molecules and between the molecule and the pore wall. This allows all the transfer coefficients at the boundary of the fluid with the wall to be calculated by the formulas presented in chapter 5 replacing parameter ε_{fg}, where the number of the pairs of sites *fg* corresponds to the site *f* in the first layer of the fluid and to the atoms on the surface of the wall *g*, by the parameter Q_1. Thus, for calculations in the LGM we should use a discrete atomistic model of the solid for the pore wall (see chapter 2). As a result, the specificity of the near-wall cells is directly taken into account in the calculation of the transport coefficients through the functions of the local densities near the walls of the pores and intermolecular interactions between the molecule and the walls, as well as between neighbouring molecules. Their values are determined by in the self-consistent manner during calculations in all cells of the calculation field.

The viscosity coefficients for the molecule–wall attraction potentials are maximum at the pore surface, so that the longitudinal velocity component is minimal (but not identically equal to zero) and the vertical component of the velocity is always zeroed because of the condition of impermeability of the fluid through the pore wall (see also section 42). In this case no conditions are imposed *a priori* on the values of the fluid density and the longitudinal component of velocity in the cells near the walls which are self-consistently determined by solving the entire system of equations. (Using the condition (30.1) eliminates the need for defining the boundary conditions for the distribution function of the velocity of the molecules.) Recall that in the gas- and hydrodynamic problems in the formulation of the boundary conditions on the walls of the pores the values of the longitudinal and vertical components of the fluid velocity are equated to zero for the border assuming that the particles of matter are so strongly attracted by the walls that they are stationary [55, 59]. But this condition is not consistent with the concepts of surface mobility and the surface flow of molecules [60, 61]. This approach eliminates this mismatch and allows to describe the effects of sliding of the dense fluid along the walls of the pores. To retain the schema of describing the hydrodynamic equations, for the second derivatives of the coordinates we formally introduce an additional layer inside the solid, in which all components of the velocity are assumed to be zero.

In addition to the near-wall boundaries, the calculation field almost always includes boundaries through which there is a stream of matter. Only in the case of the analysis of the processes of internal redistribution of the molecules within the area under consideration are all the boundaries considered impermeable. In other situations, we investigate the flows of molecules and the properties of cells at these permeable boundaries are assumed to be equal to the properties of the cells adjacent to them on the inner side of the system.

Initial conditions

The derived transport equations of the molecules require the initial conditions starting at which they describe the evolution of the system. The procedure of averaging over the local states corresponds to the minimal time equal to about 10^{-12} s. This is the characteristic time corresponding to the free flight time of the molecule through its diameter. As an upper limit of the microhydrodynamic equations we accept macroscopic times characteristic for this problem and depending on the binding energy between all system components. In problems of transport of molecules in narrow pores the processes related to the basic stages of adsorption and desorption are usually limited. Therefore, the formally constructed transport equations do not have an upper limit for the time of use and its actual value depends on the capabilities of the numerical investigation.

The question about the initial conditions is quite important, as it demonstrates the accuracy of the information about the system available to the researcher. We are talking about the reliability of defining the initial state for a full set of sites of the whole calculation field. The simplest initial conditions are the conditions of the equilibrium distribution of the molecules over a given volume of the system (the initial velocity of the molecules is excluded, and their equilibrium spatial distribution is solved by the methods set forth in chapters 2 and 3 and in Appendix 4). Then at the initial moment of time we can enter any given perturbation both for the entire calculation field and in some local area. In particular, we can introduce perturbations at the boundaries through which the exchange of molecules with external fields takes place. However, this approach is associated with a long relaxation process in the volume of the entire calculation field. With larger systems, this greatly increases the time required to the states in which we are interested, in particular the stationary solutions.

Below we used mainly condition of the equilibrium distribution of the molecules. The influence of the initial perturbation of the system at short times is specifically discussed in section 49.

Microhydrodynamic approach
The LGM was used as a basis when deriving microscopic hydrodynamics equation that considers the intrinsic volume of the molecules and intermolecular interactions between both the molecules of the fluid and between the molecules and the channel walls. The transport equations are written for each cell of the area under consideration (in terms of difference derivatives), which are identical in form with the difference form of writing differential Navier–Stokes equations.

The transfer coefficients (this follows from the connection of the coefficients with the thermal velocity of the molecules and the expression for the hopping rate of the molecules (28.8), as well as Appendix 7) account for the non-local properties of the fluid. In LGM any two molecules are always in different cells, so the lateral interactions of molecules that determine the cooperative properties of the system, and the coefficients of transfer of viscosity and thermal conductivity, reflecting the dynamics of the interaction of neighbouring molecules, are always non-local. To close the transport equations in the microdynamic approach, we use the kinetic equation for the pair distribution function, which is similar to the Smoluchowski equation in the kinetic theory of dense liquids [62–66].

In these equations, the transport coefficients depend on the local density and temperature having a strongly nonuniform distribution. This approach enables the construction of a unified system of equations describing the flow of both dense gases and liquids in micro- and mesoporous systems, including the area of capillary phenomena. The model covers the changes in the concentration of the fluid from the gaseous to a liquid state and a wide range of temperatures, including the critical region, which allows us to consider the dynamics of flows of the vapor, liquid and vapor–liquid fluids in the presence of capillary condensation. With increasing pore size the derived equations become the hydrodynamic transport equations for the flows of gas or liquid, retaining the relationship of the transport coefficients with the intermolecular potentials.

In this approach, the conditions of closure of the kinetic equations, strictly justified for dense fluids, are extended (extrapolated) to low densities. As a result of this at low densities $\theta \sim 10^{-4} \div 10^{-3}$ the expressions for the transport coefficients obtained below change to the corresponding known expressions of the elementary kinetic theory of gases [26, 27] (but not to the expression of the kinetic theory [26, 51–52]). For high densities ($\theta \sim 1$) the expressions obtained for the transport coefficients change to the corresponding well-known expressions for liquids, providing them with the correct temperature dependence [4, 26, 67].

It is well known that the transition to the low-density range, corresponding to the elementary kinetic theory, requires caution. This is due to certain limitations, approximations based on the mean free path in the gas. So, if the elementary theory "is applied to a gas cylinder in a state of uniform rotation, it gives rise to viscous stresses of the same magnitude as in the laminar flow, whereas in reality the viscous stresses are absent" [64].

In this regard, we note that the microhydrodynamic approach can be extended to obtain a closer connection with the traditional methods of kinetic theory: accounting for the spatial temporal correlation of a higher order (in particular, neglecting the single-step approach with respect to time), taking into account the velocity distribution function for rarefied phases (see the discussion of the way of such a generalization when considering thermal diffusion) and accounting for collective motions of molecules in the rarefied and dense phases (see Appendix 7). Of course, any direction leads to an increase in the complexity of the calculations.

Chapter 6 shows examples of studies of molecular flows in narrow pores using the microhydrodynamic approach: 1) taking into account the relationship of surface and volumetric flows, 2) analysis of the change of flow regimes from film to solid with volume filling of the pore during capillary condensation (in the presence of extensive redistribution of concentrations and velocities of molecules causing local velocity fluctuations), 3) consideration of the effect of structural and chemical heterogeneity of the walls on the characteristics of fluid flow in different modes. Analysis of the model variants of the porous systems is essential for understanding the correctness of the calculation procedure of the flow characteristics to predict flows in macroscopic systems of interconnected pores in real porous materials, etc.

References

1. Glauber L., // J. Math. Phys. 1963. V. 46, P. 541
2. Stanley H. E., Introduction to Phase Transitions and Critical Phenomena. – M.: Mir, 1973. [Clarendon, Oxford, 1971].
3. Eyring H. // J. Chem. Phys. 1935. V. 3. P. 107
4. Glasston S., Laidler K. J., Eyring H., The Theory of Rate Processes. – M.: Izd. inostr. lit., 1948. – 583 p.[Princeton University Press, New Jersey, 1941].
5. Tovbin Yu.K., Fedyanin V.K. // Kinetika i kataliz. 1978. T. 19, No. 4 P. 989
6. Tovbin Yu.K. // Zh. Fiz. Khimii. 1981. V. 55, No. 2. P. 284. [Russ. J. Phys. Chem., 1981 V. 55, No. 2, P. 159].
7. Tovbin Yu.K. // Dokl. AN SSSR. 1984. V. 277. No. 4. P. 917.
8. Tovbin Yu.K. Theory of physico-chemical processes at the gas–solid interface. – Moscow: Nauka, 1990. – 288 p. [CRC Press Boca Raton, FL, 1991.]
9. Tovbin Yu.K. // Progress in Surface Science. 1990. V. 34. No. 1 – 4, P. 1 – 236.
10. Tovbin Yu.K. // Poverkhnost'. Fizika. Khimiya. Mekhanika. 1989. No. 5 P. 5.
11. Tovbin Yu.K. // Thin Films and Nanostructures. V. 34. Physico-Chemical Phenomena in Thin Films and at Solid Surface / Eds. L.I. Trakhtenberg, S.H. Lin, O. J. Ilegbusi. – Amsterdam: Elsevier, 2007. – P. 347
12. Tovbin Yu.K., Cherkasov A.N. // Teor. Eksp. Khimiya. 1984. V. 20, No. 4. P. 507.
13. Pandit R., Forgacs G., August P. // Phys. Rev. B. 1982. V. 25. P. 1860.
14. Ma Shang-keng, Modern theory of critical phenomena. – Moscow: Mir, 1980. – 298 p. [Benjamin, Inc., London, 1976].
15. Nikitin E.E. Theory of elementary atomic-molecular processes in gases. – Moscow: Khimiya, 1970.
16. Tovbin Yu.K. // Kinetika i kataliz. 1984. V. 25, No. 1. P. 26.
17. Tovbin Yu.K. // Dokl. AN SSSR. 1982. V. 267, No. 6. P. 1415.
18. Tovbin Yu.K., Fedyanin V.K. // Zh. Fiz. Khimii. 1980. V. 54. No. 12. P. 3132.
19. Tovbin Yu.K., Fedyanin V.K. // Fiz. Tverd. Tela. 1980. V. 22, No. 5. P. 1599.
20. Prigogine I., Defay R. Chemical Thermodynamics. – Novosibirsk: Nauka, 1966. – 510 p. [Longmans Green, London, 1954].
21. Tovbin Yu.K. // Kinetika i kataliz. 1979. V. 20, No. 5. P. 1226.

282   Molecular Theory of Adsorption of Porous Solids

22.   Tovbin Yu.K., Votyakov E.V. // Zh. Fiz. Khimii. 1996. V. 71. No. 2. P. 271. [Russ. J. Phys. Chem. 1996. V. 71. № 2. P. 214].
23.   Tovbin Yu.K., Titov S.V. // Superkriticheskie flyuidi: teoriya i praktika. 2011. V. 6. No. 2. P. 35.
24.   Tovbin Yu.K. // Equiliblia and Dynamics of Gas Adsorption on Heterogeneous Solid Surfaces / Eds.: W. Rudzinski, W. A. Steele, G. Zgrablich. – Amsterdam: Elsevier, 1997. – P. 201
25.   Roberts M.W., McKee C.S. Chemistry of the Metal-Gas interface. – Oxford: Clarendon, 1978.
26.   Hirschfelder J. O., Curtiss C. F., Bird R. B., Molecular Theory of Gases and Liquids. – Moscow: Izd. inostr. lit., 1961. – 929 p. [Wiley, New York, 1954].
27.   Reif F., Statistical Physics. Berkeley Physics Course, Vol. 5 – Moscow: Nauka, 1977. – 352 p.
28.   Tovbin Yu.K. // Dokl. AN SSSR. 1990. V. 312. P. 1423.
29.   Tarasenko A.A., Chumak A.A. // Fiz. Tverd. Tela. 1980. V. 22. P. 2939.
30.   Chumak A.A., Tarasenko A.A. // Surface Sci. 1980. V. 91. P. 694.
31.   Hochman E.M., Tovbin Yu.K., Fedyanin V.K., Physics of interfacial phenomena. – Nalchik KBSU, 1981. – P. 145.
32.   Gvozdev V.V., Tovbin Yu.K. // Khim. Fizika. 1999. V. 18, No. 1. P. 88. [Chem. Phys. Report, 1999. T. 18. № 1. C. 179].
33.   Tovbin Yu.K., Votyakov E.V. // Poverkhnost'. Fizika. Khimiya. Mekhanika. 1991. No. 3. P. 112.
34.   Tovbin Yu.K. // Kinetika i kataliz. 1986. V. 27. No. 3. P. 655.
35.   Kiskinova M.P., Bliznakov G.M. // Surface Sci. 1982. V. 123, No. 1. P. 63.
36.   Shmachnov V.A., Malakhov V.F., Kolchin A.M., Problems of kinetics and catalysis. – Moscow: Nauka, 1978. – V. 17. – P. 170.
37.   Ertl G. // Adv. Catal. 1990. V. 37. P. 213.
38.   Imbihl R., Ertl G. // Chem. Rev. 1995. V. 95. P. 697.
39.   de Wolf C.A., Nieuwenhuys B.E. // Catal. Today. 2001. V. 70. P. 287.
40.   Tovbin Yu.K., Murovtsev A.N. // Zh. Fiz. Khimii. 1984. V. 58, No. 9. P. 2286.
41.   Tovbin Yu.K., Kolchin A.M. // Zh. Fiz. Khimii. 1988. V. 62. No. 12. P. 3306.
42.   Tovbin Yu.K., Zyskin A.G., Snagovsky Yu.S. // Kinetika i kataliz. 1991. V. 32, No. 3. P. 571.
43.   Tovbin Yu.K. Modern Chemical Physics. – Moscow: MGU, 1998. – P. 145.
44.   Tovbin Yu.K. // Khim. Fizika. 2002. V. 21, No. 1. P. 83.
45.   Tovbin Yu.K. // Zh. Fiz. Khimii. 2002. V. 76, No. 1. P. 76. [Russ. J. Phys. Chem. 2002. V. 76. № 1. P. 64].
46.   Tovbin Yu.K. // Zh. Fiz. Khimii. 1998. V. 72. P. 1446. [Russ. J. Phys. Chem. 1998. V. 72. № 8. P. 1298].
47.   Gurov K.P., Kartashkin B.A., Ugaste E.Yu., Interdiffusion in multiphase metal systems. – Moscow: Nauka, 1981.
48.   Borovsky I.B., Gurov K.P., Marchukova I. D., Ugaste Yu.E., Processes of the mutual diffusion in alloys. – Moscow: Nauka, 1973.
49.   Tovbin Yu.K. // Teor. Osnovy Khim. Tekhnol. 2005. V. 39. No. 5. P. 523. [Theor. Found. Chem. Engin. 2005. V. 39. No 5. P. 493].
50.   Bogolyubov N.N., Problems of dynamical theory in statistical physics. – Moscow: Gostekhizdat, 1946. [Wiley Interscience, New York, 1962].
5151. Ferziger J. H. H., Kaper H. G., Mathematical theory of transport processes in gases. – Moscow: Mir, 1976. – 554 p. [North_Holland, Amsterdam, The Netherlands, 1972].

52. Huang K., Statistical Mechanics. – Moscow: Mir, 1966. – 520 p. [Wiley, New York, 1963].
53. Chapman S., Cowling T. G., Mathematical Theory of Non-Uniform Gases. – Moscow: Izd. inostr. lit., 1960. – 5410 p. [Cambridge University Press, Cambridge, 1952].
54. Rumer Yu.B., Rivkin M.Sh., Thermodynamics, statistical physics and kinetics. – Moscow: Nauka, 1977. – 552 p.
55. Landau L.D., Lifshitz E.M. Theoretical Physics. VI. Hydrodynamics. – Moscow: Nauka, 1986. – 733 p.
56. Sedov L.I., Continuum Mechanics. – Moscow: Nauka, 1970. – 492 p.
57. Tovbin Yu.K. , Zhidkova L.K., Komarov V.N. // Zh. Fiz. Khimii. 2003. V. 77. No. 8. P. 1367. [Russ. J. Phys. Chem. 2003. T. 77. № 8. C. 1222].
58. Tovbin Yu.K. , Komarov V.N. // Izv. AN. Ser. khim. 2002. No. 11. P. 1451. [Russ. Chem. Bull. 2002. V. 51. № 11. P. 1871].
59. Bird R. B., Stewart W., Lightfoot E. N., Transport Phenomena. – Moscow: Khimiya, 1974. – 687 p. [John Wiley and Sons, New York, 1965].
60. Mason E. A., Malinauskas A. P., Gas Transport in Porous Media: The Dusty_Gas Model. – Moscow: Mir, 1986. - 200 p. [Elsevier, Amsterdam, 1983].
61. Satterfield Ch. N. Mass transfer in heterogeneous catalysis. – Moscow: Khimiya, 1976. – 240 p. [MIT Press, Cambridge (Mass.), 1970].
62. Kirkwood J.G. // J. Chem. Phys. 1946. V. 14. P. 180.
63. Kirkwood J.G. // J. Chem. Phys. 1947. V. 15. P. 72.
64. Kirkwood J.G., Buff F.P., Green M.P. // J. Chem. Phys. 1947. V. 17. P. 988.
65. Eisenschitz R., Statistical theory of irreversible processes. – Moscow: Izd. inostr. lit., 1963. – 128 p. [Oxford University Press, London, 1948].
66. Croxton C. A., Physics of the liquid state: A Statistical Mechanical Introduction. – Moscow: Mir, 1979. – 400 p. [Cambridge University Press, Cambridge, 1974].
67. Reid R. C., Prausnitz J. M., Sherwood T., Properties of gases and liquids. – Leningrad: Khimiya, 1982. [3rd ed. McGraw-Hill, New York, 1977].

5

Transport coefficients

33. Models of molecular transport in the bulk phase

This chapter describes the molecular models of transport of mass, momentum and energy in the vapor–liquid system, leading to the corresponding expressions for the kinetic coefficients of the equations in chapter 4. We confine ourselves to the case of spherically symmetric particles (more precisely, a monatomic fluid) to focus attention only on the concentration dependence of the transport coefficients of self-diffusion (D^*), shear (η) and bulk (ξ) viscosities and thermal conductivity (κ), to eliminate the effect of internal degrees of freedom on the coefficients of bulk viscosity and thermal conductivity.

The method for constructing molecular transport models is formulated most easily for the bulk phase. Given the wall potential in wide mesopores can be neglected in a first approximation, in fact this procedure relates to the central portion of the mesopores away from the channel walls.

From the kinetic theory it follows [1–3] that the transport coefficients characterize flows for small deviations from the equilibrium state of the system. We shall characterize the state of the fluid away from the pore walls by concentration θ and temperature T. The equilibrium distribution of the particles relative to each other will be calculated in the quasichemical approximation taking into account the direct correlations between the interacting particles. (Recall that the simpler mean-field approximation does not account for the correlation effects and does not provide a self-consistent description of the equilibrium distribution of the particles and the

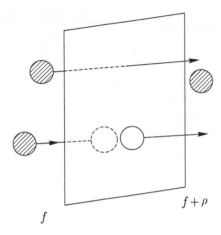

f

$f + \rho$

Fig. 33.1. Transfer of the property S through the selected plane where S: 1) the number of molecules – for calculation of self-diffusion coefficient D_i^* and mass transfer coefficients D_{ij}, 2) the number of impulses – for the calculation of the shear η and bulk viscosity ξ, and 3) the amount of energy – for the calculation of the thermal conductivity κ. There are two channels of transfer of the property S: $\eta = \eta_1 + \eta_2$, $\kappa = \kappa_1 + \kappa_2$; 1) the transfer of molecules via a separated plane – the calculation of the coefficients D_i^*, D_{ij}, η_1 and κ_1; 2) transfer of the property (momentum and energy) through the collisions – the calculation of the coefficients η_2, ξ and κ_2.

rates of elementary processes ([4, 5], see also Appendix 6), so it can not be used).

Calculation of the kinetic coefficients [6]
To calculate the kinetic coefficients as usual [3, 7], we select a plane in space 0 and consider the particle fluxes and the momentum and energy transferred by them. We use the concept of the average speed of moving particles w. We draw two planes parallel to the plane 0 (with $x = 0$) at distances $x = \pm \rho$, where ρ is the mean free path of a particle, then the properties of the particles in these planes can be written as $S(x = \pm \rho) = S(x = 0) \pm \rho dS/dx$, where the symbol S means the concentration, the momentum (in the direction y, for example) or the energy of the particles moving along the axis X. The flow of quantity S through the plane 0 consists of two oppositely directed movements of particles from the planes $x = \pm \rho$.

Two channel of transfer of momentum and energy operate in dense fluids. The first is connected with the movement of the particles, as in the rarefied phase, and the other is determined by collisions between particles. The particle in question may not cross the plane 0 if its path was blocked by other particles in the sites to the plane 0, or a particle in close proximity on the other side of the plane 0 and preventing it crossing a given plane. Both cases are not considered by the elementary kinetic theory in the gas and the kinetic theory of condensed systems must be used [4, 5].

The self-diffusion coefficient characterizes the transport of labelled particles in the equilibrium state of the system due to the thermal motion of all particles. The resulting flow of the labelled particles is composed of their direct and reverse flows through a dedicated plane perpendicular to the direction of the concentration gradient (e.g., along the axis X). We denote the concentration of labelled particles in the x plane through $\theta^*(x)$, then the concentration of unlabelled particles $\theta(x)$ is equal to $\theta - \theta^*(x)$, since for any section x the total concentration of particles is constant. Note that the intensity of interaction of particles is not dependent on whether or not they are labelled. The hopping rate of the particles $U(\rho)$ (28.8) also does not depend on the coordinates, as the vacancy concentration is everywhere constant. The flow of the labelled particles along the x axis is represented by the difference of the number of hops per unit time between two planes at a distance ρ (here S value is $\theta^*(x)$, and the defined plane is allocated between them at distance $\rho/2$)

$$J = z^* \left(\vec{U}(\rho) - \bar{U}(\rho) \right) = z^* K_\rho \left[\theta^* (-\rho/2) - \theta^* (\rho/2) \right] t_{AV}(1) \left[t_{VV}(1) \right]^{\rho-1} T_\rho, \tag{33.1}$$

wherein z^* is the number of possible hops between the adjacent layers (or neighbouring sites in an adjacent layer at the distance ρ for each site in the layer x).

Assuming that the concentration of labelled particles between the adjacent planes varies little we can write $\theta^*(\rho/2) = \theta^*(-\rho/2) + \rho d\theta^*/dx$. Comparing the resulting expression with the first Fick's law: $J = -D^* dc^*/dx$, we obtain (taking into account the link between concentrations of particles in the volume (c) and in these layers (θ): $c = \rho\theta$)

$$D^*(\rho) = \rho^2 z^* K_\rho t_{AV}(1) \left[t_{VV}(1) \right]^{\rho-1} T_\rho = \rho^2 z^* U(\rho)/\theta. \tag{33.2}$$

In the absence of lateral interactions, this formula has the form $D^*(\rho) = \rho^2 z^* K_\rho (1 - \theta)^\rho$. For small concentrations at any interaction formula (33.2) can be simplified: $D^*(\rho) = \rho^2 z^* K_\rho$. For large concentrations migration can occur only in the nearest neighbouring vacancy and $\rho = 1$. In this case, according to [4] it may be written that $(1 - \theta) \approx \exp(-\beta E_v)$, where $E_v = z\varepsilon$ is the energy required to form a vacancy in the dense phase. As a result, this model allows us to describe the exponential temperature dependence of the self-diffusion coefficient: $D^*(\rho) = \rho^2 z^* K_1 \exp(-\beta E_v)$. For a rarefied gas the formula (33.2) can be rewritten as follows: $D^*(\rho) = \rho z^* w/4$. It coincides with the known formula [3, 7] except the numerical value $1/4$, instead of $1/3$. This discrepancy appears in the elementary kinetic theory due to the use of average speeds along the preferred direction or in all directions.

Derivation of the self-diffusion coefficient equation (33.2) is similar to the conclusion of a similar factor in the elementary kinetic theory of gases. For contributions of particle fluxes to the coefficients η and κ we should repeat verbatim derivation of (33.2). In the case of viscosity the particles in the planes $x = \pm \rho$ move parallel to these planes in different directions y, and in the plane 0 the particles will not be moved in direction y. We are talking about the shear viscosity and its y-component. Then $S = m\theta w_y$, m is the mass of the particle, w_y is the y-component of the particle velocity. In the case of thermal conductivity in different planes temperature T differs and and $S = \theta C_v T$, where C_v is the specific heat per particle.

The resulting flows of values S through the plane 0 in the direction $x > 0$ lead to the following expressions:

$$\eta_1 = m\theta D^*, \quad \kappa_1 = C_v \theta D^*, \tag{33.3}$$

where the subscript 1 indicates that this is the contribution of the considered transport channel. Exactly the same relations hold for rarefied gases [3, 7]. The resulting formulas (33.2) and (33.3) consider the effect of blocking the movement of the particles along the trajectories containing other fluid particles to the selected plane 0, through the probability of formation of a vacancy trajectory. In order to accommodate the transfer of momentum also through the second channel (the collision between the particles), as well as to obtain the expressions for the bulk viscosity coefficient, it is necessary to consider the Maxwell viscosity theory.

The viscosity of the dense phase
Mathematical treatment of the phenomenon of viscosity as the reaction (reconstruction) of the medium by deformation to the mechanical impact (represented by stress) was developed by Maxwell [8]. The essence of the theory is that the reaction of the medium is not instantaneous and occurs after a certain period of time τ, called

relaxation time. The basic equations of the Maxwell theory are $\eta_2 = \tau_{sh} G_{sh}$ and $\xi = \tau_{comp} G_{comp}$, where η_2 and ξ are the shear and bulk viscosity coefficients, and τ_{sh} and τ_{comp} are the relaxation times of shear and bulk viscosity, G_{sh} and G_{comp} are the corresponding elastic moduli.

For the molecular interpretation of Maxwell's equations we use Frenkel's concept [9] that the fluid flow as the solid flow is a set of consecutive short intervals of discrete movements of individual particles. At times t, shorter than the characteristic time of the settled life of the particle τ, the fluid behaves as a solid body and can experience only elastic deformation of compression and/or shear. Describing the elementary processes of the particle displacement based on the molecular model, we can get a connection of the kinetic coefficients with the interparticle potentials. In particular, for the calculation of the shear and compression moduli we can use the results of the microscopic theory of the mechanical properties of solids [10]. The above moduli are expressed in terms of the elastic constants of the crystal lattices. We restrict ourselves to the simplest case of cubic structures which have only three independent elastic constants c_{11}, c_{12} and c_{44}; for isotropic structures the additional condition $c_{11} = c_{12} + 2c_{44}$ is satisfied, i.e., $c_{44} = (c_{11} - c_{12})/2$. For them we have $G_{sh} = c_{44}$ and $G_{comp} = (c_{11} + 2c_{12})/3$ [10].

In turn, the elastic constants can be expressed in terms of the coefficients of elasticity of concrete connections between the particles by the following formulas: $c_{11} = 2f\theta/\lambda$ and $c_{12} = c_{44} = f\theta/\lambda$, where it is considered that the stresses refer to the particles of the fluid (and not to vacancies). Thus, $G_{comp} = 4f\theta/3\lambda$ and $G_{sh} = f\theta/\lambda$. The relationship between the elasticity coefficient f with the parameters of the interparticle potential φ is listed as the simplest example of a model with the Mie potential (see (17.3)).

If we consider that the only process (other than the considered elastic reactions) responsible for the relaxation of the medium at different external perturbations is the migration of particles and vacancies, respectively, it should be assumed that $\tau_{sh} = \tau_{cmop} = \tau$, where $\tau = 1/K_{ef}$, $K_{ef} = U/\theta$. then

$$\eta_2 = f\theta^2/(\lambda U), \quad \xi = 4f\theta^2/(3\lambda U). \qquad (33.4)$$

Equations (33.4) indicate that both viscosities for the monatomic particles have similar values in the dense phase range. The total shear viscosity coefficient $\eta = \eta_1 + \eta_2$ at low densities varies linearly with θ according to (33.3), since the contribution from η_2 is small ($\eta_2 \sim$

θ^2). At high densities, the contribution from η_1 decreases and the contribution η_2 becomes decisive. Similar behaviour is shown by the bulk viscosity coefficient ξ (equal to zero at low concentrations, in accordance with the kinetic theory of monatomic particles [11]). At higher densities both viscosities, given the concentration dependence of the elastic coefficient f, depend in a complicated manner on θ. The type of the dependences $\eta_2(\theta)$ and $\xi(\theta)$ is determined by: 1) the probability of meeting of two particles, and 2) the influence of their neighbours on the activation hopping of the particles. The first factor is quadratic in θ, which is consistent with the molecular-statistical theory of liquid phases [12, 13], and the second factor increases the exponent for θ to $2(z - 1)$. It is important to note the change of the dependence of shear viscosity on temperature. At low densities, η depends as $T^{1/2}$, and at high densities we have an exponential dependence on temperature, as in the Eyring model [14].

Thermal conductivity coefficient

To construct the contribution to the thermal conductivity coefficient of energy transfer through particle collisions we used the model arguments close to those that were used by Andrade to calculate the shear viscosity [15]. Consider the energy flux carried by the vibrations of neighbouring particles. Every oscillation along the flow direction carries a certain amount of energy by contact between the particles lying in adjacent planes at a distance of λ. Transfer occurs $2v$ times per second for each of θ particles on the surface (the fact that near is a particle and not a vacancy is taken into account by function t). Each particle in the collision transfers energy $C_v T$, where the value of C_v was defined above. Assuming that the separating plane is situated between two neighbouring particles, we write the expression for the energy flux: $J = 2z^* v\lambda \, [S(x = -\lambda/2) - S(x = \lambda/2)]/6$, where we used the notion of 'property' S introduced above. As a result, we obtain

$$\kappa_2 = z^* v(\theta)\lambda^2 \theta_{AA} C_v/3. \qquad (33.5)$$

The concentration dependence of the frequency of the harmonic oscillations of the particles in the model used is expressed as

$$v(\theta)/v_1 = \left\{ \left[1+(z-1)t\right]\left(1+t\delta_m\right)^{n+2/n-m}\Big/6\left(1+t\delta_n\right)^{m+2/n-m}\right\}^{1/2}, \qquad (33.6)$$

where $v_1 = \{mnD/\mu\}^{1/2}/2\pi r_e$ is the frequency of harmonic oscillations of an isolated dimer formed by fluid particles (r_e is the position of the minimum of the potential φ at $s_{n,m} = 0$). Ratio $v(\theta)/v_1$ is obtained if for any particle we fix one of its neighbouring particles and carry out averaging for all other neighbours. According to (33.6), the ratio of the vibration frequency $v(\theta = 1)/v_1$ for a cubic lattice ($z = 6$) is 1.42 and for the face-centreed lattice ($z = 12$) -1.74. Thus, the total thermal conductivity coefficient $\kappa = \kappa_1 + \kappa_2$ at low concentrations has a linear concentration dependence, and at high concentrations of the fluid the power of the order 4 with respect to θ.

For small fillings of the bulk phase the limit values of the coefficients are equal

$$D^* = w\rho\gamma_D, \quad \eta = m\theta w\rho\gamma_\eta, \quad \kappa = \theta C_v w\rho\gamma_\kappa, \qquad (33.7)$$

where all numerical coefficients are $\gamma_D = \gamma_\eta = \gamma_\kappa = 1/3$. More accurate kinetic theories give the following values of these coefficients: $\gamma_D = 3/8$, $\gamma_\eta = 5/16$, $\gamma_\kappa = 25/32$ [3]. In the formula (33.7) we must include all the previously constructed concentration dependences of the kinetic coefficients with decreasing concentration of the adsorbate.

In concluding this section we note that the shear viscosity coefficient can be obtained by generalizing the Eyring model [14]. This path is an approximation of the equation derived above in the form of contributions from two mechanisms of the momentum transfer $\eta = \eta_1 + \eta_2$. In the Eyring model, both transfer channels are approximated by a single expression for the thermal velocity of the molecule. The original Eyring model was proposed for the cell model of the liquid phase (in the absence of vacancies). It is used for liquids, but can not be transferred to the dense and rarefied gases. The generalizations of the LGM made in [15] for the whole range of densities and temperatures, allows to use the same molecular energy parameters in the entire concentration range of the substance. The final expression for η can be written as

$$\eta = \eta_0/K_{\ni\phi}, \quad K_{\ni\phi} = U_{av}(\rho)/\theta, \qquad (33.8)$$

where $\eta_0 = (mkT/\pi)^{1/2}/(\pi\sigma^2)$ is the viscosity of the ideal rarefied gas, m is the mass of the atom, $U_{av}(\rho)$ is the average speed of the particle hopping over the distance ρ (ρ is the length of hopping of particle A to a vacant site) and is expressed by the formulas in section 28 and Appendix 7.

At low densities, the viscosity varies linearly with θ. It should be noted that the nature of the temperature dependence of the shear viscosity changes for different densities. At low densities, η depends on $T^{1/2}$, and at high densities it shows an exponential dependence on temperature, as in the traditional Eyring model [14].

The constructed kinetic coefficients are used for the self-consistent description of the fluxes of particles, momentum and energy in the whole range of concentrations, including the area of phase transitions and critical area, allowing their use for calculating flows in the capillary condensation of particles in the mesopores. In the limiting areas of concentrations these expressions become the well-known rigorous expressions of the molecular–kinetic rigorous theory of gases and liquids, as well as the correct temperature dependence of the transport coefficients in the different concentration ranges. Similarly, all the modifications of the LGM, discussed in section 9, can be taken into account.

34. The concentration dependence of the transport coefficients in the volume

The derived expressions for the coefficients of shear viscosity and thermal conductivity of the transfer were tested by experimental data. When calculating the dynamic characteristics we introduce the energy parameter of interaction between the molecules in the transition state ε^* with neighbours in the ground state. This parameter is different from the interaction of molecules in the ground state ε. All other molecular parameters that are discussed in section 11 remain unchanged.

For argon atoms [16, 17] comparison with the experimental data was performed using a modified version of the LGM, taking into account the two-body interactions at arbitrary distances and triple interactions of the nearest particles [4, 15] as well as the compressibility of the lattice structure [18] (see equations in section 9). The results are compared with similar calculations performed on the basis on the 'quasichemical' gas kinetic model [3, 16, 17, 19, 20].

According to the experimental data the value of the shear viscosity coefficient depends not only on temperature but also on pressure $\eta = \eta_0 (1 + \alpha p + ...)$, where η_0 is the coefficient of viscosity for a rarefied gas. In [19], the viscosity of the gas in which simple association of molecules can take place, i.e., the formation of dimers and quasidimers of the gas was introduced in the form $\eta = \eta_a + \eta_c$. The first term represents the viscosity of a mixture of monomers and dimers, and it is described using conventional expressions for an ideal gas mixture. For example, using the well-known Wilkie formula

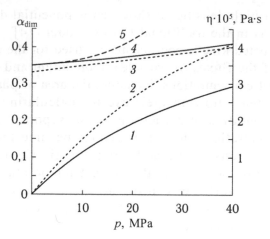

Fig. 34.1. Molar fraction of bonded dimers of argon (curves *1* and *2*) and the dynamic viscosity of argon (curves *3–5*) at $T = 500$ K. Calculations by the gas kinetic model (curves *1, 4, 5*) and LGM (lines *2* and *3*).

[20] $\eta_a = \eta_a (x_i, \eta_i^0)$, where x_i and η_i^0 are the mole fractions and viscosity coefficients of the of the mixture components. The second term η_c, due to collisions of molecules, for the hard-sphere model can be represented as [19]: $\eta c = 7b\eta^{(0)}n/(40V)$, where the value of $b = d[TB(T)]/dT$. Here $B(T)$ is the second virial coefficient, which can be presented in three parts (see section 11).

Figure 34.1 shows the curves *1* and *2* for the mole fractions of the bonded argon dimers depending on the gas pressdure at constant temperature $T = 500$ K. At low pressures, these curves are similar, but they differ by about 30% at a pressure of 400 atm.

The dependence of the viscosity coefficient on pressure is given by the curve *3* (LGM), *4* (taking into account the second and third coefficients in the virial expansion of the viscosity coefficient) [21] and *5* (for the gas-kinetic model). Calculations using the formula for the gas-kinetic model $\eta = \eta_a + \eta_c$ (curve *5*) coincide with the curve *4* only for small values of the pressure (up to 10 atm). This corresponds to an area in which the mole fraction of the bonded dimers α_{dim} is close to the analogous value for the lattice gas model. As can be seen from Fig. 34.1, the curve for the lattice model is close to the estimates by the virial coefficient of viscosity in a wide pressure range, differing only slightly.

Figure 34.2 *a* shows the effect of the ratio of the energy parameter $\alpha = \varepsilon^*/\varepsilon$ for the shear viscosity coefficient of pure argon in the bulk phase on a diluted gas to the liquid state. Experimental data [22]

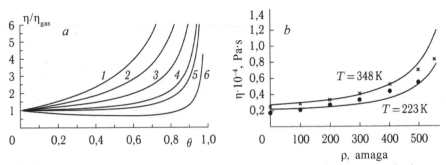

Fig. 34.2. The concentration dependence of the viscosity of argon: a – the impact of the parameter ε^* (α varies from 0.4 to 0.9, $z = 12$), b – the effect of temperature. Circles and crosses – the experimental data [16, 17, 20], the lines – calculation by the lattice gas model.

correspond to the value $\alpha = 0.85$ for $z = 6$ or $\alpha = 0.60$ for $z = 12$. It can be seen that the change of this parameter significantly changes the course of the concentration dependence of the viscosity on the density of argon.

Figure 34.2 b compares the concentration dependences of the viscosity coefficient, calculated in the lattice gas model at two temperatures, with the experimental data presented in [16, 17, 22]. The magnitude of ε was found from the equilibrium data for the compressibility factor and in the calculation of the dynamic characteristic (viscosity coefficient) its value did not vary.

Other gases [23]

Calculations of the viscosity as the compressibility coefficient (see section 11) for helium, hydrogen, methane and nitrogen were carried out in the ranges of temperatures $T_c < T < 400$ K and pressure $P_c < P < 100$ MPa, which were best represent the experimental results [24–29]. This range of pressures and temperatures relates to the supercritical range of these gases. The equations of equilibrium distribution and the parameters used are specified in section 11.

Figure 34.3 a shows the dependences of the coefficient of dynamic viscosity on the pressure of helium and hydrogen. The model allows to obtain a correct description of the viscosity in the same pressure range for helium atoms and hydrogen molecules for the two temperatures (Fig. 34.3 a), and Fig. 34.3 b shows the viscosity curves for nitrogen molecules at two temperatures as a function of pressure. Figure 34.3 b also shows the same dependences as a function of the density of the gas phase. The unit of amaga is the density of

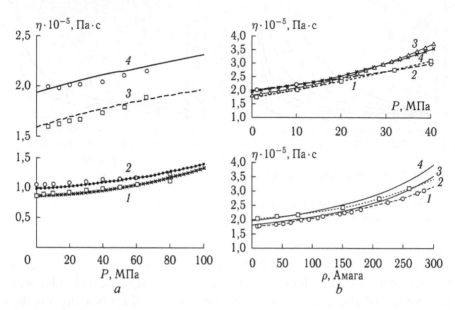

Fig. 34.3. The dependence of the dynamic viscosity coefficient on pressure, the curves – calculations, icons – the experiment (*a*): molecular hydrogen at T = 288 (*1*) and 373 K (*2*) [24, 25]; helium at T = 223 (*3*) and 298 K (*4*) [25, 26]. The dependence of the dynamic viscosity η (Pa·s), molecular nitrogen on pressure P, MPa, and the density ρ, amaga (*b*): the dashed lines *1* and *2*, marked by open circles and squares, experiment [24, 29] for T = 298 and 348 K, respectively; solid lines *3* and *4* – calculations at the same temperature.

nitrogen under standard conditions [8, 24]. The dimensionless density in the lattice gas model θ (as in Fig. 2) is directly proportional to the amaga. Table 2.4 summarizes the parameters corresponding to the curves shown.

The values of particle diameter σ in Table 2.4 were selected to match the theoretical and experimental values of the viscosity under standard conditions. The parameter α* was determined from the experimental curves of the dependence of viscosity on pressure at constant parameters found from the compressibility curves in Fig. 11.4 and viscosity at zero coverages.

The values of the energy parameters of the Lennard–Jones potential in the calculations were used as basic values. However, because of the rather large spread in the values of the experimental data, some parameters had to be matched with each other at zero coverages. Recall that in the LGM all the characteristics of the system in an ideal gas are assumed to be known. The model itself is

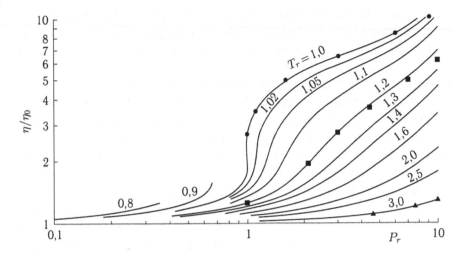

Fig. 34.4. Generalized dependence of the viscosity of the molecules, obeying the law of corresponding states [2, 3].

focused on the extrapolation of the behaviour of these characteristics in the region of large concentrations of molecules.

In describing nitrogen viscosity in the range up to 300 atm the model qualitatively ensures the proper concentration behaviour and describes the intersection of the curves for the two temperatures differing by 50°C (Fig. 34.3 b). The same viscosities as a function of the density do not cause intersections. With the variation of the temperature of the gas by a relatively low temperature (under sufficiently 'normal' variation of the gas density with pressure in the vicinity of the curves 6 and 7 in Figs. 11.2 and 11.3) this effect is not usually discussed in terms of virial equations [19]. Both dependences of viscosity on Fig. 34.3 b on pressure and density are described qualitatively correctly by this model. At higher pressures the model overestimates the data relative to the experiment. A more detailed discussion of the parameters and description other experiments can be found in [23].

Note that a large value of parameter d_e for helium physically means changing of the nature of the interaction between helium atoms at a certain value of its concentration on the opposite: the attraction of atoms at low densities is replaced by their repulsion with increasing concentration. The experimental data on the coefficient of compressibility and viscosity coefficient are described in a self-consistent manner to each other and the model can therefore be considered sufficiently reliable for a number of predictions. In particular, when the temperature is lowered by 3–5 K this results in an anomalously sharp decrease in the viscosity coefficient, which has traditionally been interpreted as the superfluidity phenomenon [30].

Generalized dependence of the viscosity coefficient

In [31] the authors compared the generalized shear viscosity coefficients of simple gases in a wide range of pressures (from rarefied states to 10 critical values) and temperatures (from subcritical to supercritical) based on the same potential functions of the interparticle interaction.

The solid lines in Fig. 34.4 show a generalized diagram of the reduced dynamic fluid viscosity $\eta^* = \eta/\eta_0$ as a function of the reduced pressure P_r for different values of the reduced temperature T_r. Calculations were carried out at the same values of the parameter of the intermolecular interaction potential (11.2) as in the calculations of the generalized compressibility factor. In addition to $\varepsilon(r)$, equation (33.8) includes the interaction parameter of the activated complex $\varepsilon^*(r)$ with neighbouring molecules. In [31] the following expressions was used for $\varepsilon^*(r)$:

$$\varepsilon^*(r) = \alpha\varepsilon(r)\left[1 - \delta_{1,r}\theta\left(d_e^* + \theta m^*(\theta)\right)\right]f^*(T), \qquad (34.1)$$

which corresponds to the usual relationship [32] between the energy parameters of the molecules in the ground and transition states, i.e. the characteristics of the potential in these two states are different (here this relates to the parameter d_e and function $f(T)$ in the formula (11.2)). Formula (34.1) contains the modified dependence of many-particle contributions – function $m^*(\theta)$, decreasing from 0.5 to –0.2 is added (compared with a constant parameter in [4, 32]). Comparison with the experimental generalized viscosity diagram shows that $\alpha = 0.5$, $d_e^* = 0.4$, $u^* = 0.45$, $u'^* = 1.0$ and the formal function $m^*(\theta)$ can be represented as $0.81 \exp[-0.47 \, P_r(\theta)] - 0.025 \, P_r(\theta)$.

The generalized diagram is only in qualitative but also quantitative agreement, with an average deviation of the modulus equal to about 6%, with the experimental values of the shear viscosity (icons in the figure), obtained in [33] (see also Fig. 139 [3]). There is reason to assume that this diagram has (similar to the compressibility diagram) a sufficiently general meaning, and can serve for qualitative estimates of the shear viscosity of other gases.

The calculations in [31] used the simplest form of relationshipis factorising the temperature dependence of the parameter $\varepsilon(r)$, although this relationship is more complex. The lattice gas model

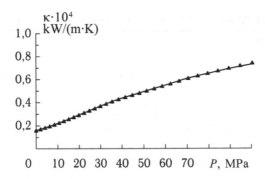

Fig. 34.5. Experimental [34, 35] and the calculated dependence of thermal conductivity on argon pressure in the bulk phase at $T = 273$ K.

also allows to obtain more stringent relationships to account for the contributions of triple and/or many-particle interactions and perform calculations of the thermodynamic parameters in both subcritical, supercritical and near-critical areas.

Thermal conductivity
As stated above in section 11, the data on the bulk phase were used to determine the parameters of the lateral interaction of the adsorbate in the model. To describe the experimental data, the authors of [34, 35] used the effective Lennard–Jones potential (34.1). The dependence of the compressibility factor on the density of the fluid for two temperatures was used to determine the parameters $d_e = 0.242$ and $u = -100$ K. The value $\varepsilon_0 = 238$ cal/mol, corresponding to the depth of the potential well of the insulated argon dimer, is the known value obtained from the analysis of the second virial coefficients. At the determined parameters d_e and u the value of ε is less than 238 cal/mol, which agrees with the results of [36].

Analysis of the concentration dependence of the coefficient of thermal conductivity was conducted under the same modifications of the interaction potential between the argon atoms at a temperature $T = 273$ K. Figure 34.5 compares the experimental data [34, 35] and the theoretical curve at a value of the dimensionless parameter $\alpha = \varepsilon^*/\varepsilon = 0.5$. The results shown on Fig. 34.5 show that the lattice model satisfactorily describes the bulk properties of inert gas (argon) to pressures of thousands of atmospheres when not only paired but also triple contributions to the interaction potential are significant.

35. Molecular models of the transport of molecules in slit-like pores

In nonuniform systems the state of the fluid is characterised by the local concentration θ_f and temperature T_f. All dissipative coefficients describe the dynamics of interaction of the neighbouring molecules in different cells, so the property of non-locality of these coefficients is principal, along with the property of non-locality when considering the intermolecular interactions using the microhydrodynamic approach in the LGM. Recall that all dissipative coefficients for molecules in the central and near-surface cells inside a pore are expressed only in terms of the parameters of the intermolecular potential between molecules and the molecule–wall potential.

As above, for the calculation of the kinetic coefficients (as usual [3, 7]) we allocate in space a plane 0 and consider the particle fluxes and the momentum and energy transferred by them. Each cell has its own degree of population, so the dedicated plane 0 should relate to the common face of the cells fg between which the local property S_f is transferred. We will use with the concept of average thermal velocity of the particle in the equilibrium state of the fluid w_{fg}, located in the cell f and moving in the direction of cell g which contains a vacancy. We confine ourselves to jumps in the next sites. The average thermal velocity of the molecule between cells f and g is $w_{fg} = \lambda U_{fg}/\theta_f$, where the value U_{fg} is the hopping probability of the molecules between cells f and g per unit time.

In this case, the expression for the hopping velocity U_{fg} is described by the following formula [4, 37, 38]:

$$W_{fg} = K_{fg} V_{fg}, \quad K_{fg} = F_f^* \exp\left(-\beta E_{fg}\right)/\left(\beta h F_f\right), \quad V_{fg} = \theta_{fg}^{AV} T_{fg},$$

$$\theta_{fg}^{AV} = \theta_{fg\,(1)}^{AV}, \quad T_{fg} = \prod_{\zeta \in z_f - 1} S_{f\zeta}^A \prod_{\zeta \in z_g - 1} S_{g\zeta}^V, \tag{35.1}$$

where K_{fg} – the hopping velocity constant of the particle from the cell f to the nearest free cell g in the empty lattice, F_f^* and F_f are the statistical sums of the states of particles in the cell f in transition and ground states respectively, E_{fg} – the hopping activation energy (for cells that are far away from the pore wall $E_{fg} = 0$, and near the walls with the adsorption potential $E_{fg} \neq 0$). The hopping velocity constants for different cells meet the condition that the the probability of the particle leaving the energy decreases with increasing depth of the well. The connection between the hopping constants and local Henry constants is described by the expressions: $a_f K_{gf}^{AV} = a_g K_{gf}^{AV}$. This takes into account that $a_f = a_f^0 \exp(\beta Q_f)$.

The concentration dependence of the migration rate of the molecule lies in the factor V_{fg}, which, in turn, consists of factors θ_{fg}^{AV} (the probability of finding a free neighbouring cell g close to the particle A in the site f) and T_{fg} (consideration of the effect of interactions between molecules located around the central pair of particles AV on sites fg, on the energy of the activated complex). The factor $S_{f\xi}^A = 1 + t_{f\xi} x^*$ included in T_{fg} belongs to $(z - 1)$ nearest neighbours ξ of the molecule A in the cell f, located in the ground state (cell g is excluded from the neighbours A); similarly, factor $S_{g\zeta}^V = 1 + t_{g\zeta}^{VA} y^*$ refers to $(z - 1)$ neighbors of the free cell g. Here $x^* = \exp[\beta(\epsilon^* - \epsilon)] - 1$, $y^* = \exp(\beta\epsilon^*) - 1$. In the absence of lateral interactions, formula (35.1) takes the form $U_{fg} = K_{fg}\theta_f(1 - \theta_g)$. Away from the from the walls of the pore the ratio F_f / F_f^* corresponds to the statistical sum of the translational degrees of freedom in the direction of motion of the particle, which is equal to $(2\pi m\beta^{-1})^{1/2}\lambda/h$. Then the hopping velocity constant in the bulk phase can be written as $K_{fg} = w/4\lambda$, where $w = (8/\pi m\beta)^{1/2}$ is the thermal motion velocity of the molecules in the gas phase (see [3]).

Inhomogeneous systems retain the same distinction in the limiting expressions for the thermal velocity of the molecules, as well as for the bulk phase [3, 16]. This discrepancy appears in the elementary kinetic theory due to the use of average speeds along the preferred direction (or in different directions from the considered cell) instead of averaging over the velocity distribution function. This question is not discussed again below for all the transport coefficients for two reasons. First – all relationships we are interested are associated with concentration dependences for which the limiting case of low concentrations serves as a point of reference. When compared with the experiment in the bulk phase the initial reference point is the experimental value for a rarefied gas, which, as shown above, is in good agreement with the measured concentration dependences of the transfer coefficients. As a reference point we use the limiting value of the appropriate transfer coefficient in the bulk phase. The second reason stems from the fact that the difference between strict and elementary kinetic theories is largely leveled if, instead of the average thermal velocity we use the average relative velocity of the molecules (Appendix 12). This replacement is crucial in mixtures for the mutual diffusion coefficient.

Self-diffusion coefficient

The self-diffusion coefficient characterizes the transfer of labeled particles in the equilibrium state of the system due to the thermal motion of all particles for any type of system. The resultant stream of the labeled particles consist of their forward and reverse flow through the selected plane perpendicular to the direction of the concentration gradient – in this case normal to the common edge between the adjacent cells fg.

The equations for the self-diffusion coefficient in inhomogeneous media were obtained in [39, 40] (section 28). Derivation of this expression is similar to that for homogeneous systems, taking into account the differences of the thermal velocity of the molecules in strong fields. As a result, due to the inhomogeneity of the equilibrium distribution of molecules in strongly inhomogeneous systems the self-diffusion coefficient essentially depends on the direction of the motion under consideration. According to these results we obtain the following expression for the local self-diffusion coefficient, which characterizes the process of redistribution of molecules between the adjacent cells:

$$D_{fg}^*(\rho) = \rho^2 z_{fg}^* U_{fg}(\rho)/\theta_f, \qquad (35.2)$$

where z_{fg}^* is the number of possible jumps between adjacent cells at a distance ρ for each cell f. As the average characteristic of the motion of the marks along the axis of the pore we have the following expression for the self-diffusion coefficient:

$$D^* = \lambda^2 \sum_{q=1}^{t} F_q \sum_{p=1}^{t} z_{qp} U_{qp} \theta_q \frac{d\theta_q}{d\theta}, \qquad (35.3)$$

where z_{qp} is the number of bonds of the site in layer q with neighbouring sites in the layer p, $d\theta_q^*/d\theta^* = d\theta_q/d\theta$. Here we take into account all the jumps of the migrating molecule in different types of neighbouring sites. Averaging is carried out using the contributions to the total flux of molecules from the sites to which the migrating molecule travels (type p site) and contributions from the sites from which it migrates (type q site).

Thermal conductivity

Expressions for the contributions of the first channel to the coefficients κ_{fg} are obtained in the same manner as for the bulk phase.

As mentioned above, two energy transfer channels operate in dense fluids. The first is connected with the movement of the particles, as in the rarefied phase, the second is determined by collisions between particles. The thermal motion of the molecules associated with the transfer of their own energy in different directions. The resulting flow of this motion of molecules in a certain direction can be expressed in terms of the self-diffusion coefficient. Of crucial importance for pore systems is the fact that they represent highly nonuniform systems in which thermal motion is strongly anisotropic due to the impact of the potential of the pore walls. This also leads to an anisotropy of the contribution of the first channel of thermal conductivity to the corresponding coefficient $\kappa_{fg}^{(1)}$. At the thermal conductivity process temperature T is different in different planes, $S = \theta_f C_v (f) T_f$, $C_v(f)$ is the specific heat capacity per particle.

As a result, we have an expression for the thermal conductivity coefficient $\kappa_{fg}^{(1)}$:

$$\kappa_{fg}^{(1)} = C_v(f)\theta_f D_{fg}^*(\rho)/v_0, \tag{35.4}$$

where the local self-diffusion coefficient in inhomogeneous media is defined in (35.2), and the specific heat capacity $C_v(f)$ per particle is defined in section 17.

To calculate the contribution to the thermal conductivity of energy transfer through collisions of particles in a dense fluid (second channel), we consider the energy flux carried by the vibrations of neighbouring particles [41]. Every oscillation along the flow direction carries a certain amount of energy by contact between the particles lying in adjacent planes f and g at a distance λ. Transfer occurs $2v_{fg}$ times per second by each of θ_f particles on the surface (the fact that a particle and not a vacancy is in the vicinity is taken into account by function t_{fg}^{AA}). It is assumed that the characteristic vibrational relaxation time is less than the characteristic time for local redistribution of molecules characterized by the evolution of the function t_{fg}^{AA}, discussed in chapter 4.

In a collision each particle transfers energy $C_v(f)T_f$, where $C_v(f)$ is the the specific heat per molecule (section 17). Assuming that the separating plane is situated between two neighbouring particles, we can write the expression for the energy flux $J = 2z^*v\lambda[S\ (x = -\lambda/2) - S\ (x = \lambda/2)]/6$, where we use the above-introduced the concept of 'property' S. As a result, we obtain

$$\kappa_{fg}^{(2)} = z_{fg}^* C_v(f)\lambda^2 v_{fg}\,\theta_{fg}^{AA}/3, \tag{35.5}$$

where v_{fg} is expressed by the formula in section 17.

The anisotropy of the distribution of neighbouring molecules θ_{fg}^{AA} causes the anisotropy of the contribution of the second channel to the energy flux. Given also the anisotropic nature of the local self-diffusion coefficients, the total thermal conductivity coefficient $\kappa_{fg} = \kappa_{fg}^{(1)} + \kappa_{fg}^{(2)}$ is anisotropic. Values of κ_{fg} for different positions of the molecules in the pore and the directions of their motion are determined by the molecular properties of the adsorbent–adsorbate system (masses of the particles and their interaction potentials). At low concentrations κ_{fg} has a linear concentration dependence, and at high concentrations of the fluid the power for θ can be about 3–4.

Viscosity coefficients

The expressions for the contributions of the first channel to the coefficients η_{fg} and for the bulk phase are obtained in a similar manner. In the case of viscosity the molecules in planes $x = \pm \lambda$ move parallel to these planes in different directions y, and in the plane 0 the particles do not move in the direction y. We are talking about the shear viscosity and its y-component; $S = m\theta_f w_{fg}(y)$, m – the mass of the particle, $w_{fg}(y)$ – the y-component of the velocity of the molecule in the direction from site f to site g. As a result, we obtain the following expression for the shear viscosity $\eta_{fg}^{(1)}$:

$$\eta_{fg}^{(1)} = m\theta_f D_{fg}^{*}(\rho)/v_0. \tag{35.6}$$

To calculate the contributions of the second channel to the shear viscosity $\eta_{fg}^{(2)}$ as well as to obtain an expression for the bulk viscosity coefficient (ξ_{fg}), we use the model of a solid, which allow us to express the local elastic constants of lattice structures if we know intermolecular potentials through the elasticity coefficients of specific bonds between molecules [41].

As above [41], we restrict ourselves to the simplest case of cubic structures and recognize that the only process (other than the considered elastic reactions), responsible for the relaxation of the medium at different external perturbations, is the migration of molecules and therefore vacancies. This implies that the relaxation times corresponding to the shear and bulk viscosity, coincide and are equal to $\tau = 1/K_{fg}(ef)$, where $K_{fg}(ef) = U_{fg}(\rho)/\theta$. As a result, we obtain expressions for the coefficients of the second component of the shear viscosity and bulk viscosity:

$$\eta_{fg}^{(2)} = M(fg)\theta_f \left[t_{fg}^{AA}\right]^k \Big/ \left(\lambda U_{fg}^{AV}(\rho)\right),$$

$$\xi_{fg} = 4\theta \left[t_{fg}^{AA}\right]^k M(fg) \Big/ \left(3\lambda U_{fg}^{AV}(\rho)\right),$$

(35.7)

where the coefficient of elasticity of the bond $M(fg)$ is expressed by the parameters of the intermolecular interaction ϕ, the value of the exponent k depends the accuracy of describing the fragment of the condensed phase containing a site f, the minimum value $k = 1$. Equations (35.7) indicate that both viscosities of monatomic particles have similar values in the region of the solid phase (this finding is consistent with experimental data for simple liquids [29]).

All the necessary molecular characteristics discussed in section 17 (elastic modulus $M(fg)$ of the bond fg, $|u|_{fg}$ – the binding energy of the particle in the condensed phase, the equilibrium distance of neighboring particles r_{fg}^{min}, the total energy of interaction of the core particle with the environment is written as ϕ_p, etc.).

Obviously, the general expression for the shear viscosity is written as follows:

$$\eta_{fg} = \eta_{fg}^{(1)} + \eta_{fg}^{(2)}.$$

(35.8)

Generalization of the Eyring model for the shear viscosity coefficient in pores

The LGM provides a generalization of the Eyring model [14] for the entire range of densities of the fluid [15]. Omitting the intermediate stage of derivation, we present the expression for the viscosity coefficient η_{fg} in shear of the fluid in the cell g relative to cell f, which is sufficient for the analysis of concentration dependences:

$$\eta_{fg} = \eta_0 \theta_f \exp\left(\beta E_{fg}(\rho)\right) \Big/ V_{fg},$$

(35.9)

where $\eta_0 = (mkT/\pi)^{1/2}/(\pi\sigma^2)$ is the viscosity of the ideal diluted gas, m is the mass of the atom, σ is the the diameter of the molecule, V_{fg} is the concentration dependence of the thermal motion of molecules between adjacent sites f and g. At low densities, the viscosity varies linearly with θ as for the ideal gas, with increasing θ the course of η (θ) depends on the ratio of the parameters ε and ε^*. At low densities η depends on temperature as $T^{1/2}$, and at high densities we have an exponential dependence on temperature, as in the traditional Eyring model [14]. Formula (35.9) is a simplified version of the expression (35.8). Nevertheless, it reflects the limiting values of the formula

(35.8). Using formula (35.8), as well as formulas for the local heat capacity (35.5) and the local reduced mass (μ (f)) is based on a simple description of the inhomogeneous fluid – taking into account the states of the cluster consisting of the nearest neighbours.

36. Concentration dependences of the diffusion coefficients and viscosity

The energy characteristics of the molecules in the pores were described in sections 4, 14, 15. When calculating the kinetic coefficients the energy characteristics are assumed to be known. A new parameter for porous systems is the activation energy of surface migration – the stage of particle hopping from one site to another on the surface of the pore. We assume that it can be specified as $E_{11} = \alpha_{11} Q_1$, α_{11} is the parameter reflecting the height of the surface relief [4], and $E_{12} = Q_1$. As above, $\alpha = \varepsilon^*/\varepsilon$. All concentration curves for kinetic coefficients are constructed in the normalized form. Normalization was carried out for the corresponding values of the coefficients of self-diffusion and viscosity at $\theta = 0$ and $Q_1 = 0$ (i.e., for the bulk phase) [37, 38, 42].

Figure 36.1 a shows the concentration dependences of the local and average (over the cross section) self-diffusion coefficients in a pore with a width of 18 monolayers. All local coefficients decrease with increasing degree of filling of the pores, as the free volume of the pores decreases. The most dramatic change is observed in the movement of molecules in the first (surface) layer (curve 1).

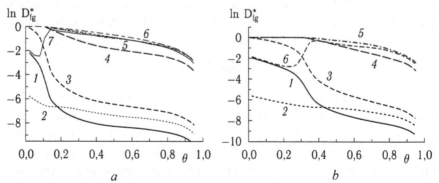

Fig. 36.1. The self-diffusion coefficients of argon slit pores in graphite 18 (a) and 6 (b) monolayers wide at $\alpha = 0.5$, $\alpha_{11} = 0.333$. The curves in a correspond local values of D^*_{fg} [cm²/s] for pairs of adjacent cells of the layers $fg = 11$ (1), 12 (2), 21 (3), 22 (4), 33 (5), 99 (6), curve 7 – average value of D^*. In b curves correspond to pairs of adjacent cells of layers: $fg = 11$ (1), 12 (2), 21 (3), 22 (4), 33 (5), the curve 6 – average value of D^*.

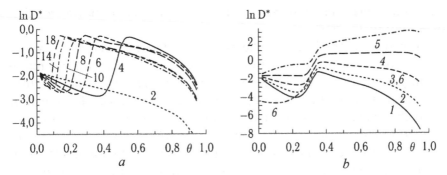

Fig. 36.2. Concentration dependence of the mean normalized values of self-diffusion coefficient of argon D^* [cm²/s]: a – in the pores of different widths (numbers of the curves correspond to the width of the pores, expressed in the number of monolayers); b – width of the pores 6 monolayers when $\alpha = -0.5$ (*1*), 0.0 (*2*), 0.5 (*3, 6*), 1.0 (*4*), 1.5 (*5*) and $\alpha_{11} = 0.33$ (*1–5*) or 1.33 (*6*).

As it is being filled there is a sharp decrease at $\theta < 0.1$, wherein the monolayer film is formed. Further reduction of D^*_{11} is connected with its densening. To move from the first layer to the second molecule it is necessary to overcome the binding energy equal to Q_1, therefore the curve *2* for D^*_{12} is below all the other curves. The reverse transition of molecules from the second to first layer corresponds to the coefficient D^*_{21} – curve *3* for it 'repeats' the curve *1* – its decrease is associated with the filling of the first layer. The movement of molecules of the second layer (curve *4*) for low coverages is the same as in the centre of the pores (curves *5* and *6*). Decrease of D^*_{22} begins at $\theta > 0.1$ after filling the surface monolayer.

Coefficient D^* reflects the contributions of filling of the individual layers, so initially it is close to D^*_{11}; at average filling it is close to the curve for the second layer, and at higher fillings it is close to the curve for the central portion of the pore. The ratio of the values D^* ($\theta = 0$) and D^* at the maximum point on the curve *7* is dependent on the activation energy of surface migration E_{11}: as E_{11} decreases $D^*(\theta = 0)$ increases.

Figure 36.1 *b* shows local self-diffusion coefficients of argon in dependence on the degree of filling of the pores in graphite with the width $H = 6$, and also their average values over the section of the pores. If $f = g$ the curves correspond to the motion of the atoms in the layer, and at $f \neq g$ – to the movement of atoms between layers. Curve *1* corresponds to the motion of the argon atoms along the surface of the pores when they overcome the activation barrier with the height E_{11}. At low fillings curve *1* is located above the curve *2*, since the transition from the first layer to the second barrier requires overcoming $E_{12} > E_{11}$.

However, with filling of the first monolayer (when $\theta > 0.4$) the proportion of free cells, enabling the movement of argon, decreases sharply and and the transition of an atom to the second layer (curve 2 is located above the curve 1) is more advantageous. The transition from the second to the first layer does not require overcoming the activation barrier, and reducing of D^*_{21} (curve 3) is also explained by the filling of the surface monolayer. The self-diffusion coefficients of argon in the second and third layers (curves 4 and 5) vary slightly at low fillings and decrease slowly with further filling of the pores. These values are sufficiently close to each other even at high fillings.

These patterns are characteristic of the pores of any width (Fig. 36.2 a) (where $\alpha = 0.5$, $\alpha_{11} = 0.333$). The case $H = 2$ is special, because here all the sites are equally, and there is a monotonic decrease of D^* in the entire region θ.

Figure 36.2 b shows the concentration dependence of the mean values of D^* over the cross section for different molecular parameters of the adsorption system. Curves 1–5 correspond to the variation of the parameter α, which characterizes the interactions between the argon atoms in the transition and ground states (with $\alpha_{11} = 0.333$). Calculations show that these interactions significantly affect the numerical values of the self-diffusion coefficients. The stronger the mutual attraction of argon atoms in the transition state, the greater the D^*. However, in all cases a significant contribution comes from the interaction with the pore wall, and the growth of θ leads to a decrease in D^*.

Increase of the activation barrier of the surface migration process (curve 6 in Fig. 36.2 b) dramatically reduces the self-diffusion coefficient at low densities. However, with filling of the surface monolayer potential the impact of the wall potential decreases and the self-diffusion coefficients in the subsequent layers cease to depend on the wall properties.

Viscosity coefficients

Calculation of the viscosity coefficients was carried out at the same molecular parameters that were used in the evaluation of the self-diffusion coefficients. Similarly, the concentration dependences wre normalized to the corresponding values of the viscosity at $\theta = 0$ and $Q_1 = 0$.

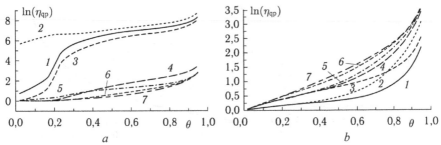

Fig. 36.3. Concentration dependences of the local shear viscosity coefficients: a – at $\alpha = 0.5$, $\alpha_{11} = 0.1$, $Q_1 = 9.24\varepsilon$; b – at $\alpha = 0.5$, $\alpha_{11} = 0.1$, $Q_1 = -1.3\varepsilon$. The curves correspond to the following pairs of adjacent cells in layers $qp = 11$ (*1*), 12 (*2*), 21 (*3*), 22 (*4*), 23 (*5*), 33 (*6*), 55 (*7*).

The concentration dependences of the local viscosity coefficients are shown in Fig. 36.3 *a* for $H = 10$. They characterize the deceleration of the flow (dissipation of momentum) as it moves through the pore. Their values depend strongly on the local flow directions and distances to the walls of the pores. This set of parameters corresponds to the argon–graphite system with strong attraction of the walls. The position and shape of the curves *1–3* in the wall region are determined by the values of Q_1 and E_{11}. In the central part of the pores (curves *4–7*) the viscosity increases as the pore is filled. Viscosity varies from the values corresponding to the gas phase to values corresponding to the liquid phase [29].

Figure 36.3 *b* shows similar curves for the pores where the walls 'repel' the adsorbate molecules. In this case, the lowest viscosity is realized near the walls and the surface layers have a minimum resistance to flow.

Figure 36.4 *a* shows the profiles of the shear viscosity coefficients of argon atoms in the pore with a width of $H = 10$ monolayers. As the distance of the layer from the pore wall decreases the coefficients η_{qq} decrease, and with increasing degree of filling of the pores they monotonically increase.

Figure 36.4 *b* shows the concentration dependences of the local viscosity in the case of weakly attractive walls in a slit pore. All the curves are relatively close to each other, since the interaction potentials of the wall and the molecules are equal. The shear viscosity of the adsorbate in the central layer depends relatively weakly on the width of the pores, although its influence can not be completely neglected. The viscosity in the surface layer at a fixed θ increases with increasing pore width H, since for $\theta = $ const increasing H increases the degree of filling of the surface layer θ_1.

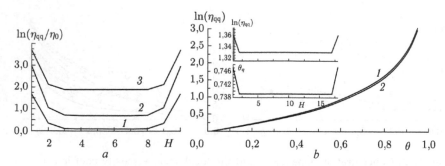

Fig. 36.4. The concentration profiles of the shear viscosity coefficient of argon atoms η_{qq}, $1 \leq q \leq 10$ (a): $\theta = 0.095$ (1), 0.485 (2), 0.905 (3) ($\alpha_{11} = 0.33$). Concentration dependences of the viscosity coefficient for a pore with a width of 18 monolayers in the case of molecules not interacting with the wall of the pores (b): $\beta\varepsilon = -0.024$, $\alpha = 0.5$, $\alpha_{11} = 0$. The inset gives concentration profiles and the shear viscosity at the filling $\theta = 0.74$.

The inset shows the profiles of concentration and viscosity at $\theta = 0.74$; they vary in a similar manner. Calculations in the framework of this lattice gas model are in good agreement with similar results for the concentration and shear viscosity profiles in the slit-shaped pores with a width of 4 and 18 monolayers [43], made by the method of non-equilibrium molecular dynamics and and on the basis of the continuum kinetic theory, which confirm the overall adequacy of the general relationships obtained in both methods.

Figure 36.5 a presents the layered coefficients of shear viscosity of argon in a pore with a width of $H = 8$. As the distance of the layer

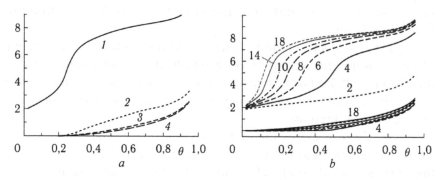

Fig. 36.5. Concentration dependences of the local normalized coefficients of the shear viscosity of argon [Pa·s] at $H = 8$, $\alpha = 0.5$, $\alpha_{11} = 0.33$, $Q_1 = 9.24\ \varepsilon$; (a) numbers of curves correspond to the number of the layer (starting from the pore wall). Concentration dependences of the viscosity of argon in the surface layer (upper group of curves) and in the central layer (lower group of curves) with different pore width; (b) the numbers of curves correspond to the width of the pores, which is expressed in the number of monolayers

from the pore wall increases the viscosity coefficients η_{qq} decrease, and with increasing degree of filling of the pores they monotonically increase. The viscosity in the surface layer shows the strongest dependence on the adsorbate concentration in the surface layer, the smallest impact of the degree of filling is exerted on the viscosity of the layer located in the centre of the pores.

The effect of pore width on the viscosity of the surface layer and in the layer at the centre of the pore can be seen from Fig. 36.5 b. The viscosity of the adsorbate in the central layer is relatively weakly dependent on the width of the pores, although its influence is completely impossible to neglect. The viscosity of the surface layer at a fixed θ increases with the width of the pores H, since at $\theta = $ const the filling of the surface layer θ_1 is greater.

The calculations show that the dynamic characteristics of the adsorbate are strongly dependent on the anisotropic distribution of the molecules over the cross section of the slit pores. Particularly large changes of the self-diffusion coefficients and shear viscosity are observed near the pore walls. In the centre of the pores these values depend on the contribution of the wall potential and the total concentration of the adsorbate. The results imply that the traditional assumptions [44–47] of the constancy of the self-diffusion and shear viscosity coefficients are generally not true. In the analysis of experimental data we must consider the relatively strong concentration dependence of the dynamic characteristics of the adsorbate in narrow pores caused by both the influence of the potential of the pore walls and intermolecular interactions.

37. Sliding friction coefficient

Background information in the microhydrodynamic approach are the potentials of interactions between the adsorbate molecules and between the adsorbate and the adsorbent walls. The interaction potential of the adsorbate with uniform walls of the slit pore with the width H monolayers is represented as a set of discrete values Q_q. This allows a molecular interpretation of the sliding friction coefficient of the fluid near the walls of the pores. This coefficient [2, 48] (β_1) is defined as the ratio of the tangential force per unit surface *relative to the flow rate* near the wall. On any solid wall, this coefficient is present in the boundary condition

$$\beta_1 u = -\eta \partial u / \partial r \big|_{r=R}. \tag{37.1}$$

The ratio $\eta/\beta_1 = \lambda$ has the dimension of length. The factor λ follows from the ratio of the dimensions because unlike β_1 the shear viscosity coefficient η is the ratio of the tangential force per unit of surface to the *velocity gradient* in the direction perpendicular to the flow direction s[48].

With increasing β_1 the speed $u = -\eta \partial u/\partial r|_{r=R}/\beta_1$ on the channel wall vanishes, which corresponds to the traditional boundary condition of the mechanics continua [1]. However, this condition should be associated with the molecular concepts of the surface mobility of the molecules and surface flows that have experimental confirmation [2, 5]. Calculations of the contributions of the 'surface' and 'bulk' transport of molecules to the total flux of labeled molecules along the pore axis (in the absence of hydrodynamic flows) considered in [37, 38], indicate the importance of the role of the surface mobility of the molecules in their thermal motion characterized by the self-diffusion coefficient. The surface flow of the molecules dominates in the case of strong attraction of the walls at low densities of filling the pores and is commensurate with the bulk transport of molecules to almost complete filling of the surface monolayer. At the repulsive potential of the walls the role of surface transport molecules increases with the degree of filling of the pores. An important role played by the activation energy of surface migration of molecules E_{11}. The smaller it is, the greater the contribution of the surface flow of molecules. Sliding effects were previously considered only for rarefied gases. In contrast to the decisive role of the specular reflection of molecules from the wall for rarefied gases, the sliding effect in dense fluids is due to the surface mobility of the molecules.

If we consider the variation of the viscosity coefficient for the surface layer according to (37.1), we obtain that

$$\beta_1 = \eta_{11}/\lambda, \qquad (37.2)$$

where the indices 11 correspond to the near-surface monolayer. As a result, the flow velocity in the subsurface layer is $u = -(\lambda \eta_{tt}/\eta_{11})\partial u/\partial r|_{r=R}$ (*tt* indices refer to the central part of the narrow pore or the bulk value of viscosity for wide channels). In the case of a strong adsorbate–adsorbent attraction or decreasing temperature the ratio η_{11}/η_{tt} sharply increases which leads to an increase in the coefficient β_1 and a decrease of the flow velocity near the wall. Formula (37.2) demonstrates the link of the above presented concentration dependences of the shear viscosity coefficient with the results of flow calculations discussed in chapter 6.

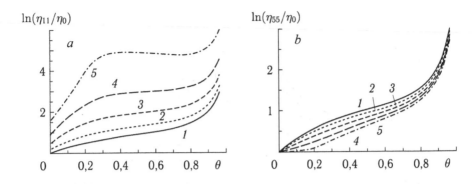

Fig. 37.1. Normalized concentration dependences of the local shear viscosity coefficients of argon atoms η_{11} (*a*) and in the central layer η_{55} (*b*) in the pores with different walls at $T = 273$ K, $H = 10$ monolayers ($\alpha_{11} = 0.33$): $Q_1/\varepsilon = 0$ (*1*), 2 (*2*) 5 (*3*) 9.24 (*4*) 16.5 (*5*).

Figure 37.1 shows the curves describing the concentration dependence of the shear viscosity coefficients on the surface [42]. Parameter α_{11} characterizes the activation energy of hopping along the surface. The non-monotonic variation of curve 5 at high degrees of filling due to the dependence $\varepsilon^*(\theta)$, found in the description of the bulk properties of the thermal conductivity coefficient.

The smaller the value of the activation barrier, the smaller the shear viscosity coefficient. Curve *1* refers to the weak adsorption, the increase of the coefficient is associated with an increase in the degree of filling the surface. Curves *3–5* refer to the strong adsorption and these curves are strongly affected by the energy of the surface barrier. In all cases, increasing the degree of filling increases the shear viscosity coefficient. The range of values of η_{11} greatly widens with increasing fluid density significantly, more than the range of variation of the viscosity coefficient η_{55} in the central monolayer.

Figure 37.1 *b* demonstrates the fact that, although primarily the surface potential directly affects the local coefficients of shear viscosity in the surface layer, it affects on the whole the state of the fluid in the entire pore through the redistribution of the molecules in layers and therefore affects the viscosity coefficients in all areas of the system. These curves correspond to the equilibrium distributions (to zero flow). In these circumstances, the local properties of the fluid vary with increase of the distance from the wall: the farther the studied region is, the smaller in the impact of the wall potential on the transfer coefficients and these properties are closer to those of the bulk phases. In the dynamics there is a more complex picture,

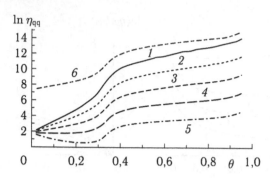

Fig. 37.2. Influence of molecular parameters of the adsorbate on the fluid viscosity in the surface layer of the pore with the width of 6 monolayers (numbers of curves and the parameters are listed in the caption to Fig. 36.2 *b*).

and the local coefficients depend not only on the distance to the wall, but also on the history of the transport process.

Figure 37.2 shows the effect of the molecular parameters of the adsorption system on the viscosity of the adsorbate in the surface layer. Increased activation barrier of argon migration along the surface increases the viscosity of the adsorbate in the surface layer, and increase in the mobility of the atoms decreases it.

Thus, the LGM provides a molecular interpretation of the coefficient of sliding friction. The effect of sliding of the flow for dense fluids is due to surface migration of molecules on solid surfaces. These results demonstrate that for the dense gases an important role is played by the effects of sliding of the dense fluid. These effects were previously considered only for rarefied gases. In dense adsorbates the sliding effect is due to the surface mobility of molecules (in contrast to the decisive role of the specular reflection of molecules from the wall for rarefied gases). Recall that in the gas and hydrodynamic problems in the formulation of the boundary conditions on the walls of the pores the values of the longitudinal and vertical components of the fluid velocity for the boundary are equated to zero on the assumption that the particles of matter are so strongly attracted to walls that they are stationary [1]. However, this condition is not consistent with the concepts of surface mobility and surface flow of the molecules [49]. Thus, the effects of sliding of dense fluids must be considered in the hydrodynamic calculations of the flow of the molecules in narrow pores

38. Concentration dependences of the thermal conductivity coefficient in slits

The most important dynamic characteristics of the adsorbate also include thermal conductivity coefficients [1–3]. Theoretical calculation of these quantities in a wide range of densities (from the gaseous to the liquid state) and temperatures is facing certain problems in the quantitative description even for the bulk phases. For porous systems the situation is more complicated. In contrast to the diffusion coefficients which are relatively reliably calculated by the molecular dynamics method [43, 50, 51], the results for the coefficients of thermal conductivity in the narrow pores are published infrequently. There are a number of papers [52–55], in which the heat flows between the walls of slit-shaped pores are modelled mainly by the molecular dynamics method (for high density) and kinetic theory for low densities [56], but the question of the concentration dependences of the thermal conductivity coefficients of dense gases and liquids in narrow pores of adsorbents to this day remains one of the toughest.

Concentration dependences of the thermal conductivity coefficient in the slit-shaped pores were studied in [56, 57]. They are represented in Fig. 38.1. The total heat transfer coefficient for the two channels is shown in Fig. 38.1 a, and the separate components corresponding to the first and second channels are shown in Figs. 38.1 b and c, respectively. The main result of comparing the contributions of both channels is that the first channel is determining at low filling degrees, but as filling of the pores continues its role is significantly reduced and at medium and large fillings the total coefficient of thermal conductivity is determined by the second channel. Figure 38.1 b shows clearly the region of high mobility of adsorbate atoms along the surface (curve 1), which passes through a maximum upon filling of the first layer (here we consider the structured wall of the surface, but the activation energy of surface migration is small). Relatively small changes in the curves 2 and 3 in Fig. 38.1 b are associated with the high energy for the stratifying of molecules in the transition from the first to the second layer and with preferential filling the first layer blocking the transitions from the second layer to the first layer. Curves 4–8 behave similarly – for them the effect of the wall is small and with filling of the pore volume the given channel conductivity makes a small contribution to the overall energy transfer. The curves in Fig. 38.1 c show that as the distance from the

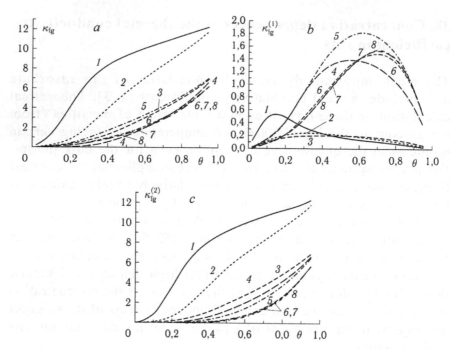

Fig. 38.1. Concentration dependences of thermal conductivity coefficient [W/m·K] for a slit pore with the width of 10 monolayers: a – the total thermal conductivity coefficient, b – the contribution of the first channel and c – the contribution of the second channel. The curves correspond to the local values κ_{fg} of the following pairs of the adjavent cells of lauers: $fg = 11$ (*1*), 12 (*2*), 21 (*3*), 22 (*4*), 23 (*5*), 33 (*6*), 34 (*7*), 55 (*8*).

wall increases the contribution of these local thermal conductivity coefficients decreases, although with increasing filling of the pores with the adsorbate they all increase.

Figure 38.1 illustrates the importance of the contribution of the adsorption potential. The total thermal conductivity coefficient has the same trend. Note the qualitative correlation between the variation of the thermal conductivity coefficient in the bulk phase with the pressure varying from 1 to 1000 atm (five times) with corresponding changes in κ_{fg} for the central portion of the slit pore (curve *8*) after forming two monolayers (filling $\theta \approx 0.4$), if we consider the effect of the coefficient d_e (its inclusion leads to different values of κ_{fg} for small and large θ, with the values differing about 6–7 times).

It should be noted that the formulas used only reflect the heat flow between the adsorbate atoms and do not include the heat transfer with the walls of the pores, so with the decrease in the degree of filling of the pores all the curves tend to zero (whereas for the diluted gas

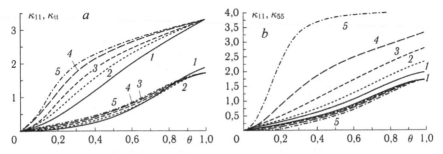

Fig. 38.2. Concentration dependences of the local thermal conductivity coefficient of argon atoms: a – κ_{11} (upper group of curves $1-5$) and κ_{tt} (lower group of curves $1-5$) in slit pores in graphite with the width of 4 (1), 6 (2), 10 (3), 20 (4), 60 (5) monolayers at $\alpha_{11} = 1/3$; b – κ_{11} (upper group of curves $1-5$) and κ_{55} (lower group of curves $1-5$) in slit-shaped pores with a width of 10 monolayers with walls of different nature with different values of $Q_1/\varepsilon = 0$ (1), 2 (2), 5 (3), 9.24 (4), 16.5 (5) when $\alpha_{11} = 1/3$.

the thermal conductivity coefficient in the bulk phase tends to the finite limiting value [2]).

For varying molecular parameters of the porous system below are the κ_{qp} curves for two extreme concentration dependences related to the surface layer $qp = 11$ and for the central layer $qp = tt$ (for pores with an even number of monolayers $t = H/2$). Figure 38.2 a shows the influence of the width of a carbon pore on the curves κ_{11} and κ_{tt} – there are two sets of curves with numbers $1-5$. With increasing pore width the curves for $\kappa_{11, tt} = $ const of both groups are consistently shifted to the left to the y-axis. This is due to the fact that the contribution of each monolayer, referring to the same value θ_q, remains, but it is achieved by increasing the width of the pores at a lower value of θ. Therefore, the limiting values of κ_{11} and κ_{tt} at $\theta \to 1$ are the same for different H (except when $t = 2$ at $H = 4$, so as the energy of the second layer is different from the energy of other central layers for $H > 4$).

Figure 38.2 b shows the impact of the energy of the adsorbent-adsorbate on the values of local coefficients κ_{11} and κ_{55} in the pore with a width of 10 monolayers. The curves numbered 1 to 5 are also split into two groups. The top group at $\theta = $ const is shifted upward, which is directly associated with an increase in the binding energy of the adsorbate with the wall in the surface layer. The lower group of curves is shifted under the same conditions down (numbers 2–4 are not listed). This behaviour is determined by the decrease in the share of the filled central layers with increasing bonding of the adsorbate with the surface. Figure 38.2 b demonstrates the significant

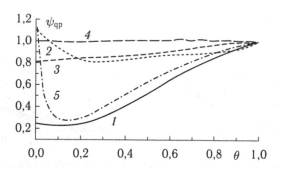

Fig. 38.3. Ratios ψ_{qp} for different layers q and directions of movement of the molecules from the layer q to layer p: $p = q + 1$ (*1, 3–5*) and $q - 1$ (*2*) $q = 1$ (*1, 4, 5*) and $q = 2$ (*2, 3*) in a pore with the width $H = 10$ monolayers when $Q_1/\varepsilon = 9.24$ (*1–3, 5*) and 0 (*4*); $\alpha_{11} = 0.33$ (*1–4*), and 1.0 (*5*).

impact of the adsorbate–adsorbent potential on the energy transfer characteristics.

The calculations show that the thermal conductivity coefficient of the adsorbate depends strongly on the anisotropic distribution of molecules in the cross section of the slit-shaped pore. Particularly large changes as compared with the homogeneous bulk gas phase or liquid are observed in the thermal conductivity coefficients near the pore walls. In the centre of the pores, these values depend relatively weakly on the wall potential and are determined the total concentration adsorbate. Figure 38.3 shows the relationship of local thermal conductivity coefficients related to different directions of motion of the molecules from the layer q: $\psi_{qp} = \kappa_{qp}/\kappa_{qq}$.

ψ_{qp} characterizes the degree of anisotropy of the tensor components of thermal conductivity associated with the energy transfer of the molecules between the adjacent layers qp, $p = q \pm 1$, and within this layer q. The largest difference is found in the components of the thermal conductivity tensor relating to the surface layer in case of a strong attraction of the molecules to the walls (curve *1*). Similar curves *2* and *3* for the second layer are much less than unity than curve *1*, and in the absence of molecular attraction to the walls the ψ_{12} value is very close to unity (curve *4*). A comparison of curves *1* and *5* shows a strong influence of the activation barrier of surface migration of molecules on the value ψ_{12}. At low fillings the energy flux to the second layer is significantly larger for the case with $\alpha_{11} = 1$ than for the system with $\alpha_{11} = 0.33$.

The fact that the heat flows mainly change in two monolayers near the walls and inside the central part of the pore the heat flow and the temperature are almost constant, creates difficulties in determining the components of the tensor of the thermal conductivity coefficient by the MD method. When calculating the normal component of the thermal conductivity coefficient the LGM and MD methods give qualitatively similar results. Thus, the ratio of the values of the normal component of the thermal conductivity coefficient between the first and second layers near the wall, obtained in the LGM (1.9) and MD (2.5), are close at the same fluid density equal to $\theta \approx 0.9$ [57]. But the task of calculating the tangential component of the thermal conductivity coefficient for the MD method is extremely difficult due to problems with a set of reliable 'statistics', whereas in the LGM such calculation is not a problem.

The thermal conductivity of the adsorbate plays an important role not only in adsorption but also in catalytic processes. In porous systems with a highly developed surface area the size factor can influence the nature of energy fluxes through the solid sorbent. This effect causes localization of heat and, as a consequence, local heating, altering the local reaction rate. Thus, changes of the conditions of heat transfer between the opposing walls of the pores in dependence on the density of the adsorbate may affect the catalytic process.

39. The mobility of the fluid in a slit-like micropore: comparison with molecular dynamics

Comparisons between different theoretical methods are sporadic, so it is important to determine the range of applicability of the LGM to describe the dynamic characteristics of the adsorbate in porous systems, as well as the accuracy of the results. For this purpose, a systematic comparison of the results of the LGM and molecular dynamics (MD) calculations was made. This problem is extremely urgent because of the large time costs required to implement the MD studies (up to 10^5 times compared to similar calculations in LGM). Furthermore, despite the study by numerical methods, namely using the LGM, it was possible to identify and explain the 'multi-dome' form of the phase diagrams in narrow-pore systems [58–60]. At the same time, an approximate calculation method based on the quasichemical approximation (CCA) in the LGM requires precision control, especially near the critical areas (hence the principle of the combined use of the CCA and accurate methods has been formulated [61]).

This section compares the profiles of distributions and mobility characteristics of the adsorbate in a slit pore for a wide temperature range and a variety of energies of the wall–fluid interactions. In a nonuniform system, the resulting flow of thermal motion of molecules in a certain direction can be expressed in terms of the local self-diffusion coefficient and the mean value of the self-diffusion coefficient in the cross section of the pore D^*. Currently, the self-diffusion coefficients D^* of the adsorbate are relatively reliably calculated by the MD method and the results are qualitatively consistent with the self-diffusion coefficients, experimentally measured by NMR methods.

The molecular model of the pores and the methodology of MD experiments are presented in [62, 63]. The equations of motion are integrated with a time step $\Delta t = 0.004$. Initially, each system was held at a high temperature ($\tau \approx 3.0 - 4.0$) until it reached an equilibrium state. The temperature was then decreased sequentially in increments $\Delta T = 0.1$ and the system relaxed to the equilibrium state in each such step. Calculations were made for the equilibrium state of the trajectory with a length of 250000–350000 Δt. Further details of the MD model of the slit pores and the technique of MD calculations and the results of its application are given in [64, 65]. The results of the LGM and the MD were compared by a special procedure for processing the MD trajectories. We emphasize that the LGM and MD use the same potential parameters (there were no matching lattice parameters ε).

Analysis (of local mobility) of the interlayer redistribution of particles [64]

In this study, the nature of the mobility of the particles is considered for the case of interlayer redistribution and average self-diffusion coefficients along the pore axis. The notion of 'layered' distribution of the particles is the starting position for the LGM (sections 14, 15). From the point of view of the MD method the layered distribution is the result of calculation of the functions of the particle distributions in the slit pore cross section. Previously it was shown that in a narrow pore with the width of 6.5σ at $\rho = 0.734$ the fluid forms a layered structure – 6 layers of approximately equal width were found, and the density of these layers depends on the interaction energy of the fluid with the wall [62, 63]. Similar layered distributions were obtained in all variants of the calculations also in this study at changes of the overall density of the adsorbate and the binding energy of the adsorbate with the walls of the pores. This allows us to directly link the average adsorbate density within each layer with the local degree of filling of the layers in the LGM, i.e. MD calculations provide the molecular justification of the layer particle distribution and partition of the pore volume into layers in the LGM.

To monitor the process of redistribution of particles in the LGM. 'a label' of unit concentration is introduced in a certain layer at the initial time. This was followed by monitoring its distribution between the layers of the pores at different time points in terms of the equilibrium distribution of particles in the system. This process, which is the exchange of labeled particles between layers, is described by a discrete system of equations of the type

$$\frac{dP_g}{dt} = \sum_{f=g\pm1} \left(D_{fg}^* P_f - D_{gf}^* P_g \right),$$ (39.1)

where P_g is the concentration of the label in layer g; here g, $f \in H$, i.e., the particle jumps are not possible inside the walls of the pores due to their impermeability. The solution of the dynamic system (39.1) was conducted with the initial conditions $P_g = 1$ for the layer in which the label was placed, and $P_g = 0$ for the remaining layers.

Similarly, in MD simulation in accordance with the obtained density profile of the fluid, the pore was divided into layers of the same width, and then the probability of particles $P_g(t)$ being at time t in layer g was calculated.

Calculations in the slit pore with $H = 6.5\sigma$ by both methods were made for temperature changes in the range $\tau = 0.6$ to 4.0. The lower boundary of the area is close to the freezing point of the fluid, the upper refers to high supercritical temperatures. The LGM calculations used a simple cubic lattice structure with the number of nearest neighbours $z = 6$, which best meets the critical parameters in the bulk phase.

Figure 39.1 illustrates the relation between MD and LGM-calculations for the temperature dependences of concentration profiles θ_f at different values of the adsorbate–wall interaction potentials. Thus, the MD calculations showed a weak temperature dependence for the walls with weak interaction with the adsorbate (the first two families of curves at $\varepsilon_{wAr} = 0$ and 1). The density of the surface layer for $\varepsilon_{wAr} = 1$ is higher than for the core layer becvause of attraction to the wall, in comparison with the case $\varepsilon_{wAr} = 0$ when the density of the core layer is lower than the density of the surface

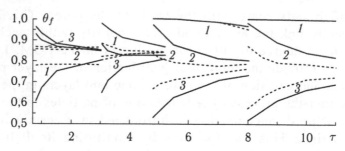

Fig. 39.1. Temperature dependences of the local fillings θ_f of the slit pore with $H = 6.5\sigma$, $\theta = 0.833$, calculated by the MD (points) and LGM (lines) methods. The number of the curve corresponds to the number of the monolayer. The first group of curves corresponds to $\tau = 0.715$–3.02, the second – $\tau = 0.609$–2.97; third – $\tau = 0.698$–4.018; fourth – $\tau = 0.996$–4.018. (Groups of the curves are shifted relative to each other so that the curves do not overlap).

layer. In both cases, the density of the second layer which occupies an intermediate position between θ_1 and θ_3 is displaced to values θ_3. The general character of the location of the curves $\theta_f(\tau)$ is the same. However, the differences between the MD and the LGM-calculations, amounting to 1 to 5% at elevated temperatures, increase with decreasing temperature. These differences are most dramatic at $\tau \leq 1.5$.

For walls with a strong attraction of the adsorbate ($\varepsilon_{wAr} = 5.16$ and 9.24) the surface layer is almost full, and the degree of filling of the third layer is much smaller. With decreasing temperature this difference increases. Qualitatively, the curves in both methods are the same. At high temperatures the difference amounts to between 1 and 10 %, which increased with $\tau \sim 1.5$ to 15 %. Further cooling of the second and third layer increases this difference and for the first layer it disappears altogether. The nature of the behaviour of the concentration profile is associated with the phenomena of condensation of molecules in different layers. The LGM equations used in this case did not include these effects, however LGM gives the correct behaviour of the curves $\theta_f(\tau)$ in a wide temperature range.

The dynamic characteristics were compared for a narrow pore $H = 6.5\sigma$ with interaction of argon atoms with different pore walls at the same intervals of temperature change.

Figure 39.2 shows typical dependences $P_g(t)$, obtained by both methods for $g = 1$–6 and $\alpha = \varepsilon^*/\varepsilon = 0.3$. The submitted time dependences of the local fillings of the argon atoms in the carbon pore refer to the case when at the initial time the entire 'label' was only in the first monolayer. During the exchange of the label

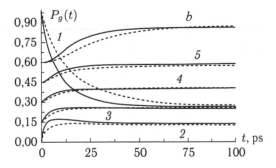

Fig. 39.2. The dynamics of the interlayer exchange of argon atoms at between six layers in the carbon slit pore ε_{wf} = 5.16, H = 6.5σ, τ = 3.0, the complete filling of the pores equals θ = 0.833. Evolution of the degree of filling layers i = 1 – 6 is described in lines LGM and points in the MD.

between the layers its concentration starts to increase gradually in the layers g = 2–6. To avoid cluttering the figure, the curves for layers 3–6 are sequentially shifted upwards on the ordinate. Even at a short distance of 6 layers we observe the lag effect of the front of the label by which the discrete description of the flow of the label differs from purely diffusion description (usually described by partial differential equations). The curves represent the typical behaviour of the dynamic curves observed in several common fillings of the pores in the temperature range where there is agreement of the concentration profiles in both methods. The greatest difference is shown for the first, second and sixth monolayer (30%). For the monolayers 3–5 the LGM and MD curves practically coincide. At times greater than 75 ps, the dynamic curves come to their steady-state values corresponding to the equilibrium distribution of the adsorbate over the cross section of the pore.

Figure 39.3 shows the dynamic curves of the interlayer exchange process of the label for α = 0.34. Number g corresponds to the number of the layer $g \leq 3$ layer, their number was reduced to simplify the diagrams. Calculations relate to ε_{wf} = 5.16 at τ = 1.0 (the first three sets of curves) and ε_{wf} = 1 and τ = 1.5. In the first case the atoms are retained by the wall and in the range up to 300 ps the label almost never comes out of the first layer (MD and LGM give concordant values of P_i). This is reflected in the first group of curves for t < 300 ps. If the label is placed in the second or third layers (second and third group of curves, respectively), then there is as almost no exchange for them with the first layer, and the exchange takes place only within the second and third layers. These

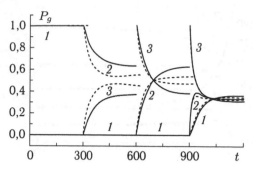

Fig. 39.3. Dynamic curves of interlayer exchange of argon atoms in a slit pore in carbon, ε_{wf} = 5.16, H = 6.5σ, τ = 1.0. For these three groups, the initial filling with the label includes the first layer ($t < 300$), the second layer ($300 < t < 600$) and the third layer ($600 < t < 900$). The fourth group refers to ε_{wf} = 1 and τ = 1.5, for this group the label in the initial moment of time was placed in the third layer ($900 < t < 1200$).

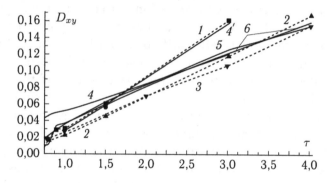

Fig. 39.4. Comparison of the temperature dependences of self-diffusion coefficients of argon atoms in a slit pore in carbon at θ = 0.833 and α = 0.34 for ε_{wf} = 0 (*1, 4, 4'*), 5 (*2, 5*), 9.24 (*3, 6*) (points – MD, lines – LGM).

curves asymptotically approach their values which coincide with the equilibrium values θ_i. The reason for the differences between the MD and the LGM-curves is the difference in the values θ_i, while the lengths of transition curves for dynamic sites are close to each other. Finally, the fourth group of the curves refers to the case of weak attraction of argon to the walls of the pores at τ = 1.5. For them the curves practically coincide during the entire time interval.

Figure 39.4 compares the temperature dependences of the average self-diffusion coefficients along the axis of the pores obtained by the MD and LGM methods for different adsorbate–adsorbent interaction potentials. The curves for strongly attracting walls are close enough to each other, but differ markedly from the case of the walls without the attracting branch of the potential. All the LGM curves *4–6* were obtained with one parameter α. In general, this parameter may be a function of temperature and density [66]. Considering this factor for the curve 4' provides a virtually exact match with the MD data with an increase of α at higher temperatures and at the same decrease at low temperature (the maximum change of α at the ends of the interval τ does not exceed 50 %).

40. Local mobility of argon: comparison with MD

Apart from the middle layered redistributions the local mobility of the molecules is also of interest. These characteristics calculated using the LGM and MD methods were compared in [65]. To do this, a special procedure for processing the MD-trajectories was developed [62, 63]. As above, the potential functions and the same calculation MD-cell and simulation conditions were used (section 39). The resultant MD-trajectories were interpreted using the following theoretical model.

The diffusion of particles in the pore is considered as a process in a three-dimensional rectangular lattice with continuous time. The pore is divided into cells along the Z axis by planes passing through the minima of the density profiles, and along the X and Y axes – into the square cells with sides equal to the diameter of the particles. Processing MD trajectories consisted of calculating the probability of the particle to move from the cell of the layer f to the cell of the layer g layer in the r-th coordination scope during time t. This probability is denoted by $P_{fg}(r|t)$. Attention was paid to all the particle displacements with respect to any of the central cell f, situated in one of the six monolayers in which the labeled particle could be placed at the initial time.

Analysis of the local mobility of the particles
In this paper, the nature of the local mobility of particles is considered for the case of redistribution of the label in the equilibrium conditions between adjacent layers and within the layers. The layered distribution is the starting position for the LGM. In MD calculations the layered distribution is the result of the calculation of the distribution functions of particles over the cross section of the slit-shaped pores. Previously it was shown that in a narrow pore with the width 6.5σ at $\theta = 0.833$ the fluid forms a layered structure with six layers of equal width, wherein the density of the layers depends on the interaction energy of the fluid with the wall [62, 63]. Similar

layered distributions were obtained in all variants of the calculations in this paper with changes and the overall density of the adsorbate and the binding energy of the adsorbate with the walls of the pores. This allows us to directly link the average adsorbate density within each layer with the local degree of filling of the layers in the LGM. Thus, the MD calculations give molecular justification of the layered particle distribution and partition of the pore volume into the layers in the LGM, and functions $P_{fg}(r|t)$ can also be determined.

In the LGM to study the process of redistribution of particles it is assumed that at the initial time a 'label' of unit concentration was introduced in the cell of layer f. This was followed by monitoring the spreading of the label between all cells in adjacent layers of the pore for different points in time. (The conditional probability $P_{fg}(r|t)$ of the cells being in layer g at time t if at zero time they were in the layer f was also calculated). This process was described by a discrete system of equations. For slit pores the algorithm of solving the LGM equations, describing the spreading of the label, is as follows: we specify the above indices f and g. We consider the volume containing layers H, $k = 1,... H$ – the number of the layer. Each layer represents a square of 21×21 cells with coordinates (i, j), $i, j = -10,... 10$ with a central cell $(0, 0)$ (here and below, the number '10' is associated with the size of the MD cell used in the calculations). Thus, in the pore volume we can defined the 'central' column that contains the label in one of the layers of k_0. Denote: $c_{i,j}^k$ – the probability of finding the label in the cell (i, j) of layer k, D_{k_1,k_2} – the local diffusion coefficient of a particle in the layer k_1 in the direction of the neighbouring cell of layer k_2. We have the equation

$$\frac{dc_{i,j}^k}{dt} = D_{kk}\left(c_{i-1,j}^k + c_{i+1,j}^k + c_{i,j-1}^k + c_{i,j+1}^k - 4c_{i,j}^k\right) + \left(D_{k-1k}c_{i,j}^{k-1} - D_{kk-1}\,c_{i,j}^k\right) +$$

$$\left(D_{k+1k}\,c_{i,j}^{k+1} - D_{kk+1}\,c_{i,j}^k\right),\ i,j = -10,...,10,\ k = 1,...H.$$
$$(40.1)$$

The right side of (40.1): 1st member – intralayer diffusion, 2nd member – exchange with the bottom layer (not for $k = 1$), 3rd member – exchange with the top layer (not when $k = H$). The initial conditions are taken as $c_{0,0}^{k_0}=1$, all other $c_{i,j}^k = 0$, i.e. the labeled particles are arranged in the central cell k_0-th layer.

The boundary conditions of equations (40.1) are as follows: $c_{-N-1,j}^k = c_{-N+1,j}^k$, $c_{N+1,j}^k = c_{N-1,j}^k$ for all j and $c_{i,-N-1}^k = c_{i,-N+1}^k$, $c_{i,N+1}^k = c_{i,N-1}^k$ for any i. This is equivalent to saying that the entire space is split to identical 'macro'-volumes (size $H \times 21 \times 21$ cells) periodically repeating each other. Using the symmetry inside the layer, we can greatly reduce the dimension of the system, leaving only the variables with indices $0 \le i \le 10$, $0 \le j \le i$.

Solution of (40.1) makes it possible determine the probability $P_{k0,k}(r)$ of finding the particle in the r-th coordination sphere of the central cell layer k_0: $P_{k_0,k_0}(0) = c_{0,0}^{k_0}$, $P_{k_0,k_0}(1) = 4c_{1,0}^{k_0}$, $P_{k_0,k_0}(2) = 4c_{1,1}^{k_0}$, as well as the likelihood that particle would be in the central cell of the adjacent layer $k_{0\pm1}$: $P_{k_0,k_0\pm1}(0) = c_{0,0}^{k_0\pm1}$. (These probabilities are shown in the figures below.)

In addition, we calculated the probability of finding a particle in a given layer $\langle c^k \rangle$ as sum of $c_{i,j}^k$ for all i and j: $\langle c^k \rangle = \sum_{i=0}^{10} \sum_{j=0}^{i} A_{i,j} c_{i,j}^k$, where coefficients $A_{i,j}$ represent the weight of the cells with the same fillings by the label: $A_{0,0} = 1$, $A_{i,j} = 4$, $j = 0$, $j = i$, $A_{i,j} = 8$, $0 < j < i$. (In this case, for $\langle c^k \rangle$ we obtain equations of the same type for the interlayer exchange of the particles if variables $c_{i,j}^k$ are replaced by $\langle c^k \rangle$, and the first term on the right side with factor D_{kk} vanishes).

Analysis of the calculations

Calculations by both methods are performed for temperature changes in the range $\tau = 0.6$ to 4.0. The lower boundary is near the region of freezing of the fluid, whereas the upper boundary corresponds to high supercritical temperatures.

Calculations indicate a different degree of agreement between the MD and the LGM, depending on the temperature and the potential of the adsorbate–wall interaction. Thus, the MD calculations showed a weak temperature dependence for the walls with weak interaction with the adsorbate (the first two families of curves at $\varepsilon_{wAr} = 0$ and 1). The density of the surface layer for $\varepsilon_{wAr} = 1$ is higher than for the core layer in the wall because of attraction to the wall, in comparison with the case $\varepsilon_{wAr} = 0$ when the density of the core layer is lower than the density of the surface layer. In both cases, the density of the second layer which occupies an intermediate position between θ_1 and θ_3 is displaced to values to θ_3. The general character of the location of the curves $\theta_j(\tau)$ is the same. However, the differences between the MD and the LGM-calculations amounting from 1 to 5% at high temperatures increase with decreasing temperature. These differences are the largest at $\tau \le 1.5$.

For walls with a strong attraction of the adsorbate ($\varepsilon_{wAr} = 5.16$ and 9.24) the surface layer is almost full, and the degree of filling of the third layer is much smaller. This difference increases with decreasing temperature. Qualitatively, the shape of the curves in both methods is the same. At high temperatures the difference is between 1 and 10 % and increases at $\tau \sim 1.5$ to 15%. Further cooling increases the difference for the second and third layers and for the first layer the difference disappears altogether. As noted above, the nature of the behaviour of the concentration profile is associated with the phenomena of condensation of molecules in different layers.

Figures 40.1–40.3 are examples of comparing the dynamic curves of spreading of the label for different conditions with different interactions with the walls of the pores, when the overall density of the adsorbate is $\theta = 0.833$ (similar curves were obtained for other degree of filling the pores). The ordinates represent the probability of the redistribution of the labeled argon atoms in a narrow pore with $H = 6.5\sigma$. Each figure contains two or three sets of curves shifted relative to each other, to avoid overlapping. On the abscissa time is measured in picoseconds. The curves, calculated by the MD method, are indicated by dots and those calculated by the LGM are indicated by lines. All the curves in Figs. 40.1–40.3 correspond to one value of $\alpha = 0.34$, i.e., there was no special selection of model parameters for the specific conditions of the redistribution process of the label.

Figure 40.1 shows the time dependences of the dynamics of redistribution of the label at $\tau = 1.5$ for weak interactions ε_{wAr}. The first group of curves ($t = 0$–50) relates to intralayer exchange in the the surface layer with $k_0 = 1$ (labeled particles were placed in the cell of the first layer). The probabilities P of detecting the label in the central cell (1), 1st (2), the 2nd (3) and 4th (4) coordination spheres, respectively, are given. The second group of the curves ($t > 50$) refers to the interlayer exchange at $k_0 = 2$ (labeled particles were placed in

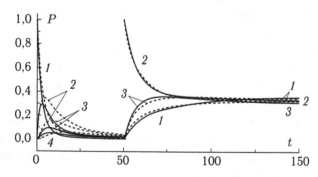

Fig. 40.1. Time dependences of the dynamics of redistribution of the label at $\tau = 1.5$ and $\varepsilon_{wAr} = 1.0$.

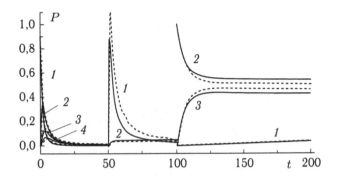

Fig. 40.2. Time dependence of the dynamics of redistribution of the label at $\tau = 3.0$ and $\varepsilon_{wAr} = 9.24$.

a cell of the second layer). The curves correspond to the probabilities P for the label to be in the 1ˢᵗ (1), 2ⁿᵈ (2) and 3ʳᵈ (3) layers.

Figure 40.2 shows the curves of the time dependences of the dynamics of the redistribution of the label as $\tau = 3.0$ for strong adsorbate–wall interactions. For the first group of curves ($t = 0$–50) there are dynamic curves for the intralayer exchange in the 2ⁿᵈ layer at $k_0 = 2$ (the label in the cell of the first layer). The numbers of the curves correspond to the probability P for the label to be in the central cell (1), 1ˢᵗ (2), 2ⁿᵈ (3) and 4ᵗʰ (4) coordination spheres, respectively. The second group of curves ($t = 50$–100) shows the exchange between the central cells of the 2ⁿᵈ and 1ˢᵗ (1) and 3ʳᵈ (2) layers at $k_0 = 2$. (Under these conditions, the interlayer exchange is strongly hindered and all the curves of the second group are enlarged 20 times.) Finally, the third group of the curves at $t > 100$ shows the evolution of the label at the interlayer exchange with $k_0 = 2$, relating to the entire second layer (and not for a single cell). The curves represent the probability P of the label to be in the 1ˢᵗ (1), 2ⁿᵈ (2) and in the 3ʳᵈ (3) layer of the pore. The differences between the curves at large times are due to the difference in the values of θ_f for the concentration profile.

Figure 40.3 shows the curves of the dynamics of redistribution of the label when $\tau = 1.5$ (here $\varepsilon_{wAr} = 5.1$). The first group of curves ($t = 0$–50) corresponds to the intralayer exchange in the 2ⁿᵈ layer at $k_0 = 2$, and the second group to the intralayer exchange in the 3ʳᵈ layer ($t = 50$–100) with $k_0 = 3$. The numbers of curves 1–4 correspond to the probability P of the label to be in the central cell, in the 1st, the 2nd, the 4th coordination spheres, respectively.

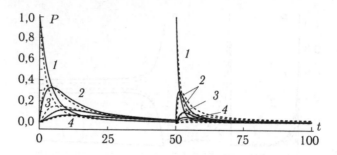

Fig. 40.3. Time dependence of the dynamics of redistribution of the label at $\tau =$ 1.5 and $\varepsilon_{wAr} = 5.1$.

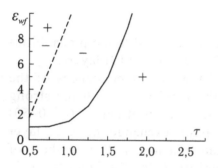

Fig. 40.4. Range of values of the parameter of the potential ε_{wf} of the adsorbate–atom interaction of the pore wall with width $H = 6.5\sigma$ and temperature in which the dynamic curves of exchange of the labeled argon atoms by the MD methods and LGM methods were compared.

Figure 40.4 summarizes the parameters of the adsorbate–wall interaction and the temperature range $\tau = 0.6$–3.0, in which both models agree and it is possible to clarify the parameters of the lattice model with the aim of achieving full quantitative agreement between the MD and LGM calculations of the dynamic curves (at $\tau > 3.0$ both models are in quantitative agreement).

The plus sign corresponds to satisfactory agreement of both methods. The minus sign corresponds to the divergence of these methods by more than 20%. The ± sign at low temperatures refers to the agreement of both methods in the time interval up to 300 ps in which the MD calculations were carried out.

Thus, at high temperatures the LGM gives a description of the dynamics which almost completely coincides with the calculations by MD. At lower temperatures, corresponding to the near-critical temperatures, this agreement deteriorates, but the LGM gives a correct qualitative description. Therefore, its use in this temperature range requires improved versions of the model [18], and not just the model outlined in this study.

It should also be noted that with decreasing temperature the results of MD simulations also becomie less perfect because of the limited cell size and complexity of a good set of statistics. Therefore, the comparison of the two models should also be made for the critical and subcritical temperatures.

The results indicate that the LGM can be a convenient and fast interpolation tool for the calculation of equilibrium and, especially, dynamic characteristics of the migration of spherical particles in strongly inhomogeneous attracting fields of the pore walls of the adsorbents. Its use can significantly expand the opportunities for using the molecular models of dynamic processes of mass and heat transfer in narrow-pore systems.

41. Dynamic characteristics of molecules in narrow cylindrical pores

Knowing the equilibrium distribution of the molecules over the cross section of the pores (chapter 2), we can calculate the probability of hopping of the molecule between different sites (related to their thermal velocities) and dynamic characteristics – self-diffusion and viscosity coefficients. Expressions for the transport coefficients (section 35) can be used for different pore geometry, with changing of the structural values of f_q and z_{qp}.

The dynamic characteristics of molecules in cylindrical pores were considered in [58]. Attention was given to the adsorption of spherically symmetric particles (atoms of argon, helium, and methane molecules in CCl_4 in cylindrical pores with the diameter of 8 to 20 cells, the walls of which model graphite atoms ($\varepsilon_{ss}/k = \varepsilon_{CC}/k = 28$ K, see [67, 68]), silica gel ($\varepsilon_{ss} = \varepsilon_{CC}/2$) and weakly binding ($\varepsilon_{ss} = \varepsilon_{CC}/8$) functional groups of polymer matrices. The potential of the walls was considered as the sum of contributions of the individual layers in the depth of the pore wall; the surface layer was approximated by structured sites with varying activation energy of hops along the surface. The lattice structure was a lattice with the number of the nearest neighbours $z = 6$, as it was repeatedly pointed out that this structure leads to the best agreement with the critical parameters of bulk fluids [3, 69]. The values of the dimensionless parameter $\alpha_{11} = E_{11}/Q_1$ were also varied; this parameter characterizes the ratio of the magnitude of the activation barrier of the surface migration of the molecule to the binding energy of the molecule with the surface ($E_{12} = Q_1$, the rest $E_{qp} = 0$). All the concentration curves presented below were plotted in a normalized form. Normalization

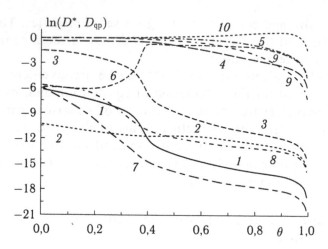

Fig. 41.1. Concentration dependences of ratios in the self-diffusion D^*_{qp} of the argon–carbon system in the pore when $\alpha_{11} = 2/3$, $d/\lambda = 8$ (*1–6*) and 10 (*7–10*), $\alpha = 0$ (*7, 9*), 0.2 (*1–6*) and 0.5 (*8, 10*), $qp = 11$ (*1, 7, 9*), 12 (*2*) 21 (*3*) 22 (*4*) 55 (*5, 8, 10*), curve *6* – average value of D^*.

is given for the corresponding values of the transport coefficients at $\theta = 0$ and $Q_1 = 0$.

In general, the calculations were carried out at $T = 198$ K, which corresponds to a dimensionless value $\beta\varepsilon = 0.6$ for argon atoms. In terms of capillary condensation the dynamic characteristics were calculated using the lever rule [4].

Figure 41.1 presents the results of calculations of the concentration dependences of the local values of self-diffusion coefficients and average value of D^* (curve *6*) of the argon–carbon system in a pore with a diameter of 4 cells. The curves show a sharp difference of the local values D_{qp}. At this activation energy of migration along the pore wall $\alpha_{11} = 2/3$ the zero value of the curve *1* is larger than for curve *2* relating to the hopping of the molecules from the first to the second surface layer. As the surface layer is being filled these curves intersect. Curve *3* is connected with hopping in the opposite direction from the second layer to the first and with its filling the value D_{21} abruptly decreases. Transport within the second layer (curve *4*) differs only slightly from the transport in the centre of the pore (curve *5*). Curve *6*, calculated by the formula (35.3), is non-monotonic due to the differences in the occupied surface and internal layers: with filling of the surface layer transport occurs mainly in the central part at a higher rate. The effect of parameter $\alpha = 0$ (curve *7*,

9), and 0.5 (curves *8*, *10*) for a pore having a diameter of 5 cells is shown for comparison.

The concentration dependences of the self-diffusion coefficients for the argon–carbon system in the case of pores with a diameter d = 8 (*1*), 10 (*2*), 20 (*3*) are shown in Fig. 41.2. With the increase in the diameter the fraction of the surface cells decreases and the increase of the self-diffusion coefficient due to the acceleration of transport in the central regions is shifted to smaller values of the degree of filling pores. The inset on the left shows the phase diagram for the pores of d = 8. The diagram is shown in coordinates τ–ρ, where $\tau = T/T_c$ (bulk), T_c (bulk) – the critical temperature in the bulk phase, and $\rho = \theta(\sigma/\lambda)^3$ – the fluid density, expressed in terms of volumes of solid spheres σ; for a rigid (incompressible) lattice $\rho = \theta/1.41$, since $\lambda = \sigma(2)^{1/6}$ (ρ units are commonly used in numerical methods [50–55, 67]). The diagram shows the condensation region in the surface layer, in the intermediate region and within the centre of the pore and – filling of the pore volume.

The inset on the right shows the average values of the diffusion coefficients for the model system for d = 8, ε_{AA}/k = 750 K, calculated

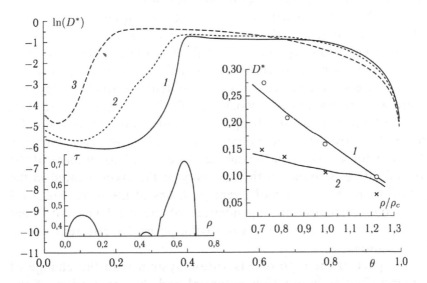

Fig. 41.2. Concentration dependences of average self-diffusion coefficients for the argon-carbon system for d/λ = 8 (*1*), 10 (*2*), 20 (*3*) at α = 0.2, α_{11} = 2/3. The inset on the left – phase diagram for the pores of d/λ = 8. The inset on the right – the average values of the diffusion coefficients for the model system with d/λ = 8, ε_{AA} = 750 K, calculated by the method of molecular dynamics (points) and in the model (lines).

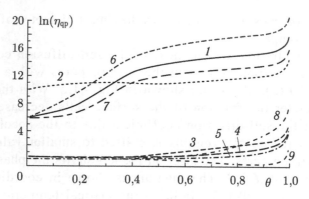

Fig. 41.3. Concentration dependences of the local coefficients of shear viscosity of the argon–carbon system at $d/\lambda = 10$, $\alpha_{11} = 2/3$, $\alpha = 0$ (6, 7), 0.3 (1–5), 0.5 (8, 9), $qp = 11$ (1, 6, 8), 12 (2), 22 (3), 32 (4), 55 (5, 7, 9).

by molecular dynamics (symbols ∘ and ×) and in the model at $T = 1.01T_c$ (pore) in the vicinity of the critical density. Curves 1 and 2 correspond to the two different conditions of the MD calculations with mirror (1) and diffuse (2) reflection of atoms from the wall. Calculation using this model was carried out taking into account the interaction in four coordination spheres at $\alpha = 0.5$ (curve 1) and 0.6 (curve 2). The comparison shows that the lattice model gives a satisfactory agreement with the MD calculations.

The concentration dependences of the local coefficients of shear viscosity of the for argon–carbon system for pores with a diameter of 10 cells are shown in Fig. 41.3. As the distance from the layer to the pore wall increases the viscosity coefficients decrease, and with increase of the degree of filling of the pores they monotonically increase. The strongest dependence on the adsorbate concentration is observed for viscosity in the surface layer, the least impact of the degree of filling is found on the viscosity of the layer located in the centre of the pores. Besides different positions of the pairs, Fig. 41.3 presents the concentration dependences of the local viscosity near the surface layer and in the centre of the pores for different values of the parameter α.

The impact of the type of adsorption system with the change of the intermolecular interaction potential and the interaction of the molecule with the pore wall is shown in Fig. 41.4. For definiteness, only two concentration dependences of the local values of the viscosity coefficient of the adsorbate near the wall (curves 1–7) and in the centre (curves 8–12) of the pore with a diameter of 10 and 20 cells were selected. Additionally, this figure shows the effect

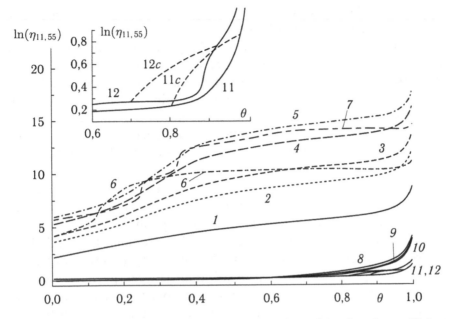

Fig. 41.4. Concentration dependences of the local values of the viscosity coefficient of the adsorbate neat the wall η_{11} (*1–7*) and in the centre η_{55} (*8–12*) of the pore with $d/\lambda = 20$ (*6, 12*) and 10 (all other lines) when $\alpha = 0.3$, $\alpha_{11} = 2/3$, $\gamma_s = \varepsilon_{cc}/\varepsilon_{ss}$ = 1/8 (*1, 8*), 1/2 (*3, 9*), 1 (*2, 4–7, 10–12*). The inset gives changes of η_{55} without considering (*11, 12*) and taking into account (*11c, 12c*) capillary condensation of argon in the centre of the graphite pore with $d/\lambda = 10$ (*11, 11c*) and 20 (*12, 12c*) at a temperature $(\beta\varepsilon_{AA})^{-1} = 0.77$ (*11, 11c*) and 1.0 (*12, 12c*).

of the factor of capillary condensation in the porous system. Curves *11* ($\beta\varepsilon_{ArAr}^{-1} = 0.77$) and *12* ($\beta\varepsilon_{ArAr}^{-1} = 1.0$) are shown in the inset: the curves calculated by the lever rule (at these temperatures there is no condensation at low filling in the surface region of the pore near the walls of the pores), marked as 11c and 12c). The Figure shows (curves *1–5*) that the molecular properties of the system, determined by the potential of the interactions between the adsorbate and with the pore wall, strongly affect the local viscosity. In the centre of the pores the normalized viscosities practically coincide (the area between the numbers 8 and 10). Curves *11* and *12*, like curves *6* and *7*, were calculated for a low temperature.

The calculations for the cylindrical pores show that, as for the slit-shaped pores, the dynamic characteristics of the adsorbate are strongly dependent on the anisotropic molecular distribution over the cross section of the pores. In both cases, the largest change of the self-diffusion and shear viscosity coefficients is observed near the pore walls. The higher the relative fraction of the surface, the more important are the contributions to the surface dynamics.

In conclusion, we briefly discuss the generalization of the model to analyze the dynamic characteristics of the spherical pores, as LGM also allows to explore the spherical pores. With the increase in the contribution of surface sites (in the transition from the cylinder to the sphere with the same value of the radius) the anisotropy of the distribution of the adsorbate inside the pore becomes greater and the non-monotonicity of the average coefficient of diffusion becomes also more distinctive. The concentration dependences of the transport coefficients of the molecules near the transition region between areas of cylindrical and spherical pores were estimated. The viscosity and self-diffusion coefficients change depending on the binding energy at the intersection of these areas. In particular, the local shear viscosity increases (decreases) with increasing (decreasing) local binding energies. The question of jointing between pores of different geometry is important for the theory of porous systems, although, in fact, it is practically not studied. Changing the width and type of the cross section of the pores changes the conditions of the phase state of the adsorbate, so the effect of taking into account different types of pores on the equilibrium and dynamic characteristics of the adsorbate is the main goal of the development of this area of work. These issues are discussed in chapter 7.

References

1.   Landau L.D., Lifshitz E.M., Theoretical Physics. VI. Hydrodynamics. – Moscow: Nauka, 1986. – 733 p.
2.   Bird R. B., Stewart W., Lightfoot E. N., Transport Phenomena. – Moscow: Khimiya, 1974. – 687 p. [John Wiley and Sons, New York, 1965].
3.   Hirschfelder J. O., Curtiss C. F., Bird R. B., Molecular Theory of Gases and Liquids. – Moscow: Izd. inostr. lit., 1961. – 929 p. [Wiley, New York, 1954].
4.   Tovbin Yu. K., Theory of physical and chemical processes at the gas-solid interface. – Moscow: Nauka, 1990. – 288 p. [CRC, Boca Raton, Florida, 1991].
5.   Tovbin Yu.K. // Progress in Surface Science. 1990. V. 34. P. 1.
6.   Tovbin Yu.K. // Zh. Fiz. Khimii. 1998. V. 72, No. 8. P. 1446.
77.   Reif F., Statistical Physics. Berkeley Physics Course, Vol. 5 – Moscow: Nauka, 1977. – 352 p.
8.   Moelwyn Hughes E.A., Physical Chemistry. – Moscow: Izd. inostr. lit., 1962. – V. 2. – P. 642. [2nd ed. Pergamon Press, London, 1961].
9.   Frenkel' Ya.I., Statistical Physics. – Moscow: Academy of Sciences of the USSR, 1948. – Ch. 12.
10.   Leibfried G., Microscopic theory of mechanical and thermal properties of crystals. – Leningrad: GIFML, 1963. – 312 p. [Handbuch der Physik, Vol. 7, Pt. 1, Ed. by S.

Flugge, Springer Verlag, Berlin, 1978].

11. Lifshitz E.M., Pitaevskii L.P., Theoretical Physics. X. Physical Kinetics. – Moscow: Nauka, 1979. – 527 p.

12. Croxton C. A., Physics of the liquid state: A Statistical Mechanical Introduction. – Moscow: Mir, 1979. – 400 p. [Cambridge University Press, Cambridge, 1974].

13. Mikhailov I.G., et al., Fundamentals of molecular acoustics. – Moscow: Nauka, 1964. – 514 p.

14. Glasston S., Laidler K. J., Eyring H., The Theory of Rate Processes. – M.: Izd. inostr. lit., 1948. – 583 p. [Princeton University Press, New Jersey, 1941].

15. Tovbin Yu.K., Fedyanin V.K. // Zh. Fiz. Khimii. 1980. V. 54. P. 3132.

16. Ferziger J. H. H., Kaper H. G., Mathematical theory of transport processes in gases. – Moscow: Mir, 1976. – 554 p. [North_Holland, Amsterdam, The Netherlands, 1972].

17. Sengers J. V. // Phys. Rev. Lett. 1965. V. 5. P. 515.

18. Tovbin Yu.K., Senyavin M.M., Zhidkova L.K. // Zh. Fiz. Khimii. 1999. V. 73. No. 2. P. 304. [Russ. J. Phys. Chem. 1999. V. 73. № 2. P. 245].

19. Mason E., Sperling T., Virial equation of state. – Moscow: Mir, 1972. – 280 p. [The Internationsl Encyclopedia of Physical Chemistry. Topic 10. ed. J.S. Rowlinson, Vol. 2].

20. The equation of state of gases and liquids / Ed. I. Novikov. – Moscow: Nauka, 1975. – 263 p.

21. Thermophysical properties of technically important gases at high temperatures and pressures. Handbook. – C. Energoatomizdat, 1987. – 232 p.

22. Sokolova I.A. // Reviews on thermophysical properties of substances. – Moscow: Publishing House of the Academy of Sciences of ITV, 1992. – No. 2 (94). – P. 36.

23. Tovbin Yu.K., Komarov V.N. // Zh. Fiz. Khimii. 2001. V. 75, No. 3. P. 562. [Russ. J. Phys. Chem., 2001. V. 75. No. 3. P. 490]

24. Tables of physical quantities. / Ed. I.K. Kikoin. – Moscow: Atomizdat, 1976. – 960 p.

25. Rivkin S.L., Thermodynamic properties of gases. Handbook. – Moscow: Energoatomizdat, 1987. – 312 p.

26. Sychev V.V., Wasserman A.A., Kozlov D., Thermodynamic properties of helium. – Moscow: Publishing House of Standards, 1984. – 240 p.

27. Sychev V.V., Wasserman A.A., Kozlov D., Thermodynamic properties of nitrogen. – Moscow: Publishing House of Standards, 1977. – 234 p.

28. Sychev V.V., Wasserman A.A., Kozlov D. Thermodynamic properties of methane. – Moscow: Publishing House of Standards, 1984. – 197 p.

29. Reid R.C., Sherwood T.K. The properties of gases and liquids. (The restimation and correlation). – N. Y.; San Francisco; Toronto, London, Sydny: McGrav-Hill Boch Comp., 1966.

30. Huang K., Statistical Mechanics. – Moscow: Mir, 1966. – 520 p. [Wiley, New York, 1963].

31. Komarov V.N., Tovbin Yu.K. // Termofizika Vysokikh Temperatur. 2003. V. 41, No. 2. P. 217. [High Temperature. 2003. V. 41. № 2. P. 181].

32. Tovbin Yu.K. // Zh. Fiz. Khimii. 1987. V. 61. P. 2711.

33. Comings E.W., Mayland B.J., Engly R.S., The Viscosity of Gases at High Pressure, Engineering Experimental Station Bulletin Ser. 354. 1944. V. 42, No. 15. Univ. of Illinois. Nov. 28.

34. Rabinovich V.A., et al., Thermophysical properties of neon, argon, krypton and xenon. – Moscow: Publishing House of Standards, 1976.

35. Anisimov M.A., Rabinovich V.A., Sychev V.V., Thermodynamics of the critical state. – Energoatomizdat, 1990. – 190 p.
36. Tovbin Yu.K. // Zh. Fiz. Khimii. 1995. V. 69. P. 118. [Russ. J. Phys. Chem. 1995. V. 69. № 1. P. 105].
37. Tovbin Yu.K., Vasyutkin N.F. // Izv. RAN. Ser. khim. 2001. No. 9. P. 1496. [Russ. Chem. Bull. 2001. V. 50. № 9. P. 1572].
38. Tovbin Yu.K., Vasyutkin N.F. // Zh. Fiz. Khimii. 2002. V. 76. No. 2. P. 319. [Russ. J. Phys. Chem. 2002. V. 76. № 2. P. 257].
39. Tovbin Yu.K. // Usp. Khimii. 1988. V. 57. P. 929.
40. Tovbin Yu.K. // Dokl.AN SSSR. 1990. V. 312, No. 6. P. 1323.
41. Tovbin Yu.K. // Khim. Fizika. 2002. V. 21, No. 1. P. 83.
42. Tovbin Yu.K., Zhidkova L.K., Gvozdeva E.E. // Inzh.-Fiz. Zh. 2003. V. 76, No. 3. P. 124. [J. Engin. Phys. Thermophys., 2003. V. 76, No. 3, P. 619].
43. Akhmatskaya E., Todd B.D., Davis P.J., Evans D.J., Gubbins K.E., Pozhar L.A. // J. Chem. Phys. 1997. V. 106. P. 4684.
44. Timofeev D.P., Adsorption kinetics. – Moscow: Publishing House of the USSR Academy of Sciences, 1962. – 252 p.
45. Kheifets L.I., Neimark A.V., Multiphase processes in porous media. – Moscow: Khimiya, 1982. – 320 p.
46. Mason E. A., Malinauskas A. P., Gas Transport in Porous Media: The Dusty_Gas Model. – Moscow: Mir, 1986. - 200 p. [Elsevier, Amsterdam, 1983].
47. Deryagin B.V., Churaev N.V., Muller V.M., Surface Forces. – Moscow: Nauka, 1985. – 400 p.
48. Lamb G.. Hydrodynamics. – M.-D. 1947.
49. Satterfield Ch. N. Mass transfer in heterogeneous catalysis. – Moscow: Khimiya, 1976. – 240 p. [MIT Press, Cambridge (Mass.), 1970].
50. Todd B.D., Evans D.J. // J. Chem. Phys. 1995. V. 103. P. 9804.
51. MacElroy J.M.D. // J. Chem. Phys. 1994. V. 101. P. 5274.
52. Todd B.D., Davis P.J., Evans D.J. // Phys. Rev. E. 1995. V. 51. P. 4362.
53. Travis K.P., Evans D.J. // Physica A. Phys. Rev. E. 1996. V. 55. P. 1566.
54. Travis K.P., Todd B.D., Evans D.J. // Physica A. 1997. V. 240. P. 315.
55. Mansour M.M., Baras F., Garsia A.L. // Physica A. 1997. V. 240. P. 255.
56. Tovbin Yu.K., Komarov V.N. // Izv. AN. Ser. khim. 2002. No. 11. P. 1871. [Russ. Chem. Bull. 2002. V. 51. № 11. P. 2026].
57. Tovbin Yu.K., Komarov V.N., Zhidkova L.K. // Zh. Fiz. Khimii. 2003. V. 77, No. 8. P. 1367. [Russ. J. Phys. Chem. 2003. T. 77. № 8. C. 1222]
58. Tovbin Yu.K., Gvozdeva E.E., Eremich D.V. // Zh. Fiz. Khimii. 2003. V. 77, No. 5. P. 878. [Russ. J. Phys. Chem. 2003. T. 77. № 5. C. 785].
59. Tovbin Yu.K., Eremich D.V. // Izv. AN. Ser. khim. 2003. V. 52, No. 11. P. 2208. [Russ. Chem. Bull. 2003. V. 52. № 11. P. 2334].
60. Eremich D.V., Tovbin Yu.K. // Zh. Fiz. Khimii. 2004. V. 78, No. 4. P. 720. [Russ. J. Phys. Chem. 2004. T. 78. № 4. C. 615].
61. Tovbin Yu.K. // Zh. Fiz. Khimii. 1998. V. 72, No. 5. P. 775. [Russ. J. Phys. Chem. 1998 V. 72, No. 5, P. 675].
62. Berlin A.A., Mazo M.A., Balabaev N.K., Tovbin Yu. K. // Fundamentals of Adsorption-7, IK International, Chiba-City, 2002. P. 402.
63. Berlin A.A., Mazo A.A., Balavaev N.K., Tovbin Yu.K. // Khim. Fizika. 2002. V. 21, No. 2. P. 3.
64. Mazo M.A., Rabinovich A.B., Tovbin Yu.K. // Zh. Fiz. Khimii. 2003. V. 77, No. 11. P. 2053. [Russ. J. Phys. Chem. , 2003. V. 77, No. 11, P. 1848].

65. Mazo M.A., Rabinovich A.B., Tovbin Yu.K. // Khim. Fizika. 2004. V. 23, No. 6. P. 47.
66. Tovbin Yu.K. // Zh. Fiz. Khimii. 2005. V. 79, No. 12. P. 2140. [Russ. J. Phys. Chem. , 2005 V. 79, No. 12, P. 1903].
67. Votyakov E.V., Tovbin Yu.K., MacElroy J.M.D., Roche A. // Langmuir. 1999. V. 15, No. 18. P. 5713.
68. Steele W.A. The Interactions of Gases with Solid Surfaces. – New York: Pergamon, 1974.
69. Batalin O., Tovbin Yu.K., Fedyanin V.K. // Zh. Fiz. Khimii. 1980. V. 53. P. 3020.

Dynamics of molecular flows

The microscopic hydrodynamics equations [1–4], constructed in section 4, constitute a new approach for the numerical investigation of molecular transport in narrow pores. The dissipative coefficients of this approach are discussed in detail in Chapter 5. The new approach was used for the calculations of fluid flows in narrow pores with a view to illustrate the possibilities of microscopic equations for the joint presence of the effects of inertia, viscosity, phase transitions, etc., to cover virtually the entire spectrum of phenomena implemented at short times in real processes in non-stationary conditions.

The following calculations are based only on information on the wall surface potential and intermolecular interactions of the adsorbate. System with argon in slit pores of different materials (carbon, silica gel, polymer matrix), which were recorded by choosing Q_1 of the adsorbent–adsorbate bond (sections 4, 14 and 21) are considered. The geometry and energy parameters of the pores are modelled by the rules set forth in section 14. In the case of delamination of the adsorbate in the pores the coexisting vapour and liquid phases were determined by the methods described in sections 21 and 25.

42. Statement of the problem and description of the calculation field

These systems simulate the real physical conditions in many adsorbents, catalysts and membranes, and their study is necessitated by the need of detailed analysis of the processes occurring in these porous bodies in a wide range of changes in various modes. Using a single molecular model provides the same accuracy of the account all the molecular characteristics and physical factors of the studied system, in contrast to the approach in section 3.

Numerical simulation was carried out of the two-dimensional flow (with velocities u_x and v_y along the x and y axes) of a monatomic fluid in a slit-like pore with width H at medium concentrations θ = 0.01–0.99 [5, 6]. Here θ is proportional to the gas-dynamic density and represents the fraction of particles in a local volume, normalized for the size of particles in the close-packed state. For rarefied gases, where the mean free path overlaps many times the width of the pore $\theta = 10^{-4}$, for dense gases $\theta \approx 10^{-3}–10^{-2}$, and for the liquid phase $\theta \approx$ 0.4–0.99. Most of the calculations were performed in the isothermal model which was tested by taking into account the energy fluxes (according to equation (30.5)).

Recall that the width of the pores H is measured in units of λ (where λ is the size of the site of the lattice structure associated with the diameter of the molecule). Pore width H was varied from 6 to 80λ. The section length of the pore L was varied depending on the type of problem from 40 to 150λ. The pore cross section is shown in Fig. 42.1: the planes BC and AD are impermeable boundaries of the pores and the planes AB and CD are cross sections perpendicular to the axis x through which the material flows. The situation in Fig. 42.1 is typical of the initial distributions in the case of two-phase states of molecules. It demonstrates the presence of the vapour–liquid delamination in the slit pore: a central dense vapour bubble surrounded by the liquid. If the vapour bubble is absent there is only the liquid, but for the low density of the fluid the pore contains the vapour instead of the fluid.

We also consider the situation with two vapour bubbles. In this case, the menisci of vapour–liquid partition are positioned so as to form the initial bubble location at a predetermined distance from each

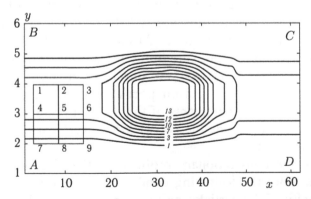

Fig. 42.1. Calculation field of slit pores. Schematically are shown 13 isolines contours of equal density and the pattern of the calculation field, consisting of 9 sites [6].

other at which there is no overlapping of the vapour–liquid transition from different bubbles. This situation is not strictly equilibrium, although each of the interfaces satisfies the equilibrium conditions of the distribution of molecules of the transition region. This is due to the fact that the total energy of the system is not minimal and depends on the distance between bubbles.

In all cases, the initial distribution was the equilibrium distribution of molecules calculated from the formulas in the chapters 2 or 3, depending upon whether or not capillary condensation takes place (or the liquid–vapour interface forms). The bubble size is determined by the width of the transition region between the vapour and the liquid and the vapour phase region defined by the conditions of the overall material balance of the substance θ in the treated area (inside the calculation field). This state of the system is sufficiently stable and may exist for a long time, if the local distributions are consistent with the macroscopic parameters of the system (T and θ).

For a given initial equilibrium distribution we calculated the local filling associated with pressure using the equation of state $P(f)$ in the cell f. The concentration profile for this section X was at a constant chemical potential in this section. Due to the interaction with the walls the given average density in the pore volume is redistributed in such a manner that the highly inhomogeneous local distribution of the fluid forms: the variable density profile over the cross section of the pore for gas and liquid phases the further complicated the gas–liquid interface (with a meniscus) in the case of capillary condensation. Calculations of the flow of the fluid in the pore at subsequent times were carried out at a constant temperature.

Initial perturbation of the states of the system was set by: 1) fluid velocity u_x along the axis in the central section of the pores, 2) increment of internal pressure $\Delta P/P_0$, where P_0 is the pressure in the initial equilibrium state, or 3) the increment of the chemical potential in terms of the pressure in the thermostat $\Delta P/P_0$ (see below).

Using the values u_x reflects the purely mechanical perturbation, constant over the cross section pores or as a Poiseuille profile ($\sim10^{-3}$–10^{-4} m/s). The increments of the internal pressure or chemical potential (pressure outside the pores) $\Delta P/P_0$ were varied over a wide range of 10^{-1} to 10^{-5}.

To solve the non-stationary problem of the viscous flow of the fluid in the pore the following conditions were formulated: 1) the initial velocity of the particles on the impermeable walls of the pores is zero; 2) at the ends of the pore we know the flow parameters at infinity u_∞, v_∞, θ_∞; 3) at the initial moment of time t_0 in addition to the boundary conditions the distribution of the flow parameters u, v, θ in the entire calculation field is also given.

The dynamic calculations were carried out for the molecular flow with the structure $z = 6$. In this case, the coefficient b in the formula (31.4) takes the value $b = 0.4$. In the process of calculation the values of the parameters at each point of the calculation field at each time are determined by the characteristics specified by the template (Fig. 42.1). Index g for any central site f ranges passes over all z of the nearest neighbouring sites f (see template of nine points in Fig. 42.1, where the central site corresponds to point 5 and the surrounding sites to the points 2, 4, 8, 6). If the pressure drop in pore is maintained constant over time then in the one of the end sections the flow parameters are fixed.

The considered flow is the movement of molecules in a strongly inhomogeneous medium. The method for solving problems of the motion of a fluid in a narrow pore, which allows to calculate the flow of a compressible viscous fluid with regard to strong concentration gradients and velocities with phase transitions taken into account and its algorithm are given in Appendix 10. Calculations by this method are carried out with high stability. The calculation procedure includes a coefficient smoothing the solution for a strong discontinuity. Discontinuities are found on the basis of density values.

Usually, in gas and hydrodynamic problems with the boundary conditions on solid surfaces the velocity components are equated to zero, based on the condition that the particles of the substance on the walls are so strongly attracted that their velocity vanishes [7]. These conditions are 'standard' for all hydrodynamic programs. In this approach, the behaviour of molecules in the cells at the border with the pore wall is different because of the influence of the wall potential from the behaviour of molecules in the inner region. This effect is taken into account when calculating the viscosity coefficients, which are functions of the local densities and intermolecular interactions. Their values are determined self-consistently in the calculation in all cells of the calculation field. For attractive potentials the viscosity coefficients are maximal on the surface, so that the longitudinal velocity component is minimal (but not identically equal to zero) and the vertical component of the velocity is always set to zero because of the condition of impermeability of the fluid through the solid surface.

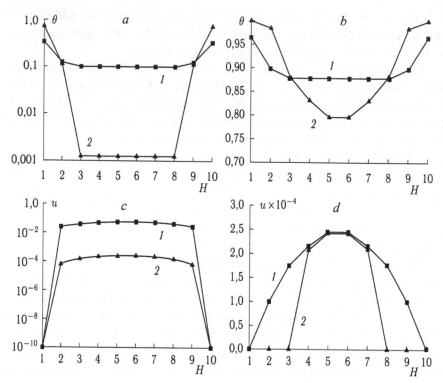

Fig. 42.2. Profiles of the distributions of argon atoms in a carbon pore with a width of 10 monolayers (*a*, *b*) at an average density of argon $\theta_{Ar} = 0.15$ (*a*) and $\theta_{Ar} = 0.9$ (*b*) and the corresponding local fluid velocity profiles along the pore axis u_x (*c*, *d*) for temperatures $T = 393.3$ K (*1*), $T = 100.8$ K (*2*) [5, 12].

The numerical method [8–11] was further adapted to accommodate the special features of the molecular potential, leading to abnormally strong gradients along the normal to the plane of the surface and near local surface roughness. In order to reflect the actual physical flow patterns on the surface of the pore wall through existing calculation schemes we introduce an additional layer formally coinciding with the surface area of a solid, for which both velocity components are equal to zero – this allows to take into account the sliding effect of the molecules in the first surface layer of the fluid relative to the wall pores.

Details of the single-phase flows are shown in Fig. 42.2 *a*, *b* which shows the density profiles of argon atoms in a pore with a width of 10 monoatomic layers for two average concentrations $\theta = 0.15$ and 0.9 [5, 12]. The first case (*a*) corresponds to the dense gas and the second (*b*) to the liquid phase. At high temperatures, the gas atoms are distributed almost uniformly over the width of the pore in its centre and with a noticeable increase of density at the pore walls. At low temperatures the distribution of atoms along the normal to the surface of the pores is highly anisotropic. The first layer is almost completely filled, and in the central part of the pore the gas is highly rarefied. (Obviously, the distributions are symmetric relative to the centre of the pores because identity both walls of the pores are identical).

For the liquid phase the local concentrations at the temperatures in question are comparable. But the differences in the conditions also change the appearance of concentration profiles. For high temperature at the wall the concentration is lower, for the bulk portion higher than at low temperature.

The density distributions discussed here are typical for both the equilibrium conditions and under weak perturbations of the distribution – in flows with velocities to $u \sim 10^{-3}$ m/s. (The same applies to distributions the viscosity and the local velocities in the gas and liquid phases respectively).

The calculated results show a significant change in the initial velocity profiles over time and a small change in the concentration profile. To obtain the stationary flow pattern the flow parameters at the left and right boundaries of the pores were gradually corrected considering their values at the central portion.

Figure 42.2 (c and d) shows the corresponding velocity profiles of the liquid and the vapour along the pore axis u_x under the same conditions. All curves are quite different from the Poiseuille profile, with curve 1 on the field (d) being similar to this profile. It corresponds to the dense phase at high temperature. With decreasing temperature there are monolayers around the wall on each side which are formed by almost frozen molecules, i.e. the molecular system itself organizes the type of boundary conditions which is determined by the temperature and interaction potentials of the molecules with the wall and with each other. The velocities of the vapour phase (field c) are significantly higher speed and they are weakly dependent on the distance from the wall, although on the surface the velocity decreases sharply.

Calculations showed that the dimensionless non-equilibrium correction of the pair distribution function (section 31) varies over a wide range from 10^{-5} and 10^{-2}, depending on the magnitude of the local flow velocity and thermal velocity u_x of the molecules. The thermal velocity of the molecules decreases with decreasing temperature and increasing local density. Values of local microhydrodynamic velocities depend on the type of flow and the coordinate of the cell. The kinetic coefficients in the flows change in a corresponding manner and because of their non-locality they depend on the entire set of neighbouring molecules affecting the thermal velocity of the molecules by the values of nonequilibrium corrections in these sites. All dissipative coefficients are calculated at each integration step in time with the corrected pair distribution function, which is somewhat different from the equilibrium value. Overall, however, the nature of the concentration and temperature dependences of the kinetic coefficients, discussed in chapter 5, is retained.

In this chapter, the time will be expressed by nanoseconds (10^{-9} s) or picoseconds (10^{-12} s).

43. Relaxation dynamics of the liquid phase

Analysis of processes at short times allows us to trace the individual stages of relaxation during external pulsed effects [13]. This kind of disturbance can be created for any impact from the outside, with a sharp overlap of the flow in the channel, under the effect of ultrasound, etc. In order to implement such a perturbation it is sufficient to create a relatively small displacement of several layers along the pore axis to produce a rarefaction (or compression) wave at the initial time. The type of local primary inhomogeneous distributions of molecules is similar to the curves in Fig. 42.2.

The isothermal flow of argon atoms at $T = 117.5$ K for three time points is shown in Fig. 43.1 [13]. In a carbon pore with a width of 10 monolayers the pulsed perturbation along the X axis from left to right was specified at the initial moment of time at point 52. Field (*a*) shows the change in the field of the values of concentrations of Ar. At the initial time the liquid was regarded as fixed for the region $X < 53$. The pulse was directed along the axis x, starting with cell $X > 53$. (Hereinafter, discrete values X represent the number of cells along the x-axis.) Under such conditions at time $t_1 = 0.04$ ns a region of the 'microbubble' type formed around the cell $X = 53$ (Fig. 43.1, *a*). The maximum density on the isolines, which is observed near the wall of the pores, corresponds to the density of about 0.9999, while the minimum value of the density isolines in the centre of the pore corresponds to $\theta = \sim 0.81$. At the interface $X = 53$ there were two decompression waves, which then began to move in opposite

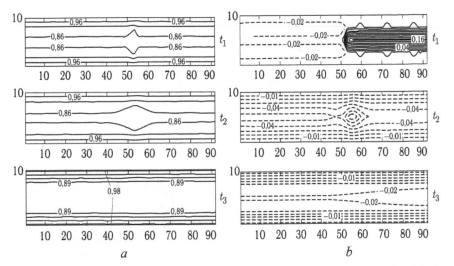

Fig. 43.1. Relaxation of the concentration field of argon atoms (*a*) and the velocity field along the axis of the pores (*b*) after a pulsed perturbation at three time points t_1, t_2, t_3 [13].

directions forming a region of transformation of the structure for the fluid flow.

At later times ($t > t_2$ = 10 ns) in accordance with the advance of the decompression waves the area of transformation of the local state of the fluid increases. The result is the formation of the liquid flow in the pore with the new distribution of the molecules in the central portion of the pores. This can be seen in Fig. 43.1 *b*, which shows the isolines of microhydrodynamic molecular velocities which give a detailed picture of the liquid microflows. (Here and below the positive values along the *x* axis of the flow velocity components correspond to the solid lines and negative components correspond to the dashed lines).

At times greater than the time t_2, the movement of the liquid flow in the left part of the pores with a small velocity $u \sim 0.01$–0.02 m/s is stabilized. For the third moment of time t_3 = 32 ns the perturbation that has been created by the initial momentum was almost completely extinguished. Thus, the numerical method allows to observe in detail all phases of the momentum relaxation in the liquid phase.

Meniscus of the liquid interlayer
In the presence of phase boundaries the initial flow parameter values were defined along the length of the pore in the form of domains with piecewise constant values corresponding to the liquid or gaseous

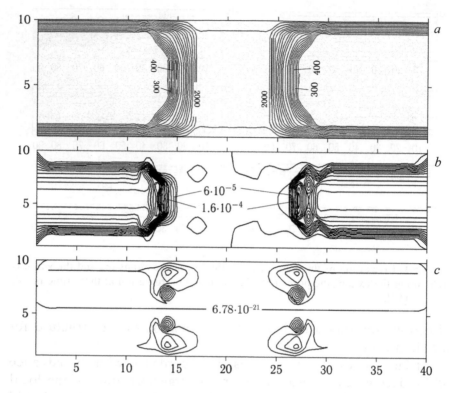

Fig. 43.2. Fields of concentration distribution (*a*), longitudinal velocity *u* (*b*) and the transverse speed *v* (*c*) of argon in a slit pore.

phase. The field $15 < X < 25$ is a liquid area between the vapour phases [5, 12].

As an example, Fig. 43.2 presents the density (concentration) field and the gas-dynamic velocity components *u* and *v* on the *X* and *Y* axes for the flow of fluid argon in the phase transition mode at temperature $T = 100$ K, the lateral interaction parameter $\beta\varepsilon = 1.18$ and the average concentration value in the cross section $\theta = 0.6$. The system parameters ($\varepsilon = 1.0$ kJ/mol, $Q_1 = 9.2\varepsilon$, $Q_2 = Q_1/8$) correspond to argon in a carbon pore [14]. In the calculations it was accepted that $\varepsilon^* = \varepsilon/2$ [15]. In this case the central part of the pore contained a liquid, and parts to the left and the right – the gas. Figure 43.1 *a* shows that after prolonged counting the one-dimensional laminar flow was established at the ends of the pore.

Figure 43.1 *b* illustrates the extent of changes of the original Poiseuille profile on the wall with $X = 1$ compared with the speed profile when $X = 5$. Near the boundary of the phase transition

($X = 15$) there is a complex two-dimensional flow. It is evident
that in addition to the initial flow vorticity in equilibrium there
are additional areas of vorticity and areas of negative values of the
velocity components.

In Fig. 43.1 c the speeds in different halves of the pore are
directed towards each other. Analysis of Figs. 43.1, b and c shows
that the gas on the left flowing into the liquid medium, is strongly

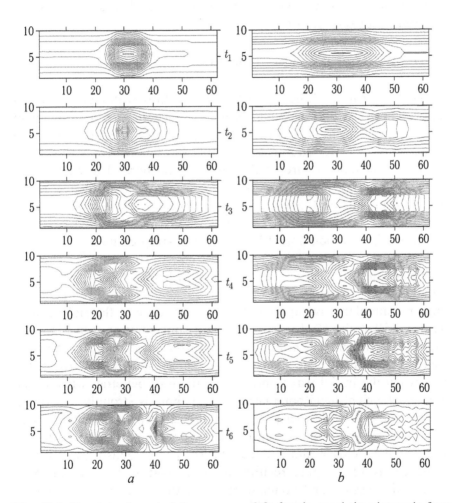

Fig. 44.1. Dynamic characteristics: $a - \tau = 0.8$; for six equal time intervals from
the start of the process until $t = 36$ ns, respectively: $\theta_{min} = 0.05, 0.2, 0.64, 0.64,$
0.60, 0.56; $\theta_{max} = 1.0, 0.95, 0.96, 0.94, 0.96, 1.0$; $\Delta\theta = 0.05, 0.05, 0.02, 0.02, 0.02,$
0.02, $b - \tau = 0.9$; for six equal time intervals from the start of the process until
$t = 60$ ns, respectively: $\theta_{min} = 0.2, 0.3, 0.54, 0.52, 0.54, 0.55$; $\theta_{max} = 1.0, 0.95, 0.92,$
0.94, 0.96, 1.0; $\Delta\theta = 0.05, 0.05, 0.02, 0.02, 0.02, 0.02$.

inhibited. The fluid itself moves slowly along the X axis from left to right.

In addition to the three types of fields shown with hydrodynamic information, the molecular theory allows to monitor the temperature in flows, the local flows of thermodynamic and thermophysical characteristics (coefficients of self-diffusion, shear and bulk viscosity, thermal conductivity), the dynamic structure of the vapour–liquid interface, along with preservation of molecular information about local thermal velocities and energies of molecules, as well as the distribution of clusters of molecules.

44. Dynamics of the vapour bubble

The complex nature of the dynamics of fluid flow with high density in the presence of capillary condensation is illustrated by the movement of a vapour bubble [12]. Figure 44.1 presented the flows for two temperatures, corresponding to $\tau = 0.8$ and 0.9, where $\tau = T/T_c$, T_c is the critical temperature of the fluid in the pore. The calculation was performed in the isothermal mode. Given the dependence $T_c(H)$, discussed in chapter 3, the second temperature is close to the critical temperature. The argon concentration fields under isothermal flow conditions in a carbon pore 10 monolayers are presented.

The flow develops from an initial equilibrium state – the gas bubble is situated in the area $28 < X < 40$, wherein X is the number of the sites along the axis of the pore, under the action of the pressure pulse rectangular wave applied at the initial time to the right end in the pores $52 < X < 62$. Lines of constant concentration θ in Fig. 44.1 a take values from the minimum θ_{min} to the maximum θ_{max} and are evenly spaced at $\Delta\theta$. The fields 2 and 3 show the flow patterns at the time when the initial wave, moving to the left, penetrates inside the bubble. The right-hand boundary of the bubble is strongly deformed, eroded, and the left border ($20 < X < 30$), as a concave lens, focuses the ascendent wave in the centre of the pores and reflects it to the right. In the fields 4 and 5 it can be seed that focusing the waves at the centre of the pore results in the formation of a region with increased pressure and concentration of the fluid which is a new boundary of the initial boundary but is concave to the other side ($25 < X < 30$). The elevated pressure, penetrating into the central part of the pore on the left, splits the original bubble into two bubbles. At the time corresponding to the field 4, the initial pressure pulse is almost completely 'eroded and 'escapes' from the calculation field, so that further development of the flow pattern occurs only under the influence of inertia, surface and intermolecular interaction forces interactions.

In field 6 it is shown that at collisions of the fluid flow moving to the right ($45 < X < 62$) and with the new boundary there appears a region of inhibition with a high fluid concentration ($35 < X < 45$), where the liquid collapses from the surface of the pore into it. The further development of the flow and deformation of the bubble are the result of the interaction of region $15 < X < 35$ and $35 < X < 45$, thereby alternately decreasing and increasing the concentrations in these areas, with the formation of 'drops' on the interface of their interactions.

Figure 44.1 b shows the concentration field of the fluid in the pore for six time points describing the isothermal flow at $\tau = 0.9$. Comparison with Fig. 44.1 a shows that at a higher temperature the influence of the initial pulse results in faster breakup of the bubble. At the initial stage the flow pattern in both cases is qualitatively similar, but at later times new bubbles formed in the interaction between them at higher temperatures do not cause those vibrational motions which form at $\tau = 0.8$. For the last time point of Fig. 44.1, b it may be seen that two films modes operate in the pore mode, combined together by the area $25 < X < 45$, wherein the transition process from one film mode to another film mode takes place.

We note the conventionality of a series of graphic images in the fields of concentrations and velocities (here and below). Because of the discrete difference equations background information is defined only on the set of the examined sites. When using standard graphics products a situation arises when the neighbouring values differ strongly enough and as a result the isolines are constructed with a resolution smaller than the size of the site. Such cases are easy to monitor and the accuracy of the images can be evaluated.

The physics of this phenomenon is confirmed by the behaviour of the longitudinal velocity of the fluid in Fig. 44.2. The velocity field corresponding to the second point in time on the field 1 in Fig. 44.2 indicates that the fluid at the initial time interval moves to the left under the influence of the pressure pulse (the dashed lines correspond to the movement of the fluid along the pores to the left). The velocities on the velocity isolines are indicated in the figure caption.

Field 2 in Fig. 44.2 at the time of reflection of the the initial compression from the left edge of the bubble and its focusing in

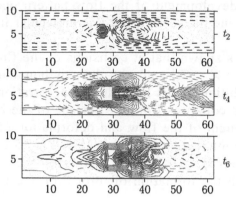

Fig. 44.2. The fields of the horizontal component of the velocity of the argon flow (in m/s) in a carbon pore with a width of 10 monolayers at $\tau = 0{,}8$ (isothermal conditions of flow) for three instants of time t_2, t_4, t_6 (as in Fig. 44.1, a), respectively: $u^+_{max} = 0.021, 0.016, 0.060$; everywhere $u^+_{min} = 0.001$; $\Delta u^+ = 0.003, 0.001, 0,005$; $u^-_{max} = -0.01$, -0.001, -0.001; everywhere $u^-_{min} = -0.15, -0.013, -0.085$; $\Delta u = 0.01, 0.001, 0.005$.

the centre of the pores is characterised by the formation of a flow ($17 < X < 32$) directed towards the main flow ($35 < X < 62$). The collision of these flows results in the formation of a zone of higher concentration where the liquid collapses from the surface of the pore into its in internal part (field 3 in Fig. 44.2). At the ends of the pores on the left and right the fluid is not moving at this moment, although the subsequent interactions in the pore generate some oscillatory movements of the fluid in its end regions.

The decay pattern of a moving bubble in the pore at the initial temperature $\tau = 0.9$ is shown in Fig. 44.3, where the variant, taking into account the limited rate of heat exchange between the fluid and the walls of the pore in fluid motion is considered. Analysis of the data in Fig. 44.3 shows that taking into account heat transfer in the problem leads to a 'softer' flow mode, where numerous waves, causing fluctuations in the liquid phase, are extinguished. In contrast to isothermal flow modes, in this case the liquid from the surfaces of the pores ($25 < X < 35$) does not collapse at the centre of the pores, since the liquid phase with the concentration θ close to unity cannot form in this mode.

The illustrations for $H = 10\lambda$ reflect basic laws of unsteady flows of fluids in narrow pores at short times. Reducing the width of the pores to 6λ (reduction of T_c (6) $\sim$ 15%) and increasing it to $H = 20\lambda$ (decrease $T_c(20) \sim 1.5\%$) qualitatively retain these features. In the case of attracting walls all the gas is adsorbed on them and there are no atoms in the bulk of the pore.

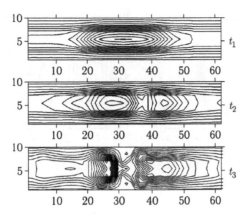

Fig. 44.3. Argon concentration fields in a carbon pore 10 monolayers wide at $\tau = 0.9$ at the initial time (non-isothermal flow conditions). For the first three time points as in Fig. 44.2 respectively: θ_{min} = 0.2, 0.3, 0.15; θ_{max} = 1.0, 0.95, 1.0, everywhere $\Delta\theta = 0.05$.

One of the main conclusions of the study is that for single-phase flows of relatively dense gases the viscous 'volume' flow (in which the main role is played by the interaction between the atoms of the gas) is almost always absent. The film flow of the fluid forms instead. In the presence of capillary condensation the pore width, of course, strongly influences the behaviour of the interface, especially across the width of the interface which at fixed interaction potentials strongly depends on the temperature of the system [16].

If we consider long periods of time, then, firstly, bubbles can be removed from the calculation field and, secondly, the ends of the pores will influence the total fluid flow. To analyze flows over long time we must transfer to a moving coordinate system.

45. A system with two bubbles

A more complex example of the flow pattern in a pore with width $H = 10$ layers at a temperature of 117.5 K from the almost equilibrium state of two coexisting equilibrium bubbles until their destruction is shown in Figs. 45.1 and 45.2 [6, 13, 17].

At the initial time (Fig. 45.1 *a*) the liquid phase corresponds to regions at $X < 12$, $X > 52$ and $X \sim 33$ in the gap between the bubbles. At a relatively low pressure drop in the form of short pulses the bubbles are deformaed in the calculated intervals deformed and almost remain in place.

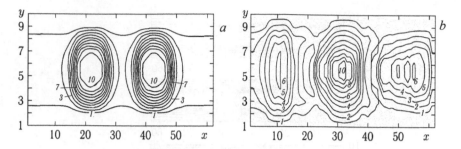

Fig. 45.1. The distribution of two bubbles in the initial moment (*a*) [6]. Evolution of the concentration field of fluid argon at process relaxation times of the order of 0.3 ns (*b*).

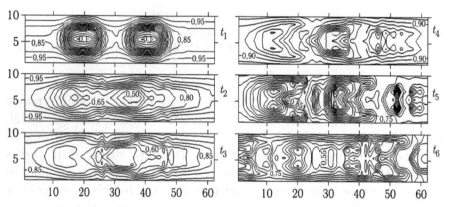

Fig. 45.2. Evolution of density distribution of the liquid, containing two vapour bubbles in a slit-like carbon pore at $T = 117.4$ K after pulsed perturbation at time points $t_1 - t_6$. The interval between the density isolines is $\Delta\theta = 0.05$. Field 1 has 18 isolines from $\theta = 0.1$ to ~1.0, for the time t_2 there are 10 isolines from 0.45 to 0.95, for the moment t_3 there are 9 isolines from 0.50 to 0.95, as well as for time t_4; at time t_5 there are 16 isolines from 0.15 to about 1.0, and finally for time t_6 there are 12 isolines from 0.35 to 0.95.

The initial state corresponds to Fig. 45.1 *a* moment t_1 in Fig. 45.2. There are two versions of the image of the process of pulsed perturbation of the overall system: a purely formal graphical representation through six equal time intervals in increments of 6 ns (total process time of 30 ns) and the state shown in Fig. 45.1 *a*, highlighted in 'manual analysis' to identify the most characteristic features of the flow structures. The field of 45.1 *b* corresponds to the case between times t_1 and t_2 in Fig. 45.2 under the influence of a pressure difference $\Delta P/P = 0.1$ at the right portion of the pores within the region $52 \leq X \leq 62$. As a result, a compression wave travelled to the calculation domain from the CD interface which resulted in the movement of the fluid. Dense gas bubbles were deformed and moved together with the whole medium. Movement of the fluid on the pore surface is accompanied by the formation of viscous forces that deform the bubbles by dragging them along the y axis. At the same time, because of the strong inhomogeneity of the medium the compression wave penetrating into the bubbles breaks them up. Thus, in particular, for the second bubble at its right boundary ($X = 37$) we observe a regular pattern of refraction when the pressure wave penetrates into the less dense gas phase. This interaction leads to the formation of a new compression wave of different intensity, which moves to the left, i.e., to the centre of the bubble, and the rarefaction wave is reflected to the right. Since the medium is inhomogeneous, the rarefaction wave in this case forms a region of a less dense gas ($45 < X < 60$) in the form of a new bubble.

The process of forming three bubbles from the two initial bubbles by crushing each initial bubbles under the influence of the initial perturbation wave is quite fast (up to 0.3 ns). (After that, the process of further fragmentation of the bubbles develops more slowly – field t_2 in Fig. 45.2 corresponds to the time 6 ns.) The initial wave moves around the pore walls along the lines of equal concentration and does not initiate similar interactions. Later, passing through the right bubble the weakened compression wave interacts with the second bubble and at its left edge ($X \sim 5$), during penetration to the liquid phase, is reflected as a compression wave inside the second bubble, increasing the density therein.

At time t_4, the initial pressure wave is completely dissipated, and flow processes within the pores in subsequent times, t_5 and t_6, are the result of the interaction of these three bubbles. The central bubble approaches the smaller bubble from the left and moves away from the bubble on the right. Two left bubbles penetrate through each other, eventually breaking up to a number of small bubbles and the liquid density between them is lowered. During this process, the diluted region, formed between the two bubbles of the right hand, is gradually filled with the liquid, which extends from both pore walls to its centre. In later stages very small bubbles form in the region of the distorted liquid interlayer $40 < X < 50$, i.e. initial bubbles are blurred and crushed. At time t_6 the first bubble has already come out from the calculation domain ABCD, and the second is strongly deformed ($X < 18$). Many small droplets formed in the pore ($20 < X < 50$) and moved behind in the second bubble to the left. After time t_6 the strongly inhomogeneous density distribution of the argon liquid is observed for the region in which the liquid phase and microscopic gas bubbles coexist. This relaxation process, associated with the redistribution of argon atoms, may continue for a long time (details are given in [6, 13]).

46. The structural inhomogeneity of the channel wall

Traditionally, in transport problems the pore walls are considered homogeneous (their properties do not vary along the length of the pores). Analysis of the real properties of the surfaces in section 6 shows that this case is a serious oversimplification, so in [18, 19] the authors considered a pore with rough walls which have steps 2λ in width and up to 2–3λ high, and the energy of the steps in a system of the argon–carbon type was approximated by the properties of the section of wall (as in section 20). The presence of the steps reduces the width of the pores with $H = 10\lambda$.

The unsteady motion of a fluid, initially located in the slit pore in the equilibrium state, occurs under the action of a velocity increment u_x, applied at all points (cells) of the pore volume (modulus u_x ranged from 0.1 to 0.5 m/s).

Figure 46.1, a gives the fields of argon concentration, and Fig. 46.1, b – a field of the axial component of the velocity of the fluid flow moving from left to right in the plane of the drawing for the four time points after the start of the fluid flow, which at the initial time moves at a speed $u = +0.5$ m/s along the x-axis. (Here and below, the positive values along the x-axis of the component of the flow velocity are described by the solid lines, and negative ones – by the dashed lines). At the initial time the distribution of the fluid in the pore with 'steps' was equilibrium – the impact of the step is spread over three monolayers along the x-axis. The initial distribution of the concentrations of the monolayers of the liquid fluid symmetrically arranged with respect to the longitudinal axis outside the area of the steps: $\theta_1 = 0.9999$, $\theta_2 = 0.977$, $\theta_3 = 0.886$, $\theta_4 = 0.85691$, $\theta_5 = 0.848$.

At short times of motion $t_1 = 1$ ns the equilibrium concentration distribution is strongly violated near the 'steps' ($45 \leq X < 60$). On the left ($0 < X < 45$) and right ($60 < X < 92$) from the step the initial equilibrium state changes as a result of movement of the entire fluid in the pore. We see that when the fluid flows around the steps two areas form behind them where the strong rarefaction wave transforms the liquid phase to the 'vapour' phase because of the strong reduction

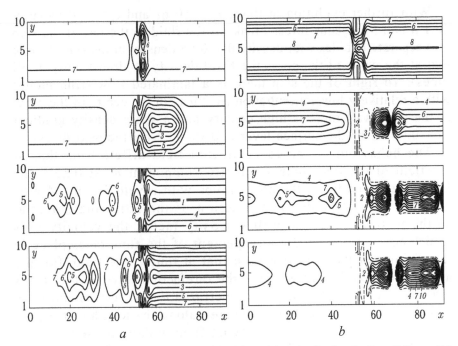

Fig. 46.1. Fields of argon concentration (*a*) and longitudinal velocity of flows (*b*) in a carbon pore with symmetrically located steps for the time intervals from the start of the process $t_1 = 1$ ns, $t_2 = 20$ ns, $t_3 = 50$ ns, $t_4 = 100$ ns.

in the density in the rarefaction wave. At $X = 54$ two microbubbles form with density ($\theta \approx 0.4$) at their centres located at opposite walls of the pore. Thus, in the passage of the step the fluid density is more than halved (the effect of 'throttling' of the flow on microscopic inhomogeneities of the pore walls), so we can talk about changing the phase state of the fluid by mechanical action. Then, the low-concentration regions relating to each of the steps are merged into one region located in the centre of the pore.

At $t_2 = 20$ ns, the perturbation from the step spreads to the region $35 \leq X \leq 75$, and at the same time on the left ($X < 34$) and right ($X > 75$) of this area changes take place in the initial equilibrium state due to the action of viscous forces on the moving fluid. Figure 46.1, *a* shows that the vapour phase, generated behind the step in the centre of the pore, begins to penetrate to the left of the step ($50 \leq X \leq 53$), which subsequently changes the nature of the fluid flow at $1 < X \leq 52$. To the right of the step two bubbles merged steps into one large bubble, which increased its size – the density in the cross section of this bubble changes from the magnitude of the almost diluted gas to 0.8.

With further evolution in times $t_3 = 50$ ns and $t_4 = 100$ ns (Fig. 46.1 a and b) viscous forces break up the fluid to the left of the step into separate bubbles with density $\theta = 0.55$ density in the centre (note that at the walls of the pores the fluid density is almost unity at all times). To right of the step there is a laminated flow (film mode) with a relatively low concentration ($\theta < 0.1$) in the centre of the pore; at the pore wall there is high density and a sharp density gradient ($\theta = 0.9999$ for the surface layer and $\theta = 0.7$ for the next layer).

The behaviour of the concentration field differs from that of the field of the longitudinal velocity component u. At time t_1 the fluid in the middle of the pore moves with a velocity of 0.4 m/s, and near the walls, due to the action of the attracting potential of the wall and the near-wall viscosity factor, the velocity falls to 0.1 m/s. Complicated rearrangement of the flow takes place in the vicinity of the step. It is seen that at time t_2 the fluid, flowing around the step, is accelerated in the rarefaction wave up to 1.1 m/s ($58 < X < 68$). At the same time, perturbation waves (compression wave) propagate to the left of the step along the length of the pore which inhibit the speed of the moving fluid initially in the range $0 < X < 47$. At $48 < X < 54$, these perturbations drag the fluid to the left.

At long times on the left of the step ($0 \leq X \leq 45$) the fluid velocity is positive, i.e. in the plane of the drawing it moves from left to right and penetrates through the neck formed by steps, and to the right of the step ($55 \leq X \leq 92$) longitudinal velocity changes its sign, the sign change occurs in the central part of the pore having a width equal to the size of the neck formed by steps. Comparison of the fields of velocity (Fig. 46.1 b) and concentration (Fig. 46.1 a) shows that, although to the right of the step the concentration field goes to a new steady-state solution, the velocity at the corresponding points of the pore continues to change, i.e., the concentration field is more conservative than the velocity field, and is mainly formed by the surface forces. To the left of the steps the fluid flow is more uniform with respect to u, although the concentration in the pore width undergoes oscillations. In other words, the individual bubbles that exist at $0 \leq X \leq 50$, interact with each other, thereby causing fluctuations in the velocity to the right of the step. As a result, to the left and right of the step the influence of surface viscous and inertial forces results in the formation of a complex interrelated, which is far from steady state. Note that the layered character of the flows clearly shown for the second time point is retained near the walls also for the following two points in time. They are not presented here

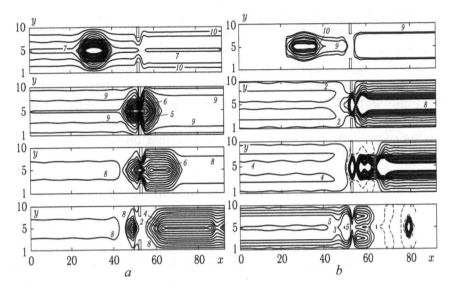

Fig. 46.2. Fields of concentration (*a*) and longitudinal flow velocity (*b*) of the vapour-liquid system in a pore with steps for the four time intervals from the beginning of the process, as shown in Fig. 46.1 *a*.

to focus on the evolution of the central part, reflecting the relaxation process on the example of isolines with $\theta = 0.1$, – over time the area covered by it, is significantly reduced.

(In Figs. 46.1 *b* and 46.2 *b* the velocity fields in the area where the modulus of the velocity approaches zero the isolines with the with negative and positive values intersect. This non-physical fact can be explained by errors in the smoothing procedure used in the standard programs of the graph plotter.)

A more complicated case of the influence of inhomogeneities in the pore walls in the form of steps on the distribution of local flows near the vapour–liquid interface when the bubble passes through a narrow part of the pore is illustrated in Figs. 46.2 *a* and 46.2 *b*. The initial distribution of the concentrations of monolayers steps of the fluid, symmetrically arranged with respect to the longitudinal axis, outside the regions of the steps and the bubbles – as in Fig. 46.1 *a*.

In this case, for the initial equilibrium distribution in the pore we selected the mode of the capillary condensation of the fluid with the liquid ($20 < X$ and $X > 40$) and gaseous ($22 \leq X \leq 40$) phases (Fig. 46.2 *a*). In the section passing through the bubble, the concentration varies from $\theta \sim 0$ for the centre of the pore to 0.9999 on the pore wall. Otherwise the situation is similar to that discussed above for Figs. 46.1 *a* and *b* for a single liquid phase. The distribution of the concentration and axial component of velocity at time $t_1 = 1$ ns is shown in the first field in Fig. 46.2, a and b. At $t_2 = 20$ ns, the vapour phase (bubble) approaches the steps and starts to penetrate through the narrow portion of the pores. At this point the bubble is subjected to the effect of surface forces from the flat walls of the pores and steps. As a result, the bubbles break up, so that part of it remains to the left of the steps ($44 \leq X \leq 52$), and the other part of it passing over steps is eroded in the flow with a uniform distribution of the concentration (for $t_4 = 100$ ns).

From the velocity field illustrated in Fig. 46.2 b at time t_1 it is seen that when the fluid having vapour and liquid phases moves at a velocity of 0.5 m/s a complex redistribution of velocities in each section of the pores takes place in the system. Thus, in the vapour phase, the velocity is doubled to about 1 m/s. The layered movement of the fluid is observed to the right of the step. At time t_2, when the gas phase permeates through the narrow neck of the pores between the steps the inhomogeneous flows, perturbed by the step and the bubble, interact. As a result, a fluid having velocity $u = 1$ m/s and a low density moves between the steps. Right in the centre of the pore there is a region with a negative velocity direction, which subsequently increases (t_3). At time t_4 when the bubble has passed through the steps, the fluid to the left of the steps moves at a low speed $u = 0.05$ m/s, and on the right a region $62 < X < 86$ with strong velocity gradients forms behind the step. This transition region, formed by the influence of steps on the movement of the fluid in the pore, inhibits the overall flow and moves with it during the relaxation of the initial perturbation.

The behaviour of the fluid as it passes through the steps, of course, depends on the initial conditions. If the initial velocity is small and/or the vapour bubble is located away from the step, the perturbations of the initial pulse, resulting in fluid movement, reach the bubble before it reaches the step or passes through it. Then, perhaps under the influence of compression waves in the area of the pore with uniform walls the bubble disintegrates earlier, similar to the previously discussed case, and the flow pattern is different from that shown in Fig. 46.2 a and b.

Comparison of the flows for two cases (Figs. 46.1 a and 46.2 a, as well as 46.1 b and 46.2 b) shows that the bubble present in the pore suppresses the process of generation of the compression wave which is formed on the left side of the step under the pulsed effect at the initial moment. As a result of this perturbation (in Figs. 46.2

a and *b*) do not extend to the left from the step (in contrast to the process in Figs. 46.1 *a* and *b*.)

The process of the fluid flow into the pore with inhomogeneities in a sense resembles the fluid flow in a pore with uniform walls under the action of the pressure difference on the ends of the pores. In both cases, in the capillary condensation mode the vapour bubble easily splits into smaller low-density areas. A transitional region from one flow mode to another forms. This is due to the fact that when the fluid moves the pore with steps includes areas of strong compression and rarefaction of the flow on opposite sides of the step which is equivalent to some extent to the effect on the bubble resting in the homogeneous pore of the moving compression wave penetrating into the bubble and then exiting from it.

47. Influence of the wall potential on molecular flows in narrow channels

In most situations in the simulation of traffic flows by molecular dynamics the values $Q_1 < 2\varepsilon$ are used. This is due to the fact that an increase in the binding energy of the molecule with the pore wall increases greatly the number of numerical problems: the fluid is difficult to 'tear' from the wall, which leads to large violations of statistics. An example of such poor statistics is clearly seen from the fact of the wrong prediction of the critical size of the diameter of the cylindrical channel in the processes of the adsorption–desorption hysteresis (section 24).

Above the calculations were carried out of the flow for the argon-carbon system with strong adsorption: the binding energy of argon with the wall is more than nine times higher than the binding energy between the argon atoms. In the microhydrodynamic approach there is not such restriction on the values of Q_1, although in the process of calculating the dynamics it is necessary to take into account the 'stiffness' of the system of equations due to the large differences in the adsorbate–adsorbate and adsorbate–adsorbent interaction potentials.

Comparative studies on the process of transition to quasistationary states for different values of the interaction potentials of molecules with the walls of the pores are considered in [20]. At the initial time $t = 0$ the flow is created by introducing a perturbation in two ways: 1) pulse perturbation of part of the volume of the system, allowing to explore the flow at short times, and 2) specifying the pressure difference pressure at both ends of the pores to explore the quasi-stationary flows. In the second case the pressure differential ΔP_x along the x-axis is defined as a proportion of the value of P. To reduce the time for solving the problem and obtain a quasi-steady flow at low values of ΔP_x, the increment of the argon concentration corresponding to the given pressure drop is distributed linearly along the pores at the initial time. The transition of the solution to its steady-state solution is controlled by the conservation of the value proportional to the fluid flow rate $q = u*\theta$ through sections 1 (AB) and 2 (CD) (Fig. 42.1). Here, in the case of the single-phase flow, $u*$ is the average speed along the axis of the pore, θ is the average value of the degree of filling of the pore across the section of the pore. As the intensity of perturbations created by the differential ΔP_x decreases the difference $\Delta q = q_1 - q_2$ tends to 0, i.e., the volume flow rate through the pore does not change in this case. A method for producing a quasi-stationary solution when we define initially the linear concentration distribution along the pore corresponds to obtaining a similar stationary solutions in the macropore (Pouisell's problem). The second method does not reflect the transient intermediate stages in the solution of the problem: the formation of waves and their interaction in the pore, what happens in reality when considering this problem, for example, in the numerical experiment.

For the entire flow pattern over time it is necessary to solve the problem by the first method – with discontinuous initial conditions. At $t = 0$ it is necessary for one half of the pores ($0 < X < L/2$) to define the pressure P_1, and in the other half – P_2 (or concentrations θ_1 and θ_2) and solve the problem of evolution of this discontinuity.

Figure 47.1 shows the flow fields generated by the first method of the perturbation of the initial state of argon atoms in the slit pore with the width $H = 16\lambda$ and length $L = 20\lambda$ at $T = 273\ K$, $\theta = 0.8$ and $Q_1/\varepsilon = 4.5$ (the argon–silica gel system). This binding energy provides a strong attraction of argon atoms to the wall surface. The flow patterns a–c relating to times $t_1 = 0.19$ ns, $t_2 = 7.1$ ns, $t_3 = 11$ ns, are given in the form of velocity isolines U_x. The pressure drop in the pore is $\Delta P_x = 3\%$.

It is seen that a fluid region forms at the discontinuity line at time t_1 in which the velocity is directed towards the positive side of the x-axis. The velocities are maximum in the centre of the pore and tend to zero near the walls.

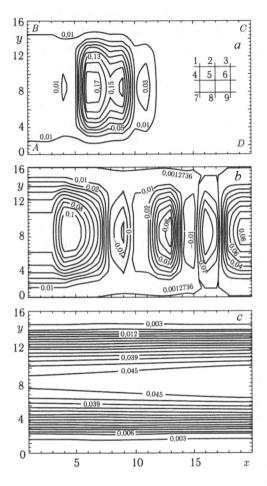

Fig. 47.1. Flow patterns of argon atoms in the pore with $H = 16\lambda$ and $L = 20\lambda$ for times $t_1 = 0.19$ ns (a), $t_2 = 7.1$ ns (b), $t_3 = 11$ ns (c), the velocity isolines U_x at $T = 273$ K, $\theta = 0.8$, $Q_1 = 4.5\varepsilon$ are shown.

If the initial pressure values are retained at the boundaries, then when the moving fluid reaches the boundaries AB and CD the fluid is reflected from them. Oscillatory motion is established in the pore. For large pressure drops ΔP_x these oscillations can last long enough on the time scale in which calculations are carried out – up to 10 ns. As a result, it is impossible to obtain the quasi-steady flow mode. Therefore, it is necessary to implement the mode of displacement of the values of the parameter of the local states of the system from internal points to the interfaces AB and CD. Since the step in the calculations is small (of the order of 10^{-2} ps), the procedure of displacement is not applied at every time step, but only after the given number of steps. Physically the procedure of transferring the parameters of the state from internal points to the boundaries denotes an increase in the length of the pores.

Figure 47.1 c shows the quasi-stationary distribution of the longitudinal velocity U_x. It is seen that the isolines of the velocity near the walls are nearly parallel to the walls of the pores, i.e., the speed is not changed along the pore. In the central portion of the pore the velocity is changing along the length of the pore, remaining at a maximum value. On the walls of the pore the velocity U_x is almost zero, that is, indeed, due to the strong attraction of the liquid flow by the wall its velocity drops to nearly zero. Analysis of the behaviour of the fields of concentration at time t_3 also shows that the flow almost reached a quasi-stationary mode. The flow rate in sections 1 and 2 is constant with a precision of 2.5 %.

To show the effect of Q_1/ε on the nature of the flow in the pore, Fig. 47.2 shows a similar pattern of development of the flow of argon in the pore with the same geometrical dimensions, but at

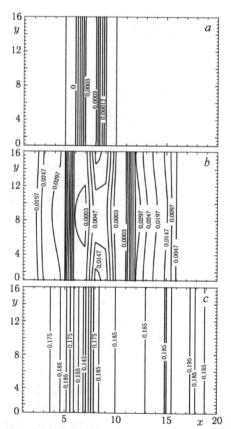

Fig. 47.2. Development of flow of argon into the same pore when $Q_1 = 0.5\ \varepsilon$, $\Delta P_x = 5\%$. Fields of longitudinal velocity U_x relate to moments $t_1 = 1.0 \cdot 10^{-3}$ ns (a), $t_2 = 0.27$ ns (b) $t_3 = 19$ ns (c).

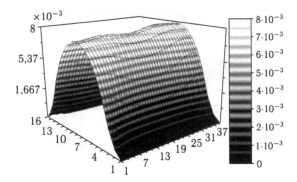

Fig. 47.3. Quasi-stationary distribution of the velocity field in the pore U_x at the moment $t = 2.9$ ns; $Q_1 = 4.5\varepsilon$, $\theta = 0.8$, $H = 16\lambda$ and $L = 40\lambda$

$Q_1/\varepsilon = 0.5$. The pressure drop in this variant is $\Delta P_x = 5$ %. The fields of longitudinal velocity U_x are given for the times: $t_1 = 1.0 \cdot 10^{-12}$ s, $t_2 = 2.7 \cdot 10^{-10}$ s, $t_3 = 1.9 \cdot 10^{-8}$ s. It is seen that waves form in the flow field and propagate the pores in both directions. The fluid flow preserves the oscillatory nature of the motion, even at times $t_3 = 19$ ns, and from the outset velocity U_x is practically independent of the y-coordinate and near the wall is not zero. Thus, when $Q_1/\varepsilon = 0.5$ the liquid slides along the wall.

Comparison of the results of calculations for two values of Q_1/ε shows that for large values of Q_1/ε the wall potential, attracting strongly the fluid flow, plays a stabilizing role in the establishment of a quasi-stationary solution.

The results of calculations of the flow in the pore with perturbation of the initial equilibrium distribution of the vapour or liquid in the second method (low pressure drop and linear law of the initial distribution of the vapour or liquid concentration along the pore axis) are shown in Figs. 47.3–47.5.

Figure 47.3 shows the pattern that describes the steady-state velocity distribution U_x in the pore at the time $t = 2.9$ ns for $Q_1/\varepsilon = 4.5$, $\theta = 0.8$, $H = 16\lambda$ and $L = 40\lambda$. As noted above, on the walls of the pores the velocity becomes equal to zero. The time to reach a quasi-stationary state in the considered short segment pores in 40λ, with a strong attraction to the walls of the pores of the molecules, is estimated at 0.1 ns.

A similar pattern of the quasi-stationary distribution of the velocity U_x is given in Fig. 47.4 for $Q_1/\varepsilon = 2.2$ at time $t = 0.94$ ns. In this case, the time to the quasi-steady state is estimated to be

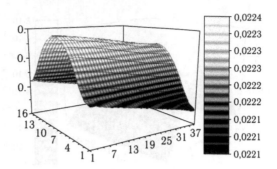

Fig. 47.4. Quasi-stationary distribution of the velocity field in the pore U_x at the moment $t = 0.94$ ns; $Q_1 = 2.2\varepsilon$, $\theta = 0.8$, $H = 16\lambda$ and $L = 40\lambda$.

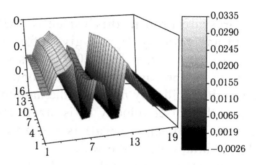

Fig. 47.5. Unsteady behaviour of the velocity field in the pore U_x at $H = 16\lambda$ and $L = 40\lambda$ for $Q_1 = 0.5\varepsilon$ and $\theta = 0.8$ at time $t = 0.7$ ns.

1.0 ns. Figure 47.4 shows a significant effect of slipping of the liquid along the surface of the pore walls.

Figure 47.5 shows the transient behavior of the velocity in a pore in the case of very weak attraction of molecules to the walls of the pore for $Q_1/\varepsilon = 0.5$, $\theta = 0.8$ at time $t = 0.7$ ns (half the length of the pore). In this case, the small value of Q_1/ε does not stabilize the flow to a quasi-stationary state, when the initial concentration values are maintained at the boundaries AB and CD. Along the normal to the surface of the pore the fluid moves almost in the same manner, so here is no sticking to the walls. The observed pattern is associated with the loss of stability of the liquid along the axis of the pore from the border AB to the border CD (in the direction of a low pressure area). At short times the liquid behaves as a system of coupled oscillators and the pulsed change in pressure along the axis of the pores leads to strong fluctuations of the velocity along this axis.

To reach the steady state flow, after the decay of these oscillations, the calculation time increases by at least 2–3 orders of magnitude. Thus, the rise time to reach the quasi-stationary state is non-linearly related to the magnitude of the ratio Q_1/ε.

Thus, the dynamic flow regimes of monatomic gas (argon) in the slit-like pores, consisting of walls of different nature, were investigated. It is shown that in the case of strong attraction of low density argon atoms to the walls of the pore the anisotropy of the flow is strong. Reducing the attraction of the molecules to the walls leads to the slipping mode of vapour or slip of the fluid along the channel wall.

48. The movement of the vapour–liquid meniscus in narrow slit pores

The question of the mechanism of movement of the meniscus is a key question in the theory of transport of molecules in narrow channels. The thermodynamic treatment of capillary flow connects it to the curvature of the meniscus of the vapour–liquid interface and the use of the Kelvin equation for the quantitative description of the driving force of the flow. As shown above (in section 26), the Kelvin equation does not hold in narrow or wide mesopores, so now we have no idea of the laws of governing the capillary flow.

In [21] the authors studied the motion of the vapour–liquid meniscus in a system with a chemical potential difference at the ends of the pores. In this case, the value of Dp reflects the change in pressure in a thermostat (outside the pore) by analogy with the traditional formulation of the problem for the dynamics of adsorption. The fluid flow in a pore with length $L = 150\lambda$ and width $H = 16$, or 40λ was considered, where λ is the linear cell size of the order of the diameter of the molecule (Fig. 48.1). Low temperatures are discussed so that the pore contains the liquid ($1 \leq X \leq 50$, $81 \leq X < 150$) and vapour phases ($50 \leq X \leq 80$).

Calculations were carried out for a liquid flow containing a vapour bubble in the pores with the pore wall material having a different nature – a different value of the interaction potential of the molecules with the wall in the surface layer of Q_1. Special attention is given the interaction of the initial shock pulse with a vapour bubble at a time when there is a transition from the equilibrium state of a substance to the non-equilibrium state with further transition to the

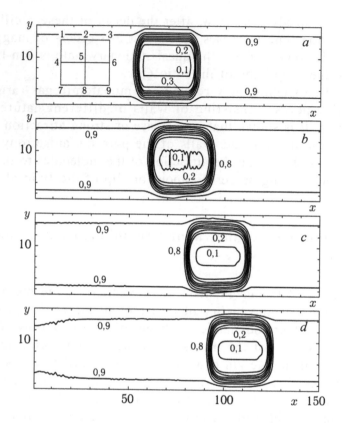

Fig. 48.1. Fluid flow pattern of argon with the vapour phase (bubble) inside the pore with $H = 16\lambda$ and $L = 150\lambda$ for times $t_1 = 0$ (*a*), $t_2 = 9.3$ ns (*b*), $t_3 = 19$ ns (*c*), $t_4 = 23$ ns (*d*). Shown are the concentration isolines at $T = 117$ K, $\theta = 0.8$, $Q_1 = 1.5\varepsilon$ and the initial pulsed perturbation $Dp = 10^{-3}P_0$, and the coordinate axes of the calculation field is a nine-point pattern used in the calculation scheme are also shown.

quasi-equilibrium flow. The physical time of studying the flow in the pore is about 100 ns.

At time $t = 0$ a perturbation appears at the left end of the pore ($1 \leq X \leq 45$). Section $1 \leq X \leq 45$ is selected to accelerate the computation time compared to the introduction of the perturbations only at the boundary $X = 1$. The perturbation results in the establishment of a concentration (perturbation is introduced by the values of the chemical potential) which provides a pressure different from the pressure in the main part of the pore ($46 \leq X \leq 150$) by some amount Dp depending on the variant of the problem. Dp changes alter the density profile of the pore which was with the equation of state (section 10) to calculate pressure changes in the equations

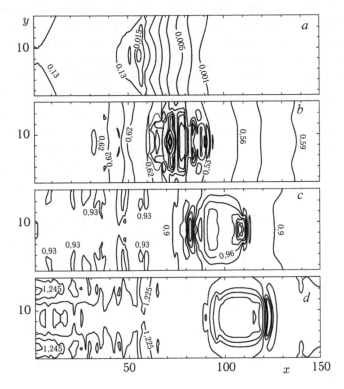

Fig. 48.2. Velocity field U_x, m/s, for the same version of the problem in the times $t_1 = 0.01$ ns (a), $t_2 = 9.3$ ns (b), $t_3 = 19$ ns (c), $t_4 = 23$ ns (d). The flow in periods $t_1 - t_3$ is unsteady, at t_4 – quasi-stationary. The average speed of the bubble at the time t_4 is $U_{bub}(t_4) = 0.6$ m/s.

of momentum transfer (section 30). Details of the calculation are listed in [21].

Dynamic problem solution requires a lot of time, which strongly depends on the parameters Q_1 and Dp. With the increase in Q_1 and Dp the calculation step with respect to time, preserving the accuracy of the calculations is reduced. This value is determined from the principle of conservation of accuracy at the maximum permissible step.

Figures 48.1 and 48.2 show the field of local concentrations θ and velocity U_x [m/s] in the form of isolines in a pore with width $H = 16\lambda$ and length $L = 150\lambda$, at $T = 117$ K, $\theta = 0.8$ and $Q_1/\varepsilon = 1.5$. Differential pressure P_0 at the ends of the pores is $Dp = 10^{-3}P_0$, where P_0 is the initial pressure. The flow of fluid occurs in the time interval $0 \leq t \leq 23$ ns.

Examination of Figs. 48.1 *a–d* shows that during the motion of the fluid in the pore the concentration, as a more conservative flow parameter, changes slightly. The vapour bubble, undergoing small deformation, is shifted along the pore by the fluid flow under the influence of the initial perturbation. The figures show only small fluctuations in the concentration around the bubble at different times. At the time t_4 (Fig. 48.1 *d*) the influence of end effects is seen at the left end of the pore: a rarefaction wave, formed on the line $x = 45$, is rebounded from the left end of the pore. The conservation of the bubble shape can be treated as the volumetric flow, with retention of the molecular distribution near the wall, as in the case the film flow regime.

More detailed information about the flow regime is given by the velocity fields (in principle, pressure fields can also be used). Analysis of the velocity fields shows that due to the rupture of the initial concentration for $t > 0$ in the region $x = 45$ the flow speed directed to the right starts to increase (Fig. 48.2 *a*). This occurs on the front and rear sides of the bubble for which the moment $t = 0$ refer to the coordinates $x = 80$ and $x = 50$, respectively. At the time t_2 (Fig. 48.2 *b*) the velocity perturbation spreads to the region occupied by the bubble. The behaviour of the velocity on the coordinate X is oscillatory.

This behaviour of velocity is dictated by the nature of the perturbations of the concentration and the pressure in the transition unsteady flow period. In the following time (Figs. 48.2 *a–d*) the flow pattern shows stabilization; the average flow is set for which in the area inside the bubble the vapour moves at its own velocity $Ux \approx$ 1.28–1.3 m/s (Fig. 48.2 *d*). The average flow velocity of the fluid in the pore is $U_x \approx 1.22$ m/s. The estimate of the average velocity of the bubble over the distance travelled by the bubble during time t_4 gives a value of $U_{x\,bub} \approx 0.6$ m/s.

Thus, in this variant of the problem the bubble moving at half the speed of the ambient flow. This means that the bubble is carried by the away along the axis of the pore at a slower speed than the speed of its environment. Analysis of the velocity and concentration shows that the molecules of the liquid 'leak' through the bubble converting into vapour and back to a liquid. In fact, it can be assumed that in the dynamic mode there is a phase transition from liquid to vapour and back again. As a result, a quasi-equilibrium motion of the bubble is established in the pore at a speed less than the average speed of the ambient fluid. Thus on the front surface of the bubble there is a

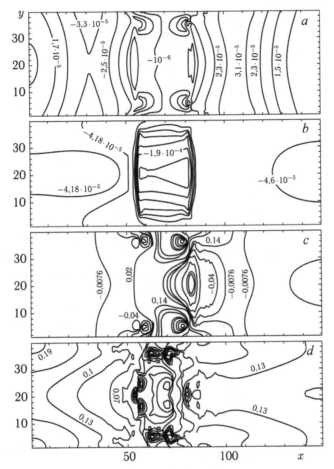

Fig. 48.3. Flow pattern of argon atoms inside the pore with $H = 40\lambda$ and $L = 150\lambda$ for times $t_1 = 0.33$ (a), $t_2 = 2.4$ (b), $t_3 = 52$ (c), $t_4 = 76$ ns (d). Velocity field at $T = 117$ K, $\theta = 0.8$, $Q_1 = 1.5\varepsilon$, and the initial pulse perturbation $Dp = 10^{-4}P_0$.

transition from liquid to vapour, and on the rear side – the vapour condenses into the liquid.

To investigate the influence of the pore width on the character of the fluid flow, containing a bubble with the same primary data as above, calculations were carried out of the molecular flows in the pore width 40λ with the value $Q_1 = 1.5\varepsilon$ and the pressure drop at the ends of the pores Dp equal to $5 \cdot 10^{-2}$ and 10^{-4} P_0. At the beginning of the flow in a wide pore the patterns of the fields of pressure and concentration are similar to those of the flow pattern in a narrow pore. But over time, small changes in the shape of the bubble lead

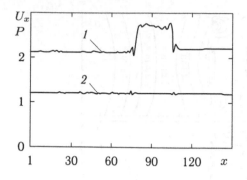

Fig. 48.4. Behaviour of $U_x(x)$ (*1*) and $P(x)$ (*2*) along the axis of the pore in the section $y = h/2$ at $t = 9.89$ ns for the variant with $Q_1 = 1.5\varepsilon$, $H = 40$, $Dp = 10^{-2}P_0$.

to a substantial restructuring of the flow. There are qualitative differences in the behaviour of the velocity.

Figure 48.3 shows the isolines of the velocity U_x in the pore for four points in time. At small Dp the flow parameters in the pore on the left and right of the bubble are almost symmetrical to each other at short times (Fig. 48.3 *a*, *b*). It is interesting that from the moment $t \approx 5$ ns the molecules at the interface of the bubble having the same speed begin to form groups (clustering). Initially, this occurs in the region of contact of the boundary of the bubble and the wall surface (Fig. 48.3 *c*), where the strong anisotropy of the flow is realized. The reason for association of the molecules with respect to velocity, or otherwise – separation of the molecules in the pore with respect to velocity is probably the fact when at the given width of the pores the molecules are bonded differently with the surface of the pore by intermolecular forces (near-surface molecules and molecules in the centre of the pores). This effect was not observed in the pore 16λ in Fig. 48.1 due to its small size.

In the behaviour of pressure for the wide pores there are also differences when compared with the previous data: the pressure along the axis of the pore reaches a constant value at t_4.

In this study the bubble motion in a wide pore at a higher value $Dp = 10^{-2} P_0$ was calculated. The analysis showed that in this case with a much larger initial pressure drop (greater speed of the fluid) there occurs clustering of the molecules with respect to velocity and pressure equalization along the length of the pore. This is shown in Fig. 48.4, which gives the profiles of the velocity and pressure in the central section of the same pore at $Dp = 10^{-2} P_0$ for time $t = 9.89$ ns. A quasi-stationary motion of the bubble along the pore is established at this moment. The vapour velocity inside the bubble, after the interaction of individual groups of molecules with each

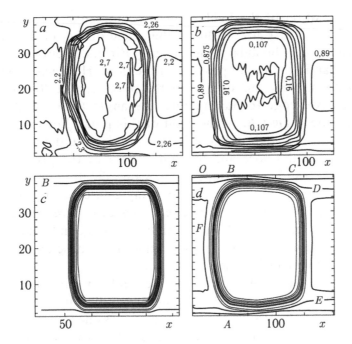

Fig. 48.5. Fragments of velocity (*a*) and concentration (*b*) fields with values of the isolines in the pore with width $H = 40$ at time $t = 100$ ns, $Dp = 10^{-2}P_0$. Concentration at times $t = 0$ and $t = 100$ ns; contact angles *OBF* and *CBD* (*c*, *d*).

other became significantly than the velocity of the liquid surrounding it. The magnitude of pressure on the scale of this figure makes a noticeable 'ripple' at the borders of the bubble. For this variant, the mass flow at the ends of the pores is constant to within 2%.

Consider in more detail the process of the evolution of the bubble shape during the motion in the pore. Figure 48.5 shows fragments of the velocity and pressure fields in the pore with width $H = 40\lambda$ and $Dp = 10^{-2} P_0$ for time $t = 100$ ns. Analysis of Fig. 48.5 *a* indicates that the velocity to the left of the bubble is less than on the right. After the transition from liquid to vapour at the right boundary there is a short burst of velocity, which gradually changes to the increase of the velocity near the boundary of vapour condensation into a liquid (see Fig. 48.4). The velocity is evenly distributed across the width of the pore. General velocity changes occur inside the bubble in the range of 2–4 %.

The behaviour of concentration (Fig. 48.5 *b*) shows an uneven distribution of concentration across the width of the pore – the denser vapour is located near the surface of the pore. To show the deformation of the bubble shape as a result of motion, Fig. 48.5 *c* shows the position of the bubble for two moments: $t = 0$ and $t = 100$ ns. It is seen that at $t = 100$ ns, the shape of the left (meniscus AB) and right (DE) boundary significantly differs. Meniscus AB is bent more than in the initial position of the pore. However, meniscus DE straightens. Near the surface of the pores (point B) there are clearly seen two edge (contact) angle: the angle OBF – left of the point B and the angle CBD – the right of this point. For this variant of the problem the angle $CBD \approx 5°$. The angles are determined by the geometry of the pore and the initial data of the problem: the velocity of the fluid flow, temperature, value Q_1. In contrast to the initial contact angle (at $t = 0$), these two angles are dependent on the velocity of the fluid. Having two edge angles of different magnitude stresses the existence of a phase transition at the boundaries AB and BD, as basic phases take part in the formation of these angles: solid surface and vapour inside the bubble.

Thus, as a result of fluid movement the fluid velocity, pressure and bubble shape are adjusted so that the quasi-equilibrium movement is realized in which the law of conservation of mass and momentum are satisfied with an accuracy of 2 and 0.5%. The flow velocity ahead of the bubble is greater than after the bubble, i.e. as a result liquid–vapour and vapour–liquid transitions the bubble whose velocity less than the velocity of the surrounding flow slows this flow down.

To analyze the effect of the energy of interaction with the wall of the molecules in the 1st and 2nd layers on the motion of the bubble, fluid motion with $Q_1 = 0.2\varepsilon$ at an initial speed of all matter in the pore with velocity $U_x = 1$ m/s was considered. Figure 48.6 shows the fields of concentration and velocity for four moments of time. Analysis shows that the sudden movement of the entire fluid at a speed $U_x = 1$ m/s, the vapour (in the form of a lens) in the pore shrinks and the velocity on the left inside the bubble greatly increases. The bubble then collapses and vapour particles settle on the pore surface. In this version of the problem, of course, the initial velocity is important and by changing this velocity it is possible to get a variety of ways of development of the flow in the pore.

It should be noted that the inclusion of energy transfer is an important contribution, but the overall picture of the examined trends remains unchanged. In this regard it should be noted that the equations in section 30 should include the contribution of heat flows between the solid and the adsorbate. The heat flows on the adsorbent are usually assumed to be faster as compared to their transport in the liquid or, especially, in the vapour phase (see also section 50).

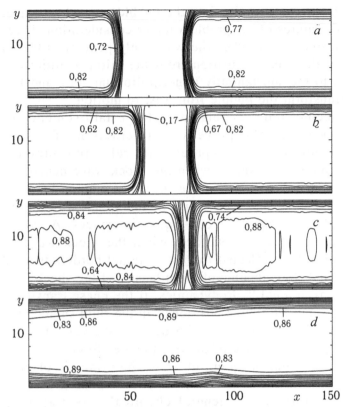

Fig. 48.6. Flow patterns (concentration fields) at time $t_1 = 0.175$ (*a*), $t_2 = 1.75$ (*b*), $t_3 = 3.5$ (*c*), $t_4 = 5.2$ ns (*d*) at the sudden movement of the entire fluid with velocity $U_0 = 1.0$ m/s. In a pore with $H = 16$ $Q_1 = 0.2\varepsilon$. At time t_3, the bubble collapse occurs and the vapour is pulled to the surface of the pore (t_4).

The studied process can be discussed from two different viewpoints: hydrodynamic and molecular. The density gradient, on the one hand, is associated with a gradient of the chemical potential, and on the other hand is associated with pressure. Adding density perturbations can be treated as a perturbation of pressure or as a perturbation of the chemical potential.

From the point of view of formal hydrodynamics collisions of a shock wave with a vapour bubble which the wave meets on its way result in multiple collisions of the shock wave with the boundaries of the bubble. As a result, the steady-state motion of the bubble is established in the pore at speeds lower than the average flow velocity of the ambient fluid. Thus on the front surface of the bubble there is a transition from liquid to vapour, and on the rear side – the vapour condenses into the liquid.

Shock wave penetration into the bubble and out of it is one of the principal features of the problem under consideration. There are two points in time. First, when the wave only penetrates the bubble. At this point there is a transition from the initial equilibrium state of the fluid to the substantially non-equilibrium state and forces form which transform the bubble from rest to motion.

The second point is related to the establishment of a quasi-stationary state. The process of interaction of a shock wave with a bubble can be compared with refractive problems in the macroscopic gas dynamics, when the shock wave penetrates through the boundary between two media (gases). Common features here are the strengthening and weakening of the shock wave when passing the liquid–vapour and vapour–liquid interfaces. As the bubble is restricted by the curved surface, when the shock wave flows into the bubble and is reflected from its internal borders there should be focusing and strengthening of the intensity of the shock wave. Recall that in conventional hydrodynamics, if the phase transitions are not considered, the air bubble positioned in the fluid should move within a channel of small diameter with the average velocity of the liquid (excluding the influence of the Archimedes force).

From the molecular point of view, the principle of the observed phenomenon is the realization of two opposing mechanisms of molecular motion. The general chemical potential difference is associated with a difference in the density of the fluid, which, in turn, is associated with the pressure drop according to the equation of state. With simultaneous action of both the pressure difference and the chemical potential difference the total flux of molecules is shifted toward lower density, but at the phase boundaries the density gradient causes intense exchange of molecules between the vapour and the liquid, which, in essence, is a phase transition in the non-equilibrium conditions. The thermal velocity of the molecules in the vapour is at least an order of magnitude higher than in the fluid (actually by three orders of magnitude). As a result, there is an intensive flow of vapour towards the low density fluid–vapour distillation. This distillation process shifts the liquid–vapour boundary in the opposite direction of the fluid flow. Partial compensation of the total liquid flow velocity takes place and the speed of displacement of the bubble is less than that of the liquid.

The analysis showed the principal possibility of describing the motion of the meniscus at the liquid–vapour interface as a whole in terms of the chemical potential gradient. This result is qualitatively different from all existing methods [22–28]. If the standard equations of hydrodynamics are used, then when considering any small element of the pores both for the vapour flow and the liquid flow in the channels of the given section, the main thing is the transfer of the momentum, which is described by the Navier–Stokes equation. At the same time, when moving to a macroscopic pore system generally the processes of the transport of the molecules are described by diffusive equations linking the mass flow with the total pressure drop (chemical potential) on either side of the solid body. This difference is in conflict with the basic equations of mass transport.

This result demonstrates the complex nature of transport of the molecules in narrow pores. This example demonstrates qualitatively the new opportunities of the microscopic equations of transport of the molecules on the basis of the microscopic hydrodynamics, which relies only on the knowledge of intermolecular interaction potentials of the molecules with each other and with the walls of the pores. These equations can simultaneously take into account any of the mechanisms of mass transfer of molecules discussed in section 3, and are not based on the phenomenological equations for the local saturated vapour pressure.

49. Molecular flows in narrow pores with small initial perturbations

In the new microscopic approach it is necessary to consider a number of provisions that are not discussed in the macroscopic formulation of the problems. In particular, it is the question of validity of the definition of the initial conditions in the generation of the molecular flow in narrow pores [29]. Traditionally, in the study of flow dynamics the initial data are given in the form of simultaneous perturbations in specific coordinates (sites) of the calculation field. To generate flows along the axis of the pore such coordinates in which the initial perturbation is specified is the 'single-point' perturbation of some section of the pores. Given that the duration of the initial perturbations, by implication, is about one integration step (about 10^{-2} ps), they lead to significant changes in the dynamics of distributions of he molecules (but not in the final state), while the relaxation of perturbations only takes place at sufficiently long times (up to hundreds of nanoseconds).

From the point of view of macroscopic time the characteristic times of the processes of transport in the nanoscale systems are extremely small. In reality, there are serious difficulties with the possibility of mechanical generation of flows at times shorter than a nanosecond. In other words, the characteristic times of mechanical perturbations are considerably greater than or comparable with the times of the molecular studies. Therefore, in such circumstances, one can not speak of an instantaneous (one-stage, as is customary in hydrodynamics) perturbation of the system. The existing techniques of perturbation with laser or electron beams, resulting in fast enough physical processes (see, e.g., [30]), allow to work with picosecond processes whose duration also exceeds by one or two orders of magnitude the duration of the perturbation considered in the molecular techniques. Perturbations of the femtosecond range, capable of transmitting more energy, are unlikely to generate the molecular transport in porous solids. The flow is formed in the course of cooperative processes with a characteristic time greater than the time of propagation of acoustic phonons, i.e., more than 1 ps.

Naturally, the question of the correctness of the simulation process of the flow arises. Thus, one of the main tasks of molecular modelling of the flows in micro- and mesoporous systems is an independent task of modelling of the initial conditions that reflect the actual mechanical perturbations, leading to the formation of molecular flows. Real processes of perturbation of the system, depending on the method of perturbations through the state parameters (temperature, density or pressure), should be given as some processes that extend not only in the spatial coordinates, but also in real time. Instead of simultaneous perturbations we should consider processes perturbing the original system within a certain time interval so that the process meets the real conditions at the microlevel. To ensure that such modelling initial conditions can be carried out we must be able to vary over a wide range the intensity of the initial perturbations, and most importantly, be able to start generating perturbations from the weakest perturbations.

To this end, attention was paid to the influence of the intensity of the initial perturbation of the equilibrium state of the vapour and liquid in a slit pore with the width ~15 nm with different interaction potentials of the molecules with the walls of the pores [29]. The question of modelling of the initial conditions for the generation of molecular flows in narrow pores was solved by analyzing the flow characteristics change with decreasing intensity of the initial perturbations δP, i.e. dimensionless quantity $\delta P = \Delta P_x / P = 10^{-2} - 10^{-5}$, where ΔP_x is the increment of pressure along the x-axis, applied on a

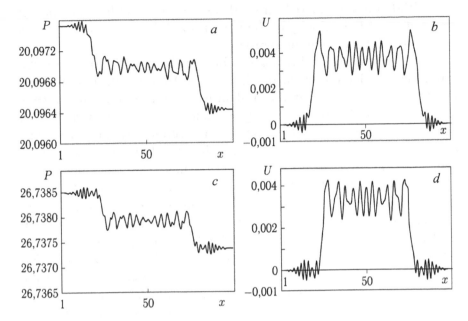

Fig. 49.1. Distribution of pressure $P(x)$ (a, c) and speed U_x along the axis of the pore (b, d) in the median section ($y = H/2$) for $\theta = 0.1$, $Q = 1.5$ (a, c), and 0.5 (b, d), $\delta P = 10^{-5}$, $t = 1.1 \cdot 10^{-10}$ s, when the initial perturbation is given as a step with a break in the middle of the pore along the axis x.

given section of a pore with length L. Studies were conducted with the value of $\delta P = 10^{-5}$, which is one (for liquid) or three (for vapour) orders of magnitude smaller than the values of δP used previously [6, 14–16].

Initial perturbations

Two methods of simultaneous perturbation were compared: 'single-point' pressure surge when the value of δP was defined at two adjacent cross sections perpendicular to the axis of the pore, and the linear pressure drop when the given differential δP was set in a distributed fashion on a segment of length L along the axis of the pore. Like the previous section, the change in concentration (pressure) was set by changing the chemical potential of the molecules.

The first method sets a perturbation as in conventional macroscopic equations. The initial perturbation was set in the middle of the calculation field at $x = 50\lambda$. The second method, in general, allows to consider more complex modes of perturbation with an arbitrary method of pressure distribution (or initial velocity). Setting the initial pressure distribution allows us to calculate the initial distribution of the velocity of the vapour or fluid using the equations of the microhydrodynamic approach.

The role of the influence of the single-point setting of the initial data is demonstrated in Fig. 49.1 at time $t = 0.11$ ns for two interaction potentials for the molecules of the dense vapour ($\theta = 0.1$) with the walls of the pore: $Q > 1$ and $Q < 1$. Point perturbation generates perturbances in the neighboring sections which spreads fast enough to a number of sections $L*$, inside which chaotic oscillations occur. The average speed in this section increases from the values of the order of 10^{-5} to 10^{-1} m/s (as Figs. 49.1 b, d), and there are small oscillations outside the inner region of size $L*$. Internal and external fluid oscillations are formed by themselves, as the system is strongly non-linear at the molecular level. These dynamic states stabilize at times $\tau* \sim 0.01$ ns for small values of $Q < 0.5$ and $\tau* \sim 0.1$ ns for large values of $Q > 1.5$. Then we can talk about the further stage in the evolution of the system and propagation of the general molecular flow at times $t > \tau*$, associated with an increase in the size of the inner region $L*$.

The distribution of pressure and velocity of the molecular flow in the middle of the pore at the axis $y = H/2$ at time $t = 0.11$ ns are represented by the oscillating curves. The small value of the initial perturbation $\delta P = 10^{-5}$, however, leads to noticeable differences in the average values of P (x) and U_x, which depend on the interaction of molecules with the walls. Increasing Q increases the average value of the average velocity and decreases the pressure inside the flow formation area. The form of the distribution of both characteristics indicates the existence of two types of oscillations inside the middle portion to the calculation field and at its borders. Oscillations in the inner region $L*$ affect 5–10 cross sections, while the 'external' oscillations occur in 1–3 sections from each side.

Increasing the width of the inner area of propagation of the perturbation is also dependent on the value Q. This effect is due to the fact that although the pore is wide enough, the majority of dense vapour molecules are attracted to the walls of the pore, so the vapour flow is largely determined by the dynamics of molecules near the pore walls. The role of Q for dense fluids was discussed in detail in the previous sections.

In the second perturbation method, the same pressure differential $\delta P = 10^{-5}$ was set linearly on a section symmetric with respect to $x = 50\lambda$ with $L = 20\lambda$ (ten sections from the centre of the pores on

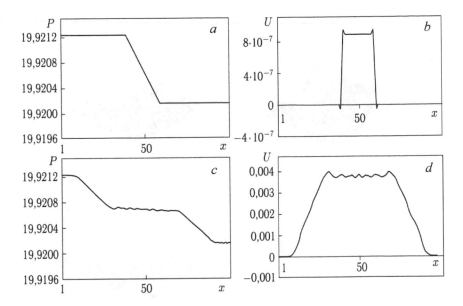

Fig. 49.2. Comparison of the initial distributions of local pressure $P(x)$ and speed U_x for x in the middle section ($y = H/2$) for the linear initial distribution of pressure on the central portion of the pore for two moments of time: $t = 4 \cdot 10^{-14}$ s (a, b), and $t = 0.11$ ns (c, d); $Q = 1.5$, $\theta = 0.1$, $\delta P = 10^{-5}$.

each side). The distribution of pressure and velocity along the axis of the pore at the time $t = 4 \cdot 10^{-14}$ s in the middle of the pores along the axis $y = H/2$ is shown in Figs. 49.2 a, b. The sharp section of the initial pulse which did not manage to spread to both sides of the section L is shown. (Recall that the integration step with respect to time in the microhydrodynamic approach is about 10^{-14} s, which is much less than the characteristic time of molecular vibrations.)

Figures 49.2 c and d show similar distributions of pressure and velocity at time $t = 0.11$ ns for $Q = 1.5$, i.e., for the time 3.5 orders later. This time corresponds exactly to the time in Fig. 49.1 b. Comparison of these distributions shows a fundamental role on the method of defining the initial perturbation. There are practically no oscillations in the inner region and the oscillations are completely absent in the outer region.

Single-point perturbations in space are highly intense and cause the oscillatory nature of the formation of the molecular flow (the nature of the oscillation is discussed below.) It should also be noted that the difference in the pressure differentials is much less than in the velocity values. This result is obvious, since the local flow

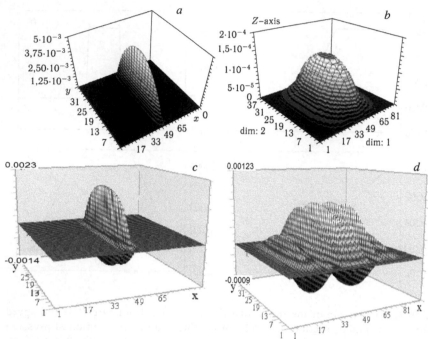

Fig. 49.3. Velocity U_x field at a point perturbation $\delta P = 10^{-5}$ of rarefied gas (a, b) with the parameters of the system $(Q = 1.5, \theta = 0.0008, T = 273$ K$)$ at time points $t = 7$ ps (a) and $t = 0.19$ ps (b) and liquid $(c\ d)$ with the system parameters $(Q = 4.5, \theta = 0.8, T = 120$ K$)$ for the moments of time $t = 20$ ps (c) and $t = 0.15$ ns (d).

velocities are much more sensitive characteristics than the local pressure. The spatial distribution of the initial perturbation leads to the realization of the strongly smoothed flow. Nevertheless, the mean values in the inner area of the flow formed in the second method shown in Figs. 49.2 c, d are sufficiently close to the average values of the pressure and velocity in Figs. 49.1 a, b (within one percent). During the relaxation of the system the average pressure profiles in two modes of perturbation generation converge.

Concentration factor
To analyze the nature of the oscillations in the single-point perturbations of dense vapour, attention was paid to similar perturbations in a rarefied vapour $(\theta = 0.0008)$ and liquid $(\theta = 0.8)$. The above-described picture of the formation of molecular flows in narrow pores for the dense vapour is modified during the transition to a dense fluid and rarefied vapour.

Perturbation of rarefied vapour

Figure 49.3 a, b shows the velocity field for two times: $t = 7$ ps (a) and $t = 0.19$ ns (b) which were calculated for an ideal gas at high temperature ($T = 273$ K); no effects of intermolecular interactions were present in the gas.

It can be seen that the initial local velocity starts to decrease from 0.007 m/s ($t = 5$ ps) to 0.00012 m/s (at $t = 0.19$ ns). Later, the average velocity continues to drop, and the velocity distribution along the section L^* expands (this also applies to the pressure field). The qualitative difference of this picture is the absence of any oscillations in the examined time range.

From a formal point of view it is also possible that in further movement of this perturbation (at greater calculation length of the pore) the 'dome' may split into separate waves of specific lengths, but such behaviour would mean a strong influence of the pore walls, which is possible only in case of very narrow pores and strong molecular attraction to the walls.

Perturbation of the liquid

The results of calculation of the evolution of the velocity field for the 'loose' liquid with an average density $\theta = 0.8$, located in the pore with a strong attraction to the walls of the molecules ($Q = 4.5$), are shown in Fig. 49.3 c, d. As above, the initial single-point perturbation is $\delta P = 10^{-5}$. Clearly shown is the presence of the oscillatory velocity profile which qualitatively coincides with the views of the curves in Figs. 49.1 b and d. In the field 49.3 c there is a blurring of the initial ridge below (negative value of the local mean velocity), while in the field 49.3 d, there are many highs and lows of local velocities. Large oscillations, much superior in their amplitude to the oscillation in Figs. 49.1 b, c, appear here. The velocities have positive and negative values, whereas in the fields 49.1 b and 49.1 d all velocities remain positive.

The initial state the form of which is close to the field 49.3 c (but without oscillations to negative velocity values) exists for a long time (almost three orders of magnitude from 10^{-2} to 10 ps). At the same time, referring to the field 49.3 c, the set of the oscillations is slightly different from the state relating to the fields 49.1 b,d. This time is characterized by the value τ^*, so there not only significant change in the maximum values of the velocities in Fig. 49.3 c and 49.3 d (almost double), but also a correlation between the total time of the start of the molecular flow, as described above, and the time of manifestation of the viscosity contribution to the flow of the fluid. In the stage to this characteristic time the viscous terms in the equations for local velocities play quite a small role.

The analysis suggests a fundamental role of the concentration factor in the oscillation behaviour of the velocity of the molecules. These oscillations exist in dense vapour and liquid (in which the amplitude of the liquid is several orders higher), but are completely absent in the diluted steam.

1. Thus, it was found that different methods of administration of the initial perturbations lead to qualitatively different situations! For single-point perturbation of the rarefied gas and the distributed method of introducing perturbations there are no oscillations in the dense vapour. In dense media (vapour and liquid) at single-point perturbation velocity oscillations appear at any density with the correlation effects of the interacting molecules. This imposes a restriction on the method of setting the initial perturbations and their compliance with the requirement for the system properties.

2. An important result of the use of microhydrodynamic approach is the ability to achieve such small initial perturbations as $\delta P = 10^{-5}$. This value is one order of magnitude less for the liquid than that used in the previous sections, and three orders of magnitude less than for the vapour used in the first calculations.

3. The principal factor is the possibility of establishing an internal correlation between the formation of the molecular flow and the presence of sufficient large viscosity contributions to the overall evolution of flows. The average flow velocities were the values when after initial stabilization of the state of the system for times τ^* after the start of the process, the velocity or the amplitude of its fluctuations started to diminish. In this time range, the initial perturbation is stabilized in its computational domain (we are talking about a certain area and not a single section) after locally chaotic 'oscillations' that are generated by (single-point) momentary perturbation in which 'hydrodynamics' appears.

Molecular flows are formed at times of the order of 50 ps for a rarefied vapour and $t = 70$ ps for a dense vapour. In fact, this range of time is the order of 10 and 100 ps, which is 'large' in terms of the molecular level, but very small in terms of macroscopic hydrodynamics and real processes. The further decrease in the velocity over time is very slow (but taking energy redistribution into account slightly accelerates it).

Table 6.1. The values of the initial velocity U_x (m/s) in dependence on Q and the intensity of the initial perturbation δP at $\theta = 0.1$ (second value in parentheses at $\delta P = 10^{-5}$ is given for $\theta = 0.0008$)

Q_1	Dp				
	10^{-2}	10^{-3}	10^{-4}	10^{-5}	
0.5	2.0	0.11	0.014	0.005	(0.0025)
1.5			0.01	0.005	(0.0015)
5.0			0.012	0.008	

Average velocities are given in Table 6.1 constructed for different values of δP and Q. The table demonstrates the strong dependence of the initial rate of formation of the dense gas flow with density $\theta = 0.1$ on the intensity of the perturbation δP. In the range of 10^{-2} to 10^{-4} the velocity is roughly proportional to the intensity of the perturbation, changing by an order of magnitude. When δP changes from 10^{-4} to 10^{-5} the velocity changes twice. (For comparison the values in the parentheses refer to the velocity of the rarefied gas with a density of $\theta = 0.0008$.) Thus, the range of values δP, which provide a correct description of the dynamics of molecular processes without introducing artificial effects due to the method of generating the initial perturbations, begins with values δP from 10^{-5} and below.

In addition, we note that a decrease in the intensity of the initial single-point perturbation the dynamic flow pattern of the vapour is qualitatively preserved: it has an oscillating character, as in the strong initial perturbations. Weak molecular attraction to the walls intensifies the oscillations in the initial phase of the fluid flow along the pore. Strong attraction of the molecules of low density to the walls of the pore produced a layer flow and the oscillations are noticeably smoothed. The intensity of the perturbation changes the absolute values of the flow parameters, including the average velocity of the emerging molecular flows. The average velocity of the hydrodynamic flow in wide pores depends weakly on the energy of interaction of molecules with the walls of the pore.

50. Wetting and spreading processes

Wetting processes are widespread in modern technology, they are implemented in the impregnation and drying processes, capillary condensation and desorption, multiphase filtration and 'spraying', in various catalytic, electrochemical, chromatographic processes, etc. [22–28, 31]. The wetting and spreading of the droplets on exposed surfaces have been investigated by various methods, such as numerical methods of molecular dynamics and the Monte Carlo method, the hydrodynamic equations and the simplest equations of molecular kinetic theory [32–40]. Although wetting problems have attracted the attention of many researchers, the very dynamics of wetting and the associated molecular motions are hitherto poorly understood. Obviously, for an understanding of the mechanism of wetting it is first necessary to formulate a molecular representation of the process.

Typically, in practice the wetting process is described by the traditional hydrodynamic equations [22–28, 31, 32]. Near the surfaces of solids, these equations reduce to the boundary conditions for considering the sharp gradients of density and velocity along the normal to the surface (a similar procedure is used in the presence of the phase boundary along the flow direction).

In this section, we examine the initial stages of contact of the liquid with a solid plate: wetting of the Wilhelmy plate [41], made of materials with different degree of hydrophilicity, and the evolution of a cylindrical droplet placed on the surface of the plate (considered is the two-dimensional dynamics of all processes in the plane xy) [42]. Attention is given to the mechanism of formation of the meniscus on the exposed surface and the influence of the nature of the plate surface, which is determined by the fluid–solid potential, depending on the type of material, and the impact of the velocity of the plate itself. The modelling of the beginning of the wetting process and the formation of a meniscus on the Wilhelmy plate surface is studied. The microhydrodynamic approach gives a detailed analysis of the formation region of the meniscus and allows to study the competition of the contributions of evaporation (flows of molecules in a gas phase) and flows of molecules around and directly on the wafer surface. The changes of the concentration fields from the beginning to the establishment of the quasistationary states are investigated; this allows to determine the velocity of the contact angle.

The analysis corresponds to the argon–the surface of graphite (carbon) real system; the molecular parameters of this system are given in [10].

Wetting problems and method of solution
Mathematically, the problem of wetting the Wilhelmy plate is reduced to the study of the evolution of the fluid flow when the liquid was previously in equilibrium with the vapour phase and at the initial moment of contact of the hydrophilic plate with the liquid

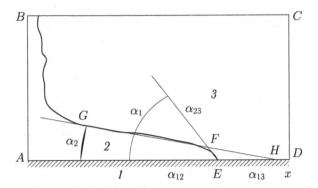

Fig. 50.1. Calculated field $ADCD$ allocation scheme and the contact angles at different molecular levels (explained in the text)

reservoir the liquid begins to move: a film is formed and flows to this plate. Fluid motion is due to the lack of equilibrium distributions of molecules in a given volume. The physical factor violating the initial equilibrium distribution of the vapour–liquid is the adsorption potential of the plate surface. Surface forces 'pull' the fluid molecule from the reservoir on the plate, and as the length of the reservoir plate is considered unlimited (both the formation of the multilayer filling or displacement of the molecules to larger distances from the reservoir can take place on it), the wetting angle forms as a result in the flow field. Due to the viscous forces the initial impulse of the introduced potential of the surface forces is scattered and the solution of the problem goes to a regime close to the quasi-stationary. Therefore, a meniscus is formed over time and its movement is quasi-stationary which allows to estimate the propagation velocity of the contact angle on a microscale, depending on the substrate material (i.e. on the binding energy of the molecule with the surface of the plate).

The two-dimensional isothermal flow of a monatomic fluid (argon) along the surface of the plate AD (Fig. 50.1) was simulated. This surface is impermeable to the fluid. Side AB is a liquid reservoir of unlimited capacity, and the sides BC and CD at the initial time relate to the vapour phase. The boundary of the formed liquid film is indicated by the line EFG. Calculations were carried out for the calculation field with sides $AD = 150\lambda$ (axis x) and $AB = 80\lambda$ (axis y), where λ – the mesh size of the order 1.12σ, and σ is the diameter of the solid sphere of the argon atom in the rectangular coordinate system x and y with spatial steps $\Delta x = \Delta y = \lambda$.

All flow calculations refer to the temperature $T = 117.4$ K (with the exception of the process of evaporation at $T = 171.4$ K) at which delamination of the liquid argon takes plate (recall that the critical temperature of argon in the bulk phase is about 150 K). The density of the liquid was of the order $\theta_{liq} = 0.96$, and the vapour density of the order of $\theta_{vap} = 0.04$. Here the density θ is measured in dimensionless numbers characterizing the probability of filling the cell with argon atoms, i.e. as the ratio of the number of molecules to the total number of sites in the system (relationship of θ with the mass density is specified in [6]).

At the initial time the vapour density in the calculation field *ABCD* and on the plate surface is assumed to be equal to θ_{vap} and microhydrodynamic velocities of all the argon atoms are zero ($v_x = v_y = 0$). Over time, as the argon atoms are transferred from the liquid reservoir the density at the boundaries *BC* and *CD* is determined by converting the material flow from the internal nearby cells (so called 'drift' procedure).

Strong adsorption field

Because of the relatively high non-equilibrium nature of the system molecular transport processes start in it at the initial moment: from the gas phase onto the surface of the plate, from the fluid reservoir into the gas phase and along the surface of the plate, as well as accompanying internal friction processes tending to match the values of the concentrations and the pulse at the liquid and plate surfaces. The wall potential causes a large difference in the velocities of the molecules along the plate and in the vapour region.

Figure 50.2 shows the evolution of the interface between the vapour and the liquid (the images of concentration fields in the form of isolines are shown) at four points in time. The last fifth moment is shown in Fig. 50.3. There are the full picture of the isolines are shown (*a*), the set of isolines related to a density greater than $\theta = 0.4$ (*b*), and a plurality of isolines related to a density lower than $\theta = 0.3$ (*c*). Finally, Fig. 50.4 presents the distribution of the density of argon atoms in two sections along the *y* axis at $X = 20$ (*a*) and 65 (*b*). These figures allow us to represent the nature of the wetting process of the plate and the formation of a liquid film with the liquid flowing from the reservoir. Due to the large differences in the densities of the fluid in the neighboring cells and, as a consequence, a strong densening of the isolines, the local densities are marked

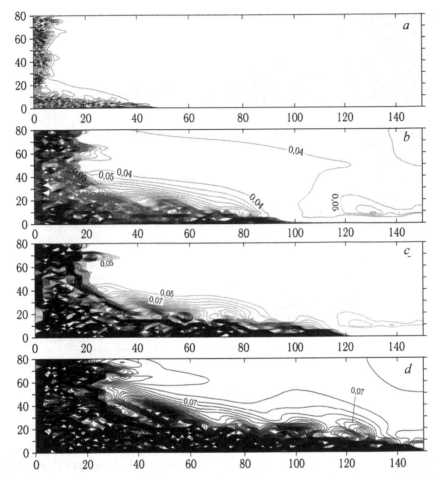

Fig. 50.2. Dynamics of the process of wetting a carbon plate by argon atoms. Concentration fields for times $t_1 = 1$ ns (a), $t_2 = 3$ ns (b), $t_3 = 4$ ns (c), $t_4 = 5$ ns (d).

by their values on only a small number of isolines (mainly for low densities).

Figure 50.2 a shows that at the initial time part of the liquid evaporates from the surface and near the plate the effect of the force of surface attraction results in the formation of a liquid film which is moving along the surface and entrains the liquid from the reservoir. Evaluations have shown that under the conditions considered in the equilibrium state the width of the vapour–liquid transition region is of the order (5–6) λ. In the dynamics the transition region width increases at least 5–6 times and becomes comparable with the width of the calculated field. This increases the size of the phase boundary

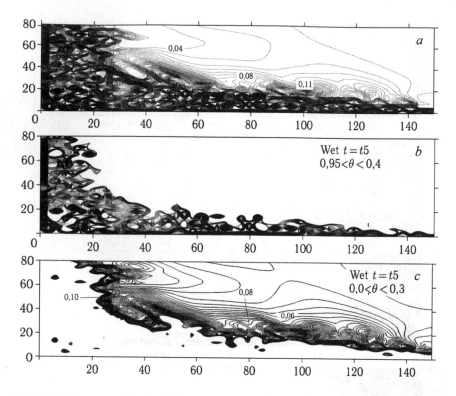

Fig. 50.3. Concentration fields for the fifth moment of time $t_s = 6$ ns: a – full set of isolines; b – isolines relating to filling $\theta > 0.4$; c – isolines related in filling $\theta < 0.3$.

surface and the extent of evaporation of the atoms to the vapour (subsequent fields in Fig. 50.2). Common to all times remains the strong irregular alternation of the local densities. The viewed isolines indicate small-scale structuring of the fluid – even at small x we observe 'light' microregions with nearly equal densities.

This can be seen clearly in Fig. 50.3, showing the highlighted areas containing portions with small (c) and high (b) densities. Comparison of fields (b) and (c) shows the distribution of the dense part of the fluid. It is concentrated near the wall and the liquid reservoir, and the less dense part of the fluid occupies a large area of the computational domain. The nature of the distribution of the fluid can be traced in more detail on the sections along one of the coordinate axes.

Figure 50.4 shows the local densities along the y axis for different values of x ($y = 1$ corresponds to the plate surface, and $y = 80$ to the line BC). Near the reservoir ($X = 20$) across the width of the

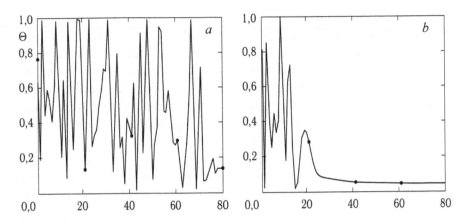

Fig. 50.4. Sections of the density of argon atoms at $x = 20$ (a) and 65 (b) for time t_5.

calculation field the local concentrations have approximately the same distribution because the liquid–vapour transition region reached this section. It is seen that the movement of the fluid is due to the transfer of a 'mixture' of particles (domains) of high and low density. For $X = 65$ we can clearly see the different character of the distribution of the fluid near the wall (strongly oscillating) and in the vapour phase from the line BC (nearly constant).

Despite the strong increase in the size of the vapour–liquid interface in the dynamic mode detailed of the local fillings allows to highlight the area of the contact angle for a moving front of the film and to consider its characteristics. Using the quasi-steady displacement of point E (in Fig. 50.1) we can estimate the velocity of the front along the plate. It is obvious that in the vicinity of the point E a concentration gradient under the action of surface forces along the normal. At smaller values of x, these density profiles have different values and their set on a selected molecular scale generates a set of isolines defining the three-phase contact angle.

The numerical results of Fig. 50.2 are schematically shown in Fig. 50.1, where two molecular scales are identified, L_1 and L_2. At the microscale (several monolayers near the surface) L_1 is equal to the order of 4–6 cells (curve EF at angle α_1) and at the high molecular scale value L_2 is about 40–50 cells (curve GH at angle α_2). On each of the scales L_1 and L_2 there is a specific contact angle, which remains almost constant in the quasi-stationary mode (at time t_5 in Fig. 50.2 the front of the film crossed the line CD). Recall that traditionally the contact angle is set in the equilibrium conditions by the Young equation through the surface tension at the boundaries of three phases, designated as α_{12}, α_{13} and α_{23} (*1* – solid, *2* – liquid, *3* – vapour), which are taken from macroscopic measurements [41]. At the microlevel the action of surface forces changes all its macroscopic quantities. In addition, in the considered microdynamics the changes of the viscosity coefficient near the surface were taken into account [6, 43, 44] (these effects are equivalent to the emergence of additional resistance forces in the surface layers).

In this case, for the time of the order of t_3 and more we can discuss the quasi-stationary motion of the front of the film along the surface of the plate; the speed of this movement was equal to ~11 m/s. It was found that for a hydrophilic plate on both molecular scales L_1 and L_2 the contact angles α_2 and α_1 are directed in the opposite direction of the front, and $\alpha_2 \approx 11°$ is significantly smaller than $\alpha_1 \approx 46°$. (Note that the accuracy of determination of angle α_1 is lower, as the difference calculation procedure of the flow parameters of the central cell uses the date for three neighbouring cells, and the lower molecular scale itself comprises 4–6 cells.) On a large scale molecular L_2 the meniscus is concave with respect to the plane of the liquid in the reservoir, which is 'pulled' in the positive direction along the *x*-axis during movement of the film.

In contrast to [6, 18] here and below we do not provide examples of the appropriate velocity fields along the *x* and *y* axes due to the extremely sharp local changes of these velocities that are difficult to be represented by the existing image editors. The complexity of the structure of local flows can be indirectly illustrated by all the drawings for the local densities with a complex pattern of distribution, taking into account the fact that the relaxation time of the local velocities is about two orders of magnitude smaller than for the local densities. Therefore, small changes in the local densities cause significant changes in the local velocities.

Weak adsorption field
The impact factor of the plate material was analyzed by similar calculation with the model system when the binding strength of an argon atom with the surface was significantly reduced and was 1.5ε; the other characteristics of the system have not changed. In this case, the plate remained hydrophilic, although the value of Q_1 was relatively little different from the interatomic interaction energy ε. However, qualitatively the observed pattern of wetting the plate in all respects remained similar to that shown in Figs. 50.2–50.4.

Quantitative changes caused by the magnitude of Q_1 led to a 25–30% decrease in the speed of the front (point E in Fig. 50.1) by comparison with the first variant of the task, as well as substantial changes in the magnitude of the contact angle – decrease to $\alpha_1 \approx 27°$, whereas the value of the contact angle on the scale L_2 remained virtually unchanged $\alpha_1 \approx 12°$. The latter indicates that the quasi-stationary properties on the system on the second molecular scale are due mainly to the nature of the intermolecular interaction and the temperature in the transition region between the wall region and fluid 'pulled' from the reservoir, and the direct contribution of the surface forces can be neglected.

Effects of motion of the plate

We have considered the initial stages of the contact of the plate with the fluid reservoir under isothermal conditions. In many situations the plate can move during of wetting, so the direction and speed of its motion may affect the characteristics of the process. The above estimates for the velocities of propagation of the wetting front of the hydrophilic plate from 8 to 11 m/s immediately allow us to estimate at what velocity the motion of the plate will affect the processes of wetting. If the velocity of the plate is substantially less than the above range of velocities, the direction of its movement with practically have no effect on the motion of the film (there will be no wetting hysteresis due to the change of direction of movement of the plate).

More complicated is the question of the legality of the isothermal analysis of the problem when there are significant contributions of the dissipation forces to the viscosity near the surface of the plate. Given that the thermal conductivity of the liquid phase is much higher than that of the gas phase attention was paid to the complete problem for variable density, velocity and temperature of the gas phase with $\theta = 0.04$ at a temperature $T = 117.4$ K and in the presence of an attractive potential of the plate $Q_1 = 9.24\varepsilon$. For comparison, three flow velocities of the gas around the plate: 10, 50 and 100 m/s were studied, taking into account the thermal conductivity of the gas. For times when the quasi-stationary flow patterns were obtained the following temperature increases were recorded near the surface of the plate: $\Delta T = 1°C$, 24°C and 135°C, respectively. Moreover, the observed changes in temperature spread to 30 monolayers from the plates, and the concentration of the substance in the gas flow in the second monolayer $\Delta \theta$ at a rate flow of 10 m/s changed by about 0.1.

This implies that the isothermal approximation is sufficiently satisfactory at velocities of the plate of up to 10 m/s (which is comparable to the velocity of the wetting front, thus completely changing the process of film movement). This fact is used in the calculation with the isothermal regime, although in this case allowing for energy transfer leads to quantitative differences. However, at velocities of the plate or gas near the plate of more than 10–20 m/s this approximation may lead to inaccuracies in the description of the redistribution of energy and distort the interpretation of the process of fluid contact with the plate.

Contact of droplets and plate

The microhydrodynamic approach allows to explore a wide range of options for the processes of contact of the liquid with the solid. One example is the variant of contact of droplets with a hydrophilic surface of the plate at $Q_1 = 9.24\varepsilon$. (For simplicity, we consider a cylindrical droplet so that we can confine ourselves to the two-dimensional geometry of the problem (Fig. 50.5 a).) Its diameter was equal to 20λ, and at the initial time it was over the plate – its lower edge at a distance of 2λ (second monolayer). Then, under the influence of the attractive potential of the plate the droplet was directed to the plate. The evolution of the system was studied for two temperatures: 117.4°C (below critical) and 171.4°C (above critical). In the first case, the droplet after contact spread over the surface of the plate, while in the second case the droplet started to evaporate rapidly.

Figure 50.5 shows the state of both systems at time $3 \cdot 10^{-9}$ s after the start of the process: b – spreading of the droplet and c – its evaporation. Figure 50.6 shows the sections of the local concentrations of the droplet centre for the three time points during its spreading. In the first case it is well evident the existence of boundary between a gas and a liquid, although the concentration profile indicates a strong heterogeneity of the distribution of atoms in a solid phase dynamics spreading. Moreover, the phase boundary itself also has a strong heterogeneity as around the central dense part of the droplet, and at the spreading areas of the drops along the surface of the plate.

In the second case, there is intensive dispersing of the substance in all directions from the centre of the droplet. Dense part of the droplet decreases near the 'evaporating' surface. It is clearly observed that the substance is removed from the droplets as small clusters (a

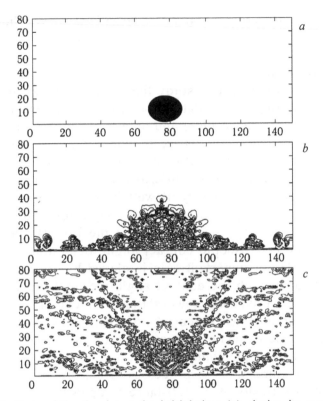

Fig. 50.5. State of the droplet at the initial time (*a*), during its spreading at $T =$ 117.4 K (*b*) and evaporation at $T = 171.4$ K (*c*) at time $t = 3$ ns.

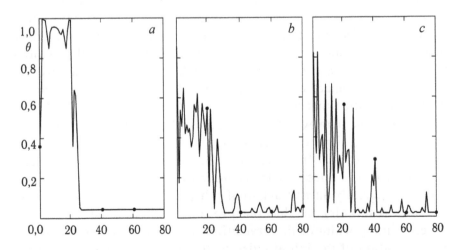

Fig. 50.6. Sections of concentration distributions of argon atoms in the course of spreading of the droplet relative to its centre at $X = 75$ at time points $t = 0.5$ ns (*a*), 1 ns (*b*) and 3 ns (*c*).

large number of different density condensations of varying shape and intensity).

Discussion of results

It should be noted that in this study there are two types of contact angles during wetting of the plate by a liquid on microscale L_1 (for a few monolayers at the surface) and on a larger molecular scale L_2 (away from the surface plane where the direct effect of the surface adsorption potential is not significant) which are different in nature from the well-known two types (micro- and macro-) of the contact angles in the macroscopic theory of wettability [41].

The latter is a consequence of the quasi-equilibrium distribution of the liquid (assuming low volatility of the liquid), and in the formation of the macroscopic contact angle, determined from the experimental data, an important role is played by the mass forces [41]. In this paper, both the contact angle are implemented at the molecular level in a dynamic quasi-stationary mode only under the influence of intermolecular potentials. So there may be differences in the values of the dynamic contact angle on the scale L_2 and the conventional macroscopic contact angle.

A common feature of these processes is the presence of strong oscillations of the density of the substance that persist for a long time (longer than the calculated time). This picture is different from the usual macroscopic concepts of the continuous flow of the film with the sharp phase boundary and the leading edge of the flowing liquid. As noted above, at the molecular level even the equilibrium phase boundary is extended and all local changes in density are reflected in detail on the considered nanoscale. All the discussed initial situations of the processes are far from equilibrium, as the procedure for enabling fluid contact with the plate leads to strong perturbations. It should be recalled that even the initial pulses that lead to fluid flow velocities in the narrow slit-shaped pores of the order of 0.1–0.5 m/s result in strong dispersion ('crushing') of the interface and identical long times (up to 10^{-6}) of the existence of the 'foam' environment. In this case, the speed of movement of the film is significantly higher (especially in contact of the highly hydrophilic plate with the fluid reservoir), so that the previously mentioned substantial increase in the width of the 'transition' region is associated with the transformation of the initial pulse to the dispersed state of the fluid (which is then maintained with time by the movement of the film on the 'free' surface of the plate). When the initial flows of matter

form the phase boundary is crushed and the system self-organizes this
process by changing the local dynamic characteristics (primarily due
to changes in viscosity strongly dependent on local concentrations).
The very fact of the long-term existence of the highly dispersed
state of the fluid, which is manifested in strong oscillations of the
density, indicates the intrinsic properties of the system under study
(which is highly nonlinear). These intrinsic properties depend only
on the temperature, the mass ratio and the interatomic potentials of
the fluid and the solid surface. In general, the microscopic process
has a macroscopic analogue in the formation and relaxation of
(vortex) perturbations in the gas macroflow around the surface (for
example, aircraft in the atmosphere), with the time of formation
of vortices of the order 10^{-3}–10^{-1} s and the relaxation time of 10^4–
10^5 s [45], i.e., there is a difference of the order of 5–8 in the times
of formation and relaxation of the highly non-equilibrium states. It is
natural to suggest that this quasi-steady development of the wetting
process will continue until such time when additional factors will
influence its evolution. In this case we can talk about the influence
of gravitational forces that were excluded when considering the
molecular processes at the microlevel. Then, by increasing the size
of the transition region between the vapour and the liquid the mass
forces will reduce the observed density oscillations (in the process of
aggregation of large clusters which represent areas of high density of
the fluid) and the picture will change to the traditional macroscopic
description. To analyze the processes at much larger time scales we
must extend calculated field, which can be done using computational
resizing of the calculation mesh.

   This section of the study was limited to the initial stage of contact
processes which takes place on the nanoscale time range. At the same
time, it should be noted that this interval may significantly exceed
the characteristic time of the elementary acts of many fast chemical
reactions, and the described complex structure of the evolution of the
distribution of matter in space (microflows) indicates the complexity
of describing the initial stages of chemical reactions in which the
the interface is perturbed.

   The calculations illustrate the fact that the microhydrodynamic
approach used for the study of flows of dense vapour or liquid in
narrow pores within the framework of Navier–Stokes equations
(with the involvement of the lattice gas model for the calculation
of the transport coefficients) allows us to describe a wide range of
phenomena associated with unsteady fluid flows, containing areas of

liquid and gaseous phases. The lattice model can be used to obtain limiting values of the shear viscosity coefficients at low densities θ, corresponding to the gas and coinciding with the literature values for the gas under normal conditions. The microhydrodynamic approach describes from the same viewpoint the flows of both vapour and liquid phases in the pores with different energies of interaction of molecules with the wall material. Moreover, the Knudsen flow regime is considered conserved in this approach, while all other types of flows, considered in section 3, are fundamentally different from the microscopic models. The type of surface potential determines the initial (quasi-equilibrium) distribution of molecules in the cross section of the pore which have a variable density. The characteristic features of pulse perturbations at the initial time of molecular states and their quasi-stationary distribution for large times were investigated. The time to the quasi-stationary distribution and the type of the relaxation phase of the evolution of this process depend on the molecule–wall (adsorbate–adsorbent) binding energ

References

1.   Tovbin Yu.K., Modern Chemical Physics. – Moscow: MGU, 1998. – P. 145.
2.   Tovbin Yu.K. // Zh. Fiz. Khimii. 2002. V. 76, No. 3. P. 488. [Russ. J. Phys. Chem. 2002. T. 76. № 3. C. 412].
3.   Tovbin Yu.K. // Khim. Fizika. 2002. V. 21, No. 1. P. 83.
4.   Tovbin Yu.K. // Teor. osnovy khim. tekhnologii. 2002. V. 36, No. 3. P. 240. [Theor. Found. Chem. Engin. 2002. V. 36. No 3. P. 214].
5.   Tovbin Yu.K., Tugazakov R.Ya., Modern Chemical Physics. – 1998. – P. 178.
6.   Tovbin Yu.K., Tugazakov R.Y. // Teor. osnovy khim. tekhnologii. 2000. V. 34, No. 2. P. 1117. [Theor. Found. Chem. Engin. 2000. V. 34. No 2. P. 99].
7.   Landau L.D., Lifshitz E.M., Theoretical Physics. VI. Hydrodynamics. – Moscow: Nauka, 1986. – 733 p.
8.   Lax P., Wendroff B. // Commun. Pure Appl. Math. 1960. V. 13. P. 217.
9.   Balwin B.S., MacCormack R.W., AJAA Paper 74-558. 1974.
10.   Rubin E.L., Burstein S.Z. // J. of Comput. Physics. 1967. No. 2. P. 243.
11.   Tugazakov R.Ya.// Izv. AN SSSR. Mekh. zhidk. i gaza. 1989. No. 2. P. 159.
12.   Tovbin Yu.K., Tugazakov R.Ya., Komarov N.V. Fundamentals of Adsorption-7. – Chiba-City: IK International, 2001. – P. 1007.
13.   Tovbin Yu.K., Tugazakov R.Ya., Komarov N.V. // Colloids and Surface. A. 2002. V. 206, No. 1–3. P. 377.
14.   Votyakov E.V., Tovbin Yu.K. // Zh. Fiz. Khimii. 1994. V. 68, No. 2. P. 287. [Russ. J. Phys. Chem., 1994. V. 68, No. 2, P. 254].
15.   Tovbin Yu. K., Theory of physical and chemical processes at the gas-solid interface. – Moscow: Nauka, 1990. – 288 p. [CRC, Boca Raton, Florida, 1991].
16.   Tovbin Yu.K., Petrova T.// Zh. Fiz. Khimii. 1995. V. 69. P. 127. [Russ. J. Phys.

Chem. 1995. V. 69. № 1. P. 114]

17. Tovbin Yu.K., Tugazakov R.Ya., Komarov V.N. // Mat. modelirovanue. 2001. No. 7. P. 73.

18. Tovbin Yu.K., Tugazakov R.Ya., Komarov V.N. // Teor. osnovy khim. tekhnologii. 2002. V. 34, No. 2. P. 115. [Theor. Found. Chem. Engin. 2002. V. 36. No 2. P. 115].

19. Tovbin Yu.K. // Appl. Surface Science. 2002. V. 196. P. 71.

20. Tovbin Yu.K., Tugazakov R.Ya., Rabinovich A.B. // Teor. osnovy khim. tekhnologii. 2008. V. 42, No. 5. P. 509. [Theor. Found. Chem. Engin. 2008. V. 42. No 5. P. 509].

21. Tovbin Yu.K., Tugazakov R.Ya. // Teor. osnovy khim. tekhnologii. 2010. V. 44, No. 1. P. 104. [Theor. Found. Chem. Engin. 2010. V. 44. No 1. P. 102].

22. Carman P. C. Flow of Gases through Porous Media. – London: Butterworths, 1956.

23. Timofeev D.P., Kinetics of adsorption. – M.: AN SSSR, 1962 – 252 p.

24. Kheifets L.I., Neimark A.V., Multiphase processes in porous media. – Moscow: Khimiya, 1982. – 320 p.

25. Mason E. A., Malinauskas A. P., Gas Transport in Porous Media: The Dusty_Gas Model. – Moscow: Mir, 1986. - 200 p. [Elsevier, Amsterdam, 1983].

26. Nikolaevskii V.N., Mechanics of porous and fractured media. – Moscow: Nedra, 1984. – 232 p.

27. Barenblatt G.I., Entov V.M., Ryzhik V.M., Movement of liquids and gases in natural formations. – Moscow: Nedra, 1984. – 208 p.

28. Lykov A.V. Heat and Mass Transfer. – Moscow: Energiya, 1978. – 480 p.

29. Tovbin Yu.K., Tugazakov R.Ya. // Teor. osnovy khim. tekhnologii. 2010. V. 44, No. 6. P. 687. [Theor. Found. Chem. Engin. 2010. V. 44. No 6. P. 902].

30. Migus A., Gauduel Y., Martin J.L., Antonetti A. // Phys. Rev. Lett. 1987. V. 58. P. 1559.

31. Deryagin B.V., Churaev N.V., Muller V.M., Surface Forces. – Moscow: Nauka, 1985. – 400 p.

32. Koplic J., Banavar J.R., Willemsen J.F. // Phys. Rev. Lett. 1988. V. 60. P. 1282–1285.

33. Thompson P.A., Robbins M.O. // Phys. Rev. Lett. 1989. V. 63. P. 766.

34. Thompson P.A., Robbins M.O. // Phys. Rev. A. 1990. V. 41. P. 6830.

35. Molecular dynamics simulation of statistical mechanics systems, eds. G. Coccotti, W.G. Hoover. – Amsterdam: North-Holland, 1986. – 610 p.

36. Molecular dynamics method in physical chemistry, ed. Yu.K. Tovbin – Moscow: Nauka, 1996. – 334 p.

37. Akhmatskaya E., Todd B.D., Davis P.J., Evans D.J., Gubbins K.E., Pozhar L.A. // J. Chem. Phys. 1997. V. 106. P. 4684.

38. Blake T.D., Clarke A., De Coninck J., Ruijter M.J. // Langmuir. 1997. V. 13. P. 2164.

39. Neogi P. // J. Chem. Phys. 1995. V. 105. P. 8909.

40. Diez J.A., Gratton R., Thomas L.P., Marino B. // J. Colloid Interface Sci. 1994. V. 168. P. 15.

41. Adamson A.W. Physical Chemisty of Surfaces. Third Ed. – N. Y.: Wiley-Interscience Publ., John Wiley, 1976.

42. Tovbin Yu.K., Tugazakov R.Ya. // Teor. osnovy khim. tekhnologii. 2002. V. 34, No. 6. P. 563. [Theor. Found. Chem. Engin. 2002. V. 36. No 6. P. 511].

43. Tovbin Yu.K., Vasyutkin N.F. // Izv. RAN. Ser. khim. 2001. No. 9. P. 1496. [Russ. Chem. Bull. 2001. V. 50. № 9. P. 1572].

44. Tovbin Yu.K., Vasyutkin N.F. // Zh. Fiz. Khimii. 2002. V. 76. No. 2. P. 319. [Russ. J. Phys. Chem. 2002. V. 76. № 2. P. 257].

45. Ishii K., Liu S.H. AIAA-87-1342. 1987. P. 1.

7

Porous systems

51. Pores of complex geometry and porous systems

The molecular theory discussed earlier for slit and cylindrical pores was extended to complex pore systems [1, 2]. Generalization was made on the basis of the LGM (lattice-gas model), the ideas of which are used twice: 1) to describe the structure of a complex porous system (supramolecular level), and 2) to account for intermolecular interactions in the quasichemical approximation (molecular level). The basis of calculation of the structure of a complex system is the procedure for separating model areas of the porous system with a simple regular geometry (slit, cylindrical, spherical and globular) of predetermined length, and also including junctions (transition areas) for various types of porous areas. The structure of the connections adjacent sections then reflects their connectivity and allows to weigh the probability of specific types of sequences using pair distribution functions, which are known analogues of the pair functions at the molecular level. Accounting the junctions allows one to describe structural defects of adjacent areas far and to further match the atomic structure of the pore walls in different areas, including a description of the inhomogeneity of the walls.

In the transition from isolated pores of ideal geometry to a complex pore system we should consider the following factors: 1) the limited length of a specific area of the pore, 2) the presence of transition regions (junctions) between different areas of pores, 3) pore size distribution in the macroscopic cross section of the porous body, 4) connectivity of different types of pores between adjacent macroscopic cross sections, 5) the changing nature of the local distribution of the molecules in the cross sections of connectios and conditions of capillary condensation of the fluid (the critical temperature and the degree of filling) as a function of the internal structure of the pore space, 6) the effect of the local structure of the connections on the dynamic characteristics of the fluid, 7) pore axis orientation relative to the direction of the macroscopic flux of molecules by averaging the local fluxes.

The first four items formulate the procedure to describe the pore space of complex porous systems, and the rest belong to the description of the calculated equilibrium and dynamic characteristics.

Supramolecular level

The real porous structure will be modeled using the areas of pores of the dispersed body of some characteristic size $L > \lambda$, where λ is the linear dimension associated with the diameter of the spherical adsorbate. We confine ourselves to a supramolecular level – the level of the grain of the adsorbent or catalyst [3, 4]. Accounting for higher supramolecular levels includes large transport pores that are many times greater than the molecular scale. Scale L refers to the supramolecular level, and the scale λ to the molecular level. The supramolecular level includes portions of the porous body with a characteristic size H of defined geometry (slit-shaped, cylindrical and spherical), where H is the slit width or diameter of the spheres and cylinders, or basic volumes of the structure of a solid in the case of globular systems defined through the dimensions of the globules. In general $H \leq L$. If $H = L$ the pore occupies the entire section. At $H = 0$ the pore does not form that allows to reflect the presence of blind pores with the adjacent portion q.

We will define the supramolecular structure by the distribution functions F_q, characterizing the fraction of sections of type q, and F_{qp}, characterizing the probability of the section of type p being close to the section of type q plot type $\sum F_{qp} = F_q$, $1 \leq q, p \leq T$, where T is the number of types of the examined sections of the porous body. We also introduce the function H_{qp} – the conditional probability of the area of the pore of type p situated around the area of the type q pore (in a selected direction) $F_{qp} = F_q H_{qp}$, and $\sum_p H_{qp} = 1$. Function H_{qp} takes into account the coupling of different types of pore areas at the supramolecular level. It allows to weigh the probability of appearance of sequences specific types of pores using pair distribution functions at the molecular level, including analogues of the radial pair functions in X-ray analysis. H_{qp} function is a complete analog of the functions d_{fg} (1) to the nearest site

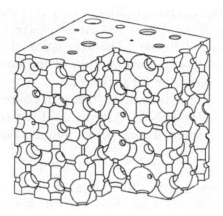

Fig. 51.1. Three-dimensional layout of the distribution of spherocylindrical pores of differing radii cylinders and spheres.

inmohogeneous systems at the molecular level [5] (section 7), which define the conditional probability of finding a site of type g next to the site of type f.

As an example of a complex porous system Fig. 51.1 shows a system consisting of spheres and cylinders of different diameters, the distance between the centres of the spheres is $L = 30\lambda$. The radii of the cylinders vary from λ to 8λ, and the radii of the spheres – from 9λ to 15λ. For both types of sections we assume the uniform pore size distribution. Figure 51.1 is a generalization of two-dimensional diagrams [6] to a three-dimensional case. Such complex porous structures can be used to model a variety of types of porous systems from zeolites cavities approximated by spheres with a relatively short length of the cylindrical parts of different diameter, to modelling junctions (intersections) of long cylindrical portions in new mesoporous materials such as MCM-41 and MCM-49 [7, 8].

The structure of the connections of the adjacent sections of the pores then reflects their coherence at the molecular level, but also allows to further match the atomic structure of the pore walls in different areas, including a description of the heterogeneity of their walls, reflecting the atomic properties of their surfaces. Taking into account the properties of connections means increasing the number of parameters of the supramolecular structure, although formally for each joint we can be confine ourselves to the indices q and p. The need to introduce additional parameters should be discussed in each case.

In complex systems, the pore space of every section of the supramolecular structure q is divided into the maximum number of molecules N_q, which can be found in it at complete filling. In general, for each site f, $1 \leq f \leq N_q$ of section q can have its own fill probability $\theta_{q,f}$. Site q, f is characterized by a local Henry constant $a_{q,f} = a_{q,f}^0 \exp(\beta Q_{q,f})$, $\beta = (kT)^{-1}$, $Q_{q,f}$ is the binding energy of the molecules with the walls of the pores, including contributions from the walls at different distances from the centre of the site, calculated in the atom–atom approximation. Stratifying of sections of the ideal structure with the same values $a_{q,f}$ and areas of the connections in which the quantities $a_{q,f}$ vary from section to section of the pores significantly reduce the range of the site types $t(q)$ within section q.

Equilibrium characteristics

Given that each site is defined by two indices q, f, we can use expressions to calculate the adsorption isotherm $\theta(P)$ and local filling of sites of different groups $\theta_{q,f}$, obtained in [5]. These expressions take into account the energy inhomogeneity of the lattice sites and the interaction between molecules at a distance R of the coordination spheres:

$$\theta(P) = \sum_q F_q \sum_{f=1}^{t(q)} F_{q,f} \theta_{q,f}(P), a_{q,f}P = \left(\frac{\theta_{q,f}}{1-\theta_{q,f}}\right)^{1+g_e} \Lambda_{q,f}, \sum_{f=1}^{t(q)} F_{q,f} = F_q, \quad (51.1)$$

$$\Lambda_{q,f} = \prod_r \prod_{p,g \in z_{q,f}(r)} \left(1 + x_{qp,fg}(r) t_{qp,fg}^{AA}(r)\right), x_{qp,fg(r)} = \exp\left(-\beta \varepsilon_{qp,fg}^{AA}(r)\right) - 1,$$

$$t_{qp,fg}^{AA}(r) = \frac{20_{p,g}}{\delta_{qp,fg}(r) + b_{qp,fg}(r)}, \delta_{qp,fg}(r) = 1 + x_{qp,fg}(r)\left(1 - \theta_{q,f} - \theta_{p,g}\right),$$

$$b_{qp,fg}(r) = \left(\delta_{qp,fg}(r)^2 + 4x_{qp,fg}(r)\theta_{q,f}\theta_{p,g}\right)^{1/2}, \theta_{q,f}^A + \theta_{q,f}^V = 1,$$

where $F_{q,f}$ is the fraction of sites of type f for the pore of type q, P is the pressure of the adsorptive, function $\Lambda_{q,f}$ takes into account the non-ideality of the adsorption system in the quasichemical approximation and the function g_e is a gauge function. In the formula (51.1) it is assumed that the lateral interaction parameter $\varepsilon_{qp,fg}^{AA}(r)$ of neighbouring molecules at a distance r of the oordination sphere can to be a function of temperature and local composition around sites q, f, and p, g. Index p, g ranges over all neighbours $z_{q,f}(r)$ of the site q, f at the distance $r \leq R$ inside the pore (R is the radius of the potential interactions, $R < L$). The molecular parameters, as defined above, are calculated in terms of the atom–atom potential functions (see section 9). Henry's constants in this case are calculated using the formulas (A4.22), (A4.30).

Equation (51.1) describes in detail the site distribution of all neighbours $z_{q,f}(r)$ of each site q, f. This detailed description can be roughened if the function $\Lambda_{q,f}$ can be rewritten in the averaged form using the function $F_{qp,fg}(r)$, which characterizes the probability of finding a pair of sites q, f and p, g at a distance r. So, if all the distributions of sites of different types can be approximated through the pair distribution function

$$H_{qp,fg}(r) = F_{qp,fg}(r)/F_{q,f}, \text{ and } \sum_{g=1}^{t(p)} H_{qp,fg}(r) = H_{qp}, \text{ then}$$

$$\Lambda_{q,f} = \prod_r \left(1 + \sum_{p,g} H_{qp,fg}(r) x_{qp,fg}(r) t_{qp,fg}^{AA}(r)\right)^{z_{q,f}(r)}, \tag{51.2}$$

here the subscript g refers to a portion of the pore of type p.

If the layer distribution of the sites is implemented, we obtain the following expression:

$$\Lambda_{q,f} = \prod_r \prod_{p,g} \left(1 + x_{qp,fg}(r) t_{qp,fg}^{AA}(r)\right)^{d_{qp,fg}(r)}, \tag{51.3}$$

wherein $d_{qp,fg}(r) = z_{q,f}(r) H_{qp,fg}(r)$.

The equilibrium distribution of particles in the sites of different types θ_k are found by solving the system of equations (51.1), setting θ, by the Newton iteration method. The accuracy of the solution of this system is less than 0.1%. The density of coexisting gas and liquid phases of the adsorbate was determined by Maxwell' equations [5, 9].

The structure of equations (51.1) remains unchanged during the transition to the porous system of any type. Changes apply only to the method of relating different types of sites g to corresponding groups related to the layers in different types of pore sections.For example, Fig. 51.2 presents isolones related to the monolayer partition of the defect cavity (including the volume of two missing globules and all neighbouring volumes of interglobular pores in a regular structure with the number of the nearest neighbours equal to 6). The section passes along the plane xy, the origin of the coordinates is in this section at the point of contact of the two missing globules. In plan view of this plane there are related fields (isolines) relating to one layer filled with the adsorbate, and the inner circles are the neighbouring globules and more distant globules in the same plane are not shown. Figure 51.2 *b* shows a perspective view of the region

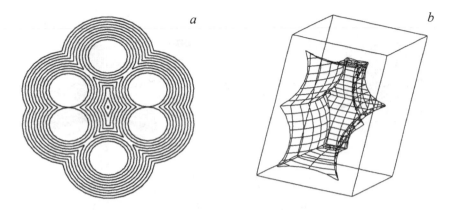

Fig. 51.2. Planar section of the globular structure $z = 6$ with a defect in the centre consisting of two adjacent remote globules (a), three-dimensional image of the discussed cavity inside the defective globular stuktura (b) [2].

of the pore space in the same globular structure containing two empty sites in the lattice supramolecular structure and contours, illustrating the layered partitioning of the space of the given pore into separate layers, filled with the adsorbate (the isolines reflect the molecular structure of the lattice).

Below we focus on the curves of vapour–liquid phase stratifying in the pores of different geometry. In section 3 it was shown that the characteristic feature of the complete phase diagrams of narrow-pore monodisperse systems of simple geometry (slit-shaped and cylindrical) is a significant fraction of the volume attributable to a few surface layers. These areas are given in the diagram in the form of domes. With an increase in the characteristic pore size the contribution of these areas in the phase diagrams rapidly decreases, and it is therefore difficult to identify them with the aid of numerical methods. This same feature is retained in the transition to a more complex geometry of the pores and porous non-monodispersed systems. This chapter illustrates the role of geometry and the influence of potential functions on the equilibrium and dynamic characteristics. This is most obvious to do for small pore sizes at which, however, due to size effects are no phase transitions.

We begin our discussion with spherocylindrical pores. In this case the system includes cylindrical and spherical portions, as well as the intermediate regions of their intersection. Additionally we take into account the pore size distribution and their relatedness. The lack of

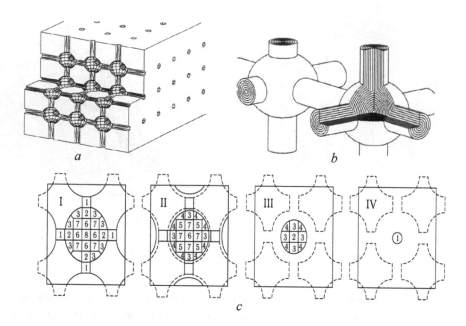

Fig. 52.1. Regular system of spherocylinders (*a*): partitioning scheme of monolayers of the pore space in spherocylinders (*b*); the lattice structure of adsorption centres (*c*).

intersections of long cylindrical pores corresponds to the system of the MCM-41 type, which was considered in chapter 3.

52. Spherocylindrical pores

Spherocylindrical systems consist of two types of elements of the free volume of pores having a finite extent, when the pore system is formed by alternating cylinders and truncated spheres [10, 11]. We denote regular spherocylindrical systems as $D_s - D_c - L_c$, which refer to: sphere diameter D_s, cylinder diameter D_c and its length L_c. Regular geometry allows to deal with with one complex periodic element.

Figure 52.1 shows a regular spherocylindrical structure. The isolines in Fig. 52.1 *b* contours reflect the lattice molecular structure for the adsorbed molecules. Figure 52.1 *c* shows the distribution of sites in the 'spherocylinder', consisting of a cylinder of unit length and radius λ, as well as a sphere of diameter 5λ. Shown are four sections from the left to right of the centre of the sphere to the cylinder, where the solid lines refer to the lines in the plane I–IV, and the dashed – lines in plane I.

For such porous systems it is necessary to analyse: 1) the impact of the ratio of the size of the simplest sites of the pores on their contributions to the phase diagrams, 2) the contributions of subsurface areas determined by the molecule–wall potential and central parts of the pores on the phase diagram, 3) the role of the transition regions and their lengths.

This example is convenient to formulate procedures describing the distribution of different types of sites, which are filled with molecules at adsorption.

Within each section q of the supramolecular scale the set of sites available to the adsorbate particles is divided into groups (type of sites) with the same binding energy with the atoms of the solid, and considering the fact that within each group the sites have the same configurations of the neighbors (up to rotation/reflection) in all considered coordination spheres r, $1 \leq r \leq R$, where R is the maximum number of coordination spheres defined by the radius of the interaction potential between the adsorbates. For example, if the pore space of the portion contains an inscribed simple cubic lattice of nearest neighbors $z = 6$ with the side λ, the distance to the sites of the first coordination sphere is λ, to the sites of the second coordination sphere $\lambda\sqrt{2}$, third $\lambda\sqrt{3}$, , the fourth 2λ, etc. In general, we have $\eta_r \lambda$, where η_r is the numerical factor for the coordination sphere r.

The structure of the section q is given by functions $F_{q,f}$ and $F_{q,fg}(r)$: $F_{q,f}$ – the fraction of sites of type f, defined as $F_{q,f} = N_{q,f}/N_q$, where $N_{q,f}$ – number of sites of type f; $F_{q,fg}(r)$ – the probability of finding the sites of type f and g at a distance of the r-th coordination sphere. It is also possible to introduce the conditional probability of finding a site type g in the r-th coordination sphere of the site of type f in the form $d_{q,fg}(r) = F_{q,fg}(r)/F_{q,f}$. Obviously, these functions must be satisfy normalizing relations $\sum_f F_{q,f} = 1, \sum_g d_{q,fg}(r) = 1,$ and $\sum_g F_{q,fg}(r) = F_{q,f}$ for any r. Within the limits of section q each site f under specified conditions corresponds to local fillings $\theta_{q,f}$ and pair probabilities, full ($\theta q,fg\ (r)$) and conditional $\left(t_{q,fg}(r) = \theta_{q,fg}(r)/\theta_{q,f}, \sum_g t_{q,fg}(r) = 1 \right)$, of finding the particles of the adsorbate in neighboring sites f and g at the distance of the r-th coordination sphere.

We denote $N_{q,fg}(r)$ – the number of pairs of sites of type f and g at the distance r, $N_{q,fg}(r) = N_{q,gf}(r)$. The number of sites of type g in the r-th coordination sphere of type f is determined as $m_{q,fg}(r) =$

Table 7.1. Example of calculation of the distribution functions

q	Supramolecular level				Molecular level												
	F_q	H_{qp}			Spherical section								Cylindrical section			
		$p=1$	2	3	f_q	d_{qp}							f_q	d_{qp}		
						$p=1$	2	3	4	5	6	7		$p=1$	2	3
1	0.125	0	1	0	0,074	0	0,8	0	0	0	0,2	0	0,444	0,4	0,4	0,2
2	0,375	0,333	0	0,667	0,296	0,25	0	0,5	0	0,25	0	0	0,444	0,5	0,5	0
3	0,5	0	0	1	0,296	0	0,667	0	0,333	0	0	0	0,111	0,667	0	0,333
4					0,099	0	0	0,5	0	0,5	0	0				
5					0,148	0	0,333	0	0,333	0	0,333	0				
6					0,074	0,167	0	0	0	0,667	0	0,167				
7					0,012	0	0	0	0	0	0	1				

$N_{q,fg}(r)/N_{q,f}$, and the conditional probability of finding a site of type g in the r-th coordination sphere of the site f is $d_{q,fg}(r) = m_{q,fg}(r)/z_{q,f}(r)$, wherein $z_{q,f}(r) = \sum m_{q,fh}(r)$ – the total number of sites in the r-th coordination sphere of site f.

An example of constructing a model of a porous body is illustrated in Table. 7.1. Suppose that the space of a rigid body is represented by the system of spherical cavities, cylindrical channels connecting them and non-porous areas. The spherical cavities have a diameter 5λ, the length of the cylindrical channels is 5λ, and their diameter 3λ and the size of the non-porous areas is 5λ. These sites are located in sites of a simple cubic lattice as follows: the coordinates of the spheres (0,0,0), cylinders (1,0,0), (0,1,0) and (0,0,1), the coordinates of non-porous portions (1,1,0), (1,0,1), (0,1,1) and (1,1,1). The method of partitioning the volume of the porous body at the supramolecular level and of porous areas at the molecular level is shown in Table 7.1.

Similarly, we introduce the distribution function $F_{qp,fg}(r)$ for particles in the sites f and g on different sections q and p at a distance r. Construction of the distribution functions of different types of site is based on the formation of the whole set of sites into which the pore space is divided, the calculation of the surface potential at each site and calculation of the local Henry constants (or setting values η_f^q). This procedure allows to select both the types of sites and the types of their neighbours in the construction of cluster and pair distribution functions. In fact, here are united procedures of partitioning the sites in a nonuniform lattice, as in Appendix 2 and the same scheme that was used in section 9 to calculate the Henry constants and their weights in a non-porous material. Here,

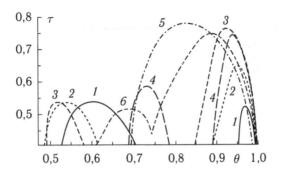

Fig. 52.2. Stratification curves for regular mesoporous spherocylindrical structures D_s–D_c–L_c: 6–4–8 (*1*), 8–4–8 (*2*) 10–4–8 (*3*), 9–6–8 (*4*), 9–8–8 (*5*), 9–6–1 (*6*).

the same principle applies to all of the pore space. If there is any symmetry, the procedure for constructing the distribution functions is greatly simplified. The isolation of simple pore portions (slits, cylinders, spheres) is a simplifying factor. The calculation algorithm is described in [11].

Recall that in the general case, the calculation uses the formula (51.1), which describes in detail the site distribution of all the neighbors $z_{q,f}(r)$ of each site q, f. This detailed description can be made more approximate if there are elements of symmetry. Then the function $\Lambda_{q,f}$ can be rewritten in the averaged form using the function $F_{qp,fg}(r)$, which characterizes the probability of finding a pair of sites q, f and p, g at a distance r (51.2) and (51.3). The distribution of different types of sites can be approximated by the pair correlation function $H_{qp,fg}(r) = F_{qp,fg}(r)/F_{q,f}$.

Regular structures of spherocylindrical pores

Curves *1–3* in Fig. 52.2 relate to systems with increasing diameter of the spherical portions at constant dimensions of the cylindrical portions. Increasing D_s increases the size of the dome relating to the area of intersection. In this case, the contributions of the intersection regions dominate over the contributions of subsurface regions (the figure does not show surface domes, located at lower temperatures τ and the degrees of filling θ). Therefore, the domes relating to filling the second layers are reduced in width at a constant value of the critical temperature and the critical temperature of the third domes increase with increasing D_s.

The curves *4* and *5* illustrate the effect of increasing diameter of the cylindrical channel with $L_c = 8$. Both domes shown in Fig. 52.2 relate to the sections of the cylinders in the vicinity of intersection

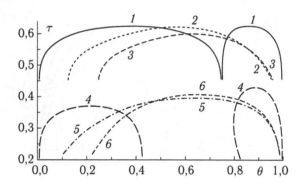

Fig. 52.3. Regular microporous spherocylindrical structures D_s–D_c–L_c: 3–2–2 (*1*), 3–1–1 (*2*), 3–1–2 (*3*), 4–2–2 (*4*), 4–1–1 (*5*) 4–1–2 (*6*), curves *1*, *2* and *3* are shifted up by 0.25.

with a sphere and to a common central parts for both types of pore, and both domes increase their critical temperatures.

Reducing the length of the cylindrical channel of the curves *4* and *6* leads to the degeneration of the surface sites inside the channels. Critical temperatures for the sites in the centre of spherical pores remain unchanged, and the proportion of these sites is significantly increased. Structure D_s–D_c–L_c = 9–6–1 for the curve *6* is strongly 'friable structure' – this greatly reduces the adsorbate–adsorbent binding energy and the mechanical stability of the adsorbent.

Microporous spherocylinders are characterized by small values of D_s–D_c–L_c. Accordingly, the entire interior of the pores is in the fields of surface potentiasl with high values of binding energies. As a result, the critical temperature of all the domes are greatly reduced. The curves in Fig. 52.3 are closer to the domes of the surface layers compared to the curves in the previous figures. According to their geometrical parameters the calculations in Fig. 52.3 comply with the following zeolite systems: D_s–D_c–L_c = 3–1–1 (Ar-chabazite and levinite with the number of windows 2, instead of 6), 3–1–2 (Ar-zeolite X – 12 windows), 4–1–1 (Ar and He – zeolite A). This range of parameters includes the system He–chabazite system – 4–2–1.

The curves *1* and *4* (and also *2* and *5*, *3* and *6*) represent the effect of increasing the diameter of the sphere. The sharp reduction in the width of the first dome for curve *4* is due to a decrease of the cylindrical portion with increasing D_s. The strong heterogeneity of the sites of the transition area leads to the fact that they are filled without forming a separate dome in the considered temperature range and, therefore, the width of the dome for the spherical portion is not increased, although its critical temperature increases. At the same time filling the sites of the transition region affects the curvature of the dome. Curves *1* and *3* (also *4* and *6*) reflect the effect of increasing the diameter of the cylindrical channel, while the curves *2* and *3* (also *5* and *6*) represent the effect of increasing its length. General trends in the properties of the phase diagrams under the effect of the geometric parameters of the structure are similar to those above for the mesoporous systems in Fig. 52.2.

Recall that in the real conditions, such as argon atoms, with values $\tau \leq 0.5$ the substance is in the solid state or in the supercooled liquid state [12]. The potential of the walls lowers the critical temperature of condensation, but its effect on the melting conditions has not been definitely defined. The experiments [13, 14] indicate that the attracting potential of the walls also lowers melting point as well as the critical condensation temperature, whereas other data suggest the possibility of increasing the melting point, which leads to a drastic reduction in the two-phase liquid–vapor coexistence region. Model (51.1) does not reflect the possibility of moving the fluid to the solid state and does not describe the two-phase liquid–solid melting region. In any case, the results obtained for stratification below $\tau < 0.4$ indicate the departure from the range of the physically admissible values that should be interpreted as the absence of a phase transition. Thus, the LGM for small pore volumes allows to get the right solutions even without the use of calibration functions, if we consider the physical meaning of the obtained parameters of the critical temperatures.

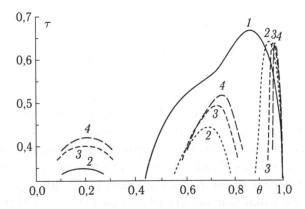

Fig. 52.4. The phase diagram for the three-dimensional system of intersecting cylinders $D_s = D_c = 5$, $L_c = 1$ (*1*), 5 (*2*), 10 (*3*), 15 (*4*).

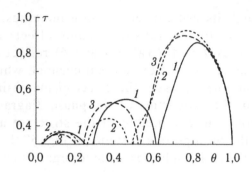

Fig. 52.5. Phase diagrams of three types of pores: cylindrical (*1*), spherical (*2*) and spherocylindrical (*3*).

Intersection of channels [10]

The special case of a regular three-dimensional structure of the spherocylinders with $D_s = D_c$ corresponds to the intersection of cylindrical channels. The intersection of the cylindrical channels appears to change the adsorption properties of the walls near their intersections. Part of the volume of the solid is occupied by the intersecting pore, so the adsorbent–adsorbate binding energy decreases. Changing the length of the cylindrical portion of the pores L_c can substantially change the energy of the adsorbent–adsorbate system; with $L_c \to \infty$ we have isolated channels. Figure 52.4 shows the phase diagram for mesoporous adsorbents. The number of channels originating from the intersection is equal to 6.

The characteristics of the surface domes for mesoporous adsorbents (Fig. 52.4) virtually little change (curves *2–4*). For curve *1*, this dome is lower, but it should be noted that the curve *1* corresponds to the unit length of the channel that leads to the open structure and its properties are very specific. In particular, such structures should have a relatively low mechanical strength.

The critical temperatures for the second domes, corresponding to the sites near the intersections of the channels, increase markedly. Finally, there are third additional domes (curves *2–4*) related to the areas with the weakest surface potential. Comparison with the phase diagram for the isolated cylinders (curves *3* and *4* in Fig. 22.1) shows that the intersections of the cylindrical channels changes the number of the domes. Thus, the effects of pore intersections may mask the properties of the porous system: three fully split domes in the case of the isolated pores are observed for $D_c > 8$ (here $D_c = 5$!) and the characteristics of the domes (their form, width, and critical parameters) vary significantly.

Table 7.2. Distribution functions for the three types of pores: cylindrical, spherical and spherocylindrical $\sum_q F_q = 1$

	Continuous description		
Pore	Cylinder	Sphere	Spherocylinder
F_1	0.331	0.259	0.275
F_2	0.264	0.210	0.232
F_3	0.198	0.166	0.191
F_4	0.207	0.365	0.302
	Discrete lattice		
Pore	Cylinder	Sphere	Spherocylinder
F_1	0.289	0.253	0.240
F_2	0.330	0.231	0.290
F_3	0.165	0.154	0.167
F_4	0.216	0.362	0.303

To compare the effect of the shape of the pore cavity on the shape of the stratification curves, Fig 52.5 shows the phase diagram for the pores with the regular contituation over the entire volume of the adsorbent: cylindrical, spherical and spherocylindrical. Here $L = 21\lambda$, the sphere diameter of the cylinder is also 21λ, and the cylinder diameter is 5.5λ. The potentials of the interactions of the molecules with the walls of different geometry of the three types of pores are given in section 5.

Table 7.2 shows the distribution function of the sites F_q for a discrete and a continuous description of the cylindrical layers and spherical shells. As explained in chapter 3, the curved surfaces describe more accurately by means of the discrete functions the distributions of the sites of the lattice system associated with the properties of the surface potential. These two methods provide several different distribution functions F_q. However, the general properties of the phase diagrams remain unchaged. There are four layers of sites in each pore, but all the phase diagrams as above contain three domes. The dome height and width correspond strictly to the contributions of surface, second and central sites in the total pore volume.

Analysis of the phase diagrams for different pore geometry reveals the following patterns.

1. There are three types of phase diagrams, depending on the adsorbate–adsorbent binding energy (changing from strong attraction

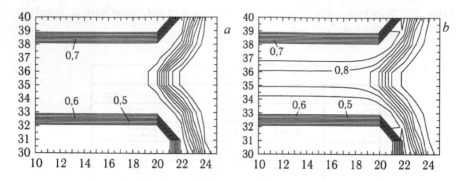

Fig. 52.6. Isolines of the density of the adsorbate in the transition region between the cylindrical and spherical portions of the mesoporous system calculated without cosnidering (*a*) and considering (*b*) the calibration function.

to its absence). In the case of strong attraction there is layer condensation of molecules, leading to additional parietal domes for the stratification curves of the adsorbate. The number of such additional domes depends on the binding energy, the long-range adsorbate–adsorbent potential, the characteristic average pore size of the system and the share of contributions of the pores with the minimum size.

2. In the absence of a strong adsorption bond the type of phase diagram is similar to the phase diagrams for the bulk phase, however, the temperature of the critical point is lower than for the bulk phase. In the intermediate case the parietal domes are missing, but the stratification curve has a bend before the 'main' stratification of volume condensation in a porous system that is associated with increased concentration of the adsorbate near the pore walls.

Influence of the calibration function [15]
Above we discussed the quantitative description of the molecular distributions in porous systems on the example of cylindrical pores. The same problems exist in the transition to more complex geometries. |For nonuniform systems the calibration function depends on the type of centre $g(f)$ [16]. The results of the, calculations showing the influence of the calibration function on the distribution of molecules in the spherocylindrical system ($D_s-D_c-L_c = 20-8-15$ at $\tau = 0.79$), are shown in Fig. 52.6.

This figure shows the density isolines corresponding to the vapour–liquid interface for the completely filled cylindrical pores and disregarding the calibration function (field *a*) and for the field *b* with the calibration function taken into account. In the latter case the adsorbate density isolines in the centre of the cylindrical pore correspond to the density of ~0.8. In the centre of field *a* the densities are equal to ~1. Thus, this example illustrates the effect of calculation accuracy of the molecular distributions on the nature of the state of the vapour-liquid system in complex surface fields. Depending on the account or ignorance of the calibration function both the equilibrium distributions and transport characteristics of the flow of molecules change quite markedly.

Note that the construction of the theory of adsorption phenomena taking into account the contribution of calibration functions, including size effects, is crucial not only for fill areas that meet the conditions for realization of capillary condensation. The behaviour of the vapour–liquid coexistence curves is sensitive to accounting the calibration functions. This changes not only the appearance of each of the domes, but also the nature of the coexistence curve, depending on the size of the pore diameter.

53. Systems of cylinders

Analysis of the properties of cylindrical pores is important for the interpretation of the phenomenon of adsorption–desorption hysteresis, as it is this geometry that is assumed in most ways of the interpretation of many of the properties of the porous systems.

For a system of cylindrical pores (Fig. 53.1), the distribution function F_q (which determines the contributions of cylinders of different diameters) can significantly change the position of the above three domes, but the general laws remain in force: there are three domes associated with the difference in the binding energies of the adsorbate in the surface monolayer, in the second monolayer and the remaining parts of the pores. Curve 5 in Fig. 53.1 *a* describes the homogeneous distribution $F_q = 1/3$. The distributions were calculated using the equations (51.1) and the condition of equality of pressure inside all pores.

Under the above conditions, the overall shape of the curves of phase diagrams is preserved, i.e., the transition to the average properties of the system still retains a simple structure of the stratification curves as for the isolated cylindrical pores. This result explains to some extent explains many intuitions about the nature of the description of porous systems, but it is the result of two traditional assumptions: 1) the process is implemented in a

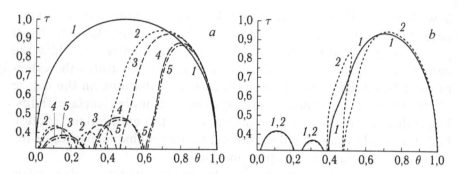

Fig. 53.1. Phase diagrams: a – bulk phase (1); systems of cylindrical pores with weighted contributions (Gaussian distribution function F_q with half-width $\sigma = 4\lambda$ and maximum) $D_m/\lambda = 20$ (2), 14 (3), 8 (4); homogeneous distribution with $F_q = 1/3$ (5); b – comparison of stratification curves in the case of ignoring (1) and considering (2) differences in the saturated vapour pressures at different stages.

porous medium with the averaged properties, as required by the thermodynamic treatment of adsorption processes, and 2) we ignore the change of the saturated vapour pressure when the width of the pore changes. The latter factor plays a crucial role.

The essence of this simplification is that the probability of filling all central sites is identical for all pore sizes. If we consider the differences in the probability of filling the central sites in different size pores, then we get a qualitatively new effect (Fig. 53.1 b). It was first discovered in [2]. Figure 53.1 b shows the phase diagrams for the porous system containing an equal ratio of two types of cylindrical pores with a diameter of 10 and 20 monolayers. The effect is that the dome related to the filling of the volume parts of the porous system is split into separate contributions corresponding to the condensation of the fluid in the pores of different sizes. As a result, under the influence of the surface potential the vapour–liquid coexistence pores for simple pore geometries then splits into contributions from the individual regions of the free pore volume.

In complex porous systems, the total stratification curve should further split into a greater number of domes because of the difference in the characteristic pore sizes included in the total pore system (number of domes in the pores of simple geometry is the minimum number of domes for the phase diagram of the complex porous systems). It should be emphasized that a significant difference in the curvature of the walls inside the cylindrical pores, changing the binding energies of molecules with the walls $Q_{q,1}$, can also lead to additional splitting of the surface domes.

This conclusion was fully confirmed by subsequent calculations [10]. Figure 53.2 shows the total stratification curves for a polydisperse system with isolated pores with the diameter D_c = 8, 9, 10, 11. The number of pores (T = 4) is selected to avoid overlapping of the curves and the contribution of each dome can be studied in

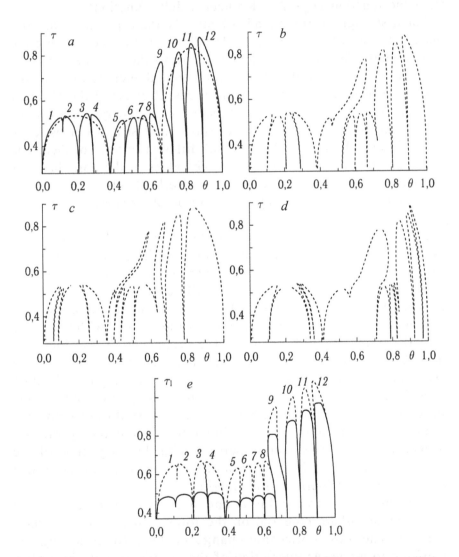

Fig. 53.2. Phase diagrams for porous systems consisting of four isolated cylinders with diameters D_c = 8, 9, 10, 11, such as Ar–C (a) and Ar–SiO$_2$ (b-d). Functions of pore size distribution: uniform (a, b), increasing (c) and decreasing (d). On the field e there is the phase diagram of the field (a) with volume calibration taken into account, but without considering the effect of calibration on the restriction of the characteristic pore size.

detail. The diameters of the pores were selected for the same reason, Fig. 53.2 (which is less than the critical pore size D^* – section 24).

The case with the uniform function of the pore size distribution corresponds to the fields a and b for which the distribution function F_q has the form $F_q = 1/T$. For a system of isolated infinitely long pores the function $D_{qp} = \Delta_{qp}$ (Kronecker delta function).

For a strongly attractive adsorbent–adsorbate potential the full phase diagram (a) consists of 12 domes: in all four pores there are three domes. The first four domes relate to the surface layers of pores of different widths. The second four domes belongs to the second subsurface layers in different pores. Because of the different wall curvature the adsorbent–adsorbate binding energy in all pores is different, so all subsurface domes split. The last four domes relate to the fillings of the central regions of different pores. For them, the splitting is due to the differences in the values of the pore diameter (the size effect) as the energy of the adsorbent–adsorbate bond is much smaller than the energy of the adsorbate–adsorbate bond and plays no role.

For comparison, field a is completed by the three domes of the phase diagram calculated under the condition that the binding energy in each of the two surface layers is the same regardless of the diameter and that the degree of filling in all central fields of the same four pores is identical. In this case, the dome of the pore system characterized by its weighted value of the width and the critical parameters with regard to the properties of the domes in the individual pores. These two conditions lead to the form of the phase diagram of the system, similar to that of the phase diagram for a single pore: there are differences in the properties between the subsurface and central domes, but there are no specifics of individual pores of different diameters. Taking into account the specificity of each pore of the system associated with the magnitude of the diameter, leads to a splitting of each of the three domes of the system to individual contributions from T pores.

Comparison of the fields a and b shows the effect of the adsorbent–adsorbate potential on the form of the complete phase diagram. The most noticeable change in these fields during the transition to a weaker interaction of the adsorbate with the wall b is due to the sharp shift to the position of the central dome of the narrowest pore in the area with the domes corresponding to the filling of the second layers: the central dome for $D_c = 8$ is located to the right of the dome of the second layer of this pore. Only after this we find the domes for the second layers in the pores with $D_c > 8$. The second difference on the diagram in the field b is due to the change in the upper parts of the domes for the surface layer due to changes in the surface potential.

The role of the function of pore size distribution F_q is illustrated by the fields b, c and d in Fig. 53.2 (exponentially increasing (c) and decreasing (d) function, so that $F_1/F_4 = 0.066/0.536$ (c) and vice versa $0.536/0.066$ (d)). The dimensions of the sections for the second and central domes in the fields b–d correlate well with the fractions of the pores. The arrangement of all the domes does not change with F_q. At the same time, the sections occupied by the first four surface domes are determined not only by the weights F_q, but also by the weights $F_{q,1}$ for surface sites that are most strongly altered by changing the diameter of the cylindrical pores. As a result, in the field c the largest section belongs to the third dome (and not to the fourth) and in the field d the largest section belongs to the the second dome (not the first). Thus, even for an isolated pore system the form of the complete phase diagram depends on both the function of the pore size distribution and the contribution from a specific type of sites in each pore.

Field e is compared with the field a by the stratification curves calculated in a calibrated QCA given volumetric calibration of the condensation parameters in the central part of the pores and on the surface values of the fluid state for the surface monolayers, but without calibration of the limitations of the characteristic pore size. The figures correspond to the same numbers of near-surface (1–4 for the first dome and 5–8 for the second dome) and central (9–12) domes. It is evident that refinement of the calculations by the calibration function significantly alters the critical temperature. From the known relation between the critical temperature and the triple point it follows that all eight surface domes for polydisperse systems obviously can not be realized because of the values of $\tau <$ 0.5. In the remaining four domes the correction of this type for the central part of the pores is not enough and, as shown in section 24, it is necessary to additionally consider the limitations of the pore volume so that there is no capillary condensation for small diameters ($D < D^*$).

Thus, in the case of strong attraction there is layer condensation of molecules, leading to additional parietal domes on the stratification curves of the adsorbate. The number of such additional domes depends on the binding energy, the characteristic average pore size of the system and the degree of contribution of the pores with the minimum. It is shown that if there is a discrete pore size distribution, we observe the splitting of the general dome relating to the filling of central sites, to a number of individual (narrow) domes, characteristic of the filling of specific pores of different diameters. Similar splitting is observed for the surface domes because of the difference in the curvature of the pore walls and the associated adsorbent–adsorbate energy, i.e. the developed method reflects the effect of the limited volume in the pores with different characteristic size.

The effect of curvature of of the dome-shaped stratification curves on the complete phase diagram of the system is associated with the simultaneous nature of filling of pores of different diameter (whereas the domes in individual pores do not have such a pronounced curvature) when the total degree of filling of the pore system for fixed external pressure changes. Note that sometimes the fact of curving of the shape of the domes was interpreted as a shortcoming of the calculation of the phase diagrams by the cluster methods. However, the new results show that the lack of curvature of the domes is the result of dealing with such situations where the condensation processes in different areas of the non-uniform system is occur completely independently from each other. In many porous systems this independence condition of different areas is not fulfilled and, conversely, increasing the pore size, their filling conditions converge.

54. Systems of slit-shaped pores

The model of slit-shaped pores is one of the most common models, along with the cylindrical pores. Formally, this model assumes the infinite extent in two directions perpendicular to the line connecting opposite pore walls. Obviously, the condition of infinite extent is unreal – it does not meet the condition of the mechanical stability of the material. In [17] the authors considered the properties of phase diagrams of stratification of the adsorbate, depending on the width of the pores for the slit-like microporous and mesoporous systems (pore width varied from 1 to 20 monolayers). These structures can be modelled by different types of clay, activated coal [18, 19]. As above, the equations (51.1) were used. The energy characteristics were the same as those in section 21, $R = 1$. The calibration function is not taken into account ($g_e = 0$).

The function of the pore size distribution F_q for the case of uniform distribution has the form $F_q = 1/T$. The main properties of the phase diagrams of the porous slit-like systems were considered for the case of a system consisting of four pores ($T = 4$) of different widths. For a system of isolated infinitely long pores $D_{qp} = \Delta_{qp}$ (Kronecker delta function). However, description of the relationship of different types of sites is much complicated by the limited length of the slit-like pore sites and how they are then connected. Below, as an example, we examine the simplest regular structure of the system slit-shaped pores.

Figure 54.1 shows the phase diagrams of the argon atoms condensed in carbon pores with width $H = 5, 7, 9, 11$ monolayers (more accurately – considering four monodisperse pore systems) (Fig. 54.1 *a*) and the total (polydispersed) system of the same pores with a uniform pore size distribution function (the curve *5* is shifted upward along the ordinate axis by an amount of 0.5). For monodisperse systems with $H \geq 5$ and the considered adsorbent–adsorbate interaction potential each complete phase diagram consists of three domes. With increasing pore size the filling area in which surface layers form is reduced, therefore, the first two domes decrease in width. It should be noted that the height of both surface domes is virtually identical, as determined by the two-dimensional character of condensation in both surface layers (the curvature of the wall has no influence, as for cylindrical pores).

The complete phase diagram of the pore system (curve *5*) consists of 7 domes. Obviously, for the non-intersecting pores it is formed as

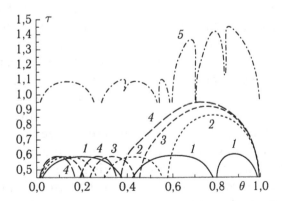

Fig. 54.1. Phase diagrams of argon in isolated slit-like carbon pores with width $H/\lambda = 5$ (*1*), 7 (*2*), 9 (*3*), 11 (*4*), and a pore system with a uniform pore size distribution (*5*). Curve *5* is shifted along the ordinate axis by an amount of 0.5.

a result of the addition of the contributions of all 12 individual domes represented by curves 1–4. The first dome refers to simultaneous filling of the first surface layers of all pores with $H = 5, 7, 9, 11$, so it width corresponds to the filling from zero to $\theta \approx 0.25$. The second dome relates to the second surface layer of the pore with minimum width $H = 5$, and the contributions from the remaining second layers for the pores with $H = 7, 9, 11$ are combined in a third dome – their the filling occurs at the same external pressure of the adsorptive. The remaining four domes correspond to the individual contribution to the total phase diagram of the central regions of each of the pores $H = 5, 7, 9, 11$. Their filling occurs at different external pressures – the smaller the width, the lower the condensing pressure.

One should emphasize the fact of noticeable curvature of the 2nd, 5th and 7th domes on the complete phase diagram of the system, while the domes in individual pores do not have such a pronounced curvature. This is the result of the influence of filling of other pores and of changes in the overall degree of filling of the porous system at a fixed external pressure. Thus, for the set of pores even isolated from each other the full phase diagram depends on the fraction of pores of different widths included in the common system.

Figures 54.2 reflect the influence on the phase diagram of the pore width range (narrow and relatively wide pores) with a uniform distribution function F_q (Fig. 54.2 a) and the variation in the proportion of pores F_q in the total pore system (Fig. 54.2 b). In the case of narrow pores for $H = 4, 5, 6, 7$ the positions of the central and the surface domes of different pores may vary with respect to the positions of the domes on the curve 5 in Fig. 54.1. First, as above, there is a common dome for the first layers. The second dome relates to a central layer in a pore with $H = 4$ – it may also be considered as a second layer in the pore. Then, the two domes sequentially related to the second layers in the pores with $H = 5$ and $H = 6, 7$. The splitting of the domes at filling of the second layers in different pores is due to the fact that the molecules of the second layer in a pore with $H = 4$ directly interact with each other, whereas at $H > 4$ is no interaction. Finally, the remaining three domes correspond to the sequential filling of the central areas of the pores with $H = 5, 6, 7$. Figure 54.2 a also shows the significant bending of the second and third domes (essential contribution of the adsorption of molecules at the same time in different parts of the porous system). In addition, the collective behaviour of the system of isolated domes makes a

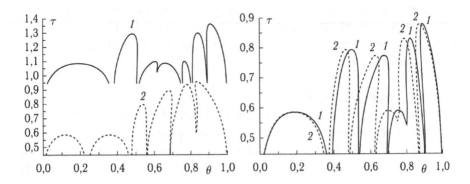

Fig. 54.2. Phase diagrams of argon: *a* – a system consisting of four isolated slits of carbon with width $H/\lambda = 4$, 5, 6, 7 (*1*), and $H/\lambda = 6$, 8, 10, 12 (*2*). Curve *1* is shifted by an amount of 0.5 on the ordinate axis; *b* – for a system of slit-like pores in clay material with width $H/\lambda = 6$, 8, 10, 12, with a pore size distribution function F_q, slightly different from the uniform function.

difference for the central domes with $H = 5$ and 7 in comparison with the same domes in Fig. 54.1 (fourth and fifth domes).

In the case of relatively wide pores (curve *2* in Fig. 54.2 *a*) the phase diagram of the system changes – splitting of the subsurface domes disappears (they are filled with the same external pressures), and the contributions of the central areas are separated and they are subsequently filled with increasing external pressure.

The variation of the function of the pore size distribution in the case of narrow pores is shown in Fig. 54.2 *b*. Here, both the distribution functions F_q are slightly different from the uniform function (shown in Fig. 54.2 *a*): the value F_q decreases linearly with increasing q for curve *1* and increases for curve *2* with a coefficient ~0.04. At a small change in F_q these changes do not affect the curvature of the dome, but at significant (order of magnitude or more) changes of F_q the curvature of the domes may change. The decrease (increase) in the proportion of pores with a fixed width causes a decrease (increase) in its contribution to the width of the respective domes. In all cases, the critical temperature of the domes does not change.

The effect of the adsorbent–adsorbate potential for narrow pores is also illustrated in Fig. 54.3. In weak interactions the role of the surface potential is small and there is no two-dimensional condensation of molecules in the surface layers. The entire pore volume takes part in the condensation process. Because of the small wall potential the filling of the pores of different widths occurs simultaneously, so the domes relating to pores with different widths are greatly curved. However, in this case, the critical temperatures of all the domes are lower than for the bulk phase, and smaller pores are filled at first.

Pores of limited length

When taking into account the limited length of the slit-like pore areas and their coupling, the method of specifying different types of sites is significantly complicated. As an example, calculation of phase diagrams is carried out for the following regular structure. It is believed that the crystal has a square shape with the side of the square L and thickness H_1 (all lengths are measured in units of λ).

Let it be that: 1) crystallites are interconnected by slit-like pores width H, and 2) they form a stack with the height $L = N_c (H + H_1)$, where N_c is the number of crystallites in a stack. This stack of the crystallites is in contact on four sides with side surfaces of the crystallites of other stacks which are perpendicularly oriented with respect to the given stack. Such contacts of the ends of the crystallites with the side surfaces of the crystallites in other stacks provide the regular structure in the volume of the entire sample and its mechanical stability. Now let all the crystallites in each stack be shifted through one in opposite directions by length L_1. The distance between the side surfaces of the adjacent stacks in one direction increases by the amount $2L_1$. Thus, one end of the crystallite (in the shift direction) is in contact with the side surface of the crystallite in the opposite stack, while the opposite end has no connection with the side surface of the crystallite in the other stack. At this point there is connection between the slit pores along the end face of the crystallite. The volume of the crystallite stack after their shift is equal to $L^2 (L + 2L_1)$ instead of the original value L^3.

Fragments of the porous regular (*a*) and irregular (*b*) structures are shown in Fig. 54.4. Contacts between the stacks for the strictly regular structure are presented in the form of point contacts (Fig. 54.4 *a*). Obviously, they are not mechanically stable. In contrast, the irregular structure (Fig. 54.4 *b*) mechanically stable, but its description requires the introduction of additional structural elements (and the corresponding geometric parameters describing contacts of the stacks).

At the ends of the crystallite this energy is less than Q_1, and at the contact angles between the adjacent crystallites is higher. The weight distribution of adsorption centers $F(Q)$ with respect to the

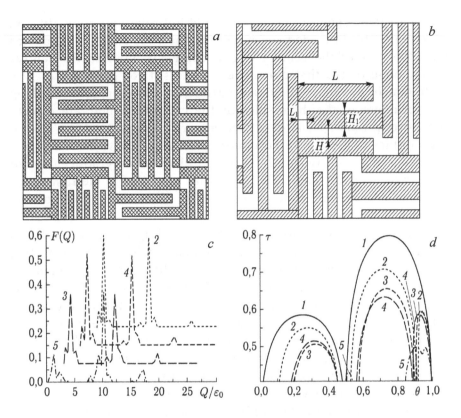

Fig. 54.4. Regular (*a*) and irregular (*b*) structures of slit-shaped pores. The distribution function of the sites of the pore system $F(Q)$ for energy Q/ε (*c*: curves are shifted along the diagonal of the figure). Effect of structural (geometrical) parameters of the system on the type of the phase diagrams of argon in a carbon pore (*d*): $L/\lambda = \infty$ (*1*), 40 (*2*), 20 (*3–5*); $H/\lambda = 4$ (*1–4*) 2 (*5*); $H_1/\lambda = \infty$ (*1*), 10 (*2, 4*), 3 (*3, 5*).

binding energies of the adsorbent–adsorbate system for four variants of the slit systems are shown in Fig. 54.4 *c*, and their phase diagrams in Fig. 54.4 *d*.

Curve *1* in the field *d* corresponds to an ideal slit of infinite length with width $H = 4\lambda$. It consists of two domes of the same width corresponding to the filling of the first and second monolayers. In the field in Fig. 54.4 *c* the curve 1 is absent: for an ideal slit $F(Q_1) = F(Q_2) = 2/H$ and $F(Q_{f>2}) = 1 - 4/H$.

The influence of the structure of the system of slit-shaped pores on the form of the phase diagram is presented in the curves *2–5* in Fig. 54.4 *d*. Due to the limited length of different types of the pores their end portions exert an effect (which is manifested in changes in binding energy) and there are changes in the local pore width in places of their connections. The limited but relatively long slits are represented by three domes (curve *2*). The first dome is connected with the area of the centres whose energy is close to the centres on the edge of the first monolayer in the slit (of the order of Q_1). Note that in the areas of contact between the two crystallites along the perimeter of the three sides of the crystallite there are centres with strong binding energy of the adsorbate (about $1.6Q_1$), and at the contact angles of three crystallites there exist the strongest centres with an energy of $2.4Q_1$ (their share is only noticeable for the narrowest gap – curve *5*). These sites are filled in the first place, but they do not form a dome at small θ, since they correspond to the quasi-one-dimensional structure.

The second dome corresponds to the filling of the central part of the slit, and the third dome is associated with the condensation of molecules in areas of the pores with width L_1 of the monolayers along the ends of the crystallite H_1. Halving L (curves *2* and *4*), as well as the reduction of the width of the crystallite H_1 (curve *3*), retains the type of the phase diagram by changing the quantitative ratios between the widths of the domes. For curve *5* the decrease of H dramatically reduces the critical temperature (if $\tau < 0.4$–0.5, then a transition to the solid state of the adsorbate, which is not taken into account by the model used, is possible).

Figure 54.4 *d* shows that unlike the isolated pores, the system of intersecting pores has a greater number of domes due to the increase in the local volume in areas of their articulation. This situation makes it difficult to determine *a priori* the total number of domes of the phase diagram for a complex porous structure based on the knowledge of the number of areas of the system with centres with very different adsorption properties.

Thus, the overall methodology can be used to study the complete phase diagrams for the vapour–liquid coexistence curves for the adsorbate in a polydisperse system of narrow slit-shaped pores. The relationship of the spatial structure of adsorption centres of different types with the multi-dome form of the phase diagrams was studied. The number of such domes depends on the binding energy, the long-range adsorbate–adsorbent potential, the average characteristic pore size of the system and the share of contributions of the pores with the minimum size.

The presence of the effect of curvature of the domes on the complate phase diagram for the pore system was also confirmed; this effect is associated with the competition in filling pores of different size which determine the overall degree of filling of the pore system at a predetermined external pressure (while the domes in the individual pores do not have such a pronounced curvature).

55. Globular porous systems

Globular system model [20]

It is believed that the molecules enter the free volume formed by the globules, but do not dissolve in the globules [35]. For porous globular systems we introduced the structure of the globules in order to describe the distribution of molecules in the bulk of the porous adsorbent, introduce the lattice for the molecules (as described above). It is assumed that all the globules have the same size (this size was designated R_g in units of λ, where λ is a constant equal to the lattice size λ for the adsorbates). This allows to pack them in a regular cubic structure with six neighbours ($z = 6$). Some part of the globules is removed from the lattice of the globules and then each site in the globular lattice can be occupied or empty. It is usually assumed that the silica gel has a chaotic structure. If φ denotes the proportion of the remote globules in the lattice for the globules, then the probability of finding in the lattice structure for the globules the two closest sites with remote globules is $\gamma = \varphi^2$.

In the general case, we have $\gamma = \varphi\psi$, where ψ is the probability of finding an empty site next to another empty site in the lattice for the globules. Function ψ depends on the specific conditions of formation of the globular structure, in the general case $\psi \neq f = \varphi$. Thus, the structure was modelled by varying the parameters φ and ψ. The maximum size of the fraction $\varphi = \varphi_M$ must be linked with the requirement of mechanical stability of the remaining portion of the globular system.

To calculate the macroscopic properties U for the complete lattice of the globules, we have to weigh any local properties $U(k, n)$ with the aid of the distribution function $F(k, n)$, which determines the probability of formation of the local structure (k, n) on the lattice of the globule, where the indices k and n denote the number of empty sites and their arrangement around the central empty site in the lattice of the globules, respectively. According to [20], the function $F(k, n)$ can be expressed in terms of the added functions φ and ψ. Function $F(k, n)$ is constructed by analogy with the quasichemical approximation for calculating the probabilities of many-particle configurations, expressed in terms of the pair distribution function. As a result

$$U = \sum_{k,n} U(k,n)F(k,n).$$

(55.1)

Equation (55.1) assumes the homogeneous distribution of empty sites across the entire globular lattice without *macroscopic* regions of sites of one type (occupied or empty). The values of φ, ψ and $F(k, n)$ allow us to calculate the structural parameters of porous bodies such as porosity (χ) and the function of the pore size distribution $\left(W_k, \sum_k W_k = 1,0\right)$. The total pore volume V may be expressed as $V = V_r W_1 + \sum_n V_n W_n$, where $n = k + 1$. Number k denotes the type of pores with volume V_k. We confined outselves to the pore size not greater than 7 remote globules. Here, $k = 1$ corresponds to the voids in a regular structure $(V_r = 0.91 \ V_g$, V_g is the globular volume in dimensionless units $R_g/\lambda)$, $k = 2$ corresponds to the isolated empty site in the lattice of the globules $(V_1 = V_g + 8V_r = 8.28V_g)$ and $k = 3–8$ corresponds to the empty central site with $(k - 2)$ removal of the globules in the first coordination sphere of the globular structure (with $V_k = V_1 + 4 (k - 2) V_r)$.

The porosity of the globular systems was evaluated using the following functions $F(k,n) = \varphi\omega(k,n)\psi^k (1-\psi)^{z-k}$, where $\omega(k, n)$ is the statistical weight of the configuration (k, n) of empty sites around the empty central site. Then, the porosity can be calculated by using the formula (55.1)

$$\chi = \chi_r \left[1 + V_g V_{1\varphi} (1 + z\psi)\right],$$

(55.2)

where $\chi_r = 0.476$ is the porosity of the regular globular structure.

To calculate the functions W_k, we must find the probability of such configurations globules which include the given pore. (This condition is limited to the value $\varphi_M = 7/27$, which is less than the percolation threshold for the cubic structure $\varphi = 0.31$). The following functions $F^*(k) = \varphi C_x^{k-2}\psi^{k-2}(1-\psi)^M$ were used for $2 \leq k \leq 8$, where $C_z^k = z!/\left[k!(z-k)!\right]$ is the number of combinations of z elements for k; $m(k) = 6, 10, 13, 15, 16, 17, 18$ for $2 \leq k \leq 8$ respectively. For $k = 1$ the function $F^*(1) = (1 - \varphi) (1 - \psi)^7$ is used. This gives $W_k = F^*(k)\mu$, where μ is defined by the value of the normalization condition $\sum W_k = 1.0$.

The globule size ranged from $R_g = 7\lambda$ to 40λ, to cover the whole spectrum of globular systems. Since the characteristic size of these globules (especially when $R_g > 10\lambda$) significantly exceeds the size

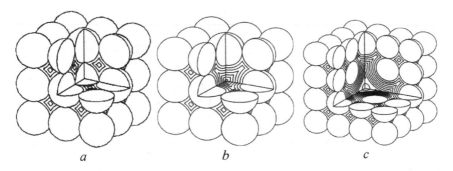

a b c

Fig. 55.1. Regular cubic structure of the globules (a) and defect structures with one (b) and four (c) remote globules.

of the adsorbate, we limited ourselves to the continuous description (in terms of geometric equations) of the share of groups of sites of various types, which determine the location of cells of different types at different distances from the surface of the globules. Figure 55.1 shows the representation of a globular structure with three orthogonal sections passing through the centre of the defects when the regular structure (a) and when one (b) and four (c) (one from the centre and three from the following three cells in a pair of orthogonal directions) globules were removed from the regular structure. Figure 55.1 shows the layered partition of the pores obtained in the globular system from which four globules in the central part of the section were removed. Complex isolines of layer of splitting, which differ significantly from flat monolayers in the slit pore, are clearly visible Sections are show areas corresponding to the isolines of the regular structure of the globules packed in a bulk cubic lattice. Each of the layers is divided into elementary sites with size v_0 from which identical group form.

Figure 55.2 a illustrates the effect of the functions φ and ψ on the porosity of the globular system. Figure 55.2 b shows as the function W_k varies for different φ. In this range φ the share all types of pores with the values $k = 2-8$ is increasing, while the share of the pores the first type ($k = 1$) decreases. (The results in Fig. 55.2 were obtained for the homogeneously distributed remote globules. Under other circumstances other distributions will be considered). The calculations were performed for a wide range of changes of R_g/λ, corresponding to the most commonly used silica gel [22–24]. The curves $1-4$ correspond to the regular structure, the curves $5-8$ – the structure with partially remote globules.

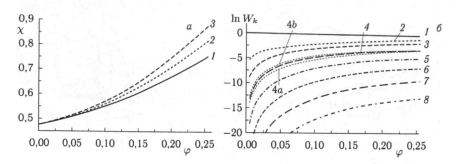

Fig. 55.2. Structural parameters globular system according to the proportion of globules remote φ: a – the porosity of the porous medium χ $\psi/\varphi = 0.7$ (*1*) 1.0 (*2*) 1.3 (*3*) b – pore distribution function W_k when $\psi/\varphi = 1.0$. *4a* and *4b* curves correspond to $k = 4$, when $\psi = 0.7$ φ and $\psi = 1.3$ φ, respectively

Distribution of molecules

It is easily to build a local distribution function f_q and m_{qp} for the equations (55.1), (55.2), where $m_{qp}(1) = zd_{qp}$, for the regular globular lattice without empty sites as well as for the irregular globular structure. Here q is the number of spherical layers around the globule having a width of λ, and wherein $p = q, q \pm 1$.

Then we have

$$f_q = \frac{V_q - V_{q-1}}{V_t - V_0}, 1 \le q \le t,$$ (55.3)

where t is the minimum value that ensures full occupation of the elementary site by spheres with radius $R + q\lambda$, placed in the centre of the closest globules; V_q is the part of the volume which belongs to the spheres of radius $R + q\lambda$; V_t is the full volume of the section (spheres of radius $R + t\lambda$ fully occupy the cell); V_0 is part of the site volume which belongs to the globules (if it exists).

Probabilities d_{qp} were calculated by the method described below. We introduced the definitions for the three types of layer states to describe these probabilities: ideal (simple flat layers), homogeneous (curved spherical layers around a single globule) and nonuniform (intersecting spherical layers in the structure of the globules). Two coefficients were also identified: the coefficient of the spherical curvature of the levels (ξ) and the coefficient of intersection of nonuniform layers (χ): $\xi_q = f_{q+1}^{homo} / f_q^{homo}$, $\chi_q = f_{q+1}^{hetero} / f_q^{hetero}$. Thus, for homogeneous and nonuniform layers ones obtains the following expression for the probabilities d_{qp}:

$$d_{qq-1}^{\text{homo}} = d_{qq-1}^{\text{ideal}}, d_{qq+1}^{\text{homo}} = d_{qq-1}^{\text{ideal}} \cdot \xi_q, d_{qq}^{\text{homo}} = 1 - d_{qq-1}^{\text{ideal}}\left(1+\xi_q\right), \quad (55.4a)$$

and

$$d_{qq-1}^{\text{hetero}} = d_{qq-1}^{\text{ideal}}, d_{qq+1}^{\text{hetero}} = d_{qq-1}^{\text{homo}} \cdot \chi_q, d_{qq}^{\text{hetero}} = 1 - d_{qq-1}^{\text{ideal}} - d_{qq-1}^{\text{ideal}} \cdot \chi_q, \quad (55.4b)$$

where the normalizing condition is satisfied for any q as $d_{qq-1} + d_{qq} + d_{qq+1} = 1$.

The value of the functions f_q and d_{qp} for the system (51.1)–(51.3) throughout the globular system was obtained by weighting functions (55.3) and (55.4) with the functions W_k. Figure 55.3 shows detailed information as the values of f_q are distributed for the globular structure (where $q = 1$ corresponds to the surface layer near the pore wall, $q = 2$ corresponds to the layer above the surface, $q = 3$ corresponds to the third layer and $q = 4$ corresponds to the rest of the 'central' parts of the volume of the pores). With increasing R_g value f_4 increases, whereas other values f_q decrease. This figure uniquely determines that, in accordance with the increase of the size of the globules, when the structure becomes more friable, the volume of the central part of the porous system increases, and the phase diagram should tend to its bulk value.

Calculation of diagrams

When calculating the energy parameters it was assumed that the surface potential influences the first two near-surface monolayers, $R = 1$. The interactions of the adsorbate with the surface is reflected through the parameter Q_1. Then, according to the Mie potential (9–3), we have that the parameter $Q_2 = Q_1/8$, and for the other layers it was assumed that $Q_{k>2} = 0$. The values of Q_1 were estimated [20]

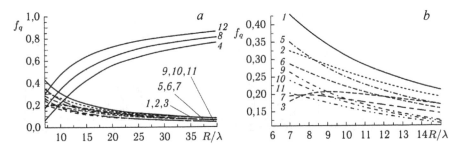

Fig. 55.3. Relationship between the functions f_q ($q = 1$ – curves 1, 5, 9; $q = 2$ – curves 2, 6, 10; $q = 3$ – curves 3, 7, 11; $q = 4$ – 4, 8, 12) and size of the globules when $\varphi = 0$ (1, 2, 3, 4), 0.11 (5, 6, 7, 8), 0.26 (9, 10, 11, 12). Figures a and b correspond to different scales R/λ

from the experimental adsorption isotherms for nitrogen and argon on silica gel [22]. This gave $Q_1 \approx 7\varepsilon$ (nitrogen) and 4ε (argon) (see section 13).

The nitrogen–silica gel system is a case of strong adsorption and the argon–silica an intermediate one. To consider the system of weak adsorption the third ratio was selected as $Q_1 = \varepsilon$ – this option corresponds to the adsorption of various gases in different polymer materials without specific interactions.

Figure 55.4 shows the effect of the restricted geometry of globular systems with different binding energy of the adsorbent–adsorbate system relative to the phase diagram in the bulk (curve 1). Phase diagrams of regular globular systems differ in the values Q_1/ε_{AA}, where Q_1 is the depth of the adsorbate–adsorbent potential, ε_{AA} is the depth of the adsorbate–adsorbate potential (globule size is 30 times larger than the adsorbate size) are shown in Fig. 55.4 a. Curves 1 correspond to the bulk phase of the fluid, 2 – the case of weak adsorption $Q_1/\varepsilon_{AA} \sim 1$, when the binding energy with the surface is of the same order as that between adsorbates, 3 – the case of strong adsorption $Q_1/\varepsilon_{AA} = 7$, when the binding energy of the adsorbate with the surface is seven times higher than their interaction (the silica gel–nitrogen system). The type of surface potential has a strong impact on the location of the critical point (critical temperature and density) and on the number of domes. However, for all cases, the critical temperature in the pores is lower than in the bulk. For weak adsorption phase diagram (curve 2) changes the form but does not contain the additional domes compared with the case of strong adsorption (curve 3). These additional domes are the result

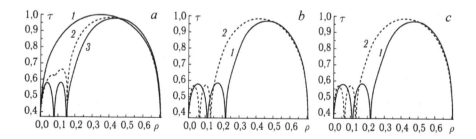

Fig. 55.4. Phase diagrams: a – for the bulk (curve 1) and for pores (curves 2 and 3) at $R_g/\lambda = 30$, $\varphi = 0$, $Q_1/\varepsilon = 1$ (curve 2), $Q_1/\varepsilon = 7$ (curve 3); b – for argon atoms $Q_1/\varepsilon = 4$ when $R_g/\lambda = 20$, $\varphi = 0$ (1) and 0.26 (2), c – for nitrogen molecules. Globular structure parameters as in the field b.

of stratified filling of the pore space in the different layers at a low temperature.

Figures 55.4 *b*, *c* show how the lattice structure for globules affects the type of phase diagrams for argon (*b*) and nitrogen (*c*) for a fixed globule size. In both cases the increase in φ gives an increase of the central dome and a decrease of the width of the domes for the surface and above-surface layers. This is due to changes in the relative number of surface atoms of the solid, which decreases when the value of φ is increased at the same predetermined size of the globules. This change in the Q_1 slightly affects both phase diagrams.

Already the first analysis of globular structures showed (see [24]), that the kind of phase diagrams of vapor–liquid stratification in porous adsorbents with regular and irregular structures is the same as in phase diagrams of other porous systems. It has been found that there are several domes (with stratified phase transitions from one layer to another) in the phase diagram depending on the type of structure. Types of phase diagrams depend on the interaction potential of molecules with the walls of the pores and lateral interactions as well as the relationship between the volumes of layers of pores in which the various contributions of interactions between molecules occur. At low temperatures, there may be two phase transitions that occur for the first and second layers, and the phase transition in the central portion of the pore volume. LGM results can be compared with calculations using the Monte Carlo method [25–29]. They are qualitatively consistent with similar calculations made by the Monte Carlo method [28] (although the biggest difference of the results refers to the type of the phase diagram for the regular structure [28]), but they are more consistent with the results of [29]. However, in both the works [28,29] the globules were smaller ($\sim 7\lambda$), therefore the free cavities which accumulate the adsorbate are sufficiently small for the formation of the phase.

56. Adsorption hysteresis and conditions of its disappearance

All the basic concepts associated with adsorption hysteresis were then introduced for the pores of the simplest geometries (slit-shaped and cylindrical pores) in secton 23 [30]. As a simple example of the existence of adsorption hysteresis in pores with complex geometries Fig. 56.1 shows the phase diagrams (solid lines) for two types of regular spherocylinders $D_s-D_c-L_c$ = 20–8–15 (*a*) and 20–15–15 (*b*)

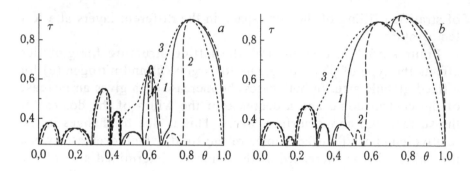

Fig. 56.1. Equilibrium diagrams and spinodal branches in spherocylindrical systems $D_s-D_c-L_c$ = 20–8–15 (*a*) and 20–15–15 (*b*) (z = 6).

(z = 6) [15, 30]. Three numbers mean the sphere diameter, the diameter and length of the cylindrical section of the pores. It is seen that both the near-surface domes as well as a 'central' dome may split into smaller domes (six and two, respectively). Additionally shown are the spinodal branches (dashed and dotted lines), characterizing the metastable areas of the vapour–liquid state. Spinodal charts also have a multi-dome structure. Full adsorption isotherms have the view similar to that discussed earlier in chapter 3 and therefore are not shown (differences appear in the number of secants).

This confirms the above conclusion regarding the qualitative agreement of only phase but also spinodal diagrams for complex porous systems with similar diagrams in the pores of the simplest geometry.

A cavity bounded on three measurements [31]

Chapter 3 investigated the conditions of extinction of adsorption hysteresis with decreasing diameter of the cylindrical pores and the side of the square pore of infinite length. In these examples the systems have the unique characteristic size: the radius or the effective radius for a rectangular cross section; the free space volume along the pore axis was not limited.

A similar situation with the disappearance of the extinction conditions of hysteresis can arise when considering spherical pores, limited in three dimensions, and more complex forms. The concept of the infinitely long channel is a conditional concept in porous bodies and natural expansion of the conditions restricting the occurrence of capillary condensation at the limited length of the channel. In fact, we are talking about an assessment of the limited cavity volume of the pores in which capillary condensation is impossible because of the small number of molecules, filling this cavity.

Spherical cavity

When inscribing a discrete lattice in a sphere with the diameter D/λ, if D/λ – odd, the lattice site fits in the centre of the sphere, if D/λ – even, then the lattice site coincides with the centre of the figure determined by the lattice basis: cube (for $z = 6$), and tetrahedron (for $z = 12$). The position of discrete lattice sites relative to the centre of the sphere is determined by the need to ensure the maximum possible number of planes of symmetry to reduce the number of types of centres with different potential and topology, and to ensure the maximum possible number of sites in the first surface layer. As above, the energy of adsorbate–adsorbent interaction is calculated by the formula (4.14), with the affiliation of the size to the pore determined by geometric feature: if the radius vector from the centre of the pore to the site is less than $D/2\lambda$, the site belongs to the pore.

In the continuum model the criterion that determines the smallest size of a spherical cavity can be written as $[(D - 2n_w)/D]^3 > \varphi^*$, where for argon atoms and nitrogen molecules with $n_w = 2$, when $\varphi^* = 0.5$, the critical dimension is $D^*/\delta = 20$.

The results of calculations of the critical values of the sphere diameter for two discrete lattices and the continuum model are given in Table 7.3 for both values of φ^*.

Cavities in the form of rectangular parallelepipeds

Investigation of cavities in the form of rectangular parallelepipeds is carried analogously to the procedure described for infinite rectangular channels. The highest binding energy in the trihedral angle is $2.36Q_1$.

As above, for a rectangular channel, consider in the continuum description the critical dimensions of the rectangular cavity in the form of a parallelepiped $h \times l \times b$ for a given φ^*. The balance of internal sites is written as

$$hlb\varphi^* = (h - 2n_w)(l - 2n_w)(b - 2n_w) - 4[(h - 2n_w - 2) + (l - 2n_w - 2) + (b - 2n_w - 2)] - 8d,$$

Here $d = 4$ for $z = 6$. Then, fixing h, l and φ^*, we find the values of b, provided that the dimensions h and l are commensurate with the minimum dimensions.

The results of calculations of the critical size of the cubic cavity are also given in Table 7.3 for both values of φ^*.

Critical length of the section of the channel

The quasi-dimensionality criterion φ is the ratio of the volume of the central and near-surface regions of the infinite channel of a constant cross section. In general, the question naturally arises about the possibility of removing the conditions of the infinite length of the channel, because the real pores are limited in extent. In this case, we should talk about the minimum length of the channel, which approximates the behaviour of an infinitely long channel in the sense that the cooperative behaviour of molecules inside the final section, defined in this section, and in the infinite channel are close enough to each other.

This formulation of the problem corresponds to the condition of the limitation in the third direction, similar to the cavities discussed above. We will construct an approximation of the cooperative behaviour of molecules in an infinite channel with the channel of the same size, but limited in the length of the area under consideration, starting from the notions of the usual mechanism of the growth process of the new phase in the condensation of molecules via the formation of a nucleus of this phase.

We define some region of the long channel and consider the contribution made by two of its borders with increase of the size of the region. For any constant section (S) the volume fraction ($2S\delta$), introduced by two boundaries in the general (SL) volume of the system is equal to $\delta_1 = 2/L$, where L − length of the channel, measured in monolayers. We assume that the channel in question approximates an infinitely long channel if $\delta_1 \leq \delta_1^*$ where δ_1^* − the numeric criterion responsible for the implementation of this approximation. The meaning of δ_1^* is completely analogous to the value of $1/D^*$, where D^* is the critical dimension of the channel in which the conditions of quasi-one-dimensional fluid behaviour in an infinitely long channel are violated and capillary condensation occurs. Therefore, as a minimum estimate to determine the critical value we assume that $\delta_1^* = 1/D^*$ then the minimum critical channel length is $L^* = 2D^*$.

Exactly the same estimate of the values δ_1^* and L^* is obtained in the channel of a constant cross section of different geometry, for example with a rectangular cross section, which corresponds to a particular pair of values of $l(h)$, linked together in a continuous or discrete description.

Comparison of the volumes belonging to the interiors with fixed quantity ϕ, for a sphere with diameter D_s and a cylinder, having a length L^* and the diameter D_c^* gives a continuum model $[(D_s - 2n_w)/2]^3 \, 4\pi/3 = 2L^*[(D_c - 2n_w)/2]^2$, where $D_c^* = 14$. Hence, for the argon atoms and nitrogen molecules $n_w = 2$ we find that the volume of a sphere equivalent in the number of internal sites to the critical length cylinder, where capillary condensation is possible, is $D_s = 20\delta$. This value coincides with the above estimate of the critical diameter of the spherical and cubic cavities $D^* = 20\delta$ in the same continuum model. For the discrete model $D_c = 13\delta$ and an equivalent volume of a sphere the value is similar $D_s = 18\delta$.

From Table 7.3, which contains critical pore sizes with various geometry, it implies that the comparison of similar porous systems – infinite cylindrical and rectangular channels (Table 3.4), as well as spherical and cubic voids demonstrates that the critical dimensions are similar.

Table 7.3 also shows that all estimates for the total number (or volume) of molecules in the areas of the critical size and different geometry at least are consistent and in qualitative agreement with each other. The qualitative agreement between the estimates of the critical pore volumes with different geometries (cavities, channels and flat 'two-dimensional' section) allows one to define the lower boundary of the implementation of the process of capillary condensation of molecules responsible for the appearance of hysteresis loops on the adsorption and desorption isotherms. When the critical volumes are reached the cooperative properties of molecules, providing fluid stratification into rarefied and dense phases become evident.

With the exception of the model of infinite slit-shaped pores (with the only restriction on the width of the pores), in most real and model porous systems we have to deal with limited pore sizes in two

Table 7.3. Minimum dimensions for spherical and cubic pores

z	n_w	Spherical cavity				Cubic pore			
		I		II		I		II	
		A	B	A	B	A	B	A	B
6	1	9^*	10^*	9	10	11	12	11	12
	2	18	20	17	20	18	21	18	21
	3	25	30	26	30	27	30	27	30
12	1	8^*	10^*	9	10	11	12	11	12
	2	17	20	17	20	18	21	18	21
	3	26	29	26	30	27	30	27	30

or three directions. The restriction in two characteristic dimensions is realized for infinitely long isolated pores of the same (cylindrical and rectangular) and variable (spherocylindrical, rectangular channels with the broadening and narrowing, globules with a regular structure) sections. The restriction is three characteristic dimensions is realized for most real porous systems.

Minimum dimensions in globular systems [32]

To analyze the conditions for the absence of capillary hysteresis we consider two cubic structures with the number of the nearest neighbours $z = 6$ (simple) and 12 (face-centered). The structure with $z = 12$ is oriented so as to provide the maximum number of sites that are near the wall pores. In Fig. 55.1 a is the diagram of the globular pores and the location of the surface layers with different binding energies. Figure 56.2 shows a diagram of the diagonal section of the pore space. The diameter of the circles D, inscribed in a continuous pore in planes passing along the normal to the axis of the open pores, periodically changes with the motion along the axis of the pore. The minimum diameter is denoted by $D_{min} = 2R(2^{1/2} - 1)$, it corresponds to the maximum narrowing of the channel, and the maximum size of the cross section of the continuous pore is denoted by D_{max} – it is in the centre of the intersection of the open pores. Distance along the axis of the pores between the centers of the circles minimum and maximum size is equal to R.

Consider the function of the dependnce of the diameter $D(l) = 2R((l^2 + 2)^{1/2} - 1)$, where $l = L/R$, L – the length of the section of minimum diameter. To compare the cavities of different shapes and variable cross-section, we enter the following average characteristics: average diameter D_{aver} of the cross section of the continuous section

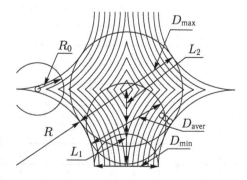

Fig. 56.2. Diagonal cross-section of the pore space at $D_{max}/\delta = 22$.

of regular packing and the linear size along the pore axis defining an area of less than $(D_{min} \leq D \leq D_{aver})$ and more than $(D_{aver} \leq D \leq D_{max})$ the average diameter of the continuous section, which is defined as $D_{aver} = \int_0^1 D(l)\,dl \approx 1.049R$. Then the first portion can be attributed to the field of connecting channels (its relative length is denoted by L_1, where $D(L_1/R) = D_{aver}$; $L_1/R \approx 0.57$), and the second – L_2 to the area intersections $L_1 + L_2 = R$.

All internal components away from the walls of the pores at distances when the binding energies $Q_{q,f}$ are about Q_q^* and less are treated as central sites of the pore for which $Q_q \sim 0$. The boundary value of Q_q^*, accounted in the calculations, is equal to $\gamma\varepsilon$, where $\gamma = 0.05$.

Analysis of the critical pore size, providing the appearance of capillary condensation, is associated with the definition of the conditions under which the volume fraction of the inner part of the pores is $\varphi > \varphi^*$, where φ^* is the critical value. The correlation of the properties of phase diagrams with possible variants of the criteria for the quantities φ showed [33] that, according to experimental data [34–36], the critical diameter corresponds to $\varphi^* = 0.5$, whereas the data [37] indicate a 'softer' condition $\varphi^* = 0.444$.

Analysis of surface areas in globular pores

The analysis technique of the minimum pore size is described in detail in chapter 3. For the given adsorbate–adsorbent interaction potential Q_1 all sites of the pore are divided to internal and near-wall pores so that $\varphi > \varphi^*$. The length of the wall potential determines the numer of subsurface domes n_w, which can be implemented at low temperature in the adsorption system. For argon atoms $n_w = 2$.

In the sites near two adjacent globules the increase in energy is largest. The sites are located in the first layer of adjacent globules, for $D_{max}/\delta = 15$ and 30 (where δ – a monolayer, for $z = 6$, $\delta = \lambda$, for $z = 12$, $\delta = \lambda\,(2/3)^{1/2}$) have energies $1.31Q_1$ and $1.58\,Q_1$, respectively. In the sites situated on the convex surface of the globule outside the region of influence of the surface of other globules (at the distance over 5λ), the value Q_1 is less than the energy on a flat uniform surface. When increasing the size of the globule the energy in the first layer tends to Q_1 (for pores with $D_{max}/\delta = 15$ and 30 the energy in the first layer is equal to $0.99Q_1$ and Q_1, respectively).

Thus, unlike in the cylindrical channel, the sites of the first few monolayers are energetically nonuniform. The assignment of sites

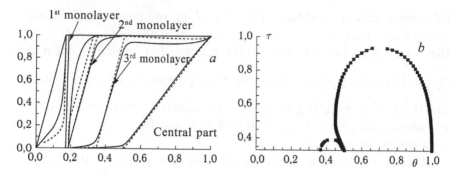

Fig. 56.3. Local filling in the pore at regular packing of globules (a): $D_{max}/\delta = 25.62$, $z = 6$, $n_w = 3$ at $\tau = 0.36$ (dashed line), 0.52 (solid lines). Phase diagram in the pore at regular packing of globules (b): $D_{max}/\delta = 25.62$, $z = 6$, $n_w = 3$.

of different types to a particular dome is made on the basis of the analysis of local adsorption isotherms. This allows one to select only those sites that are responsible for the formation of the central dome that corresponds to the actual capillary condensation. The basis of the methodology for determining the share of the central sites is the consideration of the binding energy of the adsorbate which depends on the distance from all walls of the pores, and the cooperative behaviour of the adsorbate in the central part pore outside the region of the effect of the surface potential of the walls.

Figure 56.3 a shows local isotherms θ_q for a local regular structure at $\tau = T/T_c = 0.36$ and 0.52 in dependence on the complete filling of the pore θ, T_c is the critical temperature in the bulk phase, $z = 6$. These curves show what types and how many sites contribute to the value θ. The first group of the curves refers to the angular sites having energy $Q_q \approx Q_1$, and to the sites in the first monolayer with energies $Q_q > Q_1$. Increase of the degree of filling of the pore takes place in the following order. The sites with the maximum binding energy Q_q are first to be filled. The next group of sites refers to the sites of the second monolayer. Finally, the remaining curves represent the local filling of the central sites in which the capillary condensation occurs.

At temperature $\tau = 0.52$ the redistribution of the adsorbate between the partially filled third layer and the central part of the pores takes place. This manifests itself in the phase diagram in the form of an extension of the central dome and 'creeping' on the dome of the adjacent surface dome formed by the third monolayer. At the temperature $\tau = 0.36$ the monolayers are sequentially filled, which leads to the formation of n_w subsurface domes. Clear stratifying of

Table 7.4. Characteristic pore sizes in regular globular structures (in units of δ); I – implementation of the critical criterion $\varphi^* = 0.444$, II – similar for the criterion $\varphi^* = 0.5$

z	n_w	Continuum model D_{max}/δ		Discrete model D_{max}/δ	
		I	II	I	II
6	1	8.8	8.8	10.6	11.4
	2	16.1	17.6	17.8	20.0
	3	23.4	26.4	23.4	27.1
12	1	8.8	8.8	8.2	8.5
	2	16.1	17.6	16.8	20.1
	3	23.4	26.4	23.2	26.6

the monolayers at a given temperature allows to estimate the share of the central part of the pore. The phase diagram for the considered globular system is shown in Fig. 56.3 *b*.

In addition to the discrete model of the globular pores the continuum model was considered in [32]. The volume of the κ-th layer is defined as the volume bounded by the surfaces of the point that are distant from the nearest globule at a distance $(\kappa - 1)\delta$ and $\kappa\delta$. This model takes into account only the specifics of sites on a curved surface having a lower binding energy Q_q compared with Q_1, and does not consider the enhancement of energy in the area of the effect of two adjacent globules.

Table 7.4 gives the values of the critical dimensions D_{max}/δ, which may cause capillary condensation in the regular structure of the globules for different adsorbent–adsorbate interaction potentials responsible for the number of surface domes n_w. The increase of n_w to three occurs when the globule size and Q_1 increase. The decrease of n_w to unity occurs at any pore size with decreasing Q_1. An important role in the discrete model begins to be played by the character of distribution of different types of sites within the first surface monolayer and also away from the surface.

For $n_w = 1$ with increasing θ situations may arise with large molecular rearrangements between sites of various types. Consequently, the shape of the first dome is deformed and a clear boundary between the domes disappears. The reason for this behaviour of local isotherms is the strong heterogeneity of the surface region for which it is well known that the number of domes in the phase diagrams can vary. Binding to the number of types of domes which close to the pore walls are the conditional characteristic of the system is made for clarity and convenience of assessment. The main physical factor of the system is the surface potential and the extent of its impact on the cooperative behaviour of the adsorbate, since the equilibrium capillary hysteresis is associated with only intermolecular interactions.

The obtained results allow us to formulate general conclusions for narrow-pore systems consisting of many different areas of pores that have restrictions on the cross-section and length of the comparable size:

1) in porous systems there should be no capillary condensation if the characteristic size of all the pores of the system is less than the critical value of D^*;

2) capillary condensation occurs at all pores, if their characteristic size is greater than critical;

3) in the intermediate case when a proportion of the pores have the characteristic pore size greater than the critical size D^*, the capillary condensation may occur only in the part of the system, wherein the characteristic length exceeds a certain critical length L^*, which can be taken for long enough pores; if for part of the porous system we have $D > D^*$, but $L < L^*$, then within this region there is no capillary condensation.

It follows from this analysis that for rectangular channels and parallelepipeds, which simulate in the real conditions the pores of finite size in activated carbons, with the *minimum* characteristic size of less than 6.0–7.5 nm, the appearance of hysteresis loops is not possible. These estimates are in good qualitative agreement with experimental data [22, 38–40] for the adsorption isotherms of nitrogen and argon for active carbons of different origin. Analysis revealed that they usually do not contain hysteresis loop and the capillary condensation phenomenon does not occur there. The difficulty of achieving the minimum pore size of 6–8 nm is due to the methods of formation of most active carbons. In cases where such loops appear, they relate to higher pressures $P/P_0 > 0.95$ (in which accurate measurements are very difficult) and the proportion of larger pores is small.

57. Polydisperse system and adsorption porosimetry

Most porous materials are polydispersed, so the general form of diagrams for systems consisting of pores of different sizes is

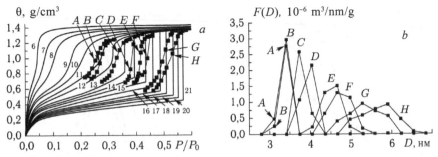

Fig. 57.1. The adsorption isotherms for isolated cylindrical pores and experimental data for the adsorption of argon in pores with different average diameters at $T = 87.3$ K (a): A, F – [37], B, C – [46], D – [47], E – [35], G, H – [36]. Functions of pore size distribution, corresponding to the experimental data, are shown in Fig. a (b).

important. Most often the experimental data on the adsorption porosimetry [22, 41], based on a comparison of the full adsorption isotherm of the system with the isotherms relating to the individual pores, are treated as a set of contributions from independent pores of different sizes, in the form

$$\theta_{exp}(P/P_0) = \int_{D_{min}}^{D_{max}} F(D)\,\theta(D \,|\, P/P_0)\,dD, \qquad (57.1)$$

where D_{min} and D_{max} are the minimum and maximum sizes of the range of the characteristic pore size under consideration; $F(D)$ is the pore size distribution function, P_0 is the saturated vapour pressure.

Figure 57.1 and 57.2 are examples of using the expression (57.1) for the evaluation of the pore size distribution function. Figure 57.1, a shows the desorption branches of the adsorption isotherms of argon atoms in the material of the MCM-41 type for isolated cylindrical pores of different diameters from 6 to 21 monolayers. There are also given the experimental data for the pores with the mean values of different diameters: A [37], B [46], C [46], D [47], E [35], F [37], G [36], H [36] at $T = 87.3$ K. Figure 57.1 b shows the determined pore size distribution functions corresponding to the experimental data. These calculations explain the slope of the experimental curves in Fig. 24.1 compared to the sharp jumps in density in the case of the pores with the monodisperse distribution.

Figure 57.2 a shows the temperature dependences of adsorption of argon atoms in the MCM-41 material for the mean size of 4.4 nm at $T = 87.3$ K (1) [36], 94 K (2) [46] K and 99.9 (3) D [47] (lines – calculations, symbols – experimental data) and the isotherms for isolated cylindrical pores. For clarity, the curves are shifted relative to each other by 0.2 along the axis of relative pressures P/P_0. The

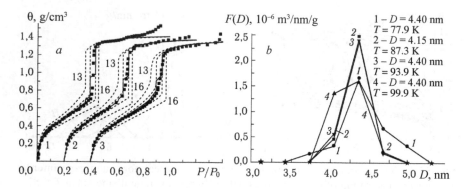

Fig. 57.2. Temperature dependence of the desorption branch of the isotherm average pore diameter $D \sim 4.4$ nm (*a*): $T = 87.3$ K (*1*) [37], 94 K (*2*) [46] K and 99.9 (*3*) D [47] (lines) and experimental data (symbols) and isotherms for isolated pores (dotted line). Pore size distribution function corresponding to the experimental curves on the *a* (*b*).

field in Fig. 57.2 *b* shows the corresponding pore size distribution function (the samples, obtained by different authors, differ markedly in the structure). Further to the field *a* the curve *1* corresponds to the temperature below the melting point of the bulk phase.

The simplest examples of the special features of the phase diagrams of polydisperse systems were shown above on model examples for cylindrical and slit-shaped pores (sections 52–54). The main features are: 1) multi-dome form of the complete stratificatioin curves associated with the difference in the characteristic areas where condensation of molecules is possible (including consideration of the surface regions with different molecule–wall interaction energy), 2) bending of the domes on the complete phase diagram of the system that is the result of the joint effect of filling some aggregate of pores (instead of a single pore) and the change in the total degree of filling of the pore system with a fixed external pressure, 3) presence of intersections and junctions of pores of different sizes and sections, which are inevitable at the limited length of individual sections of the pores, which can be approximated by the form of simple geometric channels, which lead to the appearance of additional domes compared with the set of domes for the individual pores of simple geometry (Figs. 52.4 and 54.4).

Assessing the possibility of determining the function
characterizing the connectivity of pores different sizes from
adsorption measurements

Above we stressed the importance of considering connections and the effect of adjacent areas on the character of filling the considered type of the section of the pore system. In this connection it is important to address the question of the problem of determining the fraction of the transitions between the pores of different size and type according to the adsorption measurements. Conventionally, the task can be designated as a search for the distribution function $F(D, H)$, wherein the symbol D refers to the diameter of cylindrical pores, and the symbol H to the slit pore width. The symbols D, H formally indicate the difference in the types of neighbouring pores. The distribution function $F(D, H)$ has a two-dimensional character, as it depends on two different structural parameters of the system. The solution of the function $F(D, H)$ is of fundamental importance for describing the dynamics of transport processes in porous bodies, carried out through a sequence of connected pores [43–45].

This question is directly related to the essence of the procedure of adsorption measurements and interpretation of the data on the adsorption–desorption hysteresis. As such concrete examples we consider two typical situations for the adsorption of argon in the porous material with SiO_2 functional groups (the parameters are given in [15, 32]) containing connections of various types of pores.

Figure 57.3, *a* shows the high temperature section of the phase diagram of argon in a spherocylindrical porous SiO_2-based material with a diameter of the spherical portion $D_s = 20\lambda$, the cylindrical portion $D_c = 15\lambda$ and the length of the cylindrical part $L_c/\lambda = 20$ (*1*) 50 (*2*) and 80 (*3*) (diameters D_s and D_c are in units of λ, λ is the constant of the adsorbate lattice structure, L_c is the length of the cylinder connecting truncated spheres in lattice structure with the number of neighbours $z = 6$) [10, 11]. The temperature on the ordinate axis is given in units $(\beta\varepsilon)^{-1}$, where ε is the interaction energy of argon atoms, equal to 238 cal/mol, $\beta = (RT)^{-1}$. The abscissa gives the degree of filling of the pore system θ, normalized for the total volume of the pore space.

The phase diagram shows two domes that correspond to capillary condensation in the central part of the cylindrical channel (left dome) and in the centre of the spherical part (right dome). The presence of connections increases the number of domes of the phase diagram in addition to the multi-dome structure of the phase diagram for isolated

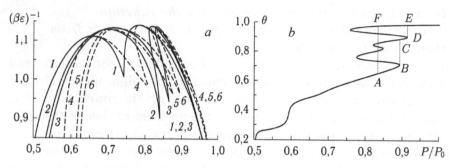

Fig. 57.3. Part of the phase diagram for the spherocylindrical system $D_s-D_c-L_c = 20-15-L_c$ (a): $L_c/\lambda = 20$ (1) 50 (2) and 80 (3), spinodal analogues of diagrams for the adsorption branch of the isotherm adsorption (curves 4–6, respectively). The adsorption isotherm of argon in the spherocylindrical porous SiO_2-based material with sizes $D_s-D_c-L_c = 11-13-20$ at $T = 92$ R (b); here P_0 is the saturated vapour pressure, $P(B)$ is the vapor pressure corresponding to the point B.

areas of pores. This fact is the first sign indicating the presence of pore connections.

The second feature of the presence of connections is demonstrated in Fig. 57.3 b. In [15, 32] it was shown that the pore connection increase the number of domes on the adsorption spinodal diagram compared with the number of domes on the equilibrium phase diagram. As a result, the adsorption isotherms form in which the adsorption branch passes through lines AB, BC, CD and DE, while the desorption branch coincides with the equilibrium density jump AF. The hysteresis loop width, equal to $P(E)/P_0 - P(F)/P_0$, increases (here by about 36%) compared with the case of the cylindrical monodisperse system $P(B)/P_0 - P(A)/P_0$. Thus, the presence of pore connections then increases the width of the adsorption isotherm branch.

Both examples point to the possibility in principle to distinguish between the pore systems with isolated areas of the same type of pores from the systems with non-isolated pores of different types. The degree of severity of both features depends on the temperature. If the temperature region is lower but close to the critical temperature, the individual adsorption isotherms may show the presence of two density jumps (first feature). This temperature range in the case of spherocylindrical pores is estimated to be approximately 15–20 % of the critical temperature for the maximum dome. The second sign of the connections of the pores is then observed over a wider temperature range.

Given both signs, we can formulate the following procedure to find the contributions from the connections between the pores of different sizes. For a system having a complex pore structure but having the periodicity (regular globular or spherocylindrical) it is enough to have measurements of the adsorption isotherms at different temperatures, including the near-critical temperature range, in order to observe two density jumps. In a polydisperse material we should analyze the difference in the angles of adsorption and desorption branches of the adsorption isotherms, which in their central parts show a linear behaviour for many experimental data.

The experimentally observed degree of filling of the adsorbent $\theta_{exp}(P/P_0)$ is expressed through the local filling of the sites of type f of the section of pores with the characteristic pore size q in the form

$$\theta_{exp}(P/P_0) = \int_{q_{min}}^{q_{max}} \sum_{q,f} F_f(q)\theta_f(q|P/P_0)dq, \qquad (57.2)$$

where q_{min} and q_{max} is the minimum and maximum sizes of the considered range of the characteristic pore size, $\theta_f(q|P/P_0)$ is the local isotherm, which determines the degree of filling of the sites of type f in the area with the pores of type q.

Since the width of the hysteresis loop $\Delta P/P_0$ for isolated pores increases monotonically with increasing D, H [30], then, using as a first step the value of the function $F(D)$ for the desorption branch, we can obtain the dependence for the adsorption branch of the isotherm $\theta_{ad}(\Delta P/P_0)$ by the formula

$$\theta_{ad}(\Delta P/P_0) = \int_{D_{min}}^{D_{max}} F(D)\theta(D|\Delta P/P_0)dD, \qquad (57.3)$$

where the value is $\Delta P/P_0$ counted relative to the desorption branch of the isotherm. Comparison of the experimental and theoretical dependences $\theta_{ad}(\Delta P/P_0)$ gives an estimate of the contribution to the connections: if $\theta_{ad}(\text{exp.}) \le \theta_{ad}(\text{theor.})$, formula (57.3) is satisfactory and the contribution of the connections is small. If we have the opposite sign relations $\theta_{ad}(\text{exp.}) > \theta_{ad}(\text{theor.})$, the contribution of the connections is large enough and the structural model of the system must include the connections between the pores. In this case, the total adsorption isotherm and the phase diagram are calculated using the formula (57.1). To calculate the adsorption branch considering connections, instead of equation (57.3) we must use the following formula:

$$\theta_{ad}\left(\Delta P/P_0\right)=\int\limits_{q_{min}}^{q_{max}}\sum_{q,f}F_f\left(q\right)\theta_f\left(q\mid\Delta P/P_0\right)dq.\qquad(57.4)$$

The ability to define a function that characterizes the coupling of the pores of different sizes and geometry, extends the capabilities of adsorption porosimetry. The system of the linked sites generates a single space of pores, and the nature of local distributions of adsorbate in any its part affects the distribution of molecules in neighbouring areas of the pores. Accounting for the contributions from regions of joining of the pores of different types to the total adsorption isotherm enables to link the structure of porous materials and the type of hysteresis curves, which are now classified by IUPAC as four types of hysteresis curves [22] (see Fig. 2.8).

58. Dynamic properties

Distribution functions
Consider the procedure of averaging of the local characteristics over the cross section of the sample [1]. The previously obtained expressions for the transfer coefficient (self-diffusion, shear viscosity and thermal conductivity) in isolated pores (section 35) formally retain their shape records. In the transition from simple pore geometry to complex systems for these factors it must be considered that the designation of the sites includes two indexes, each of which describes the type of area of the pore system and the cell number in this area, as in the expressions for the equilibrium distribution (51.1).

All the above distribution functions of different parts of the pores apply to the whole grain. To analyze the molecular transport it is necessary to define the direction of flow from one edge of the grain to the other. The coordinate along the flow is denoted by x. The grain boundaries correspond to X_1 and X_2, respectively. Then the function $F_q(x)$ will characterize the fraction of portions of type q in the cross section with coordinate x in the direction along the flow of molecules: $\int_1^2 F_q(x)dx=F_q$ (integration region $X_1\le x\le X_2$ denoted by subscripts 1 and 2), where $dx=\Delta x\sim L$. The minimum step on the supramolecular level is the value of the lattice parameter of the supramolecular structure. Respectively, instead of the pair correlation functions F_{qp} it is necessary to use average $F_{qp}(x,y)$ in the neighbouring sections x and $y=x\pm 1$, where also

$|x - y| \sim L$. We define the conditional probability $H_{qp}(x, y) = F_{qp}(x, y)/$
$F_q(x)$. The normalizing relations $\int_1^2 F_{qp}(x,y)\,dx\,dy = F_{qp}$ and $\sum_{p=1}^T H_{qp}(x,y) = 1$
are satisfied for the introduced functions.

For dynamic problems we should also extend the definition
of distribution functions at the molecular level [48, 49]: $F_{q,f}(x)$
describes the proportion of sites of type f in the area q in the cross
section x, and $F_{qp,fg}(r|xy)$ – the proportion of adjacent pairs of sites
q, f in section x and p, g in section y at distance r, with obvious
normalizing connections $\int_1^2 F_{q,f}(x)\,dx = F_{q,f}$ and $\int_1^2 F_{qp,fg}(r|xy)\,dx\,dy = F_{qp,fg}(r)$.
In particular, $H_{qp,fg}(r|xy) = F_{qp,fg}(r|xy)/F_{q,f}(x)$ describes the conditional
probability of finding the site p, g in section y at distance r from a
selected site q, f in section x. They satisfy the following relations:

$$\int_1^2 H_{qp,fg}(r|xy)\,dx\,dy = H_{qp,fg}(r) \text{ and } \sum_{g=1}^{t(p)} H_{qp,fg}(r|xy) = H_{qp}(xy).$$

Local dynamic characteristics

The most important coefficients for describing the flow of molecules
of one species are the coefficients of self-diffusion, shear viscosity
and thermal conductivity. The joint effect of the adsorption potential
of the pore walls and the intermolecular interaction creates the
equilibrium distribution of the molecules strongly nonuniform over
the cross section of the pores. In calculating these coefficients it is
assumed that the equilibrium distributions are weakly perturbed, so
to find them we use solutions of equations (51.1).

The equations for the self-diffusion coefficients in nonuniform
media were obtained earlier [5, 48, 49]. The local self-diffusion
coefficient characterizes the process of redistribution of molecules
between adjacent cells. It is calculated by the following expression:

$$D^*_{qp,fg}(\rho) = \rho^2 z^*_{qp,fg}(\rho) U_{qp,fg}(\rho)/\theta_{q,f}, \qquad (58.1)$$

wherein $z^*_{qp,fg}(\rho)$ – the number of possible jumps from site q, f to
the site p, g over the distance ρ. Here $U_{qp,fg}(\rho) = K_{qp,fg}(\rho) V_{qp,fg}(\rho)$
is the flow (the speed of hopping of the molecules per unit time)
from the site f to site g at a distance ρ, characterising the speed
of movement of the molecules in a given direction; $K_{qp,fg}(\rho)$ is the
speed constant of jumps, $V_{qp,fg}(\rho)$ is the concentration component
of the speed of jumps. This component depends on the ratio of the
interaction energy of molecules in the ground $\varepsilon_{AA}(r)$ and transition
$\varepsilon^*_{AA}(r)$ states, according to conventional notions of the theory of

absolute reaction rates. Expressions $U_{qp,fg}(\rho)$, built in the framework of the theory of absolute reaction rates in the condensed phases are described in detail in several papers [5, 48, 49]. When calculating the thermal conductivity and shear viscosity it is also necessary to use the expression of the local self-diffusion coefficient (58.1).

The local thermal conductivity coefficient at the energy transfer of molecules over the distance ρ in two channels (migration and collisional) can be written as follows:

$$K_{qp,fg}(\rho)=\theta_{q,f}C_v(q,f)\{D^*_{qp,fg}(\rho)+t^{AA}_{qp,fg}(1)\lambda^2 v_{qp,fg}/3\}, \quad (58.2)$$

where the self-diffusion coefficient $D^*_{qp,fg}(\rho)$ is defined above in formula (58.1), $C_v(q,f)$ is the expression for the specific heat given in section 17, in which the index f is replaced by q, f.

Within the framework of the generalized Eyring model the expression for the local shear viscosity coefficient in shear of a fluid in the cell g relative to the cell f over the distance ρ has the form

$$\eta_{qp,fg} = z^*_{qp,fg}(\rho)\eta_0 \frac{\exp(\beta E_{qp,fg}(\rho))}{V_{qp,fg}(\rho)}, \quad (58.3)$$

where $\eta_0 = (mkT/\pi)^{1/2}/(\pi\sigma^2)$ is the ideal dilute gas viscosity, m is the mass of the atom, σ is the diameter of the molecule, $E_{qp,fg}(\rho)$ is the activation energy of the molecule hopping between sites q, f and p, g.

Average dynamic characteristics
Averaging the dynamic characteristics over any cross section x, $X_1 \leq x \leq X_2$, is carried out by means of the distribution functions for sites of pores of different types $F_q(x)$ and their pairs $F_{qp}(x, y)$. This averaging is preceded by averaging over the site types f and their pairs fg using the functions $F_{q,f}(x)$ and $F_{qp,fg}(r|xy)$ along the section x within each pore.

As the average characteristics of the motion of labelled molecules along the pore axis we can use the following expression for the coefficient of self-diffusion:

$$D^*_{aver} = \sum_{\rho}\sum_{q}F_q(x)\sum_{f=1}^{t(q)}F_{q,f}(x)\sum_{p,g=1}^{t(p)}H_{qp,fg}(\rho\,|\,xy)\times$$

$$\times\cos(x\,|\,qp,fg)D^*_{qp,fg}(\rho)\frac{d\theta^*_{q,f}}{d\theta^*}, \quad (58.4)$$

where $\cos(x|qp, fg)$ is the cosine of the angle between the direction of flow x and the direction of connecting site q, f and p, g in adjacent sections qp. Formula (58.4) reflects the thermal motion of the labelled molecules inside the pore space at the impermeability of the pore walls (i.e., no dissolution of the molecules in the pore walls); in the state of equilibrium distribution, the relation $d\theta^*_{q,f}/d\theta^* = d\theta_{q,f}/d\theta$ is satisfied. Local constants of hopping of the molecules and local Henry constants are related by the formula $a_{qf} K_{qp,fg} = a_{p,g} K_{pq, gf}$.

Unlike the transport of the molecules, the average coefficients of thermal conductivity of the adsorbate may be introduced both along $(\kappa_\parallel)$, and normal $(\kappa_\perp)$ to the macroscopic flow of molecules. These coefficients are described by the formulas:

$$\kappa_\parallel = \sum_\rho \sum_q F_q(x) \sum_{f=1}^{t(q)} F_{q,f}(x) \sum_{p,g=1}^{t(p)} H_{qp,fg}(\rho|xy)\cos(x|qp, fg)\kappa_{qp,fg}(\rho),$$
(58.5)

$$\kappa_\perp = \sum_p \sum_q F_q(x) \sum_{f=1}^{t(q)} F_{q,f}(x) \sum_{p,g=1}^{t(p)} H_{qp,fg}(\rho|xy)\sin(x|qp, fg)\kappa_{qp,fg}(\rho).$$
(58.6)

Traditionally, in the narrow pores the average hydrodynamic flow rate v, by analogy with macrochannels, calculated by the Darcy's equation [43, 44, 50]:

$$v = -K_1 \mathrm{grad}_x P/\eta,$$
(58.7)

where K_1 is the permeability coefficient (dimension m²), $\mathrm{grad}_x P$ is the pressure gradient in the direction of x; and the mean volumetric flow rate of gas or liquid is calculated by the Hagen–Poiseuille equation:

$$Q = K_2 R^n \left(1 + md_{slip}\right)/\eta, K_2 = K_3 \mathrm{grad}_x P,$$
(58.8)

where R is the radius of the channel, K_3, n and M are the numeric constants that depend on the shape of the channel: for a cylinder $m = n = 4$ and $K_3 = \pi/8$, for an extended slit having inlet width H and length W, we have $R = H/2$, $m = 1$, $n = 3$, $K_3 = 2W/3$. In describing the transfer of momentum in the equation of Darcy's law (58.7) for the porous system and in the Hagen–Poiseuille equation for a single pore (58.8) there is a constant viscosity value η of the liquid or gas.

Calculations based on the LGM for isolated pores of different geometry showed that all transport coefficients are functions of the distance from the pore walls and of the direction of flow of molecules (i.e., have a tensor character). However, in the derivation of equations (58.7) and (58.8) the constancy of η is crucial. Therefore, to be able to compare the results obtained for the nonuniform fluid flow with the conclusions of the Darcy's and Hagen–Poiseuille equations, we must enter the average shear viscosity coefficient along the selected direction of flow in the grain. This can be done in two ways. We can formally perform averaging of (58.4) over the cross section of the pore provided impermeability of the pore walls. Then, by analogy with the formula (58.4), we obtain

$$\eta_{\text{aver}} = \sum_p \sum_q F_q(x) \sum_{f=1}^{t(q)} F_{q,f}(x) \sum_{p,g=1}^{t(p)} H_{qp,fg}(\rho \mid xy) \cos(x \mid qp, fg) \eta_{qp,fg}(\rho).$$
$$(58.9a)$$

We can additionally take into account that the density profile of the adsorbate is variable over the cross section of the pores (unlike the assumptions made in the derivation of the Hagen–Poiseuille equation) and then we obtain the following expression:

$$\eta_{\text{aver}} = \sum_p \sum_q F_q(x) \sum_{f=1}^{t(q)} F_{q,f}(x) \sum_{p,g=1}^{t(p)} H_{qp,fg}(\rho \mid xy) \cos(x \mid qp, fg) \eta_{qp,fg}(\rho) \frac{\theta_{q,f}}{\theta_q},$$
$$(58.9b)$$

where the factor $\theta_{q,f}/\theta_q$ renormalizes the contribution of each layer f of the pore in section q.

These determinations are used below to calculate average values for the dynamic characteristics of the individual pores (i.e. weighting of the section of the pore in the monodispersed pore system) and then for the system of slit pores.

Contributions of the individual pores

For certain types of pores the average dynamic coefficients (58.4)–(58.6), (58.9) can be obtained by weighting the contributions over the cross section. Self-diffusion coefficients (58.4) for the individual slit and cylindrical pores of different sizes were considered in section 30. Below are the average coefficients of thermal conductivity (58.5), (58.6) and shear viscosity (58.9) for slit-shaped pores of different widths.

The average thermal conductivity coefficients characterize the energy transfer of the molecules along and normal to the direction of their macroscopic flow without energy exchange of the molecules with the walls of the pores. With increasing degree of filling θ of

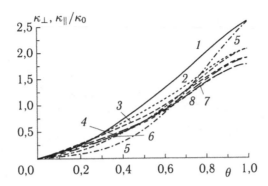

Fig. 58.1. Concentration dependences of the average thermal conductivity of argon atoms along $\kappa_{\parallel}$ (1–4) and perpendicular $\kappa_{\perp}$ (5–8) to the axis of the carbon slit pore with width $H/\lambda = 4$ (1, 5), 10 (2, 6), 20 (3 7), 60 (4, 8), $T = 273$ K.

the pores the heat conductivity increases sharply along the pores (Fig. 58.1) as the surface layers are filled. The second layer is poorly filled, so the values of $\kappa_{\perp}$ are much smaller than $\kappa_{\parallel}$ for small fillings. A further increase in θ results in gradual filling of the overlying layers and both components of the thermal conductivity tensor converge, coinciding for $\theta = 1$. The contribution of the surface layer decreases with increasing width of the pores, so the curves $\kappa_{\parallel}$ and $\kappa_{\perp}$ with increasing values of H are correspondingly lower and the difference between them decreases.

Formula (58.9) for η_{aver} gives an overview of the average coefficient of viscosity of the flows in porous systems with pores of different types and characteristic size, though the local viscosity coefficients (52.3) have a pronounced tensor character [49]. They allow one to compare the characteristics of viscosity of the flows used for macropores and wide capillaries with their counterparts for the mean values in micro- and mesoporous systems. Figure 58.2 shows the dependence (58.9a) for slit-shaped pores of the argon -carbon system of different widths. The concentration dependences of η_{aver} for pores 4 to 60 monolayers wide are shown in Fig. 58.2 a. Throughout the filling range of every pore the average viscosity coefficient decreases with increasing width of the pores, which reflects a decrease in the contribution of subsurface areas in which realized the highest viscosity is reached.

Figure 58.2 shows the curves for $H = 4\lambda$ and $H = 60\lambda$. The region between the $H = 4\lambda$–30λ covers the entire range of 'narrow pores' for spherical particles with the Lennard–Jones interaction. It can be seen that the difference between the curves depends on θ. In the range

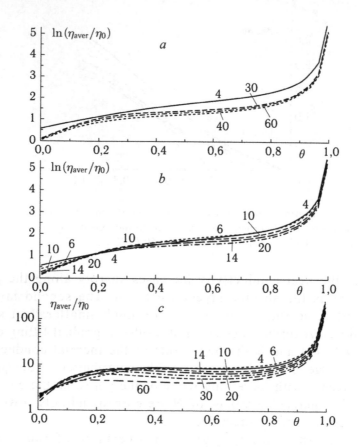

Fig. 58.2. Mean shear viscosity coefficients of individual slit pores η_{aver}: a – calculated according to formula (10a), $H/\lambda = 4, 30, 40, 60$, b – by the formula (10a), $H/\lambda = 4$, 6, 10, 14 20, c – by the formula (10b), $H/\lambda = 4, 6, 10, 14, 20, 30, 60$.

$\theta \sim 0.2$–0.3 it is minimal. More detailed analysis of the effect of the width of the pores indicates the complex nature of the dependence of $\eta_{aver}(\theta)$ on H. For small $\theta < 0.15$ and high filling values $\theta > 0.8$ the dependence of η_{aver} on H is linear. Within the region $0.15 < \theta < 0.8$ the change of θ changes results in the 'change of the curve' with the most relevant mean viscosity. And at $\theta \sim 0.22$ the viscosity of the narrowest pores $H = 4\lambda$ is the smallest. The reason for this non-linear change of the curves $\eta_{aver}(H)$ is related to the nature of layering filling of the pores with increasing pore width, as well as the fact that the dependences of η_{aver}/η_0 on θ in Fig. 58.2 are presented at $\theta = $ const.

For systems such as argon–carbon with strong attraction of the wall the highest viscosity is found in the surface layer of the adsorbate. But its maximum contribution to the average value of the viscosity of the fluid over the cross section of the pore is obtained for the overall degree of filling $\theta \sim 0.5$ for $H = 4\lambda$, whereas for $H = 10\lambda$ it corresponds to the degree of filling $\theta \sim 0.2$. Therefore, the average viscosity at $\theta \sim 0.2$ for $H = 10\lambda$ is higher than for $H = 4\lambda$. However, for large $H > 20\lambda$ the increase in the total pore volume offsets the increase in the contribution of the surface layer with relatively small θ, and is illustrated in Fig. 58.2, a. Qualitatively similar dependences for formula (58.9b) are shown in Fig. 58.2 c. However, the numerical values of both formulas (58.9) differ markedly. For formula (58.9 b) the above non-linearity of the coefficients η_{aver} with respect to H is unchanged in the range $\theta \leq 0.4$.

In the surface monolayer the local equation (58.3) gives a molecular interpretation of the so-called 'coefficient of sliding friction β_1' when calculating velocity profiles of different types of fluids in the pores (or channels). This allows us to express d_{slip} in formula (58.8) in the dimensionless form as $d_{\text{slip}} = \eta_{tt}\lambda/(\eta_{11}R)$. In the case of a strong adsorbate–adsorbent attraction or reducing temperature of (58.3), the ratio η_{11}/η_{tt} increases sharply which leads to an increase coefficient β_1 and a decrease in the flow velocity near the wall. In contrast to the decisive role of the specular reflection of molecules from the wall for rarefied gases, in dense fluids due to the slip effect is due to the surface mobility of the molecules. Its role is particularly important in narrow pores.

Porous systems

In the study of real adsorbents experiments are conducted when the external conditions change: temperature and pressure of the adsorptive, while pores of different width are filled and the observed isotherm is the average of different pore fillings at a predetermined pressure value. Similarly, all the dynamic characteristics of the system are implemented at a given external pressure. It means that the argument for the dynamic functions of the pore system as a whole should be pressure P, and not the degree of filling the considered pore. The relationship between the external pressure and fillings of different pores $\theta (H)$ is provided by the isotherms relating to the pores of different widths, with $\ln (aP) = $ const (Fig. 58.3, a). Enough visual information about the concentration dependences of the main dynamic characteristics is given in Fig. 58.3.

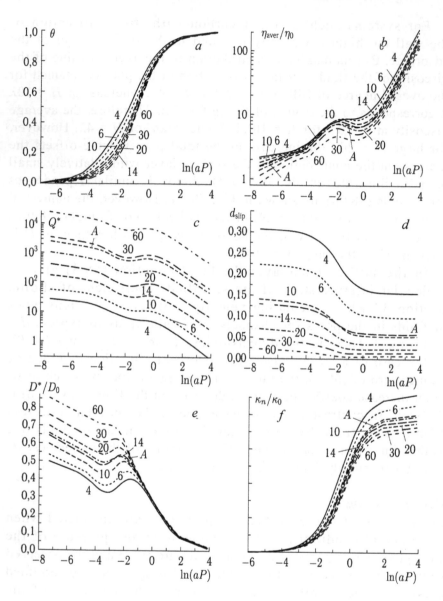

Fig. 58.3. Calculations of characteristics for slit-shaped pores of different widths (explanations in text) and for a system with slit-shaped pores (curve A).

Figure 58.3 contains the concentration dependence of the following parameters: a) adsorption isotherms, b) the average normalized values of the shear viscosity calculated by the formula (58.9 b), c) normalized steady flow $Q^* = Q/K_2$, d) function d_{slip}, characterizing the contribution of the sliding effect to the overall flow of matter,

d) average normalized values of self-diffusion coefficients, and e) the normal component of thermal conductivity.

Figure 58.3, like Fig. 58.1 and 58.2, shows the normalization volume values of the respective ratios of the gas phase. This figure shows the curves both from the region of 'narrow' pores – for fixed size H/λ = 4, 6, 10, 14, 20, 30, and for $H = 60\lambda$, which is twice the width of the 'narrow' pores, and also mean curve A, which is obtained by averaging with a uniformly distributed distribution function for all even $F_q\, H/\lambda$ from 2 to 30, further including the pores with $H/\lambda = 60$.

Figure 58.3 b shows that the non-monotonic behaviour of the average viscosity in relation to H (Fig. 58.2) is maintained in the range $H = (4-20)\lambda$, and it is difficult to construct any empirical correlations between H and η (curve A is between the $H/\lambda = 14$ and 20). With increasing pressure (density) of the adsorbate the stationary flow decreases due to the increase in viscosity (Fig. 58.3 c). In the ln (aP) range from -2 to 0 the $Q^*(\theta)$ dependence is non-monotonic. For this dependence curve A is just below $H = 30\lambda$, whereas for the dependence of d_{slip}(Fig. 58.3 d) curve A is near $H = 10\lambda$. Curves in the field 58.3 d indicate that for all the pores related to 'narrow' pores the contribution of the sliding effect is quite noticeable in the total volume flow (58.8). The calculation results for d_{slip} relate to relatively low activation energy of migration of molecules in the surface layer when $E_{11} = Q_1/3$. It is obvious that with increasing pore width the contribution of d_{slip} decreases.

Figure 58.3 f shows how the self-diffusion coefficients in the different pores changes with increasing pressure of the adsorbate. The maximum at $\ln(aP) \sim -2$, associated with the filling of the surface monolayer, is shown for the entire range of narrow pores and does not appear at $H = 60\lambda$. The self-diffusion coefficient significantly decreases with increasing pressure from values corresponding to the gas phase, to values corresponding to the liquid phase. Conversely, the coefficient of thermal conductivity normal to the direction of flow of the molecules increases monotonically with increasing pressure. In Figs. 58.3 e and 58.3, f the curves A are near $H = 14\lambda$ for the self-diffusion coefficient and between $H = 10\lambda$ and 14λ for the thermal conductivity coefficient $\kappa_\perp$.

All averaged dependences A in Fig. 58.3 pertaining to the pore system as a whole are located in different regions of H. This is due to the individual connections between the properties of the considered characteristics and wide pores. This specificity makes it difficult to search for the universal relations for the average characteristics and requires specific numerical analysis of the studied pore system.

Introduction of the averaged viscosity characteristics makes sense only for a qualitative comparison of changes in fluid viscosity in the pores of different types and characteristic dimension, since the local viscosity coefficients (58.3) have a pronounced tensor character (section 36). Furthermore, the expression (58.3) allows the molecular interpretation of the so-called sliding friction coefficient β_1 when calculating velocity profiles of different types of fluids in the pores (or channels) [51, 52].

This factor is present only in the boundary condition at the channel wall, and with its increase the speed at the channel wall vanishes and corresponds to traditional notions of continuum mechanics [28]. In dense adsorbates the sliding effect is due to the surface mobility of molecules (in contrast to the decisive role of the specular reflection of molecules from the wall for rarefied gases).

From a comparison of equation (58.3) with the expression for the boundary conditions for the rate of fluid flow along the walls of the pores [51, 52] it is easy to show that the discussed coefficient $\beta_1 = \eta_{11}/\lambda$, where the indices 11 correspond to the surface layer of the fluid, λ is the width of the monolayer or the call size $\sim 1.12\sigma$. Multiplier λ follows from the relation of the dimensions, since, unlike β_1, the shear viscosity coefficient η is the ratio of the tangential force per unit surface to the *velocity gradient* in the direction perpendicular to the flow direction [27]. In the case of a strong adsorbate–adsorbent attraction or decreasing temperature, according to (58.3), the value η_{11}, abruptly increases which leads to an increase in the coefficient β_1 or a decrease of the flow velocity near the wall.

Local dynamic characteristics in spherocylinders [11]

The joint effect of the adsorption potential of the pore walls and the intermolecular interaction creates a strongly nonuniform equilibrium distribution of the molecules in the cross sections of the pores. When calculating any dynamic coefficients it is assumed that the equilibrium distribution is weakly perturbed, so to find them we need to use equations (51.1). The most important coefficients for describing the flow of molecules of one species are the self-diffusion and shear viscosity coefficients.

Figure 58.4 shows the concentration dependence of the self-diffusion coefficients for intersecting microchannels (*a*) and a mesoporous spherocylindrical system (*b*) at different temperatures.

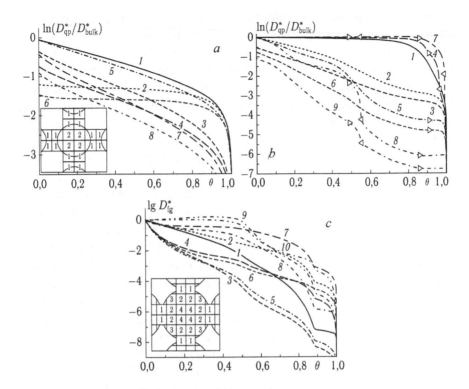

Fig. 58.4. Concentration dependences of the local self-diffusion coefficients for intersecting microchannels D_s–D_c–L_c = 2–2–2 (*a*) and a mesoporous spherocylindrical structure D_s–D_c–Lc = 10–4–8 (*b*). The same for a spherocylinder with dimensions of the diameter of the sphere 5λ, cylinder diameter 2λ, cylinder length λ (*c*). Numbers of curves 1 to 10 correspond to the following pairs of hops respectively: 11, 12, 21, 22, 23, 24, 32, 33, 42, 44 (site numbers indicated in the inset figure).

All the curves are normalized to the corresponding value for the self-diffusion coefficients for the small densities of the bulk phases at a fixed temperature. These differences are due to the coefficients at zero density.

Field *a* shows the local self-diffusion coefficients for molecules in a channel: pair of sites 11 — within the channel and 12 — from the channel to the intersection area (see inset) as well as for the two directions of movement of the molecules in the intersection region (site pair 21 — from the intersection region to the channel and 22 — between the sites of the intersection region) The inset in field *a* reflects the positions of the sites of type 1 in the channel and type 2 in the intersection region, and also differences in their adsorption properties due to the differences of energies. Calculations were performed for two temperatures. Field a is calculated with the following parameters Q_f: $(\beta\varepsilon)^{-1} = 1.23$ (solid lines $1, 2, 3, 4$), 1.00 (dashed lines $5, 6, 7, 8$); $D_{1,1}$ ($4, 8$), $D_{1,2}$ ($2, 6$), $D_{2,1}$ ($1, 5$), $D_{2,2}$ ($3, 7$). The temperature dependence of hopping in the intersection region is the weakest. To migrate along the cylindrical channel the temperature dependence is the strongest due to the high value of the activation energy of surface migration.

Figure 58.4 b shows the concentration dependences of the self-diffusion coefficient in the regular spherocylindrical system at three temperatures. The calculation in field b was performed for $(\beta\varepsilon)^{-1} = 1.23$ (solid lines $1, 2, 3$), 0.85 (dashed lines $4, 5, 6$), 0.60 (dashed lines $7, 8, 9$); $D_{1,1}$ ($3, 6, 9$), $D_{11,11}$ ($2, 5, 8$), $D_{23,23}$ ($1, 4, 7$). The solid curves 1–3 refer to the supercritical region, and the dashed curves 4–6 – to a temperature corresponding to the condensation in the central part of the system at coverages θ of about 0.90–0.96. The dotted curves 7–9 refer to the temperature at which the two domes form in the sites belonging to the surface ($\theta \sim 0.5$) and central ($\theta = 0.86$–0.98) system areas. The form of the curves in the areas of capillary condensation was calculated taking into account the Maxwell rule (the lever rule) – these areas are marked by symbols on these curves. The numbers of the pairs correspond to the following areas: $1, 1$ – sites in the cylinder surface (curves $3, 6, 9$) $11, 11$ – central sites in the cylinder (curves $2, 5, 8$), $23, 23$ – in the central sites in the sphere (curves $1, 4, 7$). The calculated results show a strong influence of the surface potential and temperature on the

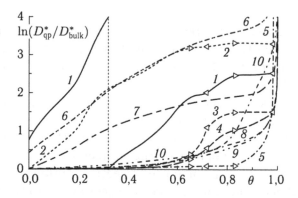

Fig. 58.5. Concentration dependence of the shear viscosity for a spherocylindrical structure D_s–D_c–L_c = 9–8–8 $(\beta\varepsilon)^{-1} = 1.23$ (solid lines $6, 7, 8, 9, 10$), 0.70 (dashed lines $1, 2, 3, 4, 5$); $\eta_{1,3}$ ($2, 7$), $\eta_{8,8}$ ($1, 6$), $\eta_{38,39}$ ($3, 8$), $\eta_{41,42}$ ($4, 9$), $\eta_{45,45}$ ($5, 10$).

migration of labelled molecules. With decreasing temperature the role of capillary condensation increases.

Fig. 58.4 c presents the local self-diffusion coefficients for a spherocylindrical system normalized to similar coefficients at $a_{q,f}^0 = 1$ (zero fill in the absence of the wall potential). The calculation is performed at $(\beta\varepsilon)^{-1} = 0.4; \alpha = \varepsilon_{AA}^*/\varepsilon_{AA} = 0.1; \alpha_{11} = E_{11}/Q_1 = 0.33$ (E_{11} – hopping activation energy in the surface layer). With the increase in the degree of filling all local coefficients decrease. The small value of the coefficients in completely filled pores is connected with the strong attraction of greatly distorted walls at these small sizes of the spherocylinder (its scheme shown in the inset of Fig. 58.4 c. Calculation was performed taking into account the capillary condensation of the adsorbate at a given temperature in the filling range from θ 0.89 to 0.99 – it leads to a 'kink' for all the curves at high coverages.

The concentration dependences of the shear viscosity coefficients for the regular structure of spherocylinders $D_s-D_c-L_c$ = 9–8–8 are shown in Fig. 58.5. The local coefficients for these pairs relate to the surface areas of the sphere $\eta_{8,8}$ (1, 6) (part of the curve 1 is shifted down by 4 units), and the cylinder $\eta_{1,3}$ (2, 7) and to the pairs oriented along the connection line between the sphere and the cylinder: $\eta_{38,39}$ (3, 8) – in the centre of the cylinder, $\eta_{41,42}$ (4, 9) – in the centre of the articulation of the cylinder and a sphere, and $\eta_{45,45}$(5, 10) – at the centre of the sphere. With increasing filling of the pore system increases the viscosity of the adsorbate. The coefficients vary significantly with increasing θ. Figure 58.5 illustrates the important role of the surface potential of the walls temperature. As also shown in Fig. 58.4 b, the graph shows areas of capillary condensation at low temperatures, when the two domes appear in the surface and in the central areas.

The considered generalization of the LGM to complex porous system allows one transfer to the self-consistent calculation of the equilibrium and dynamic characteristics of virtually all known variants of the structure of porous bodies. Formulated molecular approach allows [1, 2] to investigate the effect of the adsorbate-adsorbent and adsorbate–adsorbate potentials on the observable characteristics in a wide range of temperature and pressure of the adsorptive. Accounting for the seven factors in the transition from the idealized pores having 'infinite' length to complex porous system changes the previously obtained results for the individual pores in chapter 5. The allowance for the finite values of the three linear measurements of each local area then will significantly affect the conditions of the distribution of molecules in their 'mouths', and the size of the transition regions between the different sites then has an impact on the formation of supramolecular level: on the pore size distribution (function F_q) and their relatedness (function F_{qp}). The system of linked sites of the pores then forms a single pore space, and the nature of local distributions of the adsorbate in any part affects the distribution of molecules in neighbouring areas of the pores. As a result, the feasibility of capillary condensation and the critical conditions for its occurrence (critical temperature and the degree of filling) depend on the properties of a certain region of the intrapore space, which in turn depends on the functions F_q and F_{qp}. The need to introduce functions F_{qp} is essential to describe the dynamics, since the related sequences of through pore sections are important for molecular transport. The expressions derived for the dynamic characteristics take into account the nature of connectedness far.

In real adsorbents and catalysts with the continuous pore size distribution the character of the phase state of the adsorbate also depends on the pore size in the local area of the adsorbent and the pore size in regions surrounding the field. This leads to the fact that the calculation of the adsorbate phase distributions and their dynamic characteristics requires generalization of the introduced above discrete functions F_q and F_{qp} to continuous (which is achieved by natural replacement of sums by integrals by analogy with the transition to continuous functions of the coordinate x along the length of the grain). As a result, a significant part of the information is not only a function of the pore size distribution, but also the functions that characterize the internal structure of the pore space of the adsorbent. Such a situation should occur when describing the structure of the pore space at a higher level supramolecular compared with the grain of the adsorbent/catalyst.

References

1. Tovbin Yu.K. // Izv. AN. Ser. khim. 2003. No. 4. P. 827.
2. Tovbin Yu.K. // Zh. Fiz. Khimii. 2003. V. 77, No. 10. P. 1875.
3. Ruckenstein E., Vaidyanathan A.S., Youngquist G.R. // Chem. Eng. Sci. 1971. V. 26. P. 1305.
4. Mamleyev V.Sh., Zolotarev P.P. Gladyshev P., Heterogeneity of sorbents. – Alma-Ata: Nauka, 1989. – 287 p.

5.   Tovbin Yu.K., Theory of physical and chemical processes on the gas-solid. – Moscow: Nauka, 1990. – 288 p.
6.   Mayagoitia V., Rojas F., Kornhauser I. // Langmuir. 1993. V. 9. P. 2748.
7.   Dufau N., Llewellyn P.L., Martin C., Coulomb J. P., Grillet Y., Fundamentals of Adsorption 6. – Paris: Elsevier, 1998. – P. 63.
8.   Grun M., Schumacher K., Unger K., Fundamentals of Adsorption 6. – Paris: Elsevier, 1998. – P. 569.
9.   Hill T., Statistical Mechanics. – Moscow: Izd. inostr. lit., 1960. – 486 p.
10. Tovbin Yu.K. , Eremich D.V.// Izv. AN. Ser. khim. 2003. V. 52, No. 11. P. 2208.
11. Eremich D.V., Tovbin Yu.K. // Zh. Fiz. Khimii. 2004. V. 78, No. 4. Pp. 720.
12. Ubbelohde A.R., Molten state of matter. – Moscow: Metallurgiya. 1982.
13. Thommes M., Kohn R., Froba M. // Appl. Surf. Sci. 2002. V. 196. P. 239.
14. Thommes M., Froba M. // J. Phys. Chem. 2000. V. 104. P. 7932.
15. Tovbin Yu.K., Eremich D.V., Komarov V.N., Gvozdeva E.E. // Khim. Fizika. 2007. V. 26, No. 9. P. 84.
16. Tovbin Yu.K., Rabinovich A.B., Votyakov E.V. // Izv. AN. Ser. khim. 2002. No. 9. P. 1531.
17. Tovbin Yu.K., Rabinovich A.B., Eremich D.V. // Zh. Fiz. Khimii. 2004. V. 78, No. 3. P. 512.
18. Fenelonov V.B., Porous carbon. - Novosibirsk: IK SO RAN, 1995. – 514 p.
19. Adamson A., Physical Chemistry of Surfaces. – Moscow: Mir, 1979. – 568 p.
20. Tovbin Yu.K., Yeremich D.V. // Colloids and Surfaces. A. 2002. V. 206. P. 363.
21. Kiselev A.V., Intermolecular interactions in adsorption chromatography. – Moscow: Vysshaya shkola, 1986. – 360 p.
22. Greg S., Singh K., Adsorption, Surface porosity. – Moscow: Mir, 1984.
23. Karnaukhov A.P., Adsorption. Texture of dispersed porous materials. – Novosibirsk: Nauka, 1999. – 469 p.
24. Tovbin Yu.K. // Zh. Fiz. Khimii. 2002. V. 76, No. 3. P. 488.
25. Ford D.M., Gland E.D. // Phys. Rev. E. 1994. V. 50. P. 1280.
26. Page K.S., Monson P.A. // Phys. Rev. E. 1996. V. 54. P. 6557.
27. Kierlik E., Rosinberg M.L., Tarjus G., Monson P.A. // J. Chem. Phys. 1997. V. 106. P. 264.
28. Sarkisov L., Page K.S., Monson P.A. Fundamentals of Adsorption 6. – Paris: Elsevier, 1998. – P. 847.
29. Kierlik E., Rosinberg M.L., Tarjus G., Monson P.A. Fundamentals of Adsorption 6. – Paris: Elsevier, 1998. – P. 867.
30. Tovbin Yu.K., Petukhov A.G., Eremich D.V.// Izv. AN. Ser. khim. 2007. No. 5. P. 813.
31. Tovbin Yu.K., Petukhov A.G. // Izv. AN. Ser. khim. 2008. No. 1. P. 18.
32. Tovbin Yu.K., Petukhov A.G. // Fiziko-khimiya poverkhnosti i zashchita materialov. 2008. V. 44, No. 3. P. 255.
33. Tovbin Yu.K., Petukhov A.G., Eremich D.V. // Zh. Fiz. Khimii. 2006. V. 80, No. 3. P. 488.
34. Morishide K., Shikimi M. // J. Chem. Phys. 1998. V. 108. P. 7821.
35. Ravikovitch P.I., Wei D., Chueh W.T., Haller G.L., Neimark A.V. // J. Phys. Chem. 1997. V. 101. P. 3671.
36. Kruk M., Jaroniec M. // Chem. Mater. 2000. V. 12. P. 222.
37. Thommes M., Kohl R., Froba M. // Appl. Surf. Sci. 2002. V. 196. P. 239.
38. Everett D.H. // The Solid-Gas Interface, Ed. by E. A. Hood. – N. Y.: Dekker, 1967. – 1055 p.

39. Karnaukhov A.P., Adsorption. Texture of dispersed porous materials. – Novosibirsk: Nauka. IK SO RAN, 1999. – 469.

40. Experimental methods in molecular adsorption and chromatography, eds. A.V. Kiselev, V.P. Dreving. – Moscow State University Press, 1973. – 447 p.

41. Plachenov S.D., Kolosentsev ??? Porosimetry. – Leningrad: Khimiya, 1988. – 175 p.

42. Chizmadzhev A., Markin V.S., Tarasevich V.R., Teals G. Macrokinetics processes in porous media. – Moscow: Nauka, 1971. – 362 p.

43. Mason E., Malinauskas A., Transfer in porous media: the dusty gas model. – Moscow: Mir, 1986. – 200 p.

44. Kheifets L.I., Neimark A.V., Multiphase processes in porous media. – Moscow: Khimiya, 1982. – 320 p.

45. Tovbin Yu.K., Eremich D.V., Komarov V.N., Gvozdeva E.E. // Zh. Fiz. Khimii. 2008. V. 82, No. 12. P. 2395.

46. Neimark A.V., Ravikovitch P.I., Grun M., Schuth F., Under K.K. // J. Colloid Interface Sci. 1998. V. 207. P. 159.

47. Thommes M., Kohn R., Froba M., Studies in Surface Science and Catalysis. 2002. V. 142. P. 1695.

48. Tovbin Yu.K. // Dokl. AN SSSR. 1990. V. 312. P. 1425.

49. Tovbin Yu.K. // Zh. Fiz. Khimii. 1997. V. 73. P. 1454.

50. Sheydegger??? A.E., Physics of fluid flow through porous media. – M.: GTTI, 1960. – 196 p.

51. Bird R., Stewart W., Lightfoot E., Transport Phenomena. – Moscow: Khimiya, 1974. – 687 p.

52. Lamb G., Hydrodynamics. – Moscow–Leningrad: OGIZ 1947. – 928 p.

8

Mixtures

The equations describing the equilibrium distribution of the mixture of molecules with measurable dimensions are given in Appendices 2 and 4. The commensurability of the sizes of mixture components means that that the LGM with an average size of the lattice constant λ can be used and each component takes up one cell (site) [1–8]. Differences in the actual size of the molecules are taken into account in the lattice parameters for the Henry constants and intermolecular interactions. These models include all modifications of the LGM discussed in section 9 to account for the internal motions of the molecules in the cells. Such a formulation allows us to consider the effect of differences in the intermolecular interactions between the components of the mixture itself and with the walls of the pores on partial equilibrium and dynamic characteristics of adsorption. Sections 59–66 show the results of using the LGM in analysis of partial equilibrium and dynamic characteristics of adsorption for a mixture of comparable sizes.

The model for a mixture of components of different sizes is discussed in section 67, the results for the partial equilibrium and dynamic adsorption characteristics are given in sections 68 and 69. The microscopic hydrodynamics equations for mixtures are contained in Appendix 11. They apply equally to mixtures of molecules of different and commensurate sizes.

59. Equilibrium properties of mixtures of components of comparable size in the bulk phase

In the case of mixtures of components of commensurate size each site can be contain only one particle: either a molecule grade i (if the centre of mass of the molecule is inside the cell), or a vacancy.

Index i enumerates the number of mixture components. Through s we denote the number of different states of occupation of any site in the system, and the index s itself will be assigned to vacancies, i.e., the number of components is equal to $(s - 1)$. Finding the molecule in the cell does not indicate the fixation of its centre of mass – it has translational, rotational and vibrational degrees of freedom [9]. Typically, the concentration of molecules of type i is assumed to be equal to the number of molecules N_i per unit volume: $n_i = N_i / V$. In the LGM the concentration of the fluid component is characterised by the value $\theta_i = N_i/N$, equal to the ratio of the number of real particles in a certain volume to the maximum possible number of densely packed particles in the same volume. Then $\theta_i = n_i v_0$. Complete filling volume is defined as $\theta = \sum_{i=1}^{s-1} \theta_i$. The relationship between the total concentration of molecules $n = \sum_{i=1}^{s-1} n_i$ and the degree of filling θ is given by $n = \theta/v_0$.

The isotherms determine the relationship between the pressures in the thermostat $\{P\}$ (symbol $\{P\} \equiv P_1,..., P_{s-1}$ is a complete set of all the partial pressures of the mixture components P_i, $1 \le i \le s - 1$, in the thermostat oven) and the partial filling of the bulk of the system $\{\theta_i\}$. They have the form [7]

$$a_i P_i = \left(\frac{\theta_i}{\theta_s}\right)^{1+g_e} \Lambda_i, \quad \Lambda_i = \prod_r (S_i)^{z(r)}, \quad S_i(r) = 1 + \sum_{j=1}^{s-1} x_{ij}(r) t_{ij}(r), \quad (59.1)$$

where the function Λ_i takes into account the interaction between molecules at a distance R of the coordination spheres in the quasichemical approximation (QCA), g_e is the calibration function, which increases the accuracy of the calculation in the vicinity of the critical temperature [10–12] (see Appendix 5 and section 24), the form of the calibration function does not change at the transition from pure substances to a multicomponent mixture; $z(r)$ – the number of sites in the r-th coordination sphere, $1 \le r \le R$, $z(1) \equiv z$; $x_{ij}(r) = \exp[-\beta\varepsilon_{ij}(r)] - 1$, $\beta = (kT)^{-1}$.

Parameter $\varepsilon_{ij}(r)$ describes the interaction of the components i and j at the distance r of the coordination spheres; interactions with the vacancies $(i, j = s)$ are equal to zero. In the formula (59.1) it is assumed that the lateral interaction parameter of components i and j at a distance r of the coordination spheres is described by an effective parameter of the Lennard–Jones type $\varepsilon_{ij}(r)$, which may be a function of temperature and local composition:

$$\varepsilon_{ij}(r) = 4\varepsilon_{ij,\,ef}^0 \left[\left(\frac{\sigma_{ij}}{r} \right)^{12} \right] - \left(\frac{\sigma_{ij}}{r} \right)^6,$$

(59.2)

$$\varepsilon_{ij,\,ef}^0 = \varepsilon_{ij}^0 \left[1 - \Delta_{1r} \sum_{k=1}^{s-1} d_{ijk} t_{ijk}(1) \right] (1 - u_{ij}T).$$

Here $\varepsilon_{ij,\,ef}^0$ is the well depth of the effective pair interaction potential, taking into account in a first approximation the triple interactions with neighbouring particles and the temperature factor when averaged interparticle interactions [7–10, 13]; $\varepsilon_{ij}^0 = \sqrt{\varepsilon_{ii}^0 \varepsilon_{jj}^0}$ is the energy of the pair interaction of components i and j at low densities of the mixture, ε_{ii}^0 and ε_{jj}^0 is the depth of the potential of interaction between the two identical molecules in single-component fluids i and j, respectively; σ_{ij} is the distance of the closest approach of two hard spheres of components i and j, $\sigma_{ij} = (\sigma_{ii} + \sigma_{jj})/2$; parameter u_{ij} considers the temperature dependence of the effective pair potential; Δ_{1r} is the Kronecker symbol that distinguishes triple contributions only for the nearest neighbours; d_{ijk} is the relative share of the contribution of triple interactions in the effective pair potential $\varepsilon_{ij}(1)$; function $t_{ijk}(1)$ is the the probability of finding k molecules in the first coordination sphere of the two central molecules ij. Here $t_{ijk}(1) = t_{ik}(1)\,t_{jk}(1)/\theta_k$; functions $t_{ij}(r) = \theta_{ij}(r)/\theta_i$ describe the conditional probability of finding the neighbouring particle j at a distance r from the central particle i; $\theta_{ij}(r)$ is the probability of finding a pair of molecules ij at a distance r. All parameters of the expression (59.2) are strictly connected with the intermolecular potentials. For the bulk phase we use the full interaction potential $R = 5$.

The quantity $a_i = \beta v_0 F_i / F_{i0}$ is the ratio of the statistical sums for molecule i, situated in the in cell (F_i) and the thermostat (F_{i0}); $F_i = F_{trans}^i F_{vib}^i F_{rot}^i$ where the factors correspond to the translational, vibrational and rotational degrees of freedom, and similarly for F_{i0}. For molecules of the spherical shape of about the same size F_{rot}^i in the cell and the thermostat can be considered identical, then $a_i = \beta v_0 F_{vib}^i V_i(\theta) / F_{vib}^{i0} V_i(\theta = 0)$, where $V_i(\theta)$ is the free cell volume, in which moves the centre of mass of the molecule i moves. The gas in the thermostat is considered as ideal, so $V_i(\theta = 0) = v_0$.

Below we restrict ourselves to the vibrations in the dense phase in the framework of Einstein'model [16, 17]. The ratio of the volume of the cell in which the component i moves, to the corresponding volume at $\theta = 0$ can be written as $V_i(\theta)/V_i(\theta = 0) = \left[1 - \sum_{j=1}^{s-1} (1 - \kappa_{ij}) t_{ij}(1) \right]^3$,

where κ_{ij} is the average relative volume of the region of the centre of mass of the molecule i at thermodynamic parameters of the fluid near the triple point, $0 \leq \kappa_{ij} < 0.15$.

In equilibrium, the probability $\theta_{ij}(r)$ of finding pairs of molecules ij at sites at a distance r in the QCA is described by a system of algebraic equations (A2.6) [7]. The lattice constant λ for a compressed lattice is determined from the minimum free energy of the system (A2.5) [18]. For an isotropic bulk phase

$$\lambda = \left[2 \sum_r z(r) \sum_{ij} \sigma_{ij} \theta_{ij}(r)(\eta_r)^{-12} \Big/ \sum_r z(r) \sum_{ij} \sigma_{ij} \theta_{ij}(r)(\eta_r)^{-6} \right]^{1/6}, \text{ w h i c h}$$

determines implicit dependence of λ, and hence of the cell volume $v_0 = \lambda^3$ on density and temperature.

The equation of state (or 'expansion pressure') is defined in section 10. It is calculated as a function of the degree of complete filling of the volume $\theta(\{P\}) = \sum_{i=1}^{s-1} \theta_i(\{P\})$ at the given partial pressures of the components in a thermostat (outside the system). If the molecular model has 'purely paired' lateral interaction parameters and ignores the excluded volume, the equation of state retains the form (10.5). The compressibility factor Z is given by $P = ZnkT$ or as $Z = \beta P v_0/\theta$, where θ is the full extent of filling the system with all components of the mixture. At low densities, the equation of state changes to the equation of state of an ideal gas $\beta P v_0 = \theta$ or $\beta P = n$.

The equations of equilibrium distribution of molecules in the bulk phase allow us to estimate the compressibility factor of the mixture based on information about individual properties of gases. Corresponding calculations for the temperature $T = 273$ K at different molar compositions of the nitrogen–argon mixture ($x_1 = \theta_1/(\theta_1 + \theta_2)$ is the molar fraction of nitrogen) are shown in Fig. 59.1 a. Both gases are described by identical curves in Fig. 59.1 (in particular, the critical compressibility for agon is $Z_c = 0.292$, and for nitrogen $Z_c = 0.291$ [19]), so for the mixture we used the additive scheme of accounting the molecular interactions in which the triple contributions are described by the simplest averaging through $d_{ij} = \sum_{k=1}^{s-1} d_{ijk} t_{ijk}$. For pure gases ($i = j$) equality $d_{ii} = d_{iii}$ is satisfied, where the indices 11 and 22 refer respectively to pure nitrogen and argon. And for $i \neq j$ it

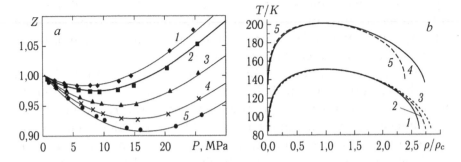

Fig. 59.1. Dependence of the compressibility factor of the nitrogen–argon mixture on pressure at a temperature $T = 273.15$ K and different ratios of mixture component concentrations (*a*): x_1 (molar fraction of nitrogen) = 0 (*1*), 0.2 (*2*) 0.4 (*3*), 0.6 (*4*) 0.8 (*5*) with $u = 0.0055$ (experiment [19] – icons, calculations – solid lines). Influence of molecular parameters d_t and κ on the behaviour of the phase diagram at $u = 0.00075$ (*b*): The first three curves refer to the change of the parameter $d_t = 0$ (*1*), 0.1 (*2*) and 0.15 (*3*) for $\kappa = 0.1$; last two curves relate to $\kappa = 0.1$ (*4*), and 0.2 (*5*) with $d_t = 0.15$

is assumed that $d_t \equiv d_{21} = d_{12}$. Similarly, the average excluded volume of the mixture is described: $\kappa = \kappa_{21} = \kappa_{12}$. Curve *1* corresponds to $x_1 = 1.000$, $d_t = 0.15$, $\kappa = 0.25$; curve *2* – $x_1 = 0.7985$, $d_t = 0,125$, $\kappa = 0.22$; curve *3* – $x_1 = 0.4845$, $d_t = 0.1$, $\kappa = 0.16$; curve *4* – $x_1 = 0.2015$, $d_t = 0.075$, $\kappa = 0.1$; curve *5* – $x_1 = 0$, $d_t = 0.075$, $\kappa = 0.1$. Everywhere x_1 refers to nitrogen: $d_{12} = d_{11} x_1 + d_{22} x_2$.

Along with the usual combination rules, $\sigma_{12} = (\sigma_{11} + \sigma_{22})/2$ and $\varepsilon_{11}^2 = \varepsilon_{11}\, \varepsilon_{22}$, it was decided that the parameter u is independent of the type of molecule. These assumptions have slashed the number of molecular parameters to two: $d(\equiv d_t)$ and κ. At the parameter values in the caption for Fig. 59.1 *a* the discrepancy between theory and experiment [19] (shown by the symbols) does not exceed 3%.

Figure 59.1 *b* shows the phase diagram for varying molecular parameters ($d_t \equiv d_{111}$, κ). Changing the parameters causes quite noticeable changes in the coexistence curve. Increasing the contribution of triple interactions (curves *1–3*) changes the liquid branch of the coexistence curve, shifting the position of the triple point to higher density. We should note the non-linear influence of d_t values on the curve. In particular, in the case of changes in the range $0.05 < d_t < 0.075$ the d_t has almost no effect. Curves *4* and *5* show the effect of the parameter of the equation of the excluded volume κ (for clarity, the last curves are shifted upwards by 50 K). Increasing κ changes significantly the position of the critical point and the two branches of the stratification curve. Therefore, the search procedure for molecular parameters from experimental data is carried out first to fit on the location of the critical point as regards density, and then determine the nature of the curvature of the coexistence curve using the parameters of the calibration function.

Varying the parameter u in the range $0 < u < 0.001$ has an insignificant effect on the behaviour of the phase diagram due to the relatively small value of the total temperature range of the coexistence curve. For values of $u > 0.001$ the calculated critical temperature T_c is less than the corresponding experimental value which does not correspond to the physical sense. The phase diagrams were calculated using the value $u = 0.00075$.

For dense gases obeying the 'law' of corresponding states (the molecules of which do not have specific interactions) we use the so-called generalized compressibility factor [4, 20]: in the coordinates Z – 'reduced' pressure for various 'reduced' temperature all substances have the same curves ('reduced' values are normalized for the corresponding values at the critical points). This leads to the fact that the generalized compressibilities for pure components retain their value also for mixtures.

Initially we calculated phase diagrams of the individual components of gas mixtures. Comparison with experimental data makes it possible to determine in the near-critical region for these components the interaction potential parameters included in the equation of state of the LGM, the size of the excluded volume and the calibration function. This is followed by the compressibility factor calculation for several values of temperature.

The results of calculation of the phase diagrams of argon and nitrogen, their comparison with experimental data and also the phase diagrams of a binary Ar–N$_2$ calculated using the LGM for different concentrations of nitrogen are shown in Fig. 59.2 (in the coordinates temperature T – reduced density $\rho_r = \rho/\rho_c$). The calculations were performed for the parameters of the intermolecular interaction potential discussed in [15]: $\varepsilon_{ii}/k = 119.8$ K, $\varepsilon_{jj}/k = 95.05$ K, $\sigma_{ii} = 3.41$ Å, $\sigma_{jj} = 3.7$ Å, $i = $ Ar, $j = $ N$_2$; $d_{ijk} = 0.005$, $u_{ij} = 0.0002$; the size of the excluded volume $\kappa_{ij} = 0.15$; coefficients of the calibration function: $A_{ij} = 0.045$, $B_{ij} = 1$. Critical temperatures and pressures of the mixture can be satisfactorily approximated by the pseudocritical parameters often used in practice and calculated by the formula $T_c^{ps} = \sum_i x_i T_i^c$, $P_c^{ps} = \sum_i x_i P_i^c$, where x_i is the molar fraction of the i-th component of a binary mixture, and T_i^c and P_i^c are the corresponding critical temperature and pressure of the

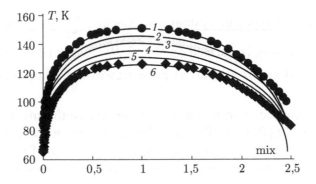

Fig. 59.2. Phase diagrams of Ar, N_2 and Ar-N_2 mixtures at different compositions with a nitrogen molar fraction xN_2 = 0 (*1*) 0.2 (*2*) 0.4 (*3*) 0.6 (*4*) 0.8 (*5*), 1.0 (*6*). Experiment – symbols, calculations – solid lines [15].

component [20]. The largest discrepancy between the experiment and the calculation near the triple point is due to the use of the linear dependence of the temperature factor. To clarify it the contribution of the vibrational motions of the molecules must be taken into account more accurately.

The results indicate that despite the simplification, the modified LGM provides a satisfactory agreement with the experimental data and the resultant model parameters have a strict physical meaning, in particular, the specificity of the interaction between the components of the mixture is taken into account. The LGM can be successfully used to describe mixtures of low-molecular substances in a wide range of densities and temperatures, if they obey the law of corresponding states. Accounting for the scaling behaviour of the thermodynamic functions of the condensed phase in the near-critical region [10–12] significantly improves the description of dense gas phases.

60. Equilibrium distributions of mixture components in the pores

The local density of particles i in the cell with number f will be denoted by $\theta_f^i, \sum\limits^{s-1} \theta_f^i + \theta_f^V = 1$. The average partial concentration of the fluid θ_i can be determined through local concentrations as

$$\theta_i = \sum_{f=1}^{t} F_f \theta_f^i,$$ where F_f is the fraction of sites of type f, $1 \leq f \leq t$, t

is the number of types of system components; $\sum_{f=1}^{t} F_f = 1$; and the total average pore filling is defined as $\theta_i = \sum_{i=1}^{s-1} \theta_i$. Symbol $\{P\} \equiv P_1,..., P_{s-1}$ denote the full set of all partial pressures of the components of the mixture P_i.

To calculate the average partial adsorption isotherms $\theta_i(\{P\})$ and local fillings $\theta_f^i(\{P\})$ on different adsorption sites we use a system of equations that takes into account the heterogeneity of the energy of the lattice sites and the interaction between molecules at a distance R of the coordination spheres [7, 21–25] :

$$\theta_i(\{P\}) = \sum_{f=1}^{t} F_f \theta_f^i(\{P\}), \quad a_f^i P_i = \frac{\theta_f^i \Lambda_f^i}{\theta_f^V},$$

$$\Lambda_f^i = \prod_r \prod_g \left(1 + \sum_{j=1}^{s-1} x_{fg}^{ij}(r) t_{fg}^{ij}(r)\right)^{z_{fg}(r)},$$ (60.1)

$$x_{fg}^{ij}(r) = \exp\left(-\beta \varepsilon_{fg}^{ij}(r)\right) - 1,$$

where the local Henry constant $a_f^i = a_f^{i0} \exp(\beta Q_f^i)$, Q_f^i – the binding energy of molecules i in the layer f with the walls of the pores f calculated as $Q_f^i = E_i(f) + E_i(H - f + 1)$, $1 \le f \le t$, and the potential of interaction of the molecule with the pore wall $E_i(f) = \varepsilon_i/f^3$ corresponds to the attracting branch of the Mie potential (9–3) [16], ε_i is the potential energy parameter.

Function Λ_f^i considers intermolecular interactions in quasichemical approximation. Index g runs through all the neighbours $z_f(r)$ of site f at a distance $r \le R$ inside the pore; R is the radius of the interaction potential. Here, the functions $t_{fg}^{ij}(r) = \theta_{fg}^{ij}(r)/\theta_f^j$, where the functions $\theta_{fg}^{ij}(r)$ are defined by equation (A2.6). In the formula (60.1) it is taken into account that the slit pores have the layered distribution of different types of sites: $z_{fg}(r)$ is the number of neighbouring sites in the layer g at a distance r from the considered site in layer f.

System (60.1) is solved by an iterative method for a given set of values $\{\theta_i\}$ or $\{P_i\}$. Its solution allows us to calculate all the equilibrium characteristics and thermal velocity in the equilibrium state, and through the thermal velocity – all transport coefficients.

Argon–krypton mixture [23]
The walls of the slit-like pores in activated carbon are composed

of carbon atoms. La0teral interactions ε are defined using the Lennard–Jones potential: $U_{ij} = 4\varepsilon_{ij} [(\sigma_{ij}/r)^{12} - (\sigma_{ij}/r) 6]$, with $r/\sigma_{ij} = 2^{1/6}$, which corresponds to the minimum of this potential. The molecular parameters of the mixture components are well known: for argon $Q_1^1 = 9.24\varepsilon_{ArAr}$ at $\varepsilon_{Ar\,Ar}/k_B = 119$ K, and for the krypton atoms $Q_1^2 = 12.17\varepsilon_{ArAr}$ and $\varepsilon_{Kr\,Kr}/\varepsilon_{Ar\,Ar} = 1.37$ (~ 326 cal/mol or 163 K) [26, 27]. The interaction parameter between particles of different varieties is assessed as $\varepsilon_{12} = (\varepsilon_{11}\,\varepsilon_{22})^{1/2}$. For the argon–krypton system the difference in the value of a solid sphere of the Lennard–Jones potential is sufficiently small: $\sigma_{Ar\,Ar} = 0.3405$ nm and $\sigma_{Kr\,Kr} = 0.363$ nm [26, 27]. For simplicity, the calculations assume that the width of the pores is commensurate with the lattice constant $\lambda = 2^{1/6}\,\sigma$, where $\sigma = (\sigma_{Ar\,Ar} + \sigma_{Kr\,Kr})/2$.

The structure of the fluid was modelled by a grid with the number of nearest neighbour cells of 12. The pore width ranged from 3 to 30 monolayers. The wall potential was taken into account to a distance corresponding to the energy lower than energy of thermal motion of the molecules $Q_f^{Ar} \sim 0.1\varepsilon_{ArAr}$, thus all the cells separated from the wall by more than 4 monolayers are considered equivalent. Calculations were limited to the isothermal conditions at $T = 1.5T_c$ (Ar), where T_c (Ar) is the critical temperature of argon, to eliminate the formation of two-phase regions.

Figure 60.1 gives an idea about the nature of equilibrium distributions of argon and krypton atoms in the cross section of a slit pore with width $H = 10$ monolayers. The figure shows the degree of local partial filling of argon atoms (solid lines) and krypton atoms (dashed lines and numbers of curves with a stroke) in different monolayers for five compositions of the mixtures $X_{Ar} = \theta_{Ar}/\theta$ at a fixed gross density of the mixture θ inside the pores. As the walls of the slit pore are identical the distribution curves of mixture components over the cross section pores are symmetrical about its centre. Figure 60.1 a refers to the case of a relatively low overall degree of filling ($\theta = 0.3$): both components are concentrated near the walls, and the filling of the central part of the pores is much smaller. Krypton atoms are attracted more than argon atoms. Therefore, for pure krypton their concentration in the surface layer is higher (curve 5′) than the concentration of pure argon (curve 1), and in the centre of the pore their concentration is lower. With decreasing proportion of the first component X_{Ar} the concentration of the first component near the wall decreases, while that of the second component increases. The influence of the wall potential

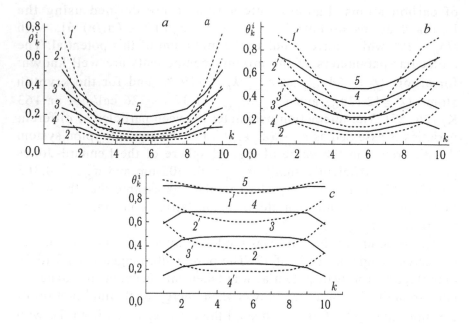

Fig. 60.1. Distribution of argon (Ar) and krypton (Kr) during adsorption in a slit-pore pore at $\theta = 0.3$ (*a*), 0.6 (*b*) and 0.9 (*c*) in a mixture of Ar and Kr for the five compositions X_{Ar}: 0 (*1*), 1/4 (*2*), 1/2 (*3*), 3/4 (*4*), 1 (*5*). The abscissa indicates the number of the monolayer, the ordinate – local degree of filling of mixture components.

is spread over three monolayer, its contribution to the fourth and fifth monolayers is small. All the curves have a similar shape with a minimum concentration in the centre of the pores.

Figure 60.1 *b* shows for the intermediate region ($\theta = 0.6$) the distribution of components which is more complicated. Krypton displaces argon from the first monolayer, so the distribution curves of argon in the cross section have maxima: Ar atoms are concentrated in the second monolayer. The distribution of the Kr atoms in the cross section is as previously with a minimum concentration at the centre of the pores, but the minimum has higher values. In Fig. 60.1 *c* for the case of filling a large pore volume ($\theta = 0.9$) the minimum concentrations of krypton atoms are observed in all the compositions in the centre of the pores. Pure argon is almost uniformly distributed over the cross section. With the decrease in the proportion of argon in the mixture argon is concentrated in the central portion of the pores, since Kr is located mainly at the walls of the pores.

The equilibrium characteristics of the binary mixture are shown in Fig. 60.2 where the abscissa gives the logarithm of the full vapour

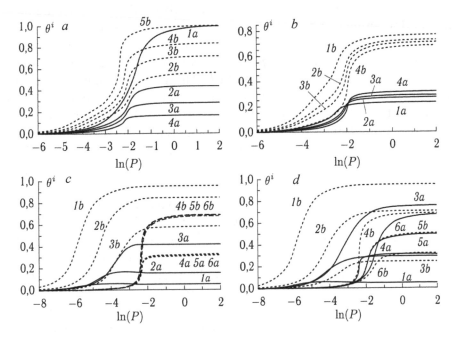

Fig. 60.2. Equilibrium adsorption characteristics $\theta_k^i(P)$ of argon ($i = 1$) and krypton ($i = 2$), where $P = P_1 + P_2$, in a symmetrical slit pore in graphite with the width of 10 monolayers (k – number of the monolayer $1 \leq k \leq 10$) for five vapour compositions: $\gamma = P_2/P = 0$ (1), 0.333 (2), 0.5 (3), 0.666 (4) and 1.0 (5). Indices a and b refer to argon and krypton, respectively. Solid lines refer to the first component, dashed – to the second. a) Average partial isotherms $\theta^i(P)$ of argon and krypton as a function of total vapor pressure P for the five compositions of the vapor mixture γ. b) Effect of the average width of the pores on the partial isotherm of argon and krypton as a function of the total vapour pressure P for $\gamma = 0.5$: $H = 5, 8, 11$ and 25 monolayers. c) Partial local isotherm. Number of curves consist of three numbers, the first number labels the binding energy value of the second component with the wall $Q_1^2/\varepsilon_{Ar\,Ar} = 12.2$ (1), 9.2 (2), 6.2 (3), the second number – the number of the component of the mixture, third number – local area: 1 – surface layer, 4 – the central region. $T = 400$ K, $R = 1$. d) Partial local isotherms. Conditions as in the field c for variation of the of interparticle interaction $\varepsilon_{22}/\varepsilon_{Ar\,Ar} = 1.37$ (1), 1.0 (2), 0.63 (3).

pressure. The fields a and b shows the average partial isotherms at variable γ and pore width H. The fields c and d are local partial fillings of the first and fourth monolayer for different values of the interaction with the wall of the second part Q_1^2 of the lateral interaction between the particles of the second component.

With increasing total pressure the filling of each monolayer increases (field a). The surface monolayers are first to be filled, then the second and so on (the count goes from the wall of the pore).

Stronger attraction of krypton leads to a shift of all partial isotherms to smaller values of the total pressure in comparison with partial argon isotherms. The influence of pore slit width ($H = 5, 8, 11$ and 25 monolayers) on the average partial isotherms of the components of the mixture is illustrates by the field b. The partial fillings are given as a function of the total vapor pressure P for $\gamma = 0.5$. Decreasing H increases the rate of filling of the pores due to the influence of the attractive potential of the wall. Further increase in the width of the pores slightly alters the conditions of filling the central part of the pores, which does not depend on the potenttial of the wall. In all cases, the degree of filling of krypton is greater than that of argon.

The effect of potential functions on the partial local isotherms is shown by the fields c and d. Given are the local degrees of filling the central and surface monolayers that reflects the overall range of variation of all local fillings at $\gamma = 0.5$. Field c shows the variation of the depth of the potential well of the second component of the mixture Q_1^2 with the wall at constant parameter ε_{22}, corresponding to krypton (the first component is argon). Reducing Q_1^2 decreases the degree of filling (of the surface layer of the pore) by the second component and increases the degree of filling by the first component. In this case the curves of local partial isotherms for the central part of the pores remain practically constant. However, the differences of the partial isotherms of the secon component $2b$ and $3b$ for the surface monolayer in comparison with similar isotherms of the first component $2a$ and $3a$ indicate the important role of intermolecular interactions in the course of filling the pore. More detail is shown by the curves in field d. Here the potential depth of interaction between the atoms of the second component of the mixture ε_{22} at $Q_1^2 = $ const, corresponding to krypton (first component is argon) changes. Reducing ε_{22} dramatically reduces the degree of filling of the surface layer by the second component and increases the degree of filling of the first component. The degree of filling of the central part of the pore by the second component decreases at the same time and the proportion of the first component increases. This result reflects the fact that the contribution of the wall potential affects the subsurface regions, and intermolecular interactions are manifested throughout the volume of the pores.

The considered isotherms were obtained under the condition of about the same dimensions of the two components. However, the width of the pores can be disproportionate to the value of the parameters σ_{ij}, so we consider the question of the proportionality

of the width of the pore and the diameter of the hard sphere of
the molecules in more detail in the example of this binary mixture
in a narrow carbon slit with the width $\sim 5\sigma_{ArAr}$. If we place five
Kr atoms in the cross section of such a pore, they can not located
be in the minima of their potential functions and this will lead to
changes in the lateral interaction parameters. This circumstance
can be solved by using a model that takes into account the change
in the lattice structure constants [18]. This is the so-called soft
lattice, its properties are determined by minimizing the free energy
of the system. A similar result can be obtained if we consider the
variation of the parameters of the lateral interaction depending on
the orientation of the neighbouring molecules.

When studying this system it was assumed that the average
distance between atoms within the five monolayers did not differ
from their bulk values, which were used in the calculations of
Fig. 60.2, and the parameters of the corresponding interactions
along the section of the pores were varied. To define them, we
selected conditions in the study [27] where the comparative data

Table 8.1. Comparison of excess adsorption of argon and krypton atoms
$\Gamma_i = \sum_{q=1}^{t} F_q [\theta_q^i - \theta_i(\text{bulk})] v_i / v_{Ar}, v_i = (1.1224\sigma_{ii})^3, i = \text{Ar, Kr}$, in the carbon pore obtained
by the MD method [27] and the QCA; θ (bulk) – dimensionless total gas density
in the bulk phase, x – Ar mole fraction in the gas bulk phase ($x = 1 - \gamma$, θ_{Ar} (bulk)
$= x\theta$ (bulk))

No.	θ (bulk)	x	Adsorption of argon atoms		Adsorption of krypton atoms	
			QCA	MD	QCA	MD
1	0.0504	0.43	0.19	0.24	1.29	1.26
2	0.0252	0.91	0.76	0.55	0.37	0.21
3	0.1030	0.89	0.86	0.86	0.53	0.38
4	0.0792	0.76	0.55	0.58	0.89	0.79
5	0.0300	0.76	0.51	0.45	0.82	0.53
6	0.0528	0.50	0.24	0.31	1.24	1.19
7	0.0132	0.55	0.24	0.18	1.00	0.63
8	0.0018	0.84	0.05	0.10	0.04	0.11
9	0.4440	0.26	−0.16	−0.14	0.51	0.51
10	0.0468	0.75	0.52	0.53	0.88	0.65
11	0.0192	0.79	0.50	0.36	0.68	0.42

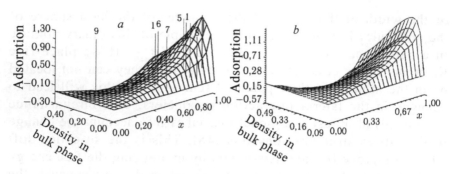

Fig. 60.3. Excess partial adsorption of argon atoms Γ_{Ar} in a binary argon–krypton mixture in a carbon pore at $T = 238\ K$: a – calculation by the density functional method [27], b – calculation using the lattice gas model. Point numbers correspond to the numbers in Table 8.1.

were obtained by the method of molecular dynamics and the density functional method, and the following parameters were evaluated: $\varepsilon_{Ar\ Ar}/k_B \approx 5.7$ K, $\varepsilon_{Ar\ Kr}/k_B \approx -5.7$ K, $\varepsilon_{Kr\ Kr}/k_B \approx -40$ K. The binding energy with the the surface was calculated according to the formulas given in section 5, where $U_{i-C}\ (r)$ is the potential function of the interaction of the atom i with a graphite surface, integration is performed over the volume of the cell $V_f = \lambda^3$ with the number f; $\lambda = 1.1224\sigma_{Ar\ Ar}$. Table 8.1 and Fig. 60.3 show the comparison between the excess (unlike Fig. 60.1, which gives complete filling values) partial filling Γ_i the pores in carbon obtained in this work and by the methods of molecular dynamics and the density functional

theory, where $\Gamma_i = \sum\limits_{q=1}^{t} F_q[\theta_q^i - \theta_i(\text{bulk})]v_i/v_{Ar}$, $v_i = (1.1224\sigma_{ii})^3$, $i = Ar$,

Kr, θ_i (bulk) – dimensionless partial density of the gas in the bulk phase.

It should be noted that the methods of molecular dynamics and density functional lead to the same degree of consistency as the calculations within the lattice gas model. Thus, the question of commensurability of the width of the pore and the diameter of the hard sphere of the molecules is important when comparing the results obtained with different methods of description.

Phase diagrams
For mixtures, the density of the coexisting vapour and liquid phases of the adsorbate was determined by Maxwell' relationships [7, 28], as for pure substanes. Recall that all phase transitions of

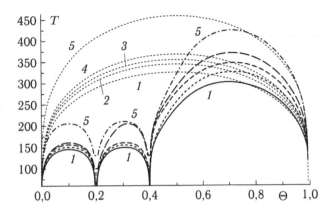

Fig. 60.4. Phase diagram (without calibration) in the slit pore width of 10 monolayers, varying the mole fraction of argon: $x_{Ar} = \theta_{Ar}/\theta = 1.0$ (*1*) 2/3 (*2*) 1/2 (*3*) 1/3 (*4*) 0 (*5*).

the first kind are stratification or condensation transitions (and the term 'stratification' is more common), in contrast to the phase transitions of the second kind in which ordering processes take place. Stratification is accompanied by the formation of the two-phase state of matter with low and high density. In complex fields of adsorption the nature of the surface potential directly determines the spatial regions in which the process of stratificatiuon to low- and high-density phases takes place. Under certain energy ratios of interactions of molecules with the pore walls and between them these spatial areas may separate from each other. A characteristic feature of the liquid–vapour stratification curves for narrow-pore systems is the presence of one or more domes, corresponds to the filling of the surface layers (chapters 3 and 7).

For mixtures, the new factor is the possibility of selective separation in one or another part of the pore volume. These spatial fields appear differently depending on the nature of the surface potential and the degree of filling of the adsorbent. Under these conditions it is important to analyse the local distribution of molecules of each species. The situation is simplified if one considers simple fluids.

An example of a binary phase diagram of an argon–krypton mixture being in a slit-like pore 10λ is shown in Fig. 60.4 for the composition of the mixture varying from one component to another. The channel walls correspond to SiO_2 groups. In the pores the phase diagrams of the mixtures retain the properties of individual substances – surface domes associated with gradual layer filling of the surface sites of the pore volume are observed. For comparison, the phase diagram for the bulk phase with the variable composition of the same mixture are presented.

Numbers of the curves relate to pure argon (*1*), mixtures (*2–4*) and pure krypton (*5*). The critical temperature of the binary mixture increases with increasing mole fraction of the component having a higher critical temperature. The same trend is seen for all the surface and central domes simultaneously. It can be seen that the critical temperature of the mixture increases with the proportion of krypton. In general, there remains the approximately linear relationship between the value of the critical point of the dome in the central part of the phase diagram and the mole fraction of the component of the binary mixture. This is due to the absence of specific bonds between the atoms of argon and krypton. These gases belong in a group of atoms and molecules obeying the law of corresponding states [4, 20] and the combination rule, preserving this law.

This example shows that all of the above patterns for a substance are transferred to multi-substance mixtures, both as regards the structure of complex heterogeneous porous structures and the use of the law of corresponding states. In the case of specific interactions, of course, the concentration dependences of equilibrium distributions and phase states will be more complicated.

61. Bulk transport characteristics of binary mixtures

The most important dynamic characteristics are the diffusion coefficients and shear viscosity coefficient. Theoretical calculation of these coefficients in a wide range of fillings (in gaseous and liquid states) and temperature is difficult. So now these dynamic characteristics of adsorbates are calculated mainly by the molecular dynamics method. Recall also that the currently available experimental methods of measurement of self-diffusion coefficients even for pure components (NMR and the labelled particles) lead to results very different from the measurements of the flow characteristics. For mixtures the situation is even more complicated.

The equations derived for the transport coefficients (Appendices 11 and 12) were tested on experimental data for the bulk phase. Calculation of the transport coefficients of individual substances was carried using parameters of the interaction potential, defined on the phase diagrams and the compressibility factor, and also the value of the interaction parameter ε^*, which gives the best agreement with the available experimental data on the transport properties (section 34). After applying the above procedure the compressibility and transport coefficients for mixtures of dense gases were calculated [15]. Clarification of the parameters required for calculating the properties of the mixture can be carried out in the presence of the corresponding experimental data for these mixtures.

The lattice-gas model (LGM) was used to calculate the thermodynamic and kinetic characteristics of the binary mixtures of molecules of the same size, the shape of which differs only slightly from spherical. The modified LGM was refine taking into account the excluded volume and the introduction of the calibration function that improves accuracy of the description near the critical temperature range. This allows the use of the LGM at temperatures both below and above the critical value.

Bulk phase

Formal generalization of the expressions for the coefficients of transfer of the single-component substance to multicomponent mixtures leads to the well-known result (Mayer formula), which is significantly different from the experiment and 'replaced' by the Stefan–Maxwell formula [16]. This characterful replacement has contributed to the development of the Chapman–Enskog kinetic theory of gases based on the use of the velocity distribution function [29–31]. Analysis of this problem from the standpoint of the theory of processes in condensed phases has shown [22,32] that the problem can be re-formulated staying within the range of the average velocities of the molecules, and get the results on the accuracy of the calculation of the mutual diffusion coefficient not worse than in the Chapman–Enskog theory. This takes into account explicitly collisions with neighbouring molecules [15, 32], which somewhat changes the form of the imperfection section (see Appendix 11).

Figure 61.1 *a* shows the concentration curves of the dependence of the coefficient of mutual diffusion in a helium–argon system for the changing mole fraction of helium atoms, calculated by different methods. The experimental data [29] (curve *5*) do not contain information about the temperature, so they are normalized to the limiting value for the mutual diffusion coefficient of the heavy component ($x_{Ar} \rightarrow 1$). This experiment demonstrates that the higher proportion of light atoms, the lower the coefficient of mutual diffusion. Limiting values of $D_{1,2}$ for $x_{Ar} \rightarrow 1$ and $x_{Ar} \rightarrow 0$ were obtained from the tabular data

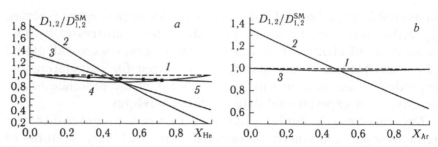

Fig. 61.1. Normalized concentration dependence of the mutual diffusion coefficient in a binary mixture: *a* – a helium–argon mixture at varying mole fraction of helium atoms [22], *b* – an argon–krypton mixture with varying mode fraction of argon atoms [23].

[29] for the average relationship of mole fractions of the two components (indicated by dots in Fig. 61.1 *a*) by linear extrapolation in accordance with the conclusions of the rigorous theory [4, 29–31]).

Curve *1* corresponds to the diffusion coefficient $D_{1,2}^{SM}$ described by the Stefan–Maxwell formula (or the rigorous kinetic theory based on the first order), curve *2* – Mayer formula, *3* – Mayer formula with a modified method of taking into account the free path: instead of the partial path the mean free path common for both components was used. Curve *4* is constructed using the modified formula in LGM [22, 32]. The crucial difference between the Stefan–Maxwell and Mayer equations is the weak dependence of the mutual diffusion coefficient on the molar composition of the mixture at a total low-density of the mixture.

It is shown that the LGM leads to the maximum deviation between $D_{1,2}$ and $D_{1,2}^{SM}$ not exceeding the magnitude $(1-1/2^{1/2}) < 0.3$ (i.e., less than 30%). For 'real' sizes and mass such as the He–Ar, the maximum value of the discussed deviations from the experimental data [29] is < 12%, which is of the same order as when using the formula in [32] (see Appendix 12).

Figure 61.1 *b* shows the normalized concentration dependences of the mutual diffusion coefficient in the binary argon–krypton gas mixture when the mole fraction of argon atoms changes. The results of calculation show that the modification of the LGM [22, 32] considerably changes the form of the dependence of the mutual diffusion coefficient in the binary mixture on the mole fraction of the light component (curve *3*) compared with the Mayer calculation formula (curve *2*).

Normalization for both systems was performed for the value $D_{1,2}^{SM}$, calculated by the Stefan–Maxwell equation for the same mixture. For the argon–krypton mixture the maximum difference in the calculations by the formulas of the new modified model differs from calculations by the Stefan–Maxwell formula (curve 1) by only 1.2%.

Dense mixtures

The diffusion of dense mixtures was studied using the same molecular parameters as in the calculation of the equilibrium density profile in Fig. 60.1. Additionally, the hopping rate was calculated using the dimensionless parameter $\alpha = \varepsilon^*/\varepsilon = \varepsilon_{fh}^{*ij}(r)/\varepsilon_{fh}^{ij}(r)$. Its value was found from a comparison with experimental data on the shear viscosity η of argon in the bulk phase [33]. According to the experiment an increase in the density of argon θ from a dilute gas to a value of 0.6 gives values of η almost twice as high. Calculations performed for a lattice structure $z = 12$ showed that $\alpha \sim 0.57$ [23]. The parameter α is sufficiently sensitive to the molecular features of the model. The same agreement with experiment was obtained earlier for a lattice with the number of nearest neighbours $z = 6$ at $\alpha = 0.85$ [34]. However, in addition to the different structural factor, previously studies used the traditional LGM which does not explicitly take into account collisions with neighbouring molecules [22, 32]. The collisions leads to a modification of the expression for the imperfection fiunction of the speed (see Appendix 7).

In practice, the so-called generalized diagrams of the transport properties of dense gases are used quite often. A generalized diagram of the self-diffusion coefficients [20] was obtained by averaging the experimental data whose amount is extremely limited (inert gas, air components, etc.), and using the Enskog kinetic theory for dense gases. The generalized diagram is the dependence of the reduced self-diffusion coefficient $PD_{ii'}/(PD_{ii'})_0$ on reduced pressure P/P_c ($0.1 < P/P_c < 4$) for certain values of the reduced temperature $\tau = T/T_c$ (in [20] $\tau = 1, 1.1, 1.2, 1.3, 1.4, 1.6, 1.8, 2.0, 3.0$). This generalized diagram has hitherto been used for quick interpolation evaluation of the values of the self-diffusion coefficients of many single-component gases in the given region of changes in pressure and temperature. Furthermore, the generalized diagram is used for determining the coefficients of mutual (mass) diffusion.

The LGM was used to describe a generalized diagram [20] and its extrapolation to high density (see Fig. 61.2). Figure 61.2 a gives the dependences $PD_{ii'}/(P\,D_{ii'})_0$ for different values of τ versus reduced

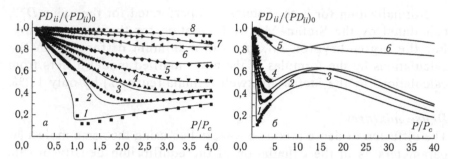

Fig. 61.2. Generalized diagram of the self-diffusion coefficient for several values of τ (calculation – solid lines, the experiment [20] – icons): a) τ = 1 (1), 1.1 (2), 1.2 (3), 1.3 (4), 1.4 (5), 1.6 (6), 2 (7), and 3 (8) for a range of pressures $P/P_c < 4$; b) τ = 1 (1), 1.1 (2), 1.2 (3), 1.3 (4), 2 (5), 3 (6) for a wide range of pressures $P/P_c < 40$.

pressure $P/P_c \leq 4$. The theoretical curves were calculated with the following parameters of the LGM : $u_{ij} = 0$, $d_{ij} = 0.1$, $0.1 \leq \kappa_{ij} \leq 0.15$, i.e. parameters u_{ij} and d_{ij} for these calculations were united, and the value κ_{ij} was recorded in the aforementioned narrow range. The value $\alpha = \varepsilon^*/\varepsilon$ was chosen from the range $0.55 \leq \alpha \leq 0.68$ and had the following form (linear smoothing): $\alpha = \alpha_1$ at $0 < \theta < \theta_1$, $\alpha = \alpha_2$ when $\theta_2 < \theta < 1$, $\alpha = \alpha_1 + (\alpha_2 - \alpha_1)(\theta - \theta_1)/(\theta_2 - \theta_1)$ with $\theta_1 < \theta < \theta_2$. The values of θ_1 and θ_2 correspond to the boundaries of the zone of the maximum change in curvature of the dependence $p(\theta)$. The study did not attempt to achieve accurate quantitative agreement between theoretical and experimental diagrams because the experimental diagram itself was approximate, although such a coincidence is achieved after a small correction of some parameters of the LGM and the introduction of non-linear smoothing $\alpha(\theta)$ (non-linearity of the value α follows from its molecular nature [35]).

Figure 61.2 b shows the generalized dependence of the self-diffusion coefficient in the pressure range an order of magnitude greater than the pressure at which this value was measured $(0 < P < 200$ MPa). As shown above, this characteristic corresponds to the region of the applicability LGM values found at relatively low densities.

The mutual diffusion coefficient of binary mixtures
As is known, the data on the diffusion coefficients for gas mixtures are even more limited when compared with the data for one-component substances and are not very accurate. One of the most common ways of the theoretical evaluation of the mutual diffusion

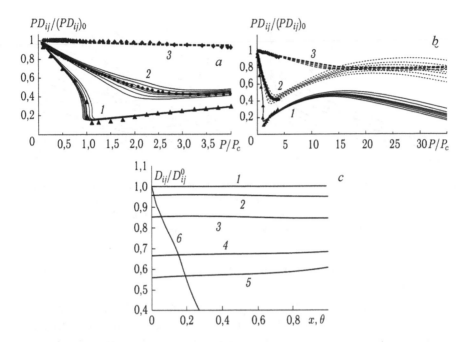

Fig. 61.3. Mutual diffusion coefficients of Ar–N$_2$ mixtures of different composition: *a* – generalized diagrams of the coefficients of the mixture for the mole fractions of nitrogen x_{N_2} = 0, 0.2, 0.4, 0.5, 0.6, 0.8, 1.0 at τ = 1 (*1*), 1.3 (*2*) and 3 (*3*) in the pressure range $P / P_c <$ 4; *b* – Analogous dependence for a range of pressures P/P_c < 40; *c* – dependence of the reduced mutual diffusion coefficient D_{ij}/D_{ij}^0 of the the the mole fraction of nitrogen x_{N_2} at the gross density θ = 0.001 (*1*), 0.1 (*2*), 0.2 (*3*), 0.5 (*4*), 0.7 (*5*), and the dependence on the gross density θ (*6*) for τ = 1.

coefficient is the method described in [20]. As the critical parameters of the gas mixture P_c and T_c are taken above pseudocritical parameters. The mutual diffusion coefficient is calculated by the generalized diffusion coefficient shown in Fig. 61.2 *a*, according to the method described in [20]. This empirical approach was used for comparisons with the results of the LGM.

As an example, Figure 61.3 *a* shows by bundles of curves shows the dependences (calculated by the LGM) of the reduced coefficient of mutual diffusion $PD_{ij}/(PD_{ij})_0$ of the Ar–N$_2$ binary gas mixture versus reduced pressure $P/P_c \leq$ 4 for the varying molar concentration of nitrogen (as x_{N_2} decreases the curve in the bundle also comes lower) for τ = 1.0, 1.3 and 3.0. Figure 61.3 *b* shows the same graphics with the scale for P/P_c magnified by an order of magnitude (0 < P < 200 MPa). Figure 61.3 *c* shows for the the case τ = 1 the dependence of D_{ij}/D_{ij}^0 on the molar composition of the Ar–N$_2$

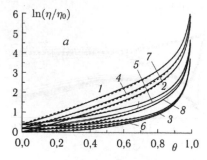 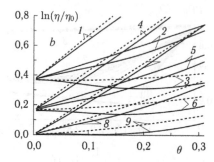

Fig. 61.4. Concentration dependences of the shear viscosity coefficient calculated for the bulk phase at $T = 15\ T_{Ar}$ (crit) for $\alpha = \varepsilon^*_{ij}/\varepsilon_{ij} = 0.3$ (1–3), 0.55 (4–6), 0.7 (7–9) and $X_{Ar} = 1.0$ ($3,\ 6,\ 9$), 0.5 ($2,\ 5,\ 8$), 0 ($1,\ 4,\ 7$): a) the full range of density of the argon–krypton mixture; b) low densities of the mixture.

mixture at different fixed values of θ. The dependence of D_{ij}/D^0_{ij} on θ is shown here.

The graphs confirm the known experimental fact of the weak dependence of the mutual diffusion coefficient on the mixture composition and its strong decrease with increasing density. The calculated curves in the bundle for each τ are aligned from the bottom up in the order of increasing nitrogen concentration. The symbols are the values of the generalized diagram of mutual diffusion obtained by the recalculation of the experimental data in Fig. 61.2 *a* according to the rules specified in [20].

Coefficient of viscosity of binary mixtures

As described in Chapter 5, to calculate the transport coefficients we must use the parameter of interaction of molecules with the activated complex defined by dimensionless parameter $\alpha = \varepsilon^*/\varepsilon = \varepsilon^{ij}_{fh}(r)/\varepsilon^{ij}_{fh}(r)$. Its value is determined from comparison with experimental data on the shear viscosity η of argon in the bulk phase [34]. The same parameters are used for mixtures.

Figure 61.4 illustrates the effect of the model used for the displacement velocity of the molecules on the form of the concentration dependences of the shear viscosity coefficient in a wide range of density of the argon–krypton mixture. The solid lines show the dependences calculated by the formulas derived in [22, 32], and the dotted lines are the curves calculated according to the formulas ignoring the type of colliding molecules. At low fillings both expressions lead to markedly different curves. For clarity, the region of θ to $\theta = 0.3$ is shown in Fig. 61.4 on a large

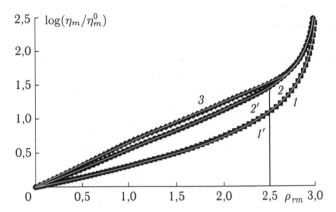

Fig. 61.5. Logarithmic dependenced η_m/η^0_m as a function of reduced density ρ_{rm} of the Ar–N$_2$ binary mixture for $\tau = 0.75$ (*1*), 1 (*2*) and 3 (*3*). Symbols – calculations, thin lines – generalized experimental dependences [20].

scale. However, the basic change of the values of the coefficient of viscosity with increasing density of the mixture remain the same. For values of $\theta > 0.4$ both models give approximately the same value (Fig. 61.4 *a*).

As above, all the dynamic characteristics are normalized to the corresponding values for the coefficients for argon in the dilute gas phase.

One of the most popular methods for calculating the viscosity of mixtures of non-polar dense gases was developed by Dean and Steele (cited in [20]). Based on the processing of large amounts of experimental data for binary mixtures of light hydrocarbons, hydrocarbons with inert gases and air components an universal correlation formula was obtained for the calculation of the so-called residual viscosity $\left(\eta_m - \eta^0_m\right)$ at high pressure: $\left(\eta_m - \eta^0_m\right) = f\left(\rho_{rm}\right) =$ 1.08 [exp (1.439 ρ_{rm}) – exp (–1.11 $\rho_{rm}^{1.858}$)], where η^0_m is the viscosity at low pressure, $\rho_{rm} = \rho_m/\rho_{cm} = \theta/\theta_c$ is the pseudoreduce density of the binary mixture, $\xi_m = T_{cm}^{1/6}/M_{cm}^{1/2} p_{cm}^{2/3}$. There are a number of methods for calculating η^0_m, as detailed in the reference literature (for example, see [20]).

Figure 61.5 shows that the dependence of the residual viscosity fully described by the LGM if η^0_m is determined by the approximation proposed by Todos et al and Herning and Zipper (see [20]). Compared are the dependences of η_m/η^0_m on the reduced density ρ_{rm}, obtained for a binary Ar–N$_2$ mixture at values of $\tau = 0.75$, 1.00 and 3.00 from the empirical formula and the LGM [15]. These relationships were plotted used a logarithmic scale as on this scale these relationships are close to linear.

The discrepancy in the results does not exceed the error of the generalized correlation based on the experimental data. Calculations were carried out by the LGM with the following parameters: $u_{ij} = 0$, $d_{ij} = 0.1$, $\kappa_{ij} = 0.15$, i.e. parameters u_{ij}, d_{ij} and κ_{ij} for these calculations are the same. The value $\alpha = \varepsilon^*/\varepsilon$ was chosen from the range $0.35 < \alpha < 0.5$. Just as in the calculation of the mutual diffusion coefficient, the linear smoothing was performed but in a substantially wider range of density $\theta_1 < \theta < \theta_2$. As can be seen in Fig. 61.5, in the region $\rho_{rm} > 2.4$ there is a sharp increase in viscosity. This fact corresponds well to the liquid state of dense systems.

62. Diffusion coefficients in the slit-like pores

The specificity of the narrow-pore systems consists in the fact that the walls of the majority of porous materials attract adsorbate molecules. The most common situation occurs when the pure adsorbates or mixtures thereof are in the central part of the pores in the rarefied state and liquid-like state on the walls of the pores (see previous chapters). As a result, a film flow of the adsorbate forms along the walls of the pores and, consequently, a sharp gradient of density and flow rates is established. Formation of the adsorbed film suggests the 'multiphase' state of the mixture in the cross section of the pores. In multicomponent systems, the differences in the diffusion mobility of the components at any density both along the pore axis and normal to the walls of the pores create the spatial anisotropy of the flow of molecules of different types and the gradients of density and molar composition form.

The question of the mutual diffusion coefficient in the mixtures is important when considering the transport processes in narrow pores (10–15 nm wide). This factor is present in the equations of mass transfer in the bulk phase, and, naturally, its presence is essential for the mass transfer in the porous bodies. The equations for calculating this coefficient are discussed in detail in the Appendices 11 and 12. In [32] it was shown that despite the difference in the mechanisms of transport of particles in an ideal gas and a solid alloy the mutual diffusion coefficient in the moving reference frame is expressed in the same way through the diffusion coefficients of the individual components D_i in the fixed frame of reference. The modification of the LGM equations are also explained there, which allows to build

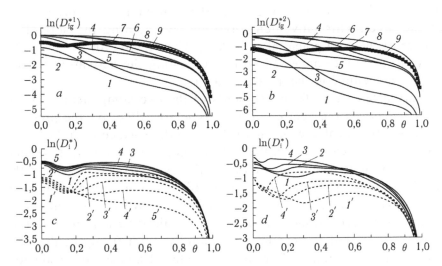

Fig. 62.1. Concentration dependences of the transport coefficients of the label in an argon–krypton mixture at $H = 10$, $T = 1.5\ T$ (crit), $R = 1$, $\alpha = 0.5$, $\alpha_{11} = 1/3$, $Q_1^{Ar} = 9.24\ \varepsilon$. Local coefficients for the following pairs of adjacent cells in layers $f_g = 11$ (*1*), 12 (*2*), 21 (*3*), 22 (*4*), 23 (*5*), 32 (*6*), 33 (*7*), 55 (*8*), curve *9* – average value of D_i^* at $X_{Ar} = 0.5$; Figure *a* relates to the argon label and Figure *b* to the krypton label. Average values D_i^* on *c* are given for $X_{Ar} = 0.005$ (*1*), 1/4 (*2*), 1/2 (*3*), 3/4 (*4*) and 0.995 (*5*). Average values in *d* are given for $X_{Ar} = 0.5$ in pores of different width $H = 6$ (*1*), 10 (*2*), 14 (*3*), 30 (*4*). In the fields *c* and *d* the numbers without primes refer to argon, those with the primes to krypton.

a unified expression for the mutual diffusion coefficient, which can be used for any phase.

Transfer coefficients of the label in the pores

The diffusion coefficient characterizes the motion of a label or self-diffusion in one-component substances subject to the equilibrium distribution of molecules in the entire considered area. Figures 62.1 *c* and *d* shows the local concentration dependences of the transport coefficients of the label of argon atoms (solid lines) and krypton (dotted line) which depend on the distance from the pore wall and the direction of migration, as well as average values of the partial transfer coefficients of the label. Curves *1–8* sequentially number the concentration curves relating to all pairs of sites f_g, where $g = f$, $f \pm 1$. The movement of the mixture occurs inside the layer f at $g = f$ and normal to the surface of the slit pore with $g = f \pm 1$. For $g = f + 1$, the motion is carried to the centre of the pore, which for the attractive potential of the wall corresponds to a decrease of the binding energy with the wall with increasing number of the layer. This movement involves overcoming the activation barrier of the wall potential $(Q_f^i - Q_{f-1}^i)$. Conversely, for $g = f - 1$ the movement is not connected with overcoming the activation barrier of the molecule i in the adjacent layers.

Partial label transfer coefficients of the two components (a) and (b) decrease as the centres of each type are filled. As the pressure increases the local coefficients related to the surface monolayer are the first to decrease, and in the last instance – the coefficients related to the central area. Average values of the transport coefficients of the labels (curves 9) vary non-monotonically. They have a maximum in the region of filling the surface monolayer, and when it is almost full, filling of the second monolayer starts. In this situation, migration in the second monolayer occurs quite rapidly. With filling of the pore volume the fraction of free sites is reduced and all the transport coefficients of labels decrease.

Figure 62.1 c shows the average values D_1^* for different compositions of the mixture. With the decrease in the proportion of krypton and increase of the proportion of argon the initial values of the transport coefficients of argon and krypton labels increase, as their share in the surface monolayer changes. It can be seen how krypton displaces argon to the central part of the pore, as Kr is adsorbed more strongly by the pore surface.

Figure 62.1 d shows the concentration dependence of the mean mass transfer coefficients of the argon and krypton atoms for pores of different width $H = 6$ (1), 10 (2), 14 (3), 30 (4). The increase of H increases the contribution of the central part of the pores, so the area of the non-monotonic variation of the coefficients is shifted to lower densities, and the curve itself (after the minimum) is shifted to a similar curve for the bulk phase.

The mutual diffusion coefficient

Figure 62.2 shows the concentrational dependences of the mutual diffusion coefficients in slit-shaped pores. The general form of the curves in Fig. 62.2 a is largely similar to the above-examined concentration dependences of the transport coefficients of the labels of both components. The same strong anisotropy with respect to the direction of local transport and the distance from the wall is observed. However, quantitative differences are quite noticeable. Figure 62.2 b shows the dependence of the average mutual diffusion coefficients at different mole compositions of the mixture in the slit pore width of 10 monolayers. Unlike the label transfer coefficient the average value of the mutual diffusion coefficient is weakly dependent on the mole fractions of the components of the binary mixture. This can be noted by the well-known analogy with the behaviour of the mutual diffusion coefficient in gases.

The calculations show that the dynamic characteristics of the adsorbate are strongly dependent on the anisotropic molecular distribution over the cross section of the slit pores. The self-diffusion coefficients vary particularly strongly near the pore walls. In the centre of the pores, these values depend on the contribution of the wall potential and the total concentration of the mixture of the adsorbate. In the analysis of experimental data we must consider the sufficiently strong concentration dependence of the partial dynamic characteristics of a mixture of adsorbates in narrow pores caused by both the influence of the potential of the pore walls and their intermolecular interactions.

63. The shear viscosity coefficients

The concentration dependence of the shear viscosity coefficients for dense gases and liquids in narrow pores are unknown. This determines the importance of molecular modelling of transport processes.

Calculation of the viscosity coefficients in the slit-shaped pores was carried out at the same molecular parameters as those used to estimate the diffusion coefficients (Fig. 62.1). Figure 63.1 a shows the components of the shear viscosity tensor in a slit pore with the width of 10 monolayers at an equal ratio of argon and krypton. As in the above examples, all of the initial values of the local shear viscosity coefficients are determined by the surface potential of the wall.

Movement along the surface layer due to overcome the surface potential barrier, which is smaller than the 'desorption' $(Q_2^i - Q_1^i)$, so for zero values $\eta_{11}(0) < \eta_{12}(0)$. Quantity $\eta_{12}(0)$ was the highest, as it is connected with the maximum change of the surface potential of the wall. (This same potential determines $\eta_{ff}(0) < \eta_{ff+1}(0)$ within the range of its effect.) As the filling of the pore increases the first layer is initially filed, and this is determined by a rapid increase in the values of $\eta_{11}(\theta)$, which is superior to $\eta_{12}(\theta)$ with further fillings (the second monolayer sites are blocked increases and the attraction effect of the surrounding molecules becomes stronger).

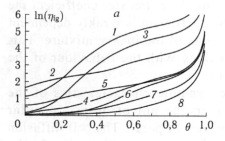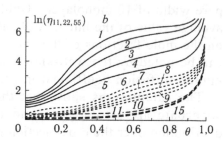

Fig. 63.1. Concentration dependences of the local shear viscosity coefficients of the argon–krypton mixture. Calculation conditions as for Fig. 62.1. *a* – local curves correspond at X_{Ar} = 0.5 to the following pairs of adjacent cells in layers *fg* = 11 (*1*); 12 (*2*); 21 (*3*); 22 (*4*); 23 (*5*); 32 (*6*); 33 (*7*); 55 (*8*); *b* – the values of the coefficients η_{ff}, *f* = 1, 2 and 5, for the compositions γ = 0 (*1, 6, 11*), 1/4 (*2, 7, 12*), 1/2 (*3, 8, 13*), 3/4 (*4 9, 14*) and 1.0 (*5, 10, 15*).

Variations of X_{Ar} are shown in Fig. 63.1 *b* for the three layers *f* = 1, 2 and 5. The greatest impact of the mixture composition can be traced to the surface layer, and the least effect was observed for the central layers – η_{55} curves almost merge into a line.

The concentration dependences of the local viscosity coefficients and the corresponding profiles along the section of the slit pore with the width of 10 monolayers are shown in Fig. 63.2. The viscosity coefficients characterize inhibition of the flux (momentum dissipation) during its movement through the pore, and their values depend strongly on the local flow directions and distances to the walls of the pores.

Figure 63.2 *b* and *c* show the profiles of the shear viscosity coefficients of the argon–krypton mixture in a pore with width H = 10 monolayers at different molar composition (corresponding to the profiles of the partial degrees of filling in Fig. 60.1). As the distance from the pore wall layer increases the coefficients η_{ff} decrease, and with the growth of the overall degree of filling of the pores they monotonically increase.

Influence of pore widths to the local viscosity of the first surface layer (curves *1–3*), the second surface layer (*4–6*) and the centre of the pores (*7–9*) can be seen from Fig. 63.3. The curve 10 corresponding to the bulk viscosity of the mixture of equimolar composition X_{Ar} = 0.5 is shown for comparison. General patterns for the mixture of fixed composition are close to the laws previously obtained for the one-component fluid [36, 37] and chapter 5. The shear viscosity of the adsorbate in the central layer is relatively

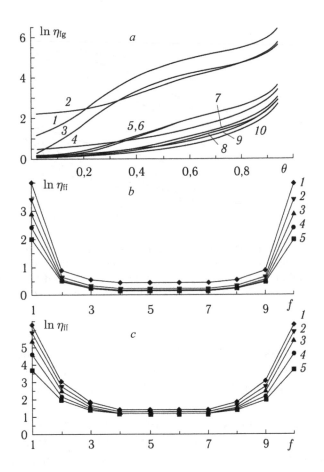

Fig. 63.2. Normalized concentration dependences of the local shear viscosity coefficients η_{fg} of the equimolar argon–krypton mixture $X_{Ar} = 0.5$ in the carbon pore with a width of 10 monolayers, depending on the total density of the mixture at $\alpha_{11} = 0.1$ (*a*): for pairs of adjacent cells in the layers $fg = 11$ (*1*), 12 (*2*), 21 (*3*), 22 (*4*), 23 (*5*), 32 (*6*), 33 (*7*), 34 (*8*), 43 (*9*), 55 (*10*). The shear viscosity coefficient profiles η_{ff} along the section of the pore at $\theta = 0{,}25$ (b): $X_{Ar} = 0$ (*1*), 0.333 (*2*), 0.5 (*3*), 0.666 (*4*) and 1.0 (*5*); for $\theta = 0{.}75$ (*a*), the rest as shown in Figure *b*.

weakly dependent on the width of the pores, although its influence is completely impossible to neglect. The viscosity of the surface layer increases with increasing pore width H. This is due to the increasing degree of filling of the surface layer θ_1 with increasing H at a fixed value θ of the overall density of the mixture inside the pores. (Recall that the calculations in the framework of this lattice gas model are in good agreement with similar results for the profiles

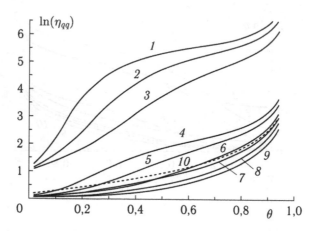

Fig. 63.3. Concentration dependences of the viscosity coefficient of the argon–krypton mixture at $X_{Ar} = 0.5$ in the surface layer η_{11} (curves *1–3*), the second surface layer (curves *4–6*) and the centre of the pore η_{55} (curves *7–9*) with different pore widths $H/\lambda = 20$ (*1, 4, 6*), 10 (*2, 5, 8*) 6 (*3, 6, 9*). The dashed curve *10* corresponds to the bulk phase.

of concentrations and shear viscosity of one-component fluids in slit-shaped pores with a width of 4 and 18 monolayers [38], made obtained by non-equilibrium molecular dynamics and the continuum kinetic theory.)

The dynamic characteristics of the mixture components are strongly dependent on the anisotropic molecular distribution over the cross section of the slit pores. The transfer coefficients vary particularly strongly near the pore walls due to the maximum influence of the surface potential. In the centre of the pores, these coefficients depend less strongly on the contribution of the potential of the wall. The important role is played by the total concentration of a mixture of molecules. The denser the system, the lower the rate of migration and all the associated mass transfer coefficients, while the viscosity coefficient increases significantly

For the transport of dense gases in the adsorbents which are adsorbed efficiently by the pore walls (for example, active charcoal) an important role is played by the sliding effect of the dense fluid. These effects are usually considered only for rarefied gases. In dense adsorbates the sliding effect is due to the surface mobility of the molecules and not a mirror reflection of molecules from the walls, as in the case of rarefied gases.

Traditional assumptions about the constancy of the self-diffusion and shear viscosity coefficients [39–43] of the mixtures are generally incorrect. In the analysis of experimental data we must consider the sufficiently strong concentration dependence of the partial dynamic characteristics of a mixture of adsorbates in narrow pores caused by the influence of the potential of the pore walls and their intermolecular interactions. The molar composition of the mixture and the nature of intermolecular interactions with the pore wall components and other components determine the degree of separation of the components in each section of the pores, which leads to strong non-linear effects for the transport coefficients when the total density of the mixture changes .

64. The sliding friction coefficient for the mixture

The molecular approach for the calculation of the shear viscosity coefficients allows us to give a molecular interpretation of the so-called sliding friction coefficient β_1 for mixtures used in the hydrodynamic equations in calculating the velocity profiles of the fluids near the surfaces [20, 44]. Recall that the coefficient β_1 is defined as the ratio of the tangential force per unit surface to *the relative velocity* of the flow near a solid wall: $\beta_1 u = -\eta \partial u/\partial r|_{r=R}$. The ratio $\eta/\beta_1 = \lambda$ has units of length. Factor λ follows from the relationship dimension, because unlike β_1, the shear viscosity coefficient η is the ratio of the tangential force, attributable per unit surface, to the *velocity gradient* in the direction perpendicular to the flow direction [21].

   In the case of mixtures the expressions for the sliding friction coefficient are retained, as all molecular information is stored in the calculation of local shear viscosity coefficients.

   According to [21, 24], $\beta_1 = \eta_{11}/\lambda$, where the indices 11 correspond to the near-surface monolayer. Therefore, the normalized values of η_{11} presented above characterize with the accuracy to the constant size of the monolayer λ the sliding friction coefficient. As a result, the flow rate in the surface layer is $u = -(\lambda\beta_v)\partial u/\partial r|_{r=R}$, where $\beta_v = \eta_{tt}/\eta_{11}$ (indexes *tt* relate to a narrow central part of the narrow pore of the volume value of the viscosity for wide channels). In the case of the strong adsorbate–adsorbent attraction or reduction of temperature the ratio η_{11}/η_{tt} sharply increases which leads to an increase of the coefficient β_1 and a decrease of the flow velocity near the wall.

   In the above figures, e.g. Figs. 63.1 and 63.3, we can clearly see the dependences corresponding to the local viscosities in the

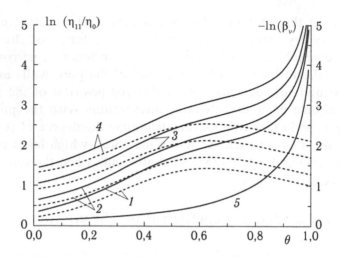

Fig. 64.1. Influence of the activation energy of surface migration of molecules on the viscosity values in the surface layer η_{11} at $X_{Ar} = 0.5$ and $\alpha = 0.60$ for $\alpha_{11} = E_{11}^{iV}/Q_1^i = 0.1$ (*1*), 0.333 (*2*) 0.666 (*3*), 1.0 (*4*) – the solid lines, curve *5* corresponds to the shear viscosity coefficient η_{55} for the centre of the pore. Dotted lines *1–4* – corresponding ratios η_{11}/η_{55} (right axis).

surface monolayer. Their behaviour is affected by both the bond of the molecule with the surface and the magnitude of the activation barrier of surface migration.

An additional important role for a mixture is played by the relationship between the values of the activation energy of surface migration E_{11}^{iV} of both components. Figure 64.1 shows the local viscosity of the mixture of the equimolar composition in the surface layer for different activation energy of surface migration of components of the mixture (the ratio between E_{11}^{iV} for argon and krypton remains constant). The viscosity of the mixture in the centre of the pores is shown for comparison. The difference between the logarithms of the local viscosity at the surface and in the centre of the pore (i.e., the ratio $\beta_v = \eta_{11}/\eta_{55}$), represented by the corresponding dashed lines, shows the non-monotonic behaviour with the increase of the total density of the mixture. However, the most important feature of the dependence $\beta_v (\theta)$ is that with an increase in the degree of filling of the pore with the liquid phase $\theta \to 1$ (in this case, of course, the densities corresponding to a mixture of solid state are ignored) it remains finite, which actually characterizes the intensity of slipping of the fluid near a solid wall. The smaller the value E_{11}^{iv}, the greater the contribution of the surface flow of molecules of each component to the overall 'sliding' flow of molecules. Case $\alpha_{11} =$ 0.1 corresponds to low activation energies of migration (which are characteristic of the migration of the metal atoms at their ideal faces of single crystals), whereas the values of $\alpha_{11} = 1$ correspond to the high activation energies when the molecules must break the bond with the substrate to jump to neighbouring sites. Accordingly, with increasing α_{11} the curves $\beta_\nu(\theta)$ are shifted to higher values.

The maximum on the curves $\beta_\nu(\theta)$ arises due to the different rates of increase of the shear viscosity coefficient near the wall compared to the same change in the bulk phase. Upon reaching a sufficiently high density in the bulk phase the rates of increase of viscosity with increasing θ are equalized in both systems. But at the same θ value in the pore the central portion is more rarefied (due to increased density on the walls) than in the bulk phase, thus the viscosity in the volume increases more rapidly than in the pore. (Note that previously no account was made not only of the possibility of such a non-monotonic behaviour $\beta_\nu(\theta)$, but also of the effect of slip in the liquid phase was not considered, as there was no unified approach to the study of gaseous and liquid phases).

65. Spherocylindrical system

As noted above for single-component substances, the theory of transport of mixtures in narrow porous systems was developed in [21, 45] (see chapter 7). In these studied the authors formulated differences in description of equilibrium and kinetic characteristics occurring at the transition from individual pores to complex pore systems. They are related to the need to consider the following factors: 1) the limited length of the specific section of the pore, 2) transition regions (junctions) between different adjacent areas of the pores and so on.

As a simple example of porous systems with variable sections we consider spherocylindrical systems in which the porous space is formed by alternating cylinders, with finite length, and truncated spheres [46]. Partition of the system to two types of levels (molecular and supramolecular) simplifies this task. The main task of the supramolecular level of the theory of processes in porous bodies is the construction of the cumulative distribution functions, which allows sufficiently accurate description of the structure of porous systems. As before, the real porous structure will be modeled using the sites of the pores of the dispersed body of some characteristic

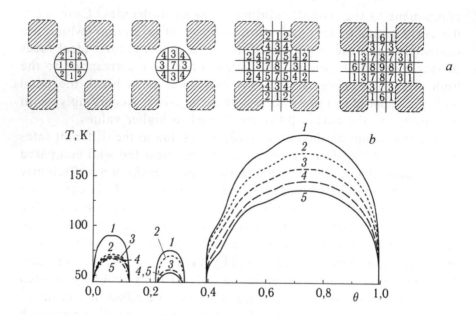

Fig. 65.1. Partitioning the pore spaces in the spherocylindrical system D_s–D_c–L_c = 5–3–1; (*a*) phase diagram (without calibration) in a regular spherocylindrical structure D_s–D_c–L_c = 20–14–6 (*b*) [47].

size $L > \lambda$, where λ is the linear size associated with the diameter of the spherical adsorbate.

Figure 65.1 *a* shows the partition into layers of the section of the spherocylindrical system consisting of a truncated sphere with a diameter 5λ and a cylindrical with channel length 1λ and diameter 3λ. Such structures are conveniently denoted as D_s–D_c–L_c = 5–3–1, where the symbols D_s and D_c relate to the intersecting sphere and the cylinder, and the symbol L_c refers to the length of the cylindrical part of the channel, all dimensions are given in units of λ. Equivalent sites are numbered identically. Section 1 corresponds to the cross section of the narrowest portion of the system – the window between two cavities. Section 4 corresponds to the central part of the sphere (for simplicity we used here the cubic lattice for the introduction of sites with the number of nearest neighbours z = 6). The sites with different binding energy of the molecules with the walls of the pores have different numbers. The interaction energy with the wall was calculated in the framework of the averaged model of a solid [47]. For this kind of lattice structures it is easy to construct unary and binary distribution functions of system sites [21].

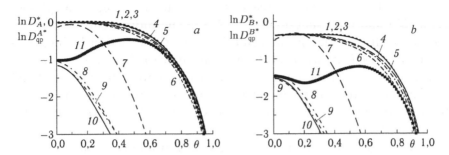

Fig. 65.2. Transfer coefficients of the label: a – for argon in a regular spherocylindrical structure $D_s–D_c–L_c$ = 20–08–20 fo the equimolar composition of the binary Ar–Kr mixture; b – for krypton in a regular spherocylindrical structure $D_s–D_c–L_c$ = 20–08–20 for the equimolar composition of Ar–Kr binary mixture (numbers of curves are the same as on the field a).

Figure 65.1 b shows analogues of the phase diagrams, as in Fig. 60.4, by varying the mole fraction of argon: $x_{Ar} = \theta_{Ar}/\theta = 1.0$ (*1*); 2/3 (*2*); 1/2 (*3*); 1/3 (*4*); 0 (*5*). As above, the complex porous structures retain in the first approximation the relationships that are typical for mixtures of molecules obeying the 'law' of the corresponding states.

Local dynamic characteristics
In the transition from simple pore geometries to complex systems we must take into account that the previously obtained expressions for the transport coefficients for mixtures in isolated pores [21, 45] formally retain their shape in recording when replacing one index for the site number by two indices, each of which describes the type of section of the porous system and the cell number in this area (as in expressions for the equilibrium distribution (see chapter 7)).

Transfer coefficients of the label [47]
The transfer coefficients of the label for all components depend strongly on the local distributions of mixture components in the cross section of the pores and on the direction of movement between the sections under consideration q and p. These coefficients are also dependent on the total local density of the system and intermolecular interactions. The specificity of the transport coefficients of the label are demonstrated by Figs. 65.2 a and b for the atoms of argon and krypton, respectively. The calculations were performed for the system $D_s–D_c–L_c$ = 20–08–20 with partition of the volume of the pores of the cubic lattice $z = 6$. The parameter $\alpha = \varepsilon_{ij}^*/\varepsilon_{ij} = 0.4$.

The system consists of three sections: a sphere (2, 4, 8), a cylinder (1, 6, 10) and their junction (3, 5, 7, 9). The movement of the label parallel to the surface in the central part of each section is described by the curves 1, 2 and 3, in the second layer of the section – curves 4, 5 and 6, and in the surface layer of the section (curves 8, 9 and 10); movement to the wall at the junction from the second layer to the surface layer – curve 7, curve 11 – the average transfer coefficient of the label.

The initial values of local transport coefficients of the label are determined by interactions with the walls of the pore. Increasing density of the mixture results in a decrease of the local diffusion coefficient. The character of the changes of the concentration dependences depends largely on the degree of reduction of the local share of the free sites. In the same figures there are average concentration dependences of the partial transfer coefficients of the label. The average values are obtained by averaging over all pairs of sites between which hopping of the molecules within the unit cell of the supramolecular structure is possible. These dependences are characterized by the non-monotonic behaviour associated with the filling of the surface region of the cell and the transition to a decrease in the free volume in the central portions of the pores. The dependences on the coefficients of the argon and krypton atoms are close enough – they differ in the energy of interaction of atoms with each other and with the walls of the pores.

The shear viscosity coefficient [47]

Calculation of the viscosity coefficients was carried out at the same molecular parameters as those used to estimate the diffusion coefficients (Fig. 65.2). Figure 65.3 shows the components of the

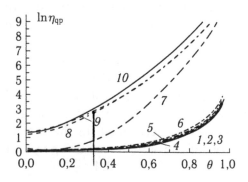

Fig. 65.3. The shear viscosity coefficients in the regular spherocylindrical structure $D_s–D_c–L_c = 20–8–20$ for equimolar composition of the Ar–Kr binary mixture (numbers of curves are the same as in Fig. 65.2).

shear viscosity tensor depending on the position of the molecules and their direction of movement at an equal ratio of argon and krypton. As in the above examples, all of the initial values of the local shear viscosity coefficients are determined by the surface potential of the wall. Movement along the surface layer is due to overcoming the surface potential barrier, which is smaller than the 'desorption' $(Q_2^i - Q_1^i)$, so the zero values $\eta_{11}(0) < \eta_{12}(0)$. Value $\eta_{12}(0)$ is the highest as it is connected with the maximum change of the surface potential of the wall. (This same potential determines the ratio $\eta_{ff}(0) < \eta_{ff+1}(0)$ within its scope.) With increasing filling of the pore the first layer is initially filled, and this determines the rapid increase in the values of $\eta_{11}(\theta)$, which are greater than $\eta_{12}(\theta)$ when filling more than one monolayer (sites of the second monolayer are blocked and the attraction effect of the surrounding molecules becomes stronger).

Average dynamic characteristics

Averaging of the dynamic characteristics over any cross section x, $X_1 \leq x \leq X_2$, was carried out by means of the distribution functions for sections of pores of different types $F_q(x)$ and their pairs $F_{qp}(x, y)$, which was preceded by averaging over the types of sites f and their pairs f_g using functions $F_{q, f}(x)$ and $F_{qp, fg}(r|xy)$ along the section x within each pore. These coefficients are determined for some macroregion comprising a plurality of local regions. The question is discussed in detail for single-component substance and the averaging procedure for a mixture is the same. As an example we present average values of the label transfer coefficient in spherocylindrical systems.

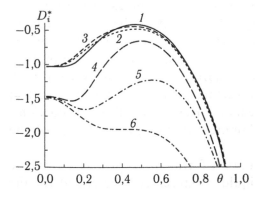

Fig. 65.4. Variation of the composition of the Ar–Kr binary mixture. Average transfer coefficients of the argon label (curves 1, 2 and 3) and krypton label (curves 4, 5 and 6) in a regular spherocylindrical structure D_s–D_c–L_c = 20–08–20: mole fraction of argon X_{Ar} = 0.1 (curves 1 and 4), 0.5 (curves 2 and 5) and 0.9 (curves 3 and 6).

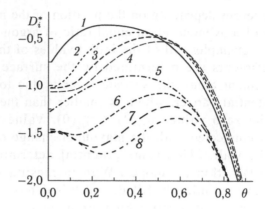

Fig. 65.5. The average transfer coefficients of the argon (curves *1*, *2*, *3* and *4*) and krypton labels (curves *5*, *6*, *7* and *8*) in different regular spherocylindrical structures: $D_s-D_c-L_c$ = 20–14–06 (curves *1* and *5*), 20–14–60 (curves *2* and *6*), 20–08–20 (curves *3* and *7*) and 20–08–60 (curves *4* and *8*) for the equimolar composition of the Ar–Kr binary mixture.

Figure 65.4 shows the average transfer coefficient of argon (curves *1*, *2* and *3*) and krypton (curves *4*, *5* and *6*) labels in different regular spherocylindrical structure $D_s-D_c-L_c$ = 20–08–20 at varying molar composition of the binary Ar–Kr mixture. Here we deal with the concentration dependence of the average of the average partial transfer coefficients. The curves fall into two groups for each atom. Argon, as a component with higher mobility, has higher transfer coefficients than krypton. With an increase in the mole fraction of krypton the krypton label transfer coefficient decreases sharply.

Figure 65.5 shows the concentration dependences of the average transfer coefficients of argon and krypton labels in different regular spherocylindrical structures for the equimolar composition of the binary Ar–Kr mixtures. This figure shows the effect of the porous structure on the macroscopic characteristics of the transfer. Increasing the length of the cylindrical part of the system reduces the transfer coefficient due to increasing proportion of the surface areas. Increasing the diameter of the cylinder increases the transfer coefficient due to the increasing share of internal areas with weak coupling of molecules with the surface. This character of the effect of the structure on the coefficients is observed in all other situations, such as an increase in the diameter of the spherical part of the system.

The molecular-kinetic theory of transport processes of the components of gaseous and liquid mixtures is generalized for the case of narrow pores of the complex structure. The theory is based on the lattice gas model, which allows to provide a self-consistent way to calculate the equilibrium and transport properties of inert gases and liquids. This model is used for both the description of the supramolecular structure of highly dispersed bodies and to calculate the distributions of mixture components within the pore space. The supramolecular structure is modeled using slit-shaped, cylindrical, spherical and globular regions, as well as their joints. The influence of the molecular properties of the mixture components, the physical and chemical properties of the pore walls and the width of the cylindrical and complex sections of the variable diameter of the spherocylindrical pores on the concentration dependences of the local transport coefficients of labels and viscosity for mixtures of spherical molecules in the supercritical temperature range is analysed. We discuss the correlation between the partial fillings of the components of the mixture and the local dynamic characteristics of the mixture in spherocylindrical systems. Increase of the binding energy increases the local viscosity of the flow and reduces the diffusion coefficients of labeled components in the mixture.

In the supercritical temperature range we can distinguish the relationship of local equilibrium partial distributions of the molecules with the local transport coefficients for complex areas of the porous structure. This is due to the important role of the interaction potential of molecules with the walls of the pores. This is most clearly seen for the transport coefficients of the label. Increase of the binding energy increases the local viscosity of the flow and reduces the diffusion coefficients of labelled components in the mixture and mutual diffusion coefficients. It should be noted that at temperatures below the critical this correlation is more complicated by contributions from the areas of the phase boundaries. To identify correlations in the latter case it is necessary to define separate contributions of interfacial regions, which in itself is a rather complicated task.

66. Distribution of microimpurities in equilibrium and flows

Adsorption and membrane technologies [42, 48] are the basis of many processes of purification and separation of liquid and gas mixtures. The effectiveness of deep cleaning of the main component to remove impurities is characterized by the separation factor. It is determined from experimental data of adsorption of mixtures of the 'microimpurity–macrocomponent' type on the studied adsorbent or membrane. However, the experimental determination of these coefficients is a laborious task, therefore the theoretical calculation methods become important here [49–51]. Most adsorption processes occur in porous systems [42, 48, 52, 53]. For such systems there are no experimental methods to control the nature of the distribution of microimpurities. Calculation methods are, apparently, the only possibility to obtain information on the distribution of microimpurities in the pores which can be of great practical importance for many applications of adsorption, chromatography and colloid chemistry.

The problem in molecular dynamics (MD) for the microimpurities is that it is necessary to consider that a very large number of molecules (about 10^4) of the main component, and a small number of impurity molecules. This makes it difficult to obtain reliable results in the calculation of the distributions of microimpurities. The LGM is an approximation, but provides a quick solution, and it has no restrictions on the amount of microimpurities. The accuracy of application of the LGM is largely determined by the accuracy of describing the properties of the solvent, which in turn strongly influence the nature of the distribution of microimpurities. The above described comparison of the MD and LGM shows that qualitatively both approaches always give the same result and modifying LGM LGM and/or 'adjusting' its energy parameters, we can get a quantitative agreement with the almost exact numerical results of the Monte Carlo method and molecular dynamics. This allows the use of the LGM for many practical situations.

Impurity at the vapour–liquid interface
To analyze the properties of the interface, the phase diagram for a fixed width of a cylindrical pore was calculated in the initial stage [54]. Further, the distributions of rarefied and solid phases at the ends of the field L were considered given and the calculation of the density distribution of the argon atoms in the transition region was carried out by formulas as described in section 26. In this study, calculations are conducted for different lattice structures $z = 6$ and 8 and different lateral interaction radii $R = 1\lambda$ and 4λ. The distributions of the pure fluid (main component) and microimpurities with $\varepsilon^* = 0.5$ (weakly interacting impurity) and 2.0 (strongly interacting impurity) were investigated. The change in ε^* simultaneously changes the adsorbate–adsorbate lateral interaction energy and the adsorbate–adsorbent interaction energy. The results are presented for the case of a fixed pore radius $R_p = 4\lambda$ (these calculations are demonstrative,

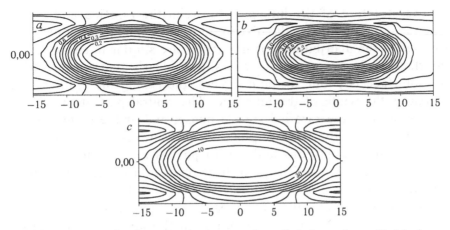

Fig. 66.1. The interface between the dense and rarefied phases in a cylindrical pore $z = 6$ at $T/T_{crit} = 0.8$: a – the distribution of the main component, b – distribution of 'weak' microimpurity $\varepsilon^* = 0.5$ (with respect to the solvent) c – distribution of the 'strong' microimpurity $\varepsilon^* = 2.0$ (with respect to solvent).

given dimensional restrictions on phase transitions in narrow pores).

Calculation of the distribution of molecules at the interface of dense and rarefied phase of the adsorbate in a cylindrical pore at $T/T_c = 0.8$, where T_c refers to the bulk gas phase, is shown in Fig. 66.1 for the lattice structures $z = 6$. In the porous systems the adsorption potential of the walls changes the course of the phase diagram and, in particular, reduces the magnitude of the critical temperature in comparison with the bulk phase. The value T_c depends on the type of lattice structure z, as it determines the number of neighbouring molecules, 'compensating' the kinetic energy of the molecules.

Figure 66.1 a shows a pure component, and Figs. 66.1 b and c refer to the weakly and strongly interacting microimpurities. The figures show the isolines of the equal density of the molecules. For clarity of images, instead of the discrete changes in local densities θ_q^i in different sites of the structure Fig. 66.1 shows the smoothed values. They are obtained by designing θ_q^i along the radius of the pore to a certain preferred direction followed by 'smoothing' by the quadratic procedure of computer graphics (for more detail see [54]). For the pure solvent the isolines represent real densities of the molecules in dimensionless variables θ. These curves are similar to curves of the distribution of molecules in the slit-shaped pores. The density values for the microimpurities were normalized to the density of the gas outside the pores.

The structure of the interface between the rarefied and dense phases is determined by the distribution of the main component, so isolines of the distributions of impurities are similar in shape to the interface of the argon atoms. It is shown clearly that the length of the interface is about six monolayers. Increasing the temperature increases the width of the interface. The central part of the figure corresponds to the rarefied ('gas') phase, and at the edges of the picture there is the 'liquid' phase. The sections of the volumes occupied by the gaseous and liquid phases reflect the nature of gradual changes of the fillings in these phases. In the gas phase near the walls there is a dense 'film' (of the order of two monolayers) of the adsorbed molecules. The weakly interacting impurity is less concentrated on the walls compared to the strongly interacting impurity. It is easily 'displaced' from the walls by the argon atoms, while the strongly interacting impurity is held at the wall also at a high density of the main component. The qualitative nature of the distributions of microimpurities for both lattice structures is similar. The quantitative differences are primarily related to different critical temperatures for both structures in the bulk gas phases when T/T_c = const. Furthermore, at a small radius of the cylinder and hence a small number of sites in the section of the pore an important role is played by the distribution of sites between the boundary area of the pore and its central part. However, due to the smooth form of the curves shown in Fig. 66.1, the role of the second factor is less important than that of the first one.

Selectivity factor inside the pore

The process of redistribution of the microimpurity between the wall and the central part of the pores is characterized quantitatively by selectivity factor

$$S = \theta_{2\,center} / \theta_{2\,wall}. \tag{66.1}$$

In [54] the authors analyzed the estimates for function S at low $(\theta_1 \sim 0)$, critical (θ_{crit}) and large $(\theta_2 \sim 1)$ fillings of the pores resulting in the following conclusions.

1. The greater the value ε^* the smaller the selectivity factor at zero filling.

2. In the near-critical conditions $S(\theta_{crit}) > S(\theta_1)$, and the value of S decreases with an increase in the value of ε^*. In the critical region at sufficiently strong attraction of the atoms to the pore walls the excess can not exceed ten.

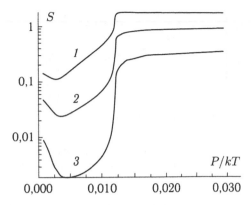

Fig. 66.2. The coefficient of distribution of the microimpurity between the centre and wall of the pore for the case $R = 1\lambda$, $z = 6$, depending on the degree of filling of the pore volume by the solvent (the abscissa gives the values of P/kT for the volume of the gas phase).

3. In the case of high densities the selectivity increases with increasing, and this growth becomes greater with increasing ε^*. This implies that even at complete filling of the pore the value of ε^* plays an important role and determines the limiting values of functions $S(\theta_2)$.

These qualitative estimates are confirmed by calculation of the selectivity factor of the microimpurity between the centre and the wall of the pore shown in Fig. 66.2.

The curves of the dependence of ln S on the volume filling of the pores show that qualitatively their form is the same for any impurity (the nature of the impurities is reflected through the parameter ε^*). With increasing degree of filling of the pores by the main component the filling of the first layer increases initially and the neighbouring adsorbed molecules hold stronger, due to lateral interactions, the impurity near the wall, so the curve ln S – (P/kT) shows a minimum. However, a further increase in filling the pores displaces the impurities from the walls to the central part of the pores due to the large amount of the main component. When the surface area of the pore is completely filled (approximately two surface monolayers), the value of S increases sharply. Typically, this behaviour manifests itself in the near-critical region, when the surface of the pore is almost completely filled (for the considered strong attraction of the molecules by the walls) and the central part of the pore is about half filled with the main component. A further increase in the filling of the central part of the pore has little effect

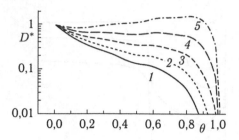

Fig. 66.3. Concentration dependence of the self-diffusion coefficient of the main component at $z = 6$, $R = 1\lambda$ and varying the parameter α: $1 - 0$, $2 - 0.05$, $3 - 0.10$, $4 - 0.15$ $5 - 0.20$

on the values of S. The stronger the interaction of the impurity with the walls andmolecules of the main component, the stronger it is held on the wall surface and the lower the selectivity factor for all the pores filled.

Self-diffusion coefficients

To calculate the derivative $d\theta_q^{*i}/d\theta_i^*$ in the presence of microimpurities we take into account that $\theta_i^* = \sum_{q=1}^{t} f_q \theta_q^{*i}$, $u_f^i = \dfrac{a_f^i \Lambda_f^i P_f^i}{a_f^A \Lambda_f^A P_f^A}$,

$$\frac{d\theta_q^{*i}}{d\theta_i^*} = \left[\left(\frac{du_q^i}{d\theta} \right) \theta_q^i + u_q^i \left(\frac{d\theta_q^i}{d\theta} \right) \right] \bigg/ \sum_q f_q \left[\left(\frac{du_q^i}{d\theta} \right) \theta_q^i + u_q^i \left(\frac{d\theta_q^i}{d\theta} \right) \right]. \tag{66.2}$$

Hopping rate constants for different sites correspond to a simple condition: the greater the depth of the well, the less likely is the particle to exit the well. The connection between the hopping constants and local Henry constants has the following form: $a_q^i K_{qp}^{iV} = a_p^i K_{pq}^{iV}$, where we have $a_q^i = a_q^{0i} \exp\left(\beta Q_q^i \right)$, as well as $E_{qp}^{iV} = \gamma Q_q^i$, $\gamma = 1.1$, if $Q_q^i \geq Q_p^i$. In the calculations it was assumed that $\varepsilon_{qh}^{*ij}(r) = \varepsilon_{ij}^*(r)$ and $\varepsilon_{qh}^{ij}(r) = \varepsilon_{ij}(r)$ for all types of sites in the structure.

The general trend of decreasing self-diffusion coefficient of molecules with increasing degree of filling of the pores is due to the decrease of the free space available to the molecules (Fig. 66.3). The nature of the concentration dependence $D_i^*(\theta)$ for the main component is strongly dependent on the ratio between the lateral interactions of molecules in the ground and transition states. If differences in the interactions of molecules in the ground and transition states are neglected, there is a monotonic decrease in the

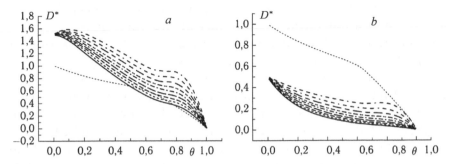

Fig. 66.4. The concentration dependence of the self-diffusion coefficient of weakly- (*a*) and strongly (*b*) interacting impurities for $z = 8$ and $R = 4$ for α (solvent) $= 0.6$: α (impurity) varies from 0.4 for the lower curve to 0.8 for the upper curve with a step of 0.05 (dotted line – curve for the solvent) [54].

concentration dependence of the self-diffusion coefficient. The slope of the curve $D_i^*(\theta)$ at low fillings is much larger than in filling in the near-critical region. This behavior of $D_i^*(\theta)$ at low filling is due to the increase in the proportion of molecules in the central portion of the pores wherein the molecules migrate faster than around the pore wall. Reducing $D_i^*(\theta)$ at higher fillings is due to the sharp decrease in the number of vacancies. For $z = 6$ and $R = 1\lambda$ the relationship $\alpha = \varepsilon_{AA}^*(1)/\varepsilon_{AA}(1) > 0.1$ is sufficient to change the nature of the monotonic decrease of $D_i^*(\theta)$ in the filling region close to critical due to the increase of the lateral attraction in the transition state. For other values of z and R the value of α, corresponding to the non-monotonous course of the dependence $D_i^*(\theta)$ is more than 0.1, but qualitatively the specified feature is retained.

Figure 66.4 shows similar relationships for the self-diffusion coefficient of weakly (*a*) and strongly (*b*) interacting microimpurities. (The curves are normalized to the value of the self-diffusion coefficient of the solvent label at zero filling). The general character of the concentration curves for the self-diffusion coefficients remained unchanged. For weakly (strongly) interacting impurities the family of curves shifted to larger (smaller) values of the self-diffusion coefficient with respect to the solvent label. But in all cases an increase in the magnitude of α leads to the non-monotonic dependence $D_i^*(\theta)$.

For the weakly interacting impurity at low fillings there appears the first local maximum, and near the critical filling – the second local maximum. For the strongly interacting impurity at low fillings the maximum does not form – the impurity is located near the walls and the migration in the central part of the pores makes a small contribution. For the weakly interacting impurity the fraction of molecules in the central part of the pores is much larger than their fraction around the pore wall. The fraction of this impurity in the centre of the pore increases with increasing concentration of the main component – this explains the presence of the first local maximum.

Figure 66.4 shows that the maximum for large θ is linked to two factors: 1) the difference in interactions of the migrating particle with its surroundings in the ground and transition states, and 2) with the concentration of the main component increasing to near-critical values. With increasing concentration of the main component the impurity is displaced from the walls to the central part of the pores where its movement is faster but with the maximum $D_i^*(\theta)$ is shown only at a certain ratio α (impurity) $> \alpha$ (solvent). Thus, the concentration dependences of the self-diffusion coefficients are very sensitive to the potential of intermolecular interactions of the molecules of the main component and is strongly dependent on its total concentration. A further increase in the density of the main component reduces the mobility of the main component and any impurities.

Distribution of the clusters [54]

In porous systems the concentration of clusters is also dependent on the parameters of the interaction of particles with the wall and the geometric dimensions of the micropores, as the effect of the pore walls changes the bulk phase diagram and therefore there are dramatic changes in the region of existence of the supersaturated state of the fluid compared with the bulk phase. It was shown above that the distribution of microimpurities is sensitive to the surface potential. The above examples of the distributions of clusters can be directly transferred to the impurity systems. The ability to calculate the concentrations of clusters at different fillings of the cylindrical pore is illustrated below by the example of dimer clusters containing impurity atoms of different types.

Probability analysis of the formation of clusters of the main component (argon) of different sizes depending on the position of the centre of mass of clusters relative to the walls of slit and cylindrical pores with different degrees of filling of micropores (section 16) led to the following conclusions.

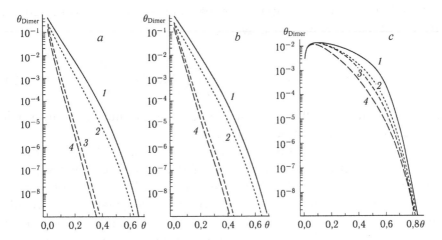

Fig. 66.5. Concentration dependence of the dimer cluster distributions within the cylindrical pore (tempearature $T = 1.01 \; T_{crit}$): a, b – different kinds of impurities, forming dimers, located in pairs of sites (12) and (34): 1 – (w–w), 2 – (s–w), 3 – (w–s), 4 – (s–s); c – the dimer formed by the solvent and weak impurity in the pair of sites (fg): 1 – (12), 2 – (23), 3 – (34), 4 – (46) [54].

1. The form of these relationships is strongly affected by the character of the interaction of molecules with the walls of the pores. In the case of non-wetting walls (when the pore walls 'repel' the argon atoms) the stability of the cluster within the pore is determined by interatomic interactions, as in the bulk phase, and in the case of well wetted walls (when the walls of the pores strongly 'attract' the argon atoms) all cluster atoms tend to stay close to the wall and have a flat structure.

2. The equilibrium cluster concentrations in the pores change dramatically when changing the position of the clusters in the volume of pores and orientations with respect to the walls of the pores.

3. With increasing degree of filling of the pore volume the concentrations of different clusters change dramatically.

The quantitative differences between the slit and cylindrical pores of the same width are due to the increased proportion of the sites in the near-wall region and decreasing proportion of the sites in the central part of the pore in the case of cylindrical pores.

Figure 66.5 examines the concentration dependence of the formation of dimer clusters with the orientations with respect to pore walls defined by the numbers of the central sites (fg). The existence probability of the dimer clusters, formed from the monomers i and

j, which are located in adjacent sites f and g, can be expressed as: $\theta_{clus}^{ij}(2|fg) = \theta_{ij}(2|fg)\Phi(2|fg)$, where

$$\theta_{ij}(2|fg) = \theta_f^i t_{fg}^{ij}; \quad \Phi(2|fg) = \prod_{h \in z_{fg}(1)} t_{fgh}^{ijV}, \quad t_{fgh}^{ijV} = \frac{\theta_{fh}^{iV}\theta_{gh}^{jV}}{\theta_f^i \theta_g^j \theta_h^V}, \quad (66.3)$$

here site h denotes all nearest neighbour sites around the central pair of the sites fg, the number of such sites is denoted by $z_{fg}(1)$.

Figure 66.5 shows the dependence of the probability of dimer formation on the degree of filling of the cylindrical pores in the case of well wetted walls (lattice structure with $z = 8$). The system components have the following notation: B – the main, W – weakly interacting impurity, S – strongly interacting impurity. In Fig. 66.5 a and b the curves belong to different classes of dimer molecules in sites (fg) in the central part of the pore. In general, the concentration of clusters decreases with increasing volumetric filling of the pores. However, the range of concentrations of the main component, in which the concentration of the dimers differs from practically zero, depends strongly on the nature of the monomer and the orientation of the cluster. The curves relating to the dimers formed from weakly-interacting monomers, decrease slowly in comparison with the curves for the same dimers formed from the strongly-interacting impurities. This is due to the displacement of the 'weak' dimers in the central part of the pores, which is filled only after filling the surface region of the pore.

In Fig. 66.5 c the curves refer to the 'main component – weakly interacting impurity' dimer molecule, located in different positions within the pore – site pairs (1–2), (2–3), (3–4) and (4–6). In this case, there is a non-monotonic dependence: first, the concentration of dimers increases and then decreases. The closer the dimers to the wall, the faster their concentration decreases due to the preferred filling of the wall region.

The analysis showed that the general relationships of the distribution of isolated clusters at various points within narrow pores found earlier [54, 55] for the main component molecules are retained in the study of clusters formed by the impurity molecules, but the energy of the impurity molecule can strongly influence the likelihood of formation of different types of mixed clusters and their orientation relative to the walls of the pores.

Microimpurity in flows

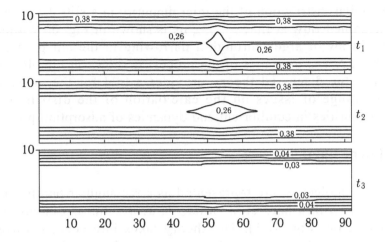

Fig. 66.6. Distribution of heavy microimpurity in the relaxation process of states of argon after a pulsed disturbance in the cell $X = 53$ in the three points in time: $t_1 = 40$ ps, $t_2 = 2.8$ ns, $t_3 = 10$ ns [56].

A small amount of the impurity can not strongly disturb the distribution of the microhydrodynamic velocity and, therefore, in a first approximation the velocity distribution is maintained. This takes into account only the local thermal redistribution of molecules in the flow that explicitly takes into account the difference in the thermal velocities of the mixture components. Below are two examples of the flows discussed in Chapter 6, for which the distributions of microimpurities at a particular time were calculated.

One of the simplest situations is associated with the relaxation of internal disturbance in the liquid. Figure 66.6 shows the distribution of atoms of a heavy impurity at different times. The density field and the field of the microhydrodynamic velocity along the axis of the slit pore are shown in Fig. 43.1.

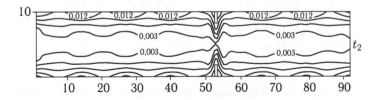

Fig. 66.7. Field of the distribution of the microimpurity at $t_2 = 20$ ns.

A similar distribution of the microimpurity in a slit-like pore with a constrained flow at time $t_2 = 32$ ns is shown in Fig. 66.7 after the start of flow of argon through a narrowing in the slit pore in the cell $X = 52$. The dynamics of the process is shown in Fig. 46.1 [57].

Thus, the discussed theory provides a tool for the analysis of a wide range of issues on the calculation of the distribution of microimpurities in equilibrium and dynamics of adsorption processes.

67. Large particles

Adsorbed molecules are represented as a rectangular parallelepiped $b \times d \times n$ with a hard core, where b, d, n – linear dimensions of the molecule (Fig. 67.1). The theory takes into account different possible orientations of the adsorbate in the plane of the surface. An important feature of the adsorption of large molecules is the ability to change the horizontal orientation of the long axes of the molecules by vertical when the degree of filling of the exposed surface and the pore walls changes [58–61]. Under certain conditions, with increasing concentration of the adsorbate the ordered distributions of the molecules with the same orientation of their long axes becomes more advantageous: an analogue of the three-dimensional phase transitions of he disordered phase – nematic (or smectic) type appears [62, 63].

The slit pore space is divided into elementary volumes. The lattice structure with the number of the nearest neighbours equal to

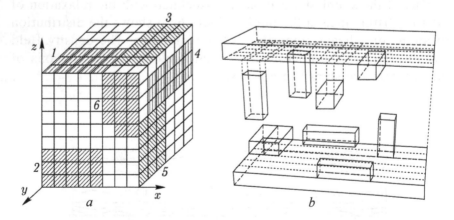

Fig. 67.1. Six types of orientations of the 'hard sphere' of the parallelepiped considered in describing the adsorption of large molecules in the slit pores at $z = 6$ $L = 3$, and (a). The arrangement of molecules in the slit pore (b).

z forms. The lattice constant is equal to the size of the adsorption centre (site), which is determined by the area of a local minimum of the adsorbent–adsorbate potential energy for one of the components of the mixture. Large molecules occupy more than one site. Figure 67.1 a shows six types of orientations of one side of the 'hard-sphere' of the parallelepiped 5×3 to be considered when describing the adsorption of large molecules in the slit-shaped pores at $z = 6$; the axes x and y are parallel to the walls of the pores, the pore width is measured along the axis z (the pore scheme is shown in Fig. 67.1 b).

The equations of the equilibrium distribution and kinetic equations for the transfer of large particles are presented in Appendix 11. The thermal velocity of large molecules is described in Appendix 7. Below are the description of the structural model with the participation of large particles and the calculation results of their equilibrium and dynamic characteristics.

The model

Let all the components of the mixture in the pore volume occupy an integer number of sites equal to M_m, $1 \leq m \leq \psi$, ψ is the number of components in the mixture. The orientation of each of the adsorbed molecules m is characterized by its energy characteristics, allowing in the statistical description of adsorption to treat the molecules in a particular orientation as a separate type of particle, so that considering different orientations of the molecules, even for a one-component system, is reduced to the problem of adsorption of a mixture of molecules of different size, each of which has a strongly fixed orientation. It is assumed that the orientations of the molecule are defined relative to the selected direction in the plane of the surface layer of the adsorbent. We denote by L_m the number of possible directions of the axes of the orientations of the adsorbed molecules; index λ, $1 \leq \lambda \leq L_m$, enumerates the specific orientation axis of the molecule. For simplicity, the number L_m is considered the same for all surface centres.

Molecule m in orientation λ will be called particle i (m, $\lambda \leftrightarrow i$, i.e. the set of pairs of indices m and λ corresponds to one index i), $M_i \equiv M_m$. The pair of indices m and λ is ordered as follows. It is assumed that the first index value i (from 1 to L_1) belongs to the first component, the following values of the index i (from $L_1 + 1$ to L_2) belong to the second component and so forth, then $1 \leq i \leq S$, where $S = \sum_{m=1}^{\psi} s_m$. Here s_m is the number of distinct orientations of the molecule m, when its long axis coincides with one of the axes of the possible directions. For asymmetric molecules we should distinguish situations where the direction of the long axis of the molecule is identical or opposite to the direction of the axis of orientation. In this case, $s_m = 2L_m$. For symmetric molecules both of these cases correspond to the same state, then $s_m = L_m$.

Consider the lattice structure with $z = 6$ and $L_m = 3$. For molecules in the form of a rectangular parallelepiped $(b \times d \times n)_m$ the number of 'kinds' $s_m = 6$ since for each of the three fixed directions of the long axis b_m of the molecule m we have two options of the location of the molecule on the other two axes at $d_m \neq n_m$. Figure 67.1 shows the different orientations of the face $(b \times d)_m$, the third axis of the molecule m is directed along the third axis. The value n_m can be any, but if $n_m = d_m$ or $n_m = b_m$, then it corresponds to the states of the molecule degenerate in space and, as for the bar, $s_m = 3$. For the parallelepiped $(b \times d \times n)_m$ we have the following sizes $b_i(\alpha)$ of the particle i along the axes x, y, z:

$$
\begin{array}{ccccccc}
x = & b_m & b_m & d_m & d_m & n_m & n_m \\
y = & d_m & n_m & b_m & n_m & b_m & d_m \\
z = & n_m & d_m & n_m & b_m & d_m & b_m
\end{array}
$$

Depending on the orientation of the given molecule m different numbers on sites will be blocked on the walls of the pores and in other layers. The number of sites occupied by particle i $(1 \leq i \leq S)$ in each layer of a slit-like pore is denoted by I_i, it is equal to $I_i \equiv I_{m,\lambda} = b_i(x) b_i(y)$ (where $I_{m1} = I_{m3} = (bd)_m$, $I_{m2} = I_{m5} = (bn)_m$, $I_{m4} = I_{m6} = (dn)_m$), and let P_i denote the perimeter of the particle i in the layer: $P_i = P_m \lambda = 2 [b_i(x) + b_i(y)]$ (where $P_{m1} = P_{m3} = 2(b + d)_m$, $P_{m2} = P_{m5} = 2(b + n)_m$, $P_{m4} = P_{m6} = 2(d + n)_m$).

We formulate a rule for determining the position of a large molecule that blocks M_m neighbouring sites of the lattice structure, in a specific site of the lattice structure. We choose one of the segments of the particle i, for example an angular segment for the parallelepiped (plate or rod) and will count from it other sites occupied by the same particle. It is assumed that the given particle occupies the site with the number f ($1 \leq f \leq N$, N is the number of sites of the lattice structure in the pore volume), which contains the selected segment of the particle i, and the section of the lattice occupied by a given particle i is denoted by $\{f\}$. (In the slit pore the number of the site f corresponds to the number of the layer k and the number of the site in the given layer l: $f \Leftrightarrow (k, l)$; the portion $\{f\}$ corresponds to its set of values $\{k, l\}$).

Energy

Each adsorption centre f in a heterogeneous lattice structure can be characterized by the local Henry constant $a_{\{f\}}^i$ in adsorption of a particle i on it: and $a_{\{f\}}^i = a_{\{f\}}^{0i} \exp \{\beta E_{\{f\}}^i\}$, $\beta = (k_B T)^{-1}$, where $a_{\{f\}}^{0i}$ is its pre-exponential factor, $a_{\{f\}}^{0i} = F_{\{f\}}^i \beta / F_i^0$; $F_{\{f\}}^i$ and F_i^0 are the statistical sums of the adsorbate in orientation i on the local fragment $\{f\}$ and in the gas phase (the case of adsorption without dissociation of the adsorptive is considered here); $E_{\{f\}}^i$ is the binding energy of the particle i with the site $\{f\}$.

We take into account the lateral interactions between the nearest neighbours. Intermolecular interactions of the neighbouring particles i and j, located in the sites f and g, will be characterized by the energy parameters $\varepsilon_{\{f\}\{g\}}^{ij}$. It is assumed that the positive values of the interaction parameters correspond to attraction. If it is assumed that the energy of intermolecular interactions is calculated in the atom–atom approximation, the values $E_{\{f\}}^i$ and $\varepsilon_{\{f\}\{g\}}^{ij}$ consist of the sum of the contributions of the pair interactions.

In addition, we can introduce the concepts of energy contacts of the molecules and the total value of the potential energy of interaction of neighbouring particles can be represented by the sum of contributions from these contacts. Measurements are taken of the surface area of the molecule in single contacts the area of which is considered to be equal to the square of the lattice constant. The area of the particle i is denoted by Q_i (it is expressed in terms of the previously introduced values P_i and I_i as: $Q_i = P_i b_i (z) + 2I_i$). The number of contacts of class φ of particle i is denote Q_i^φ, $Q_i = \sum_{\varphi=1}^{\tau_i} Q_i^\varphi$, where the sum over φ from 1 to τ_i denotes the summation over all classes of contacts of particle i, and σ_{fg}^{ij} denotes the contact area of adjacent particles i and j stuated in the sites f and g. For any *complete particular* set of neighbouring particles $\{j\}$, taking into account the sequence of their location and orientation relative to the central particles (denoted by the symbol $\alpha (j)$), we have

$Q_i = \sum \sigma_{fg}^{ij}$, where the sum over j is taken over all neighbouring particles around particle i. Then the values of the parameter $\varepsilon_{\{f\}\{g\}}^{ij}(\varphi\xi)$ can be represented as the sum of contributions $\varepsilon_{\{f\}\{g\}}^{ij}(\varphi\xi)$ between adjacent contacts φ and ξ, related to the area of the single contact: $\varepsilon_{\{f\}\{g\}}^{ij} = \sum_{(\varphi\xi)} \varepsilon_{\{f\}\{g\}}^{ij}(\varphi\xi)$, where the sum over pairs of adjacent contacts $(\varphi\xi)$ is carried out over a site with area of σ_{fg}^{ij}. Here contact φ refers to a particle i, occupying section $\{f\}$, and contact ξ belongs to particle j, occupying section $\{g\}$.

Each site in the lattice structure can be blocked by the adsorbate or be free. Free site v is counted as a particle of the kind $S + 1$, all its contacts are equivalent and $Q_v \equiv Q_{S+1} = z$; $m_{S+1} = 1$. Interactions of the particles with the free sites are equal to zero. Lateral adsorbate-adsorbate interactions between the nearest neighbours are taken into account in the quasichemical approximation preserving effects of direct correlations, and in the mean field approximation in the absence of correlation effects.

Macroscopic systems

The population of individual sites can be described only for small fragments of porous systems containing about 10^3 sites. The properties of porous macro-objects may be transmitted appropriately by describing the adsorption process on small fragments only if they have uniform walls or the strictly regular arrangement of non-uniform adsorption sites on the pore walls. In general, we should consider all possible variants of mutual arrangement of various adsorption sites. Therefore, to describe macroscopic porous systems with heterogeneous walls we must enter the distribution functions of different types of sites and use them to average the obtained solutions. In turn, to account for a wide spectrum of arrangements of different adsorption sites it is necessary to use the averaging procedure of contributions to local isotherms from each adsorption centre of the various types of local structures. For this it is necessary to formulate a method for allocating partial contributions of heterogeneous adsorption sites on the lattice structure. The same need arises in the case of single-component adsorption of large molecules. The specified procedure is also required to match theory and the numerical results obtained by using the Monte Carlo and molecular dynamics methods.

As a result, for the full adsorption isotherm we obtain the well known expression [58–61]

$$\theta(P) = \sum_{i=1}^{S}\sum_{q=1}^{T} f_q \theta_q^i, \quad \sum_{q=1}^{T} f_q = 1. \qquad (67.1)$$

Here f_q is the unary function of the distribution of sites of the lattice structure as regards the adsorption capacity, T is the number of site types on this structure. For large values of T the sum over site types q in equation (67.1) is often replaced by the integral over the energy state of the molecules. If there is no summation over the types of molecules, formula (67.1) gives an expression for the partial adsorption isotherm. The values of local fillings θ_q^i in (67.1) are determined from the system of equations in Appendix 11 using the procedure outlined below.

Let the site with number f is a site of type q. Around this site there is its own specific set of sites. For another site type q, for example a site with number h, there is another adjacent site environment. Function f_q in equation (67.1) takes into account the fraction of sites of type q, but in no way characterizes the types of sites situated around the sites f and h, i.e., the function f_q does not reflect the structure of the surface and does not specify which neighbouring sites can be blocked by large particles. In the slit pore, when the particle i is in the site of type q, the neighbouring sites of different types, blocked by this particle, may be in different layers.

For a particular type of particle i when deriving expression θ_q^i it is necessary to average over the probability of realization of the different ways of distribution of the adjacent $(M_i - 1)$ sites around a site of type q, blocked by the particle i. As a result, the function θ_q^i is defined as

$$\theta_q^i = \sum_{q_1=1}^{T}\cdots\sum_{q_{M-1}=1}^{T} F_{q_1,\ldots,q_{M-1}}(f)\theta_f^i, \qquad (67.2)$$

here q is the type of site from which we count the region blocked by the particle i; teh sums are gradually taken over all types of sites of the blocked region from $q = 1$ to $q = M - 1$; $F_{q_1,\ldots,q_{M-1}}$ is the conditional probability of sites of type $q_1,\ldots,q_{M-1}$ situated in the blocked region, the index f numerates the position of the 'beginning' of the molecule in the site of type q. Function θ_f^i relates to the corresponding local probability of finding the particle i at the site f, as defined above in equations (67.1).

We define the so-called binary separation factor [49–51] for an ideal gas phase in the form $\alpha_{ij} = \theta_i P_j / (\theta_j P_i)$. This ratio allows one to characterize the effectiveness of the proccess of cleaning liquid and gas mixtures. It links the equilibrium concentrations of the components i and j in the gas–adsorbate system for multicomponent mixtures. By analogy with (67.2) we obtain a generalization of previously published formulas for the case of adsorption of a mixture of large molecules:

$$\alpha_{ij} = \sum_{q=1}^{T} f_q U_q^i \Big/ \sum_{q=1}^{T} f_q U_q^j, \quad U_q^i = \sum_{q_1=1}^{T} \cdots \sum_{q_{M-1}=1}^{T} F_{q_1 \cdots q_{M-1}}(f) a_f^i \frac{\theta_{\{f\}}^{M_{iv}}}{\Lambda_{\{f\}}^i}, \quad (67.3)$$

where U_q^i is the partial contribution of particle i, located on the site of type q, to the binary separation factor. It is expressed through the local values of Henry's constants a_f^i, the probability of finding an empty area on the surface $\theta_{\{f\}}^{M_{iv}}$ in which adsorption of the given molecule i is possible, and also through the non-ideal adsorption system $\Lambda_{\{f\}}^i$.

68. Equilibrium properties of large molecules

Binary mixture of planar molecules in two-layer pores
The general equations in Appendix 11 for the equilibrium distribution of molecules can be used in some situations to derive simpler equations that reflect the specific properties of the adsorbents studied. As a simple example, consider a model situation with the adsorption of molecules i ($i = A, B$) in a binary solution in a microporous adsorbent with pores with the width of about two monolayers. The difference in the size of the molecules M_i is due to differences in their areas $S_i \equiv I_i$ for the unit height of the monolayer $h = 1$ ($M_i = I_i = S_i h$). In this case there are no molecular reorientation effects. If we confine ourselves to considering lateral interactions in the molecular field approximation, we can obtain one non-linear equation describing the filling of the pore with molecules A [61].

Parametric analysis of this equation allows us to reflect all the molecular features of the binary mixture. This model is used to describe the experimental data for the adsorption of benzene molecules in the micropores of the activated carbon AG-1 from a C_6H_6–CCl_4 solution at 300 K [64] (see Fig. 68.1). Data on the excess values of adsorption were converted to full concentrations. The model allows us to give a qualitatively correct description of the experiment in a wide range of concentrations. However, at low fillings of benzene ($\theta_{benzene} < 0.2$) the model does not provide quantitative agreement. In order to improve the description of the experiment, it is necessary to abandon the condition of homogeneity of the pore walls and enter strongly adsorbing centres for molecules A of low concentration (about 2%) with a binding energy 10 kT higher than for the rest of the surface sites. This result is natural, since from the quantitative description we must transfer to correct mathematical models with a large number of local concentrations.

The constructed theory of adsorption of a mixture of large molecules [61] allows, on the one hand, to obtain a sufficiently detailed description of the molecules in the slit-shaped pores with uneven walls, as is done by the numerical Monte Carlo and molecular dynamics metrods, and on the other hand, allows us to offer different versions of simplified models with clearly formulated assumptions at the molecular level to describe the experimental data of multicomponent adsorption.

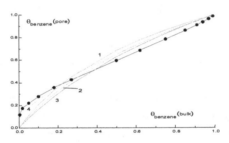

Fig. 68.1. Description of the experimental data for the adsorption of benzene molecules from the benzene–carbon tetrachloride solution in the microporous active coal AC-1 at 300 K (solid points), $L_{benzene}/L_{CCl4} = 1.04$; all energy parameters are given in kT units: uniform surface, curve 1 – $E_A = 11$, $E_B = 8.8$, $\varepsilon_{AA} = 1.0$, $\varepsilon_{BB} = 0.76$; curve 2 – $E_A = 17$, $E_B = 14$, $\varepsilon_{AA} = 1.5$, $\varepsilon_{BB} = 0.76$; curve 3 – $E_A = 17$, $E_B = 14.5$, $\varepsilon_{AA} = 1.5$, $\varepsilon_{BB} = 2.0$; curve 4 – taking into account the nonuniformity of the surface with the parameters of curve 2.

Calculations for the rods

Rods of equal length L were considered; the lateral interaction parameter of the adjacent contacts ε was also regarded as one and the same in contact of the rods and zero in contact with a vacancy. Special attention was paid to the difference of distributions of rod-shaped molecules in comparison with the distributions of spherically symmetric particles. The main characteristics of the distributions were volume fractions of rods: the partial densities ω^X, ω^Y and ω^Z, representing the probability of the space segment belonging to the rod oriented in the appropriate direction. Unless otherwise indicated, the abscissa gives the value $\omega = \omega^X + \omega^Y + \omega^Z$ – the probability that the space segment is occupied by a segment of some rod, called the gross density. Note that in the case of single-site particles the bulk and molar fractions are identical.

The values of L and ε were varied. The absolute values of ε, attributable to one contact, did not exceed 300 cal/mol, which corresponds to the average values of the interaction energies of the dispersion type for systems without the formation of specific bonds. The system temperature was 300 K. We considered a cubic lattice with $z = 6$ and three orientations of the rods along the Cartesian coordinate axes. This is fully consistent with the well-known Di Marzio model [65], which is actively used in the bulk phase.

The main difference between the system containing rods or multi-site particles and the system of single-site particles is that for sufficiently large values of the parameters L and/or ε it has ordered additional solutions, beginning with some density θ_{bif}, which is called the bifurcation point. In the case of the non-uniqueness of the solutions we must choose the initial approximation in such a manner as to ensure that the optimization process would lead to the desired solution of the equilibrium system of equations. Therefore, the field is oriented in a certain direction. This is achieved as follows: slightly changing the Henry's law constant for this direction, we obtain a non-isotropic solution of the modified system. We use this solution as the initial approximation for the system being solved. Returning to their previous values Henry constants, we solve our system. If the values of parameters L and/or ε are not large enough or density has not yet reached the critical value ω_{bif}, then the solution is unique and nothing but the isotropic solution can be obtained. But if the density $\omega \geq \omega_{bif}$, then the solution is 'stratified' and the density of particles, oriented in the direction of the field, is very different from the densities of particles oriented in other directions. Physically, the outcome is clear: for the rod-shaped particles to fill the space with a high density, one of the solutions should be that they are all oriented in one direction. And the longer the bars, the sooner this process should begin. This 'ordering' occurs abruptly, as also demonstrated by the following calculations [66].

Bulk phase
The ordering effect of rod-shaped molecules in the bulk phase in the absence of lateral interactions is shown in the inset of Fig. 68.2 *a*. The abscissa gives the density in mole fractions $\omega = \omega^X + 2\omega^Y$, the ordinate – filling by rods oriented along the axes: ω^X – the upper

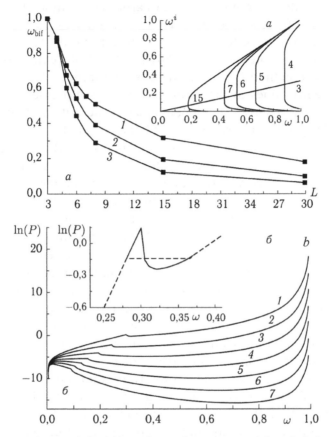

Fig. 68.2. Dependence of the bifurcation point ω_{bif} on rod length L for different parameters of intermolecular interaction (*a*): $\varepsilon = -840$ (*1*), 0 (*2*), 840 (*3*) J/mol in the bulk phase. Dependence of the logarithm of pressure ln (*p*) on the density of the molecules ω at different parameters of lateral interaction (*b*): $\varepsilon = 0$ (*1*), 210 (*2*), 420 (*3*), 630 (*4*), 840 (*5*), 1050 (*6*) 1260 (*7*) J/mol in the bulk phase.

branches of the curves, and $\omega^Y = \omega^Z$ – the lower branches of the curves.

The distributions of concentrations for different values of the rod length L from 3 to 8 are presented. When $L \le 3$ there is only one isotropic solution $\omega^X = \omega^Y = \omega^Z$. When $L > 3$, starting with a certain density value $\omega = \omega_{bif}$, there appears a non-isotropic ordered solution of the system, i.e. the curves ω^X and $\omega^Y = \omega^Z$ are sharply divided: the curve ω^X goes up ($\omega^X \to 1$ for $\omega \to 1$), the curve corresponding to $\omega^Y = \omega^Z$ goes down ($\omega^Y = \omega^Z \to 0$, $\omega \to 1$). In this case the greater the length of the rod L, the lower is the critical value ω_{bif}, as the molecules try to order themselves to 'avoid' crossing with each other.

These results are consistent with earlier calculations by the Monte Carlo method [67]. Moreover, even a small increase to $L \geq 3.05$ leads to ordering, of course for values ω_{bif} close to unity. This fact for $\varepsilon = 0$ was previously mentioned in [68] for $L = 3.06$ (calculation by the Monte Carlo method for a two-dimensional lattice), but the situation with non-zero values of the parameter ε was not previously analyzed.

Figure 68.2 a shows the dependence of the critical density ω_{bif} on the rod length L for three values of the parameter $\varepsilon = -840$ (1), 0 (2), 840 (3). With increasing length of the rod L the critical density ω_{bif} is shifted to lower values of density ω. If the point $\{L, \omega\}$ for the predetermined parameter value ε is below the appropriate curve, only one isotropic solution is possible. If, on the contrary, this point would be above the curve, an ordered solution can be obtained. The inset shows the dependence of the distributions of ω^X (upper branches) and ω^Y (lower branches) on the gross density ω for rods of different lengths $L = 3, 4, 5, 6, 7, 15$ in the bulk phase.

Calculation of the isotherms for the rod $L = 5$ in the inset of Fig. 68.2 b shows that even when $\varepsilon = 0$ a phase transition of the first kind takes place in the bulk phase, which is manifested in the formation of the non-monotonic isotherm similar to the van der Waals loop in attraction of spherically symmetric molecules [62]. With increasing repulsion the position of the bifurcation point shifts to the right to higher densities, and with increasing attraction the bifurcation point is shifted to the left to small density values (Fig. 68.2 b). A change of lateral interactions has quite a noticeable effect on the critical values of density ω.

Monolayer adsorption [66]
The surface adsorption potential influences the distribution of the rod axes. Since the horizontal axes X and Y are parallel to the surface and the Z axis is directed perpendicular to the surface, the wall potential wall removes the degeneration along the axes Y and Z. Therefore, the situation becomes more complicated. This is most easily seen for monolayer adsorption.

For simplicity, we assume that the binding energy of the rod in the horizontal orientation is proportional to the number of contacts, i.e. it is L times larger than in the vertical orientation. This results in that the Henry's law constant in this case can be expressed as $a_i = \exp(\beta L_i \Delta Q)$, where $L_i = L$, if $i = X, Y$, and $L_i = 1$ if $i = Z$.

Figure 68.3 a shows the dependence of the distributions of rods of various orientations on the gross density ω for $L = 4, 5$ and 6,

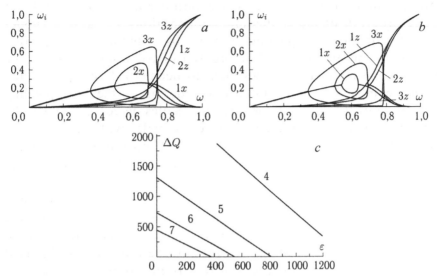

Fig. 68.3. Adsorption of rod-shaped molecules in the surface monolayer. The curves with letter z correspond to ω^z, the curves with letter x to ω^x (upper branch) and ω^y (lower branch). a – distribution of ω^i depending on the density ω for rods of different length $L = 4$ (*1*), 5 (*2*), 6 (*3*) at $\Delta Q = 200$; b – distribution of ω^i depending on the density ω for different values of the parameter of lateral interaction $\varepsilon = 75$ (*1*), 100 (*2*), 125 (*3*) for rods of length $L = 5$ at $\Delta Q = 200$; c – the chart of conditions of realization of ordering of molecules in the monolayer (the abscissa gives the values of the lateral interaction parameter ε, the vertical axis – the smallest value ΔQ at which the ordering can take place; numbers of curves correspond to the length of the rods L).

with $\varepsilon = 100$ and $\Delta Q = 200$. As for the bulk phase, ordering of the rods in realized in the *XY* plane, i.e. similar patterns. The probability of vertical orientation depends strongly on the length of the rod. Although the energy of the rod in the horizontal orientation in L times larger than in the vertical orientation, with filling of the surface the tendency towards the vertical orientation of the rods increases and becomes dominant at $\omega \rightarrow 1$.

If Fig. 68.3 a shows the effect of rod length L on the distribution of filling by the rods, oriented along axes, then Fig. 68.3 b shows the results of similar calculations for $\varepsilon = 75$, 100, 125, showing the effect of the lateral interaction ε (cal/mol) on this distribution at $L = 5$ and $\Delta Q = 220$.

The wall potential plays a major role for weak interactions. As the density increases the proportion of molecules orientated vertically increases monotonously, as the rods in the plane of the monolayer start to block each other. With the increase in the intermolecular interaction its role becomes more important. In the monolayer plane ordering of molecules takes place along the axis X (upper branches of the corresponding curves), by reducing the proportion of rods in the direction Y. The ordered distribution of rods in the plane of the monolayer allows them to occupy a larger part of the surface compared with the absence of ordering, so the corresponding proportion of the vertically oriented molecules decreases. Physically, the outcome is clear: for the rod-shaped particles with a high density to fill almost all the space they need to be almost all oriented in the vertical direction, because each such particle occupies just a single cell of the monolayer.

From these results it is clear that the bifurcation in the monolayer of the rods of length L does not occur for any values of ε and ΔQ. But if for a rod of length L for some values of the wall potential ΔQ and the potential of the lateral interaction ε in the interval $[\omega'; \omega'']$ bifurcation takes place, it occurs in this interval also with an increase in any of the parameters: ΔQ, ε or L.

The diagram of the conditions of realization of ordering of molecules in the monolayer for rods from $L = 4$ to 7 in the coordinates (ε, ΔQ) is shown in Fig. 68.3 c. An almost linear relationship exists between the binding energies of the adsorbate-adsorbent and adsorbate–adsorbate defining the line between regions with and without the ordering of molecules. Above the corresponding curve the ordering possible and below – no. With increasing L the possibility of ordering increases – the separating curve shifts to lower values of ε and ΔQ due to an increase of the number of contacts per molecule.

Slit-like pore

The distribution of rods in slit pores additionally depends on the combined influence of both walls. This is particularly noticeable in the narrow pores. In the slit pore we consider the density ω_q^i – the probability that a given cell situated in the layer q is a segment of the rod oriented in the direction i.

Figure 68.4 presents the distribution of rods of length $L = 4$ in a pore with width $H = 9$ as a function of gross density ω for the values of the wall potential $\Delta Q = 0$ (a, b) and 200 (c, d) with $\varepsilon = 0$. Moreover, because of the symmetry of the pore $\omega_q^i = \omega_{H-q+1}^i$ the distributions are given only for $q \leq t = [(H + 1)/2]$. The numbers of the curves correspond to the number of the layer q. The field b shows the densities of horizontally oriented rods ω_q^X and ω_q^Y. As

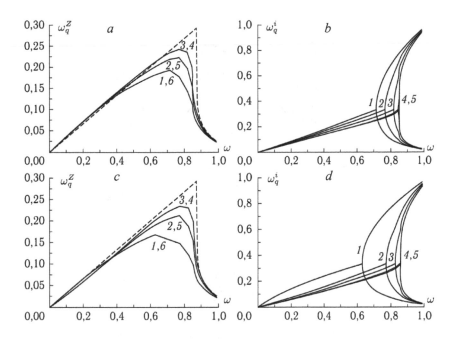

Fig. 68.4. Distribution ω_q^X and ω_q^Y as a function of gross density ω for rods with length $L = 4$ in a pore with width $H = 9$ at $\varepsilon = 0$, $\Delta Q = 0$ (a, b) and $\Delta Q = 840$ J/mol (c, d). Numbers of curves correspond to the number of the layer q. Upper branches of the curves correspond to ω_q^X, lower to ω_q^Y.

previously, the upper branches of the curves correspond to the density of molecules oriented in the direction X, the lower branches to the density of molecules oriented in the direction Y. The field a shows the densities of vertically oriented rods ω_q^Z. A typical feature is the presence of a maximum in the density ω_q^Z associated with a sharp decrease of ω_q^Z due to reorientation of the rods in a plane parallel to the plane of the pore wall at high occupancies of the pore volume. For comparison, the dotted line in the field a also shows density ω^Z in the bulk phase.

In the surface layers there is a strong effect of the wall potential, leading to earlier ordering. This influence gradually wanes, and for the horizontally oriented rods the distribution ω_q^X and ω_q^Y in the central layers practically coincides with the volume distribution. The presence of the wall exerts an effect for the vertically oriented rods. Increase of density ω_q^Z with increasing layer number q is due to the fact that if the 1st layer contains only segments of vertically oriented rods, terminating in this layer, the subsequent layers also contain segments of rods, ending in earlier layers, differing from the number

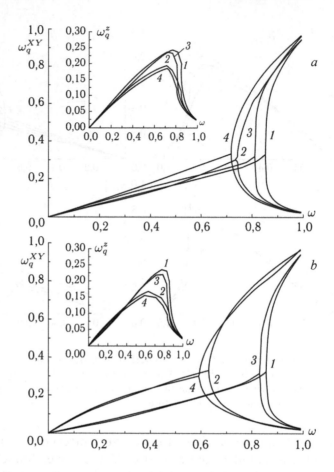

Fig. 68.5. Influence of parameters ε and ΔQ on ordering of the system $L = 4$ and $H = 9$. Dependence of distributions ω_q^i on density ω for rods of length $L = 4$ in the central (*1, 3*) and the surface (*2, 4*) layers at $\Delta Q = 0$ (*a*) and 840 (*b*) kJ/mol; ε = 0 (*1, 2*) and 840 (*2, 4*) kJ/mol. Distributions of ω_q^x (upper branches of the curves) and ω_q^y (lower branch of the curves) are presented in the main figure, the distribution of ω_q^z – in the insets.

of the layer q by no more than $L - 1$. Recall that the final segment of the rod i is the segment with the lowest value of the coordinate i.

The influence of the wall potential ΔQ and the lateral interaction parameter ε on the ordering of the system is shown in Fig. 68.5. As usual, the abscissa gives the gross density ω, the ordinate – the distributions of the horizontally oriented bars in the central layer (curves denoted *1* and *3*) and in the surface layer (curves marked *2* and *4*). Figures *1–4* correspond to the following parameter values: ε = 0, $\Delta Q = 0$ (*1*), ε = 0, $\Delta Q = 200$ (*2*), ε = 200, $\Delta Q = 0$ (*3*), ε = 200, $\Delta Q = 200$ (*4*). As in the case of Fig. 68.4, $L = 4$ and $H = 9$.

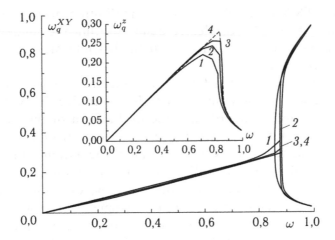

Fig. 68.6. Dependence of the distribution of ω_q^i on density ω for rods of length $L = 4$ in the central layer of the pore with width $H = 9$ (*1*), 13 (*3*), 15 (*3*) and in the bulk phase (*4*) with $\Delta Q = 840$, $\varepsilon = 420$ J/mol. Distributions of ω_q^x (upper branches of the curves) and ω_q^y (lower branches of the curves) are shown in the main figure, the distribution of ω_q^z – in the inset. The dotted line in the inset shows ω^z for the bulk phase.

In the system under consideration $H > 2L$. In the central layer the effect of the wall is weakened and the main role is played by the lateral interaction parameter ε, and the distribution of horizontally oriented rods in the central layer is close to the volume distribution at the corresponding value of parameter ε.

Note that in the absence of the wall potential the increase in the parameter ε narrows the ordering area. In the presence of interaction with the wall the increase in ε 'widens' the ordering area, pushing the start of stratification in the surface layer in the region of lower density.

Increased of the wall potential ΔQ enhances the effect of ordering in the surface layer and slows down ordering in the central layers.

Nature of the distribution of molecules in narrow pores is determined by all the molecular parameters : lateral interactions and potential wall, and rod length and width of the pores.

It is clear that with increasing pore width H the partial density distribution in the central layer should tend to the volume distribution. For the horizontally oriented rods this fact was noted above. Therefore, the criterion of the closeness of the distribution in the central layers of the pore to the volume distribution is the proximity ω_q^z and ω^z.

Figure 68.6 shows for $L = 4$ the distribution of ω_q^z in the central layer at $H = 9, 13, 15$, clearly demonstrating the tendency of the distribution of ω_q^z to reach the volume distribution of ω^z, indicated by a dotted line.

The equilibrium distribution of molecules reflects features of the cooperative behaviour of rod-shaped molecules with specific interactions and their ordering due to the form of the 'hard-cores'. Effects of blocking of adjacent layers by vertically oriented molecules and the tendency of rods to the vertical orientation with increasing degree of filling of the wall surface for a monolayer are replaced by the effect of limiting the width of the pores, which reverses the ordering of molecules throughout the volume of the pores. Increasing length of the rod increases the width of the pore in which the effect of the wall on the distribution of molecules in the surface region is evident. The situation is largely dependent on the ratio between the energy of intermolecular interaction and the interaction with the pore wall. For the rods for the concept of 'narrow' pores is usually greater than the 10 nm region, therefore, a detailed analysis of the effect of the energy characteristics of the adsorption system on the nature of its phase state should be carried out.

69. Transfer coefficients of rod-shaped molecules

Calculation conditions and the bulk phase
Equilibrium distributions of rod-shaped molecules are necessary for the calculation of the dynamic characteristics, considered in [66]. Recall that in addition to the isotropic distribution the rod-shaped molecules have the ordered distributions in the direction of the preferred axis i ($i = X, Y$ or Z). The ordering effect is observed for rods of length $L > 3$, and for small values of the volume fraction (density) ω of the molecules the system is isotropic and has a unique solution, and when passing through the value ω_{bif} in a jump in the form of a bifurcation we obtain additional solutions that meet the orderly distributions of molecules. The value ω_{bif} depends on the length of the rod: the longer the rod, the earlier the bifurcation forms, and it also depends on the lateral interaction parameter ε: increase of ε leads to earlier ordering. The resultant distributions indicate the important role of the effect of ordering of the molecules along their long axes on the distribution of the molecules.

Below we consider the ordered distributions in the direction of the axis X. All calculations were carried out at a temperature

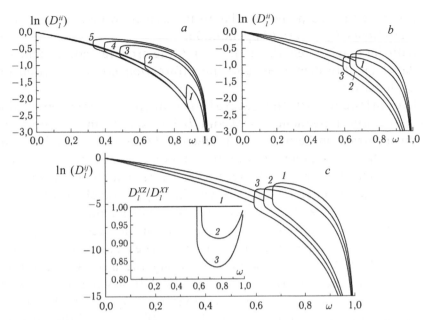

Fig. 69.1. Dependence $\ln D_l^{ij}$ for movement of the rod from the volume fraction of the fluid ω: a – the movement along the long axis of the rod with length $L = 4$ (*1*), 5 (*2*), 6 (*3*), 7 (*4*), 8 (*5*) at $\varepsilon = 420$ kJ/mol; b – movement along the long axis at varying parameter: $\varepsilon = 0$ (*1*), 420 (*2*), 840 (*3*) kJ/mol for the rod with length $L = 5$, and c – movement across the long axis at varying parameter: $\varepsilon = 0$ (*1*), 420 (*2*), 840 (*3*) kJ/mol for the rod with length $L = 5$.

$T = 300$ K. The interaction energy of adjacent contacts ε ranged from zero to 1260 J/mol, which is a typical area in the absence of specific contributions to intermolecular interactions. The ratio of the energy in the transition state to the ground state energy equals $\alpha = \varepsilon^*/\varepsilon = 0.55$ (all contacts of the rod are energetically identical). All the concentration dependences of the dynamic characteristics are normalized to their value for a rarefied gas in the bulk phase. In all the graphs below the abscissa gives the volume fraction (density) of the molecules ω.

Figure 69.1 shows the transfer coefficients in the bulk phase of labeled rods of different length L, and different orientations. The influence of the size of the rod is demonstrated in Fig. 69.1 a. The ordering effect is observed for all sizes of rods – at the bifurcation point the concentration dependence of the coefficient is divided to two branches: the upper branch corresponds to the motion of the long axis of the rod along the axis of ordering X, which corresponds to coefficient D_l^{XX}. The lower branch corresponds to the motion of the long axis of the rod perpendicular to the axis X, which corresponds to coefficients $D_I^{YY} = D_I^{ZZ}$ (Appendix 7).

With increasing length of the rod L the transfer coefficient of the label quickly decreases, and the bifurcation point is shifted to lower densities.

The influence of the lateral interaction parameter ε on concentration dependences D_I^{XX} and D_I^{YY} is shown in Fig. 69.1 b. Stronger attraction between the molecules holds them together and reduces the coefficient of transfer of labeled molecules (in the general case, the shape of the curves is also influenced by the values of the parameter α). As above, the upper branch corresponds to coefficient D_I^{XX}, and the lower branch – coefficients $D_I^{YY} = D_I^{ZZ}$.

Figure 69.1 c shows similar curves for the second type of movement. The upper branches of the curves correspond to the coefficients $D_I^{YY} = D_I^{ZX}$ of movement along the axis X, the lower branches – to the coefficients of movement transverse to the axis X. Note that at zero lateral interactions ($\varepsilon = 0$) $D_I^{XY} = D_I^{ZY}$. With increasing concentration the transfer coefficients of the second type decrease significantly faster than the coefficients of movement of the first type. Since the coefficients of movement across the X-axis differ little from each other, Fig. 69.1 c shows the curves corresponding to $D_I^{ZY} = D_I^{YZ}$, and the inset to Fig. 69.1 c – the ratio of the coefficients D_I^{XZ}/D_I^{YZ}. Note that the coefficient of the second type of motion of the rod, located along the axis of ordering X, is large than the coefficient of the corresponding motion of the rod, located perpendicular to the axis X.

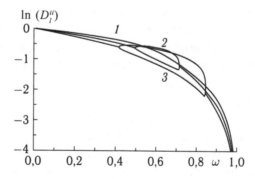

Fig. 69.2. Dependence ln D_I^{ij} in diffusion of the molecules by the first type of movement (along the long axis) in the surface monolayer on the volume fraction of fluid ω while varying the parameter of lateral interactions $\varepsilon = 0$ (*1*), 420 (*2*), 840 (*3*) kJ/mol for the rod length $L = 6$. Upper branches of the curves correspond to the coefficient D_I^{XX}, lower branches to the the coefficient $D_I^{YY} = D_I^{ZZ}$.

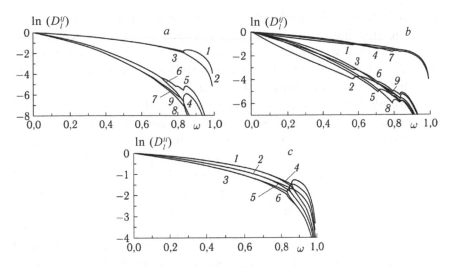

Fig. 69.3. Transfer coefficients of the label D_I^y in a slit pore at $\alpha = 0.55$, $L = 4$, $H = 9$: a – rectilinear motion of the rods in the cental layer of the pore for the lateral interaction $\varepsilon = 840$ kJ/mol; b – transfer coefficients of molecules D_I^{iX} in the X direction in different layers of the pore; $\varepsilon = 840$ kJ/mol; c – varying lateral interactions $\varepsilon = 0$, 420 and 840 kJ/mol for the first type of movement in the central layer.

Monolayer system

In the presence of a homogeneous solid surface an important molecular factor is the difference in the binding energies ΔQ between the horizontally and vertically oriented rods. The calculations of monolayer adsorption used the value $\Delta Q = 840$ J/mol, referred to one contact. The surface was assumed to be almost completely smooth, so that the activation energy along the surface was considered small. Figure 69.2 shows the transport coefficients of the label for rods with $L = 6$ for the first type of movement at different intermolecular interactions ε. Under these conditions, in the absence of lateral interaction between the molecules the ordered arrangement of the molecules is not realized. The greater the attraction between the molecules, the higher the rate of decrease of the transfer coefficient of the label with increasing surface filling. The upper branches of the curves *2* and *3* correspond to the movement of the rod along the axis of ordering and the bottom branches – across this axis. At high fillingsthe vertical orientation of the rods becomes dominant and both branches coincide.

Slit-like pores

In the slit-like pores the potential of both walls affects the nature of the distribution of the molecules with different orientations. As the total concentration of the molecules changes they can be reoriented. The calculations were performed for $H = 9$ monolayers and the rod length $L = 4$. The potential of each wall was considered the same as for the open surface with $\Delta Q = 840$ J/mol.

Figure 69.3 a shows the achieved variants of various components of the tensor of the label transfer coefficient for a fixed layer ($q = 2$) of the first type of movement (dependence of $\ln D_l^{ij}$ on the volume fraction of the fluid ω in the second layer ($q = 2$) of the slit-like pore). The curves decay in the direction of motion of the rod, and further splitting occurs with respect to the axis of ordering in the ordering region. Coefficients D_l^{ii} in movement of the first type are represented by curves 1–3, corresponding to the movement of the rod along the longer axis : $\ln D_l^{ii}$, $i = X$ (1), Y (2), Z (3). Movement of the rod transversely to the long axis D_l^{ii} is described by the curves 4–9; in the direction $j = X$ the curves (4) for $i = Y$ and (5) ($i = Z$); in the direction $j = Y$ – curves 6 ($i = Z$), and 7 ($i = X$); in the direction $j = Z$ – 8, curves ($i = X$) and 9 ($i = Y$).

The influence of the position of the molecule and the direction of its motion for three tensor components is shown in Fig. 69.3 b, which shows the dependence of $\ln D_l^{ii}$ on the volume fraction of the fluid ω in different layers of the slit pores. Curves 1–3 describe the transfer coefficients D_l^{iX} in the 1st surface layer, 4–6 – coefficients D_l^{iX} in the 2nd layer, 7–9 – coefficients D_l^{iX} in the 3rd pore layer (when $i = X$ the molecules are transferred by the first type of molecular motion, $i \neq X$ – by the second type). (The lower branches of the curves in Fig. 69.3 b are omitted) We note that with increasing number of the layers the bifurcation point moves monotonically from θ_{bif} values in the 1st layer to the value θ_{bif} in the central layer, close to the bulk phase.

Finally, the dependence of $\ln D_l^{ij}$ for molecular motion along its long axis in the central layer of the slit pore on the volume fraction of fluid at ω $\alpha = 0.55$, $L = 4$, $H = 9$ is shown in Fig. 69.3 c. Curves 1–3 represent the coefficients of motion in the tangential direction relative to the axis of the pore, with the upper branch corresponding to D_l^{XX}, the lower branch to D_l^{YY}. Curves 4–6 represent the coefficients of motion D_l^{ZZ} in the transverse direction relative to the axis of the pore at the values $\varepsilon = 0$ (1, 4), 420 (2, 5) and 840 (3, 6) J/mol. Figure 69.3 c shows that in addition to the effect of the direction of motion and the direction of the axis of ordering, the concentration dependences of the transfer coefficient of the label are affected by the lateral interaction parameters. Increased of the intensity of the

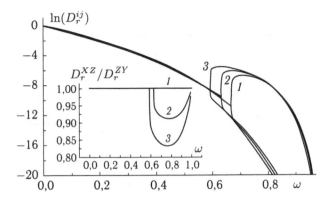

Fig. 69.4. Dependence on the coefficients of rotational diffusion D_r^{ij} of rod-shaped molecules $L = 5$ on the volume fraction of the fluid ω in the bulk phase by varying the lateral interaction coefficient $\varepsilon = 0$ (*1*), 420 (*2*), 840 (*3*) J/mol.

interaction of the molecules strongly inhibits the transfer of the label and the coefficient decreases sharply.

In general, Fig. 69.3 shows that the concentration dependences of the transfer coefficients of the rod-shaped molecules are in nature much more complex than similar dependences for small spherical molecules, which in narrow pores also have the tensor nature and depend on both the position of the molecules in the pore volume and the direction of motion.

Coefficient of rotational movement of the label
The average angular velocity of rotary motion (or frequency of rotation) is denoted by w_f^i – it is associated with the average hopping speed $W_f(i \rightarrow k)$ by the expression $w_f^i = \varphi W_f(i \rightarrow k)/\theta_f^i$.

In the transition of the molecules from the orientation state i to the 'nearest' state k the reference plane 0 should be considered the plane that is perpendicular to the axial rotation vector from the state i to state k which is centered between the orientational states i and k. For rotary motion we can introduce a similar diffusion coefficient of rotational diffusion of the label. This coefficient characterizes the thermal velocity of reorientation of the particles between different orientational states under equilibrium conditions. By analogy with (35.2) we obtain the following expression for the rotational diffusion coefficient of the label:

$$D_f^{i*} = z_f^*(i \rightarrow k)\varphi^2 W_f(i \rightarrow k)/\theta_f^i, \qquad (69.1)$$

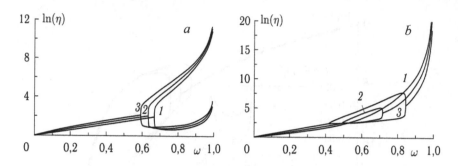

Fig. 69.5. The dependence of shear viscosity of ln (η) of the fluid volume fraction ω: a – in the bulk phase for rods with $L = 5$ for values of $\varepsilon = 0$ (*1*) 210 (*2*) 420 (*3*) kJ/mol, b – in the surface monolayer of rods with $L = 6$ for values of $\varepsilon = 0$ (*1*) 210 (*2*) 420 (*3*) J/mol.

where $z_f^*(i \to k)$ is the number of possible rotations between states $i \to k$ for cell f. This number depends on the specifics of the potential relief of the system. If 'direct' rotations take place, then by definition adopted above for the types of particles $z_f^*(i \to k) = 1$. The dimension of the rotational diffusion coefficient is [rad²/s].

We agree to denote the transfer coefficient of the label in rotational motion through D_r^{ij} with superscripts i and j, characterizing the initial and final orientation of the rod. In rotation we consider only cases where $i \neq j$.

Figure 69.4 shows the rotational diffusion coefficients of the rod-shaped molecules in the bulk phase. The lateral interaction parameter ε was varied. With increasing attraction of the molecules the rotational diffusion coefficient rapidly decreases. Since rotation is possible only if there is a large vacancy area, the attraction of neighbouring molecules makes the process less likely.

The rotational diffusion coefficients greatly differ when the molecules rotate in the direction of ordering $D_r^{YX} = D_r^{ZX}$ (upper branches of the curves) in comparison with the rotational diffusion coefficients across this direction $D_r^{YZ} = D_r^{ZY}$ (lower branches of the curves). Note that the rotation coefficients $D_r^{XZ} = D_r^{XY}$ of the rods lying along the axis X differ so little from the respective rotation coefficients $D_r^{YZ} = D_r^{ZY}$ of the rods situated transversely to the axis X, that in the above graph this difference is almost imperceptible (see the inset which shows the relationship D_r^{XZ}/D_r^{YZ}).

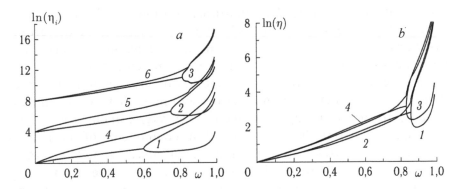

Fig. 69.6. The dependence of shear viscosity of $\ln(\eta)$ of the fluid volume fraction in the slit pore ω: $a - H = 9$, for $L = 4$, $\varepsilon = 420$ kJ/mol, the curves are shifted sequentially by the unit 4, *1, 4* – the first monolayer, *2, 5* – the second and *3, 6* belong to the third monolayer; $b - H = 9$ for $L = 4$, curves *1–2* relate to the parameter $\varepsilon = 0$, and curves *3–4* – to $\varepsilon = 840$ J/mol.

The shear viscosity coefficient

The expression for the local viscosity coefficient η_{fg} in shear of the mixture of molecules in a cell g relative to f (here $\chi = 1$, so the index is omitted) has the form

$$\eta_{fg} = \left[\sum_{j-1}^{s-1} x_f^j (\eta_{fg}^j)^{-1}\right]^{-1}, \quad \eta_{fg}^j = \frac{\theta_f^j}{U_{fg}^{jV}}, \quad x_f^j = \frac{\theta_f^j}{\theta_f}, \quad \theta_f = \sum_{i=1}^{s-1} \theta_f^j \quad (69.2)$$

where x_f^i is the mole fraction of component i in site f, θ_f is the complete filling of the site f, the expression for the thermal migration velocity $U_{fg}^{iV}(\chi = 1)$ was defined in Appendix 7.

The shear viscosity coefficient in the bulk phase for rods of length $L = 5$ shown in Fig. 69.5 *a*. The coefficient increases with increasing fluid density. The ordering effect is strongly dependent on the interaction between the molecules ε. The shear viscosity coeffcient in the ordered phase is lower than in the disordered phase.

A similar dependence for the viscosity coefficient of monolayer adsorption is shown in Fig. 69.5 *b* for rods with $L = 6$. The upper branches of the curves correspond to the viscosity in the direction X, the lower branch – to the viscosity in direction Y.

Slit-shaped pores

As above, the calculations for the slit-shaped pores were made for the gap at $H = 9$ monolayers and the rod length $L = 4$. In Fig. 69.6 the numbers of the curves *4, 5, 6* relate to the vertical movement of motion Z, the remaining motions in the plane XY. Figure 69.6 a represents the viscosity coefficient in different directions of movement of the rod and in various monolayers. The curve number i indicates the direction of movement. Curves 1, 2 and 3 describe the coefficients for motion of the component in the plane XY. The ordering of molecules within each layer separates these components, the upper branch $i = X$ axis coincides with the direction of ordering. The corresponding components of motion in the Z direction correspond to the number of curves 4, 5, 6. Overall, the curves match, however it is obvious that the ordering is first implemented in the first layer, then the second and then the third. As the layer number increases the difference between the branches X and Y also increases. The upper branches of the curves 1–3 represent the viscosity in the direction X, the lower branch – the viscosity in the direction Y.

Figure 69.6 b shows the effect of intermolecular interaction on the concentration dependence of the viscosity coefficient for different directions of motion in the layer $q = 3$. The upper branches of the curves 1 are the viscosity in the direction X, the lower branch – the viscosity in direction Y. Curves 2, 4 represent the viscosity in direction Z. The stronger the attraction between the molecules, the faster is the increase of the viscosity coefficient with increasing volume filling of the pores. Here the difference between the viscosity coefficient on different branches within the layer decreases with increasing parameter ε.

References

1.   Guggenheim E.A., Mixtures. – Oxford: Clarendon Press, 1952. – 271 p.
2.   Shakhparonov M.I., Introduction to molecular theory of solutions. – Moscow: Gostekhizdat, 1956. – 480 p.
3.   Prigogine I.R., Molecular theory of solutions. – Moscow: Metallurgiya, 1990.
4.   Hirschfelder J., Curtis H., Bird R., Molecular theory of gases and liquids. – Moscow: Izd. inostr. lit., 1961.
5.   Smirnova N.A., Molecular theory of solutions. – Leningrad: Khimiya, 1982. – 334 p.
6.   Morachevsky A.G., Smirnova N.A., Piotrovskaya E.M., et al., Thermodynamics of the liquid–vapor equilibrium, Ed. A.G. Morachevsky. – Leningrad: Khimiya, 1989.
7.   Tovbin Yu.K., Theory of physical and chemical processes at the gas-solid interface. – Moscow: Nauka, 1990. – 288.
8.   Tovbin Yu. K. // Progress in Surface Science. 1990. V. 34. P. 1.
9.   Tovbin Yu.K. // Zh. Fiz. Khimii. 1995. V. 69, No. 1. Pp. 118.
10.   Tovbin Yu.K. // Zh. Fiz. Khimii. 1998. V. 72. P. 775.
11.   Tovbin Yu. K., Rabinovich A.V. // Langmuir. 2004. V. 20. P. 6041.

12. Tovbin Yu.K., Rabinovich A.V., Votyakov E.V. // Izv. AN. Ser. khim. 2002. No. 9. P. 1531.
13. Tovbin Yu.K. // Zh. Fiz. Khimii. 2005. Volume 79, No. 12. P. 2040.
14. Tovbin Yu.K., Komarov V.N. // Zh. Fiz. Khimii. 2005. Volume 79, No. 11. P. 2031.
15. Komarov V.N., Rabinovich A.V., Tovbin Yu.K. // TVT. 2007. V. 45, No. 4. P. 518.
16. Melvin Hughes E.A., Physical Chemistry. – Moscow: Izd. inostr. lit., 1962. – Pr. 2. – P. 642.
17. Leibfried G., Microscopic theory of mechanical and thermal properties of crystals. – Leningrad: GIFML, 1963. – 312 p.
18. Tovbin Yu.K., Senyavin M.M., Zhidkova L.K. // Zh. Fiz. Khimii. 1999. V. 73. P. 304.
19. Tables of physical quantities, Ed. Kikoin. – Moscow: Atomizdat, 1976. – 960 p.
20. Bird R., Stewart W., Lightfoot E., Transport Phenomena. – Moscow: Khimiya, 1974. – 687 p.
21. Tovbin Yu.K. // Izv. AN. Ser. khim. 2003. V. 52, No. 4. P. 827.
22. Tovbin Yu.K. // Izv. AN. Ser. khim. 2005. No. 8. P. 1717.
23. Tovbin Yu.K., Rabinovich AB // Izv. AN. Ser. khim. 2005. No. 8. P. 1726.
24. Tovbin Yu.K., Gvozdeva E.E., Zhidkova L.K. // Engineering Physics journal. 2006. Volume 79, No. 1. P. 121.
25. Tovbin Yu.K. // Adsorption. 2005. V. 11. P. 245.
26. Steele W. A. The Interactions of Gases with Solid Surfaces. – N.Y.: Pergamon, 1974.
27. Sokolowsky S., Fischer J. // Molec. Phys. 1990. V. 71. P. 393.
28. Hill T., Statistical Mechanics. – Moscow: Izd. inostr. lit., 1960. – 486 p.
29. Chapman S., Cowling T., Mathematical Theory of Nonuniform Gases. – Moscow: Izd. lit., 1960 (Experimental data Lonius-Schmidt from Table 24).
30. Ferziger G. and Kaper G., Mathematical theory of transport processes in gases. – Moscow: Mir, 1976.
31. Kogan M.N. Dynamics of rarefied gases. – Moscow: Nauka, 1967.
32. Tovbin Yu.K. // Teor. osnovy khim. tekhnologii. 2005. V. 39, No. 5. P. 523.
33. Sokolova I.A. // Reviews on thermophysical properties of substances. – Moscow: Publishing House of the Academy of Sciences of ITV, 1992. – No. 2 (94). – P. 36.
34. Egorov B.V., Komarov V.N., Markachev Yu.E., Tovbin Yu.K. // Zh. Fiz. Khimii. 2000. V. 74. No. 4.
35. Tovbin Yu. K. // Rus. J. Phys. Chem. Suppl. 1. 2005. V. 79. P. S1.
36. Tovbin Yu.K., Vasyutkin N.F. // Izv. AN. Ser. khim. 2001. No. 9. P. 1496.
37. Tovbin Yu.K., Vasyutkin N.F. // Zh. Fiz. Khimii. 2002. V. 76, No. 2. P. 319.
38. Akhmatskaya E., Todd B.D., Davis P.J., Evans D.J., Gubbins K.E., Pozhar L.A. // J. Chem. Phys. 1997. V. 106, No. 11. P. 4684.
39. Timofeev D.P., Adsorption kinetics. – Moscow: Publishing House of the USSR Academy of Sciences, 1962.
40. Summerfield Ch.N. Mass transfer in heterogeneous catalysis. – Moscow: Khimiya, 1976.
41. Mason E., Malinauskas A., Transfer in porous media: the dusty gas model. – Moscow: Mir, 1986.
42. Ruthven D.M., Principles of Adsorption and Adsorption Processes. – N. Y.: John Wiley, 1984.
43. Lykov A.V., Heat and Mass Transfer. – Moscow: Energiya, 1978. – 480 p.
44. Lamb G., Hydrodynamics. – M.-L. : OGIZ 1947. – 928 p.
45. Tovbin Yu.K. // Zh. Fiz. Khimii. 2003. V. 77, No. 10. P. 1875.
46. Eremich D.V., Tovbin Yu.K. // Zh. Fiz. Khimii. 2004. V. 78, No. 4. P. 720.

47.   Tovbin Yu.K., Eremich D.V., Gvozdeva E.E. // Khim. Fizika. 2007. V. 26, No. 4. P. 88.
48.   Timashev S.F., Physical chemistry of membrane processes. – Moscow: Khimiya, 1988. – 237 p.
49.   Kuznetsova T.A., Sigaeva A.E., Tolmachev A.M. // Vestn. Mosk. Univ. Ser. Khimiya. 1988. V. 29, No. 1. P. 43.
50.   Gilyazov M.F., Lopatkin A.A., Tolmachev A.M. // Vysokochistye veshchestva. 1987. No. 6. Pp. 40, 48.
51.   Tovbin Yu.K., Gilyazov M.F., Tolmachev A.M. // Vysokochistye veshchestva. 1990 No. 1. Pp. 76, 83.
52.   Greg C., Singh K., Adsorption. Specific surface. Porosity. – Moscow: Mir, 1984. – 310 p.
53.   Kiselev A.V., Intermolecular interactions in adsorption chromatography. – M.: Vysshaya shkola, 1986. – 360 p.
54.   Tovbin Yu.K., Votyakov E.V. // Izv. AN. Ser. khim. 2000. No. 4. P. 605.
55.   Tovbin Yu.K., Komarov V.N., Vasyutkin N.F. // Zh. Fiz. Khimii. 1999. V. 73, No. 3. P. 500.
56.   Tovbin Yu. K., Tugazakov R.Ya., Komarov N.V. // Colloids and Surface A. 2002. V. 206, No. 1-3. P. 377.
57.   Tovbin Yu.K., Tugazakov R.Ya., Komarov V.N. // Teor. osnovy khim. tekhnologii. 2002. V. 34, No. 2. P. 115.
58.   Tovbin Yu.K. // Izv. AN. Ser. khim. 1997. No. 3. P. 458.
59.   Tovbin Yu.K. // Khim. Fizika. 1998. V. 17. No. 6. P. 140.
60.   Tovbin Yu.K. // Izv. AN. Ser. khim. 1999. No. 8. P. 1467.
61.   Tovbin Yu.K., Zhidkova L.K., Komarov V.N. // Izv. AN. Ser. khim. 2001. No. 5. P. 752.
62.   Chandrasekhar S., Liquid Crystals. – Moscow: Mir, 1980. – 344 p.
63.   Bazarov I.P., Gevorgyan E.V., Statistical theory of solid and liquid crystals. – Moscow: MGU, 1983. – 262 p.
64.   Larionov O.G., Dissertation, Dr. Chem. Sciences. – M.: Institute of Physical Chemistry, Academy of Sciences of the USSR, 1975.
65.   Di Marzio E.A. // J. Chem. Phys. 1961. V. 35. P. 658.
66.   Tovbin Yu.K., Rabinovich A.V. // Izv. AN. Ser. khim. 2006. No. 9. P. 1476.
67.   Yoon K. et al. // J. Chem. Phys. 1981. V. 72, No. 2. P. 1412.
68.   Herzfeld J. // J. Chem. Phys. 1982. V. 76, No. 8. P. 4185.
69.   Tovbin Yu.K., Rabinovich A.V. // Izv. AN. Ser. khim. 2006. No. 9. P. 1485.

9

Conclusion

The molecular theory provides a new look at the traditional problems in the theory of adsorption. This primarily relates to the question of the classification of the pore size, which is briefly discussed in Sec. 1, the assessment of the volume of micropores, the applicability of the thermodynamic approaches when considering the equilibrium adsorption and flows of molecules in porous solids, and the prospects of development of the theory.

70. Classification of the pore size

Analysis of the vapour–liquid stratification curves showed that the current classification, built on the basis of measurements of the adsorption isotherms, needs to be clarified. The phase diagrams in Fig. 23.1*a* with strong adsorption show a dramatic change in their behaviour in the range of the transition dimensions from micro- to mesopores, for which there is a non-monotonic variation of the critical density at pore width from 3 to 10 molecular diameters. The critical temperature decreases monotonically with decreasing width of the pores. This decrease of the critical temperature becomes greater with the increase of the intensity of the interaction of the adsorbate with the pore walls. In the case of weak adsorption the critical density has a single maximum of θ_c in the range $H = 8–10$.

This fact is of fundamental importance in the discussion of the classification of the pore size and determining the volume of micropores. As stated in section 1, the boundaries of linear mesopore sizes are in the range of 1.5–1.6 to 100–200 nm, which corresponds to the limit of applicability of the Kelvin capillary condensation equation. This classification has been built on the basis of adsorption measurements and different mechanisms of sorption phenomena occurring in the pores, differing in size, in which the process of capillary condensation takes place. In the case of small dimensions the surface potential of the walls is dominant which suppresses the influence of lateral interactions and the filling of micropores is independent of the intermolecular interaction.

The recently obtained new data on the relationship of the geometry and size of the mesopores with the possibility of occurrence of capillary condensation indicate that the traditional mesopore range from 3 to 100 nm should be divided in more detail depending on the cooperative properties of molecules. The position of the boundary between micropores and mesopores is experimentally fixed by the disappearance/appearance of the hysteresis loop in strictly equilibrium measurements of the adsorption–desorption isotherms with a decrease/increase in the pore size. The molecular theory allowed to also formulate the criteria for the characteristic values of pore sizes of different geometry related to small volumes, in which there are no phase transitions of the first kind (sections 23, 24 and 56) [1, 2].

The size range of mesoporous systems, traditionally studied by the methods of adsorption porosimetry, with a typical size of up to 100 nm, should be divided into the following sections.

The first portion – pore sizes up to ~4 nm. There is no hysteresis loop and capillary condensation is due to the small size of the cylindrical channels. This does not exclude the presence of the so-called diffuse phase transitions, manifested in the concentration and temperature dependences of the specific heat of the adsorbate. However, such experiments are extremely complex and unknown in the literature. To date, the only evidence is direct experiments on adsorption measurements. Numerical experiments do not provide sufficient accuracy and reliability to describe the actual experiments.

In the case of long slit-shaped pores, which correspond to the model of infinitely long pores, stratification can take place for small values of the width of the pores (section 21). However, such systems are unstable mechanically. Real carbon adsorbents correspond to the model of pores of bounded cavities (section 54), and for them this classification is valid. For globular and other systems with a complex shape of cavities, including limited volume slit pores, the said value of 4 nm should relate to the 'through' channel portion.

When the pore walls are intersecting at right or other angles, then these portions are not included in the specified dimensions.

The second section – the pore sizes from about ~4 nm to 15 nm. This is an area where dimensional constraints so strongly affect the critical parameters of the central dome that they can not be neglected. In this area of the so-called narrow pores there are indirect interactions between the walls due to the cooperative behaviour of the adsorbate. The dependence of the critical temperature of stratification on the pore size is confirmed by all the existing theoretical methods and adsorption measurements.

The third section of the sizes – from 15 nm to 100 nm, includes relatively wide pores in which there is not only no overlapping of the surface potentials, but the limited pore volume leads to small (less than 2%) differences in critical parameters of the process of stratification of the adsorbate. In this area there is complete weakening of indirect interactions of the walls through the phase state of the adsorbate.

This classification reflects the nature of the cooperative behaviour of the adsorbate in limited volumes of porous systems. In the presence of a long-range contribution to the adsorbate–wall interaction potential these areas correspondingly increase in proportion to the radius of the interaction potential between the adsorbate molecules.

Cooperative processes play a key role in the stratification of the adsorbate in porous systems. The non-uniform distribution of the surface potential is responsible for the appearance in porous media of a plurality of local areas with their coexisting vapour and liquid phases. The stratification curves ranging from isolated pores of different geometry to complex porous systems have a multidome structure, depending on the energy of the interaction of molecules with each other and with the walls of the pore. Cooperative processes 'independently' isolate subdomains of the total pore volume in which the phase transformations take place. Main phase transformations are of the quasi-two dimensional and quasi-three dimensional nature, although they involve in the change in the local degrees of filling numerous other neighbouring areas. The new concept of 'narrow' mesopores is also associated with the determining role of the cooperative behaviour of molecules in limited volumes of the pores in which phase transitions can take place.

Taking stratification of the adsorbate into account is important for the interpretation of experimental adsorption isotherms for micro- and mesoporous sorbents, as well as for the interpretation of dynamic data (from the coefficients of self-diffusion and mass transfer). In real adsorbents with a continuous distribution of the pore sizes the character of the adsorbate phase state depends greatly on the size of the pores in a given local area of the adsorbent, and taking into account the intersections and junctions of pores of different types then also leads to the dependence on the pore size in the areas surrounding this region. This leads to the fact that the calculation of phase distributions of the adsorbate requires information not only on the functions of the pore size distribution, but also on the structure of the adsorbent.

The observed non-monotonic dependence of the critical density on the width of the pores makes such calculations more complex, and this complicates the task of molecular interpretation of the experimental data on the equilibrium adsorption and transport characteristics, as it requires more information on the nature of the porous structure of the adsorbent. The described features allow to analyze the processes of adsorption in any porous systems, thereby extending the range of practical application of adsorption techniques, in particular adsorption porosimetry. Currently, the molecular theory has come out of the field of representations that focus only on carbon adsorbents, to arbitrary types of porous materials.

71. Micropore volume

The concept proposed by M.M. Dubinin that molecules in the micropores are a special state of matter, was launched on the basis of numerous adsorption measurements for a wide variety of adsorbents in a wide range of temperatures and pressures. This concept has been repeatedly affirmed in all the discussions in the interpretation of numerous experimental data [3–8]. The reason for this was the notion that the size of micropores are commensurate with the dimensions of the adsorbed molecules, so the microporous adsorbent + adsorbate system can be viewed as a single-phase system. This assumption allows the use of thermodynamic concepts that any real pore system is identified with the homogenized state of the dispersed phase. For such a system the representations of layer-by-layer filling of the pore surface naturally lose the physical meaning. At any nature of the adsorbate interactions with the surface it is believed that the adsorption field is created in the entire micropore volume and adsorption occurs therein by the volume filling mechanism. Another characteristic feature of the microporous system was the firmly established experimental fact that there is no capillary condensation in the micropores. This suggests that there are qualitative differences between the states of matter in the micropores in comparison with the traditional vapour properties in the bulk phase.

The complex nature of the real structure of microporous materials indicates the need to the Dubinin–Radushkevich equation [3] by the Dubinin–Astakhov equation and later versions of the theory of volume filling [9, 10], as well as the need for prior separation of the contributions of adsorption in mesopores [7].

The definition of the micropore size by M.M. Dubinin (to submicropores, actually micropores and supermicropores) finds its natural explanation in the framework of the molecular examination, which leads to a drastic change in the adsorption properties with the change of the width of slit-shaped pores of the activated carbons for which this classification was proposed. Calculations in section 21 show the discrete nature of the change of the adsorption curves that is directly related to the number of molecules that form the complete filling of the pores up to 2 nm wide. With further increase in the width of the pores these changes become far smoother.

Systematic study of the phase diagrams of adsorbed molecules in porous systems allowed to answer the question to what extent the molecules in the micropores are a special state of matter. The overall heterogeneity characteristic of activated carbons (structural roughness, chemical heterogeneity, etc.) suggests that the pore volume consists of many small local interconnected cavities. To analyze the properties of the adsorbate in narrow pores, in addition to the stratification curves it is necessary to know the temperature dependence of the saturated vapour pressure at which there is complete (volume) filling the pores [11] (see section 21).

P–T-section

Figure 71.1 schematically shows the change in the position of the saturated vapour curve on the section $P–T$ of the phase diagram for the bulk phase and a slit pore with the pore size H (a similar situation holds for cylindrical pores, for them H – pore diameter). Line AB refers to the bulk phase (point A – triple point, B – critical), and the line $A^H B^H$ – to the pore with size H [12].

With decreasing pore size H the differences between the line $A^H B^H$ and the line AB increases. Simultaneously, with the decrease of the critical temperature point B changes to point B^H. Similarly, the

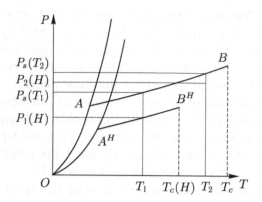

Fig. 71.1. Sections of $P-T$ phase diagrams of the adsorptive in the bulk phase and the adsorbate in a slit-like pore of width H; T_c and $T_c(H)$ – the critical temperature; $P_s(T)$ – saturated vapour pressure at temperature T; $P_{1,2}(H)$ – filling pressure of the volume of micropores with width H at temperatures T_1 and T_2, respectively [12].

vapour pressure at the critical point decreases, and the entire line of the saturated vapour is displaced downward. Figure 71.1 shows that in the narrow pores there may form two situations depending on the temperature of the experiment : $T_1 < T_c(H)$ and $T_2 > T_c(H)$, where $T_c(H)$ – the critical temperature of the adsorbate to the pore width H. At temperatures T_1, increasing the pressure to the value of $P_1(H)$ we cross the line $A^H B^H$ and adsorbate condensation occurs in the pore. If the system would have been closed, the stratification phase transition would take place in the pore, but the real experiment was conducted in an open system, so the pore is completely filled with the 'liquid' adsorbate.

This value $P_1(H)$ is an analogue of the saturated vapour pressure of the bulk phase. At temperatures T_2 the adsorbate in the pore is above the critical state and volumetric filling of the pores occurs without a phase transition. The pressure under which the micropores are filled at temperatures T_2 is denoted $P_2(H)$. In both cases, the micropore filling ends at some pressure values $P_{1,2}(H)$, which are lower than saturated vapour pressure at the same temperature P_s (see section 21).

The micropore volume

Molecular analysis shows: to take into account the real pressure of filling of micropores with a characteristic size H, using equation (6.6) it is necessary to change the standard way of defining W_0, where W_0 is the value of the linear extrapolation of the dependence ln W $- [\varepsilon(x)]^2$ to the values $x = 1$. In this case, the value $W_0(H)$ should correspond to the appropriate value of $P(H)$, so instead of (6.6) we use the equation [12]

$$\ln\frac{W}{W_0(H)} = -k\left(\frac{RT}{\gamma}\right)^2\left(\ln\frac{P_{1,2}(H)}{P}\right)^2. \qquad (71.1)$$

It differs from the equation (6.6) only by replacing the numerator P_s at $P_{1,2}(H)$. The relationship with the traditional method of determining the micropore volume is shown in Fig. 71.2.

If the experimental curve is rectified in the coordinates of equation (71.1), its intersection with the ordinate determines the micropore volume $W_0(H)$ with a characteristic size H. Note that the relation (71.1) can not be extended to higher pressures and thus we cut off all the uncertainties in the interpretation associated with the deviation of the experimental curve from the linear dependence at higher pressures. (These curves are traditionally interpreted as the presence of micropores with a different value of the parameter k in equation (6.6) or other characteristic size, although they are observed at relatively high fillings). If the experimental curve in the coordinates of (6.6) deviates from the linear dependence, this indicates the strong heterogeneity of the microporous system. The presence of bends at relatively high pressures in the coordinates of (6.6) and their absence in the coordinates of equation (71.1) indicates

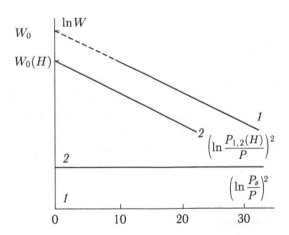

Fig. 71.2. Curves of adsorption isotherms in the coordinates of (6.6) and (71.4) with the corresponding abscissa axes *1* and *2* at pressures lower than the pressure $P(H)$ – volume filling of micropores of width H; dotted portion of the curve *1* – standard procedure for determining the micropore volume W_0

the existence of a distribution function of the volumes markedly different from the function (6.7). Typically, these deviations are observed for sufficiently large values of P/P_s, which have no relation to the micropores of the studied system. Figure 71.2 shows that the micropore volume $W_0(H)$ is smaller than W_0. With an increase in H the differences between $W_0(H)$ and W_0 decrease.

If we use the above estimates for the values $b_{1,2}(H)$, then the change in the volume of micropores values depends on the angle of the rectified isotherm in the coordinates (6.6) or (71.1). These differences for the experimental data may vary from 10 to 50%, which indicates the need to consider the actual value of the pressure corresponding to the filling of micropores. In this regard, it is much more correct to use the value $0.2P/P_s$ [7], which was first chosen as a measure of filling of the micropores. This value $0.2\ P/P_s$ is in full agreement with the molecular calculations, similar to the results in Fig. 21.7. (Later, the micropore volume W_0 was the extrapolated value at $P = P_s$, that does not meet the physical sense.)

The estimates obtained provide the correct pressure dependence of the volume filling of micropores on the size of the adsorbate: increasing size of the adsorbate decreases the number of monolayers H. This leads to an increase in the values $b_{1,2}(H)$ and, correspondingly, a greater reduction in the filling pressure of the pores. As a result, if in the equation (6.6) the curves must intersect at one point W_0, now the isotherm curves must end with the same value of W_0, but at different pressures $P_{1,2}(H)$.

72. Microheterogeneity of porous systems

The molecular theory is based only on information about the intermolecular interaction potentials between the adsorbate molecules and the atoms of the adsorbate and the adsorbent. It is shown that all specific features of the adsorption in porous materials are connected with the heterogeneity of the molecular distribution in the potentials of the pore walls that depend on the structure of the material that limits the volume available for the adsorbate, and on the intermolecular interaction of the adsorbate.

The observed multi-dome structure of the adsorbate stratification phase diagrams showed the importance of considering the microheterogeneity of porous materials. The cooperative behaviour of molecules itself forms stratification regions of substances and mixtures according to the properties of the adsorbent. This fact raises the question of the correctness of the applicability of traditional thermodynamic models in the interpretation of adsorption measurements. Recall that classical thermodynamics reflects 'only' the law of conservation of energy and is not related in any way to the molecular properties of adsorption systems.

In particular, thermodynamics can not give a recipe for constructing the equation of state in the bulk phase, nor, especially, within the pore cavities. In thermodynamics, there is no concept of the size of molecules, their orientation or structure of a certain region (e.g. the surface) since the thermodynamics applies to the sciences of a continuous medium. In addition, the thermodynamics can not be applied to small objects (which is especially important in the study of the behaviour of the adsorbate in small cavities of isolated pores) and the thermodynamics completely neglects any effects of fluctuations [13, 14].

The most comprehensive questions of application of classical thermodynamics to the adsorption systems were outlined by Hill [15]. Naturally, the macroscopic description gives one the opportunity to reflect only the general laws of the adsorbate material balance between the two phases: the adsorbent (porous body) and the external gas or liquid phase. In a sense, the thermodynamic description homogenizes the state of the system and completely ignores the real microheterogeneous properties of the adsorbate–adsorbent system. This view was largely forced by virtue of great uncertainty about the microheterogeneous structure of the adsorbent. However, during the long-term use of thermodynamic approaches this view was so widely employed so that the currently available experimental information and opportunities that far surpass the possibility of the last century, are largely unused.

Above we mentioned the incorrect use of the Kelvin equation to calculate the saturated vapour pressure in porous systems and the Poliani potential adsorption theory [16], which was previously used to estimate the micropore volume. The molecular theory allowed to determine the region of applicability of these thermodynamic approaches. The central result of the comparison of the molecular calculations with the thermodynamic models is that for highly nonuniform systems, which include the porous system, they shows that it is completely incorrect to use the thermodynamic methods. In particular, the Dubinin–Radushkevich equation may significantly overestimate the pore volumes, and the Kelvin equation gives acceptable accuracy assessment only for pores larger than 25 nm.

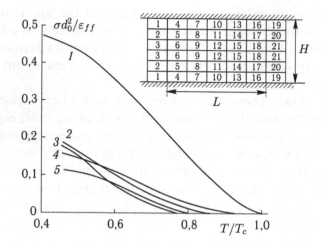

Fig. 72.1. Interfacial tension at the gas–liquid interface in the free volume (*1*) and in slits with a width of H/λ = 12 (*4*), 8 (*2*), 6 (*5*), 4 (*3*). The diagram shows the numeration of the sites of the adsorption lattice in the transition layer.

These results are not surprising, since the thermodynamic approach, being the method of the theory of continuous media, basically gives the averaged characteristic that refers to the state of the material smoothed with respect to the inhomogeneities of the material.

Additionally, we should also note other limitations of the thermodynamic approaches in adsorption, particularly problems with determining the surface tension inside the free volume of the pores.

Figure 72.1 shows the temperature dependence of the surface tension at the liquid–vapour boundary inside the slit-shaped pores of different widths. Curve *1* corresponds to the bulk phase. For any temperature and the slit dimensions the surface tension inside the pores is considerably less than in the bulk phase.

With increasing temperature, the surface tension decreases and vanishes at the critical point. Similar curves for the substance in the pore also vanish: the smaller the pore width, the lower the critical temperature. By lowering the temperature the curves for the surface tension may overlap, reflecting the rather complicated nature of the distribution of molecules at interfaces under the influence of the surface potential. This figure confirms once again the impossibility of using the Kelvin equation, which is commonly used with the bulk value for the surface tension.

Figure 72.2 shows the separating surface at the vapour–liquid interface in a spherocylindrical system [17]. It is seen that at this

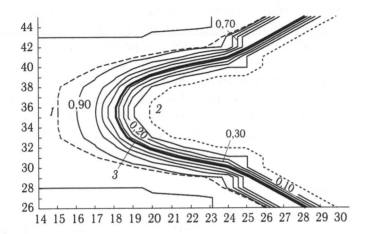

Fig. 72.2. The liquid–vapour pore transition region in a spherocylindrical pore D_s–D_c–L_c = 20–15–15 when τ = 0.75. Tension surfaces (3) and the interfaces of the liquid (1) and vapour (2) are shown.

temperature there is complete filling of the cylindrical part of the system and a relatively small filling of the spherical regions. The numbers mark the degree of filling at the density isolines. Menisci form in the mouths of the cylindrical pores. The density isolines are presented: it is clear that the shape of the phase boundary is different from spherical. In order to be able to calculate the surface tension in narrow pores, using equation (A2.5), we first need to determine the boundary of the vapour–liquid transition region and establish the form of the tension surface.

The dashed line shows the line between the transition region from the liquid phase, and the dotted lines shows the line separating the transition region from the gas phase (which is almost at the same level with the line θ = 0.1). Both lines define a complete transition region between the two phases (phase boundary). The thick line between the isolines is the section of the tension surface, the plane passing through the centre of the cylindrical channel. This line terminates at the exit from the mouth of the cylindrical channel, since the inner walls of the spherical pores are also filled with the adsorption film for which there is no surface tension of liquid–vapour.

In the general case, the calculation of the surface tension within the pores is a big problem in view of a strong influence of the surface potential, which causes that the surface tension value becomes a complicated function (having the tensor nature) of the coordinate of the dividing surface, which varies not only with temperature but with all the molecular properties of the system: starting with the heterogeneity of the adsorbent – its composition and structure (see [17]). Therefore, in such situations, the very concept of surface tension becomes purely formal, since the rules for its unambiguous calculation are becoming so complex that it eliminates the practical value of the concept. The situation is further complicated in the polydisperse porous systems when the local states of the adsorbate may be affected by the neighboring porous areas.

73. Transport of molecules

The molecular–kinetic theory based on the LGM provides the opportunity for modelling using a single method of calculating the dynamic molecular distributions in all three states of matter and all three types of phase boundaries. The theory of condensed phases allows to calculate the velocity of the multistage physico-chemical processes in a wide range of characteristic sizes of systems with a resolution of one atom/molecule to the macroscale. This transition is achieved by averaging the elementary processes at the microscopic level using the distribution functions of different dimensions and defined on different spatial scales. The theory allows us to keep the specifics of the elementary processes in the whole time range and use common definitions for the kinetic coefficients for all phases and their interfaces.

Testing the theory and comparison with similar calculations stochastic methods: Monte Carlo, Brownian and molecular dynamics, as well as the use of hydrodynamic equations and molecular kinetic theory, showed that the theory has significant advantages associated with accuracy and speed of calculations. Stochastic MD methods and the Monte Carlo method can, in principle, can be used to describe many processes in condensed phases, but the main problem in their use is obtaining reliable statistics which is time consuming (due to the need to control).

Calculations based on the LGM for pores of different geometry showed that all transport coefficients are functions of the distance from the pore walls and of the direction of flow of molecules (i.e. have a tensor character). They vary especially much near the pore walls. It was found that the traditional assumptions [18–22] of the constancy of the dynamic characteristics of the molecules in narrow pores are not true. In the analysis of experimental data we must consider the sufficiently strong concentration dependence of the dynamic characteristics of the adsorbate in narrow pores caused by both the influence of the potential of the pore walls and by intermolecular interactions.

Comparison of the calculated self-diffusion coefficients and shear viscosity coefficient of the fluid in a wide range of densities in narrow pores, obtained in the framework of this theory (see chapter 5) and by the method of molecular dynamics, as well as a comparison with the experimental data on the shear viscosity of simple gases in the bulk phase in the entire range of the densities on the rarefied gas to the dense fluid demonstrates the effectiveness of this approach. Finally, as described in chapter 2, the new theory allows to quantitatively describe the experimental data on the fluid compressibility factor in the bulk phase in the whole range of densities from the dilute gas to a liquid.

The proposed approach completely replaces the procedure proposed by Kirkwood for calculating the dissipative coefficients in the liquid phase (Appendix 9). Recall that his approach [23–27] is based on the introduction of the so-called friction coefficients. For flows near the solids the construction of such expressions, the solution of the corresponding equations and even the formulation of the boundary conditions are such complex tasks that they have not been fulfilled so far. So different approaches were proposed to circumvent many difficulties and extend the kinetic theory to a wide range of densities. Recently, emphasis of the theoretical studies has shifted significantly to molecular dynamics simulations which also have significant limitations on the time intervals and the low accuracy of the calculation of a number of characteristics (e.g., the tangential component of the thermal conductivity).

The formulated microhydrodynamic approach to the description of dense flows of molecules allowed to investigate the types of dynamic distributions of molecules and their velocities depending on the adsorbate–adsorbate and adsorbate–pore wall interaction potentials. The microhydrodynamic approach describes from the same viewpoint flows of both the vapour and liquid phases in the pores with different energies of interaction of molecules with the wall material. The characteristics of pulsed disturbances of the states of molecules at the initial time and the quasi-stationary distribution for large times were demonstrated. The time to the quasi-stationary distribution and the type of the relaxation stage of the evolution of this process depend on the molecule–wall (adsorbate–adsorbent) binding energy.

The attracting surface potential leads to the formation of a strongly anisotropic distribution of the molecules perpendicular to the surface of the pore walls at relatively low densities, which leads to the film flow. By lowering the temperature and/or increasing the density of the fluid results in capillary condensation (fluid density varies little in the pore cross section) and the volumetric fluid flow is observed. The flow profile is different from the quadratic Poiseuille profile. Under the same conditions, the slip of the surface layer molecules takes place in the rarefied phase. In narrow pores there is no viscous flow of the gaseous phase, if the pore walls attract the molecules. Instead, the film flow occurs and the contribution of the surface flow is commensurate with the flow through the middle of the pores. The transitional regime between the film and bulk liquid phase flow strongly depends on the initial conditions and requires more detailed study because of the complex nature of the redistribution of velocities along and normal to the walls of the pores in the transitional regime.

Pulsed perturbation of the vapour–liquid phase boundaries lead to their fragmentation followed by a prolonged phase of relaxation of the system. Numerical results show that as the temperature and density of the fluid approach the critical conditions for the investigated pore the bubbles present in the initial period quickly disintegrate under the influence of the applied perturbation. In the subcritical region, in particular at temperatures below the critical temperature the bubbles existing in a state close to equilibrium also remain at relatively strong external perturbations. They behave like large molecules, making swinging motions from their original position. With further increase of the perturbation the bubbles are deformed and begin to move. This may be accompanied by their fragmentation.

The proposed numerical analysis method allows to control the flows of the molecules in narrow pores and analyze: 1) the relationship of surface and volumetric flows, 2) the change of the flow regimes from film to solid with the volume filling of the pores in capillary condensation, 3) the stability of the nature of the flow near phase transitions and the flow characteristics in different modes.

At a microscopic level it is shown that the heterogeneities on the walls of the pores in the form of steps can lead to a significant change of the local densities of the fluid as it moves along the pore i.e. mechanical perturbations cause phase transformations in the stream. So, limiting the size of the pore section, the steps, depending on the initial conditions of the problem, may represent the interface between two (or more) flow regimes with significantly different fields of concentrations and velocity. The lifetime of such phase fluctuations requires a special investigation. In this situation, consider the process of evolution at aftereffect times 4–5 orders of magnitude higher than the timing of the pulse. Thus, the study of the influence and the role of changes in the phase state of the systems at mechanical effects in various nanotechnology processes can be of practical importance.

For dynamic processes, using the Kelvin equation in assessing the driving forces for the curvature of the meniscus leads to even larger differences than for the equilibrium characteristics. The molecular theory showed that in the motion of molecules in the pore, when phase transitions take place in the pore, a vapour bubble moves in the liquid at its own speed, which is different from the fluid velocity. On the moving phase boundary there are breaks in the concentration and speed. By this the given process is different from the movement of the bubble in the macropores when phase transitions are not considered: the air bubble, positioned in the fluid moving within a small-diameter channel, moves with the average velocity of the liquid.

Analysis of the equations of microscopic hydrodynamics raised for the first time the question of the correct assignment in molecular approaches of the initial conditions for the formation of flows in narrow pores. It was demonstrated that taking into account the differences in different ways to create a one-time initial perturbation, namely one-point and distributed ways of generating the initial perturbations δP, leads to different results. Reached the smallness of δP we can transfer to modeling of almost any kind of external perturbations without introducing artificial effects. The method allows to start the development of models of perturbations of equilibrium distributions that have not even been discussed in order to avoid artificial effects in the generation of flows. Using the models of the initial perturbations as time processes rather than one-time information makes it possible in future to consider the dynamic processes in the real world. Although the studies were limited to the initial stages of contact processes, which takes place on the nanoscale time interval, this interval may significantly exceed the characteristic time of the elementary acts of many fast chemical reactions and the demonstrated complex structure of the evolution of the distribution of matter in space (microflows) indicates the complexity of describing the initial stages of chemical reactions during which the phase boundary is perturbed.

These examples illustrate the possibilities microhydrodynamic approach to the analysis of local non-equilibrium concentration distributions of molecules and their velocities in narrow pores. This approach can also be used for studies of wetting processes at the molecular level of the liquid in contact with a solid plate, which can be of various nature (from highly hydrophilic to hydrophobic). The early stages in the process of wetting of the plate from a fluid tank, the effects of movement of the plate, spreading droplets and evaporation were considered. In all processes there is competition between the contributions of evaporation (flows of molecules in a gas phase) and flows of molecules directly around and along the surface. The approach allows to check the validity and applicability of the different views on the nature of transport (flows) of the molecules under the influence of strong adsorption fields to calculate the flows near the surfaces of solids.

It should be noted that the microhydrodynamic approach allows to describe a wide range of phenomena associated with the non-stationary flow of a fluid containing liquid and gaseous phases, for different types of inhomogeneities (structural and chemical) of the surfaces of the plates. Its application allows to naturally incorporate mass forces in the analysis of the dynamics of processes at large spatial scales, since the equations of this approach become ordinary hydrodynamic equations on large spatial scales. In this case, the theory can be used to obtain limiting values of the transfer coefficients in the bulk phase at all densities.

The new approach opens the possibility of detailed study flows of the molecules in complex porous systems, since the calculations for single pores can be extended to far more complex geometries of the pores, including the pores having a variable cross-section and in merger of several pores into a single pore or, alternatively, the separation of a pore in several smaller pores. The development of this direction allows to explore details of the surface processes of spreading of films and the related dynamic characteristics of viscosity, surface tension and heat transfer in open (non-porous) surfaces.

The generalization of the microhydrodynamic approach to mixtures can significantly extend the scope of its application. It is admissible to generalize the theory for multicomponent systems without loss of molecular information. The proposed method for constructing the transport equations and calculation of the dissipative coefficients of the mixture of adsorbates of spherical shape of approximately the same size in narrow pores is generalized to the explicit consideration of the rotational and vibrational motions of the mixture components and the case of large differences in the size of different types of molecules. The structure of the equations describing the non-equilibrium distribution function amendments, remains unchanged. Reducing the characteristic time scale, we can consistently analyze all contributions of faster processes. This allows to include in consideration the dynamics of changes in the internal states of different molecules at arbitrary densities of the mixtures of fluids.

74. Prospects

We should also briefly discuss the prospects of using the molecular theory in more complex tasks of modeling processes in disperse phases, which require further development of the theory. The microhydrodynamic approach is an alternative method of molecular dynamics. The new method is comparable to the accuracy with the MD method and in the speed of calculations is at least two orders faster. In the future this approach will offer great benefits associated with the ability to adapt to a variable discrete size scale in view of the strict binding to molecular information when defining the probability of the elementary stages. These properties of the LGM indicate the great prospects for improving the accuracy of calculations taking into account the fluctuation effects.

Both approaches operate only with one and the same information about the potential functions of the interaction between components of the mixture and the walls of the pores. The LGM gives full information about all the local flows and concentrations of the molecules, the vectors of microhydrodynamic speeds and temperature, on the dynamic structure of the phase boundaries, the local coefficients of self-diffusion, shear and bulk viscosity, thermal conductivity, while maintaining information about the local thermal velocities and energies of molecules, and also the cluster distributions of molecules. This information helps to identify the impact of structural and chemical inhomogeneities of solid surfaces on the nature of unsteady flows of fluid containing liquid and gaseous phases, in the transport of molecules in the pores, in the wetting of surfaces and the participation of molecules in the formation of new phases.

The LGM differs the method of molecular dynamics by the degree of details of distributions in space. In LGM the unit of length is the characteristic linear size of the molecules, whereas in the molecular dynamics method this size is typically ~20 times smaller. Accordingly, in those situations where a detailed description of distributions is required, the molecular dynamics method can, in principle, give new information. However, in the problems of the phase distribution of the components at the molecular level such a detail is superfluous. This confirms the earlier comparison of both approaches when calculating both the equilibrium and dynamic characteristics discussed in the monograph. An important factor in this study is the calculation time. It is substantially less in the LGM. For example, estimates of the time of calculation of phase diagrams in narrow pore systems indicate a reduction in the time in the LGM by 4–5 orders of magnitude compared with the molecular dynamics method.

In practical use of this approach we always consider a limited amount of the system being studied, since the dimension of the system of equations is limited. Formally, this means that, as in any numerical molecular method of the Monte Carlo type or molecular dynamics, a limited number of molecules is studied. The width of narrow pores specifies one of the characteristic dimensions of the system. The characteristic dimensions in other directions should also be limited, so the numerical flow analysis is possible in a limited area along the axis of the pore. For the slit-shaped pores (in the case of two-dimensional formulation of the problem) it is natural to introduce periodic conditions along an axis perpendicular to the direction of the flow as well as to the width of the pores. This allows one to consider formally the infinite extent in the depth pores, and the distribution functions adequately describes the conditions. In analyzing the three-dimensional flows in slit-shaped pores as well as cylindrical and more complex pore structures such a simplification is not possible. They require a specific formulation of the problem to study the flow characteristics over relatively short distances of the length of the pores (the specific length of the pores is determined by the technical possibilities of solving the corresponding system of equations of the desired dimension). The LGM can be used to increase the volume of the system being studied by the transition to mesocells of different scale within which the fluid properties are obtained by averaging over groups of cells, and to describe the dynamics for which we can construct a procedure for calculating the kinetic coefficients related to mesocells in general. This effect in addition to the simultaneous spatial non-locality of the microscopic hydrodynamics enables molecular interpretation of so-called 'memory function' [28, 29].

An important property of the molecular theory is its self-consistency in the description of the equilibrium and kinetic characteristics, which is associated with the use of the full range of configurations of the molecules in the phase space, compared with the uncontrolled generated variety of configurations in stochastic methods. These features can significantly extend the range of objects to which the molecular theory is applicable, due to the transition to processes multi-scale in time and space in the processes in complex porous bodies. In particular, these include the following processes.

1. With decreasing temperature, the liquid can pass to the solid state. Accounting for processes of crystallization and glass transition in the field of the surface potential is necessary to describe the low temperature. Obviously, the characteristic times of the transport of molecules in the gas and solid are very different, however, the proper understanding of the correctness of using transfer equations, containing or not containing the effects of viscosity, is important for modeling of transport phenomena. All existing numerical techniques of molecular dynamics and Monte Carlo, in fact, leave open the question of the presence of the solid phase, as the methods of assessment of the state of phases used in them are indirect (by analysis of pair distribution functions or time estimates). Especially necessary is the correct assessment of possible two-phase states of the mixture at low temperatures.

2. The equilibrium distributions and dynamic characteristics of large molecules with large differences in the characteristic times of reorientation and phase transformations, must depend on the molecular structure and flexibility of the segments of linear molecules. It allows one to simulate many practical processes in porous solids.

3. Taking into account the influence of the permeability of the walls of the pores in secondary absorption processes of the adsorbate molecules into the depth of the porous material, which inevitably changes the properties of the overall adsorption system, etc. We can also mention the possibility to explore encapsulation cases (when the adsorbate penetrates very narrow cavities, for example,

in zeolites with the size comparable with interstitial cavities in the metal lattices, which actually include absorption concepts), as well as the variants of deformable adsorbents. However, in the latter case it is important to take strictly into account the relationship between the boundaries/grains of the porous material where different rheological processes, up to flows, can take place under severe deformation.

4. Transition to long-range intermolecular interaction potentials significantly expand the range of substances, solutions and simulated situations that reflect the constructed equations for the equilibrium distribution of molecules and processes for the dynamics of their transfer. This will give a unified approach at the molecular level to the description of ion exchange processes in ion exchangers and membranes. This area today is far behind the level of interpretation of the experimental data compared to the adsorption of simple low-molecular molecules.

5. It will be recalled that the original development of the method of discrete distribution functions used as a basis for developing this molecular theory, has been focused on the adsorption and catalytic processes on open surfaces, so this approach allows us to include in the studied range of processes also catalytic processes in porous catalysts, as well as and various chemical reactions in various dispersed phases, most of which are used in practice.

Investigation of these situations require little or no revision of the already discussed equations – here we are just discussing the method of partition of the constructed equations for the equilibrium and dynamic distribution of molecules for different types of nonuniform subsystems with their energies and characteristic times of the elementary processes, and ways of solving systems of equations of high dimension.

It is also necessary to develop this approach to solve the problems that have traditionally been studied sufficiently.

6. Analysis of the role of fluctuations in small finite systems, and improvement of methods for calculating the vicinity of the critical point of the system, etc.

7. The role of amendments of the contribution from the decomposition of the collision integral for the low-density vapour. The kinetic theory of dense phases in the limit of low densities should change to the kinetic theory of gases, which is sufficiently well developed. However, the physical principles of the theory of liquid phases differ significantly from the initial assumptions of the theory of gas, so this transition is not currently possible. The

main factor in the kinetic theory of gases is the account of the velocity dispersion of the molecules, whereas in the dense phase the main factor is the spatially non-uniform distribution of the mixture components. However, in highly nonuniform porous systems the non-equilibrium amendments should be taken into account on the basis of the formulated microhydrodynamic approach in the LGM.

8. The important direction of studies generalizing the results of this communication may be studies on the use of many-particle interaction potentials. It is well known for the bulk phase and uniform surface that the contributions of triple interactions significantly alter the concentration dependences of the thermodynamic characteristics compared to considering only pairwise interactions. It is natural to expect similar changes for nonuniform surfaces and inside the pores, especially for a mixture of molecules of different sizes.

9. Further development of the theory for multicomponent mixtures will have practical applications in the description of the transport of molecules in porous solids and near solid surfaces. These processes play a crucial role in the implementation of a wide range of modern technologies: catalytic, sorption, membrane, electrochemical, chromatographic, purification and separation of liquid and gas mixtures, capillary condensation and desorption, multiphase flow and 'spraying', wetting, impregnation and drying a wide range of disperse systems, etc. It is also important to note that microhydrodynamics allows us to describe a wide range of phenomena associated with the non-stationary flow of the fluid containing liquid and gaseous phase in the presence of structural and chemical inhomogeneities of solid surfaces and pore walls. In addition to this, we note that the theory allows us to consider at the molecular level processes of adhesion, lubrication and friction, as well as the formation of new phases of thin films and nanomaterials for microelectronics. Its use will answer many questions that arise in the interpretation of experimental data in tunneling and atomic force microscopy.

References

1. Tovbin Yu.K., Petukhov A.G. // Izv. RAN. Ser. khim. 2008. No. 1. P. 18.
2. Tovbin Yu.K., Petukhov A.G. // Fiziko-khimiya poverkhnosti i zashchita materialov. 2008. V. 44, No. 3. P. 255.
3. Dubinin M.M., Radushkevich L.V. // Dokl. AN SSSR. 1947. V. 55. P. 331.
4. Dubinin M.M., Zaverina E.V., Radushkevich L.V. // Zh. Fiz. khimii. 1947. V. 21. P. 1410.
5. Dubinin M.M., Zaverina E.V. // Zh. Fiz. khimii. 1949. V. 23. P. 1129.

6.   Radushkevich L.V. // Zh. Fiz. khimii. 1949. V. 23. P. 1410.
7.   Dubinin M.M. // Usp Khimii. 1955. V. 24. P. 3.
8.   Dubinin M.M. // Zh. Fiz. khimii. 1965. V. 23. P. 1410.
9.   Plachenov T.G., Kolosentsev S.D., Porometry. – Leningrad: Khimiya, 1988. – 175 p.
10.   Greg S., Singh K., Adsorption, specific surface, porosity. – Moscow: Mir, 1984. – 310 p.
11.   Tovbin Yu.K., Votyakov E.V. // Izv. RAN. Ser. khim. 2001. No. 1. P. 48.
12.   Tovbin Yu.K. // Izv. RAN. Ser. khim. 1998. No. 4. P. 659.
13.   Landau L.D., Lifshitz E.M., Theoretical Physics. V. 5. Statistical Physics. – Moscow: Nauka, 1964. – 567 p.
14.   Bazarov I.P., Thermodynamics. – Moscow: Vysshaya shkola, 1991. – 376 p.
15.   Hill T.L., Catalysis, theory and research methods. – Moscow: Izd. inostr. lit., 1955. – P. 276.
16.   Adamson A., Physical Chemistry of Surfaces. – Moscow: Mir, 1979.
17.   Tovbin Yu.K., Eremich D.V., Komarov V.N., Gvozdeva E.E. // Khim. Fizika. 2007. V. 26, No. 9. P. 98.
18.   Timofeev D.P., Adsorption kinetics. – Moscow: Publishing House of the USSR Academy of Sciences, 1962. – 252 p.
19.   Raczynski V.V., Introduction to the general theory of sorption dynamics and chromatography. – Moscow: Nauka, 1964. – 134.
20.   Satterfield Ch.N., Mass transfer in heterogeneous catalysis. – Moscow: Khimiya, 1976. – 240 p.
21.   Ruthven D.M. Principles of Adsorption and Adsorption Processes. – N. Y.: J.Willey & Sons, 1984.
22.   Kheifets L.I., Neimark A.V., Multiphase processes in porous media. – Moscow: Khimiya, 1982. – 320 p.
23.   Kirkwood J.G. // J. Chem. Phys. 1946. V. 14. P. 180.
24.   Kirkwood J.G. // J. Chem. Phys. 1947. V. 15. P. 72.
25.   Kirkwood J.G., Buff F.P., Green M.S. // J. Chem. Phys. 1947. V. 17. P. 988.
26.   Eisenschitz R.K., Statistical theory of irreversible processes. – Moscow: Izd. inostr. lit., 1963.
27.   Croxton K., Physics of the liquid state. – Moscow: Mir, 1979.
28.   Zubarev D.N., Morozov V.G., Röpke G., Statistical mechanics of nonequilibrium processes. Vols 1 and 2. – Moscow: Fizmatlit 2002.
29.   Ottinger N.S., Beyond equilibrium thermodynamics. – Hobeken, New Jersey: Wiley, 2005.

Appendix 1

Stastistical justification of the lattice-gas model

Equilibrium systems play an important role in the theory of condensed phases. The most general description of equilibrium systems gives the thermodynamic theory [1–3]. It is applicable to all systems in different phases and states of aggregation, including the interface between coexisting phases. However, thermodynamic consideration is the method of the theory of the continuous medium and it is very limited in the accounting of the molecular properties of the phases.

The molecular–statistical theory of equilibrium systems is based on the Gibbs distribution for microscopic states for a fixed selected set of macroscopic variables (specifying the type of ensemble), describing the observed state of the system [4–7]. For a canonical ensemble belonging to N particles of one kind situated in volume V at temperature T, this distribution is written as

$$\theta\left(p_1, r_1, ..., p_N, r_N\right) = \frac{1}{N! h^{3N} Q_N} \exp(-\beta H), \quad H = K_N + U_N, \quad (A1.1)$$

$$K_N = \sum_{1 \le i \le N} \frac{p_i^2}{2m}, \quad U_N = \sum_{1 \le i \le j \le N} \varepsilon\left(\left|r_i - r_j\right|\right), \quad (A1.2)$$

$$Q_N = \frac{1}{N! h^{3N}} \int_V ... \int_V dp_1 \, dr_1 ... dp_N \, dr_N \exp(-\beta H), \quad (A1.3)$$

where $\theta(p_1, r_1, ..., p_N, r_N)$ is the full distribution function of N

molecules; p_i – momentum and r_i – coordinate of the centre of mass of the molecule i, $1 \leq i \leq N$, m – mass, h – Planck constant, Q_N – normalizing factor (or the statistical sum over states of the system), $\beta = (k_B T)^{-1}$, k_B – Boltzmann constant; H – total energy of the system, K_N and U_N – its kinetic and potential contributions, $\varepsilon(|r_i - r_j|)$ – the potential function of the interaction of molecules i and j, M – partial function θ $(r_1,..., r_M)$, where $M \leq N$, characterizing the probability of various configurations of molecules [8–12], is determined by partial averaging of the full distribution function (A1.1)

$$\theta(r_1,...,r_M) = \frac{\lambda^{*N}}{(N-M)!} \int_V ... \int_V dr_{M+1} ... dr_N \exp\left[\beta(F_N - U_N)\right], \text{(A1.4)}$$

where $\lambda^* = (2\pi mkT/h^2)^{3/2}$, $F_N = -kT \ln (Q_N)$ – the free energy of the system.

The question of strict theoretical justification of the lattice models of the liquid state and the lattice gas model has been debated in the literature [13]. It is believed that the first such a justification was given by Kirkwood [14, 15] (see also [9, 16, 17]). He started from the initial partition of the volume of the system by the lattice to unit cells as small as the size of the molecule. Inside the cell, a molecule moves in the field of its neighbours, having fixed positions in the centres of their cells. Under these assumptions integral equations of the distribution of the centre of mass of the molecules inside the cell in the mean field approximation were derived. Also an expression was derived for the mean free volume per one molecule as a function of intermolecular interactions.

The main objection to lattice models was caused by the very existence of the lattice defining the long-range order in the liquid, which is absent in reality [13]. Studies [14, 15] did not remove the objection. In a review [10] the authors discussed several forms of averaging the position of the centre of mass of molecules inside the cells, which could eliminate the formal presence of the lattice, but the relevant papers were not implemented. Perhaps one of the main reasons for this result was associated with the use of equations in the molecular field approximation, while it was clear that when considering the liquid state the correlation effects must be taken into account.

We should note the study [18], which uses a Gaussian distribution function of the molecules inside the cell. Later it was shown [19] that the formal presence of the lattice does not prevent the construction of a chain of equations of the BBGKY type (Bogolyubov–Born–

Kirkwood–Yvon) for conditional correlative distribution functions. Finally, in [20, 21] the author showed a direct link between the BBGKY chain and LGM equations (see also [22]). To prove this connection, the cluster approach is used [23]. This approach was developed to address the problem of describing the kinetic processes in systems with strong intermolecular interactions [24]. It also allowed to generalize the traditional equation for the equilibrium distribution of molecules for homogeneous phases to arbitrary inhomogeneous systems [25].

The cluster approach
The essence of the cluster approach is to replace the calculation of the statistical sum of the studied system by the solution of the system of equations for the cluster distribution functions that characterize the probability of different local configurations of molecules $\theta(1,..., M)$ with the coordinates $r_1,..., r_M$.

To build this system of equations, we choose m 'central' sites in the centre of the region M. The width of the remaining area containing $(M - m)$ sites must be equal to the radius of the interaction potential between particles R. In this case, the central sites can not communicate with the sites located outside the area under consideration M. We denote such distribution function as $\theta(\{m\}|M)$, where the symbol $\{m\}$ is a list of γ_f^i, $1 \le f \le m$ and $1 \le i \le s$, of the occupation states m of the central states. If we fix the type of particles in the central sites and consider the relationship between the functions $\theta(\{m\}|M)$ with different states of occupation of the central sites, we obtain relationships of the type $\theta(i|M)/\theta(j|M)$ with particles of type i and j, for example $m = 1$. These relationships, on the one hand, are expressed in terms of elementary cluster Hamiltonians $h(i|M)$ for the region containing M sites and the particle i in the centre as

$$\frac{\theta(i\,|\,M)}{\theta(j\,|\,M)} = \exp\left\{\beta[h(j\,|\,M) - h(i\,|\,M)]\right\}, \qquad (A1.5)$$

Using the principle of inclusion–exclusion of probabilities and normalizing relations, the functions $\theta(i|M)$ can be expressed through a sequence of smaller correlators (A1.4).

Accounting for nearest-neighbour interactions
Below is a specific example of the construction of cluster equations

in the case of considering the interactions between the nearest neighbours. Consider the homogeneous lattice system in which any site can be occupied by a particle of type i, $1 \leq i \leq s$, s is the number of components.

We assume that the internal degrees of freedom of the particles are weakly dependent on the energy change of the interparticle interaction of neighbouring particles by replacing particles of one type by another. If ε_{ij} denotes parameters of pair interaction of neighbouring particles of type i and j, then the potential energy of the system will consist of all possible pairwise contributions of the interacting particles and the effective Hamiltonian can be written as [23, 25]

$$H = \sum_{f,i} \nu_i \gamma_f^i - \frac{1}{2} \sum_{f,g} \sum_{i,j} \varepsilon_{ij} \gamma_f^i \gamma_g^j \qquad (A1.6)$$

where the sum over f is taken over all lattice sites, and the sum of g over all z neighbors of site f; the sum over i and j is taken over all the occupation states of the lattice sites. Factor $1/2$ takes into account that each pair is calculated twice.

In the cluster approach, the original lattice system is shown as a set of clusters. We confine ourselves to clusters with one central site. The original lattice is displayed on clusters consisting of $z + 1$ sites: a central site and its neighbours z. For each cluster, we can enter values h_f, characterizing the contributions to the total energy of the system of central particles (cluster Hamiltonians). When accounting for the nearest-neighbour interaction the cluster Hamiltonian for a cluster with one central site has the form

$$h_f = \sum_{i=1}^{s} h_f^i, \quad h_f^i = \left(\nu_i - \sum_g \sum_{j=1}^{s} \varepsilon_{ij} \gamma_g^j \right) \gamma_f^i. \qquad (A1.7)$$

We define two types of cluster correlation. Correlators of the first type can be represented as follows:

$$\theta_{fg_1 \ldots g_z}^{ij_1 \ldots j_z} = \frac{1}{Q} \sum_{i_1=1}^{s} \ldots \sum_{i_N=1}^{s} \gamma_f^i \prod_{g=1}^{s} \gamma_g^i \exp(-\beta H) =$$

$$= \frac{1}{Q} \sum_{i_1=1}^{s} \ldots \sum_{i_N=1}^{s} \gamma_f^i \prod_{j=1}^{s} (\gamma_g^j)^{n_j} \exp[-\beta(h_f + G_f)] \equiv \theta_{\{[n]\}}^i, \qquad (A1.8)$$

where $G_f = H - h_f$. The second equation determines the transition from defining the type of particle in site g, where g is any of the coordination sphere sites, to the number of energy bonds of the central particle i with the neighbouring particles j in the coordination sphere of the cluster. The energy of the central particle in the uniform lattice does not depend on the numbers of sites on which the particle and its neighbours are situated, so for our purposes, the subscripts f, $g_1 \ldots g_z$ omitted. The straight brackets in functions $\theta^i_{\{[n]\}}$, where $\{[n]\} = [n_1 \ldots n_s]$, indicate that the coordination sphere of the cluster contains exactly n_1 particles of type 1, n_2 particles of exactly type 2, etc. To uniquely specify the state of occupation of the sites of the coordination sphere it is enough to set the $(s - 1)$ value n_j, as

$$n_s = z - \sum_{j=1}^{s-1} n_j, \quad 0 \le n_j \le z, \tag{A1.9}$$

i.e. type s particles may be considered as addition to the complete filling of the sites of the coordination sphere of the cluster. The curly brackets denote a complete set of values n_j.

Correlators of the first type (A1.8) characterizes the probability of finding in the cluster centre of the particle i and n_j the particles of the type j, $1 \le j \le s$, at the sites of its coordination sphere.

The expression for G_f in (A1.8) does not depend on the state of occupation of the site f, but depends on the state of occupation of the sites of its coordination sphere, so that for any particular set of particles $\{[n]\}$ formula (A1.8) can be rewritten as

$$\theta^i_{\{[n]\}} = \Lambda \exp\left[\beta\left(-v_i + \sum_{j=1}^{s} \varepsilon_{ij} n_j\right)\right], \quad \Lambda = \frac{\mathrm{const}\left(\{[n]\}\right)}{Q}, \tag{A1.10}$$

where the unknown constant const $(\{[n]\})$ takes into account the energy contributions from all lattice sites, except for the central site of the cluster with the particle i.

Correlators of the second type are obtained from the correlators (A1.8) when averaged over the states of occupation of the sites of the coordination sphere. Using a number of energy bonds of the central particles with its neighbours, these correlators can be written as

$$\theta^i_{\{n\}} = \left\langle \gamma^i_f \prod_{j=1}^{s-1} \left(\gamma^j_g\right)^{n_j} \right\rangle, \quad 0 \le n_j \le z, \tag{A1.11}$$

Functions (A1.11) characterize the probability that the cluster centre there is particle i, and in the coordination sphere of the cluster there are not less than n_1 particles of type 1, n_2 particles of type 2, etc., at least n_{s-1} particles of type $s - 1$. In this case (A1.9) is not satisfied, and in the notation (A1.11) the straight brackets are omitted.

Correlators (A1.11) express all the thermodynamic characteristics of the system. Thus, according to the general definition of correlators of the m-th order $1 \le m \le N$, N is the total number of sites in the system:

$$\theta_{f_1 \dots f_m}^{i_1 \dots i_m} = \left\langle \gamma_{f_1}^{i_1} \dots \gamma_{f_m}^{i_m} \right\rangle = \sum_{i_1=1}^{s} \dots \sum_{i_N=1}^{s} \prod_{n=1}^{m} \gamma_{f_n}^{i_n} P(\{\gamma_f^i\}), \qquad (A1.12)$$

where the sums are taken over all states of occupation for all sites in the system, i_f is the type of particle in the site with the number f; $P(\{\gamma_f^i\})$ is the normalized distribution function of the particles in the system. In equilibrium, it is an expression $P(\{\gamma_f^i\}) = \exp[-\beta H(\{\gamma_f^i\})]/Q$, where Q is the normalization factor, called the statistical sum of the system, or the sum over states. Here H is determined by the expression (A1.6).

Formula (A1.11) is a special case of formula (A1.12), which characterizes the probability of finding an arbitrary configuration of particles on a lattice: the site f_1 has the particle of type i_1, in the site f_2 there is the particle of type i_2 and so on, up to m-th particle inclusive. The occupation states of other $(N - m)$ sites are of no interest – averaging is carried out over them.

$\theta_{\{0\}}^i = \theta_{0 \dots 0}^i = \left\langle \gamma_f^i \right\rangle = \theta_i$ is the probability of finding a particle i at any site of the lattice; fixing θ_i values in the equations of sorption isotherms (adsorption or absorption) determines the external pressure of the molecules of the gas phase or vice versa.

Functions $\theta_{0 \dots j \dots 0}^i = \left\langle \gamma_f^i \gamma_g^j \right\rangle = \theta_{ij}$ (with the j-th index equal to unity) are the probabilities of finding the particles i and j at neighbouring sites of the uniform lattice. Knowing these averages, we can find the heat of sorption or mixing, heat capacity, etc.

Introduction of the correlators of two types allows us to reduce the whole process of constructing systems of equations for the correlation functions only to work with probability relations between them. Detailed stages of this work will be considered for a binary system.

We construct equations describing the local distribution of particles A and B of a binary solution in equilibrium [23, 25]. For clarity, we restrict ourselves to the pairwise interactions between the central particle and its neighbours, according to equation (A1.6). At the first stage we use the expression (A1.10) and consider the ratio of the first type of correlators, differing in the type of the central particle, but having the same state of occupation of the sites of the coordination sphere:

$$\theta_{[n]}^{B} = M_{n}\theta_{[n]}^{A}, \quad M_{n} = \exp[-\beta(v + n\omega)],$$

$$v = v_{B} - v_{A} + z(\varepsilon_{AB} - \varepsilon_{BB}), \quad \omega = \varepsilon_{AA} + \varepsilon_{BB} - 2\varepsilon_{AB},$$

(A1.13)

where n is the number of particles A in the coordination sphere of the cluster. The ratio of the functions (A1.10) excludes unknown Q and const $(\{[n]\})$.

In a second step we consider the connection of the first and second correlator types. By definition, the correlators of the second type containing no less than n particles A in the coordination sphere of the cluster, we represent

$$\theta_{n}^{A} = \sum_{k=0}^{z-n} C_{z-n}^{k} \theta_{[n+k]}^{A}, \quad 0 \le n \le z,$$

(A1.14)

The coefficient of the ratio C_{z}^{k} (A1.14) can be easily obtained from the following considerations. Adding to the n particles another particle A, we proceed to a configuration containing $n + 1$ particles A and characterized by the correlator $\theta_{[n+1]}^{A}$. Such a transition is possible in $(z - n)$ ways. A similar shift to $(n + 2)$ particles is possible by adding two particles A by $(z - n)(z - n - 1)/2$ methods, etc. The coefficients C_{m}^{k} are the number of combinations z of elements in n: $C_{m}^{k} = m!/(k!(m-k)!)$.

Inverting the linear system of equations (A1.14) with respect to the first type of correlators, we find

$$\theta_{[n]}^{A} = \sum_{k=0}^{z-n} (-1)^{k} C_{z-n}^{k} \theta_{n+k}^{A}.$$

(A1.15)

A similar expression holds for the correlation of the first type to the core particle B.

In the third stage, we substitute the expression (A1.15) for the central particle A and B in relation (A1.13):

$$\sum_{k=0}^{z-n} (-1)^{k} C_{z-n}^{k} \theta_{n+k}^{B} = M_{n} \sum_{k=0}^{z-n} (-1)^{k} C_{z-n}^{k} \theta_{n+k}^{A}.$$

(A1.16)

This linear system of equations connects the unknown functions θ_{n}^{A} and θ_{n}^{B} together. It is solved sequentially, starting with $n = z$. Its solution has the form

$$\theta_{n}^{B} = \sum_{k=0}^{z-n} C_{z-n}^{k} M_{n+k} \sum_{r=0}^{z-n-k} C_{z-n-k}^{r} \theta_{n+k+r}^{A} = \exp[-\beta(v + n\omega)] \sum_{k=0}^{z-n} C_{z-n}^{k} x^{k} \theta_{n+k}^{A},$$

(A1.17)

where $x = \exp(-\beta\omega) - 1$. In the second equality the terms are regrouped so that the first sum changes the correlation dimension and the energy factors are in the inner sum, and the explicit form of M_n (A1.13) is taken into account.

The system of equations (A1.17) is the starting point for calculating the cluster correlators. With proper sequential search of all lattice sites the system (A1.17) allows, in principle, an exact solution. This is done for the one-dimensional lattice [26]. For two- and three-dimensional lattices the number of equations of the system (A1.17) is less than the number of unknowns θ_n^A and θ_n^B (taking into account the normalization condition and the condition of invariance of filling the central and adjacent sites in the cluster), so it must be closed. Depending on how the system of equations (A1.17) is closed, by approximating higher correlators through the correlators of lower dimension we obtain different approximations characterized by the accuracy of taking into account the correlation effects. The structure of the equations is such that it is necessary to formulate a general rule for the approximation of all higher correlators through the lower ones, which defines the functional dependence of the closed system for uncoupled correlators. This, in turn, allows after finding the uncoupled correlators to calculate self-consistently all higher correlators, both first and second type in this approximation.

Continuous distribution of molecules

To describe the distribution of the molecules we use a continuous analogue of the cluster approach [20–22]. For a group with the same number of molecules M but different coordinates, we consider the ratio of the functions (A1.4)

$$\theta(r_1,...,r_M) = \theta(r_1^*,...,r_M^*)\xi\exp\left\{\beta\sum_{1\leq i\leq j\leq M}[\varepsilon(|r_i^* - r_j^*|) - \varepsilon(|r_i - r_j|)]\right\},$$

$$\xi = \frac{\Lambda}{\Lambda^*}, \quad \Lambda = \int_V...\int_V dr_{M+1}...dr_N \exp\left\{-\beta\left[\sum_{1\leq i\leq M;M+1\leq j\leq N}\varepsilon(|r_i - r_j|) + \right.\right. \quad (\text{A}1.18)$$

$$\left.\left. + \sum_{M+1\leq i\leq j\leq N}\varepsilon(|r_i - r_j|)\right]\right\},$$

where the coordinates of the molecules in the different groups are denoted by r_i and r_i^* respectively (coordinates with an asterisk correspond to Λ^*).

From M molecules we separate a group containing m molecules in volume ω, which will be considered central. The volume ω is surrounded by a sphere ω_R whose radius is not less than $\omega^{1/3} + R$, where R is the radius of the interaction potential between the molecules. In the volume $(\omega_R - \omega)$ there are the remaining $(M - m)$ molecules. If we fix their coordinates, then regardless of the position m of the central molecules $\xi = 1$ and the formula (A1.18) can be rewritten as

$$\theta(r_1,...,r_m \mid r_{m+1},...,r_M) = \theta(r_1^*,...,r_m^* \mid r_{m+1},...,r_M) \times$$

$$\times \exp\left\{\beta\left(\sum_{1 \le i \le j \le m} [\epsilon(\mid r_i^* - r_j^* \mid) - \epsilon(\mid r_i - r_j \mid)] + \right.\right.$$

$$\left.\left. + \sum_{1 \le i \le m; m+1 \le j \le M} [\epsilon(\mid r_i^* - r_j \mid) - \epsilon(\mid r_i - r_j \mid)]\right)\right\} \cdot \text{(A1.19)}$$

Expressions (A1.19) are sets of relations, based on different locations of both central molecules and molecules in the surrounding area $(\omega_R - \omega)$.

In the case of discrete distribution of molecules, these relations are used in the cluster approach for lattice structures with $m = 1$ and 2 [23, 25]. These relations can be used to describe systems with a continuous distribution of the molecules. To prove the latter, consider a system (A1.18) in which M is replaced by m, and the value of N by M, and putting $r_i = r_i^*$, $2 \le i \le m$, $r_1^* = r_1 + dr_1$. Expanding the expression for dr_1 (A1.18), we obtain

$$\frac{\partial}{\partial r_1} \ln \theta(r_1,...,r_m) + \beta \frac{\partial}{\partial r_1} U_m + \frac{\partial}{\partial r_1} \xi = 0, \qquad \text{(A1.20)}$$

where U_m is the energy of a group of m molecules, defined in (A1.2), and the derivative with respect to ξ, according to (A1.1), (A1.4), (A1.18), leads to the following expression:

$$\frac{\partial}{\partial r_1} \xi = \frac{\beta(M-m)}{V\theta(r_1,...,r_m)} \int_V dr_{m+1} \theta(r_1,...,r_{m+1}) \frac{\partial}{\partial r_1} \epsilon(\mid r_1 - r_{m+1} \mid). \quad \text{(P1.21)}$$

Thus, equations (A1.20) and (A1.21) represent a system of integrodifferential BBGKY equations [8–11]. Equations (A1.18) determine the relationship between different configurations of groups consisting of the same number of molecules that differ in their coordinates. Differential changes in these probabilities change in coordinate-wise change of the position of each group of molecules obey the BBGKY system of equations.

Equations (A1.19) represent one of the specific ways of the distribution of the molecules obeying the integral relations (A1.18). The enumeration of the distribution of the molecules in areas ω and $(\omega_R - \omega)$ can be arranged arbitrarily. We divide the spaces of these areas by some grid with step λ_n: $n\lambda_n = 2^{1/6}D$, where D is the diameter of the molecule, and n is the number of grid sites, falling to its diameter. (We restrict ourselves to a uniform grid with the same pitch λ_n in all directions in space, in general, steps in different directions may differ). Such a statement of the problem reduces it to the problem of distributing molecules occupying more than one lattice site, taking into account the interactions between them.

Denoting the event of the centre of the molecule located at a site of the grid f through the state of occupation by component A, and the absence of the molecule in a site through a vacancy (component V), system (A1.19) can be rewritten in the following discrete form:

$$\theta^V_{\delta,\{n\}}(f) = \exp\left[-\beta\left(v_f + \sum_{r=1}^{R} \varepsilon_{fg}(r)n_r\right)\right] \prod_{r=1}^{R}\left\{\sum_{k_r=1}^{z_r-n_r} C^{k_r}_{z_r-n_r} x^{k_r}_{fg}(r)\right\}\theta^A_{\delta,\{n+k\}}(f),$$

$$(A1.22)$$

where $v_f = v^V_f - v^A_f$; all curly brackets in (A1.22) represent a complete list of coordination spheres within the radius of the interaction potential R; $x_{fg}(r) = \exp(-\beta\varepsilon_{fg}(r)) - 1$; $\varepsilon_{fg}(r)$ is the energy of interaction of the molecules at distance r, the potential between the molecules $\varepsilon_{fg}(r)$ is considered to be long-range. In general, this energy may depend on the coordinates of the molecules (as it occurs at the nonuniform surfaces). Symbol δ refers to the number of the sublattice inside the area blocked by the hard sphere of the molecule, as in Fig. 5.4. List $\{n_r\} \equiv n_1,..., n_R$ refers to the coordination spheres outside the blocked area. Here the combinatorial factor C^k_n is the number of combinations of n by k – written for different coordination spheres r. For any site f, $1 \le f \le N$, we obtain an algebraic system of equations relative to the local cluster distribution functions $\theta^i_{\{n\}}$ (f), describing the probability of finding n_r the particle A in the r-th coordination sphere on the sites $g_r \in z_r$ around the central particle i at site f. Similarly we construct equations for multicomponent systems, replacing the binomial coefficients C^k_z by polynomials.

When $\theta^A_{\delta,\{0\}}$ = const we have a liquid state. If there is preferred filling of one of the sublattices δ, then we can talk about the implementation of the long-range order. In the liquid state the condition of constant density allows to exclude exp (βv) from the first equation of the chain and the reciprocal distributions of the molecules are described by equations (A1.22). In Appendix 2 the full set of equations (A1.22) for all sites in the lattice describes the molecular distribution in the sites of the nonuniform lattices. However, the condition $\theta^A_{\delta,\{0\}}$ = const is inapplicable for the nonuniform systems.

The system of equations (A1.22) was derived for lattice structures with variable lattice parameter λ_n. Its change changes the values of the number of coordination spheres within the potential radius R and, respectively, sets of numbers z_r, but the structure of the system of equations remains unchanged. Using the variable parameter λ_n we can construct successive approximations to describe the structure of the liquid state. In this sequence the first approximation ($\lambda \equiv \lambda_1$) satisfies the condition that each molecule occupies one lattice site – the traditional position of the lattice models [16]. To describe the thermodynamic characteristics of liquid systems we often need only a first approximation, and to analyze the structure we require a fine grid [17].

Fine grid

The idea of a fine grid in the LGM can be implemented as a task for investigating the distribution of large particles using equations (A1.22), or as deriving integrated continuum equations on a given lattice structure. Correctness of the first method follows directly from the analysis of the distribution of multicentre molecules on the one-dimensional structure [27–29]. The adsorption isotherm of the multicentre particle of size n in the absence of interaction is described by the equation $aP = \dfrac{\theta[n-\theta(n-1)]^{n-1}}{[n(1-\theta)]^n}$, $\theta = nN/L$, where L – the number of lattice sites, a – Henry's constant, P – pressure of the gas phase. Reducing the grid spacing λ_n leads to the asymptotic behavior $(n \sim \lambda_n^{-1})$ of isotherms $aP = b \exp(b)$, where $b = \theta/(1 - \theta)$, and using the relation between the expansion pressure in the LGM π and the adsorption isotherm $\beta\pi\lambda = \int_0^\theta \theta d\ln P$, we obtain the Tonks equation of state [13, 30, 31]: $\beta_p\lambda = \theta/(1 - \theta)$.

A similar transition for the same problem from a discrete description of the pair distribution function, expressed as $\theta_{AA}(r) = \theta t(r) = \theta t_{AV}(t_{VV})^{r-1} t_{VA}$, where $t_{AV} = t_{VV} = n(1 - \theta)\sigma$, $t_{VA} = \theta\sigma$, $\sigma = [n - \theta(n - 1)]^{-1}$, to the limit at $\sigma \to 0$, leads to an expression for the pair distribution function of the continuum [32]: $\theta_{AA}(r) = \theta^2$ exp $[-\theta(r - D)/D(1 - \theta)])/(1 - \theta)$, where D is the diameter of the molecule [20, 21].

The idea of splitting the cell volume was used in the theory of melting [33, 34] and in the development of the theory of the liquid state based on the equations for the polymer molecules [35].

This same idea has been applied in the search for effective algorithms in the Monte Carlo method [36, 37]. It is technically much simpler to implement a stochastic process on a discrete set of sites than in the continuous space. The idea of splitting the cell was also used to describe the adsorption of spherically symmetric molecules (which are approximated by a sphere or square) on amorphous surfaces [38, 39]. Simplification is achieved by a special measure in which we use for each sublattice the equation for single-site particles, which can be solved quite quickly, instead of solving the equations for large molecules. On an example of calculating the adsorption of argon atoms on the amorphous surface of rutile this procedure was shown to be highly effective [38, 39].

As a result, the statistical justification of lattice models of the liquid state takes place without the use of a priori concepts of the lattice. The difference system of equations for discrete distribution functions in the case of the fine grid gives a description of the structure of the liquid state, as in the theory of integrodifferential or integral equations of the liquid. In fact, in [14, 15] the authors proposed a procedure for obtaining the average values of effective parameters of the interparticle interaction using the continuous potential curves in the mean field approximation. This procedure, which reveals the meaning of the effective parameters, retains its value also in the new procedure for constructing successive approximations for different λ_n. Development of these ideas with a more accurate quasi-chemical approximation is discussed in Appendix 4 and section 9.

Structural justification of the LGM is a key opportunity for a unified description of all three phases. The initial formulation of the LGM, strictly tied to the crystal lattice of a solid, restricted its use for solids and their surfaces. Removing this limitation provides a general approach to liquids and steam, as well as all of their phase boundaries.

References

1 1.   Gibbs J.W., Selected Works. Thermodynamics. – Moscow: Nauka, 1982. – P. 9.
2.   Kubo R., Thermodynamics. – New York: Wiley, 1970.
3.   Prigogine I., Defay R., Chemical Thermodynamics. – Novosibirsk: Nauka, 1966. – 510 p.
4.   Gibbs J.W., Selected Works. Statistical mechanics. – Moscow: Nauka, 1982. – P. 350.
5.   Fowler R., Guggenheim E., Statistical Thermodynamics. – Moscow: Izd. inostr. lit., 1948.
6.   Kubo R., Statistical Mechanics. – Moscow: Mir, 1967.
7.   Huang K., Statistical Mechanics. – Moscow: Mir, 1966. – 520 p.
8.   Bogolyubov N.N., Problems of dynamical theory in statistical physics. – Moscow: Gostekhizdat, 1946.
9.   Hirschfelder J., Curtis Ch., Bird R., Molecular Theory of Gases and Liquids. – Moscow: Izd. inostr. lit., 1961. – 929 p.
10.   Hill T.L. Statistical Mechanics. Principles and Selected Applications. – N. Y.: McGraw-Hill Book Comp. InP., 1956.
11.   Croxton K., Physics of the liquid state. – Moscow: Mir, 1979. – 400 p.
12.   Maeyr J., Goeppert-Maeyr M., Statistical mechanics. – Moscow: Mir, 1980.
13.   Fisher I.Z., Statistical Theory of Liquids. – Moscow: Fizmatgiz, 1961. – 200 p.
14.   Kirkwood J.G. // J. Chem. Phys. 1950. V. 18. P. 380.
15.   Salsburg Z.W., Kirkwood J.G. // J. Chem. Phys. 1952. V. 20. P. 1538.
16.   Barker J.A. Lattice theories of the liquid state. – Oxford: Pergamon Press, 1963.
17.   Barker J.A., Henderson D. // Rev. Mod. Phys. 1976. V. 46, No. 4. P. 587.
18.   Mayer J.E., Careri G. // J. Chem. Phys. 1952. V. 20. P. 1001.
19.   Rott L.A., Statistical theory of molecular systems. – Moscow: Nauka, 1979. – 280 p.
20.   Tovbin Yu.K., Dissertation, Karpov Institute. Moscow. – 1985. – 380.
21.   Tovbin Yu.K. // Theoretical methods for describing the properties of solutions: Intercollegiate Collection of Scientific Works. – Ivanovo, 1987. – p. 44.
22.   Tovbin Yu.K. // Zh. Fiz. Khimii. 2005. V. 79, No. 12. P. 2140.
23.   Tovbin Yu.K. // Zh. Fiz. Khimii. 1981. V. 55, No. 2. P. 273.
24.   Tovbin Yu.K. // Zh. Fiz. Khimii. 1981. V. 55, No. 2. P. 284.
25.   Tovbin Yu.K.m Theory of physico-chemical processes at the gas–solid interface. – Moscow: Nauka, 1990. – 288 p.
26.   Fedyanin V.K., Statistical physics and quantum field theory. – Moscow: Mir, 1973. – P. 241.
27.   Tovbin Yu.K. // Zh. Fiz. Khimii. 1974. V. 46, No. 5. P. 1239.
28.   Tovbin Yu.K. // Teor. i eksp. khimiya. 1974. V. 10, No. 3. P. 258.
29.   Tovbin Yu.K. // Teor. i eksp. khimiya. 1976. V. 12. No. 2. P. 196.
30.   Tonks L. // Phys. Rev. 1936. V. 50, No. 9. P. 955.
31.   Heer K., Statistical mechanics, kinetic theory and stochastic processes. – Moscow: Mir, 1976.
32.   Verhagen A.M.W. // J. Chem. Phys. 1970. V. 53, No. 3. P. 2223.
33.   Cowley E.R. // J. Chem. Phys. 1979. V. 71. P. 458.
34.   Cowley E.R., Barker J.A. // J. Chem. Phys. 1980. V. 73. P. 3452.
35.   Sanchez I.C., Lacombe R.H. // J. Phys. Chem. 1976. V. 80. P. 2353.
36.   Panagiotopoulos A.Z. // J. Chem. Phys. 2000. V. 112. P. 7132.
37.   Floriano M.A., et al. // Langmuir. 1999. V. 15. P. 3143.
38.   Tovbin Yu. K. // Langmuir. 1999. V. 15. P. 6107.
39.   Gvozdev V.V., Tovbin Yu.K. // Izv. AN. Ser. khim. 1997. No. 6. P. 1109.

S

Appendix 2

The lattice gas model for non-uniform systems

Concentration dependences of the thermodynamic characteristics of adsorption systems are determined by the combined influence of surface non-uniformity and lateral interaction between adsorbed particles. In this appendix we consider the hierarchy of adsorption equilibrium models associated with differences in the spatial scales of the considered part of the surface and with the use of different distribution functions of different types of adsorption sites in the description of the nature of surface non-uniformity. Expressions are obtained for the isotherms of adsorption equilibrium and the equilibrium constant of adsorption displacement of molecules with each other in the adsorption of gas mixtures.

In presentation the emphasis is placed on the two-dimensional surface systems, for which it is easy to demonstrate the emerging relationship between the distribution functions of different types of sires. Similar issues arise when considering the complex porous materials. In addition to the presentation, we require the relationships between sites belonging to different monolayers within the pore volume.

Model
To describe the adsorption system we will use the lattice model. Each lattice site is an adsorption centre which contains a particle of any sort i, including a vacancy, $1 \leq i \leq s_f$, s_f is the number of system components that may situated in the site with the number f, $1 \leq f \leq N$, N is the number of sites. Since the sites are different in

nature, then they may contain particles of different sorts. (the last s_f index will be assigned to the free sites). The state of occupation of each site is characterized by the random variable $\gamma^i_f, \gamma^i_f = 1$, if the site f contains a particle of sort i, and $\gamma^i_f = 0$ if the site contains a particle of other sort. Random variables are subject to the following relations: $\sum\limits_{i=1}^{sf} \gamma^i_f = 1, \gamma^i_f \gamma^i_f = \Delta_{ij}\gamma^i_f$, where Δ_{ij} is the Kronecker symbol, which means that any site is necessarily occupied by some, but only one, particle.

The site type with number f will characterize by the value of η^q_f which assumed to be known and constant during the adsorption process (non-reconstructed surface) $1 \leq q \leq t$, t is the the number of types of sites; $\eta^q_f = 1$ if the site f is a site of type q, and $\eta^q_f = 0$ otherwise. The complete set of values $\{\eta^q_f\}$ $1 \leq f \leq N$, uniquely determines the surface composition and structure of the surface, which can be arbitrary.

We assume that the lateral interaction is described by pair potentials with the interaction radius R, R is any given number. Distances will be counted in numbers of coordination spheres (c.s.) (see Fig. 7.1 b). The number of sites in the r-th c.s. of the site with the number f of type q is denoted by $z_f(r)$ (or $z_q(r)$), $1 \leq r \leq R$. The interaction parameter of the particles i and j, located on the sites of type q and p with numbers f and g at a distance r, denoted by $\varepsilon^{ij}_{fg}(r)$ (or $\varepsilon^{ij}_{qp}(r)$, as the site number and its type are uniquely linked by the values η^q_f).

The total energy of the in the grand canonical ensemble can be written as follows [1, 2]:

$$H = \sum_{f,i} v^i_q \gamma^i_f \eta^q_f - \frac{1}{2}\sum_{r=1}^{R}\sum_{f,g_r}\sum_{i,j} \varepsilon^{ij}_{qp}(r)\gamma^i_f \gamma^j_{g_r} \eta^i_f \eta^j_{g_r}, v^i_q = -\beta^{-1}\ln y^i_q,$$

$$y^i_q = \left(a^i_q P_i\right)^{1/m_i}, a^i_q = \left(F^i_q\right)^{m_i}\beta\exp\left(\beta Q^i_q\right)/F^0_i, \beta = 1/(kT), \qquad (A2.1)$$

where the index g labels the sites that are in the r-th c.s. of the site f; a^i_q is the coefficient of adsorption of particle i at site q, F^i_q and F_i are the statistical sums of the adsorbed particle i on the site q and the molecule i in the gas phase, P_i and m_i are the partial pressure and the degree of dissociation of the molecule i_m. If $m_i = 1$ then $Q^i_q = \varepsilon^i_q$ where ε^i_q is the energy of bonding of the particle i with site q, if $m_i = 2$, then $Q^i_q = \varepsilon^i_q - D_p$, where D_i is the dissociation energy of the molecules in the gas phase i_m. In dissociation of the molecule of the AB type to particles A and B the equilibrium of the process $AB_g + 2Z_q = Z_q A + Z_q B$, where Z_q and Z'_q is the free site and the suite q occupied by the particle i, characterized by constant $a_q^{AB} = F_q^A F_q^B \exp\left[\beta\left(\varepsilon_q^A + \varepsilon_q^B - D_{AB}\right)\right]/F_{AB}^0$, where D_{AB} is the dissociation energy of the molecule in the gas.

Discrete (two-dimensional) model

The task of the statistical theory is to construct closed expressions describing the distribution of adsorbed particles on the surface. The interaction between the particles leads to the dependence of the states of occupation of each site on the state of occupation of its neighbours. To describe the distribution of interacting particles on non-uniform lattices it is most convenient to use the so-called cluster approach, which was first formulated in [3]. Its essence is as follows.

1. The original lattice is shown on the set of clusters. Each cluster consists of central sites and their R coordination spheres. The central sites may be any number of sites. Usually only one or two are enough. If there is one central site, the lattice is displayed on the N clusters, if two central sites and the distance between them is ρ, 1 $\leq \rho \leq R$, then the lattice displayed on $zq(\rho)N/2$ clusters. The the order of the distribution of clusters is 'remembered'. Sites with indices g, h, l are the sites that are closest to the site with the number f, are in first, second and third coordination sphere. Figure 7.2 shows an example of such a map. For the averaged model discussed below that is a simplification of the cluster description – it is the continuation of the scheme, the lattice is 'split' to isolated pairs of sites at a distance of the first and second neighbours.

2. For each cluster we construct a precise system of equations for the probabilities of the various configurations of particles, which can exist on it. The resulting systems of equations are finite by the finiteness of the number of sites in the cluster, but not closed.

3. We introduced the method of closing which is the same for all systems of equations belonging to different clusters which takes into account the location of the real sequence of clusters on the original lattice. The result is a single closed system of equations for the entire lattice as a whole.

The procedure for constructing equations for the distributions of molecules is completely identical with the procedure in Appendix 1 for an uniform surface.

If we only consider the direct correlations between interacting particles, the closed system of equations for θ_f^i – the probability of finding the particle i at a site with the number f and θ_{fg}^{ij} – the probability of finding the particles i and j at sites f and g at a distance r, using the results of [4] and skipping intermediate transformations, written in final form for a fully distributed model: 1) local fillings θ^i_f characterizing the probability of filling the site f with the particle i:

$$\exp\left[\beta\left(v^s_f - v^i_f\right)\right] = a^i_f P_i = \frac{\theta^i_f \Lambda^i_f}{\theta^s_f},$$

$$\Lambda^i_f = \prod_{g\in z(f)} \prod_{j=1}^{s} {}^*\theta^{ij}_{fg} \frac{\exp\left(\beta\varepsilon^{sj}_{fg}\right)}{\theta^i_f},$$

(A2.2)

2) for the chemical potential:

$$\mu^i_f = \mu_i(\text{Thermostat}) = M^i_f - M^s_f,$$

$$M^i_f = v^i_f + \sum_{r=1}^{R}\sum_{g_r} \varepsilon^{ii}_{fg}(r) z^{ii}_{fg}(r)/2 - \beta^{-1}\ln\theta^i_f - \frac{1}{2}\beta^{-1}\sum_{r=1}^{R}\sum_{g_r} \ln\left(\frac{\theta^{ii}_{fg}(r)\theta^{ik}_{fg}(r)}{\left(\theta^i_f\right)^2 \theta^{ki}_{fg}(r)}\right).$$

(A2.3)

Index k plays the role of the reference varieties, its choice is determined by the convenience of calculation, and the numerical values do not depend on it.

3) for the local expansion pressure π_f in the LGM determined through derivative $\mu^s_f = \partial F/\partial N_f$, as $\pi_f v^0_f = -\mu^s_f$ (here free or vacant sites correspond to the s-th component of the mixture)

$$\beta\pi_f v^0_f = -\ln\theta^s_f - \frac{1}{2}\sum_{g\in z(f)} \ln\left(\frac{\theta^{ss}_{fg}}{\theta^s_f \theta^s_g}\right).$$

(A2.4)

In (A2.2)–(A2.4) we use the notation ${}^*\theta^{ij}_{fg} = \theta^{ij}_{fg}\exp\left(-\beta\varepsilon^{ij}_{fg}\right)$, which are connected by a system of equations ${}^*\theta^{ij}_{fg}{}^*\theta^{km}_{fg} = {}^*\theta^{im}_{fg}{}^*\theta^{kj}_{fg}$, where θ^{ij}_{fg} is the probability of finding the particle pairs ij on sites fg of the type qp.

Free energy
The free energy of a multicomponent mixture F in equilibrium in a quasi-chemical approach is given as follows:

$$F = \sum_{f,i}\theta^i_f\, kT\ln J^i_f + \frac{1}{2}\sum_{r}\sum_{f,g}\sum_{i,j}\varepsilon^{i,i}_{fg}(r)\theta^{ij}_{fg}(r) - kT\sum_{f,i}\theta^i_f\ln\theta^i_f -$$

$$-\frac{kT}{2}\sum_{r}\sum_{f,g}\sum_{i,j}\left[\theta^{ij}_{fg}(r)\ln\theta^{ij}_{fg}(r) - \theta^i_f\theta^j_g\ln\theta^i_f\theta^j_g\right]$$

or

$$F = E - TS = \frac{1}{N} \sum_{f=1}^{N} \sum_{i} M_f^i \theta_f^i = \sum_{f=1}^{t} \sum_{i} F_f M_f^i \theta_f^i, \qquad (A2.5)$$

where $E = -\langle H \rangle$ is the internal energy and S is the entropy of the system. In formula (A2.2) a_f^i is the local Henry coefficient. All local partial filling θ_f^i are functions of the complete set of partial pressures $\{P_i\}$, $1 \leq i \leq (s-1)$. The non-ideality function Λ_f^i is taken into account in the quasi-chemical approximation only by direct correlation between the interacting particles. The system of equations (A2.2) subject to the normalizations functions θ_f^i and θ_{fg}^{ij} is closed. It allows one to calculate the required equilibrium distributions of the molecules.

In particular, the local isotherm equations (A2.2) have the form

$$\theta_f^s y_f^s = \theta_f^i \prod_{r=1}^{R} \prod_{g \in z_f(r)} S_{fg}^i(r),$$

$$S_{fg}^i(r) = \sum_{j=1}^{s} \theta_{fg}^{ij}(r) \exp\left\{ \beta \left[\varepsilon_{fg}^{sj}(r) - \varepsilon_{fg}^{ij}(r) \right] \right\} / \theta_f^i, \qquad (A2.2a)$$

or

$$S_{fg}^i = 1 + \sum_{j=1}^{s-1} \theta_{fg}^{ij}(r) x_{fg}^{ij}(r) / \theta_f^i,$$

$$x_{fg}^{ij}(r) = \exp\left\{ \beta \left[\varepsilon_{fg}^{sj}(r) - \varepsilon_{fg}^{ij}(r) \right] \right\} - 1 = \exp\left[-\beta \omega_{fg}^{ijss}(r) \right] - 1,$$

$$\theta_{fg}^{ij}(r) \theta_{fg}^{kl}(r) = \theta_{fg}^{il}(r) \theta_{fg}^{kj}(r) \exp\left[\beta \omega_{fg}^{ijkl}(r) \right], \qquad (A2.6)$$

$$\omega_{fg}^{ijkl}(r) = \varepsilon_{fg}^{ij}(r) + \varepsilon_{fg}^{kl}(r) - \varepsilon_{fg}^{il}(r) - \varepsilon_{fg}^{kj}(r),$$

$$\sum_{i=1}^{s} \theta_{fg}^{ij}(r) = \theta_g^j, \quad \sum_{j=1}^{s} \theta_{fg}^{ij}(r) = \theta_f^i, \quad \sum_{i=1}^{s} \theta_f^i = 1. \qquad (A2.7)$$

For adsorption of particles of one type ($s = 2$) of equation (A2.6) allowed explicitly:

$$\theta_{fg}^{AA}(r) = \frac{2\theta_f^A \theta_g^A}{\delta_{fg}(r) + b_{fg}(r)}, \quad \delta_{fg}(r) = 1 + x_{fg}^{AA}(r)\left(1 - \theta_f^A - \theta_g^A\right),$$

$$b_{fg}(r) = \left\{ \left[\delta_{fg}(r) \right]^2 + 4x_{fg}^{AA}(r) \theta_f^A \theta_g^A \right\}^{1/2}. \qquad (A2.8)$$

The remaining pair functions for $s = 2$ are determined from the normalization of relations (A2.7). For $s > 2$ for small fillings $\left(\theta_f^s \sim 1,\right.$ the remaining functions $\theta_f^i \sim 0,$) paired functions expressed in terms of density as $\theta_{fg}^{ij}(r) = \theta_f^i \theta_g^j \exp\left[\beta\varepsilon_{fg}^{ij}(r)\right],$ where i and $j \neq s$. In general, a solution of non-linear equations (A2.6) and (A2.7) is carried out numerically. The method of this significant reduction of the dimension of the system of equations is discussed below.

The constructed system of equations describes the distribution of particles is more detailed manner. It operates with the filling of sites of the considered fragment of the surface. After approximate closing of the system of equations, all clusters 'returned' to their sites, so that the original lattice is restored completely, and the engagement of all the sites in the system with each other. The possibility of practical use of the system (A2.2a), (A2.6), (A2.7) for large N considering lowering its dimensions is determined by the computer capabilities.

The question naturally arises of constructing other cruder models, such that on the one hand, we transfer to the description of a larger size of the surface area, and on the other hand, rely on the same molecular adsorption system parameters used in (A2.2a), (A2.6), (A2.7). Such a transition is carried out using the average distribution of sites on separate sections of the surface.

Average (point) models
Point models are obtained if, instead of 'returning' the clusters to their places in the starting lattice, we group them so that all clusters are identical regardless of the number of the central site are in the same group. The identical clusters and those for which only types of central sites and all sites in the coordination spheres coincide. The number of identical clusters is the statistical weight of this type of cluster. The problem displaying the original lattice on the clusters is that the clusters reflect the actual structure of the original non-uniform surface. In the two-dimensional model such a problem does not arise, as it displays the specific fragment of the surface.

Grouping the clusters by types violates the connection of lattice sites, and to preserve the local structure of the original surface, we display it on two groups of clusters. The first group – cluster with one central site on which we 'mentally' display the lattice in the construction of the equations describing the distribution of the particles (and now this display is 'real').

The second group are the clusters with two central sites located at distances $\rho \leq R$, and with all its neighbouring sites located at distances $r \leq R$, to at least one of the central sites. These clusters provide a balanced account of the influence of neighbouring sites, surrounding each of the central sites f and g, on which there are the particles i and j, characterized by the function $\theta_{fg}^{ij}(\rho)$. For convenience of numbering the neighbours, we introduce the concept of a single c.s. with radius r of the dimeric molecule occupying central sites representing the totality of sites at the same distance r from the nearest of the central sites [1, 2].

$\pi_r(\rho)$ denotes the number of different orientations of the positions of neighbouring sites in the r-th c.s. relative to the fixed position of the dimeric molecule. Orientation $\omega_r(\rho)$ is characterized by the angle between the line l_1, connecting central sites, and the line l_2, connecting the neighbouring site and the middle of the distance between the central sites, $1 \leq \omega_r(\rho) \leq \pi_r(\rho)$. The number of sites in the r-th c.s. with the orientation $\omega_r(\rho)$, through which the line l_2 passes when rotating around the line l_1, is denoted by $\kappa(\omega_r|\rho)$.

In the second equality (A2.2a) for the non-ideality function $S_{fg}^i(r)$ in the apparent form we consider zero particle interactions with free sites (these particles have index s_g). In order not to complicate the notation for the symbol s_g, we omitted the site number to which it relates. This is clear in the meaning: when summing the number refers to the considered in the sum, and in the notations θ_f^s and $\theta_{fg}^{is}(r)$ it refers to the corresponding lower index f or g.

Figure 7.1 b shows the distribution of neighbouring sites around the central sites f and g at distances $\rho = 1$ and 2 on a square lattice for $R = 2$. When $\rho = 1$ the first coordination sphere includes sites 1–6 ($\pi_1(1) = 4$), the second – the sites 7–10 ($\pi_2(1) = 2$): $\omega_1(1) = 1$ corresponds to site 1, $\omega_1(1) = 2 - 2, 6, \omega_1(1) = 3 - 3, 5, \omega_1(1) = 4 - 4$; $\omega_2 (1) = 1 - 7, 8$; $\omega_2 (1) = 2 - 9, 10$. If $\rho = 2$, the first coordination sphere includes sites 1–6 ($\pi_1(2) = 3$), the second – the sites 7–12 ($\pi_2(2) = 4$): $\omega_1(2) = 1$ correspond to the sites 1 and 2; $\omega_1(2) = 2 - 3, 6; \omega_1(2) = 3 - 4, 5; \omega_2(2) = 1 - 7; \omega_2(2) = 2 - 8, 12; \omega_2(2) = 3 - 9, 11; \omega_2(2) = 4 - 10$.

On a non-uniform lattice the spacing between c.s. may depend on the types of sites, however, in addition to the structural characteristics shall further indicate the types of central sites and their neighbours. If the central sites located at a distance ρ, are the sites are of type q, p, then the distribution of the neighbouring sites will be given values $\lambda_{qp\xi}(\omega_r|\rho)$ and $\kappa_{qp}(\omega_r|\rho)$, denoting the number of sites of type ξ and the total number of sites in the r-th c.s. with the orientation

$$\omega_r(\rho); \sum_{\xi=1}^{t} \lambda_{qp\xi}(\omega_r|\rho) = \kappa_{qp}(\omega_r|\rho).$$

To construct the average model is necessary to determine the following distribution functions of the sites, pairs of sites and clusters of different types. (for them we need to specify clearly the type of site, not its number, as in a distributed model, where the type and site number are uniquely linked by the values η_f^q).

Function $f_q = N_q / N$ characterizes the surface composition (mole fraction of sites of type q), $N_q = \sum\limits_{f=1}^{t} \eta_f^p$ is the number of sites of type q. Function $f_{qp}(r) = N_{qp}(r)/(Nz_q(r))$ describes the surface structure, $N_{qp}(r) = \sum\limits_{f=1}^{t} \sum\limits_{g \in z_f(r)} \eta_f^q \eta_{gr}^p$ is the number of pairs of sites of type qp, at a distance r. Normalizing relations for the above functions have the form $\sum\limits_{p=1}^{t} f_{qp}(r) = f_q$ and $\sum\limits_{q=1}^{t} f_q = 1$.

We also define the distribution function of clusters on the lattice structure $f(q\{\lambda\}R)$ and $f(qp\{\lambda\}R|\rho)$. The former characterizes the probability of finding a cluster on the lattice with a central site of type q and its neighbours, denoted by a set of numbers $\lambda_{qp}(r)$, $1 \le p \le t$, $1 \le r \le R$, $\lambda_{qp}(r)$ is the the number of sites of type p, at distance r from the site q; symbol $\{\lambda\}R$ is the compact notation means of the entire set of numbers. The number of different types c.s. of the clusters with one central site q is denoted by σ_q.

The second functions characterize the probability of finding clusters on a lattice with two types of sites p and q, and all their neighbosrs at distances $r \le R$. For them $\{\lambda\}R$ means a complete set of numbers $\lambda_{qp\xi}(\omega_r|\rho)$. The number of different types of the c.s. with two central sites p and q at a distance ρ is denoted by $\sigma_{qp}(\rho)$. These functions relate to normalizing connections

$$\sum_{\sigma_q} f\left(q\{\lambda\}R\right) = f_q \text{ and } \sum_{\sigma_{qp}(\rho)} f\left(qp\{\lambda\}R|\rho\right) = f_{qp}(\rho). \quad \text{(A2.9)}$$

We also introduce the conditional distribution function

$$d_{qp}(r) = \frac{f_{qp}(r)}{f_q}, \quad d\left(qp\{\lambda\}R|\rho\right) = \frac{f\left(qp\{\lambda\}R|\rho\right)}{f_{qp}(\rho)},$$

$$d\left(q\{\lambda\}R\right) = \frac{f\left(q\{\lambda\}R\right)}{f_q}, \quad d\left(u\left(qp\{\lambda\}R|\rho\right)\right) = \frac{f\left(qp\{\lambda\}R|\rho\right)}{f\left(q\{\lambda\}R\right)}.$$

$$\text{(A2.10)}$$

In the latter definition $u(qp\{\lambda\}R|\rho)$ denotes the set of sites remaining after separation of the cluster with two central sites p and q at a distance ρ, i.e. the sites of the truncated cluster $p\{\lambda\}R$. For $\rho = 1$ (Fig. 7.1 b) this set of sites includes sites 9, 4, 10, for $\rho = 2 - 9$, 4, 10, 5, 11. The number of types of truncated clusters $p\{\lambda\}R$ is denoted by σ'_p.

Particle distribution on the considered non-uniform lattices is described by the functions $\theta^i_q(\{\lambda\}R)$ and $\theta^{ij}_{qp}(\{\lambda\}R|\rho)$ which are derived from functions θ^i_f and $\theta^{ij}_{fg}(\rho)$ by averaging over the sites of various types. The index f runs over all lattice sites, and the index g over the neighbouring sites of p at a distance r in accordance with the local structure of clusters within R c.s.

As a result, from the equations (A2.2a)–(A2.7) we go to the following equations, which differ in the accuracy of the description of the non-uniform lattice structure and the influence of the local state of the lattice on the probability of filling molecules.

A. The model that takes into account the effect of the local structure of the lattice on the occupation of the central site:

$$y^i_q \theta^s_q(\{\lambda\}R) = \theta^i_q(\{\lambda\}R) \prod_{r=1}^{R} \prod_{l=1}^{z_q(r)} \sum_{\sigma'_q} d\left(u\left(qp\{\lambda\}R|r\right)\right) S^i_{qp_l}(r),$$

$$S^i_{qp_l}(r) = 1 + \sum_{j=1}^{s-1} \frac{\theta^{ij}_{qp}(\{\lambda\}R|r) x^{ij}_{qp}(r)}{\theta^i_q(\{\lambda\}R)}. \qquad (A2.11)$$

Index l in p_l indicates the site number in the r-th c.s. of the central site of type q, here p-type site with index l. Equations for $\theta^{ij}_{qp}(\{\lambda\}R|\rho)$ have the form similar to equations (A2.6) where they are instead of the function $\theta^{ij}_{fg}(r)$. The normalizing relations (A2.7) and equation (A2.8) are modified in the same manner.

This model allows to take into account the real structure of the non-uniform surface in local areas of the size $z_q(R)R/2$. For the model the main characteristic of filling of lattice sites is $\theta^i_q(\{\lambda\}R)$ (and not a function θ^i_q), whose value depends on the type of the central site and all its neighbours. The system of equations with respect to $\theta^i_q(\{\lambda\}R)$ and $\theta^{ij}_{qp}(\{\lambda\}R|\rho)$ allows to calculate the average filling with particles i of the sites of type q and the number of pairs of particles ij, located at the sites of type qp at distance r:

$$\theta_q^i = \sum_{\sigma_q} d\big(q\{\lambda\}R\big)\theta_q^i\big(\{\lambda\}R\big),$$

$$\theta_{qp}^{ij}(r) = \sum_{\sigma_{qp}(r)} d\big(qp\{\lambda\}R|r\big)\theta_{qp}^{ij}\big(\{\lambda\}R|r\big). \qquad (A2.12)$$

B. The model does not take account the effect of types of centres of neighbouring sites on the cluster occupation of the central site:

$$y_q^i\theta_q^s = \theta_q^i\sum_{\sigma_q} d\big(q\{\lambda\}R\big)\prod_{r=1}^{R}\prod_{p=1}^{t}\big[S_{qp}^i(r)\big]^{\lambda_{qp}(r)},$$

$$S_{qp}^i(r) = 1 + \sum_{j=1}^{s-1}\frac{\theta_{qp}^{ij}(r)x_{qp}^{ij}(r)}{\theta_q^i}. \qquad (A2.13)$$

Nevertheless, this model reflects the local structure of the lattice through the influence of neighbouring molecules, located inside the cluster, im the local areas of size $z_q(R)$.

C. The model with the approximation of the distribution function of clusters of the distribution functions of pairs of sites of different types:

$$y_q^i\theta_q^s = \theta_q^i\Psi_q^i, \quad \Psi_q^i = \prod_{r=1}^{R}\big[D_q^i(r)\big]^{z_q(r)}, \quad D_q^i(r) = 1 + \sum_{p=1}^{t}d_{qp}(r)\sum_{j=1}^{s-1}\frac{\theta_{qp}^{ij}(r)x_{qp}^{ij}(r)}{\theta_q^i}.$$

$$(A2.14)$$

This model is reflects most approximately the structure of the surface. It coincides with the model B in the case of the equilibrium distribution of different types of sites [6, 7], and in particular cases of surfaces with a spot, random distribution of sites and some cases their orderly distribution.

In models B and C, the equation for the pair distribution functions of particles have the form of equations (A2.6), in which the functions $\theta_{fg}^{ij}(r)$ replaced by $\theta_{qp}^{ij}(r)$, and in (A2.7) and (A2.8) instead $\theta_{fg}^{ij}(r)$ and θ_f^i there must be $\theta_{qp}^{ij}(r)$ and θ_q^i. Hence, the replacement of variables, discussed below in order to reduce the dimensionality of the system of equations is fully applicable to the models A–C.

Hierarchy of models

Having considered the equilibrium adsorption model for specific fragments of the surface and its averaged description, we move on to discuss the complete set of models and their relationships (see Fig. A2.1).

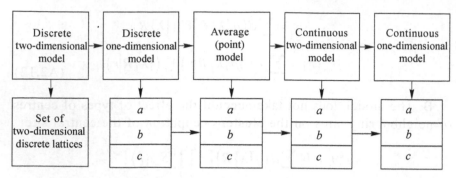

Fig. A2.1. Scheme of hierarchical relationships of adsorption equilibrium models: a – the influence of neighbouring types of cluster sites on the central site occupation f_q, $f(q\{\lambda\}R)$, $\lambda_{qp}(r)$, $f(qp\{\lambda\}R|\rho)$ $\lambda_{qp}(\omega_r|\rho)$; b – ignoring the impact of the types of adjacent cluster sites on the central site occupation f_q, $f(\{\lambda\}R)$, $\lambda_{qp}(r)$; c – approximation of the distribution function of the clusters of the pair function distribution of of sites of different types f_q, $f_{qp}(r)$, $z_q(r)$.

The original lattice represents at the atomic and molecular level the two-dimensional or three-dimensional discrete set of sites. It corresponds to a two- or three-dimensional discrete model, which may reflect any of its fragment (fragment shape is arbitrary). The use of two- or three-dimensional discrete model to describe the macroscopic characteristics of the adsorption system correspond to the macroscopic lattice on which the considered fragment translates in two or three directions and fills it completely. In some cases, this model is an accurate reflection of the real situation: 1) stepped surfaces of simple crystal faces when steps are located at the same distance from each other, 2) the surface of strictly ordered alloys, 3) stepped surfaces of strictly ordered alloys, etc., when the fragment size is not smaller than the regularly repeated 'pattern' formed by other imperfections of the surface [7], as well as cases corresponding to the isolated imperfections on uniform areas, if the number of sites of the fragment is large enough. The corresponding three-dimensional models reflect the situation with multilayer adsorption.

A close but more complex case corresponds to the situation when there is a set of independent surface fragments, such as the different faces of the ordered crystal or vicinal faces with different angles of cleavage, etc. Filling each fragment occurs in an independent way and to describe the overall process it is necessary to introduce a distribution function of fragments $f(f)$, which characterizes the fraction of the surface attributable on the fragment of type q, $1 \le q \le t(f)$, where $t(f)$ is the number of types of fragments. This function is similar to the function f_q (where each fragment is characterized by its surface composition and structure).

One-dimensional discrete models are obtained during (simultaneously for all rows) averaging over sites, pairs of sites and clusters of different types within each rows of the two-dimensional lattice. We represent the number of sites of the two-dimensional lattice f as $f = (m, l)$, m – rows number, l – site number in the rows. The ratio $f_q(m) = N_q(m) / N(m)$, where $N_q(m)$ and $N(m)$ is the number of sites of type q in a rows of type m and the total number of sites of the rows m, characterized by the mole fraction of sites of type q in rows m. Function $f_{qp}(mk|r)$ characterizes probability of finding sites of type q and p at a distance r in rows m and k, $f_{qp}(mk|r) = N_{qp}(mk|r)$ / $(N_q(m)\, \kappa_q(mk|r))$, where $N_{qp}(mk|r)$ is the number of pairs of sites of type qp at distance r in layers m and k, $\kappa_q(mk|r)$ is the number of sites located in layer k at a distance r from the site q in layer q.

Normalizing relations for them are written as functions

$$\sum_{q=1}^{t} f_{qp}\left(mk\,|r\right) = f_q\left(k\right),\ \sum_{p=1}^{t} f_{qp}\left(mk\,|r\right) = f_q\left(m\right),\ \sum_{q=1}^{t} f_q\left(m\right) = 1.$$
$$(A2.15)$$

For a more detailed account of the local structure of the surfacewe define the distribution function of clusters $f_m(q\{\lambda\}R)$ and $f_{mk}(qp\{\lambda\}R|r)$. The normalization condition for the first takes the form

$$\sum_{\sigma_q(m)} f_m\left(q\{\lambda\}R\right) = f_q\left(m\right),\qquad (A2.16)$$

where $\lambda_{qp}(mk|r)$ is the number of sites of type p in layer k, located at a distance r from the central site of type q in layer m. The compact record of the entire list of the values $\lambda_{qp}(mk|r)$, $1 \le q$, $p \le t$, $1 \le r \le R$, denoted as above by $\{\lambda\}R; \sum_{p=1}^{t}\lambda_{qp}\left(mk\,|r\right) = \kappa_q\left(mk\,|r\right)$, the index k enumerates the rows corresponding to each value of $r: \sum_{k}\kappa_q\left(mk\,|r\right) = z_q\left(r\right)$.

Function $f_m(q\{\lambda\}R)$ (A2.16) characterize the probability of the centre of the cluster $q\{\lambda\}R$ being in rows. The number of different types of all R c.s. of the cluster with the centre q in rows m is denoted by $\sigma_q(m)$. Similarly, we define the function $f_{mk}(qp\{\lambda\}R|r)$ as generalized functions $f(qp\{\lambda\}R)$. For them, $\{\lambda\}R$ is a complete list of values $\lambda_{qp\xi}(mkn|\omega_r(\rho))$, defining the number of sites of type ξ, located in layer n with orientation $\omega_r(\rho)$ in the r-th c.s. of the central pair of sites qp in layers mk at a distance ρ. The normalization condition for them is $\sum\limits_{\sigma_{qp}(mk|\rho)} f_{mk}\left(qp\{\lambda\}R\big|\omega_r(\rho)\right)=f_{qp}\left(mk|\rho\right)$, where $\sigma_{qp}\left(mk|\rho\right)$ is the number of different types of all R of the c.s. of the cluster with two central sites of type qp, located at a distance ρ in layers m and k.

The transition to one-dimensional models automatically means the uniform distribution of sites, site pairs and clusters of different types within the same rows, which naturally 'roughens' the original real situation. The equations describing the distribution of the particles in the discrete one-dimensional models with varying accuracy accounting for the distribution of different types of sites, have a form similar to equations (A2.11)–(A2.14), in which instead of functions θ_q^i and $\theta_{qp}^{ij}(r)$ there are functions $\theta_q^i(m)$ and $\theta_{qp}^{ij}\left(mk|r\right)$, i.e., there are additional indices of the numers of rows, as in formulas (A2.15) and (A2.16) compared with (A2.9).

The point models are obtained by averaging the particle distribution across all sites of a discrete two-dimensional lattice or all rows in a discrete one-dimensional model.

In describing the macroscopic systems of size L, there are two alternatives. Either properties of the system are such that the above models are satisfied, i.e., there is a certain uniformity of the distribution of sites of different types (uniform–non-uniform surfaces [8]) or the properties of the system change in a certain way when moving from one small enough macroportion to another. In the second case we must take into account the changing nature of the non-uniformity of macroscopic sites with a characteristic size Δx, where $\Delta x \ll L$ but $\Delta x \gg R$, i.e. to move towards continuous models.

Each region of the order Δx^k, k is the dimension of the area, is characterized by its functions of the distribution of the sites that represent the abover introduced distribution functions of the point model, relating not to the entire surface, and its macroscopic area with coordinate x. For $k = 2$ we must specify the law of variation of the distribution functions for both coordinates, for $k = 1$ along one coordinate there exists a uniform distribution of sites, which changes in the direction of x. Depending on the required accuracy of description in the continuous models we can use any of the models a, b, c (the scheme in Fig. A2.1). The distribution functions for them additionally contain macroscopic coordinate $\bar{x}: f_q\left(\bar{x}\right), f_{qp}\left(r|\bar{x}\right),$ $f\left(q\{\lambda\}R|\bar{x}\right),$ $f\left(qp\{\lambda\}R|r|\bar{x}\right)$; in the same way the additional index $\bar{x}$ is present in the functions describing the probability of filling sites and their pairs: $\theta_q\left(\bar{x}\right), \theta_{qp}\left(r|\bar{x}\right)$ etc.

Adsorption isotherms

Adsorption isotherms link the external partial pressures of the gas phase molecules i_m and the partial degree of filling the surfaces θ_i by them. Depending on the type of model of adsorption equilibrium: sets of fragments of lattices, discrete one-dimensional, point and continuous, defined as

$$\theta_i = \sum_{q=1}^{t(\phi)} f_q(\phi)\frac{1}{\tilde{N}_q}\sum_{f=1}^{\tilde{N}_q}\theta_{f}^i, \; \theta_i = \frac{1}{\kappa}\sum_{m=1}^{\kappa}\sum_{q=1}^{t} f_q(m)\theta_q^i, \; \theta_i = \sum_{q=1}^{t} f_q\theta_q^i,$$

$$\theta_i = \int_{\bar{x}_1}^{\bar{x}_2} f_q(x)\theta_q^i(x)\,d\bar{x}/S,$$

(A2.17)

where $\tilde{N}_q$ is the number of sites of the fragment q, κ are the numbers of rows of one-dimensional systems, $\bar{x}_1$ and $\bar{x}_2$ are the values, determining the dimensions of the surface S. The total surface coverage θ_Σ and the fraction of the vacancy sites $\theta_s \equiv \theta_V$ can be written as $\theta_\Sigma = \sum_{i}^{s-1}\theta_i$ and $\theta_V = 1-\theta_\Sigma$.

The adsorption equilibrium constant of displacement of molecules in the adsorption of the mixture of molecules characterizes the competition between the adsorbed molecules of different sort. In the absence of dissociation of the molecules the scheme process can be written as $B_g + ZA \leftrightarrow A_g + ZB$, where Z_i is adsorbed state of the molecule i, $i = A, B$; subscript 'g' refers to gas phase. If the molecules A and B have about the same volume, which is reflected in the given recording of the scheme of the process, the adsorption equilibrium constant of displacement $K_{AB} = \theta_A P_B / (\theta_B P_A)$ coincides with the constant of adsorption separation of gases [9]. The above adsorption equilibrium equations for the mixture allow us to calculate the adsorption isotherms (A2.17), as well as constants K_{ij}, $1 \le i, j \le s - 1$ (defined $K_{ji} = K_{ij}^{-1}$).

Note that the adsorption of the mixture of gases has been studied both experimentally and theoretically far less extensively than the adsorption of individual gases. Basically, this included the calculations of the the adsorption isotherms in the absence of interaction based on ideal models for uniform and non-uniform surfaces or consideration of the interaction of the nearest-neighbour interactions on uniform surfaces in the molecular field and quasi-chemical approximations.

The expressions for the equilibrium constant of adsorption displacement of the adsorbed molecules by each other are easily obtained from the above equations. Using the definition (A2.1) for y_q^i we write the formula (A2.14), such as $a_q^i \theta_q^s / \Psi_q^i = \theta_q^i / P_i$. Multiplying this equation on the left and right on f_q and taking the sum over q, we obtain $\theta_i P_i = \sum\limits_{q=1}^{t} f_q a_q^i \theta_q^s / \Psi_q^i$. Writing a similar expression for the molecule j, we obtain

$$K_{ij} = \sum_{q=1}^{t} f_q \theta_q^s a_q^i \left(\Psi_q^i\right)^{-1} \bigg/ \sum_{q=1}^{t} f_q \theta_q^s a_q^j \left(\Psi_q^j\right)^{-1}. \qquad (A2.18)$$

When accounting for nearest-neighbour interactions ($R = 1$), the expression (A2.18) becomes the expression in [10]. In [11, 12] the authors investigated the concentration dependences of the equilibrium adsorption coefficient of displacement in a binary mixture of micro- and macrocomponents ($\theta = 3$), taking into account the combined influence of surface irregularities (the case of $t = 2$ was discussed in detail) and interparticle interaction. For a uniform surface of type q the formula (A2.18) is greatly simplified (at $R = 1$ it changes to the formula of [13]):

$$K_{ij} = \frac{a_q^i}{a_q^j} \prod_{r=1}^{R} \left[\sum_{l=1}^{s} t_{qq}^{il}(r) \exp\left\{ \beta \left(\varepsilon_{qq}^{jl} - \varepsilon_{qq}^{il} \right) \right\} \right]^{z_q(r)}. \qquad (A2.19)$$

Heat capacity

To calculate the thermal conductivity coefficient we require the local partial coefficients of thermal conductivity. In formula (A2.5) it is indicated that $E = -\langle H \rangle-$ is the internal energy of the system, calculated in terms of molecular distribution, the equations for which are given above:

$$E = \sum_{f=1}^{t}\sum_{i} F_f E_f^i = \sum_{f=1}^{t}\sum_{i} F_f \theta_f^i \left[\ln J_f^i + \frac{1}{2}\sum_{r}\sum_{g}\sum_{j} \varepsilon_{fq}^{i,j}(r) t_{fg}^{ij}(r) \right]. \qquad (P2.20)$$

If we divide the contributions to the internal energy E to the kinetic E_1 and potential E_2, then the heat capacity of the system can be written as $C_v = C_{v1} + C_{v2}$, where $C_{v1} = \dfrac{d}{dT} E_1$ and $C_{v2} = \dfrac{d}{dT} E_2$ and each of the terms defines its heat capacity contribution. When calculating the temperature derivative, the total density of molecules in the system is considered to be constant.

From expression (A2.20) for any non-uniform system we can determine the local internal energy E_f^i related to the specific type of sites occupied by molecules of sort i: this is the term in square brackets (A2.20). On the basis of this expression we calculate the contributions for local heat capacities of the mixture components:

$$C_v^i(f) = C_{v1}^i(f) + C_{v2}^i(f), \text{where } C_{v1}(f) = \frac{d}{dT}E_f^i(1) \text{ and } C_{v2}(f) = \frac{d}{dT}E_f^i(2).$$
(A2.21)

Many-body configurations

Expressions for the probabilities of many-body configurations through smaller correlators are written in the quasichemical approximation as

$$\theta_{f(g_1\cdots g_z)R}^{i(j_1\cdots j_z)R} = \theta_f^i \prod_{r=1}^{R}\prod_{g=1}^{z}\prod_{j=1}^{s}\left[t_{fg}^{ij}(r)\right]^{n_{fg}^j(r)},$$
(A2.22)

where the functions θ_f^i and $t_{fg}^{ij}(r) = \theta_{fg}^{ij}(r)/\theta_j^i$ determined by solving the system of equations (A2.2a), (A2.6), (A2.7), here $1 \leq i$ and $j \leq s$, $1 \leq r \leq R$; lower symbol $f(g_1\cdots g_z)R$ denotes the complete list of sites on a non-uniform lattice with fixed locations of all sites within R coordination spheres around a central site f (cluster $f(g_1\cdots g_z)R$); superscript $i(j_1\cdots j_z)R$ is the full list of particles in this cluster $f(g_1\cdots g_z)R$ of the non-uniform lattice around the central particle i; $n_{fg}^i(r)$ is the number of particles species j on pairs of sites of type fg at the distance r of the coordination sphere around the centre of the particles i, where $\sum_{j=1}^{s} n_{fg}^j(r) = z_{fg}(r)$; in the mean field approximation

$$\theta_{f(g_1\cdots g_z)R}^{i(j_1\cdots j_z)R} = \theta_f^i \prod_{r=1}^{R}\prod_{g=1}^{z}\prod_{j=1}^{s}\left[\theta_g^j\right]^{n_{fg}^j(r)}.$$
(A2.23)

On the role of describing the nature of surface non-uniformity [14]

The models differ in how they describe the nature of the distribution of sites on a non-uniform surface. To get a clear idea about the role of accounting the structure of the surface, we compare the distributions of particles of one kind ($s = 2$) on the surface sites, consisting of two types of sites obtained by the models (A2.11) and (A2.13) with allowance for the interaction of the nearest neighbours.

The results of calculations with a random distribution of sites of the first and second type are shown in Fig. A2.2. The graph shows

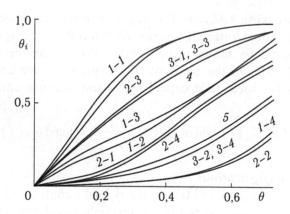

Fig. A2.2. The occupation of the central cluster sites and the average values θ_q for the surface $f_1 = 1/2$, $f_{12} = 0.25$, $z = 4$, $a_1/a_2 = 15$. Number of the curves $m-n$ indicate: $m = 1$ – cluster 1[4.0], $m = 2$ – cluster 1[0.4], for $n = 1$ or 3; $m = 1$ – cluster 2[4.0], $m = 2$ – cluster 2[0.4], for $n = 2$ and 4, the values $m = 3$ correspond to the average filling $q = 1$ if $n = 1$ and 3 and $q = 2$ when $n = 2$ and 4. For $n = 1$ and 2 values $\beta\varepsilon$ (1) = 1.2, $n = 3$ to 4 values $\beta\varepsilon$ (1) = –1.2. Curves 4 (θ_1) and 5 (θ_2) were calculated by the formula (A2.13).

the occupation of the central cluster sites with different c.s. and averages θ_q, $q = 1$ and 2, obtainable by (A2.12), depending on the degree of surface coverage θ. (If $\theta > 0.7$, all the curves are almost linear to the point (1.0, 1.0).) Average fillings θ_1 in the models (A2.11) and (A2.13) are given by the curves 3–1, 3–3 (for different interaction parameters) and 4. Differences between them do not exceed about 10%.

However, the analysis of the occupations of the central sites of the clusters 1[4, 0] and 1[0, 4], (curves 1–3 and 2–3 for repulsive and 1–1, 2–1 for attracting particles) indicates their sharp contrast both between themselves and in relation to the average values. (Curves of cluster occupations 1[3, 1] 1[2, 2], 1[1, 3] are in the area between the curves for 1[4, 0] and 1[0, 4] and are not shown in the Figure). The same applies to the occupation of the second type of sites for clusters 2[4, 0] and 2[0, 4], which correspond to curves 1–4 and 2–4 for the repulsive and 1–2, 2–2 for attracting particles, when comparing them with the average values 3–2, 3–4.

Figure A2.3 shows the occupation of the sites of the first type for surfaces whose sites tend to stratify (curves 1, 2, 3, 7) and become ordered (curves 4, 5, 6, 8). Average values θ_1 of the occupations by models (A2.11) and (A2.13) are sufficiently close to each other (curves 3 and 7), and the local occupations (curve 1 – cluster

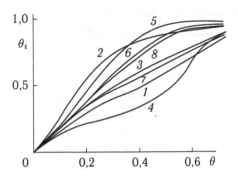

Fig. A2.3. Occupation of sites of the first type at $\beta\varepsilon(1) = -1.2$ (strictly ordered lattice). Curves *1–6* are calculated by formulas (A2.11), curves *7, 8* – the formula (A2.13), $f_{12} = 0.125$ (*1–3, 7*), $f_{12} = 0.375$ (*4–6, 8*). Curves *1* and *4* correspond to cluster 1[4, 0], curves *2* and *5* – cluster 1[0, 4], and the rest – the average filling.

1[4.0] and curve *2* – 1[0.4]) are substantially different. Even greater differences of local occupations are realized for the same cluster on a more ordered surface (curves *4* and *5*). That is, the surface structure has a great impact on local occupations of sites of different types and their differences can be up to 3–5 times. Note also that the local occupations, in contrast to the average ones, strongly depend on the sign of the interaction parameters. With increasing modulus of the interaction parameter these differences increase.

Thus, if we are talking about such 'rough' characteristics as adsorption isotherms, then in some situations it is enough to use equations (A2.13) for them. If we consider more 'accurate' characteristics (e.g., heat, heat capacity, adsorption kinetics), in some situations the use of equations (A2.13) can lead to significant deviations from the results of applying equations (A2.11).

From (A2.12) it follows that the functions θ^i_q and $\theta^i_q(\{\lambda\}R)$ coincide only if $d(q\{\lambda\}R) = 1$. For systems in which $d(q\{\lambda\}R) = 1$, include systems with ordering of the particles, systems with the ordering of sites of different types, transition regions of phase boundaries and several others. If $d(q\{\lambda\}R) \neq 1$ we need to consider the whole range of different states of the c.s. of the clusters which significantly increases the dimension of the system (A2.11) compared with (A2.13) and the correct use of the functions θ^i_q should be checked in each case.

Questions regarding a more accurate calculation of molecular distributions in non-uniform systems than under the QCA, are discussed in Appendix 3.

Lowering the dimensionality of the system of equations in the quasichemical approximation [2, 15]

The closed system of equations in the quasi-chemical approximation for any form of recording contains the equations (A2.6) with respect to functions $\theta_{fg}^{ij}(r)$, where $1 \leq i$ and $j \leq s$, and equation (A2.7) with respect to normalization of the functions θ_f^i. Increasing the number of components increases dimensionality of the system of equations. The number of equations for θ_f^i increases linearly with increasing s, whilst for the equations of pair functions $\theta_{fg}^{ij}(r)$, even for a uniform lattice with $R = 1$, the increase occurs as $s(s - 1)/2$. Increase of the interaction radius R and the number of types of lattice sites t increases evenmore the dimensionality of the system of equations.

To reduce the dimensionality of the system of equations we should used (A2.6): $\theta_{fg}^{ij}(r)\theta_{fg}^{ss}(r) = \theta_{fg}^{is}(r)\theta_{fg}^{sj}(r)\exp\left(\beta\varepsilon_{fg}^{ij}(r)\right)$, determining all pairs functions via functions $\theta_{fg}^{is}(r)$ or $\theta_{fg}^{sj}(r)$. Using these functions as independent variables, it is possible to express all the functions $\theta_{fg}^{ij}(r)$ according to (A2.6) and substitute into (A2.2a). Finally, using full balance of the pairs $\sum\limits_{i,j=1}^{s}\theta_{fg}^{ij}(r) = 1$, which holds for any interaction radius r, we define through unknown $\theta_{fg}^{is}(r)$ or $\theta_{fg}^{sj}(r)$ the function $\theta_{fg}^{ss}(r)$. As a result, equation (A2.6) transforms to the identities $1 \equiv 1$ and the dimension of the system of equations is greatly reduced.

References

1.   Tovbin Yu.K. // Zh. Fiz. khim. 1990. V. 62, No. 4. P. 865.
2.   Tovbin Yu.K., Theory of physico-chemical processes at the gas-solid interface. Moscow: Nauka, 1990. – 288 p.
3.   Tovbin Yu.K. // Zh. Fiz. khim. 1981. V. 55, No. 2. P. 273.
4.   Tovbin Yu.K. // Zh. Fiz. khim. 1992. V. 66, No. 5. P. 1395.
5.   Tovbin Yu.K. // Poverkhnost'. 1982. No. 9. P. 15.
6.   Tovbin Yu.K. // Kinetika i kataliz. 1983. V. 24, No. 2. P. 308, 317.
7.   Tovbin Yu.K. // Zh. Fiz. khim. 1982. V. 56, No. 7. P. 1689.
8.   Tovbin Yu.K. / / Kinetika i kataliz. 1980. V. 21, No. 5. P. 1165.
9.   Tolmachev A.M. // Usp. khimii. 1981. V. 50, No. 5. P. 769.
10.   Barrer R.M., Klinowski J. // JCS Farad. Trans. Part II. 1978. V. 74. No. 904.
11.   Tovbin Yu.K., Pak V.M. / / Zh. Fiz. khim. 1989. V. 63, No. 3. P. 788.
12.   Tovbin Yu.K., Gilyazov M.F., Tolmachev A.M. // Vysokochistye veshch. 1989. No. 6. P. 76.
13.   Gilyazov M.F., Tovbin Yu.K., Tolmachev A.M. / / Vysokochistye veshch. 1989. No. 6. P. 83.
14.   Tovbin Yu.K. // Dokl. AN SSSR. 1989. V. 306, No. 4. P. 888.
15.   Tovbin Yu. K. // Progress in Surface Science. 1990. V. 34. No. 1–4. P. 1.

Appendix 3

New approaches in the lattice gas model for non-uniform systems

This appendix provides new approaches in the LGM to improve the accuracy of calculation of molecular distributions in the theory of adsorption, compared with the quasi-chemical approximation: the expansion of the possibilities of the matrix method of calculation and the new fragment method to obtain accurate results, as well as the combined use of the cluster approach and the stochastic numerical methods. The total energy of the adsorption system in the grand canonical ensemble is defined in Appendix 2 which also examines the case of the approximation of many-body configurations of molecules in the quasi-chemical approximation. Appendix 4 discusses methods for modifying the LGM by taking into account the internal motions of the molecules.

The exact solution for describing the distribution of interacting molecules even on an uniform surface is possible only in rare situations [1–3], so basically we use approximate methods. Within the framework of the lattice gas model the main methods of calculating the characteristics of adsorption are the cluster method (CM) [4–6], which allows to obtain closed systems of equations in various approximations (quasi-chemical, mean-field, etc.), and the numerical Monte Carlo method [7, 8]. The Monte Carlo method allows us to investigate surface areas of up to 10^6–10^8 sites. This number is large enough for the results obtained by this method to be considered accurate. Naturally, these values of N allow us to study many types of topographies of the non-uniform surfaces. However, this method may lead to artefacts. For example, study [6] discusses the case detection of the secondary peak of the thermal desorption spectrum due to the lack of achieving the 'true' equilibrium surface distribution of the molecules in the phase transition of ordering implemented in the desorption process, and other cases. Such a situation may occur in numerical studies of any phase transitions even on uniform surfaces and special techniques must be used to control [7, 8] the lack of metastable states. On non-uniform surfaces the probability of the existence of metastable states increases due to local inhomogeneities and this makes it difficult to conduct numerical study (note that the CM provides a regular method of constructing phase diagrams on non-uniform surfaces [9]). Alternative exact methods of calculating the adsorption characteristics are needed to assess the reliability of the results of the Monte Carlo and the CM methods. The exact methods include the matrix method [2, 3, 10–12] and the fragment method [13, 14].

Matrix method (MM)
The matrix method [10–12] is based on constructing a transition matrix linking the probability of all realizable configurations of particles on a lattice structure, and determining the eigenvalues of this matrix through which the equilibrium distributions are calculated. It is this method that was used to obtain exact solutions describing the adsorption for some parameter values on uniform surfaces [2, 3]. The ability to use the MM is associated with the existence on the surface of the periodically repeating in one of directions of the surface structure of the surface fragment of size N, forming the strip (and the entire two-dimensional surface is represented as a set of such strips). This method allows one to test the results of the Monte-Carlo method.

For simplicity, consider a single-component adsorption of $s = 2$ ($i = A$ – the site is occupied and $i = v$ – the site is free for which $a^v p_v = 1$). In the MM the sum over states can be written as

$$Q = \sum_{\delta_1} \left(P_{\delta_1 \delta_1} \right)^m = \mathrm{Sp}\left(P_{\delta_1 \delta_1} \right)^m, \quad P_{\delta_k \delta_{k+1}} = \exp\left[-\beta \left(E_{\delta_k} + E_{\delta_k \delta_{k+1}} \right) \right],$$

$$E_{\delta_k} = -\sum_{f,i} \beta^{-1} \ln\left(a_f^i p_i \right)^{1/m_i} - \sum_{f,g} \sum_{i,j} \frac{\varepsilon_{fg}^{ij}(r)}{2}, \quad E_{\delta_k \delta_{k+1}} = -\sum_{f,g} \sum_{i,j} \frac{\varepsilon_{fg}^{ij}(r)}{2}, \tag{A3.1}$$

where δ_i is the number of configuration of the k-th fragment; $P_{\delta_k \delta_{k+1}}$

is the matrix element for the transition between configurations of the repeated fragment, each includes $L = N/m$ rows (m is the number of repeating fragments in the strip consisting of N sites; at the macroscopic limit $m \to \infty$). For $s = 2$, the number configurations is equal to $B = 2^N$, $1 \le \delta_k \le B$, and $\delta_{m+1} = \delta_1$. Symbol Sp P^m denotes the sum of the diagonal elements of the transition matrix raised to the power m, E_{δ_k} is the total energy of the k-th fragment, having a configuration δ_k, subscripts f and g number the sites of the k-th fragment; $E_{\delta_k \delta_{k+1}}$ is the lateral interaction energy between molecules on neighbouring cells k (subscript f) and $k + 1$ (subscript g) in the configurations δ_k and δ_{k+1}. The matrix method also takes into account the lateral interactions and the non-uniformity of the grid. Ways to solve the problem with this method are discussed in [14], since the number of configurations B increases rapidly with increasing fragment size that complicates the calculation of the maximum eigenvalue of the transition matrix.

To improve the accuracy of the calculations of adsorption on non-uniform surfaces with different topography, a number of improvements have been proposed for the matrix method [14]. This allows: a) extend the range of topographies adequately described by the MM, and b) to determine how accurately the correlation effects are taken into account in the method fragments. Early estimates of the desorption rate were performed using the MM on the assumption that the interaction of the activated complex with neighbors absent. In fact, this corresponds to the collision model for the model of the desorption rate $\varepsilon_{fg}^{ij*}(r) = 0$. The modification of the MM extends the description of non-uniform systems at removing all the previously mentioned limitations of $\varepsilon_{fg}^{ij*}(r) = 0$ with increasing size of the fragment by combining the MM and the cluster method [13, 14].

This feature of the MM allows one to analyze some adsorption systems with the locally non-uniform distribution of molecules: a) the orderly distribution of adsorbate on a uniform surface [15], b) stepped surfaces and other systems containing the repeating element of the structure. MM calculations carried out for a uniform surface, show [16] that the strip of four rows ($N = 4$) describes quite well the thermodynamic characteristics of an infinitely wide strip, in particular differences from the strip with $N = 8$ are of the order of 1%. Simultaneous consideration of the surface non-uniformity and lateral adsorbate interactions was performed in [17] for a regular sequence of two types of sites in a chain and in [18] for three adjacent chains, each of which consists of sites of one type, modeling the step on the surface.

The number of different configurations of the fragment increases rapidly with an increase in its size, and thus increase the mathematical problems in the MM. Ways to solve this problem are discussed in studies [19–22]. The maximum number of sites N, used in calculations [23] of the heat of adsorption and desorption rates at a linear surface heating rate (thermal desorption spectra) on the faces of single crystals is 12. In addition, the calculation of the rate of desorption was carried out under the condition of neglecting the interaction of the activated complex with the neighbouring molecules (in CM there is no such restriction [4, 6]).

In studies [13, 14]: 1) the scope of MM is expanded to a wider class of non-uniform systems, 2) restrictions are removed on the values of the parameters of interaction of the activated complex with the neighbouring molecules; 3) it is proposed to use both the MM and CM in order to increase the accuracy of describing the distribution of molecules in comparison to the CM, and to increase the size of the fragment compared with the MM. The first generalization concerns the application of the MM to a more complex non-uniform surface topography. Instead of the periodically repeating fragment we can consider the totality of repeating fragments having the same size N, but different 'internal' topography. This situation is a generalization of [24], in which the one-dimensional chain is studied, for an arbitrary number of series in the band. In this case, instead of finding the eigenvalues of the transition matrix we should use the direct matrix multiplication and determine the asymptotic behaviour of their products. The second generalization is associated with the calculation of the probabilities of many configurations and follows directly from the use of methods of direct products of transition matrices. Finally, to describe the combined distribution of molecules (on one part of the sites we used the MM, while on the other part of the sites – CM) the expressions for the energy contributions at the transition matrix elements responsible for the 'coupling' of the two surface portions were modified, which leads to the necessity of the self-consistent solution of equations. The latter requires use of an iterative procedure for finding the solutions (compared with linear non-iteration operations in the standard MM).

The fragment method (FM)

The FM [13, 14] allows us to study the topography of the fragment of the non-uniform surface larger than the fragment size of the MM, on the other hand, the accuracy of describing the distribution of molecules is higher than in the CM. Its essence is the calculation of the statistical sum for all configurations of the fragment without any approximations. The fragment in question is a periodic section of the macroscopic surface. Its internal topography can be arbitrary. If we compare each configuration of molecules on a fragment with some sort of quasi-particle, the full range of configurations can be interpreted as a mixture of quasi-particles. This 'mix' is ideal, and all the thermodynamic functions are obtained from the formulas for ideal mixtures [25].

The FM was used to eliminate the need to use approximations with the closure of a set of equations for the cluster functions (A1.10). When const($\{[n]\}$) = 1 in equation (A1.10) the probability of finding the i-th configuration of molecules on a fragment takes the form $\theta_i = \exp(-\beta E_k)/Q$, where E_k is the total energy of the molecules to be calculated from the relation (A1.2) where f and g numerate the sites of the fragment. For const = 1, two situations are possible: a) the fragment has a fixed environment and, in particular, it breaks all ties linking it with the lattice, and b) adsorption states of the surrounding sites are connected with those of the central sites by periodic boundary conditions. In case b the fragment simulates the macroscopic surface with a periodic topology consisting of N/N_f fragments on the surface, where N_f – number of sites on the fragment. The fragment can have an arbitrary inner topology. If each configuration of molecules on the surface of the fragment is assigned to a certain type of quasi-particles that can be denoted as $k\left(k \leftrightarrow \{\gamma^i_f\}\right)$, then the full range of configurations can be considered as a mixture of quasi-particles (the number of types of quasi-particles $B = 2^{N_f}$).

The concentration of the quasi-particles can be defined as

$$\theta_k = \left[\sum_{n=1}^{B} \exp\{\beta[E_n - E_k]\}\right]^{-1}, \quad \sum_{k=1}^{B} \theta_k = 1. \qquad (A3.2)$$

This implies that the mixture of quasi-particles is ideal and all thermodynamic and kinetic functions are computed by the equation for the ideal mixtures. The probabilities of any local configurations of adsorbed molecules $\theta\left(\left[\gamma^i_{\{f\}}\right]\right)$ (where $i_1 \ldots i_m$ are the adsorption states of m sites $f_1 \ldots f_m$, of the fragment; $m \leq N_f$) are calculated from the total probabilities (A3.2) by summing the adsorption states of other sites (h_1... h_p, $m + p = N_f$) of the fragment. We denote the number of occupied sites with the k-th configuration n_k, then the probability of complete filling of the surface takes the form

$$\theta = \sum_{k=1}^{B} n_k \theta_k. \tag{A3.3}$$

The method leads to the exact solution of the problem of the distribution of the adsorbed molecules by a finite number of sites of the fragment. The maximum number of surface sites is bounded from above and is equal to the number of sites of the fragment. The full set of the topology of the given fragment consists of all possible locations of various sites.

Like in the MM, the number of types of quasi-particles increases rapidly with increasing fragment size. However, this method allows us to consider the fragments that are larger than the size of the element in the MM. Even the simplest computers allow one to consider the fragments of up to 24 sites, on better computers, the fragment size canbe increased up to 32 sites. Comparison of the calculations by the MM and the FM [26, 27] for uniform stepped surfaces showed that both methods give almost identical results. Thus, differences for the adsorption isotherms are less than 0.1% outside the phase transition region and 0.5% in the vicinity of phase transitions. (In the critical region, both methods are approximate due to the limited size of the fragments.) Calculations in the FM can be performed on a larger number of sites than in the MM, so this method allows one to explore a greater number of topographies of non-uniform surfaces. Considering also that the calculations based on the FM equations are much easier than in the MM, this method can be considered as most convenient for the study of adsorption.

Analysis of isotherms and heats of adsorption and desorption-adsorption rates for different types of non-uniform surfaces [26–28] showed that the QCA with the fragment description of the structure of the surface (see Appendix 2) gives the results that differ from the exact calculations by the FM by no more than 5%. These maximum deflection are observed in early and late phase transitions. For ranges of small, medium and large fillings (out of phase transitions) differences were not more than 1.5%. It should be noted that with increasing degree of non-uniformity of the surface the differences between the CM and accurate methods are reduced, so fragmentary description of the structure in the CM is the most promising for theoretical calculations. For applied problems, the equations of the cluster method can be used to obtain more simpler equations, providing good agreement with experimental data on mono- and multilayer adsorption of individual gases and their mixtures [29].

Combined use of cluster and numerical methods
It is difficult to use accurate numerical methods in the calculation of the kinetics and dynamics of adsorption processes in non-isothermal conditions requiring multiple definitions of the coexisting concentrations of dense and diluted adsorbate phases and types of phase states (i.e., full knowledge of the phase diagram of the system studied is required). Such calculations are not carried out by numerical methods due to time-consuming. In this situation, it is desirable to have methods that provide fast calculation of phase diagrams and equilibrium characteristics, as in the case of approximate methods, and the accuracy close to the accuracy of numerical methods (Monte Carlo and molecular dynamics). The joint application of these two groups of methods is naturally considered [30]. The essence of the combined method lies in the use of lattice models as the 'interpolation' system: lattice models provide calculations in the 'intermediate' regions between known solutions obtained by the accurate numerical methods. To ensure that these solutions coincide with the approximate methods at the points of exact solutions, we introduce the calibration (trial) function (see section 24 and Appendix 5). Such a formulation leads to the need to study the effectiveness of procedures for joint use of the exact and approximate methods. Analysis of the various approximations shows [30] that in the first approximation it is sufficient to consider only accurate information at the critical point and use the QCA as an interpolation procedure: it allows one to quickly perform calculations, reflect the effects of direct correlations, gives accurate results in the field of small and strong interactions, and provides a qualitatively correct description in the critical region (better than the mean-field approximation). Additionally, the consent with the critical exponents [31] in the vicinity of the critical region may be required.

References

1.   Hill T.L., Statistical Mechanics. Principles and Selected Applications. N. Y.: Mc-Graw-Hill Book Comp. Inc., 1956.
2.   Onsager L. // Phys. Rev. 1944. V. 65. P. 117.
3.   Baxter R.J. Exactly Solved Model in Statistical Mechanics. London: Acad. Press, 1982.

4.   Tovbin Yu.K. Theory of physical and chemical processes on the gas-solid interface. Moscow: Nauka, 1990. – 288 p.
5.   Tovbin Yu.K. // Poverkhnost'. Fizika. Khimiya. Mekhanika. 1982. No. 9. P. 15 ; No. 10. P. 45.
6.   Tovbin Yu. K. // Progress in Surface Sci. 1990. V. 34. P. 1.
7.   Nicolson D., Parsonage N.G., Computer Simulation and The Statistical Mechanics of Adsorption. N. Y.: Acad. Press, 1982.
8.   Monte Carlo Methods in Statistical Mechanics. Topics in Currene Physics / Ed. by K. Binder. – Berlin: Springer-Verlag. V. 7. 1979 ; V. 36. 1984.
9.   Votyakov E.V., Tovbin Yu.K. // Zh. Fiz. Khimii. 1993. V. 67. P. 391.
10.   Kramers H., Wannier G.H. // Phys. Rev. 1941. V. 60. P. 252.
11.   Montroll E.W. // J. Chem. Phys. 1941. V. 9. P. 117.
12.   Montroll E.W., Stability and phase transitions. – Moscow: Mir, 1973. – P. 92.
13.   Tovbin Yu.K. // Khim. Fiz., 1996. V. 15. P. 75.
14.   Tovbin Yu.K. // Zh. Fiz. Khimii. 1996. V. 70. P. 700.
15.   Kinzel W., Selke W., Binder K. // Surface Sci. 1982. V. 121. P. 13.
16.   Derrida B., de Seze L., Vannimenus J. // Lecture Notes in Physics. Berlin: Springer-Verlag, 1981. V. 149. P. 46.
17.   Hill T.L. // Polymer Sci. 1957. V. 23. P. 549.
18.   Kiselev A.V., Lopatkin A.A., Lourie B.B. // Teor. eksp. khimya. 1974. V. 10. P. 254.
19.   Domb C. // Proc. Roy. Soc. 1949. V. A196. P. 36.
20.   Domb C. // Adv. Phys. 1960. V. 9. P. 149.
21.   Rikvold P. A., Kaski K., Gunton J. D., Yalabik M. C. // Phys. Rev. B. 1984. V. 29. P. 6285.
22.   Fush N. H. // Phys. Rev. B. 1990. V. 41. P. 2173.
23.   Payne S. H., Zhang J., Kreuzer H. J. // Surface Sci. 1992. V. 264. P. 185.
24.   Balagurov B.Ya., Sveshnikov V.G., Zaitsev O.V. // Fiz. tverd. tela. 1974. V. 16. P. 2302.
25.   Prigogine I., Defay R. Chemical Thermodynamics. Novosibirsk: Nauka, 1966. – 510 p/.
26.   Tovbin Yu.K., Petrova T. // Zh. Fiz. khimii. 1996. V. 70. P. 870.
27.   Tovbin Yu.K., Petrova T. // Zh. Fiz. khimii. 1997. V. 71. P. 123.
28.   Tovbin Yu.K., Petrova T., Votyakov E.V. // Zh. Fiz. khimii. 1997. V. 71, No. 3. P. 548.
29.   Tovbin Yu. K. // Langmuir. 1997. V. 13, No. 5. P. 979.
30.   Tovbin Yu.K. // Zh. Fiz. khimii. 1998. V. 72, No. 12. P. 2254.
31.   Stanley G., Phase transitions and critical phenomena. Moscow: Mir, 1973. – 400 p.

Appendix 4

Movement of molecules inside cells

This appendix sets out the approach based on the LGM, taking into account the motion of the molecules inside the cells. For simplicity we consider the case where all the components of the solution are comparable in size. This makes it simple to formulate a general procedure for the preparation of the desired expressions describing the equilibrium characteristics of the system. The basic framework of the proposed methodology will remain when considering arbitrary volume ratios of the solution components.

Solution model [1]

To describe the liquid phase we will use the lattice model. As usual, the lattice site is understood to be some cell in space, which has the characteristic dimensions of the order of the diameter of the molecule. Each site is no more than one molecule. For the methodological reasons stated above, it is assumed that all the molecules are close in size. Recall that the lattice models have been successfully used to calculate the thermodynamic characteristics of solutions of oligomers and polymers [2–5], but for the calculation of reaction rates we require knowledge of the complete set of cluster distribution functions [6, 7].

We believe that the sites form a periodic structure. Distances between the sites will be measured in numbers of coordination spheres r, $1 \le r \le R$, R is the radius of the interaction potential, the highest value of the radii of the potentials of interactions between molecules i and j, $1 \le i, j \le s$, s is the number of solution components, including vacancies (free sites), denoted by the last index s. The number of sites of the r-th coordination sphere is denoted by $z(r)$. The state of occupation of the site with the number f ($1 \leq f \leq N$, $N = V/V_0$, V – volume of the system, V_0 – volume unit) is reported as γ_f^i: $\gamma_f^i = 1$ denotes that the site contains a molecule of variety i, and $\gamma_f^i = 0$ otherwise. Obvious condition $\sum_{i=1}^{s} \gamma_f^i = 1$ means that any site of the structure contains a molecule of any sort.

To analyze the distribution of polyatomic molecules, in addition to specifying the coordinates of their centres of mass (via the site number f), it is necessary to specify their orientations. Imagine a vector from the centre of mass of a polyatomic molecule directed to one of its atoms, and we characterize the orientation of a rigid molecule by the angles between this vector and the vector of the external (hypothetical) field. In a spherical coordinate system the orientation is defined by the angle η_x, $0 \leq \eta_x \leq 2\pi$, determining the projection of the vector on the plane $X0Y$, measured from the axis X (the vector of the external field coincides with the axis z), and the angle η_z, $0 \leq \eta_z \leq \pi$, between the vector and the axis Z (for linear molecules), as well as by an additional angle η, $0 \leq \eta \leq 2\pi$, describing the rotation about this vector, for non-linear molecules. For brevity, we denote the set of values of angles η_x, η_z, η of the molecule i by one symbol φ_i.

By analogy with the transition from a continuous set of values of the coordinates of the centre of mass of molecules to the discrete one used in lattice models, we introduce a set of discrete orientations of the molecules relative to their centres of mass instead of continuous orientations. In this case, the value of φ_i correspond to discrete angles η_x, η_z, η. The number of different orientations of the molecule i is denoted by Φ_i. This approach was used to account for the orientation of intermolecular bonds in [8, 9]. In the lattice models discrete molecular orientations are naturally bonded with the directions along the bonds with neighbouring molecules ($\Phi_i = z$). This model explains the existence of the upper and lower critical points in the phase diagrams of stratification of the binary solutions [9]. In more recent studies [10–13], the number of orientations was considered as a model parameter that is not associated with the lattice structure.

To describe the orientation of the molecules, we introduce the random variable γ_i^φ, which characterizes the orientation φ_i of the molecule i: $\gamma_i^\varphi = 1$, if a molecule i has an orientation φ_i, and

$\gamma_i^{\varphi} = 0$ otherwise. For the value γ_i^{φ} the equality $\sum\limits_{\phi=1}^{\Phi_i} \gamma_i^{\phi} = 1$ is satisfied. There are no orientations for the vacancies, $\Phi_s = 1$. With these values the total energy of the mixture of polyatomic molecules in a large canonical ensemble can be written as

$$H = \sum_{f=1}^{N}\sum_{i=1}^{s}\sum_{\phi=1}^{\Phi_i} v_i^{\phi}\gamma_f^i\gamma_i^{\phi} - \frac{1}{2}\sum_{r=1}^{R}\sum_{f,gr}^{N}\sum_{i,j}^{s}\sum_{\phi,\psi_j=1}^{\Phi_i\Phi_j} \varepsilon_{ij}^{\phi\psi}\gamma_f^i\gamma_i^{\phi}\gamma_{gr}^j\gamma_j^{\psi},$$

$$v_f^i = -\beta^{-1}\ln(J_f^i) - \mu_i,$$

(A4.1)

where J_f^i is the statistical sum of the internal motions of the molecule i with orientation φ_i, μ_i is the chemical potential of the molecule i; the sum of f is taken over all lattice sites, the sum of g_r – over $z(r)$ sites of the coordination sphere; $\varepsilon_{ij}^{\varphi\psi}(f,g\,|\,r)$ is the interaction parameter between the molecule i with orientation φ_i and molecule j with orientation ψ_j, located at the sites f and g_r at distance r. Interactions with vacancies vanish. For the polyatomic molecules generally $\varepsilon_{ij}^{\varphi\psi}(f,g\,|\,r) \neq \varepsilon_{ji}^{\psi\varphi}(f,g\,|\,r)$, as molecular rearrangement changes their contact areas (i.e., atoms of different molecules at the smallest distance at a given value of r). It should also be borne in mind that for the fixed indices r, φ_i and ψ_j the change in the position of the neighbouring molecule (var g_r) changes the values of the energy parameters. This is the main difference between the potentials considered, reflecting the direction of bonds, and the spherical potentials. However, it is clear that the sequence of indices does not itself change values of the energy parameters $\varepsilon_{ij}^{\varphi\psi}(f,g\,|\,r) = \varepsilon_{ji}^{\psi\varphi}(g,f\,|\,r)$.

Equation (A4.1) can be rewritten in the equivalent form, if we enter a value $\gamma_i^{\varphi}(f) = \gamma_f^i\gamma_i^{\varphi}$, characterizing the complex event: in site f is the molecule i and it has an orientation φ_i. For these values the relations: $\gamma_i^{\varphi}(f)\gamma_i^{\psi}(f) = \Delta_{ij}\Delta_{\varphi\psi}\gamma_i^{\varphi}(f)$, where Δ is the Kronecker delta, $\sum\limits_{i=1}^{s}\sum\limits_{\varphi_i=1}^{\Phi_i}\gamma_i^{\varphi}(f) = 1$, that is, each site contains only one molecule of one discrete orientation. This allows one to formally introduce new concept of the sort of the particles λ, located at site f, if the sort of the molecule i and its orientation φ_i are related to one index λ ($\lambda \leftrightarrow i, \varphi_i$), then $\gamma_f^{\lambda} = \gamma_i^{\varphi}(f)$, where the values of the index λ vary from unit to $\Phi = \sum\limits_{i=1}^{s}\Phi_i$. (It is agreed that the last index Φ refers to vacancies.) As a result the equation (A4.1) can be rewritten as

$$H = \sum_{f=1}^{N}\sum_{\ell=1}^{\Phi} \nu_{\ell}\gamma_f^{\ell} - \frac{1}{2}\sum_{r=1}^{R}\sum_{f,gr}\sum_{\ell,m}^{N} \varepsilon_{fg}^{\ell m}(r)\gamma_f^{\ell}\gamma_{gr}^{m}. \qquad (A4.2)$$

Equation (A4.2) is similar to the expression for the total energy of a multicomponent mixture [7], but now the intermolecular interaction parameter depends on the relative positions of the particles λ and m at the sites f and g_r. This leads to the necessity of accounting for the bulk isotropic lattice of local orientation in the distribution of the molecules. The latter fact is described by the cluster method [7], developed for locally non-uniform lattices.

Equilibrium distribution of the molecules

In the cluster approach [7] the original lattice is shown on the set of clusters. For each cluster we construct a precise system of equations for the probabilities of the various configurations of particles that can exist on it. We consider the clusters consisting of a central site f and its R coordination spheres. For each cluster, we construct two types of cluster distribution function (CDF):

$$\theta_{f\{g\}}^{l(\{m\})} = \left\langle \gamma_i^{\phi}(f)\prod_{r=1}^{R} \prod_{g\in z_f(r)} \gamma_j^{\psi}(g_r) \right\rangle = \left\langle \gamma_i^{\phi}(f)\prod_{r=1}^{R} \prod_{g\in z_f(r)} \prod_{m=1}^{\Phi}(\gamma_{gr}^{m})^{n_{gr}^{m}} \right\rangle =$$

$$= \theta_f^{l}(\{n_{gr}^{m}\}). \qquad (A4.3)$$

Symbol $\{[m]\}$ for CDF of the first type is a list of all particles m (molecules j and their orientations ψ), located on all g_r sites of each of R cluster coordination spheres. The angle brackets indicate that the states of the occupation of all sites outside the cluster were used for averaging using the distribution function $\exp(-\beta H)/Q$, where H is the total energy of the system (A4.1) or (A4.2), Q is the statistical sum of the system. Symbol $\{m\}$ for CDF of the second type means that the list of particles m may not be comprehensive: for some sites g_r averaging was carried out over the states of occupation. In this list the number of the sort is replaced by the index 'zero'.

The second equality in (A4.3) gives the equivalent entry through the number of fillings $n_{gr}^{m} = 0$ of the sites g_r with the particles of type m: $n_{gr}^{m} = 1$ if the particle is present, and $n_{gr}^{m} = 0$ if the particle is absent. The CDF number of the first type coincides with the CDF number of the second type, and they are connected to each other by linear equations:

$$\theta'_f(\{n^m_{gr}\}) = \prod_{r=1}^{R} \prod_{g \in z_f(r)} \left\{ \sum_{k^1_{gr}=0}^{1-\tilde{n}} \cdots \sum_{k^{\Phi-1}_{gr}=0}^{1-\tilde{n}+k^+} \right\} \theta'_f(\{[n^m_{gr} + k^m_{gr}]\}),$$

$$\tilde{n} = \sum_{m=1}^{\Phi-1} n^m_{gr}, \quad k^+ = \tilde{k} - k^{\Phi-1}_{gr}, \qquad \text{(A4.4)}$$

$$\theta'_f(\{n^m_{gr}\}) = \prod_{r=1}^{R} \prod_{g \in z_f(r)} \left\{ \sum_{k^1_{gr}=0}^{1-\tilde{n}} \cdots \sum_{k^{\Phi-1}_{gr}=0}^{1-\tilde{n}+k^+} C_1^{\{k^m_{gr}\}} (-1)^{\tilde{k}} \right\} \theta'_f(\{[n^m_{gr} + k^m_{gr}]\}),$$

where $C_1^{\{k^m_{gr}\}}$ is the number of combinations of sets of numbers k^m_{gr} for each site g_r; numerical values of this coefficient are equal to unity, but the coefficient is placed behind the values of the occupation numbers for easy tracking. The signs of the products with respect to R and g_r indicate the compact entry of the corresponding numbers of the sums for different sites in the cluster coordination spheres.

Central particles do not interact with the particles located outside the cluster so the CDF of the first type is

$$\theta'_f(\{n^m_{gr}\}) = \text{const}\,(\{n^m_{gr}\}) \frac{\exp\left[-\beta h'_{gr}(\{n^m_{gr}\})\right]}{Q},$$

$$h'_{gr}(\{n^m_{gr}\}) = \left(v'_f - \sum_{r=1}^{R} \sum_{g \in z_f(r)} \varepsilon^{lm}_{fg}(r) \gamma^m_{gr} \right) \gamma'_f \equiv v'_f - \sum_{r=1}^{R} \sum_{g \in z_f(r)} \sum_m \varepsilon^{lm}_{fg}(r) n^m_{gr}, \qquad \text{(A4.5)}$$

where h'_f is the energy of the central particles l in site f, the value of const $(\{n^m_{gr}\})$ does not depend on the state of employment the central site. The equations describing the distribution of the particles were obtained in three steps: 1) The relationships of the CDF of the first type with different central particles but with identical conditions of all coordination spheres exclude unknown Q and const $(\{n^m_{gr}\})$:

$$\theta^b_f(\{[n^m_{gr}]\}) =$$

$$= \theta'_f(\{[n^m_{gr}]\}) \exp \beta \left(v'_f - v^b_f + \sum_{r=1}^{R} \sum_{g \in z_f(r)} \sum_m [\varepsilon^{bm}_{fg}(r) - \varepsilon^{lm}_{fg}(r)] n^m_{gr} \right), \qquad \text{(A4.6)}$$

2) in (A4.6) the CDF of the first type are replaced by the CDF of the second type, according to (A4.4); and 3) we solve the system of equations of the second type with the central particle b through the CDF of the second type with the central particle l. As a result, we

obtain

$$\theta_f^b(\{[n_{gr}^m]\})\exp\beta\left(v_f^l - v_f^b + \sum_{r=1}^{R}\sum_{g\in z_f(r)}\sum_{m}[\varepsilon_{fg}^{bm}(r) - \varepsilon_{fg}^{lm}(r)]n_{gr}^m\right)\times$$

$$\times\prod_{r=1}^{R}\prod_{g\in z_f(r)}\left\{\sum_{k_{gr}^1=0}^{1-\tilde{n}}\cdots\sum_{k_{gr}^{\Phi-1}=0}^{1-\tilde{n}+k^+}\prod_{m=1}^{\Phi-1}[x_{fg}^{lmb}(r)]^{k_{gr}^m}C_1^{\{k_{gr}^m\}}\right\}\theta_f^l(\{n_{gr}^m + k_{gr}^m\}), \quad \text{(A4.7)}$$

$$x_{fg}^{lmb}(r) = \exp[-\beta\omega_{fg}^{lmb\Phi}(r)] - 1,$$

$$\omega_{fg}^{lmb\Phi}(r) = \varepsilon_{fg}^{lm}(r) + \varepsilon_{fg}^{b\Phi}(r) - \varepsilon_{fg}^{l\Phi}(r) - \varepsilon_{fg}^{bm}(r).$$

These systems of equations, due to the finite number of cluster sites, are finite, but not closed.

Quasichemical approximation (QCA)

The constructed system of equations describe the equilibrium distribution of the molecules. For their input circuit common to all cluster approximation. The result is a single closed system of equations for the entire array as a whole. To close the system in the quasi-chemical approximation is necessary to use an approximation

$$\theta_f^l(\{n_{gr}^m\}) = \theta_f^l\prod_{r=1}^{R}\prod_{g\in z_f(r)}\prod_{m=1}^{\Phi-1}[t_{fg}^{lm}(r)]^{n_{gr}^m}, \quad t_{fg}^{lm}(r) = \frac{\theta_{fg}^{lm}(r)}{\theta_f^l}, \quad \text{(A4.8)}$$

where $t_{fg}^{lm}(r)$ is the conditional probability of finding the particle m at a site g_r close to the particle l at site f. The probability of many-body configuration ($\{n_{gr}^m\}$) expressed as the product of the unary and pair distribution functions.

The final form of the QCA system of equations is given by the following formulas, with the proviso that all the clusters are equivalent to each other:

$$\theta_f^{\Phi} = \theta_f^l \exp\beta(v_f^l - v_f^{\Phi})\prod_{r=1}^{R}\prod_{g\in z_f(r)}S_{fg}^l(r), \quad \sum_{l=1}^{\Phi}\theta_f^l = 1,$$

$$\text{(A4.9)}$$

$$S_{fg}^l(r) = 1 + \sum_{\lambda=1}^{\Phi-1}t_{fg}^{l\lambda}(r)x_{fg}^{l\lambda}(r), \quad x_{fg}^{lm}(r) = \exp[-\beta\varepsilon_{fg}^{lm}(r)] - 1,$$

$$\theta_{fg}^{lm}(r)\theta_{fg}^{\Phi\xi}(r) = \theta_{fg}^{l\xi}(r)\theta_{fg}^{\Phi m}(r)\exp[\beta\omega_{fg}^{lm\Phi\xi}(r)],$$

$$\sum_{\lambda=1}^{\Phi}\theta_{fg}^{l\lambda}(r) = \theta_f^l, \quad \sum_{l=1}^{\Phi}\theta_{fg}^{l\lambda}(r) = \theta_g^{\lambda}, \quad \text{(A4.10)}$$

where the parameter $\omega_{fg}^{lm\,\Phi\xi}(r)$ was defined above. Equations (A4.9) are the equations for the mole fractions of particles l $\left(\theta_l = \langle \gamma_f^l \rangle\right)$. They describe the vapor–liquid equilibrium by substituting $v_f' = v_l$ in the expression for the values of the chemical potentials v_l of the molecules in the gas phase.

In the theory of liquid solutions it is usually assumed that the mole fraction θ_i of component i is known, $\theta_i = \sum\limits_{\phi=1}^{\Phi_i} \theta_i^\phi$. In this case, equation (A4.9) can determine the mole fraction of molecules i with a specific orientation φ_i, and also allow the use of experimental data on the vapor–liquid equilibrium and other thermodynamic characteristics of solutions to determine the parameters of intermolecular interaction.

Equations (A4.10) – a system of equations for the pair distribution function $\theta_{fg}^{l\lambda}(r) = \langle \gamma_f^l \gamma_{gr}^\lambda \rangle$, describing the probability of finding the particles l and λ (i.e., molecule i with the orientation φ_i in the site f and molecule j with the orientation ψ_j in site g_r) on the relative distance r.

Equations (A4.9) and (A4.10) are a generalization of the equations given in [10–13], previously used to study the phase diagrams of binary solutions of molecules with specific interactions. They reflect the contributions of all possible orientations of the molecules of each species. The orientations are 'selected' through specific values of energy parameters, which in the case of specific interactions 'cut out' locally oriented arrangements of neighbouring molecules, given the tendency of molecules to the organization of quasicrystalline structures (specific solvation). Typically, specific solvation is characterized by the relatively strong interaction that leads to a change in the internal motion of the molecules (and change in parameter v_l). This change depends on the specific environment of each molecule by its neighbours, i.e., it is the many-body effect. Simple approximate description of changes in internal states of molecules is achieved following the pair approximation:

$$v_f^l(\{\lambda\}) = v_l^0 + \sum_{r=1}^{R}\sum_{\lambda=1}^{\Phi} \delta v_{fg}^{l\lambda}(r) n_g^\lambda(r), \qquad (A4.11)$$

wherein $\delta v_{fg}^{l\lambda}(r)$ is the change in the contribution to the statistical sum of the central particle l in the site f, provided by the neighbouring particle λ, located in the r-th coordination sphere of the site f, $n_g(r)$ is the number of particles λ in site g_r, equal to zero or one; v_l^0 is the statistical sum of the particle l in a standard condition (relating, for example, to the pure component). Such an approximation leads to the replacement value $\varepsilon_{fg}^{l\lambda}(r)$ by $\tilde{\varepsilon}_{fg}^{l\lambda}(r) = \varepsilon_{fg}^{l\lambda}(r) + \delta v_{fg}^{l\lambda}(r)$.

The size of the algebraic system of equations (A4.10) and (A4.11) can be reduced by introducing new variables (see Appendix 2).

Lattice parameters of bulk fluid [14]

We compare the free energies of the system in a discrete and a continuous version of the recording of the spatial distribution of the molecules. The lattice model for a mixture of polyatomic molecules based on the pair interaction potential describing the orientation direction of the bonds, was stated above.

The values of the energy parameters $\varepsilon_{fg}^{lm}(r)$ depend on the types of particles l and m, and also the mutual arrangement of the sites f and g at distance r, as with the change of the position of polyatomic molecules (var g_r) at constant l, m and r the closest atoms of these molecules also change. In formula (A4.1), the index f runs over all N lattice sites, and the index g_r – all $z_f(r)$ adjacent sites in the r-th coordination sphere of the central site f.

When calculating the statistical sum Q_{lattice} summation is carried out over all states of occupation γ_f^l of all N lattice sites. In order to 'separate' from the expression (A4.1) one- or two-particle contributions to the statistical sum of the system, we use [7]: 1) 'projection' properties of the values γ_f^l (5.1) and 2) define the local correlators of the lattice system (Appendix 1).

Denoting these contributions by B_f^l and $B_{fg}^{lm}(r)$ respectively, we obtain

$$B_f^l = \langle \gamma_f^l Q_{\text{lattice}} \rangle = \exp(-\beta v_f^l) = F_f^l,$$

$$B_{fg}^{lm}(r) = \langle \gamma_f^l \gamma_{gr}^m Q_{\text{lattice}} \rangle = F_f^l F_{gr}^m \exp[\beta \varepsilon_{fg}^{lm}(r)]. \qquad (A4.12)$$

The statistical sum of the same system in the continuum model can be written as

$$Q_{\text{cont}} = \frac{1}{\prod_{i=1}^{s-2}(h^{p_i}\sigma_i)^{N_i}} \int\int d\bar{q}\, d\bar{q}\, \exp[-\beta H(\bar{q},\bar{p})], \qquad (A4.13)$$

$$H(\bar{q},\bar{p}) = \sum_{i=1}^{s-1}\sum_{n_i=1}^{N_i}\left[H_{\text{int}}^i(n_i) + \sum_\alpha \frac{p_{i\alpha}^2}{2m_i} + \frac{1}{2}\sum_{j=1}^{s-1}\sum_{n_j=1}^{N_j}\varepsilon(|q(n_i)-q(n_j)|) \right],$$

where $H_{\text{int}}^i(n_i)$ is the energy of the internal states of the molecule i, n_i is the number of molecules of type i, $1 \le n_i \le N_i$; $p_{i\alpha}$ is the components of the momentum of the translational movement of the centre of mass of the molecule i, $\alpha = x, y, z$; $\bar{q}$ is the vector of the coordinates of all molecules $q_i = (r_i, \Omega_i^\delta)$, δ labels the Euler angles defining the orientation of the molecule; $\bar{p}$ is the vector of all the momenta of molecules, ρ_i is the number of degrees of freedom and σ_i is the number of symmetry of the molecule i; ε is the potential function of the interaction of molecules i and j. Integrating over all momenta of the molecules, we can rewrite (A4.13) in the following form:

$$Q_{\text{cont}} = \prod_{i=1}^{s-1} \frac{Z_{\{N\}}}{(v_i^0)^{N_i}}, \quad v_i^0 = \lambda_i^3, \quad \lambda_i = h / (2\pi M_i \beta^{-1})^{1/2},$$

$$Z_{\{N\}} = \int d\bar{q} \exp \left\{ \sum_{i=1}^{s-1} \sum_{n_i=1}^{N_i} \left[\ln F_i^0 - \frac{1}{2} \beta \sum_{j=1}^{s-1} \sum_{n_j=1}^{N_j} \varepsilon(| q(n_i) - q(n_j) |) \right] \right\},$$

$$(A4.14)$$

where F_i^0 is the statistical sum of the internal states of the molecule i; v_1^0 is the volume per thermal de Broglie waves of the molecule i. In formulas (A4.13) and (A4.14) the molecules each class are numbered by indices n_j. This is done to highlight the contributions of specific molecules in a fixed volume of space. Numbering molecules makes them distinguishable so the formulas (A4.13), (A4.14) do not contain factor $\left(\prod_{i=1}^{s-1} N_i! \right)^{-1}$.

We divide the volume V of the system to the same elementary volumes $v(f)$ which appear in the lattice model, where f is the cell number (site); $V = \sum_{f=1}^{N} v(f)$, N is the number of sites of the lattice. (In general, no conditions are imposed on the size of each site.)

Similarly, the 'volume' of the space of rotation of the molecule i relative to the centre of mass is divided into Φ_i sectors with the 'volume' Ω_i^φ, each of which contains 'its own' orientation φ_i, defined in the lattice model: $\sum_{\varphi_i=1}^{\Phi_i} \Omega_i^\varphi = 8\pi^2 = V_\Omega$ is the 'volume' of the space of rotations of polyatomic molecules [15, 16]. For simplicity, in the formulas here and below, the factor $\tilde{V}_i = \sigma_i h^3 / \prod_{i=1}^{3} (2\pi J_i^\alpha \beta^{-1})^{1/2}$, where J_i^α is the main central moment of inertia of the molecule i ($\alpha = 1\text{--}3$), normalizing the space of 'rotation' in the preparation of statistical sum of the rotational states $F_{rot} = V_\Omega / \tilde{V}_i$ obtained from the formula (A4.13) in the integration of the rotational momentum in the classical approximation, is attributed to the expression for F_i^0.

The integral in (A4.14) is a multiple integral, its multiplicity is the number of molecules $\sum_{i=1}^{s-1} N_i$. Each of these integrals over $dq(n_i)$ may be represented as a sum of integrals over $dv(f)$, where $dv(f)$ is the element of the three-dimensional space of the cell f, and the sums of integrals over $d\Omega_i^\varphi$; $d\Omega_i^\varphi$ is the element of the three-dimensional space of rotation Ω_i^φ:

$$\int_{V,V_\Omega} (...)dq(n_i) = \int_V \int_{V_\Omega} (...)dv(n_i)d\Omega(n_i) = \sum_{f=1}^{N}\sum_{\varphi_i=1}^{\Phi_i} \int_V \int_{V_\Omega} (...)dv(f)$$

As a result the formula (A4.14) can be rewritten as

$$Q_{cont} = \prod_{i=1}^{s-1}\prod_{n_i=1}^{N_i}\sum_{f=1}^{N}\sum_{\varphi_i=1}^{\Phi_i}\frac{1}{v_1^0}\int_{v(f)}\int_{\Omega^\varphi}\exp\left\{\ln F_i^0 - \frac{1}{2}\beta\sum_{j=1}^{s-1}\sum_{n_j=1}^{N_j}\varepsilon(|q(n_i)-q(n_j)|)\right\}\times$$

$$\times dv(f)d\Omega_i^\varphi.$$
$$(A4.15)$$

Here in each cell $v(f)$ one molecule of type i with the number n_i and orientation φ_i is allowed to exist. Every other molecule is in its cell. Choosing the site number g_r and placing in it a molecule of sort j with the number n_j and orientation ψ_j, we get the unambiguous correspondence between these indices. The notion of finding a molecule in a cell means finding in it the centre of mass of the molecule.

Formula (A4.15) is a lattice analogue of the continuum model. Directly from it we can 'highlight' local simple one- and two-particle contributions to the Q_{cont}. The one-particle contribution for the molecule i with the orientation φ_i located at site f, corresponds to

$$B_f^l = \int_{v(f)}\int_{\Omega_i^\varphi}\frac{F_i^0 dv(f)d\Omega_i^\phi}{v_i^0} = \frac{F_i^0 dv\Omega_i^\phi v(f)}{v_i^0}.$$
$$(A4.16)$$

The second equality is a consequence of the independence of F_i^0 on the position of the centre of mass in the volume of the cell and the orientation in the sector Ω_i^ϕ; ratio $v(f)/v_i^0$ is the statistical sum for the translation movement of the centre of mass of the molecule. Comparison of formulas (A4.12) and (A4.16) gives the relation between the one-particle lattice parameter F_f^l and its molecular interpretation in terms of sums over the states of various motions of the molecule: orientation, translational and internal (electronic, vibrational).

The two-particle contribution for the molecule i with the orientation φ_i at site f and molecule j with orientation ψ_j, located in the site g, can be written as

$$B_{fg}^{lm}(r) = \frac{1}{v_i^0 v_j^0} \int\limits_{v(f)} \int\limits_{v(gr)} \int\limits_{\Omega_i^\phi} \int\limits_{\Omega_j^\psi} F_i^0(q_i) F_j^0(q_j) \times$$

$$\times \exp[-\beta\varepsilon(|q_i - q_j|)]dv(f)dv(g_r)d\Omega_i^\phi d\Omega_i^\psi$$
(A4.17)

(when recording the coordinates of the molecules q_i and q_j their numbers are omitted). In general, the value F_i^0 and F_j^0 depend on the current coordinate values $q_{i,j}$ and may not be brought outside the integral. However, if the 'external' field of the intermolecular interaction ε changes only slightly the internal state of the molecules i and j, then the values $F_{i,j}^0$ can be taken outside the integral sign.

The energy interaction parameter, according to (A4.12) and (A4.17) is expressed as follows:

$$\varepsilon_{fg}^{lm}(r) = \beta^{-1}\ln D_{fg}^{lm}(r),$$

$$D_{fg}^{lm}(r) = \frac{B_{fg}^{lm}(r)}{B_f^l B_{g_r}^m} = \frac{1}{F_i^0 F_j^0 v_i^0 v_j^0 \Omega_i^\phi \Omega_j^\psi} \int\limits_{v(f)} \int\limits_{v(gr)} \int\limits_{\Omega_i^\phi} \int\limits_{\Omega_j^\psi} F_i^0(q_i) F_j^0(q_j) \times$$

$$\times \exp[-\beta\varepsilon(|q_i - q_j|)]dv(f)dv(g_r)d\Omega_i^\phi d\Omega_i^\psi.$$
(A4.18)

In the energy expression (A4.1), the sign of the interaction parameter is chosen so that positive values correspond to the attraction and negative ones to repulsion. Such a choice of sign corresponds to an increase of the heat of adsorption with increasing filling for the attracting molecules in the case of physical adsorption and to a reduction – to repulsion of the molecules during chemisorption. The attraction regions of the potential in the continuum model correspond to the negative values of the potential function. The lattice model parameters for different r replace the continuous potential curve for a discrete set of average values of potential energy.

Lattice parameters of the adsorption system [14]
Analysis of adsorption systems is performed similarly. The previously

constructed lattice parameters used the condition of constant density of the isotropic gas and/or liquid phase. For adsorption systems the lattice parameters will differ from the bulk phase of the solution by the presence of formulas for the additional potential interaction with the solid surface. For discrete and continuum models we have

$$\tilde{H} = H - \sum_{f,l} \varepsilon_f^l \gamma_f^l, \quad \tilde{H}(\bar{q}, \bar{p}) = H(\bar{q}, \bar{p}) - \sum_{i=1}^{s-1} \sum_{n_j=1}^{N_i} U(q(n_i)), \quad \text{(A4.19)}$$

where ε_f^l is the bonding energy of the particle l, located at site f, with the surface of the adsorbent, $U(q(n_i))$ is the potential of interaction of the molecule of sort i with the number n_i and orientation φ_i with the same surface.

In both models, this interaction is included in the single-particle contribution of the adsorbed molecules. As a result, we obtain

$$\tilde{B}_f^l = \tilde{F}_f^l \exp(\beta \varepsilon_f^l) = \int_{v(f)} \int_{\Omega_i^\phi} \tilde{F}_i^0(q_i) \exp[-\beta U(q_i)] dv(f) d\Omega_i^\phi / v_i^0. \quad \text{(A4.20)}$$

In general, the potential of interaction with the adsorbent can affect the internal state of the adsorbate and the value of $\tilde{F}_i^0$ can not be placed behind the integral sign, i.e., the lattice parameters $\tilde{F}_f^l$ and ε_f^l can not be expressed in terms of molecular parameters (A4.13) and (A4.19) separately, without additional assumptions. This fact is usually 'bypassed' introducing modified mean values of $\widehat{F}_i^0$ ($\neq F_i^0$ in the phase bulk), which are placed outside the integral sign (A4.17)–(A4.19).

In weak interaction of molecules with surfaces that is realized in physical adsorption, we can assume $\tilde{F}_f^l = \widehat{F}_f^l$ and

$$\varepsilon_f^l = \beta^{-1} \ln \left\{ \int_{v(f)} \int_{\Omega_i^\phi} \exp[-\beta U(q_i) dv(f) d\Omega_i^\phi / v_i^0] \right\}. \quad \text{(A4.21)}$$

The ratio of one-particle contributions $\tilde{a}_f^l = \tilde{B}_f^l / B_f^l$ is a local Henry's constant (it is the full content of molecules)

$$\tilde{a}_f^l = \frac{\tilde{F}_f^l \exp(\beta \varepsilon_f^l)}{F_f^l} = \int_{v(f)} \int_{\Omega_i^\phi} \frac{\tilde{F}_i^0(q_i) \exp[-\beta U(q_i)] dv(f) d\Omega_i^\phi}{v_i^0 F_i^0 \Omega_i^\phi} \quad \text{(A4.22)}$$

linking the concentration of molecules in the adsorbed state and in that same volume of the site $v(f)$ in the absence of the influence of the potential of the adsorbent (in the bulk phase). If the concentration of molecules in the bulk (gas) phase is expressed in terms of pressure, the local Henry constant a_f^l is defined as $a_f^l = \beta \tilde{a}_f^l$.

Separation of two-particle contributions of the adsorption system allows us to express the energy parameters of the lattice model, by analogy with formula (A4.18)

$$\tilde{\varepsilon}_{fg}^{lm}(r) = \beta^{-1} \ln \tilde{D}_{fg}^{lm}(r), \qquad (A4.23)$$

where $\tilde{D}_{fg}^{lm}(r) = \int\limits_{v(f)} \int\limits_{v(gr)} \int\limits_{\Omega_i^\phi} \int\limits_{\Omega_j^\psi} \tilde{F}_i^0(q_i)\tilde{F}_j^0(q_j)\exp\{-\beta[U(q_i)+U(q_j)+\varepsilon(|q_i-q_j|)]\}\times$

$\times dv(f)dv(gr)d\Omega_i^\phi d\Omega_j^\psi$.

Thus, in general, the energy parameters of the lattice model depend not only on the intermolecular interaction potential ε ($|q_i - q_j|$) for the bulk phase, but also on the nature of the interaction of molecules with the surface of the adsorbent.

Continuum lattice model [17]

The translational degree of freedom of the molecule inside the cell can be considered explicitly. In this case, the continuum cell model provides more stringent values of lateral interactions ε (θ, T, D). It summarizes the LGM equation describing the distribution of particles in the studied system by taking into account shifts of the centre of mass of the particle from the centre of the cell (as in cell theories of liquids [16,18,19]). We write the total energy of the system as follows:

$$H = \sum_{f,i} \int\limits_{\Delta(f)} v_f^i(r)\gamma_f^i(r)dr_i - \frac{1}{2}\sum_{r=1}^{R}\sum_{f,g}\sum_{i,j} \int\limits_{\Delta(f)}\int\limits_{\Delta(g)} \varepsilon(r_{fg})\gamma_f^i(r)\gamma_g^j(r)dr_i\,dr_j,$$

$$(A4.24)$$

where the value of $v_f^i(r)$ reflects the internal degrees of freedom of the particle i and the potential of interaction of the particle with the solid in the cell f, *the* cell volume f is denoted by $\Delta(f)$, R is the radius of the interaction potential. Values $\gamma_f^i(r)$ define the event: the centre of mass of the particle i is at the point r within the cell f, $r\in\Delta$ (f). Since the particle i may be located at any point of the volume $V = \Sigma_f \Delta(f)$, in this case $\Sigma_{f,i} \gamma_f^i(r) = N_i$ is the number of particles type i and $\Sigma_r \Sigma_{f,i} \Sigma_{g,j} \gamma_f^i(r_i)\gamma_g^j(r_j) = N_{ij}(r)$ is the average number of particle pairs ij at a distance $r = |r_f^i - r_g^j|$.

The unknown functions in the continuum lattice model are distribution functions $\theta_f^i(r)$ and $\theta_{fg}^{ij}(r_ir_j)$. The first functions characterize the probability of finding the particle i at the point

r inside the cell f. Functions $\theta^{ij}_{fg}(r_ir_j)$ characterize the probability of finding the centres of mass of particles i and j at the points r_i and r_j inside the cells f and g at a distance r. For them, there are normalizing relations similar to equations

$$\int_{\Delta(f)} \theta^i_f(r)dr = \theta^i_f, \quad \int_{\Delta(f)} \int_{\Delta(g)} \theta^{ij}_{fg}(r_ir_j)dr_idr_j = \theta^{ij}_{fg}, \quad \sum_{j=1}^{s}\theta^{ij}_{fg} = \theta^i_f. \quad (A4.25)$$

Recall that the functions θ^i_f and $\theta^{ij}_{fg}(r)$ are the usual functions of the discrete lattice problem. The cluster approach gives the following system of equations for the unknown functions, taking into account direct correlations between the interacting particles:

$$\exp(-\beta v^i_f(r))\theta^s_f(r) = \theta^i_f(r)\Lambda^i_f(r),$$

$$\Lambda^i_f(r_i) = \prod_{p=1}^{R} \prod_{g \in z_f(\rho)} \sum_{j=1}^{s} \int_{\Delta(g)} \theta^{ij}_{fg}(r_ir_j)\exp[-\beta\varepsilon^{ij}_{fg}(r_ir_j)]/\theta^i_f(r_i)dr_j.$$

$$(A4.26)$$

As above, the function $\Lambda^i_f(r_i)$ takes into account the contributions of intermolecular interactions. To close the system of equations, in addition to equations (A4.25) and (A4.26) we require equations relating the pair distribution functions (continuum quasichemical approximation):

$$^*\theta^{ij}_{fg}(r_ir_j)^*\theta^{kn}_{fg}(r_ir_j) = {}^*\theta^{in}_{fg}(r_ir_j)^*\theta^{kj}_{fg}(r_ir_j),$$

$$^*\theta^{ij}_{fg}(r_ir_j) = \theta^{ij}_{fg}(r_ir_j)\exp[-\beta\varepsilon^{ij}_{fg}(r_ir_j)]. \quad (A4.27)$$

The constructed system of equations (A4.25)–(A4.27) contains the cell dimensions $\Delta(f)$ as the parameters. To eliminate the arbitrariness in the values of these parameters we should minimize the free energy of the system F. In this approximation, the free energy is expressed as

$$F = \sum_{f,i} \int_{\Delta(f)} \left\{ v^i_f(r_i) + \beta^{-1}\ln\theta^i_f(r_i) \right\}\theta^i_f(r_i) + \frac{F_f\beta^{-1}}{2} \right\}dr_i,$$

$$F_f = \sum_{p=1}^{R} \sum_{g \in z_f(\rho)} \sum_{j=1}^{s} \int_{\Delta(g)} [\theta^{ij}_{fg}(r_ir_j)\ln{}^*\theta^{ij}_{fg}(r_ir_j) - \theta^i_f(r_i)\theta^j_g(r_j)\ln\theta^i_f(r_i)\theta^j_g(r_j)]dr_j.$$

$$(A4.28)$$

Formula (A4.28) generalizes the previously obtained expression for the free energy of a non-uniform lattice system [20].

The solution of the system of integral equations (A4.26)–(A4.28) allows one to find the unknown distribution functions and use them to calculate the lattice parameters

$$\varepsilon_{ij}(\rho) = \frac{1}{\theta_{fg}^{ij}(\rho)} \sum_{f,g} \int_{A(f)} \int_{A(g)} \varepsilon_{fg}^{ij}(r_i r_j)\theta_{fg}^{ij}(r_i r_j) dr_i dr_j, \qquad (A4.29)$$

where the sum over (fg) is taken over all pairs of sites at a distance ρ of the coordination spheres. Thus, defining the information about the structure of the solid and intermolecular interaction potentials, in a lattice model, as in the other methods of the molecular level, it is possible to obtain a complete description of the thermodynamic and kinetic characteristics of adsorption.

On the basis of the continuum approach we can construct various approximate expressions simplifying the calculation of molecular distributions.

To find the lattice parameters from the interaction potentials, in [21,22] the authors proposed assessments to determine their minimum and maximum values. For the lattice parameter of the lateral interaction ε we have

$$\varepsilon(\text{min}) = \beta^{-1}\ln(D_{qp}), \quad D_{qp} = \int_{V_q}\int_{Vp} \exp[-\beta U_{AA}(r)]\frac{dV_q dV_p}{V_q V_p},$$

$$\varepsilon(\text{max}) = \frac{\displaystyle\int_{V_q}\int_{V_p} E\exp[-\beta E(r)]dV_q dV_p}{\displaystyle\int_{V_q}\int_{V_p}\exp[-\beta E(r)]dV_q dV_p};$$

$$\qquad (A4.30)$$

$$Q_q = \frac{\displaystyle\int_{V_q} U(r)\exp[-\beta U(r)]dV_q}{\displaystyle\int_{V_q}\exp[-\beta U(r)]dV_q}, \quad A_q = \int_{V_q}\exp[-\beta U(r)]dV_q.$$

There are also expressions for the lattice parameters Q_q and A_q. In formula (A4.30) V_q is the volume of the site with index q above the surface of the adsorbent, $U(r)$ is the adsorbate–adsorbent interaction potential obtained by summing all paired contributions $U_{A-i}(r)$ from the atom–atom interaction potentials of the adsorbate atom (A) with solid ions (i) within a sphere of radius R. The upper estimate is a weighted average (with the Boltzmann weight) energy of the pair interaction of two particles in neighbouring cells. The lower estimate is the RMS value of the same energy. Model calculations carried out with the Lennard–Jones potential give the following estimates: $\varepsilon(\min) \sim 0.4$–$0.5D$ and $\varepsilon(\max) \sim 0.9$–$0.95D$ (which almost coincides with the qualitative estimate of $\varepsilon(\max) \approx D$). For contributions of the second neighbours have respectively the estimates of the maximum and minimum values of the lattice parameters: $\varepsilon(\max) \sim 0.10$–$0.125D$ and $\varepsilon(\min) \sim 0.10$–$0.15D$. Differences in the estimates for the nearest neighbours are quite significant.

This approach was used to construct the distribution functions for different types of sites using the example of adsorption of CH_4, CO_2 molecules and Kr atoms in a glassy matrix of polycarbonate based on the minimum estimate of the lattice parameter values [22] and the adsorption of pure gases Xe, Kr, N_2, CH_4 and mixtures thereof Xe–Kr, CH_4–N_2 in zeolites NaX based on the maximum estimates of the values of the lattice parameters [21]. Earlier in the lattice models the parameters ε, Q_q and A_q were traditionally considered as 'the adjustable' parameters determined by comparing the calculated and experimental values. In the case of (A4.30) and (A4.29), these values are not independent parameters – they are determined by the parameters of atom–atom potential curves, which can be found in the literature. If the parameters of the potential curves are not available, can be found by (A4.30) and (A4.29) from the comparison of the calculated and experimental adsorption characteristics.

The main advantage of the lattice model of adsorption is the ability to take into account the main molecular properties of the adsorption system: non-uniformity of the surface, lateral interactions, the effects of the orientation of the adsorbate, etc. The use of the QCA provides fast calculations. Introduction of the calibration functions can be used in this approximation in the area of the phase transition, including critical areas, and dramatically reduce the time of calculation of phase diagrams. The continuum lattice theory allows to relate the lattice parameters and the potentials of intermolecular interactions. It is natural to assume that the combined use of the calibrated QCA and the continuum lattice theory will not be inferior in the accuracy of description to the numerical Monte Carlo and molecular dynamics methods, and the speed of calculations wull still be their main advantage and will significantly exceed that of the the numerical methods.

References

1.	Tovbin Yu.K., // Zh. Fiz.. khimii. 1995. V. 69, No. 2. P. 220.
2.	Guggenheim E.A., Mixtures. – Oxford: Univ. Press, 1952. – 270 p.
3.	Miller A.R., The Theory of Solutions of High Polymers. – Oxford: Clarendon Press, 1948.
4.	Smirnova N.A. // Kolloid. Zh. 1979. V. 41. P. 1152.
5.	Tovbin YuK., Batalin O. // Zh. Fiz.. khimii. 1987. V. 61. P. 2535.
6.	Tovbin Yu.K., Cluster approach to the theory of atomic and molecular processes in condensed phase. Dissertation. Zh. Fiz.. khimii. Moscow: Karpov Institute. 1985.
7.	Tovbin Yu.K., Theory of physico-chemical processes at the gas–solid body interface. – Moscow: Nauka, 1990..
8.	Smirnova N.A., Chemistry and Thermodynamics of solutions. – Leningrad: Leningrad State University, 1968. – Issue. 2. P. 8, 1982. P. 87.
9.	Barker J.A., Fock W. // Disc. Farad. Soc. 1953. No. 15. P. 188.
10.	Anderson G.A., Wheeler J.C. // J. Chem. Phys. 1978 . V. 69. P. 2082 .
11.	Walker J.C., Vause C.A. // Phys. Lett. A. 1980 . V. 79. P. 421.
12.	Goldstein R.E., Walker J.C. // J. Chem . Phys. 1983 . V. 78. P. 1492 .
13.	Walker J.C., Vause C.A. // Ibid. 1983 . V. 78. P. 2660 .
14.	Tovbin Yu.K., // Zh. Fiz. khimii . 1995. V. 69, No. 1. P. 118 .
15.	Godnev I. Calculations of thermodynamic functions using the molecular data. – Moscow: GITTL, 1956 .
16.	Hirschfelder J., Curtis H., Bird R. Molecular theory of gases and liquids . – Moscow: Izd. inostr. lit. 1961.
17.	Tovbin Yu.K., // Zh. Fiz.. khimii. 1998. V. 72. No. 5 . P. 775.
18.	Lennard-Jones J.E., Devonshire A.F. // Proc. Roy. Soc. 1937. V. 163A. P. 53.
19.	Barker J.A. Lattice theories of the liquid state. – Oxford: Pergamon Press, 1963.
20.	Tovbin Yu.K. // Zh. Fiz. khimii. 1992 . V. 66. P. 1395 .
21.	Gilyazov M.F., Kuznetsova T.A ., Tovbin Yu.K. // Zh. Fiz. khimii. 1992. V. 66. P. 305.
22.	Tovbin Yu.K.// Zh. Fiz.. khimii. 1995. V. 69. 1990. P. 288.

Appendix 5

Calibration functions in calculation of equilibrium characteristics

The accuracy of calculations of phase diagrams by approximate clusters methods in the LGM for uniform and non-uniform systems was improved using calibration functions, which are constructed on the basis of information about the exact solutions or experimental data in the vicinity of the critical points. This allows for a more accurate description of molecular distributions than in QCA.

In the study of equilibrium thermodynamic characteristics of adsorption (isotherms, heats, specific heats) experiments are usually carried out in a wide range of the temperature of the adsorption system and the pressure of adsorbed compounds [1–3]. In areas of medium and large fillings there are important effects of interparticle interactions that determine the aggregate states and type of ordered structures of condensed phases of the adsorbate, which affect the concentration dependences of the adsorption characteristics [4]. Exact solutions for the description of adsorption exist only for one-dimensional ($d = 1$, d is the dimension of the system) uniform and non-uniform lattices and for a small number of special cases for two-dimensional uniform arrays [5–9]. This has stimulated the development of active approximate [1, 4, 6, 8] and numerical (Monte Carlo and molecular dynamics) [10–13] methods for calculating the equilibrium characteristics. Numerical methods provide a detailed description of the distribution of molecules and, in principle, make it possible to obtain almost exact (from a statistical point of view) results for $d = 2$ and 3. However, they require two to five orders of magnitude more time for their implementation than the approximate methods based on the lattice gas model [4, 8]. Therefore, in the literature has addressed the issue of such modifications of approximate methods of calculation, which would allow to improve the accuracy of the description of the adsorption characteristics without significantly increasing the time these calculations [14–16]. The basis of these modifications was the idea of using the exact values of the critical parameters to improve the quasi-chemical approximation (QCA). The quasichemical approach has several advantages: 1) it allows one to quickly perform calculations, 2) reflects the effects of direct correlations, i.e., consider the effect of direct interaction between particles on their distribution (effects of indirect correlations when the particles are at a distance greater than the radius of the potential interaction, this approximation is not taken into account), 3) gives accurate results in the field of small and strong interactions, and 4) provides a qualitatively correct description in the critical region (better than the mean-field approximation) [4, 8, 16]. Using additional data on the exact solutions, we can construct a calibration (fitted) function, which in the region of existence of exact solutions would match them with the approximate calculations and thus greatly improved the accuracy of the calculations in the entire region of the phase diagram.

Justification of introducing calibration functions [17, 18]
Consider a uniform lattice with the number of nearest neighbours z, containing N_A particles A on N sites, for which the customary relationships of the LGM are satisfied: the number of vacancies is $N_v = N - N_A$ and $2N_{ii} + N_{ij} = zN_i$ are the relations between pairs of the nearest neighbours. The configuration part of the statistical sum for a fixed value N_A can be written as [6, 8]

$$Q_{\text{conf}} = \sum_{N_{ij}} W\left(N_{ij}\right)\exp\left[-\beta N_{ij}\varepsilon_{ij}\right],$$ (A5.1)

where the sum is taken over all pairs N_{ij}, ε_{ij} is the energy of the pair interaction, W is the statistical weight of a given value of N_{ij}, $\beta = (kT)^{-1}$, k is the Boltzmann constant, T is temperature.

In the quasi-chemical approximation (QCA) is assumed that $W^* = (zN/2)!/\{N_{AA}!N_{vv}!(N_{Av}!)^2\}$. This overestimates the exact value of W, since all pairs are considered independent from each other. To compensate for this overestimation of values W, we introduce the normalizing factor $W = hW^*$, where h is independent of N_{ij}. Factor h

is selected so that the value W in the absence of interaction passed into the well-known result for an ideal system $W = N!/(N_A! \, N_v!)$ [6, 8]. This gives

$$W = \frac{N!}{N_A! N_v!} \prod_{ij} \frac{N_{ij}^*!}{N_{ij}!}, \quad N_{ij}^* = \frac{z N_i N_j}{N}, \quad (A5.2)$$

where N_{ij}^* is the mean number of pairs ij in the case of a random distribution of molecules. Here the function h plays the role of a calibration function providing an exact solution for the special case $\beta \varepsilon_{ij} = 0$, and correcting the value of W for all other situations.

This fact was first pointed out in [14] (see also [8]), in which it was proposed to introduce a common link between functions W and W^* in the form $W = \gamma W^* + \delta$, where both functions γ and δ are independent of N_{ij}. Two conditions must be met when defining the functions γ and δ [14]: 1) the critical temperature in the modified QCA should coincide with the exact value, and 2) the statistical weight W coincides with the exact value at $\beta \varepsilon_{ij} \to \infty$ or dimensionless temperature $(\beta \varepsilon)^{-1} \to 0$ (as usual, $\varepsilon = \varepsilon_{AA} + \varepsilon_{vv} - 2\varepsilon_{Av}$, where for the vacancies $\varepsilon_{iV} = \varepsilon_{Vi} = 0$, $i = A, V$, so that $\varepsilon = \varepsilon_{AA}$). Such a modification resulted in an almost exact agreement with the Onsager solution at low temperatures. However, for high temperatures this modification was unsuccessful.

Differences in the behaviour of the exact solution and in the QCA are shown for the stratification curve shown in Fig. 18.2 and for the temperature dependence of the pair function in Fig. 5.3. The curves differ most strongly near the critical points. For example, the critical exponents β of the stratification curve $(\rho L - \rho G) \sim (1 - T/T_c)^\beta$ (18.2), where θ_L and θ_G are densities of the liquid and the vapour, differ as $\beta(\text{exact}) = 0.125$ and $\beta = 0.5$ for all the mean-field approximations [19]. For the pair function at high densities and high temperatures the approximate solution based on the phase (ordered) state of the particles moves closer to the exact solution. Therefore, the focus should be given to the intermediate temperature range.

Obviously, the introduction of calibration functions can be executed in various ways. The type of calibration function should enable the transition of the equation of calibrated QCA (CQCA) to equations of the conventional QCA for large ($\theta \to 1$) and small ($\theta \to 0$) fillings, as well as for low ($\beta \varepsilon \to \infty$) and high ($\beta \varepsilon \to 0$) temperatures, and their parameters must satisfy the equations describing the position of the critical point $d \ln P/d\theta|_T = d^2 \ln P/d\theta^2|_T = 0$, P is the adsorbate pressure.

Such functions can be constructed explicitly by means of the condition of the free energy $F = -kT \ln Q_{conf}$, which leads to cumbersome expressions [14, 15], or by using adjustable parameters in the calibration functions, which bring together the exact and approximate solutions. In [16, 20] the author discussed approaches using effective parameters of interactions and entropic contributions. Below we use the second way.

Uniform lattice

Imagine the calibrated equation for the adsorption isotherm as [18]

$$aP = \left[\frac{\theta}{1-\theta}\right]^{1+g} \Lambda, \ \Lambda = (1+xt)^z, \ x = \exp(-\beta\varepsilon)-1,$$

$$t = \frac{\theta_{AA}}{\theta_A} = 1 - \frac{2(1-\theta)}{(1+\delta)}, \ \delta = \left\{1+4\theta(1-\theta)\left[\exp(\beta\varepsilon)-1\right]\right\}^{1/2}, \text{(A5.3)}$$

where a is the Henry constant; g is the calibration function, modifying the entropic component of free energy; function Λ allows for lateral interactions between the adsorbate, with $g = 0$ we have the conventional QCA; $t = \theta_{AA}/\theta_A$ ($\theta_A \equiv \theta$), for the following relations are satisfied for the pair functions: $\theta_{AA}\theta_{vv} = \theta_{Av}^2 \exp(\beta\varepsilon)$, where θ_{ij} is the probability of finding the next pair of particles $ij(i, j = A, v;$ here A is the adsorbate, v is the vacancy) and $\theta_{ii} + \theta_{ij} = \theta_{i}$, and $\theta_A + \theta_v = 1$. Equation (A5.3) defines all the other equilibrium adsorption characteristics.

In general, the problem of constructing a calibration function is ambiguous and it is advisable to take into account the specific nature of the differences between the exact and approximate solutions. For a uniform lattice current and QCA values θ_c coincide, so it suffices to consider the case when the calibration function depends on the temperature and does not depend on θ. This automatically satisfies the second condition for the critical parameters $d^2 \ln P/d\theta^2|_T = 0$.

The critical temperature is determined by the condition $d \ln P/d\theta|_T = 0$, so differentiating (A5.3) with respect to filling θ we have

$$\frac{d \ln(P)}{d\theta} = \frac{1+g(\tau)}{\theta(1-\theta)} + \frac{d \ln(\Lambda)}{d\theta}, \qquad \text{(A5.4)}$$

where $\tau = (\beta\varepsilon)_{exact}/(\beta\varepsilon)$ is the reduced temperature; $(\beta\varepsilon)_{exact}$ corresponds to the exact value of the critical temperature in the model. Then the critical value of the reduced temperature in the QCA is denoted as $\tau = (\beta\varepsilon)_{exact}/(\beta\varepsilon)_c^{QCA}$.

To modify the QCA, the function $g(\tau)$ must be different from zero in the neighborhood of the interval $(1, \tau_c)$ and ensure the uniqueness of the solution of the equation $d\ln (P)/d\theta = 0$ for $\tau = 1$ and the critical value of the degree of filling of $\theta_c = 0.5$.

In accordance with these requirements for the function $g(\tau)$, according to (A5.4), we can construct an expression

$$g_e = -1 - \theta_c (1 - \theta_c) \frac{d\ln(\Lambda)}{d\theta}\bigg|_{\theta_c=0,5;\tau=\tau_e}, \qquad \text{(A5.5)}$$

where the following expressions have been used for different temperature ranges [17]:

$$g(\tau) = g_e \begin{cases} (1-A)\exp(-B(\tau_e - \tau)), & \tau_* \leq \tau < \tau_e, \\ 1 - A(\tau - \tau_e)/(\tau_c - \tau_e), & \tau_e \leq \tau \leq \tau_c, \\ (1-A)\exp(-B_1(\tau - \tau_c)), & \tau_c \leq \tau < \tau^*. \end{cases} \qquad \text{(A5.6)}$$

Here A and B are the parameters that determine the curvature of the stratification curve in the critical region (parameter A determines the jump of the function at τ_e: the higher A, the more shallow the dome, the parameter B determines the rate of decrease of the value of the calibration function below the critical temperature); τ_* and τ^* are the parameters defining the temperature region in which calibration function 'works': $\tau_* = (0.7-0.85)\tau_e$ and $\tau^* = 2\tau_c - \tau_e$. Parameter B_1 determines the nature of the transition of the calibrated QCA in the supercritical region to the usual QCA: $B_1 \approx 5/(\tau_c - \tau_e)$ at τ^*. The parameters A and B in function (A5.6) must be obtained from a comparison with the exact solution: analytical, Monte Carlo or molecular dynamics, if any, or from the experimental data on the phase diagrams. The reliability of the parameters can be checked by the critical exponent β (see Table A5.1).

Curve 4 with a high degree of accuracy corresponds to the Onsager curve. The determined parameters will be used in the future for other calculations. As an example we note the description of

Table A5.1. Critical exponent β depending on the parameters of the calibration function

Variant	1	2	3	4	5	6	7
A	0.20	0.25	0.30	0.35	0.40	0.35	0.35
B	45	45	45	45	45	20	60
β	0.199	0.169	0.146	0.127	0.111	0.117	0.140

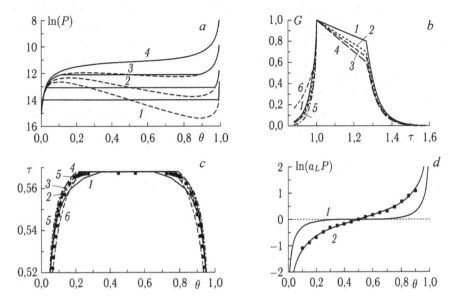

Fig. A5.1. Effect of the calibration function on the adsorption isotherm (*a*): $g = 0$ (*1*), 0.1 (*2*), 0.2 (*3*), 0.3 (*4*). The above calibration function $G = g/g_e$ (*b*) and the stratification curves (*c*) in dependence on temperature for different values of the parameters of the calibration function; the numbers of curves *1–6* correspond to the variants 1, 3–7 in Table. 1. Isotherms of adsorption (*d*): in the critical region (*1*) and at a temperature of $T = 2T_c$ (*2*).

the experimental data on the adsorption of argon atoms at a second monolayer on the surface of $CdCl_2$ (see Fig. 18.1 *b*).

Figure A5.1 shows the effect of the calibration function [17] on the adsorption characteristics.

Figure A5.1 *d* shows the effect of the calibration function [17] on the adsorption characteristics in the supercritical region $T = 2T_c$. Points on the curve *2* correspond to the exact values obtained in [21].

Another example of uniform systems is the bulk phase. In this case, the 'exact' values are the experimental data. The properties of the bulk phase were described using LGM modifications assuming the full interaction radius $R = 4$, the softness of the lattice, the excluded volume and the contribution of triple interactions (see Fig. 18.1 *a*) [17].

Thus, the introduction of the calibration function can significantly improve the traditional QCA and calculate the equilibrium characteristics with a high degree of accuracy in the entire region of the phase diagram. According to its 'action' the calibration function makes a refinement similar to taking into account large-scale fluctuations in the scaling theory of phase transitions [19] (absent in the QCA) in the vicinity of the critical region in addition to local correlations in the QCA. Therefore, the basic properties of the calibration function $g(\tau)$, established for uniform systems, must also be retained for non-uniform systems, if the size of non-uniformity areas is small, i.e., there are no two-dimensional macroscopic phases in different parts of the surface, in particular a phase transition of stratification takes place [22].

Inhomogeneous systems

Direct generalization of (A5.3) for local isotherms based on the model [17, 18], taking into account the interaction in the quasi-chemical approximation for non-uniform lattices, can be written as

$$\theta(P) = \sum_{f=1}^{N} \frac{\theta_f(P)}{N}, \quad a_f P = \left(\frac{\theta_f}{1-\theta_f}\right)^{1+gk} \Lambda_f, \quad \Lambda_f = \prod_{h \in z}\left[1 + t_{fh} x_{fh}\right], \tag{A5.7}$$

where θ_f is the degree of filling of the site f, $1 \le f \le N$, N is the number of lattice sites, a_f is the local Henry constant for the site f with the adsorbate–adsorbent interaction energy Q_f. The index h denotes the sites around the central site f. Function Λ_f describes intermolecular interactions at a distance R (Appendix 2–4). In formula (A5.7) we assume a detailed description of the expected degree of filling of all sites of the non-uniform lattice. Expressions for the conditional probabilities t_{fh} of finding the neighbouring particle on site h relative to the 'central' particle at a site f in (A5.7) are as follows $t_{fh} = \theta_{fh}^{AA}/\theta_f$, where θ_{fh}^{AA} is the probability of finding two particles in the neighbouring nodes f and h: $t_{fh} = 2\theta_h/[\delta_{fh} + b_{fh}]$, $\delta_{fh}(r) = 1 + x_{fh}(1 - \theta_f - \theta h)$, $x_{fh} = \exp(-\beta\varepsilon_{fh}) - 1$, $b_{fh} = \{[\delta_{fh}]^2 + 4x_{fh}\theta_f \theta_h\}^{1/2}$. Additionally we take into account the following normalization relations: $\theta_{fh}^{AA}(r) + \theta_{fh}^{AV}(r) = \theta_f \equiv \theta_f^{A}$ and $\theta_{fh}^{VA}(r) + \theta_{fh}^{VV}(r) = \theta_f^{V} = 1 - \theta_f$ (as above the index v denotes a vacancy – a free site) as well as $\theta_f^{A} + \theta_f^{V} = 1$.

Index 'k' in the calibration functions g_k refers to the number of the dome implemented in a non-uniform system. Therefore, for non-uniform systems it is not necessary to have a direct connection between the indices of the site type q and the number of the dome, as the number of site types t does not coincide with the number of domes on the complete stratification curve.

In chapter 3 it is shown that the number of phase transitions depends on the composition and structure of the considered two- and three-dimensional gratings [4]. In particular, in the case of the same order of magnitude of all values f_q, $1 \le q \le t$, it was identified in [4, 22] (see chapters 3 and 7) that the number of phase transitions of the first kind can vary from 1 to t, depending on the method of location of the centres of adsorption of different types (i.e., on the functions d_{qp} at fixed values of f_q). The unity corresponds to the case where all sites of different types of lattice form a periodically repeated 'pattern' and all are involved in the formation of dense and diluted phases. Number t corresponds to the case when each section, consisting of the centres of the same type, forms its dense and dilute phases.

The independence of the function g in (A5.7) on the indices f or q reflects the macroscopic nature of the cooperative (and not local) behaviour of interacting particles in the vicinity of the critical point. Consequently, in non-uniform lattices there is a set of values of pairs of critical densities θ_c^k and temperature T_c^k where the number of phase transitions is $1 \le k \le K(f_q, d_{qp})$, here $K(f_q, d_{qp})$ is the number of phase transitions for these fixed values of functions f_q and d_{qp}, describing the composition and structure of the non-uniform lattice.

When using the calibration functions (A5.7) it is considered that the regions of existence of different phase transitions are described in the QCA accurately enough and for each phase transition the functions (A5.7) depend only on the temperature, as well as for uniform lattices. The physical basis of this assertion is the fact that at low temperatures the QCA provides almost the exact solution of the problem of the distribution of molecules over the different types of sites [4]. For each phase transition the values of critical fillings θ_c^k are determined from the preliminary calculation of the phase diagrams in the QCA, like values of the local densities θ_{qc}, the critical temperature T_c^k (QCA) and derivatives $d \ln (\Lambda q)/d\theta$ for fixed θ_c^k. These values are then used in the calibration functions g_k, $1 \le k \le K(f_q, d_{qp})$. Additionally we need to know the exact values of the critical temperature T_c^k (exact) or obtain their estimates. In the absence of independent data on the exact solutions for all phase transitions of non-uniform systems, such estimates can be obtained on the basis of the available data on the exact T_c(exact, d) and approximate T_c(QCA, d) critical temperatures in homogeneous lattices of dimension d with appropriate amendments Δ_d, $d = 2$ or 3. Therefore, the parameters of the calibration functions determined

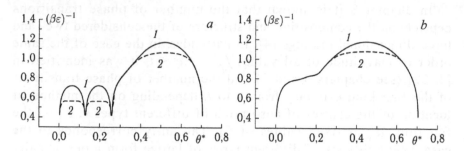

Fig. A5.2. Phase diagrams of argon in a slit pore carbon (*a*): in QCA (*1*) and CQCA (*2*); $Q_1 = 9.246\varepsilon$, $\varepsilon = 0.238$ kcal / mole, $Q_2 = Q_1/8$, $z = 6$, $H = 11$. Calibrated phase diagrams of argon in slit pores with a width of 11 monolayers for $Q_1 = 1.0\varepsilon$, $Q_2 = Q_1/8$, $z = 6$ (*b*).

for two- and three-dimensional systems can be used to predict phase transitions in complex non-uniform systems. This path is used for the analysis of the stratification curves in slit pores in which a central portion of the pores can be seen as a three-dimensional region with critical parameters close to the bulk phase, and subsurface monolayers – as quasi-two-dimensional subsystems, in which we can use the two-dimensional system parameters.

As an example, Fig. A5.2 shows the effect of calibration functions for pores with different binding energy of the adsorbate with the walls [18].

Figure A5.2 *a* shows phase diagrams for the adsorption of argon atoms in a carbon slit pore with a width of 11 monolayers in the QCA and CQCA. The molecular parameters of the system correspond to the literature data [32]: $Q_1 = 9.24\varepsilon$, $Q_2 = Q_1/8$, the remaining $Q_{q>2} = 0$; ε is the interaction parameter between the argon atoms, 0.238 kcal/mol. The initial phase diagram of the QCA consists of three domes. The first two domes relate to the first two near-

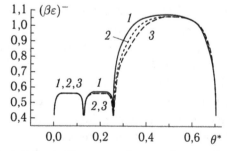

Fig. A5.3. Calibrated phase diagrams of argon in slit pores with a width of 11 monolayers for $Q_1 = 9.246\varepsilon$ (*1*), 4.0ε (*2*), and 2.0ε, $Q_2 = Q_1/8$, $z = 6$.

surface layers, and the third one refers to the remaining part of pores including its central part, starting from the third monolayer (the layers are numbered from the pore walls).

For the slit pore of the same width, but with the weak adsorbent-adsorbate interaction the influence of the calibration function is shown in Fig. A5.2 *b*. For the molecules it is not advantageous to form a two-dimensional flat structure near the pore wall. They form clusters (associates) of a new phase in several neighbouring layers. In this situation, there are no near-surface domes and only one dome, referring to the bulk filling of the pore, forms. As above, in CQCA the dome flattens and the critical temperature decreases.

Figure A5.3 shows the effect of the wall potential on the phase diagram for the above model example of a pore with a width of 11 monolayers. For different wall potentials exceeding the the interaction between the argon molecules, there is clear layered condensation. The curves show that a more accurate calculation using additional information about the exact solutions for two- and three-dimensional systems, fully confirms the qualitative picture obtained earlier in the QCA, regarding the presence of near-wall phases and allows us to explain their lack in numerical methods.

Usually, the numerical methods have been used to study the range $0.6–0.65 < \tau \leq 1$: decreasing τ increases significantly the computation time and reduced the accuracy of results. The data in Fig. A5.3 show that for the study of wall phases we must consider the temperatures corresponding to $\tau < 0.5$, but this area is not easily accessible for numerical techniques.

Thus, the methodology used for the calibration functions in the QCA not only allows to quickly obtain more reliable information, but also to check the calculations performed by numerical methods, as it is based on accurate information.

References

1. Avgul' N.N., Kiselev A.V., Poshkus D.P., Adsorption of gases and vapors on homogeneous surfaces. — Moscow: Khimiya, 1975. — 384 p.
2. Roberts M., McKee Ch., Chemistry of the gas–solid interface. — Moscow: Mir, 1981. — 539 p.
3. Experimental Methods in Catalytic Research, Ed. R. Anderson. — Moscow: Mir, 1972. — 420 p. [Academic, New York, 1968].]
4. Tovbin Yu. K., Theory of physical and chemical processes at the gas-solid interface. – Moscow: Nauka, 1990. – 288 p. [CRC, Boca Raton, Florida, 1991].
5. Onsager L. // Phys. Rev. 1944. V. 65. P. 117.
6. Domb C. // Adv. Phys. 1960. V. 9. P. 149; 245.

7.   Baxter R. J., Exactly solvable models in statistical mechanics. – Moscow: Mir, 1985. – 486 p. [Academic] Press, London, 1982].
8.   Hill T.L., Statistical Mechanics. Principles and Selected Applications. – Moscow: Izd. Inostr. lit., 1960. – 486 p. [N.Y.: McGraw–Hill Book Comp. Inc., 1956].
9.   Tovbin Yu.K. // Khim. Fizika. 1996. V. 15. No. 2, P. 75. [Chem. Phys. Report, 1996. V. 15. No. 2. P. 231].
10.   Nicolson D., Parsonage N.G., Computer Simulation and the Statistical Mechanics of Adsorption. — N. Y.: Academic Press, 1982.
11.   Molecular dynamics simulation of statistical mechanics systems, Eds. G. Coccotti, W.G. Hoover. — Amsterdam: North-Holland, 1986. — 610 p.
12.   Steele W. A. // Chem. Rev. 1993. V. 93. P. 2355.
13.   Tovbin Yu.K., Method of molecular dynamics in physical chemistry, Ed. Yu.K. Tovbin. — Moscow: Nauka, 1996. — P. 128.
14.   Prigogine I.P., Mathot-Sarolea L., van Hove L. // Trans. Farad. Soc. 1952. V. 48. P. 485.
15.   Cavalloti P. // Surface Sci. 1979. V. 83. P. 325.
16.   Tovbin Yu.K.// Zh. Fiz. khimii. 1998. V. 72, No. 5. P. 775 [Russ. J. Phys. Chem. , 1998 V. 72, No. 5, P. 675].
17.   Tovbin Yu.K., Rabinovich A.B. // Langmuir. 2004. V. 20, No. 12. P. 6041.
18.   Tovbin Yu.K., Rabinovich A.B., Votyakov E.V.// Izv. AN. Ser. khim. 2002. No. 9. P. 1531 [Russ. Chem. Bull. 2002. V. 51. № 9. P. 1667].
19.   Stanley H. E., Introduction to Phase transitions and critical phenomena. – Moscow: Mir, 1973. – P. 42. [Clarendon, Oxford, 1971].
20.   Tovbin YuK. // Zh. Fiz. khimii. 1998. V. 72, No. 12. P. 2254 [Russ. J. Phys. Chem. 1998. V. 72. № 12. P. 2053].
21.   Fisher M.E. // J. Math. Phys. 1963. V. 4. P. 278.
22.   Tovbin Yu.K. // Dokl. AN SSSR. 1981. V. 260. P. 679.

Appendix 6

About self-consistent description of elementary reaction rates and the equilibrium state of reaction systems

The key problem in the kinetic theory for all phases is the self-consistency description of the expressions for the rates of elementary reactions (stages) and the equilibrium state of the reaction system. The essence of this statement is that by equating the expressions for the reaction rates of any of the stages that occur in the forward and reverse directions, we obtain the equations describing the equilibrium distribution of the molecules of the system. In the LGM we can find the conditions under which these self-consistency conditions of the description of reaction rates and equilibrium of the system are satisfied [1, 2]. This issue is discussed for elementary processes occurring on one and two sites of the non-uniform lattice system with the interaction in the QCA for any range of interaction between the neighbours R taken into account. The self-consistency conditions are fulfilled for the entire surface if they occur for each site and for each pair of sites. Therefore, we consider the expressions for the distributed model (see Appendix 2), in which we will operate with specific individual surface centers, i.e., before the averaging procedure with different distribution functions in the composition and structure of the non-uniform surface [2, 3]. The averaged models are obtained from the expressions for the discrete model by weighting with the functions discussed in the chapters 1 and 2.

Single-site reaction

First, consider the single-site reaction between the adsorbed particle A and gas particle B: $ZA + B \leftrightarrow ZC + D$, taking place on the site of type q with the number f. The reaction proceeds in the forward and reverse directions. The equation for the velocity of the single-site stage is:

$$U_{AB} = \frac{1}{N}\sum_{f=1}^{N} U_f^{AB}, U_f^{AB} = \widehat{K}_f^{AB}\theta_f^A S_f^A, S_f^i = \prod_{r=1}^{R}\prod_{h\in z_f(r)} S_{fh}^i(r),$$

$$S_{fh}^i(r) = \sum_{j=1}^{s}\theta_{fh}^{ij}(r)\exp\left[\beta\delta\varepsilon_{fh}^{ij}(r)\right]/\theta_f^i = 1 + \sum_{j=1}^{s-1}\theta_{fh}^{ij}(r)x_{fh}^{ij}/\theta_f^i,$$

$$x_{fh}^{ij}(r) = \exp\left[\beta\delta\varepsilon_{fh}^{ij}(r)\right] - 1, \delta\varepsilon_{fh}^{ij}(r) = \varepsilon_{fh}^{ij*}(r) - \varepsilon_{fh}^{ij}(r). \tag{A6.1}$$

Here N is the number of sites of a distributed system, $\widehat{K}_f^{AB}$ is the rate constant of the elementary reaction between the adsorbed particle A and gas particle B; $\varepsilon_{fh}^{ij*}(r)$ is the interaction parameter of neighbouring particle j, located in the ground state at a distance r, with the activated complex of the reacting molecule i; the upper symbol $\wedge$ means that the expression for the velocity includes the factor of pressure for a gas phase particle (which is not reflected in the symbol of the site f); $\varepsilon_{fh}^{ij}(r)$ is the interaction parameter of the two neighbouring particles at a distance r, in the ground state.

In the equilibrium conditions the equation for the reaction rates in the forward and reverse directions $U_f^{AB} = U_f^{CD}$ gives the following relationship:

$$\widehat{K}_f^1 = \frac{\widehat{K}_f^{AB}}{\widehat{K}_f^{CD}} = \frac{\theta_f^C \psi_f^{CA}}{\theta_f^A}, \quad \psi_f^{CA} = \prod_{r=1}^{R}\prod_{h\in z_f(r)} \frac{S_{fh}^C(r)}{S_{fh}^A(r)}. \tag{A6.2}$$

In the absence of interaction the function $\psi_f^{CA} = 1$ and $\widehat{K}_f^1$ is the effective equilibrium constant for the site f of the ideal adsorption system; $\widehat{K}_f^1 = K_f^1 P_B/P_D$ where K_f^1 is the true equilibrium constant.

For the non-uniform lattice in the QCA we have the following relations for the pair distribution function [2,3]:

$$\theta_f^i\theta_{fh}^{jk}(r) = \theta_f^j\theta_{fh}^{ik}(r)\exp\left[\beta\{\varepsilon_{fh}^{jk}(r) - \varepsilon_{fh}^{ik}(r)\}\right]/S_{fh}^i(j|r),$$

$$\text{where } S_{fh}^i(j|r) = \sum_{k=1}^{s} t_{fh}^{ik}(r)\exp\left[\beta\left(\varepsilon_{fh}^{jk}(r) - \varepsilon_{fh}^{ik}(r)\right)\right]. \tag{A6.3}$$

and

$$\theta_{fh}^{kl}(r) = \theta_{fh}^{mn}(r)\exp\left[\beta\{\varepsilon_{fh}^{kl}(r) - \varepsilon_{fh}^{mn}(r)\}\right]S_{fg}^{k}(m|r)S_{fg}^{l}(n|r)\frac{\theta_{f}^{k}\theta_{h}^{l}}{\theta_{f}^{m}\theta_{h}^{n}}.$$

(A6.4)

Using the fact that the properties of the activated complex (AC) are not dependent on the direction of the reaction ($\varepsilon_{fh}^{Aj*}(r) = \varepsilon_{fh}^{Cj*}(r)$ for any index j), from (A6.3) and the equations for the QCA it follows that

$$\frac{S_{fh}^{C}(r)}{S_{fh}^{A}(r)} = \left[S_{fh}^{A}(C|r)\right]^{-1}$$

(A6.5)

and therefore, $\psi_{f}^{CA}(r) = \left[S_{fh}^{A}(C|r)\right]^{-1}$, i.e., the right side of expression (A6.2) does not depend on the interaction of AC with its neighbours and the effective equilibrium constant is expressed only through the parameters of the interaction of particles in the ground state (rather than transient) and the equilibrium concentration of the particles. This result is entirely consistent with the concept of the equilibrium distribution of the particles, and the resulting expression (A6.2) is the equilibrium constant in a non-uniform system.

In other words, the equations for the equilibrium distribution are obtained regardless of the method of their construction: from kinetic analysis or directly for the equilibrium state of the molecules of the mixture.

Two-site reaction
A similar conclusion can be obtained from the relations (A6.3)–(A6.5) if we equate the velocity of the forward and reversed directions of two-site reaction stage ($ZA + ZB + C \leftrightarrow ZE + ZD + F$), taking place in the two neighbouring sites f and g [1–3]. These velocities are written as

$$U_{fg}^{ABC}(1) = \widehat{K}_{fg}^{ABC}(fg)\theta_{fg}^{AB}(1)\exp\left[-\beta\varepsilon_{fg}^{AB}(1)\right]\prod_{r=1}^{R}\prod_{\omega_{r}=1}^{\pi(r|qp)}\prod_{h\in z(\omega_{r}|qp)}S_{fgh}^{AB}(\omega_{r}),$$

$$S_{fgh}^{AB}(\omega_{r}) = \sum_{j=1}^{s}t_{fgh}^{ABj}(\omega_{r})\exp\left[\beta\delta\varepsilon_{fgh}^{ABj}(\omega_{r})\right],$$

$$\delta\varepsilon_{fgh}^{ABj}(\omega_{r}) = \varepsilon_{fgh}^{ABj*}(\omega_{r}) - \varepsilon_{fgh}^{ABj}(\omega_{r}),$$

$$t_{fgh}^{ABj}(\omega_{r}) = \theta_{fh}^{Aj}(r_{1})\theta_{gh}^{Bj}(r_{2})/\theta_{f}^{A}\theta_{g}^{B}\theta_{h}^{j}.$$

(A6.6)

Moreover, for large distances $\theta_{fh}^{ij}(r > R) = \theta_f^i \, \theta_h^j$, that consequently gives $S_{fgh}^{AB}(\omega_r)\big|_{r_1 > R} = S_{gh}^B(r_2)$.

To prove the self-consistency condition, it is necessary to prove the following relation:

$$\frac{S_{fgh}^{ED}(\omega_r)}{S_{fgh}^{AB}(\omega_r)} = \frac{S_{fh}^E(r_1) S_{gh}^D(r_2)}{S_{fh}^A(r_1) S_{gh}^B(r_2)}, \qquad (A6.7)$$

where the values of ω_r and r_1, r_2 are uniquely linked, $\varepsilon_{fgh}^{ABj}(\omega_r) = \varepsilon_{fh}^{Aj}(r_1) + \varepsilon_{gh}^{Bj}(r_2)$. Then the equilibrium distribution of the molecules does not depend on the method of achieving it through the kinetics of the forward and reverse directions or direct examination of only equilibrium configurations.

To prove (A6.7), we must remember that $\varepsilon_{fh}^{Aj*}(r_1) = \varepsilon_{fh}^{Ej*}(r_1)$ and $\varepsilon_{gh}^{Bj*}(r_2) = \varepsilon_{gh}^{Dj*}(r_2)$ because of the independence of the properties of the activated complex on the direction of the process. We introduce the notation $\tilde{\theta}_{fh}^{ij}(r) = \theta_{fh}^{ij}(r) \exp\left[-\beta \varepsilon_{fh}^{ij}(r)\right]$ and rewrite the left side of the formula (A6.7) as

$$\frac{\theta_f^A \theta_g^B \sum\limits_{j=1}^{s} F_j \tilde{\theta}_{fh}^{Ej}(r_1) \tilde{\theta}_{gh}^{Dj}(r_2)}{\theta_f^E \theta_g^D \sum\limits_{j=1}^{s} F_j \tilde{\theta}_{fh}^{Aj}(r_1) \tilde{\theta}_{gh}^{Bj}(r_2)},$$

where $F_j = \exp\left[\beta\left\{\varepsilon_{fh}^{Aj*}(r_1) - \varepsilon_{gh}^{Bj*}(r_2)\right\}\right] \big/ \theta_h^j$, and we have to show that the ratio

$$\frac{M}{L} = \frac{\tilde{\theta}_{fh}^{Ej}(r_1) \tilde{\theta}_{gh}^{Dj}(r_2)}{\tilde{\theta}_{fh}^{Aj}(r_1) \tilde{\theta}_{gh}^{Bj}(r_2)}$$

does not depend on the index j.

To do this, we express the functions $\tilde{\theta}_{fh}^{Ej}(r_1)$ and $\tilde{\theta}_{gh}^{Dj}(r_2)$ through the functions $\tilde{\theta}_{fh}^{Aj}(r_1)$ and $\tilde{\theta}_{gh}^{Bj}(r_2)$ respectively, using general QCA relationships for the pair function (e.g., $\tilde{\theta}_{fh}^{Ej}(r_1) = \tilde{\theta}_{fh}^{Aj}(r_1) \, \tilde{\theta}_{fh}^{EA}(r_1) \big/ \tilde{\theta}_{fh}^{AA}(r_1)$ and so forth).

Consequently

$$\frac{M}{L} = \Lambda_{fh}^{EA}(r_1) \Lambda_{gh}^{DB}(r_2),$$

where $\Lambda_{fh}^{ik}(r) = \left\{\tilde{\theta}_{fh}^{ii}(r) \tilde{\theta}_{fh}^{ik}(r) \big/ \left[\tilde{\theta}_{fh}^{ki}(r) \tilde{\theta}_{fh}^{kk}(r)\right]\right\}^{1/2}$.

Then, considering the obvious equality: $\sum\limits_{j=1}^{s} F_h^j M \Big/ \sum\limits_{j=1}^{s} F_h^j L = M/L$,
you can see that the left side of relations (P6.7) is

$$\frac{\theta_f^A \theta_g^B \Lambda_{fh}^{EA}(r_1) \Lambda_{gh}^{DB}(r_2)}{\theta_f^E \theta_g^D}.$$

On the other hand, the formulas $\dfrac{S_{fh}^E(r_1)}{S_{fh}^A(r_1)} = \Lambda_{fh}^{EA}(r_1)\dfrac{\theta_f^A}{\theta_f^E}$ and $\dfrac{S_{gh}^D(r_2)}{S_{gh}^B(r_2)} = \Lambda_{gh}^{DB}(r_2)\dfrac{\theta_g^B}{\theta_g^D}$ can be proved in the same way, so we have proved the legality of (A6.7).

Using the approximations, which do not include the effects of spatial correlation, such as the mean field or random approximation, does not meet the conditions of self-consistency, because they lead to a renormalization of the activation energy of the process by the value $\varepsilon_{fg}^{ij}(1)$ (see [1–3]).

References

1.   Tovbin Yu.K. // Zh. Fiz. khimii. 1981. V. 57. P. 284 [Russ. J. Phys. Chem. 1981. V. 55. I₂. P. 159].
2.   Tovbin Yu. K., Theory of physical and chemical processes at the gas-solid interface. – Moscow: Nauka, 1990. – 288 p. [CRC, Boca Raton, Florida, 1991].
3.   Tovbin Yu. K. // Progress in Surface Sci. 1990. V. 34, No. 1–4. P. 1–236.

Appendix 7

Thermal velocity of the molecules

Vacancy area for single-site particles
The need to describe simultaneously the transport of molecules within the adsorbent in rarefied gas and dense liquid phases makes it difficult to use the concept of the mean free path of the molecules, changing from $10^4\lambda$ for the gas to λ for the fluid. In the lattice model, as in the kinetic theory of liquids, the main concept is the concept of the hopping probability (or displacement) of the molecule $W(\chi)$ over the considered distance χ, rather than the mean free path.

All neighbours located in sites 1–25 and site ξ simultaneously interact with a molecule moving from site f to an adjacent site g (hopping length $\chi = 1$) (Fig. A7.1). In general, $\chi > 1$ for movement

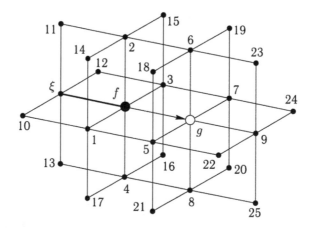

Fig. A7.1. Hopping scheme of the molecule from site f to a free neighboring site g in the first two coordination spheres of the lattice structure with $z = 6$.

within the condensed phase requires a vacancy trajectory through which the molecule moves with thermal velocity $W(\chi)$. Such a trajectory is created in a fluctuation manner. The longer the trajectory χ, the more neighbours are involved around it in the formation of the environment, affecting the hopping probability. The task of the theory is to calculate the probability of its formation so that the molecule could move, and consider the impact of neighbours on such transfer. This requires a self-consistent way to calculate the probability of many-body configurations forming the vacancy trajectory.

The probability of a molecule hopping between sites is calculated using the Eyring transition state model [1–3], which deals with the displacement of molecules as the activation process of overcoming the barrier. The formulation of this model for non-ideal reaction systems when the overcome barrier is created by the potentials of neighbouring particles and the solid surface is given in [4, 5].

As follows from the ratio of the dimensions, the average thermal velocity of the molecules is expressed in terms of the rate of hopping of the molecule $U_{fg}(\chi)$ between the sites f and g over the distance χ in the form

$$W_{fg} = \frac{\chi U_{fg}(\chi)}{\theta_f}, \tag{A7.1}$$

Section 28 gives the formula for the hopping of the molecule $U_{fg}(\chi = 1)$ to the next available site when the effect of only the nearest neighbours is taken into account. In general, in the state of thermal equilibrium the hopping probability of molecules between the sites is described by

$$U_{fg}^{AV}(\chi) = K_{fg}^{AV}(\chi)V_{fg}^{AV}(\chi), \tag{A7.2}$$

where $K_{fg}^{AV}(\chi)$ is the rate constant of the molecule hopping from site f to a free site g at a distance χ, and $V_{fg}(\chi)$ is the concentration dependence of the rate of hopping of the molecule;

$$K_{fg}^{AV}(\chi) = K_{fg}^{*AV}(\chi)\exp[-\beta E_{fg}^{AV}(\chi)], \tag{A7.3}$$

$K_{fg}^{*AV}(\chi)$ is the pre-exponential factor of the rate constant, $E_{fg}^{AV}(\chi)$ is the hopping activation energy determined by the potential of the pore walls (the influence of interparticle interactions on the magnitude of the activation hopping is defined by the imperfection function $\Lambda_{fg}^{AV}(\chi)$);

$$V_{fg}^{AV}(\chi) = \theta_{fg}^{AV}(\chi)\Lambda_{fg}^{AV}(\chi), \tag{A7.4}$$

$$\theta_{fg}^{AV}(\chi) = \theta_{fg_1}^{AV}(1) \prod_{k=1}^{\chi-1} t_{g_k g_{k+1}}^{VV}(1). \qquad (A7.5)$$

The concentration dependence of the migration rate of the molecule $V_{fg}(\chi)$ is expressed by the product of two factors:

1) $\theta_{fg}^{AV}(\chi)$ – the probability of formation of a vacancy trajectory of χ free sites from the cell f with the length χ through a sequence of cells $g(1)$, $g(2)$ and so on until the cell $g \equiv g(\chi)$, for $\chi = 1$ the cell $g(1)$ is finite;

2) $\Lambda_{fg}^{AV}(\chi)$ – the imperfection function, which takes into account the effect of interactions of molecules that are around a given trajectory, on the probability of hopping inside the trajectory:

$$\Lambda_{\xi fg}^{AV}(\chi) = \prod_{r=1}^{R} \prod_{\omega_r=1}^{\pi_r} \prod_{h \in m(\omega_r)} \sum_{k=1}^{v} t_{fgh}^{ivk}(\omega_r \mid \chi) E_{fgh}^{ivk}(\omega_r \mid \chi),$$

$$t_{fgh}^{ivk}(\omega_r \mid \chi) = \frac{\theta_{fh}^{ik}(r_1)\theta_{gh}^{vk}(r_2)}{\theta_f^i \theta_g^v \theta_h^k}. \qquad (A7.6)$$

The co-factor $t_{fgh}^{ivk}(\omega_r \mid \chi)$ describes the probability of finding the neighbouring particle k in site h at a distance of r_1 from reagent A and r_2 from vacancy V. The particles A and V themselves are situated at a distance χ. The setting of the particles around the activated complex is conveniently numbered using the numbers of sites $m(\omega_r \mid \chi)$, located in orientations $\omega_r(1 \leq \omega_r \leq \pi_r)$, π_r is the number of orientations in the r-th single coordination sphere around a dimer AV on central sites f and g at a distance χ (single coordination sphere of radius r, $1 \leq r \leq R$ is defined as a plurality of sites at a distance r from the site or f, or from the site g); R is the radius of the interaction potential. Orientation are measured from the centre of the dimer AV: the point of intersection line linking central sites f_g, and a line connecting the site h with the centre of dimer AV.

In the hopping process (middle of the connection fg, see Fig. A7.1) the moving molecule A is subjected to the influence of neighbouring particles, the energy of this interaction is denoted by the parameters $\varepsilon_{fh}^{*Aj}(r)$ which differ from identical energy parameters for the particles in the ground state $\varepsilon_{fh}^{Aj}(r)$ (i.e. in the lattice). The co-actor $E_{fgh}^{AVi}(\omega_r)$ reflects the difference in the interaction of the activated complex with its neighbours via the difference of the pair interactions $(\varepsilon_{fh}^{*Aj}(r) - \varepsilon_{fh}^{Aj}(r))$:

$$E_{fgh}^{AVi}(\omega_r \mid \chi) = \exp \beta [\delta\varepsilon_{fh}^{Ai}(r_1) + \delta\varepsilon_{gh}^{Vi}(r_2)], \quad \delta\varepsilon_{fg}^{ij}(r) = \varepsilon_{fg}^{*ij}(r) - \varepsilon_{fg}^{ij}(r).$$
$$(A7.7)$$

Traditionally, the pre-exponential factor of the rate constant $K_{fg}^{*AV}(\chi)$ is expressed through the ratio of the statistical sums of reactants and the activated complex as [1–3]

$$K_{fg}^{*AV}(\chi) = \frac{kT}{h} \frac{F_{fg}^{*AV}}{F_f^A F_g^V},$$

where F_{fg}^{*AV} and F_f^A are the statistical sums of the molecule in both transient and ground states, for the vacancy $F_V = 1$, h is Planck's constant.

In the absence of lateral interactions formula (A7.2) can be written as

$$U_{fg}(\chi) = K_{fg}\theta_f(1-\theta_g)^\chi.$$

Away from the walls of the pores ratio F_A/F_{AV}^* represents the translational degree of freedom in the direction of particle motion and equals $(2\pi m\beta^{-1})^{1/2}\chi/h$, then $K_\chi = [(2\pi m\beta)^{1/2}\chi]^{-1}$ or $K_\chi = w/4\chi$, where $w = (8/\pi m\beta)^{1/2}$ [6].

Knowing the hopping speed $U_{fg}(\chi)$, we can calculate the self-diffusion coefficients and other transport coefficients.

Kinetic parameter

All dynamic characteristics (including transfer coefficients) are determined by the energy parameters ε^* and ε. The quantity ε is the averaged value of the interaction potential between the particles and averaging is carried out over the locally equilibrium distribution of the particles in the momenta inside the given cell (section 9 and Appendix 4):

$$\varepsilon_{fg}^{ij}(\rho) = \frac{1}{\theta_{fg}^{ij}(\rho)} \int_{\Delta(f)} \int_{\Delta(g)} \varepsilon_{fg}^{ij}(r_1 r_2 \mid \rho)\theta_{fg}^{ij}(r_1 r_2 \mid \rho)dr_1 dr_2,$$

where $\theta_{fg}^{ij}(\rho) = \int_{\Delta(f)} \int_{\Delta(g)} \theta_{fg}^{ij}(r_1 r_2 \mid \rho)dr_1 dr_2$, the average probability of finding molecules i and j in cells f and g at a distance ρ; here the value $\Delta(f)$ is the volume of the cell f.

To calculate the average interaction ε^* of the migrating molecule (or AC) in the given cell with the surrounding molecules it is necessary to carry out averaging over all possible 'non-equilibrium' states of the moving particle with the molecules in neighbouring cells that affect the potential of the migrating particle. As regards the meaning this value is the only value used in the calculation of all dissipative characteristics. The most natural assumption is that of the equal probability of all directions of motion of the particle. In this case, the average interaction between the migrating particle i and its neighbours j at a distance ρ is defined as [7]

$$\varepsilon_{fg}^{*ij}(\rho) = \int\limits_{\Delta^*(f)} \int\limits_{\Delta^*(g)} \varepsilon_{fg}^{*ij}(r_1 r_2 \mid \rho)\, dr_1 dr_2. \qquad (A7.8)$$

The difference between $\Delta^*(f)$ and $\Delta(f)$ consists in the fact that $\Delta^*(f)$ is a sector inside cell f in which the reacting particle moves considering the excluded volume by blocking the neighbouring molecules. If there is no full blocking, the sector size is determined by the size of AC and the neighbouring particles – there is a partial blocking. In fact, the averaging takes place over the section on the edge connecting two sites. Here r_1 and r_2 relate to the saddle point. Fluctuations of AC refer to directions perpendicular to the lines between the sites f and h, between which the hopping $g \neq h$ takes place. Therefore, when calculating ε_{fg}^{*ij} we must take into account the real allowed trajectories of displacements of the molecules – a complete or partial blockage of these shifts should also affect the probability of the hopping taking place.

Relative thermal velocity [8–10]

Equations (A7.2) are the traditional interpretation of the LGM, when the process of hopping of the molecule through the activation barrier is influenced by the excitations generated by all the surrounding molecules and solids (thermostat). To consider the impact of the contributions of the individual neighbours on the probability of overcoming the activation barrier, they must be separated from the overall impact of the thermostat. In this case, the overall process of gathering the excess energy of the reaction system is also influenced by the thermostat, but the very act of overcoming the activation barrier depends on the direction of movement and the masses of the colliding particles before the final passage of the barrier. This circumstance affects the rate of relative motion of the migrating (or reactive) molecule.

Modification of the formula for the rate of hopping of the molecule i with explicit allowance for the type of molecule j with which it collided last time, when calculating its contribution to traffic flow leads to the following expression:

$$U_{\xi fg}^{iv}(\chi) = \sum_{j=1}^{s-1} U_{\xi fg}^{(j)iv}(\chi), \ U_{\xi fg}^{(j)iv}(\chi) = K_{\xi fg}^{(j)iv}(\chi) V_{\xi fg}^{(j)iv}(\chi),$$

$$V_{\xi fg}^{(j)iv}(\chi) = \left\langle \theta_{fg}^{iv}(\chi) \right\rangle \left\langle t_{f\xi}^{ij} \right\rangle \Lambda_{\xi fg}^{AV}(\chi), \qquad (A7.9)$$

$$\left\langle \theta_{fg}^{iv}(\chi) \right\rangle = \theta_{fg(1)}^{iv}(1) \prod_{\psi} t_{\psi\psi+1}^{vv}(1),$$

where the index ξ belongs to the neighboring site on the ρ-scale relative to the site f, containing molecule j, which determines the magnitude and direction of the momentum of the molecule i hopping from site f to a free site g after a 'head-on' collision with molecule j (site ξ is aligned on a straight line with the sites f and g on the other side from the site g); $K_{\xi fg}^{(j)iv}(\chi) = w_{i(j)}/\chi$ is the hopping rate constant, expressed by the average thermal velocity $w_{i(j)} = (2\pi\mu_{ij}\beta)^{-1/2}$ of the molecule i, with respect to the molecule j; $\mu_{ij} = m_i m_j/(m_i + m_j)$ is the reduced mass of the colliding particles i and j, m_i is the mass of component i.

The concentration dependence of the migration rate of molecules is expressed by the co-factor $V_{\xi fg}^{(j)iv}(\chi)$. It takes into account three factors: 1) $\left\langle \theta_{fg}^{iv}(\chi) \right\rangle$ – the probability of realization of a free path from the cell f to the cell g with length χ (that was not blocking the trajectory of other molecules); 2) $\Lambda_{\xi fg}^{i}(\chi)$ – influence of lateral interactions of neighbouring molecules located around a given trajectory, on the probability of hopping on it, and 3) $\left\langle t_{f\xi}^{ij}(\rho) \right\rangle$ – the probability of finding the neighbouring particle j at site ξ (on the ρ-scale); it is expressed as

$$\left\langle t_{f\xi}^{ij}(\rho) \right\rangle = t_{f\xi}^{ij}(\rho) \exp[\beta\delta\varepsilon_{f\xi}^{ij}(\rho)]. \qquad (A7.10)$$

Normalization (A7.10) on θ_ξ gives in the absence of interaction $\left\langle t_{f\xi}^{ij}(\rho) \right\rangle / \theta_\xi = x_\xi^j$ – mole fraction of particle j at site ξ (see Appendix 12).

Functions $t_{fg}^{ij}(r)$ in (A7.9) are calculated by the equations of the equilibrium state of the fluid.

Function $\Lambda_{\xi fg}^{AV}(\chi)$, which takes into account the lateral interactions of the surrounding molecules in the quasichemical approximation, is written as

$$\Lambda_{\xi fg}^{AV}(\chi) = \prod_{r=1}^{R}\prod_{\omega_r=1}^{\pi_r}\prod_{h \in m^*(\omega_r)} \sum_{k=1}^{v} t_{fgh}^{ivk}(\omega_r \mid \chi) E_{fgh}^{ivk}(\omega_r \mid \chi). \quad (A7.11)$$

Functions $t_{fgh}^{ivk}(\omega_r \mid \chi)$ and $E_{fgh}^{ivk}(\omega_r \mid \chi)$ are defined above (A7.6) and (A7.7): the distance from site h to site f is r_1, and to site g it is r_2. Symbol ω_r characterizes the position of the site h: through the angle formed by a line connecting a pair of 'central' sites fg, and the line connecting site h with the midpoint of the line fg, and through the set of sites at distance r from one of the two central sites f and g; $m^*(\omega_r)$ – the set of neighbouring sites with fixed values of r and ω_r, from which the site ξ is excluded if it meets the given values of r and ω_r.

In the absence of lateral interactions formula (A7.9) can be written as $U_{\xi fg}^{iv}(\chi) = \sum_{j \in \xi} K_{\xi fg}^{(j)iv}(\chi) x_j \theta_f (1-\theta_g)^x$. For a rarefied gas ($\chi \to \infty$) we find that the average relative velocity w_i of component i is rewritten as follows: $w_i = \sum_{j=1}^{s-1} w_{i(j)} x_j$.

Large particles [11–13]
Consider the expressions for the thermal velocities of the molecules, occupying more than one site, using the example of rod-shaped molecules of length L. Elementary movement of the rods has the form of displacement of the centre of gravity of the mass and the rotation about the centre of mass. It is believed that the displacement of the centre of the mass of the rod i without changing its orientation λ_i is its translational movement (hopping) between the adjacent cells f and g.

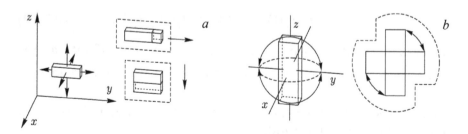

Fig. A7.2. Directions of movement of the rods (indicated by arrows) and the corresponding vacancy fields: a – translational movement, b – rotational movement; dotted line shows schematically the nearest neighbours surrounding the trajectory and affecting the hopping probability of the molecule.

For single-site migrating particles the vacancy region consists of at least one neighbouring vacancy into which the molecule moves. Movement over longer distances $\chi > 1$ requires the vacancy region, providing the trajectory hopping the molecule consisting of a linked sequence χ of the vacancies. For single-site rotating particles the change of orientation is considered as a change in the internal state of the particles without the need for vacancies to take part.

Different elementary movements of the rod require different vacancy regions corresponding to the motion of the given type (Fig. A7.2). In displacement of the the centre of inertia by one site the vacancy regions coincide with $S_{m\gamma}$: in the direction of movement of the long axis of the rod should be one vacancy ($S_{m\gamma} = 1$) and in the direction of the side face of the rod there must be L vacancies ($S_{m\gamma} = L$). If the displacement of the rod occurs over distance $\chi > 1$, then the vacancy area covers the trajectory consisting of a sequence of vacancies with length χ, and the number of vacancies in the section of this trajectory should be equal to $S_{m\gamma}$.

Translational motion of the rods [11]
For the average forward velocity of the rods from the site f to the free site g over the distance χ we obtain

$$U_{fg}^{iv}(\chi) = K_{fg}^{iv}(\chi)V_{fg}^{iv}(\chi),$$
$$K_{fg}^{iv}(\chi) = (2\pi m_i \beta)^{-1/2} \exp(-\beta E_{fg}^{iv}(\chi))/\chi,$$
$$V_{fg}^{iv}(\chi) = \left\langle \theta_{fg}^{iv} \right\rangle \Lambda_{fg}^i, \quad \left\langle \theta_{fg}^{iv} \right\rangle = \theta_{f(1)}^{iv*} \prod_\ell t_{\ell\ell+1}^{vv*}, \tag{A7.12}$$

where $K_{fg}^{iv}(\chi)$ is the hopping rate constant, expressed by the average thermal velocity of the particle i; m_i is the mass of the particle i, $E_{fg}^{iv}(\chi)$ is the hopping activation energy, which depends on the distance from the wall and the direction of motion of the rod. Away from the wall potential of the pore the activation energy is zero.

The concentration dependence of the migration velocity of the particle is expressed by factor $V_{fg}^{iv}(\chi)$. It takes into account two factors.

1. $\left\langle \theta_{fg}^{iv} \right\rangle$ – the probability of the realization of the free path which eliminates blocking the path from cell f to cell g by other particles, this trajectory consists of sites $f, f+1, f+2$, etc. to the site g. The trajectory has $S_{i\gamma}$ vacancies in the section (which are marked with *). The direction of movement from site f to site g for a given orientation of the rod i uniquely determines the full size

of the vacancy region, which consists of a portion of the trajectory with length $\chi - 1$ (described by co-factor with respect to l in the expression for $\left\langle \theta_{fg}^{iv} \right\rangle$) and the vacancy region for shift of the rod by one site. If $\chi = 1$, then $g = (1)$ and the product along l is absent in the expression for $\left\langle \theta_{fg}^{iv} \right\rangle$.

2. Λ_{fg}^{i} – the function taking into account the effect of lateral interactions of the surrounding particles disposed around the entire path from the site f to g with the section $S_{i\gamma}$, at which hopping of the particle i takes place. According to the model contacts [11], when taking into account the quasi-chemical approximation for contact pairs we have (Π_i is the number of contacts of the molecule of type i):

$$\Lambda_{fg}^{i} = \prod_{k=1}^{\Pi_i} {}^{*}\sum_{j=1}^{\Phi+1}\sum_{n=1}^{\Pi_j} t_{ij}^{kn}(fh)\exp(\beta\delta\varepsilon_{ij}^{kn})\prod_{d=1}^{\Pi_v} {}^{*}\sum_{j=1}^{\Phi+1}\sum_{n=1}^{\Pi_j} t_{vj}^{dn}(gh)\exp(\beta\delta\varepsilon_{vj}^{dn}),$$
(A7.13)

where $\delta\varepsilon_{ij}^{kn} = \varepsilon_{ij}^{*kn} - \varepsilon_{ij}^{kn}$, ε_{ij}^{*kn} is the energy parameter of the interaction of contact k of molecule i in the activated state in migration with the adjacent contact n of molecule j, located in the site h. In the first product the symbol h refers to the neighbours j of the 'central' particle i. The asterisk at the first sign of the product means that the from the set of contacts $k \in \Pi_i$ of the molecule i we exclude contacts with the neighbouring vacancy region into which the particle i moves in the direction of site g. The asterisk at the second mark work means that from the set of contacts $d \in \Pi_v$ of the vacancy region we excluded only the contacts by the migrating particle i in the direction of the site f (in the second product index h refers to the neighbours surrounding vacancy region).

The transition from formula (A7.12) to the relative velocity of the rods (or any other large molecule) is carried out similarly to the transition from formula (A7.2) to (A7.9).

Rotation of the particles relative to the centre of inertia [11–13]

We find expressions for the velocity of the rotational motions, using the theory of the transition state that the number of activated complexes (AC) per unit volume θ^{*} depends on the rate of the elementary process $U = \theta^{*}\nu$, where ν – the frequency of crossing the activation barrier (s^{-1}). The magnitude of ν for rotational motion $\nu = u_r/\varphi$, where u_r is the average angular velocity of rotation when passing the activation barrier, which has the dimension [rad/s], φ

is the length of the arc in radians. The average velocity for rotary motion in a given direction is expressed as $u_r = (kT/2\pi I^*)^{1/2}$, where I^* is the moment of inertia of the AC. The concentration of the AC at the top of the barrier can be written as $\theta^* = \theta F_r$, where θ – concentration of the AC after replacing one degree of freedom of 'swinging' motion, similar to oscillatory motion, by rotational motion, $F_r = (2\pi I^* kT)^{1/2} \varphi/h$. This implies that $U = \theta\, kT/h$. Introducing in the conventional manner the specific rate of the elementary process $K_i = U/\theta_i$ for movement of the rod I, we have $K_i = \theta kT/(h\theta_i)$. This conclusion is completely analogous to the derivation of the formula for translational motion [14] for which v is equal to the ratio u_t/δ, where u_t is the average velocity of movement of the AC with the mass m^*, which is equal to $u_t = (kT/2\pi m^*)^{1/2}$, δ is the length of the activation barrier. On the other hand, for the translational motion $\theta^* = \theta F_r$, $F_t = (2\pi m^* kT)^{1/2}\delta/h$, so $K_i = \theta kT/(h\theta_i)$.

Ratio $\theta/\theta_i = F^*/F_i$ is expressed in terms of sums over the states for AC (F^*) and for the molecule in the ground state (F_i); this allows us to express the rate constant of the elementary motion of the gas phase in the usual form $K_i = kTF^*/(F_i h)$. As a result the type of motion of the molecule (translational or rotational) is specified only in the expressions for the sum over the states of AC and the molecule in the ground state. For a rarefied phase $F^*/F_i = F_r$. For non-ideal reaction systems this relationship is complicated because it takes account of the influence of neighbouring molecules. Their influence is seen through the lateral interactions and through the vacancy areas in the dense phase, permitting implementation of the given type of elemental movement.

The vacancy area for rotation $i \rightarrow k$ consists of the total number of free sites which the rod crosses in the transition from the state i to state k. We denote it by $S(i \rightarrow k)$. In the initial and final positions the rod blocks L vacancies, so $S(i \rightarrow k) > L$. At rotation of the molecules relative to the pore wall the vacancy area should consist of $S_w(i \rightarrow k)$ vacancies, providing rotation of the rod on some arc from state i to k. In general, $S(i \rightarrow k) > S_{my}$, and $S_w(i \rightarrow k) > S(i \rightarrow k)$.

Processes the reorientation of particle i proceed differently with respect to its rotation relative to the centre of inertia and the pore wall. In the first case, the rotation can be treated independently from the translational motion of the rod, if the origin of the moving coordinate system coincides with the centre of inertia of the rod. In the second case, both types of motion are interconnected.

Upon rotation of the rods we will only consider 'next' discrete orientational states between which 'direct' rotations (transitions) are possible.

For example, let f be the central site of the rod of type i, and g the site closest to it, such that the direction of the fg coincides with the direction of orientation of the rod. When turning the central site of the rod remains unchanged. Only the orientation of the rod and, therefore, its variety change. It is denoted by k. Through g_1 we denote the site nearest to site f, such that the direction of bond fg_1 coincides with the new direction of the rod. Direct transitions are the discrete transitions from bond fg to bond fg_1, such that the angle φ between the vectors of the original and new directions does not exceed 90°.

All rotation of the rod i will be divided into rotations with angles greater than and less than the angle $\varphi/2$.

In the first case the rotation process relates to the 'internal' processes, the index i and orientation λ_i are retained. There is a 'swing' of the molecule with respect to the centre of mass for small deviations from the orientation λ_i. In the second case the rotational process is 'external' in which the index i of the molecule is changed to k, and its orientation λ_i – to λ_k. $W_f(i \rightarrow k)$ denotes the average speed of the elementary process of the change of the orientation of the rod i, located at site f, to the rod k whose centre of mass remains in the same site f. Obviously, in general, the geometric centre coordinate of polyatomic molecules does not coincide with the coordinate of the centre of gravity, which makes indexing sites in describing the dynamics of rods more difficult. (In the description of the equilibrium distribution the real position of the numerated segment of the molecule does not play any role.)

The presence of impermeable pore walls affects the nature of the molecules in two ways: first, the wall restricts rotation angles around the centre of mass of the molecules near the walls, and secondly, it is possible to change the molecular orientation state by rotating around the atoms (or functional groups) of the surface. At the same the molecules are characterized by moments of inertia calculated from the contact with the surface. This kind of rotational motion simultaneously with the orientation of the molecules change the position of the centre of inertia, and their contribution must be taken into account when analyzing the flows moving centres of mass.

Vacancy area for rotations of large particles
Let *the* particle *i* rotates around its main axis of rotation passing through the centre of mass at the site *f*. In the condensed phase the particle rotates after it has acquired by thermofluctuational perurbations the medium reaches the activation energy of rotation and gets the momentum in the right direction. In addition, the rotation of the molecule is affected the full external torque acting on a given particle *i* due to the adsorption potential. The rate of the reorientation process can be described by analogy with the traditional way without detailing the specific contribution of collisions with neighboring particles. In this case, the reorientation rate $i \to k$ at the site of the rod *f* is expressed as

$$W_f(i \to k) = K_f^i(i \to k)V_f^i(i \to k),$$

$$K_f^i(i \to k) = (2\pi I_{ik}\beta)^{-1/2} \exp(-\beta E_f^i(i \to k))/\varphi,$$

$$V_f^i(i \to k) = \left\langle \theta_f^i(i \to k) \right\rangle \Lambda_f^i(i \to k), \qquad \text{(A7.14)}$$

$$\left\langle \theta_f^i(i \to k) \right\rangle = \theta_f^{iv*} \prod_\ell t_{\ell\ell+1}^{vv*}.$$

Here $K_f^i(i \to k)$ is the reorientation rate constant of the molecule *m* from state *i* ($i \leftrightarrow m$, λ_i) to state *k* ($k \leftrightarrow m$, λ_k) and $E_f^i(i \to k)$ is its activation energy; I_{ik} is the principal moment of inertia of the particle *i*, corresponding to its rotation about the centre of inertia with transition to state *k*, for the rod $I_{ik} = I$, where $I = ML/12$ in the continuum approximation, *M* is the mass of the rod, *L* is its length, or $I = \sum_{A \neq B} m_A m_B L_{AB}^2/M$, L_{AB} is the distance between the atoms *A* and *B* in the atomically discrete model of the rod.

Here, as above, $V_f^i(i \to k)$ is the concentration dependence of the reorientation rate $i \to k$, consisting of two factors:

1) $\left\langle \theta_f^i(i \to k) \right\rangle$ is the probability of formation of the vacancy region $S(i \to k)$, the size and shape of which are sufficient to implement the reorientation process $i \to k$. Formally, the equation for the given probability in the expression (A7.14) differs from (A7.12) by the lack of the index of site *g* and the symbol of the vacancy in it, since the rotation is relative to the site *f*;

2) $\Lambda_f^i(i \to k)$ is the function of the effect of ambient particles on the reorientation process of the particle *i*. For rotary motion the expression (A7.13) is replaced by a similar equation

$$\Lambda^i_j(i \to k) = \prod_{k=1}^{\Pi_i}{}^* \sum_{j=1}^{\Phi+1} \sum_{n=1}^{\Pi_J} t^{kn}_{ij}(fh)\exp(\beta\delta\varepsilon^{kn}_{ij}) \prod_{d=1}^{\Pi_v}{}^* \sum_{j=1}^{\Phi+1} \sum_{n=1}^{\Pi_J} t^{dn}_{vj}(fh)\exp(\beta\delta\varepsilon^{dn}_{vj}),$$

(A.7.16)

As above, in the first product symbol h refers to the neighbours j of the 'central' particle i at site f, whereas in the second product symbol h refers to the neighbours at the end of 'trajectory' of rotation surrounding the vacancy area in which there is a particle k. The asterisk in the first sign of the product means that from the set of contacts $k \in \Pi_i$ of the molecule i we excluded contacts with the neighbouring vacancy area $S(i \to k)$, in which rotation from particle i to particle k takes place. The asterisk at the second sign of the product means that from the set of contacts $d \in \Pi_v$ of the vacancy area $S(i \to k)$ we excluded contacts of the central particle k, facing the particle i.

Rotation of the particles relative to the pore wall

Upon rotation of the molecules relative to the wall its centre of inertia is displaced, i.e., the processes of displacement and rotation of the rod are inseparable. In describing the rotation of the rods relative to the pore wall the formulas (A7.14)–(A7.15) for the velocities of the elementary rotations modified. And only the form of the vacancy area $S_w(i \to k)$ changes, which provides rotation of the rod along some arc from state i to k, and in the expression for the reorientation rate constant $K^i_{fw}(i \to k)$ (A7.14) I_{ik} is replaced by the moment of inertia I^w_{ik} of the molecule relative to the atoms of the wall.

The average thermal velocity of the translational motion of the site f to the free site g is expressed through the rotational rate relative to the wall:

$$w^i_{fg} = \frac{\chi^i_{fg}(i \to k)w_{fw}(i \to k)}{\theta^i_f},$$

(A7.16)

where $\chi_{fg}(i \to k)$ is the projection of the centre of mass of the particle i in the direction of the bond fg during rotation $i \to k$; function $W_{fw}(i \to k)$ can be described modified formulas (A7.14)–(A7.15) using $K^i_{fw}(i \to k)$ and $S_w(i \to k)$.

Taking into account the restrictions on the angles of rotation in the near-wall layers, resulting from impermeability of the pore wall, changes the shape and reduces the size of the free volume, consisting of vacancies required for this type of rotation. In this case, some of the factors from the contacts of adjacent particles in the equation

(A7.12) which are 'replaced' by atoms of the wall are absent. Their contribution is replaced by the wall potential, which is accounted for in the expression for the activation energy for $E_{fg}^{iv}(\chi)$.

Vacancy region and collective motion [15–18]

Th expressions derived for the thermal velocity of the molecules are oriented on any aggregate state. They are based on individual displacements of the molecules in the field of neighbouring molecules, which remain in their places, although they may oscillate, with reciprocating rotational displacement within their cells. Strictly speaking, the motion of individual molecules occurs in the gas phase. Individual hops of atoms in a solid are always accompanied by oscillatory movements of neighbours and for them the individual hops are a convenient form of the description. For the liquid phase the model of individual hops is a convenient approximation which is well supported by comparisons with molecular dynamics studies [19] and sections 39 and 40.

The concept of the collective motion of the molecules in the liquid phase was first formulated by Fisher [15] and further developed in [16–18] and other studies. The gist of it is that, because of the strong correlation between the molecules of the liquid their motion is more complicated than originally anticipated in the Frenkel model which introduced the concept of the time for the molecule living at fixed position and its vibrations in the localized state [20]. This concept leads to a relatively small correction in the calculation of the coefficient of self-diffusion in pure liquids [16–18], however, the possibility of this type of motion can also be considered in the LGM.

The idea of collective motion can be formulated as a shift of clusters of varying size in the direction of a more rarefied local area fluid. Such a path can easily be visualized on the basis of constructed expressions for the motion of large molecules, as indicated above. A second alternative of this concept is to consider the molecular motion on a finer grid (in accordance with the conclusions of Appendix 1) for conventional 'single' molecules that are displaced over distances smaller than their diameter. This path also leads to equations similar to the case of large molecules on the grid of a variable size. Both of these features lead to the expressions described above and they can be easily studied in the framwork of the LGM.

References

1. Glasston S., Laidler K. J., Eyring H., The Theory of Rate Processes. – M.: Izd. inostr. lit., 1948. – 583 p. [Princeton University Press, New Jersey, 1941].
2. Entelis C.G., Tiger R.P., Kinetics of the reactions in the liquid phase. – Moscow: Khimiya, 1973. – 416 p.
3. Moelwyn Hughes E.A., Equilibrium and kinetics of reactions in solutions. – Moscow: Mir, 1975. – 470 p.
4. Tovbin Yu.K., Theory of physico-chemical processes at the gas-solid interface. – Moscow: Nauka, 1990. – 288 p. [CRC Press Boca Raton, FL, 1991.]
5. Tovbin Yu.K. // Progress in Surface Science. 1990. V. 34. P. 1 - 236.
6. Hirschfelder J. O., Curtiss C. F., Bird R. B., Molecular Theory of Gases and Liquids. – Moscow: Izd. inostr. lit., 1961. – 929 p. [Wiley, New York, 1954].
7. Tovbin Yu.K. // Zh. Fiz. khimii. Suppl. 1. 2005. V. 69, No. 1. P. S1.
8. Tovbin Yu.K. // Teor. osnovy khim. tekhnol. 2005. V. 39. No 5. P. 523. [Theor. Found. Chem. Engin. 2005. V. 39. No 5. P. 493]
9. Tovbin Yu.K. // Teor. osnovy khim. tekhnol. 2005. V. 39. No 6. P. 613 [Theor. Found. Chem. Engin. 2013. V. 39. No 6. P. 579].
10. Tovbin Yu.K. // Izv. RAN. Ser. khim. 2005. No. 8. P. 1717. [Russ. Chem. Bull., 2005. T. 54. № 8. C. 1768]
11. Tovbin Yu.K.// Zh. Fiz. khimii. 2006. V. 80, No. 6. P. 1134 [Russ. J. Phys. Chem., 2006. T. 80. № 6. C. 995].
12. Tovbin Yu.K. // Khim. Fiz. 2005. V. 24, No. 9. P. 91.
13. Tovbin Yu.K. // Khim. Fiz. 2006. V. 26, No.12. P. 43.
14. Glasston S., Laidler K. J., Eyring H., The Theory of Rate Processes. – M.: Izd. inostr. lit., 1948. – 583 p. [Princeton University Press, New Jersey, 1941].
15. Fisher I.Z. / / Zh. eksp. teor. fiz. 1971. V. 61, No. 4. P. 1647.
16. Fisher I.Z., Zatovsky A.V., Malomuzh N.P. // Zh. eksp. teor. fiz. 1973. V. 65, No. 1. P. 297.
17. Malomuzh N.P., Fisher I.Z. // Strukturnaya khimiya. 1973. V. 14, No. 4. P. 1105.
18. Malomuzh N.P., Troyanovskii V.S. // Zh. Fiz. khim. 1983. V. 57, No. 12. P. 2967.
19. Grivtsov A.G., Molecular dynamics method in physical chemistry. – Moscow: Nauka, 1996. – P. 16.
20. Frenkel' Ya.I., Kinetic Theory of Liquids. – Moscow: Publishing House of the USSR Academy of Sciences, 1945 [Oxford University Press, Oxford, 1946].

Appendix 8

Methods for modeling the dynamics of processes in condensed phases

This appendix discusses the basic concepts of the non-equilibrium theory, which are realized in the form of kinetic equations, and our goal is to see how the basic concepts of the kinetic theory are reflected in the different numerical techniques for modeling physical and chemical processes.

In general, the physical meaning of the process (characteristic times and sizes of local areas) is contained in the probability values of the realization of elementary processes or stages W_α (section 28). If W_α is derived using intermolecular potentials of [1–9], these are model of a molecular level, in other cases it is the effective characteristics of the elementary process that should be evaluated from experimental data. In particular, the last case includes the rate constants of the macroscopic or atomic level. Transfer of molecular information between the different levels can be carried out only through intermolecular potentials.

The starting point of the theory of non-equilibrium processes is the fact that the non-equilibrium distribution functions enable us to obtain the transport equation and calculate the kinetic coefficients through correlation functions and for the equilibrium case they are transferred to the Gibbs distribution. The main concept was proposed by N.N. Bogolyubov regarding the hierarchy of relaxation times in non-equilibrium statistical mechanics [10], which was formulated for rarefied gases and is as follows. If the initial distribution is arbitrary, the state of the system in the initial stage may be very different from the equilibrium and to describe we need to set a large number of distribution functions: not only the single- and two-particle, but also higher-order functions that are rapidly changing with time according to the Liouville equation. This stage is called the stage of 'initial chaotization'.

Next comes the 'kinetic stage' of 'synchronization' of the distribution functions when all distribution functions are completely determined by specifying the single-particle distribution function. For example, for low-density gases or weak interaction this is achieved during the time approximately equal to the collision time [11–13]. However, for dense systems the knowledge of only the single-particle distribution function is insufficient and requires knowledge of the pair distribution function [14–21]. Formally the kinetic stage is mentioned in the kinetic theory of gases but little attention has been paid to it and the hydrodynamic regime of the processes is then considered. It should be noted that the kinetic equations of the kinetic stage are rarely discussed in the literature because of the complexity of the self-consistent derivation. But for the solids and especially for processes at the interfaces of the solid body with the vapour and liquid this step is important because its range extends to the macroscopic time.

For even larger time scales (for gases – much longer free path time) the number of parameters needed to describe the state of the system is reduced and the hydrodynamic stage starts, which can be described by the hydrodynamic equations (together with the heat equation), i.e. by only a few moments of the distribution function (average number of particles, average energy and average speed). The distribution function begins to depend on time only through these parameters.

Recall that the hydrodynamic stage is characterized by a small deviation of particle distributions from the local equilibrium distribution of components through which all dissipative coefficients of the hydrodynamic equations are derived. If deviations from the equilibrium distributions are not small, it is necessary to use more sophisticated systems of equations containing the kinetic equations, at least for the pair distribution functions, since the knowledge of only the local densities is insufficient to fully characterize the state of the system

The general methodology of the non-equilibrium theory is extremely useful [22, 23]. However, its practical implementation

Table A8.1. Methods of modeling the kinetic physical and chemical processes and the characteristic times of their use

№	Method	Range time with
1	Molecular dynamics (MD)	$10^{-13}-10^{-7}$
2	Kinetic Monte Carlo (KMC)	$10^{-12}-10^{0}$
3	Brownian dynamics	$10^{-7}-10^{-5}$
4	Boltzmann equation (for gases, continuum)	$10^{-8}-10^{-4}$
5	Boltzmann equation (discrete)	$10^{-9}-10^{0}$
6	Lattice gas model (LGM)	$10^{-12}-10^{0}$
7	Microscopic hydrodynamics	$10^{-13}-10^{-5}$
8	Lattice automata	$10^{-5}-10^{0}$
9	Hydrodynamic equations	$10^{-5}-10^{0}$

meets many problems. Note the problematic use of kinetic approaches based on the statistical theory of liquids [1, 20–24]. This is due to the need to consider an instantaneous location of the neighbours for those reactants that participate in the reactions or transport (see Appendix 7). In the continual description of the state of the mixture components we must be able to self-consistently calculate the probability of many-particle high-dimensional configuration. This requirement virtually eliminates the possibility of using a theory based on integral equations [25–28]. Thus, their use even for the flow of small molecules in narrow channels (i.e., in strongly inhomogeneous systems) ended only in the equations in the mean field approximation [29, 30] that can not be considered satisfactory for the microscopic theories.

Table A8.1 is a list of basic kinetic methods used to calculate the dynamics of the processes and their time range. For practical calculations using strict integral equations it was necessary to transfer to the simplest variant, in which the correlation effects are absent – the density functional method (and in fact, the 'spoiled' version of the theory of integral equations) [31–33]; at the same time as indicated by Appendix 6, the self-consistency conditions are violated), so they are not listed in Table A8.1.

The following discussion of these methods applies only to the principle of construction of the functions W_α, which are key in the multiscale modeling of processes and the method of accounting for intermolecular interactions and correlation effects. Obviously, each method contains a lot of different techniques and algorithms, adapted to the specific types of processes and materials (their discussion is beyond the scope of this application.)

Molecular dynamics (MD)

The method of molecular dynamics (MD) [34–39] is based on the numerical solution of Newton's equations for all molecules of the system. It starts from the time of the order of 10^{-15} s in steps of $1\text{--}2 \cdot 10^{-15}$ s, to take account in the differential equations of the sharp change in the potential at a distance of 10^{-13} cm – characteristic distance of the sharp increase in the repulsive branch. Accordingly, the passage time of a molecule of its diameter at room temperature is of the order of $10^{-13}\text{--}10^{-12}$.

In the course of solving the system of equations we accumulate and analyze all the temporal and spatial data on thr positions and velocities of the molecules in different parts of the area under consideration. The method operates with the potential functions of intermolecular interactions and allows us to investigate both the velocity and spatial distributions of the molecules used to calculate all the dynamic and equilibrium characteristics of the system. This method is currently serving as the primary method of research of many molecular systems.

The principal drawback of the method is the dependence on the progress of the trajectory in two positions. 1. Any averaging in the method is carried out over some interval of time. But the total length of the trajectory in the method is not determined so with increasing length of the trajectory is necessary to change the weight of the already passed time intervals. This fact affects the overall normalization condition in any averaging procedure. 2. Although any trajectory, starting from a fixed initial state, 'forgets' after a certain time its initial position, however, in situations with strong intermolecular interactions in many situations, the system can not get far from the initial state. This means violation of the ergodicity condition used when replacing the averaging over ensembles copies by time averaging as required by the Gibbs statistical mechanics [40]. These constraints lead to the fact that the method of the MD, being in principle a statistically accurate method, may in practice give equivocal results, especially in the case of highly heterogeneous systems (see chapter 3).

The MD method usually works until the time of $10^{-8}\text{--}10^{-7}$ s. Next we go to the Brownian or Langevin dynamics [41, 42].

Kinetic (or dynamic) Monte Carlo method (KMC)

The kinetic Monte Carlo method (KMC) is based on the use of equation (27.1) for the total distribution function [43–48], in the same way as in the LGM. This method is stochastic. To use it we need to formulate rules for the formation of the trajectory of stepwise motion in time. Each step involves enumeration of all particles in the system and estimation of the probability to change the occupation status of each site. These rules are directly related to the probability of realization of the elementary stage W_α depending on the conditions of occupation of all the nearest neighbours. The meaning of the transition probabilities W_α in the KMC is the same as in the LGM. Therefore, all the constructions discussed above in the equations in section 27 are important for them. Dynamics of the total system is composed of the rates of elementary processes involving all components of the system. Depending on how the expressions for the elementary stages are derived, the theory may start with the different minimum evolution time and take into account these or other microscopic features of the molecular system.

The method was used in the study of adsorption–desorption processes, surface reactions, stages of transport of substances inside the crystals and along the surface, during crystal growth, the formation of a new phase and for multistage physico-chemical processes, etc. Currently the KMC method is the traditional numerical method of statistical physics. This method seriously competes with the theory of kinetic equations, since the difficulty of their solutions increases with the size of the system. The KMC algorithms are fairly simple and allow to obtain solutions in a standard way, so the method has been used successfully in all areas, including critical. In many cases, the differences between the obtained and the known exact solutions are insignificant. This has led some researchers to adopt this method as accurate. Unfortunately, this assessment is overly optimistic. As the MD method, the KMC method is approximate, and the accuracy of its calculations is strongly dependent on the properties of the system. The problem is that in this method the weight of the calculated averages varies consistently with increasing path length. The 'quality' of the trajectory is determined by the physical principles laid down in the transition probabilities W_α, and depends on the account of the local state of the considered site. In case of strong intermolecular interactions local condensation of molecules takes place quite easily (subdomains with different states of occupation) and the exchange between them is difficult, i.e. the ergodicity hypothesis is violated.

In KMC we are usually limited only to the implementation of requirements (27.2) on the probability W_α without reference to the absolute reaction rate theory (see, e.g., [49–51]). The parameters frequently used in the functions W_α have no physical meaning, although they may qualitatively reproduce quite complicated dynamic situations in catalytic processes. Such a situation in modeling has developed on the basis of studies describing the process of oxidation of CO molecules on the basis of the law of acting masses [52, 53]; 'piecewise' dependences are constructed for the rates of elementary reactions which hold for individual concentration regions, and 'switches' are introduced to change these relationships with changes in the degree of filling of the reagents.

Formally, the KMC and LGM can work in a wide range of time with the characteristic value of this time determined by the rate constants of the elementary processes, which play the role of kinetic parameters. These constants acquire physical meaning only when lined with the intermolecular potentials.

Brownian (or Langevin) dynamics [41, 42]

At large times of the molecular dynamics simulation the small particles tend to local equilibrium. If the system additionally contains large particles, their characteristic times of transition to local equilibrium have significantly higher values. Therefore, the system can be quasi-equilibrium for rapid (mobile) components, but to non-equilibrium for large (slow) molecules. During further movement along the trajectory there is very slow overall relaxation of the system to its equilibrium state. This process can be accelerated if the original system of the equations of motion of all components is reduced and we retain only the slow components. From Newton's equations we transfer to Langevin equations containing in the right-hand part the random forces and friction coefficients.

This transition is usually associated with the loss of the initial molecular information on the atom–atom potentials, which should be partially stored in the form of friction coefficients reflecting the molecular features of the system being studied. In general, the number of coefficients can be quite large (it depends on the purpose of the study), but the primary information about the structure of the system is lost, and the friction coefficients already introduced are secondary characteristics and depend on the method of their introduction. In this regard, the most consistent way of introducing the friction coefficient is a direct numerical calculation of the initial MD task and determination of the basis of this trajectory of equilibrium and quasi-equilibrium correlation functions of pulses, forces, etc., through which they are defined [41, 42].

Boltzmann equation (for gases, continuum)

The Boltzmann equation has a lower limit for the characteristic time for gas of the order of 10^{-9} s [54, 55]. Real times of using the Boltzmann equation for gas are longer than 10^{-9} s. It is believed that the process of averaging in time captures the interval of the order of up to 100-fold the lower value for the characteristic time interval, which corresponds to times from 10^{-7} and higher. At numerical implementation of solutions of the Boltzmann equation an important role is played by the methods of averages right-hand sides in the space velocities of the molecules of which the most practical interest is the Monte Carlo method. This approach has been very widely used for gas mixtures (polyatomic and chemical reaction).

Boltzmann equation (discrete)

The discrete Boltzmann equation has been used widely in recent years [56, 57]. Its appearance was associated with the desire to simplify the full Boltzmann equation to the maximum extent. Later this method was transferred to the related problems in dense phases.

The system consists of a plurality of cells between which there is a transfer of molecules and in which chemical reactions are allowed. Formally, this technique allows us to consider the majority of atomic and molecular processes as its basis reflects the multistage nature of real processes: molecular collisions (with or without reaction) and the transfer of molecules from one system of particles to another. Unlike the continuum approach, the process inside the cells is artificially made stochastic that allows to model denser systems than the gas phase in the continuum approach. At the same time the unambiguous relationship between the dynamics of movement and distribution of molecules in space is lost. Accordingly, the connection with with the intermolecular potentials in construction of W_a is also lost. The principal disadvantage of this approach is the *a priori* loss of spatial correlation effects of molecules that are essential for the description of the dense phases at the microscopic level.

The lattice gas model (LGM)

The kinetic equations in the LGM described above are based on discrete distribution functions (section 27). They form the basis of methods focused on the condensed phases. This molecular theory is based on the use of atom–atom potentials, so the theory, as well as all existing stochastic numerical MD and MC methods, use a conventional set of energy parameters of the atom–atom potentials. These potentials are now considered to be known. By its opportunities this approach most closely fits the kinetic Monte Carlo: it allows you to reflect all the stages of evolution of dynamic systems from short times of the order of 10^{-12} s, including the kinetic stage, up to macroscopic times. The whole dynamics of the system consists of the dynamics of elementary processes involving all components of the system. The rates of the elementary stages are built in a similar way, so the theory can start with the minimal evolution time and reflects the microscopic features of the system.

Microscopic hydrodynamics

Microscopic hydrodynamics is based on the LGM equations [58, 59]. It uses only the intermolecular potentials and is adapted to calculate the flow of gases and liquids near solid surfaces. At its core, this approach describes the state of the system close to the quasi-equilibrium distribution of molecules in strong surface fields of the solids. This method is described in more detail in chapter 4.

Lattice automata

The lattice automata method has been proposed to simplify the stochastic calculations in the dynamic MC about three decades ago [60–65]. It is based on the idea of numerical analysis techniques plurality of discrete cells, each of which can be in the discrete state of occupation. All cells are treated at each time step in a random sequence. Changing the conditions of occupation of each cell is controlled by the selection rules that depend on the states of occupation of the neighbouring cells of employment. With regard to the selection of rules for the implementation of elementary events, this method is similar to the kinetic Monte Carlo or LGM method. However, in this approach the probabilities themselves W_{α} are not associated with the intermolecular interactions or other molecular properties of the system.

The main objective of this approach is in tracking the spatial distribution of the particles and their spatial structure. Each cell may refer to the micro-, meso- or macro-area, depending on the meaning that is attributed to the values of W_α. The values of W_α themselves W_α may be deterministic or stochastic. In the latter case the 'noise' from the operation of the method of lattice automata plays the role of temperature in the equilibrium system and the method itself can serve as a tool for analyzing the phase formation in the non-equilibrium conditions [66].

Most often the main goal of such work is the temporal analysis of structural changes of materials at the macroscopic level. The method is widely used in studying the dynamics of the kinetics of phase transitions in terms of the theory of dynamic systems to describe the growth dendrites, the dynamic of biological and other systems [67–69]. In more detail the current state of the method is described in [65].

Hydrodynamic equations
These equations, as noted above, include the equations of hydrodynamic flow regimes in gases and liquids, describing the dynamics of the system in terms of the concepts of the theory of continuous media. More generally, this type of equations includes the following types of kinetic equations: 1) actual hydrodynamic Navier–Stokes equations or analogues for complex molecular systems [70–75], 2) equations of chemical kinetics under ideal models (law of mass action) which operate only with the single-particle distribution functions (concentrations of reactants) [76–80], 3) the equations of classical thermodynamics to simplify the calculations of the nucleation processes [81–84], 4) equations of the thermodynamics of irreversible processes, which include items 1–3 [85, 86], and 5) equations of the dynamics and equilibrium in mechanics for *deformation states* of solids [87–91].

In fact, all of these equations now form the basis of thermodynamics of irreversible processes, which includes items 1–3 and all types of molecular mechanisms of transport processes in addition to the convective transport. The equations of chemical kinetics in the ideal models, i.e. using the law of acting masses and the information only about the concentrations of the reactants are also the equations at the hydrodynamic level at any value of the system volume. The equations of classical thermodynamics are often used for the calculation of the nucleation process to simplify and omit the calculation of the stages of condensation and desorption of single molecules to the emerging phase. The new phase (drop) itself is described by the excess free energy functions (through surface tension).

Actually the hydrodynamic equations are well known for any macroscopic times and macroscopic systems [1, 70–75]. In practice, the lower limit is in the range of from 10^{-5} to 10^{-4} s because of the need to link with the observed experimental data. In recent years, very often these equations have been formally transferred to arbitrary small times and sizes of areas, not tracking the physical meaning of the input parameters, which are traditionally taken from macroscopic experiments and applies on the scale of 1–2 nm [92, 93].

For hydrodynamic modeling level there is a very large number of specific algorithms for specific processes [94, 95]. Among them we can mention the finite element method [95–97], which is focused on calculations of systems with complex geometric configuration and irregular physical structure. In the finite element method the problem of finding the function is replaced by the problem of finding a finite number of its approximate values at specific points – sites. This may explain the view that the finite element method is the grid method designed for solving microlevel problems, for which the model of the object is defined by a system of differential equations in partial derivatives with given boundary conditions.

A detailed discussion of different hydrodynamic techniques is beyond the scope of this section. The meaning of this discussion is the separate models of the molecular level and show the extent to which the macroscopic equations can be helpful or misleading at the transition to the nanosized objects such as the porous system.

The dynamics and equilibrium equations in mechanics for *deformation states* of solids are traditionally discussed in terms of the equations of elasticity theory [87–91] which use the same algorithms as in hydrodynamics. The general nature of these hydrodynamic equations follows from the fact that all the equations of thermodynamics, hydrodynamics and mechanics belong to the science of continuous media and their application conditions are interrelated. The transition out of the continuum approximation and accounting the discrete structure of matter in the description of small systems are discussed in sections 73 and 74.

References

1. J. Hirschfelder, Curtis H., R. Bird, Molecular Theory of Gases and Liquids. — Moscow: Izd. lit., 1961.
2. Kaplan GN Introduction to the theory of intermolecular interactions. — Moscow: Nauka, 1982. — 311.
3. Steele W. A. The Interactions of Gases with Solid Surfaces. — N.Y.: Pergamon, 1974.
4. Kiselev AV Intermolecular interactions in adsorption chromatography. — M.: Higher School, 1986. — 360.
5. Intermolecular Forces // Adv. Chem. Phys. / Ed. by J. O. Hirschfelder. — N. Y.: Interscience Publ., 1967. — V. 12. — 643 p.
6. Intermolecular interactions: From diatomic molecules to biopolymers / Ed. B. Pullman. — New York: Wiley, 1981. — 592 p.
7. Molecular interactions / Ed. H. Ratajczak and Orville H. Thomas. — New York: Wiley, 1984. — 600.
8. Bakhshiev NG Spectroscopy of intermolecular interactions. — Leningrad: Nauka, 1972. — 264.
9. Kitaygorodsky AI Molecular crystals. — Moscow: Nauka, 1971. — 424.
10. NN Bogolyubov, Problems of dynamical theory in statistical physics. — Moscow: Gostekhizdat 1946. — 96.
11. NS Krylov Works on the foundations of statistical physics. — Moscow: Publishing House of the USSR Academy of Sciences, 1950. — 207.
12. Klimontovich Yu L. Statistical Physics. — Moscow: Nauka, 1982. — 608.
13. DN Zubarev, static -equilibrium thermodynamics. — Moscow: Nauka, 1971. — 416.
14. Tovbin JK Theory of physical and chemical processes at the gas -solid. — Moscow: Nauka, 1990. — 288.
15. KP Gurov, Kartashkin BA Ugaste EY Interdiffusion in multiphase metallic systems. — Moscow: Nauka, 1981. — 350.
16. Tovbin JK // Dokl. USSR Academy of Sciences. 1984. T. 277, № 4. S. 917.
17. Saito Y., Kubo R. // J. Stat. Phys. 1976. V. 15. P. 233.
18. Vlasov LS, Schneider VE // Zh. exp. and theor. physics. 1977. T. 73. S. 1493.
19. Binder K., Staiffer D. // Adv. Phys. 1976. V. 25. P. 343.
20. Eyzenshits P. Statistical theory of irreversible processes. — Moscow: Izd. lit., 1963.
21. Croxton K. Physics of the liquid state. — New York: Wiley, 1979 (Croxton CA Liquid State Physics — A Statistical Mechanical Introduction. — Cambridge: Cambridge Univ. Press, 1974).
22. DN Zubarev, VG Morozov, G. Röpke, Statistical Mechanics of nonequilibrium processes. Vols 1 and 2. — Moscow: Fizmatlit 2002.
23. Ottinger H. C. Beyond Equilibrium Thermodynamics. — Hobeken New Jersey, Wiley, 2005.
24. Rowlinson, B. Widom, Molecular Theory of Capillarity. — New York: Wiley, 1986.
25. Mayer, J., M. Goeppert-Mayer, Statistical Mechanics: Per. from English. / Ed. DN 3ubareva. 2nd ed. — New York: Wiley, 1980. — 544.
26. Oray C. O., Gubbins K. E. Theory of Molecular Fluids. — Oxford: Clarendon Press, 1984. — V. 1. — 626 p.
27. Fisher IZ Statistical Theory of Liquids. — Moscow: Fizmatgiz, 1961. — 200 c.
28. TL Hill, Statistical Mechanics: Per. from English. / Ed. SV Tyablikov. — Moscow: Izd. lit., 1960. — 485.

29.   Pozhar L. A., Gubbins K. E. // J. Chem. Phys. 1991. V. 94. P. 1367.
30.   Pozhar L. A., Gubbins K. E. // J. Chem. Phys. 1993. V. 99. P. 8970.
31.   Sokolowski S., Fischer J. // Molec. Phys. 1990. V. 71. P. 393.
32.   Evans R., Marconi U. M. B., Tarazona P. // J. Chem. Phys. 1986. V. 84. P. 2376.
33.   Evans R. // J. Phys.: Condens. Matter. 1990. V. 46. P. 8989.
34.   Grivtsov AG molecular dynamics method in Physical Chemistry / Ed. YK Tovbin. — Moscow: Nauka, 1996. — S. 16.
35.   Molecular dynamics method in Physical Chemistry / Ed. YK Tovbin. — Moscow: Nauka, 1996. — 334.
36.   Allen M. P., Tildesley D. J. Computer Simulation of Liquids. — Oxford: Claredon Press, 2002.
37.   Haile J. M. Molecular Dynamics Simulation: Elementary Methods. — N. Y.: Wiley, 1992.
38.   Molecular dynamics simulation of statistical mechanics systems / Eds. G. Coccotti, W. G. Hoover. — Amsterdam: North-Holland, 1986. — 610 p.
39.   Evans D. J., Morriss G. P. Statistical Mechanics of Nonequilibrium Liquids. Second Edition. — Cambridge: Cambridg Univ. Press, 2008.
40.   Gibbs JW Thermodynamics. Statistical mechanics. — Moscow: Nauka, 1982. — 584 p.
41.   Lemak A. S., Balabaev N. K. // Molecular Simulation. 1995. V. 15. P. 223.
42.   Lemak A. S., Balabaev N. K. // J. Comp. Chemistry. 1996. V. 17. P. 1685.
43.   Monte Carlo methods in statistical physics / Ed. K. Binder. — M.: Mir, 1982. — 400.
44.   Nicolson D., Parsonage N. G. Computer Simulation and The Statistical Mechanics of Adsorption. — N. Y.: Acad. Press, 1982.
45.   Allen M. P. Introduction to Monte Carlo simulations // In Observation, Prediction and Simulation of Phase Transitions in Complex Fluids / Eds.: M. Baus, L. F. Rull, J.-P. Ryckaert. — Boston: Kluwer Acad. Publishers, 1995. — 339 c.
46.   Jorgensen W. L. Monte Carlo simulations for liquids // Encyclopedia of Computational Chemistry / Ed. P. V. R. Schleyer. — N. Y.: Wiley, 1998. — P. 1754.
47.   Jorgensen W. L., Tirado-Rives J. // J. Phys. Chem. 1996. V. 100. P. 14508.
48.   Binder K., Landau D. P. // J. Chem. Phys. 1992. V. 96. P. 1444.
49.   Gorodetskii V. V., Elokhin V. I., Bakker J. W., Nieuwenhuys B. E. // Catalysis Today. 2005. V. 105. P. 183.
50.   Erohin VI, EI Lukin, Matveeva. V. Gorodetsky VV // Kinetics and Catalysis. 2003. T. 44. S. 692.
51.   Matveev A. V., Latkin E. I., Elokhin V. I., Gorodetskii V. V. // Chem. Engin. J. 2005. V. 107. P. 181.
52.   Imbihl R., Cox M. P., Ertl G. et al. // J. Chem. Phys. 1985. V. 83. P. 1578.
53.   Moller P., Weizl K., Eiswirth M. et al. // J. Chem. Phys. 1986. V. 85. P. 5328.
54.   Bird G. Molecular Gas Dynamics. — New York: Wiley, 1981. — 320.
55.   Nonequilibrium phenomena: the Boltzmann equation / Ed.: J. Liebowitz, E. W. Montroll. — New York: Wiley, 1986. — 269 with.
56.   Gardiner KV Stochastic methods in the natural sciences. — New York: Wiley, 1986. — 526 p.
57.   Succi S. The lattice Boltzmann equation for fluid dynamics and beyond. — Oxford: Oxford Univ. Press, 2001.
58.   Tovbin JK // Chem. physics. 2002. T. 21, № 1. C. 83.
59.   Tovbin JK // Zh. nat. Chemistry. 2002. T. 76, № 1. Pp. 76.
60.   Kier L. B., Seybold P. G., Cheng C.-K. Cellular Automata Modeling of Chemical

Systems. — Dordrecht: Springer, 2005.
61. Wolfram S. // Cellular Automata. Los Alamos Sci. 1983. V. 9. P. 2.
62. von Neumann J. // Theory of Self-Replicating Automata / Ed. A. Burks. — Urbana: University of Illinois Press, 1966.
63. Zuse K. The Computing Universe // Intern. J. Theor. Phys. 1982. V. 21. P. 589.
64. Toffoli T., Margolus N. Cellular Automata Machines. — Cambridge, MA: The MIT Press, 1987.
65. Wolfram S. A New Kind of Science. — Wolfram Media, Champaign, IL, 2002.
66. Bak P. How Nature works. The Science of Self-Organized Criticality. — Oxford: Oxford University Press, 1997.
67. Langer J. S. // Rev. Mod. Phys. 1980. V. 52. P. 1.
68. Rosen R. // Prog. Theor. Biol. 1981. V. 6. P. 161.
69. Cellular Automata / Eds.: D. Farmer, T. Toffoli, S. Wolfram S. — N. Y.: North-Holland, 1984.
70. Landau LD, Lifshitz EM Theoretical Physics. VI. Hydrodynamics. — Moscow: Nauka, 1986. — 733 p.
71. R. Bird, W. Stewart, E. Lightfoot, Transport Phenomena. — Moscow: Khimiya, 1974. — 687 c.
72. Collins R. fluid flow through porous materials. — New York: Wiley, 1964. — 350.
73. Sheydegger AE physics of fluid flow through porous media. — M.: Gostopizdat, 1960. — 250.
74. Nigmatulin RI Fundamentals of mechanics of heterogeneous media. — Moscow: Nauka, 1973. — 336.
75. Nicholas V. Mechanics of porous and fractured media. — Moscow: Nedra, 1984. — 232.
76. Glasstone S., Leydler K., Eyring, Theory of absolute reaction rates. — Moscow: Izd. Lita. 1948. — 583.
77. Entelis C. G., Tiger RP kinetics of the reactions in the liquid phase. — Moscow: Khimiya, 1973. — 416.
78. Melvin Hughes EA Equilibrium and kinetics of reactions in solution. — New York: Wiley, 1975. — 470.
79. Eremin EN Fundamentals of chemical kinetics. — Moscow: Higher School, 1976. — 374 p.
80. Kiperman SL Fundamentals of chemical kinetics in heterogeneous catalysis. — Moscow: Khimiya, 1979. — 350.
81. M. Volmer, Flood N. // Z. phys. Chem. A. 1934. V. 170. P. 273.
82. M. Volmer kinetics of formation of a new phase. — Moscow: Nauka, 1986.
83. Fuchs NA Mechanic aerosols. — Moscow: Khimiya, 1959. — 500.
84. Lushnikov AA Sutugin AG // Usp. 1976. T. 45. S. 385.
85. Groot C., P. Mazur, Non-equilibrium thermodynamics. — New York: Wiley, 1964.
86. Haase R. Thermodynamics of irreversible processes. — New York: Wiley, 1967.
87. LI Sedov, Mechanics of Continua. T. 1. — Moscow: Nauka, 1970. — 492.
88. Landau LD, Lifshitz EM Theoretical Physics. VI. Theory of Elasticity. — Moscow: Nauka, 1987. — 246.
89. Introduction to micromechanics / Ed. M. Onami. — Moscow: Metallurgy, 1987. — 280.
90. Theodosiou K. Elastic models of crystal defects. — New York: Wiley, 1985. — 352.
91. Ions VN, Selivanov VV Dynamics destruction deformable body. — M.: Mechanical Engineering, 1987. — 270.

92.   Persson B. N. J. // Phys. Rev. B. 1994. V. 50. P. 5590.
93.   Persson B. N. J. // J. Chem. Phys. 1995. V. 103. P. 3849.
94.   Fletcher K. Methods in Computational fluid dynamics. Vols 1 and 2. — New York: Wiley, 1991.
95.   Hoffman K. A., Steve T. C. Computation Fluid Dynamics. V. 1–3. — Wichite, Kansas: Engineering Education System, 2000.
96.   G. Strang, G. Fix theory of finite element method. — New York: Wiley, 1977. — 349 p.
97.   Zinkevych O., Morgan K. Finite elements and approximation. — New York: Wiley, 1986. — 318.

Appendix 9

Kinetic equations of gas and liquid

The transport of molecules through porous materials without chemical reactions between molecules or with the walls of the channel is much like the fluid flow inside macroscopic channels or in flow around macroscopic bodies. The starting position for constructing transport equations are the Liouville equations which operate with the full distribution function of the molecules of the system, and to describe the state of each molecule they use six variables that describe its spatial coordinates and velocities [1–4]. The same principle is used for discrete distribution functions.

The state of the system at time t is described by the full distribution function $\theta_{(N)}\left(\{i,f,\mathbf{r}_f^i,\mathbf{v}_f^i\},t\right)$, which characterizes the probability of a particle of species i (state of occupation of the site f), of being in a cell with number f, $1 \le f \le N$ (full list of the cells in given in the braces $\{\ \}$), at point $\mathbf{r}_f^i$ and having velocity $\mathbf{v}_f^i$. This designation corresponds to the total distribution function $P(\{I\},\, t)$ of section 27 for which velocities were not determined $\mathbf{v}_f^i$.

The local distribution functions in space $\theta_{(N)}\left(\{i,f,\mathbf{r}_f^i,\mathbf{v}_f^i\},t\right)$ used here differ from the continual distribution functions by their normalization in the cell volume (not the volume of the system). They are not determined outside the cells under consideration: $\theta_{(N)}\left(\{i,f,\mathbf{r}_f^i,\mathbf{v}_f^i\},t\right)_{\mathbf{r}_{f\alpha}\notin\omega_f} = 0$, where $1 \le f \le N$, $\alpha = x, y, z$; $\mathbf{r}_f \in \omega_f$ (here symbol ω_f is used for the volume of the cell number with number f instead of v_f, to distinguish this symbol from the particle velocity). The boundary conditions in the space of velocities coincide with the usual conditions: these functions vanish for aspiration module infinity velocity $\theta_{(N)}\left(\left\{i,f,\mathbf{r}_f^i,\mathbf{v}_f^i\right\},t\right)_{\mathbf{v}_{f,\alpha}=\pm\infty}=0$

To describe the dynamics of the system we should go to a chain of coupled BBGKY equation (Bogoliubov–Born–Green–Kirkwood–Yvon) for the reduced functions $\theta_{(s)}\left(\left\{i,f,\mathbf{r}_f^i,\mathbf{v}_f^i\right\},t\right)$, s is the order of the distribution function, which connects all of the distribution functions $1 \leq s \leq N$ [1–4].

The chain equation for the function $\theta_{(s)}$ can be written as $L^0_{x_1,\dots,x_s}\theta_{(s)}$

$$-\sum_{1\leq i,j\leq s}B_{ij}\theta_{(s)}=n\sum_{1\leq i\leq s}\int B_{i,s+1}\theta_{(s+1)}dx_{s+1},$$ where the following designations

are used: $L^0_{x_1,\dots,x_s}=\dfrac{\partial}{\partial t}+\sum_{1\leq i\leq s}\left(\mathbf{v}_i\dfrac{\partial}{\partial r_i}+F_0\dfrac{\partial}{\partial p_i}\right)$, $F_0=\dfrac{\partial u(r_i)}{\partial r_i}$ is the external

force, $u(\mathbf{r}_i)$ is the potential energy of the molecule i in the external field, $x_i = (\mathbf{r}_i, \mathbf{v}_i)$, $\mathbf{v}_i$ is the velocity of the molecule i, $\mathbf{p}_i$ is its momentum; $B_{ij}=\dfrac{\partial\varepsilon_{ij}}{\partial r_i}\dfrac{\partial}{\partial p_i}+\dfrac{\partial\varepsilon_{ij}}{\partial r_j}\dfrac{\partial}{\partial p_j}$, ε_{ij} is the intermolecular interaction

potential between the molecules of type i and j.

Due to the extreme complexity of the chain of equations for the sequence of the distribution functions $\theta_{(1)}$, $\theta_{(2)}$,... we naturally desire to obtain the approximate closed equations for the simplest distribution functions which is achieved by 'a coarser' description of the processes in the system.

We will use the quasichemical approximation for considering the intermolecular interactions, which allows the probability of any configuration of the molecules, i.e., the full distribution function, to be represented through local unary $\theta_{(1)}\left(\mathbf{r}_f^i,\mathbf{v}_f^i,t\right)$ and paired $\theta_{(2)}\left(\mathbf{r}_f^i,\mathbf{v}_f^i,\mathbf{r}_g^j,\mathbf{v}_g^j,t\right)$ distribution functions, $1\leq f, g \leq N$, in the form [5–7]:

$$\theta_{(N)}\left(\left\{i,f,\mathbf{r}_f^i,\mathbf{v}_f^i\right\},t\right)=\prod_{f=1}^{N}\theta_{(1)}\left(\mathbf{r}_f^i,\mathbf{v}_f^i,t\right)\prod_{g\in z_f}\left\{\xi_{fg}^{ij}\left(\mathbf{r}_f^i,\mathbf{v}_f^i,\mathbf{r}_g^j,\mathbf{v}_g^j,t\right)\right\}^{1/2},$$

$$(A9.1)$$

where the exponent 1/2 takes into account that the pairs of cells are listed twice, the index g runs through all z neighbours of site f; $\xi_{fg}^{ij}\left(\mathbf{r}_f^i,\mathbf{v}_f^i,\mathbf{r}_g^j,\mathbf{v}_g^j,t\right)=\theta_{(2)}\left(\mathbf{r}_f^i,\mathbf{v}_f^i,\mathbf{r}_g^j,\mathbf{v}_g^j,t\right)/\left[\theta_{(1)}\left(\mathbf{r}_f^i,\mathbf{v}_f^i,t\right)\theta_{(1)}\left(\mathbf{r}_g^j,\mathbf{v}_g^j,t\right)\right]$ — the pair correlation function (below the time argument t is omitted for simplicity).

For the complete set of these local distribution functions we derive the kinetic equations. Given the normalization relations for the single-component substance it is sufficient to construct the kinetic equations for local unary $\theta_{(1)}(x_f)$ and pair $\theta_{(2)}(x_f, x_g)$ distribution functions, which have the 'usual' form [1–4].

The kinetic equations for the unary functions can be written as [5–7]

$$\left(\frac{\partial}{\partial t} + v_f \frac{\partial}{\partial \mathbf{r}_f} + \frac{F(f)}{m} \frac{\partial}{\partial \mathbf{v}_f} - \sum_h \int \frac{\partial \varepsilon_{fh}}{\partial \mathbf{r}_f} \frac{\partial t_{fh}(x_f, x_h)}{m \partial \mathbf{v}_f} dx_h \right) \theta_{(1)}(x_f) =$$

$$= J = \int \frac{\partial \varepsilon_{fg}}{\partial \mathbf{r}_f} \frac{\partial \theta_{(2)}(x_f, x_g)}{m \partial \mathbf{v}_f} dx_g, \quad (A9.2)$$

where ε_{fg} is the interaction potential function of the molecules in the cells f and g; m is the mass of the molecule; $\mathbf{F}(f)$ is the vector of the external conservative force in cell f (in narrow pores the main contribution comes from the potential of the walls, and the gravitational field can be neglected). Here we use the numerical density θ instead of the traditional mass density $\rho (\rho = m\theta/v_0)$.

The sum over h is taken over all neighbours of the site f and describes the terms created by the interactions of the neighbouring molecules in the neighbourhood of the site f with which the molecule in the site f does not collide in the given time period (so-called Vlasov contributions). The value of this time period is determined by the change of time on the left side of the kinetic equation in the derivative $\partial/\partial t$. The structure of these members is completely similar to the collision integral on the right. The intermolecular interaction potential ε_{fg} corresponds to the locations of the molecules at the sites h at a distance to R coordination spheres. For rarefied gases, this term is absent, then the formula (A9.2) becomes the Boltzmann equation. The presence of neighbours greatly complicates the form of the kinetic equation and to solve we must know the pair distribution functions. Usually in the theory of solids and plasma [4, 8], these terms are considered in the mean field approximation, in which the closure occurs at the level of the unary distribution functions, i.e., instead of the functions $\partial \theta_{(2)}(x_f, x_g)/\partial \mathbf{v}_f$ we consider the derivatives $\partial [\theta_{(1)}(x_f)\theta_{(1)}(x_g)]/\partial \mathbf{v}_f$. If we neglect the contribution of the collision integral J, we obtain the Vlasov equation [9, 10]. In this case, for the liquid–vapour system we considered the short-range potential LD, and therefore it is necessary to keep both types of terms. As the density increases the role of the Vlasov contribution also increases.

However, as shown above, the unary distribution functions do not provide a self-consistent description of systems with a wide range of density changes, so we need to keep the pair distribution functions.

The kinetic equations for the pair distribution functions are written in a similar manner [5–7]

$$\left(\frac{\partial}{\partial t} + v_f \frac{\partial}{\partial q_f} + \frac{F(f)}{m} \frac{\partial}{\partial v_f} + v_g \frac{\partial}{\partial r_g} + \frac{F(g)}{m} \frac{\partial}{\partial v_g} - \frac{\partial \varepsilon_{fg}}{\partial r_f} \frac{\partial}{m \partial v_f} - \frac{\partial \varepsilon_{fg}}{\partial r_g} \frac{\partial}{m \partial v_g} \right) \times$$

$$\times \theta_{(2)} \left(x_f, x_g \right) - \frac{1}{m} \sum_{\xi} \int \left\{ \frac{\partial \varepsilon_{f\xi}}{\partial r_f} \frac{\partial}{\partial v_f} + \frac{\partial \varepsilon_{g\xi}}{\partial r_g} \frac{\partial}{\partial v_g} \right\} \theta_{(3)} \left(x_f, x_g, x_\xi \right) dx_\xi =$$

$$= \frac{1}{m} \int \left\{ \frac{\partial \varepsilon_{fh}}{\partial r_f} \frac{\partial}{\partial v_f} + \frac{\partial \varepsilon_{gh}}{\partial r_g} \frac{\partial}{\partial v_g} \right\} \theta_{(3)} \left(x_f, x_g, x_h \right) dx_h, \qquad (A9.3)$$

Potential function ε_{fg} refers to a pair of the considered molecules in the sites f and g. In the sites ξ there are neighbouring molecules that interact with (simultaneously or separately, depending on the distance) the molecules of the sites f and g, but do not interfere with them during the considered time period. As above, the role of Vlasov contributions increases with increasing density of the molecules.

The system of equations (A9.2) and (A9.3) closes the system kinetic equations at the level of the pair functions in the superposition approximation $\theta_{(3)}(x_f, x_g, x_\xi) = \theta_{(2)}(x_f, x_g)\theta_{(2)}(x_f, x_\xi)\theta_{(2)}(x_g, x_\xi)/\theta_{(1)}(x_f)\theta_{(1)}(x_g)\theta_{(1)}(x_\xi)$ and describes the dynamics of the non-uniform system. This system differs from the system (27.6) and (27.7) in the LGM by taking molecular velocities into account.

We discuss the conditions under which the solutions of these equations are obtained for a rarefied gas, using the example of the Boltzmann equation, and the Kirkwood approach for a liquid [11].

Rarefied gas
In the Boltzmann equation the right-hand side of (A9.2), which determines the collision integral, can be written as

$$J = \int \int \int (\theta_1' \theta_2' - \theta_1 \theta_2) |p_1 - p_2| \sigma \, dp_2 dp_1' dp_2', \qquad (A9.2a)$$

where σ is the collision cross section, the subscripts 1 and 2 correspond to the numbers of the colliding particles and the upper symbol 'prime' means that the function refers to a particle after the collision. The molecules are considered as hard spheres, and the integral takes into account only the repulsive branch of the interaction potential of the colliding molecules. The momenta of the scattered molecules are linked with the momenta of the colliding molecules by the mechanics equations. The formula (A9.2a) suggests that the time scale is much larger than the transit time of the colliding molecules in the sites f and g of the distance at which the molecules interact with each other. It is also believed that the radius of interaction between the molecules is much smaller than the mean free path of the colliding molecules. Values $\mathbf{v}_f$ change rapidly in collisions in magnitude and direction, and the momentum and energy of the particles also change. The derivation of equation (A9.2a) is based on the molecular chaos hypothesis, allowing to neglect the correlations in the distribution of molecular velocities and providing irreversible relaxation processes.

A number of methods have been developed to solve the Boltzmann equation; they are base on the expansion of the collision integral $J\left(\theta_{(1)}(\mathbf{v})-\theta^0\right)$ for small deviations of the non-equilibrium distribution function $\theta_{(1)}(\mathbf{v})$ relative to the equilibrium Maxwell distribution function $f^0(v)=n\left(m/2\pi k_B T\right)^{1/2} \times\exp\left[-\beta m\left(v_\alpha-u_\alpha\right)^2/2\right], \beta=\left(k_B T\right)^{-1}$, here n is the concentration of the molecules, v_α is the component of the thermal velocity, u_α is the component of the hydrodynamic velocity $u_\alpha=\langle v_\alpha\rangle$.

In the Chapman–Enskog equations the non-equilibrium distribution function is expanded in terms of small deviations from the equilibrium distribution function with respect to the gas-dynamic variables: density θ, macroscopic velocity u_i (three components) and temperature T through the function of the coordinate r and time t.

Much attention has been paid to the development of the theory for dense gases, generalizing the Enskog method. This question became dependent on the concept of the principle of total attenuation of correlations at the initial time which, as shown by Bogolyubov, is necessary to justify the Boltzmann equation. Generalization to the paired functions was performed in [12]. It was believed that successively increasing the number of correction terms in the expansion of the density of subsequent terms of the series, we can obtain the kinetic equations for gas in any approximation in density. At each step we use the assumptions about the complete attenuation of the initial correlations at times much shorter than the collision time τ_{col}, and about the continuity of the process of collisions. However, the implementation of this program resulted in fundamental difficulties (see [13]): it appears that the contributions to the collision integrals from higher approximations of density contain divergent (in time) integrals. This showed that the construction of kinetic equations for dense gases by the directly use of the method of successive approximations in powers of the density is impossible.

It was stated that a condition of the total attenuation of the initial correlations is approximate [14, 15] and that the task of constructing kinetic equations for dense gases can be solved if we abandon the condition of complete attenuation of the initial correlations at times much smaller than τ_{col}. In general, there is only partial weakening of the correlations. Large-scale fluctuations are not damped quickly enough and should therefore be taken into account in the construction of the kinetic theory. Their role is seen in particular in that the distribution function $\theta_{(1)}$, for which the kinetic equation is written, is not strictly deterministic. When using the condition of partial easing of the pair correlations the Liouville equation does not lead to the Boltzmann transport equation. Simplification of the original Liouville equation consists only in the fact that instead of the equation for the exact distribution function $\theta_{(N)}$ we obtain the approximate equation for the smoothed distribution function $^*\theta_{(N)}$, for which an analogue of the BBGKY chain is constructed, but for the smoothed distribution functions. This chain of equations differs from the BBGKY chain by the fact that it already takes into account the dissipation due to binary pair collisions, and all functions are slowly changing.

To account for the large-scale fluctuations in gases we use two approaches. In the first approach the starting equations are the kinetic equations, which are treated as Langevin equations with random sources [16–18]. The second approach is based on the approximate solution of the chain of equations for the distribution functions for the smoothed distribution functions in which the dissipation caused by binary collisions is taken into account from the onset [19]. Also widely used is the method for obtaining generalized kinetic equations [20–23] in which the most general kinetic equations, taking into account the arbitrary particle interactions, are taken into account. The generalized kinetic equation is much simpler than the Liouville equation, as it is the equation for the first distribution function. However, it is still difficult for practical use.

Problems using the kinetic theory for dense gases showed that the direct way of using the kinetic equations with direct molecular collisions (generalized kinetic Boltzmann equations [2], accounting for the Vlasov terms [24], the inclusion of different kinds of large-scale fluctuations [4], etc.) encounters great difficulties, so for liquid systems another variant of the kinetic theory is used.

Liquid

In the kinetic theory of liquids we consider high-density systems consisting of particles with a continuously decreasing potential of the interparticle interaction. In such systems, the concept of purely pair collisions is meaningless [3, 11, 25]. Although in the case of solid spheres of high density we can assume that the evolution of the system occurs due to frequent pair collisions of the same type, as alleged in the derivation of the Boltzmann equation. In dense systems, the change in the momentum in the interaction is usually very small, i.e., $\Delta p \ll p$, so here the hypothesis of molecular chaos is rejected.

Moreover, if in the case of strong binary collisions both colliding particles move along trajectories determined by the conservation of total energy and momentum, in the case of high density and smooth interaction potential the movement of the given particle can be considered as stochastic motion with correlation time τ, so that for long times its movement does not correlate with the initial movement.

The characteristic of the process is the autocorrelation function of the force acting on the particle, obtained by averaging over the states of the remaining $N - 1$ particles, which must vanish for times $s > \tau$, i.e.

$$\Xi(s) = \langle F_1(t) \cdot F_1(t+s) \rangle = 0, \quad s > \tau. \tag{A9.4}$$

Here $F_1(t) = -\partial U/\partial r_j$ is the force acting on the specified particle 1. The condition (A9.4) means that the forces acting on the particle at different times are statistically independent if the characteristic correlation time τ is exceeded. This condition introduces the dissipation in the dynamic description of the system and in this sense replaces the hypothesis of molecular chaos in the case of a rarefied gas, ensuring the irreversibility of the evolution of dense gas or liquid to the equilibrium state. The 'memory loss', alleged by the equality $\Xi(s) = 0$ for $s > \tau$, does not make it possible to reproduce the phase trajectory by turning the time in the dynamically reversible Liouville equation.

For a given initial distribution of $\theta_{(1)}(p, t)$ due the continuous change of the pulses we have

$$\theta_{(1)}\left(\mathbf{p}+\Delta\mathbf{p},t+\Delta t\right)=-\frac{1}{\tau}\int_0^\tau\!\!\int\!\!\int\frac{1}{m}\sum_{i=1}^N\left(\mathbf{F}_i\frac{\partial\theta_{(N)}}{\partial p_i}\right)\prod_{j=2}^N dp_i\,dr_j\,ds=$$

$$=\int W_{\Delta\mathbf{p}}\theta_{(1)}\left(\mathbf{p},t\right)d(\Delta\mathbf{p}),$$

(A9.5)

where $\theta_{(N)}$ is the full N-particle distribution function, $W_{\Delta\mathbf{p}}$ is the transition probability per time Δt of particle 1 from the state $(\mathbf{p},\,t)$ to the state $(\mathbf{p}+\Delta\mathbf{p},\,t+\Delta t)$. The second equality formally defines the value $W_{\Delta\mathbf{p}}$, which is considered to be independent of time. This condition is a key item in the considered approach. According to (A9.5) distribution $\theta_{(1)}$ $(\mathbf{p}+\Delta\mathbf{p},\,t+\Delta t)$ is defined as the integral over the momentum increment $\Delta\mathbf{p}$. If this increment is assumed to be infinitely small, the integrand (A9.5) can be expanded in a Taylor series and the evolution equation takes the form

$$\frac{\partial\theta_{(1)}}{\partial t}+\frac{\mathbf{p}_1}{m}\frac{\partial\theta_{(1)}}{\partial\mathbf{q}_1}=\frac{\partial}{\partial\mathbf{p}_1}\left\{\frac{\langle\Delta\mathbf{p}_1\rangle}{\Delta t}\theta_{(1)}+\frac{1}{2}\frac{\partial}{\partial\mathbf{p}_1}\frac{\langle\Delta\mathbf{p}_1\Delta\mathbf{p}_1\rangle}{\Delta t}\theta_{(1)}\right\},$$

(A9.6)

where the averages are defined as

$$\langle\Delta\mathbf{p}_1\rangle=\int W_\mathbf{p}^{\mathbf{p}+\Delta\mathbf{p}}\Delta\mathbf{p}_1 d\left(\Delta\mathbf{p}_1\right),$$

$$\langle\Delta\mathbf{p}_1\Delta\mathbf{p}_1\rangle=\int W_\mathbf{p}^{\mathbf{p}+\Delta\mathbf{p}}\Delta\mathbf{p}_1\Delta\mathbf{p}_1 d\left(\Delta\mathbf{p}_1\right).$$

(A9.7)

The expression in the curly brackets on the right side of equation (A9.6) describes the change in the distribution function at small momentum increments similar to the collision integral of the Boltzmann equation. (If we add to equation (A9.6) the term taking into account the influence of external force $\mathbf{X}_1$, then this equation is well known as the Fokker–Planck equation. This equation is used heavily in physical kinetics, in particular, it is the fundamental equation in the theory of Brownian motion).

To use equation (A9.6), we must obtain explicit expressions for $\langle\Delta\mathbf{p}_1\rangle$ and $\langle\Delta\mathbf{p}_1\Delta\mathbf{p}_1\rangle$, i.e. it is necessary to determine the probability of transition $W_\mathbf{p}^{\mathbf{p}+\Delta\mathbf{p}}$.

For this purpose we use the expression [25]

$$\theta_{(N)}\left(r_1...\mathbf{p}_N\right)=\theta_{(1)}\left(r_1,\mathbf{p}_I\right)\theta_{(N-1;1)}\left(r_2...r_N,\mathbf{p}_2...\mathbf{p}_N;r_1,\mathbf{p}_I\right),$$

(A9.8)

where the second factor is the conditional distribution of the dynamic variables $(N-1)$ for the particles at a fixed value of the coordinates and the momentum of the first particle. It is assumed that the conditional probability (second factor) is weakly dependent on the values of $\mathbf{p}_1$ and $\mathbf{r}_1$, then the distribution function will have the form of the equilibrium distribution, which depends on time only through the time dependence $\mathbf{p}_1$:

$$\theta_{(N-1;1)}\left(\mathbf{r}_2...\mathbf{r}_N,\mathbf{p}_2...\mathbf{p}_N;r_1,\mathbf{p}_I\right)=\theta_{(N-1;1)}=\exp\left(\frac{2p_1^2-p_1'^2}{2mkT}\right)\theta_{(N)}^0=$$

$$=\exp\left(\frac{p_1^2}{2mkT}\right)\left(1-\frac{\mathbf{p}_1\cdot\Delta\mathbf{p}_1}{mkT}\right)\theta_{(N)}^0.$$

It can be shown [25] that $\int W\,d(\Delta\mathbf{p}_1)=1$, and this helps to avoids the direct use of the transition probabilities W, using the corresponding mean values for the functions introduced above. For the momentum increment $\Delta\mathbf{p}_1 = \int_{t-r}^{t}\mathbf{F}_1\,ds$ we write the mean and the square of the mean as

$$\left\langle\Delta\mathbf{p}_1\right\rangle_{p_2...r_N} = -b\mathbf{p}_1\tau, \quad \left\langle\Delta\mathbf{p}_1\Delta\mathbf{p}_1\right\rangle_{p_2...r_N} = 2bmkT\tau, \quad \text{(A9.7a)}$$

where, using the fact that the time integral of the autocorrelation function (A9.4) does not depend on the time, we introduce the friction constant b by the following expression

$$b = \frac{1}{mkT}\int_{t-r}^{t}\left\langle\mathbf{F}_1\left(t\right)\cdot\mathbf{F}_1\left(t+s\right)\right\rangle_{p_2...r_N}\,ds. \quad \text{(A9.9)}$$

As a result, equation (A9.6) can be rewritten as

$$\frac{\partial\theta_{(1)}}{\partial t}+\frac{\mathbf{P}_1}{m}\frac{\partial\theta_{(1)}}{\partial q_1}=b\frac{\partial}{\partial\mathbf{P}_1}\left\{\mathbf{P}_1\theta_{(1)}+\frac{1}{2}\frac{\partial}{\partial\mathbf{P}_1}\left(kT\theta_{(1)}\right)\right\}. \quad \text{(A9.10)}$$

According to the first equation (A9.7), the term 'friction constant' indicates that the value of b characterizes the force acting on the particle proportional to the particle velocity and directed in a direction opposite to the direction of motion of the particle.

For the kinetic theory of liquids it is more important to derive a similar Fokker–Planck equation for the pair distribution function, since it is necessary to take into account the spatial and temporal correlations between the particles in the liquid. The binary distribution function is defined as

$$\theta_{(N)}\left(\mathbf{r}_1...\mathbf{p}_N\right)=$$

$$\theta_{(2)}\left(\mathbf{r}_+,\mathbf{r}_-,\mathbf{p}_+,\mathbf{p}_-\right)\theta_{(N-2;1,2)}\left(\mathbf{r}_3...\mathbf{r}_N,\mathbf{p}_3...\mathbf{p}_N;\mathbf{r}_1,\mathbf{p}_1,\mathbf{r}_2,\mathbf{p}_2\right).$$
$$\text{(A9.11)}$$

Here, the following variables of the coordinates and momenta are used for two fixed particles: $r_+ = (r_1 + r_2)/2$, $r_- = (r_1 - r_2)/2$; $p_+ = p_1 + p_2$, $p_- = p_1 - p_2$. The change of variables reflects the motion of the centre of mass of the pair of particles and their relative displacement. Then we can construct a finite differential equation for the evolution of the pair distribution function, using the tensor quantities for 'friction constants'

$$b_{\mu\nu} = \frac{2}{mkT} \int_{t-r}^{t} \langle F(t) \cdot F(t+s) \rangle \, ds, \qquad (A9.12)$$

where instead of the symbols μ and ν we have the signs 'plus' and 'minus'. As above, the friction tensors do not depend on time, but depend on the coordinates and momenta of the particles 1 and 2

$$\left[\frac{\partial}{\partial t} + \frac{1}{2m} \left(\sum_{\mu} p_{\mu} \cdot \frac{\partial}{\partial r_{\mu}} \right) + 2 \langle F_- \rangle \frac{\partial}{\partial p_-} \right] \theta_{(2)} =$$

$$= \sum_{\mu\nu} \frac{\partial}{\partial p_{\mu}} \cdot \left\{ b_{\mu\nu} \cdot \left[p_{\nu} + \frac{\partial}{\partial p_{\nu}} \cdot (2mkT \, b_{\mu\nu}) \right] \right\} \theta_{(2)} + O, \qquad (A9.13)$$

where the symbol O denotes two terms, for which it is assumed that they cancel each other [25]. Included in the equation (A9.13) the mean and standard momentum increments and average products of increments are expressed in terms of the friction tensors

$$\langle \Delta p_+ \rangle = -\tau \sum_{\nu} b_{+\nu} \cdot p_{\nu},$$

$$\langle \Delta p_- \rangle = -\tau \sum_{\nu} b_{-\nu} \cdot p_{\nu} - 2 \int_{t-\tau}^{t} \langle F \rangle \, ds, \qquad (A9.14)$$

$$\langle \Delta p_{\mu} \Delta p_{\nu} \rangle = 4mkT \, b_{\mu\nu} \tau.$$

These kinetic Fokker–Planck equation for the unary function, and even more so for the pair distribution function, are so complex that their direct use is difficult. They are presented here to illustrate the real problems that exist in a strict sequence of averages in dense phases, if we start with the Liouville equation, and the formulation of assumptions in the course of such averages. In 'practice' in the kinetic theory of liquids these equations are replaced by their analogues averaged over the momenta.

Smoluchowski equation

In dense systems the locally equilibrium distribution of momenta is established much faster than the spatially equilibrium configuration. Therefore, in a dense real system the evolution of the momentum distribution is carried out mainly by the quasi–Brownian dissipative motion of the molecules in the field of smoothly varying forces, while the configuration of their evolution is influenced by the strong short-term Boltzmann interaction. In this case, for the dense phase it is considered that the evolution of the momentum distribution in the characteristic time interval ∂t is almost complete and only configuration relaxation occurs in the system. Under these assumptions the Fokker–Planck equation describing the motion in the phase space becomes the Smoluchowski equation for the motion in the configuration space.

Obviously, the liquid itself can not be considered as an ensemble of Brownian particles, primarily due to the fact that a strong intermolecular interaction takes place between the liquid molecules; to take this interaction into account we examine the evolution of a selected pair of molecules. To define the various transport coefficients linking flows with corresponding generalized forces in the liquid – temperature or velocity gradients – it is required to solve the corresponding kinetic equation. This solution was obtained [3, 11] for the two-particle Smoluchowski equation. The transition to this equation is performed by introducing two flux vectors:

$$\mathbf{j}_+ = \frac{1}{2m}\int \mathbf{p}_+ \theta_{(2)}\,d\mathbf{p}_+ d\mathbf{p}_-, \quad \mathbf{j}_- = \frac{1}{2m}\int \mathbf{p}_- \theta_{(2)}\,d\mathbf{p}_+ d\mathbf{p}_-.$$

After multiplying equation (A9.13) by $d\mathbf{p}_+\,d\mathbf{p}_-$ we integrate it over momenta. The following equation is obtained:

$$\frac{\partial \theta_{(2)}}{\partial t} = -\left(\frac{\partial}{\partial \mathbf{r}_+}\right)\cdot \mathbf{j}_+ - \left(\frac{\partial}{\partial \mathbf{r}_-}\right)\cdot \mathbf{j}_-, \tag{A9.15}$$

where the pair function $\theta_{(2)}$ depends only on the relative distance between a pair of particles.

In addition to the equation for $\theta_{(2)}$ we obtain the equations for the mean values of $\mathbf{p}_+ d\mathbf{p}_+ d\mathbf{p}_-/2$ and $\mathbf{p}_- d\mathbf{p}_+ d\mathbf{p}_-/2$, multiplying by these values equation (A9.13) on the right and the left and integrating over the momenta. Integration gives new unknown functions in averaging the squares and cross-products of the components of the momenta. To simplify the construction and avoid the appearance of these new unknown non-equilibrium functions, these functions are replaced with the corresponding equilibrium averages. As a result, we can construct explicit expressions for the unknown flows $\mathbf{j}_+$ and $\mathbf{j}_-$. Collecting all the terms together, we obtain an integral equation for the pair distribution function $\theta_{(2)}$, which is called the Smoluchowski equation:

$$\frac{\partial \theta_{(2)}}{\partial t} = -\frac{\partial}{\partial \mathbf{r}_+}\cdot\left[\left(\frac{\partial}{\partial \mathbf{r}_+}\right)\left(\frac{kT\theta_{(2)}}{bm}\right)\right] - \frac{\partial}{\partial \mathbf{r}_-}\cdot\left[\left(\frac{\partial}{\partial \mathbf{r}_-}\right)\left(\frac{kT\theta_{(2)}}{bm}\right) - \frac{\langle \mathbf{F}_-\rangle\theta_{(2)}}{bm}\right],$$

(A9.16)

where the terms with the accuracy to $1/b^2$ are retained. For simplicity, instead of the friction constant tensor we used its counterpart — the scalar friction constant. Function $\langle \mathbf{F}_-\rangle$ in (A9.16) describes the relative force of the indirect influence of the particles of the pair in the medium, including the interactions between the particles and with all the surrounding particles

$$\mathbf{F}_- = -\frac{\partial U(r_{12})}{\partial \mathbf{r}_-} - \sum_{j=3}^{N}\left(\frac{\partial U(r_{2j})}{\partial(\mathbf{r}_j-\mathbf{r}_2)} - \frac{\partial U(r_{1j})}{\partial(\mathbf{r}_j-\mathbf{r}_1)}\right). \quad (A9.17)$$

Equation (A9.16) is solved assuming $\theta_{(2)} = \theta^0_{(2)}(1+w(r))$, $\theta^0_{(2)}$ where is the equilibrium pair distribution function and $w(r)$ is a small non-equilibrium correction that is sought by solving the Smoluchowski equation with the appropriate boundary conditions.

To find the transport coefficients in a stationary (time-independent) non-equilibrium case, the Smoluchowski equation (A9.16) will contain additional terms corresponding to the momentum and energy fluxes, respectively, due to the velocity and temperature gradients.

As an illustration, we present an expression for the non-equilibrium correction

$$\frac{\partial w}{\partial t} + \nabla^2 w + \frac{1}{kT}(\mathbf{F}_-\cdot\nabla w) = \left(\frac{1}{kT}\right)^2 bm\mathbf{F}_-\cdot\mathbf{a}\cdot\mathbf{r}_-, \quad (A9.18)$$

where only the relative motion of the particle pair is considered and vector $\mathbf{a}$ is the shear rate vector. This equation shows that the relative motion of a pair of molecules in a viscous flow has the form of forced diffusion imposed on the free diffusion movement that occurs due to the presence of the drift velocity which is supported by the external conditions.

Using the solution of this equation we can determine the shear viscosity coefficient as $\eta = \frac{2\pi\theta_0^2}{15}b\int_0^\infty \frac{d\varepsilon(r)}{dr}\xi_2^0(r)u(r)u^3dr$, where θ_0

is the density of the fluid, ξ is the correlation function, and the friction coefficient b must be known. Function $u(r)$, included in the expression η, is linearly related to the non-equilibrium correction $w = Au(r)$, where the coefficient A reflects the relationship with the shear flow rate and it must be determined from the formula:

$$u''(r) + \left(\frac{2}{r} - \frac{1}{kT}\frac{d\varepsilon(r)}{dr} \right)u' - \frac{6u}{r^2} = -\frac{m\beta_{(2)}}{(kT)^2}\alpha r \frac{d\varepsilon(r)}{dr}, \quad \text{(A9.19)}$$

with the appropriate boundary conditions.

Equation (A9.19) must be satisfied for the two boundary conditions. One of them is obvious, namely

$$u(r) \to 0 \quad \text{at} \quad r \to \infty. \quad \text{(A9.20)}$$

Two conditions were proposed as a second boundary condition: 'weak' [26] (it implies that the function $u(r)$ is finite everywhere, but it implies a strong anisotropy, not detected experimentally [25, 27])

$$u(r) \to 0 \quad \text{at} \quad r \to 0 \quad \text{(A9.21)}$$

and 'strong'

$$r^3 u(r) = 0 \quad \text{at} \quad r \to \infty, \quad \text{(A9.22)}$$

which is also placed at infinity instead of the natural condition for $r \to 0$.

(To find the bulk viscosity and thermal conductivity coefficients, it is necessary to solve the equations of the auxiliary functions appearing instead of the function $u(r)$ for the shear viscosity.)

The described foundations of kinetic equations for gas and liquid show that the two theories are in quite a poor agreement with the conditions necessary to describe the flows in porous bodies. Recall that small typical pore sizes and pore wall non-uniformity impose severe restrictions on the existence of all the derivatives in space for both gas and liquid. The materials of the previous chapters clearly shows the sharp changes in the density of matter in the surface potential field on non-uniform surfaces and inside the pore volume of different geometry. Similar drastic changes in density are observed at the liquid–vapor interfaces in capillary condensation of substances in the pores.

In general, we can state the following problems in the kinetic theory of vapour and liquid.

1. The central problem is that the need for joint consideration of the vapor–liquid system requires the use of a single kinetic equation,

while the description of the evolution of gas is based on the functions of the velocity distribution of the molecules, and the description of the evolution of the liquid is based on the distribution functions of the molecules in space.

2. Both types of approaches have developed methods of constructing equations for uniform and weakly non-uniform systems. In both methods, the minimum area of the phase under consideration exceeds the radius of the intermolecular interaction potential. The radius of the intermolecular potential for the LD potential is estimated as $\sim 2.5\sigma$, which corresponds to $R = 5$ coordination spheres for the lattice $z = 12$. As follows from the preceding chapters, the specificity of porous bodies is the strong spatial non-uniformity at the atomic and molecular level with a typical size commensurate with the size of the molecules, which prevents the use of these approaches.

This region provides a qualitative estimate the density at which the region of high density starts, when the neighbouring molecules affect the trajectory of motion of the molecules. A sphere with such a radius occupies a volume equal to $4\pi R^3/3\sigma^3$ or $\approx 108\sigma^3$, i.e. about 10^2 sites. Therefore, if the numerical density θ is less than 10^{-2}, then we can speak of a free path of the molecules and the region of applicability of methods for describing the dynamics of rarefied gas molecules. If $\theta > 10^{-2}$, then the molecule is almost always in the region of the effect of potential neighbours and we should consider their impact, as in the theory of liquids.

3. The friction coefficient in the theory of liquids is the average complex characteristic of the system, not directly related to the intermolecular interaction potential of the molecules and elementary motions of the molecules. Only in the simplest case, when using a number of simplifying assumptions by Kirkwood [3, 26–30] it was possible to obtain a relationship between the pair correlation function $g_{(2)}(r)$ and the pair potential function $\varepsilon(r)$ of the interaction of molecules: $\beta_{(1)} = (4\pi m/3)\int g_{(2)}(r)\nabla^2\varepsilon(r)r^2 dr$. It is quite difficult to construct such relationships in more realistic conditions and the determination of the friction coefficients is an extremely complex task. Especially, the construction of such a characteristic looks problematic in highly non-uniform porous systems. First of all, it refers to large differences in the characteristic times for the motion of the molecules in different monolayers near any surfaces.

4. Each approach also has its own internal problems to be solved in each case. Thus, we note that there are two problems in finding the the correction $u(r)$ in the liquid theory to calculate the shear

viscosity: 1) the uncertainty of the boundary conditions at small distances between the particles, which is natural because of the neglect of the discreteness of the spatial distribution of molecules, and 2) the condition for the equation (A9.19), which is obtained for an incompressible fluid; this is not suitable for a wide range of fluid density changes. For a compressible fluid there should be further corrections of its compressibility.

In the case of the gas a constant problem is the self-consistency of the derived kinetic equations with the expressions for the thermodynamic non-dissipative characteristics. Another issue is to consider a sharp anisotropy of density near the wall even for a rarefied vapour. In this direction, there has been progress compared with the works discussed in section 3 both for the exposed surface [31] and for the open surface [32, 33] at low vapour density, which relates to the Knudsen flow regime.

The problems for dense gas and liquid are solved using the method of molecular dynamics and other dynamic techniques (see Appendix 8). Their development – a reaction to the modern development of the non-equilibrium theory (see, e.g., [34, 35]), where strict development can not be brought down to the simplest practical problems due to their mathematical complexity. As a result, studies of the kinetic theory on the basis of integral equations in external fields have almost completely disappeared, such as [24], when the short-range order dynamics is not used in practice, although the results of section 28 points to the need to integrate it at the microlevel. Applying rigorous approaches in porous systems resulted only in the development of equations of the mean field approximation type [36–38], which is insufficient for a self-consistent description of the equilibrium and dynamic properties of the dense phases.

In contrast to the kinetic theory of gases and liquids, in the solid state theory and of the solid–vapour or liquid interface (see sections 27 and 28) the role of the Vlasov terms (which affect by their interaction potential the course process, but do not participate in the elementary stage) was taken into account immediately with the development of the kinetic theory (see review [39]). So, the dense vapour and liquid should be considered in the framework of the same approaches that were taken into account in the formulation of equations (A9.2) and (A9.3). In addition, in the development of stochastic methods in the kinetic theory special attention was given to the concept sampling space, which is very similar to the LGM, but it does not fix the cell volume [40, 41]. As part of this review the transitions of the molecules from one cell to another were recorded; this is completely equivalent to the system of equations in the LGM (section 27). To harmonize the system of equations (A9.2) and (A9.3) with the LGM it is also necessary to introduce additional terms I_f^i and I_{fg}^{ij} which in addition to the collisional and Vlasov terms reflect the balance of departure and arrival of the molecules in the cell. In such constructions one should monitor the physical nature of the flows of the molecules: although the hydrodynamic velocity is determined by averaging the thermal velocities of the molecules, the condition fundamentally important but in the discrete version of the spatial distributions of molecules is that the microscopic velocity **u** is calculated from the equations of momentum transfer (i.e., through the equation of the Navier–Stokes type), and not through the hopping of molecules in a stationary thermostat, as it sometimes is assumed in constructing 'lattice hydrodynamics' (see compilation [42]). Such constructions replace the notion of flux of the momentum by the concentration flux of molecules, which is similar to the diffusion flux, rather than hydrodynamic (this aspect is not specifically analyzed).

References to Annex 9

1.   Bogolyubov N.N., Problems of dynamical theory in statistical physics. – Moscow: Gostekhizdat, 1946 [Wiley Interscience, New York, 1962].
2.   Ferziger J. H. H., Kaper H. G., Mathematical theory of transport processes in gases. – Moscow: Mir, 1976. [North_Holland, Amsterdam, The Netherlands, 1972].
3.   Croxton C. A., Physics of the liquid state: A Statistical Mechanical Introduction. – Moscow: Mir, 1979. [Cambridge University Press, Cambridge, 1974].
4.   Klimontovich Yu.L. Kinetic theory of non-ideal gas and non-ideal plasma. — Moscow: Nauka, 1975.
5.   Tovbin Yu.K., Modern Chemical Physics. — Moscow: Moscow State University Press, 1998. — P. 145.
6.   Tovbin Yu.K. // Khim. fiz.. 2002. V. 21, No. 1. P. 83.
7.   Tovbin Yu.K. // Zh. Fiz. khimii. 2002. V. 76, No. 1. P. 76 [Russ. J. Phys. Chem. 2002. V. 76. № 1. P. 64].
8.   Bazarov I.P., Statistical theory of the crystalline state. — Moscow: Moscow State University Press, 1972. — 118 p.
9.   Vlasov A.A., Statistical distribution functions. — Moscow: Nauka, 1966. — 356 p.
10.   Vlasov A.A. // Zh. eksp. teor. fiz.. 1938. V. 8. P. 291.
11.   Kirkwood J.G. // J. Chem. Phys. 1947. V. 15. P. 72.
12.   Uhlenbeck G., Ford G., Lectures on statistical mechanics. — Moscow: Mir, 1965.
13.   Cohen E., Statistical Mechanics at the Turn of the Decade. — N. Y., 1971.
14.   Sandri G. // Ann. Phys. 1963. V. 24. P. 332.
15.   Hofield J., Batin A. // Phys. Rev. 1968. V. 168. P. 193.
16.   Kadomtsev B.B. // Zh. eksp. teor. fiz. 1957. V. 32. P. 943.

17. Kogan Sh.M., Schul'man A.Ya. // Zh. eksp. teor. fiz. 1969. V. 56. P. 862.
18. Kogan Sh.M., Schul'man A.Ya. // Fiz. Tverdogo Tela. 1969. V. 11. P. 308.
19. Klimontovich Yu.L. // Usp. Fiz. Nauk. 1973. V. 110. P. 573.
20. Prigogine I., Goerge G., Rae J. // Physica. 1971. V. 56. P. 25.
21. Balescu R., Brenig J., Wallenborn J. // Physica. 1971. V. 52. P. P10.
22. Zubarev D.N., Kalashnikov V.P. // Teor. Mat. Fizika. 1971. V. 7. P. 372.
23. Zubarev D.N., Novikov M.Yu. // Teor. Mat. Fizika. 1972. V. 13. P. 403.
24. Davis H.T. // J. Chem. Phys. 1987. V. 96. P. 1474.
25. Eisenschitz R., Statistical theory of irreversible processes. – Moscow: Izd. inostr. lit., 1963. [Oxford University Press, London, 1948].
26. Kirkwood J.G., Buff F.P., Green M.S. // J. Chem. Phys. 1947. V. 17. P. 988.
27. Eisenschitz R. // Proc. Phys. Soc. 1957. V. 59. P. 1030.
28. Kirkwood J.G. // J. Chem. Phys. 1946. V. 14. P. 180.
29. Kirkwood J.G. // J. Chem. Phys. 1947. V. 15. P. 72.
30. Zwanzig R.W., Kirkwood J.G., Stripp K., Oppenhneim I. // J. Chem. Phys. 1953. V. 21. P. 2050.
31. Borisov S.F., Balakhonov N.F., Gubanov V.A., Interactions of gases with the surface of the solid.—Moscow: Nauka, 1988. — 200 p.
32. Borman V.D., Krylov S.Yu., Prosyanov A.V. // Zh. eksp. teor. fiz. 1990. V. 97. P. 1795.
33. Borman V.D., Krylov S.Yu., Prosyanov A.V. // Zh. eksp. teor. fiz.. 1986. V. 90. P. 76.
34. Zubarev D.N., Morozov V.G., Röpke H., Statistical Mechanics of non-equilibrium processes. Vols 1 and 2. — Moscow: Fizmatlit, 2002.
35. Ottinger H.S., Beyond equilibrium thermodynamics. — Hobeken, New Jersey, Wiley, 2005.
36. Pozhar L.A., Gubbins K.E. // J. Chem. Phys. 1991. V. 94. P.1367.
37. Pozhar L.A., Gubbins K.E. // J. Chem. Phys. 1993. V. 99. P. 8970.
38. Pozhar L.A., Transport Theory of Inhomogeneous Fluids. — Singapore: World Scientific, 1994.
39. Tovbin Yu.K. // Progress in Surface Science., 1990. V. 34. No. 1 – 4, P. 1 – 236.
40. Gardiner K.V., Stochastic methods in natural sciences. — Moscow: Mir, 1986. — 526 p. [Gardiner C. W., Handbook of Stochastic Methods for Physics, Chemistry and Natural Sciences, Springer – Verlag, Berlin, 1983]
41. Succi S., The lattice Boltzmann equation for fluid dynamics and beyond. — Oxford: Oxford Univ. Press, 2001.
42. Pattern Formation and Lattice-Gas Automata / Eds. A.T. Lawniczak, R. Kapral // Fields Institute Communications. 1996. V. 6.

Appendix 10

Algorithm for calculating molecular flows

The kinetic equations based on the LGM lead to discrete transport equations that are written for each cell of the region under consideration, and which are identical in form with the different form of writing differential Navier–Stokes equations [1–4].

All cells inside the considered region or on its boundary are completely equal: the transport equations for the boundary cells have the same form as inside the lattice system.

In general, the equations of unsteady motion of the fluid in the pore under isothermal conditions, reflecting the laws of conservation of mass and momentum, have the form

$$U_t + F_x + G_y = E,$$

where U, F, G, E are the vectors with components

$$U = (\rho,\, \rho u,\, \rho v,\, e)$$

$$F = \left(\rho u,\, \rho u^2 + \sigma_{xy},\, \rho uv + \tau_{xy},\, (e + \sigma_x)u + \tau_{xy}v + \kappa_x\right),$$

$$G = \left(\rho v,\, \rho uv + \tau_{xy},\, \rho v^2 + \sigma_{yk} + Q_2,\, (e + \sigma_{yk})v + \tau_{xy}u + \kappa_y\right),$$

$$E = (0,\, 0,\, 0,\, \varphi_\kappa).$$

Here, ρ is the density per unit volume (in kg/m^3), which is associated with the concentration $\theta = \rho v_0/(30M)$, v_0 is the volume of the molecule, M is the molecular weight; $e = E + \rho(u^2 + v^2)/2$ is the total energy, $E = C_v T$ is the internal energy, C_v is the specific heat;

u and v are the components of the gas-dynamic velocity along the x and y axes, $\varphi_k = (Q_k - Q_{k+1})/\lambda$ is the the potential created by the wall in a layer k, κ_x and κ_y are the coefficients of thermal conductivity along the x and y axes.

The components of the vectors F and G are expressed through shear (μ) and bulk (η) viscosity:

$$\sigma_x = P + \left(\frac{2\mu}{3} - \eta\right)\left(u_x + v_y\right) - 2\mu u_x; \quad \tau_{xy} = -\mu\left(u_x + v_y\right);$$

$$\sigma_y = P + \left(\frac{2\mu}{3} - \eta\right)\left(u_x + v_y\right) - 2\mu v_y,$$

and the coefficients of thermal conductivity.

The expressions of the dissipative coefficients via the local density and fluid velocity are defined in the works that are discussed in chapter 5. The expressions for the dissipative coefficients were derived using the model of the transition state of the condensed phases [5], which uses the intermolecular interactions of the particles in the ground (ε) and transition (ε^*) states (in the calculations it was assumed $\varepsilon^* = \varepsilon/2$ – this ratio is used most often when describing the various experimental data in other problems for surface processes).

To solve the problems of the motion of the fluid in a narrow pore we need a method that would allow to calculate the flow of a compressible viscous fluid taking into account the strong concentration gradients due to phase transitions. We use a two-step explicit finite-difference method of second order accuracy, which is a two-step time variant of the Lax–Wendroff method [6]. Similar modifications of the Lax–Wendroff method are used in many studies of gas dynamics, such as the splitting of the spatial coordinates in [7]. In order to use the method [6] on the strong discontinuities, the 'weight' factor β_1 is introduced in the finite difference scheme smoothing strong discontinuities whilst preserving the second-order accuracy for smooth solutions [8].

Calculations are performed in a rectangular coordinate system x, y, where $\Delta x \sim \Delta y \sim \lambda$. Let us consider the equations of motion of the fluid in the space x, y, t. The difference scheme for solving the equations is presented in time t with nine points (3×3) with the parameters known at these points (the pattern is shown in Fig. 6.1 in chapter 6).

Finding the solutions in the next moment $t_1 = t + \Delta t$ at the point (i, j) is carried out in two stages. First, the flow parameters are searched in the four middle points using the formulas

$$U_{k,l}(t+\Delta t) = \left(U_{k+1/2,l+1/2} + U_{k+1/2,l-1/2} + U_{k-1/2,l+1/2} + U_{k-1/2,l-1/2}\right)/4 +$$

$$+ \left(F_{k+1/2,l+1/2} - F_{k+1/2,l-1/2} + F_{k-1/2,l+1/2} - F_{k-1/2,l-1/2}\right)\Delta t/2\Delta x +$$

$$+ \left(G_{k+1/2,l+1/2} - G_{k+1/2,l-1/2} + G_{k-1/2,l+1/2} - G_{k-1/2,l-1/2}\right)\Delta t/2\Delta y +$$

$$+ \left(E_{k+1/2,l+1/2} + E_{k+1/2,l-1/2} + E_{k-1/2,l+1/2} + E_{k-1/2,l-1/2}\right)\Delta t/4,$$

$$(A10.1)$$

where $k = i \pm 1/2$, $l = j \pm 1/2$.

The final result at the central point (i, j) is searched as

$$U_{i,j}(t+\Delta t) =$$

$$= \beta_1 U_{i,j}(t) + \left(U_{i+1,j+1} + U_{i-1,j+1} + U_{i+1,j-1} + U_{i-1,j-1}\right)(1-\beta_1)/4 +$$

$$+ \left(F_{i+1,j} - F_{i-1,j} + F_x(t+\Delta t)\right)\Delta t/2\Delta x +$$

$$+ \left(G_{i,j+1} - G_{i,j-1} + G_y(t+\Delta t)\right)\Delta t/4\Delta y +$$

$$+ \left(E_{i,j}(t) + E_{k,l}(t+\Delta t)\right)\Delta t/2,$$

$$(A10.2)$$

where $F_x(t + \Delta t)$ and $G_y(t + \Delta t)$ are the central differences F and G at $t_1 = t + \Delta t$ and $E_{k,l} = 1/4\Sigma(E_{i \pm 1/2, j \pm 1/2}(t))$.

Calculations by formulas (A10.1) are conducted with high stability. The results obtained by the formula (A10.1) describe the solution of the problem at the central point (i, j) with the first order accuracy of the coordinate and time (since this formula is the finite-difference scheme of the first order). To simplify the calculations of the fluid flow in the pore we can confine ourselves to this solution, but in the calculation of the flow in a region with strong density gradients this solution gives strong distortions: the first-order schemes blur contact discontinuities (in this case this refers to the interphase boundaries). Therefore, to accurately describe the fluid flow in the pore we apply the second step in time in the form of formula (A10.2).

The final formula (A10.2) provides the second-order accuracy for smooth solutions. Factor β_1, entered in the scheme, smoothes the solution at a strong discontinuity. The discontinuities are found on the basis of density. Before calculating the parameters of the calculation point by formulas (A10.1) and (A10.2) we analyze the density gradient along the axes x, y, and depending on its value for this type of task we select coefficient β_1.

This two-step method is stable at integration step $\Delta t \sim \lambda^2 \rho / \mu$. Due to the smallness of this value, to solve the problem we require a large number of steps, to $\sim 10^6 - 10^7$ [9, 10]. In the absence in the flow of phase transitions, the stratified one-dimensional flow is established quickly enough in the pore (about 10^4 steps) layered set-dimensional flow. However, when phase transitions are realized in the flow field, the flow at the phase boundaries is calculated by 'smearing' the boundary into multiple cells. This is achieved by considering the influence on the parameter of the central calculation point of the points surrounding it.

The contribution of this influence is determined from the condition of satisfying two mutually exclusive factors. First, it is necessary to 'smear' the border to a few cells to ensure that the asecond-order scheme did not lose stability at discontinuities occurring in the flow field. Second, the 'smearing' should not be strong enough to avoid losing the accuracy of calculation. Applying this calculation method greatly accelerates the process of establishing the quasi-stationary flow regime.

References

1.   Tovbin Yu.K., Modern chemical physics, Moscow, Moscow State University,1998, 145.
2.   Tovbin Yu.K., Zh. Fiz. Khim., 2002 V. 76, No. 3, 488 [Russ. J. Phys. Chem. 2002. T. 76. № 3. P. 412].
3.   Tovbin Yu.K., Khim. Fiz., 2002, V. 21, No. 1, 83.
4.   Tovbin Yu.K., Teor. Osnovy Khim. Tekhnol., 2002, V. 36, No. 3. 240 [Theor. Found. Chem. Engin. 2002. V. 36. No 3. P. 214].
5.   Tovbin Yu.K., Theory of physico-chemical processes at the gas-solid interface. – Moscow: Nauka, 1990. – 288 p. [CRC Press Boca Raton, FL, 1991].
6.   Lax P., Wendroff B., Pure Appl. Math., 1960, V. 13, 217.
7.   Balwin B.S., MacCormack R.W., AJAA Paper 74–558. 1974.
8.   Tugazakov R.Ya., Izv. AN SSSR. Mekh. Zhidk. Gaza, 1989, No. 2, 159.
9.   Tovbin Yu.K., Tugazakov R.Ya., Modern chemical physics, Moscow, Moscow State University,1998, 178.
10.   Tovbin Yu.K., Tugazakov R.Ya., Teor. Osnovy Khim. Tekhnol., 2000. V. 34, No. 2, 117 [Theor. Found. Chem. Engin. 2000. V. 34. No 2. P. 99].

Appendix 11

Microhydrodynamic equations of flows of dense mixtures of molecules of different size in pores

The equations describing the equilibrium and dynamic states of mixtures have a common structure for the components of comparable and different sizes. This application collects the equation to describe these states of the mixtures.

Kinetic equations

As in Appendix 9, we formulate kinetic equations forming the basis of the transition to the microscopic hydrodynamics equations [1, 2]. To describe the dynamics of the mixture components we consider the full time distribution function $\theta\left(\left\{i, f, \mathbf{v}_f^i, \Omega_f^i\right\}, t\right)$, which characterizes the probability of finding in a cell with the number f, $1 \leq f \leq N$, a particle i (i.e. the state of occupation of site f), having a translational velocity $\mathbf{v}_f^i$ and rotational Ω_f^i movements (as above, the complete list of the cells is indicated by braces {}), where t is time. We retain the same way to describe the dynamics of particles in each cell of the lattice structure, considering that we are talking about the centre of mass of particles i, although the particles can occupy multiple cells. This makes it relatively easy to reformulate the equations, previously constructed for single-site particles, for a mixture of components of larger sizes (see section 67).

In the quasi-chemical approximation the full distribution function is released through local unary $\theta^i_f\left(\mathbf{v}^i_f,\Omega^i_f,t\right)$ and paired $\theta^{ij}_{fg}\left(\mathbf{v}^i_f,\Omega^i_f,\mathbf{v}^j_g,\Omega^j_g,t\right)$ distribution function, $1 \leq f,\ g \leq N$, as

$$\theta\left(\left\{i,f,\mathbf{v}^i_f,\Omega^i_f\right\},t\right)=\prod_{f=1}^{N}\theta^i_j\left(\mathbf{v}^i_f,\Omega^i_f,t\right)\prod_{g\in\Pi_i}\left\{\frac{\theta^{ij}_{fg}\left(\mathbf{v}^i_f,\Omega^i_f,\mathbf{v}^j_g,\Omega^j_g,t\right)}{\theta^i_f\left(\mathbf{v}^i_f,\Omega^i_f,t\right)\theta^j_g\left(\mathbf{v}^j_g,\Omega^j_g,t\right)}\right\}^{1/2},$$

$$(\text{A11.1})$$

where the exponent 1/2 takes into account that pairs of cells are listed twice; index g ranges over all z neighbours of site f, the index j – the sort of particle in the site g (the time argument t will be omitted). The local distribution functions used here in space differ from normal distribution functions by their normalization for the cell volume (not the volume of the system). They are not defined outside of the examined cells. The boundary conditions in the space of velocities coincide with the usual conditions: these functions vanish when the velocity modulus tends to infinity.

To describe the evolution of the system it is sufficient to construct the kinetic equation for the local unary $\theta^i_f\left(\mathbf{v}^i_f,\Omega^i_f\right)$ and paired $\theta^{ij}_{fg}\left(\mathbf{v}^i_f,\Omega^i_f,\mathbf{v}^j_g,\Omega^j_g\right)$ distribution functions, in which the spatial coordinate is replaced by the number of the cell f. Unary distribution functions obey equation: $\mathbf{D}^i_j\theta^i_f\left(\mathbf{v}^i_f,\Omega^i_f,t\right)=St^i_f$, where $\mathbf{D}^i_f$ is the differential operator and $St^i_f=\sum_{g,j}St^{ij}_{fg}$ is the collision integral for a particle i, located in the site f, $g\in\Pi_i$ (Π_i is the number of contacts of the molecule of type i), $1\leq i\leq\Phi$, which has the form

$$\mathbf{D}^i_f=\frac{\partial}{\partial t}+v^i_f\frac{\Delta}{\Delta q_f}+\frac{F_i(f)}{m_i}\frac{\partial}{\partial v^i_f}+\Omega^i_f\frac{\Delta}{\Delta\lambda_i}+\frac{M_i(f)}{I_i}\frac{\partial}{\partial\Omega^i_f},$$

$$(\text{A11.2})$$

$$St^{ij}_{fg}=\int\left(F^{ij}_{fg}\frac{\partial\theta^{ij}_{fg}\left(\mathbf{v}^i_f,\Omega^i_f,\mathbf{v}^j_g,\Omega^j_g\right)}{m_i\partial v^i_f}+M^{ij}_{fg}\frac{\partial\theta^{ij}_{fg}\left(\mathbf{v}^i_f,\Omega^i_f,\mathbf{v}^j_g,\Omega^j_g\right)}{I_i\partial\Omega^i_f}\right)dq_g\,dv^j_g\,d\Omega^j_g.$$

Here in the left-hand side are $\Delta/\Delta\mathbf{q}_f$ – difference derivative in the spatial variable $\mathbf{q}$ for site f, reflecting the discrete nature of the variation of the coordinate of the particles of the mixture, as well as the difference derivative $\Delta/\Delta\lambda_i$ for the change of the orientation of the particle i (for simplicity of notation the symbol discrete orientations λ is retained as differential variable λ_i). Note that, although the indexes i and j take into account the change of the orientation of adjacent discrete particles, however, at short times (and at small angles of rotation) the particles may have their angular velocities, which correspond to the dynamics of the oscillations of the particles in a dense phase. Here $\mathbf{F}_i(f)$ is the external force vector in the cell f, acting on the particle i, comprising contribution $-\partial U_f^i / \partial \mathbf{q}_f$ from the conservative field, U_f^i is the potential energy of particle i in cell f and $-\partial U_{fh}^{ij} / \partial \mathbf{q}_f$ for the dynamic field from neighbouring particles j in cells h (Vlasov terms); m_i is the mass of the particle i; $\mathbf{M}_i(f) = -\partial U_f^i / \partial \Omega_f^i$ is the angular momentum in the cell f, generated by the external forces and causing the rotation of the particle i with the tensor of the momentum inertia $\mathbf{I}_i$ (and similar Vlasov term, see below). In narrow pores and near solid surfaces a major role is played by the surface potential and contribution of gravity can be ignored.

The right side of the equations under the integral sign retains the differential derivatives with respect to the spatial coordinate $\mathbf{q}_f$, and the orientation coordinate λ_i relative to the potential function of interaction of the particle (this result in accurate accounting of the spatial properties potential functions); vector $\mathbf{F}_{fg}^{ij} = -\partial U_{fg}^{ij} / \partial \mathbf{q}_{fg}$ is the force on particle i by particle j. Angular velocities Ω_g^j are included in the integration variables at fixed values of the index j. Value $\mathbf{M}_{fg}^{ij}$ is the momentum of a pair of intermolecular interaction forces applied to the particle i by particle j, which is equal to the vector product $\mathbf{M}_{fg}^{ij} = \left[\lambda_i \times \partial U_{fg}^{ij} / \partial \lambda_i \right]$. Relation between the force $\mathbf{F}_{fg}^{ij}$ and the force pairs is expressed as $\mathbf{M}_{fg}^{ij} + \mathbf{M}_{gf}^{ij} + \left[\mathbf{q}_{fg} \times \mathbf{F}_{fg}^{ij} \right] = 0$ (index order is essential for $\mathbf{F}_{fg}^{ij}$ and $\mathbf{M}_{fg}^{ij}$, as for $\mathbf{M}_{fg}^{ij} \neq -\mathbf{M}_{gf}^{ij}$) [3–5]. This ratio follows from the invariance of the potential energy in rotation of the system consisting of two particles, as a whole, by an infinitesimal angle.

The pair distribution functions for a mixture of large molecules are described the following equations:

$$\mathbf{D}_{fg}^{ij} \, \theta_{fg}^{ij} \left(v_f^i, \Omega_f^i, v_g^j, \Omega_g^j \right) = \mathrm{St}_{fg}^{ij} + \mathrm{St}_{gf}^{ij}, \quad \mathrm{St}_{fg}^{ij} = \sum_{h,k} \mathrm{St}_{fgh}^{ijk}, h \in \Pi_i, 1 \leq k \leq \Phi,$$

$$\text{where } \mathbf{D}_{fg}^{ij} = \mathbf{D}_f^i + \mathbf{D}_g^j - \frac{\partial}{\partial t} - \left(\frac{F_{fg}^{ij}}{m_i} \frac{\partial}{\partial v_f^i} + \frac{F_{fg}^{ij}}{m_j} \frac{\partial}{\partial v_g^j} + \frac{M_{fg}^{ij}}{I_i} \frac{\partial}{\partial \Omega_f^i} + \frac{M_{gf}^{ij}}{I_j} \frac{\partial}{\partial \Omega_g^j} \right), \quad (A11.3)$$

$$\mathrm{St}_{fgh}^{ijk} = -\int \left(F_{fh}^{ik} \frac{\partial}{m_i \partial v_f^i} + M_{fh}^{ik} \frac{\partial}{I_i \partial \Omega_f^i} \right) \theta_{fh}^{ik} \left(v_f^i, \Omega_f^i, v_h^k, \Omega_h^k \right) \times$$

$$\times \frac{\theta_{gh}^{jk} \left(v_g^j, \Omega_g^j, v_h^k, \Omega_h^k \right)}{\theta_h^k \left(v_h^k, \Omega_h^k \right)} dq_h \, dv_h^k \, d\Omega_h^k.$$

here $\mathbf{D}_f^i$ and $\mathbf{D}_g^i$ are defined in (A11.2), and the third term removes the 'extra' time derivative. The bracket in (A11.3) reflects the contribution of the intermolecular interaction of a pair of particles ij to the evolution of this pair at the translational and rotational motions of its constituent particles. The second term St_{gh}^{jk} in the collision integral is similar to the first term St_{fg}^{ij} – it is obtained by simultaneous permutation of pairs of indices (i, f) and (j, g).

Additional types of rotations (or simultaneous shift–rotation) near the pore wall in this section and below are not mentioned: from a formal point of view, only the values of $L_{m,1}$ and corresponding moments of inertia $\mathbf{I}_i$ change.

Conservation equations

To examine the flows of uncharged mixtures it is sufficient to consider the properties S, where S is the mass, momentum, total angular momentum and energy of the particles i, $1 \leq i \leq \Phi$ [6]. To construct microhydrodynamic equations we construct the conservation equations of the properties S on the basis of kinetic equations for the unary and pair distribution functions. The equations of conservation of properties S in each cell are obtained by averaging (A11.2) and (A11.3) with respect to the coordinates inside the cell, the forward velocity of the particles and the angular velocities of each particle i. Since we do not consider changes in the internal degrees of freedom at any particle collisions (elastic collisions), the collision integrals vanish when identical particles collide, and have non-zero values for different particle collisions.

Averaging the kinetic equation (A11.2) gives the following conservation equation for each cell in the system:

$$\frac{\partial \langle \theta_f^i S \rangle}{\partial t} + \sum_{j=1}^{3} \left[\frac{\Delta \langle \theta_f^i v_{fj}^i S \rangle}{\Delta q_{fj}} - \theta_f^i \left\langle \frac{v_{fj}^i \Delta S}{\Delta q_{fj}} \right\rangle + \frac{\Delta \langle \theta_f^i \Omega_{fj}^i S \rangle}{\Delta \lambda_{ij}} - \right.$$

$$- \theta_f^i \left\langle \frac{\Omega_{fj}^i \Delta S}{\Delta \lambda_{ij}} \right\rangle - \theta_f^i \left\langle \frac{F_i(f)_j \partial S}{\partial v_{fj}^i} \right\rangle / m_i - \theta_f^i \left\langle \frac{S \partial F_i(f)_j}{\partial v_{fj}^i} \right\rangle / m_i - $$

$$\left. - \theta_f^i \left\langle \frac{M_i(f)_j \partial S}{\partial \Omega_{fj}^i} \right\rangle / I_i - \theta_f^i \left\langle \frac{S \partial M_i(f)_j}{\partial \Omega_{fj}^i} \right\rangle / I_i \right\} = \langle St_f^i S \rangle,$$

$$\text{(A11.4)}$$

where the index $j = 1$–3 relates to vectors directed along the axes x, y, z from the cell f; v_{fj}^i and Ω_{fj}^i is the component j of the velocity vector $\mathbf{v}_f^i$ and the angular rotational velocity $\mathbf{\Omega}_f^i$ of the particle i in cell f (for the vacancy $v^v_f = \Omega^i_{fj} = 0$ and to simplify the arguments we omitted the velocities in unary functions). If in averaging in $\langle St^i_f S\rangle$ we use the equilibrium distribution functions for the translational and rotational velocities of all particles then $\langle St^i_f S\rangle = 0$.

By analogy with (A11.4), we construct the equation of conservation of the properties S belonging to two cells f and g. To do this, we average equation (A11.3) over the velocities of the molecules in these cells. This leads to the following equation (for compactness herein and below we omit the velocity arguments in the pair functions):

$$\frac{\partial \langle \theta^{ij}_{fg} S\rangle}{\partial t} + \sum_{k-1}^{3}\left[\frac{\Delta\langle \theta^{ij}_{fg} v^i_{fk} S\rangle}{\Delta q_{fk}} - \theta^{ij}_{fg}\left\langle\frac{v^i_{fk}\Delta S}{\Delta q_{fk}}\right\rangle + \frac{\Delta\langle \theta^{ij}_{fg}\Omega^i_{fk} S\rangle}{\Delta \lambda_{ik}} -\right.$$

$$-\theta^{ij}_{fg}\left\langle\frac{\Omega^i_{fk} S}{\Delta\lambda_{ik}}\right\rangle - \theta^{ij}_{fg}\left\langle\frac{F_i(f)_k\partial S}{\partial v^i_{fk}}\right\rangle / m_i - \theta^{ij}_{fg}\left\langle\frac{S\partial F_i(f)_k}{\partial v^i_{fk}}\right\rangle / m_i -$$

$$-\theta^{ij}_{fg}\left\langle\frac{M^i_{fk}\partial S}{\partial \Omega^i_{fk}}\right\rangle / I_i - \theta^{ij}_{fg}\left\langle\frac{S\partial M^i_{fk}}{\partial \Omega^i_{fk}}\right\rangle / I_i - \theta^{ij}_{fg}\left\langle\frac{F^{ij}_{fgk}\partial S}{\partial v^i_{fk}}\right\rangle / m_i +$$

$$+\theta^{ij}_{fg}\left\langle\frac{S\partial F^{ij}_{fgk}}{\partial v^i_{fk}}\right\rangle / m_i - \theta^{ij}_{fg}\left\langle\frac{M^{ij}_{fgk}\partial S}{\partial \Omega^i_{fk}}\right\rangle / I_i + \theta^{ij}_{fg}\left\langle\frac{S\partial M^{ij}_{fgk}}{\partial \Omega^i_{fk}}\right\rangle / I_i \left.\right] + \{_g\} =$$

$$= \langle St^{ij}_{fg}S\rangle, \quad (A11.5)$$

where the sum over k is built with respect to the contributions from the cell variables f, $\{_g\}$ denotes the same sum constructed for the cell g (it is obtained by replacing the indices f and i by g and j). In deriving (A11.5) it is considered than we can neglect the contribution of simultaneous triple collisions of particles, and the property S is retained in pair collisions of identical particles which are implemented in the potential field created by a third particle, located next (i.e. in collisions there are no changes in the internal states of molecules or in their orientations). In general, the collision of different components $\langle St^{ij}_{fg} S\rangle \neq 0$ but using the equilibrium distribution function $\langle St^{ij}_{fg} S\rangle = 0$.

If S corresponds to the mass of the two molecules, equation (A11.5) yields an analogue of the continuity equation for the flow of particle pairs ij (i.e. for all sorts of molecules in different orientations)

$$\frac{\partial \theta^{ij}_{fg}}{\partial t} + J^{ij}_{fg}(t) + J^{ij}_{fg}(r) = I^{ij}_{fg}(t) + I^{ij}_{fg}(r), \quad (A11.6)$$

$$I_{fg}^{ij}(t) = \sum_{h \in z_g^*} \left\{ U_{fgh}^{(iv)j} - U_{fgh}^{(vi)j} \right\} + \sum_{h \in z_g} \left\{ U_{hfg}^{i(vj)} - U_{fgh}^{i(jv)} \right\},$$

$$I_{gf}^{ij}(r) = \sum_{\lambda} \left\{ \left[W_{gf}^i(\lambda \to i) - W_{gf}^i(i \to \lambda) \right] + \left[W_{gf}^i(\lambda \to j) - W_{gf}^i(j \to \lambda) \right] \right\},$$

$$J_{fg}^{ij}(t) = \sum_{\kappa} \left[\left(u_k(f) + u_k(g) \right) \frac{\Delta \theta_{fg}^{ij}}{\Delta q_k} + \theta_{fg}^{ij} \frac{\Delta \left(u_i(f) + u_j(g) \right)}{\Delta q_k} \right], \qquad \text{(A11.7)}$$

$$J_{fg}^{ij}(r) = \left(\Omega(f) + \Omega(g) \right) \frac{\Delta \theta_{fg}^{ij}}{\Delta \lambda} + \theta_{fg}^{ij} \frac{\Delta \left(\Omega(f) + \Omega(g) \right)}{\Delta \lambda},$$

here the functions $U_{hfg}^{(vi)j}$ have the form $U_{hfg}^{(vi)j} = U_{hf}^{vi} \Psi_{hfg}^{ij}$, U_{hf}^{vi} is the thermal velocity of the particles of type i, moving from site f to free site h, the function Ψ_{hfg}^{ij} describes the influence of neighbouring particles j on this process (all functions defined below); $W_{\xi fg}^i(i \to \lambda) = W_{\xi f}^i(i \to \lambda) \Psi_{hfg}^{ij}$, $W_{\xi f}^i(i \to \lambda)$ is the thermal velocity of the change of orientation of the particle i at site f to particle λ; function Ψ_{fg}^{ij} takes into account the effect on the rotation of the particle i in the presence of a neighboring site f particles j in the site g, $\Omega(f)$ is the average angular velocity of rotation of the mixture in site f.

Formula (A11.6) explicitly takes into account that the evolution of pairs in adjacent cells is realized in two ways. There is a fast relaxation mechanism of changes in the pair distribution function due to the thermal motion of the molecules and the slow mechanism by molecular flows (movement of the thermostat). The fast mechanism is responsible for the contributions of elementary displacements $I_{fg}^{ij}(t)$ and rotations $I_{fg}^{ij}(r)$, and the slow mechanism – contributions $J_{fg}^{ij}(t)$ and $J_{fg}^{ij}(r)$, defined by (A11.7). They have the same structure as the continuity equation for the unary functions. Because of the diversity of the local values of translational and rotational velocities in neighbouring cells local microflows may affect the neighbouring pairs of particles (disconnect and form these pairs). Here, as in section 31, it is assumed that there is no correlation between the average local translational ($\mathbf{u}(f)$ and $\mathbf{u}(g)$) and rotational ($\Omega(f)$ and $\Omega(g)$) velocities in the neighbouring cells. Equations (A11.6) and (A11.7) are equivalent to the Smoluchowski equation [7, 8]:$I_{fg}^{ij}(t)$ – a discrete form of writing in the LGM the momentum-averaged collision integral of the mixture particles in the Fokker–Planck equation for the translational motion (the latter is obtained in the framework of a strictly kinetic approach to dense liquids [7, 8]);

$I_{fg}^{ij}(t)$ – a similar expression for rotational movement. Recall that in dense fluids, unlike rarefied gas, the non-equilibrium flow of momentum and energy due to the translational motion depends on the coordinates of the particles, and not on their momentum distribution. The same applies to the orientation coordinates of the molecules in the dense phases.

Since the molecular flows in the narrow pores are small in relation to the average thermal velocity of the molecules, the structure of (A11.6) corresponds to the hierarchy of the Bogolyubov time for the evolution of unary and pairwise distribution functions [9, 10]. Imagine θ_{fg}^{ij} as $\theta_{fg}^{ij} = \theta_{fg}^{*ij}\left[1 + \delta_{fg}^{ij}\right]$, where the asterisk denotes the affiliation of the pair function to the equilibrium distribution (see below), and its absence – to the non-equilibrium distribution, δ_{fg}^{ij} is the sought correction to the equilibrium distribution. It is determined from the condition that the explicit time dependence of the pair is equal to zero, according to the hierarchy of the time for the establishment of the equilibrium state for the unary and pair distribution functions [10]. After expansion in the small parameter of the equations written above, in the first approximation we obtain a solution

$$\delta_{fg}^{ij} = -\frac{J_{fg}^{ij}(t) + J_{fg}^{ij}(r)}{b_1 U_{hfg}^{(vi)j} + b_2 W_{\xi fg}^{j}(i \to \lambda)}, \tag{A11.8}$$

where $b_{1,2}$ are the numerical coefficients of the approximation functions, depending on the lattice structure and the size and shape of the particles. Formula (A11.8) has a very simple structure – it is the ratio of the microscopic characteristics of the velocities of the and thermal velocities of the particles in the cell f. Formula (A11.8) is obtained in one-step approximation in time and space coordinates for the central pair of the sites fg. Upon closer examination, by analogy with [10], it is necessary to solve a linear system of equations for the complete set of corrections δ_{fg}^{ij} for all spatial and orientation coordinates, but the structure of solutions (A11.8) through the ratio of the local microhydrodynamic and thermal velocities is retained. Everywhere below in the formulas for local characteristics, we use the non-equilibrium function θ_{fg}^{ij}.

Thus, the equations of the local distribution functions contain the equilibrium distribution functions and non-equilibrium amendments that are expressed through the local hydrodynamic and the average thermal velocities of the molecules. It is usually assumed that the characteristic size of the hydrodynamic scale is not smaller than the maximum value of b_m, d_m, n_m. Since the values of b_m, d_m, $n_m \geq \lambda_c$ (in this application λ_c is the lattice constant), then going to the hydrodynamic level raises the question of the need for subsequent coarsening of the spatial description of the particle distribution in comparison with the above procedure for constructing the equations. This procedure is discussed in [2].

Transport equations
We will consider a non-equilibrium state of the system when a pressure gradient forms along the axis of the pore, leading to the flow of the mixture. For single-site particles the volume element is the cell volume v_0, each cell has z faces. In fact, the equation [1, 2] for the single-site particles belong to the movement of the point (centre of mass) and they do not change, as they reflect the translational movement. Consideration of the size of the large particles is associated with the abandonment of the point review and the description of molecular motion becomes similar to the description of the motion of the 'bulk' solid body [11], therefore additionally we should have equations for the angular velocities of the particles.

Let us assume that the amount of the component of the mixture i flows through the surface of the cell f per unit time; this amount of is equal to $\theta_f^j \mathbf{u}_i(f)$, where $\mathbf{u}_i(f) \equiv \int \mathbf{v}_f^i \theta_f^j (\mathbf{v}_f^i) d\mathbf{v}_f^i$ is the vector of the average local flow velocity of the molecules i in the cell f, and the average is taken across all directions and the absolute values of the velocity $\mathbf{v}_f^i$ of the molecule i in the cell. As usual, we assume that the flow is positive, if the mixture comes out of the cell, and negative if it flows into the cell. Additionally, the component i in the mixture flow may rotate in the vicinity of the cell f with the average angular speed $\Omega_i(f)$.

$(\nabla \cdot \alpha) \equiv \mathbf{div}(\alpha) = \Delta(\alpha)/\Delta x + \Delta(\alpha)/\Delta y + \Delta(\alpha)/\Delta z$ denotes the difference derivative, where the differences Δx, Δy, Δz are equal to the linear dimension of the cell λ_c. Then the local equation of continuity for the centres of mass of the particles i in the mixture of Φ components can be written as

$$\frac{\partial \rho_f^i}{\partial \tau} = -(\nabla \cdot \mathbf{J}_f^i) + \mathbf{I}_f(i), \quad \mathbf{J}_f^i = \rho_f^i \mathbf{u}(f) + \mathbf{j}_i(f), \quad (A11.9)$$

where $\mathbf{J}_f^i$ is the mass flow of component i, consisting of convective $(\rho_f^i \mathbf{u}(f))$ and diffusion $(j_i(f))$ flows. Here we use the notation

$\rho_f^i = \dfrac{m_i \theta_f^i}{v_0}$ – average density of the mixture, and $\mathbf{u}(f) = \sum\limits_{i=1}^{\Phi} y_f^i \mathbf{u}_i(f)$ – the average flow rate in cell f, $y_f^i = \dfrac{\rho_f^i}{\rho_f}$, $\rho_f = \sum\limits_{i=1}^{\Phi} \rho_f^i$. The average speed of the convective flow $\mathbf{u}(f)$ is measured from the stationary wall, and the diffusion flow refers to the coordinate system moving with the average flow velocity. Local thermal motion of the molecules is controlled by local gradients of variables describing the equilibrium state of the mixture of fluids: the partial concentrations of the components, the total pressure and temperature [12, 13], so that the total flow of each component can be written as $\mathbf{j}_i(f) = \mathbf{j}_j^{(1)}(f) + \mathbf{j}_j^{(2)}(f) + \mathbf{j}_j^{(3)}(f)$, where $\mathbf{j}_j^{(1)}(f)$ is the molecular diffusion under the influence of a concentration gradient, $\mathbf{j}_i^{(2)}(f)$ is the diffusion due to the pressure gradient, $\mathbf{j}_i^{(3)}(f)$ is the thermal diffusion under the influence of the temperature gradient. There is no so-called contribution of forced diffusion under the influence of external potentials [13], since the system has no external electric or magnetic fields, and the relatively short-range adsorption potential of the pore walls is taken into account in the values of the activation energy of hopping of the particles between the cells.

Term $I_f(i)$ is the contribution of the reorientation of particles i with the centre of mass of these particles at the site f. The reorientation of the particle is considered as a 'reaction' with the change of value λ. In the absence of the flow of the mixture the discrete change of orientation of the particle i is described by: $I_f(i) = \sum\left[W_{\xi f}(k \to i) - W_{\xi f}(i \to k)\right]$ where $W_{\xi f}(i \to k)$ is the rate of transition of particle i to particle k, which is the same molecule m, but with different orientation λ, $1 \leq \lambda \leq L$. In the case of a large number of possible orientations L we use the continuum description $I_f(i)$ using the rotational diffusion coefficient [4, 14, 15].

In the presence of the flow of the mixture the change of the orientations of the components may be caused by rotation of the local volume with the average angular velocity $\Omega_f = \sum\limits_{i=1}^{\Phi} z_f^i \Omega_i(f)$, where $z_f^i(\gamma\phi)$ is the average proportion of component $(\gamma\phi)$ of the tensor of the momentum of inertia of particle i at site f, which is defined as $\mathbf{z}_f^i = x_f^i \mathbf{I}_i / \mathbf{I}(f)$, $\mathbf{I}_i$ is the tensor of the momentum of inertia of the particle i, and $\mathbf{I}(f) = \sum\limits_{i=1}^{\Phi} x_f^i \mathbf{I}_i$ is the average momentum of inertia of the mixture in the vicinity of the site $f\left(x_f^i = \theta_f^i / \theta_f, \theta_f = \sum_{i=1}^{\Phi} \theta_f^i\right)$.

Then, by analogy with the convective and diffusive fluxes of the translational motion of the centre of mass, the term $I_f(i)$ in equation (A11.11) can be expressed in terms of the local convective ($\mathbf{I}(f) \cdot \Omega_f$) and diffusion ($\mathbf{H}_i(f)$) flows of rotational motion:

$$I_f(i) = \sum_k \left[W_{gf}(k \rightarrow i) - W_{gf}(i \rightarrow k)\right] + \rho_f^i I_i \Omega_f = -\left(\nabla \cdot \mathbf{I}_f^i\right),$$

$$\mathbf{I}_f^i = \left(\mathbf{I}(f) \cdot \Omega_f\right) + \mathbf{H}_i(f). \tag{A11.10}$$

As above, we represent the contribution of rotational diffusion in the form of $\mathbf{H}_i(f) = \mathbf{H}_i^{(1)}(f) + \mathbf{H}_i^{(2)}(f) + \mathbf{H}_i^{(3)}(f)$, where $\mathbf{H}_i^{(1)}(f)$ is the rotational diffusion under the influence of the concentration gradient, $\mathbf{H}_i^{(2)}(f)$ is the rotational diffusion due to the hydrostatic pressure gradient, $\mathbf{H}_i^{(3)}(f)$ is the rotational thermal diffusion under the effect of the temperature gradient. The short-term adsorption potential of the pore walls is taken into account in the activation energies of the elementary processes of reorientation of components i in the expressions of the velocity $W_f(i \rightarrow k)$.

As a result, equation (A11.9) remains the same as the continuity equation in continuum mechanics [16] and includes the account of the microstructure of the medium. In the case of the single-site particles it relates to the lower limit of the volumes for the transport equations applicable to randomly selected small macrovolumes. Therefore, it contains the difference derivatives of the local values of density and average velocities of the molecules over distances defined on the sites of the lattice structure. However, for larger particles the equation (A11.9) describes the transfer of particles at a very shallow λ_c-scale which reflects the molecular but not hydrodynamic level of description. To go to the hydrodynamic level of description we must smear the molecular level.

Local rotational motions of the medium are modified by the equations of transfer of the momentum and energy, and they also lead to an additional equation for the average angular velocity of the mixture Ω_f.

The local equation of translational motion of the mixture is written as

$$\frac{\partial(\rho(f)u(f))}{\partial t} = -(\nabla \cdot \Phi_u(f)) + \sum_{i=1}^{\Phi} \rho_f^i F_i(f), \tag{A11.11}$$

$$\Phi_u(f) = \rho(f)\mathbf{u}(f)\mathbf{u}(f) + \pi(f),$$

where $\Phi_u(f)$ is the flow of momentum in the cell f, $\rho(f) = \sum_{i=1}^{\Phi} \rho_f^i$, $\pi(f) = P(f)\delta - \sigma(f)$, $\pi(f)$ is the pressure tensor and $P = \sum^{s-1} P_f^i$ is the total (static) pressure in the cell f, δ is the unit tensor, $F_i(f)$ is the external force acting on the particle i (see (30.3)), $\sigma(f)$ is the component of the momentum flux due to viscosity. The relationship of $\sigma(f)$ with the velocity gradient has the usual form [16] of the components of the viscous stress tensor $\sigma_{\gamma\phi}(f)$: $\sigma_{\gamma\phi}(f) = \sigma_{\gamma\phi}^1(f) + \sigma_{\gamma\phi}^2(f) + \sigma_{\gamma\phi}^3(f)$, where $\sigma_{\gamma\phi}^1(f) = \eta_{\gamma\phi}(f)\Delta u_\gamma(f)/\Delta\phi$, $\sigma_{\gamma\gamma}^2(f) = \xi_{\gamma\gamma}(f)\Delta u_\gamma(f)/\Delta\gamma$. Due to the need to consider the anisotropy of the mixture near the walls of the pores the first two components are presented in the non-symmetrized form (unlike traditional symmetrized form [16]), here $\eta_{\gamma\phi}(f)$ and $\xi_{\gamma\phi}(f)$ are the local components (ik) of the tensors of shear and bulk viscosity in the cell f.

The third component of the viscous stress tensor describes the rotational viscosity: $\sigma_{\gamma\phi}^3(f) = \eta_{\gamma\phi}^\gamma(f)\left[2\Omega_f - \mathrm{rot}(\mathbf{u}(f))\right]$, where $\eta_{\gamma\phi}^\gamma(f)$ is the component ($\gamma\phi$) of the tensor $\boldsymbol{\eta}_r(f)$ of the rotational viscosity of the mixture in the cell f.

The local equation of the average angular velocity Ω_f reflects the continuation of the internal angular momentum of the mixture $d(\mathbf{I}\Omega)_f/dt = -(\nabla \cdot \mathbf{L}_f)$, where $\mathbf{L}_f = \langle n\mathbf{CI}\Omega\rangle_f$ the tensor of the momentum flux in the vicinity of the cell f. As usual the symbol of the average $\langle\ \rangle$ denotes averaging using the non-equilibrium velocity distribution function $\mathbf{C}$ of the *translational* motion of all components of the mixture [3, 4]. Quantity $\mathbf{L}_f$ plays exactly the same role when considering the contributions of rotational motion as the stress tensor $\pi(f) = \langle\rho\mathbf{CC}\rangle_f$ in the neighborhood of site f in the derivation of equations for the momentum transfer. Since our microhydrodynamic approach uses the equilibrium function, then all average characteristics are obtained through changes in the populations of particles by different sorts of sites in the elementary processes of displacement and rotation of the components of the mixture. Therefore, the equation for the conservation of angular momentum is obtained similarly to the derivation of the momentum conservation of the translational motion

$$\mathbf{I}_f \frac{d\Omega_f}{dt} = \mathrm{grad}\left(\eta_{rs}(f)\,\mathrm{div}\langle\Omega\rangle_f\right) - \mathrm{rot}\left(\eta_{rs}(f)\,\mathrm{rot}\,\Omega_f\right) - \quad \text{(P11.12)}$$

$$-2\eta_r(f)\left[2\Omega_f - \mathrm{rot}(\mathbf{u}(f))\right] + \rho_f\mathbf{K}(f),$$

where $\mathbf{I}_f$ is the tensor of the average inertia moment introduced above (if the average moment of inertia relates to per unit mass, $\mathbf{I}(f)$ on the left side of (P11.12) should be replaced by $\mathbf{I}_f \rho_f$). In the near-wall region, this characteristic is determined by taking into account the contribution of the rotation of the molecules relative to contact with the wall. Thus for the first ($f = 1$) layer $\mathbf{I}_1$ is determined only by the moment of inertia with respect to this contact. For the second ($f = 2$) layer $\mathbf{I}_2$ is defined via the moments of inertia with respect to this contact, and also through the moment restricted with respect to the rotation angles relative to the centre of mass located in the second layer (if this is permitted by the dimensions of the molecules). Symbol $\mathbf{K}(f)$ denotes the mass moments in the site f, which may occur due to the orienting influence of the external field (electrical, magnetic, etc.).

The structure of (A11.12) is consistent with the constructions of [17, 18], if we assume that the first two terms on the right are determined by additional types of rotation of the particles relative to the pore wall. Symbol $\boldsymbol{\eta}_{rs}(f)$ indicates viscosity coefficients additional to the tensor $\boldsymbol{\eta}_r(f)$, which are associated with the specifics of the elementary movements of molecules near a solid wall (they are present in the boundary conditions [17, 18]). However, one should distinguish between different types of elementary movements under the influence of the surface potential. For example, in [19] it was shown that the additional flow characteristics, such as the coefficient of sliding friction, which usually appears near the interface during the formal constructions of continuum mechanics [20], can be expressed at the molecular level through the same basic movements and distributions function, like the 'normal' shear viscosity coefficient. Therefore, this factor becomes a tensor due to the influence of the surface potential, even for single-site particles. In the absence of the first two terms the equation (A11.12) changes to the appropriate equation of non-equilibrium thermodynamics [21].

With increasing differences in the size and in the implementation of ordered state of the mixture components the rotational motion fundamentally changes its contribution to the transport characteristics and, accordingly, their description is complicated: instead of one coefficient of rotational viscosity $\eta_r(f)$ the equation for the isotropic bulk mixtures has additional viscosity coefficients [21–24]. For the bulk phase it was shown in [15] that the total set of the non-zero components of the viscosity tensor is determined by two independent vectors: the polar vector of the translational velocity and the axial angular momentum vector, in contrast to earlier studies [25–27]. In the presence of the pore walls there appear further viscosity coefficients reflecting the effects of limited rotation and rotation of the molecules with respect to the contact with the wall. The specific set of viscosities $\eta_r(f)$ and $\eta_{rs}(f)$ is defined by the rules of tensor transformation of the flows reflecting the properties of the molecules and the pore walls at which their symmetry properties remain unchanged.

The energy equation for a mixture of components with regard to their rotation includes the energy of translational and rotational motions:

$$\frac{\partial \rho(f)\left\{U(f)+0.5(f)^2+0.5I_f\Omega_f^2\right\}}{\partial t}=$$

$$=-(\nabla\cdot e(f))+\sum_{i=1}^{\Phi}\left(J_f^i\cdot F_i(f)\right)+\rho(f)\left(K(f)\cdot\Omega_f\right),\ \text{(A11.13)}$$

$$\mathbf{e}(f)=\rho(f)\left\{U(f)+P(f)v_0+\frac{u(f)^2}{2}+\frac{I_f\Omega_f^2}{2}\right\}\mathbf{u}(f)+\mathbf{q}(f)-$$

$$-\left[\left(\sigma^1(f)+\sigma^2(f)\right)\cdot\mathbf{u}(f)\right]+\rho(f)\left(\Omega_f\cdot\sigma^3(f)\right),$$

where $e(f)$ and $U(f)$ is the total energy flux and the internal energy of the fluid in the cell f, $q(f) = q^{(1)}(f) + q^{(2)}(f) + q^{(3)}(f)$ is the energy flux relative to the average mass velocity $u(f)$ and the average speed of rotation Ω_f, comprising of $q^{(1)}(f)$ – the energy flux due to thermal conductivity, $q^{(2)}(f)$ – the energy flux induced by mutual diffusion, $q^{(3)}(f)$ – the flow due to the concentration gradient (Dufour effect) [21]. Here, the average moment of inertia I_f is related to the unit of mass – the specificity of its definitions associated with different elementary rotations explained above.

The dynamics equations contain dissipative coefficients. Before considering these factors we construct the expressions for the equilibrium distribution of the components that are needed, according to (A11.8), to calculate the dissipative coefficients.

The equilibrium distribution of molecules in the pore

Intermolecular interactions of all components of the mixture will be considered in the quasi-chemical approximation. To use we need the local functions $\theta^i_{\{f\}}$ and $\theta^{ij}_{\{f\}\{g\}} = \left\langle\Gamma_f^i\Gamma_g^i\right\rangle$, describing the distribution of particles in a lattice: $\theta^i_{\{f\}}$ is the probability of filling the site f by particle i; function $\theta_{\{f\}}^{M_i v}$ is the probability that the portion $\{f\}$ with size M_i is free because the symbol M_{iv} is a free site of the lattice with size M_i in which can the particle i can be placed; $\theta_{\{f\}\{g\}}^{ij}$ is the probability of finding the particle i in the area $\{f\}$ and the particle j in the area $\{g\}$, where $1 \leq j \leq (\Phi + 1)$. The local partial isotherm equations relate the probability that a given portion $\{f\}$ is filled with particle i or is free. In this model the adsorption isotherms for particle i can be written as follows:

$$a_{\{f\}}^i p_i \theta_{\{f\}}^{M_i v} = \theta_{\{f\}}^i \Lambda_{\{f\}}^i, \quad \Lambda_{\{f\}}^i = \sum_{\alpha(j)} \prod_j t_{\{f\}\{g\}}^{ij} \exp\left[-\beta \varepsilon_{\{f\}\{g\}}^{ij}\right], \quad (A11.14)$$

where $p \equiv p_m$ is the partial pressure of the adsorptive m. Function $\Lambda_{\{f\}}^i$ is a function that takes into account the effect of lateral interactions between the molecules. Summation over $\alpha(j)$ means the sum over all possible locations of neighboring molecules j.

Function $t_{\{f\}\{g\}}^{ij} = \theta_{\{f\}\{g\}}^{ij} / \theta_{\{f\}}^i$ is the conditional probability of finding the particle i near the area $\{f\}$ close to particle j in area $\{g\}$, where $1 \leq j \leq (\Phi + 1)$. They are determined by solving the following system of algebraic equations:

$$\theta_{\{f\}\{g\}}^{ij} \theta_{\{f\}\{g\}}^{M_i v, M_i v} = \theta_{\{f\}\{g\}}^{i, M_j v} \theta_{\{f\}\{g\}}^{M_i v, j} \exp\left[-\beta \varepsilon_{\{f\}\{g\}}^{ij}\right], \quad \sum_j \theta_{\{f\}\{g\}}^{ij} = \theta_{\{f\}}^i. \quad (A11.15)$$

Equations (A11.14) and (A11.15) are closed by the normalization condition for the local filling of the sites and by the method of calculating functions $\theta_{\{f\}}^{M_i v}$:

$$\theta_{\{f\}}^{M_i v} = \theta_f^v \prod_h t_{hh+1}^{vv}, \quad (A11.16)$$

where the index h numerates $(M_i - 1)$ sites of the blocked particle i with size M_i, θ_f^v is the probability that site f is free, t_{hh+1}^{vv} is the conditional probability finding a free site with the number $(h + 1)$ close to the free site with index h.

Particular cases of the equations for rigid rods of length n and for plates with the size $b \times d$ are obtained from the equations considered for the three-dimensional case $b \times d \times n$ by simplifications $b = d = 1$ and $n = 1$, reducing the size and number of measurements of the molecules, respectively, in the first and second cases.

The system of equations (A11.14)–(A11.16), with respect to $\theta_{\{f\}}^i$ defines all the partial isotherms $\omega_m(\{p\}) = \sum_{i=1}^{s_m} \sum_{f=1}^{N} \omega_{\{f\}}^i(\{p\})/N$, where

$i \in (m, \lambda)$, the total adsorption isotherm of $\omega(\{p\}) = \sum\limits_{m=1}^{\psi} \omega_m(\{p\})$ and all local equilibrium characteristics of the mixture required to calculate the flows.

The local equation of state (or 'expansion pressure' in the terminology of LGM) in a dense mixture of different monolayers is written as [28]

$$\beta P(f)v_0 = -\ln(\theta_f^v) - \sum_{g \in \Pi_l} \ln\left[\frac{\theta_{\{f\}\{g\}}^{vv}}{(\theta_f^v)^2}\right]/2, \qquad (A11.17)$$

where θ_f^v is the proportion of vacant sites and $\theta_{\{f\}\{g\}}^{vv}$ is the probability of finding two vacant sites in the sites f and g; here the sum over g denotes the summation over the neighbouring sites around the site f. At low densities, the bulk phase equation (A11.17) gives the equation of state of an ideal gas $\beta Pv_0 = -\ln(\theta_v) = \theta$ or $\beta P = n$.

Thermal velocity of the particles
The velocities of the thermal translational and rotational motion of the mixture components commensurate in size and in the case of large components are given in Appendix 7. Therefore, here we present only the factors that are present in the equation (A11.7)

translational velocity

$$\Psi_{hfg}^{ij} = t_{fgh}^{ivj}(\omega_r) E_{fgh}^{ivj}(\omega_r) \Big/ \sum_{k=1}^{\Phi} t_{fgh}^{ivk}(\omega_r) E_{fgh}^{ivk}(\omega_r). \quad (A11.18)$$

rotational velocity

$$\Psi_{fg}^{ij} = t_{fg}^{ivj}(\omega_r) E_{fg}^{ivj}(\omega_r) \Big/ \sum_{\ell=1}^{\Phi} t_{fg}^{iv\ell}(\omega_r) E_{fg}^{iv\ell}(\omega_r). \quad (A11.19)$$

The expressions (A11.18) and (A11.19) are directly reformulated for elementary rotations relative to contact with the wall and with constraints on the rotation angles of the near-wall layers due to the impermeability of the pore wall: only the form of the free volume required for this type of rotation changes.

Dissipative coefficients [1, 2, 29]
All dissipative coefficients are constructed by calculating the amount of the transferred property in forward and reverse directions through the allocated reference plane 0. The first channel is connected with the movement of particles as in a rarefied gas. The second channel

is determined by collisions between particles when the particle in question may not cross this plane 0 if its trajectory is blocked by other particles situated in the sites to plane 0, or a particle in close proximity on the other side of the plane 0 and preventing its crossing of the plane.

Mass transfer is possible only for the first channel. The additivity of the contribution of two channels for the transfer of momentum and energy is a direct consequence of the two possibilities for the examined particle to cross or not to cross the selected plane. For the shear viscosity and thermal conductivity coefficients we have the following structure of the expressions: $\eta_{\xi fg} = \eta_{\xi fg}^{(1)} + \eta_{fg}^{(2)}$ and $\kappa_{\xi fg} = \kappa_{\xi fg}^{(1)} + \eta_{fg}^{(2)}$, where the superscript 1 or 2 represents the number of the transfer channel.

For large particles besides translational motion we must consider contributions of rotational motion which changes the orientation of the particles. As for the translational movement, rotational diffusion takes place in the first channel, characterized by the presence of free space (except for the spherical particles, in which the centre of mass coincides with the geometrical centre). The transfer of the angular momentum and energy is possible in both channels. In the transition from the orientation state i to the 'nearest' state k the reference plane 0 is the plane that is perpendicular to the axial vector of rotation from state i to state k and is situated in the middle of these links.

Translational movement is characterized by the following diffusion coefficients: transfer of the labels, mass transfer under the influence of a concentration gradient, pressure, external field and temperature. The first coefficient in a multicomponent mixture characterizes the process of the thermal motion of labeled particles of type i, $1 \le i \le \Phi$, under the equilibrium conditions between neighbouring sites f and g, located in neighbouring planes at a distance χ. The remaining coefficients characterize the process of mass transfer when the relevant parameters of the system change.

The local transfer coefficient of the label depends on the mixture density and intermolecular interaction of the label with all its neighbours. For a labeled particle of type i we have an expression

$$D_{\xi fg}^{i*}(\chi) = z_{fg}^{*}(\chi)\chi^2 U_{\xi fg}^{iv}(\chi)/\theta_f^i, \tag{A11.20}$$

where $z_{fg}^{*}(\chi)$ is the number of possible hops between sites over the distance χ for each cell f, the speed of hopping $U_{\xi fg}^{iv}(\chi)$ is defined in Appendix 7.

The local mass transfer coefficients are obtained from a consideration of the flow J_i of molecules of type i as the difference in hopping velocities in free sites in the forward and backward directions along the axis of the flow. We expand the flow with respect to the independent parameters of the state of the system A_k, where A_k are the mole fractions $x_k (1 \leq k \leq \Phi)$, pressure and temperature:

$$J_{fg}^i = z_{fg}^* \chi \frac{U_{fg\xi}^{vi}(\chi) - U_{\xi fg}^{iv}(\chi)}{v_0} =$$

$$= z_{fg}^* \chi \sum_{j=1}^{\Phi} \frac{U_{fg\xi}^{vi(j)}(\chi) - U_{\xi fg}^{(j)iv}(\chi)}{v_0} = \sum_{k=1}^{\Phi} D_{\xi fg}^{ik}(\chi) \operatorname{grad}(A_k),$$

$$\tag{A11.21}$$

where we introduced the 'partial' diffusion coefficients of mass transfer in a multicomponent mixture

$$D_{\xi fg}^{ik}(\chi) = \sum_{j=1}^{\Phi} D_{\xi fg}^{(j)ik}(\chi),$$

$$D_{\xi fg}^{(j)ik}(\chi) = z_{fg}^*(\chi) \chi^2 U_{\xi fg}^{(j)iv}(\chi) \frac{d}{dA_k} \frac{\ln\left\{ Y_{fg}^i(\chi) Y_{f\xi}^{ij}(\rho) a_f^{i*} p_i \right\}}{\theta_f v_0},$$

$$Y_{fg}^i(\chi) = \frac{\theta_{fg}^{vv}(\chi)}{t_{fg}^{iv}(\chi) t_{gf}^{iv}(\chi)}, \quad Y_{fg\xi}^{ij}(\rho) = \frac{\langle t_{f\xi}^{ij}(\rho) \rangle}{S_{f\xi}^i(\rho)}. \tag{A11.22}$$

The expression under the logarithm (A11.22) depends only on the functions of the equilibrium distribution of components. With increasing distance χ the factor $Y_{fg}(\chi)$ tends rapidly to unity. Value $(a_f^{i*} p_i)$ is included in the isotherm equation, which relates the difference between the chemical potentials of component i and vacancies (see before). The values $(a_f^{i*} p_i)$ in (A11.22) are determined from the equations for the partial isotherms (A11.14). Here, $a_i^* = a^i/ a_i^c$ where a_i^* is the component that does not depend on the density of the function a_i, a factor a_i^c considers all concentration contributions. In particular, if we assume that the vibrations do not depend on density, then $a^{*i} = F_{vib}^i / F_{vib}^{i0}$ and $a_i^c = V_i(\theta)/V_i(\theta = 0)$. Formula (A11.22) for medium mass transfer coefficients of the molecules i when the mole fractions x_j change transfers to the well-known expression [10, 21], which relates the mass transfer coefficient with the gradient of the chemical potential of component i.

The mutual diffusion coefficient is expressed by the formulas (A11.22) as

$$D_{\xi fg}^{i,k}(\chi) = D_{\xi fg}^{ik}(\chi) - x_f^i \sum_{j=1}^{\Phi} D_{\xi fg}^{ij}(\chi). \qquad (A11.23)$$

From this expression it follows that unlike the gas the mutual diffusion coefficient in the solid phase depends not only on the individual properties of the colliding molecules, but also on the intermolecular interactions. As a consequence, a significant role is played by the character of mutual arrangement of the components, i.e., the configuration effects (short- and long-range orders) and temperature have a strong influence on the mutual diffusion coefficients. Similarly, we obtain expressions for the coefficients of barodiffusion and forced diffusion [30]. The formula for the coefficient of thermal diffusion requires a more detailed analysis [30].

For rotary motion we can introduce similar coefficients of diffusion of labels and change the orientation. The diffusion coefficient of the label for rotary motion characterizes the thermal velocity of the reorientation of the particles between different orientational states under the equilibrium conditions, and the rotational diffusion coefficient characterizes the process of redistribution of molecules between different orientational states under the non-equilibrium conditions when the relevant system parameters (concentration, pressure, external field and temperature) change.

The rotational diffusion coefficient of the label has the form

$$D_{\xi f}^{i*} = z_{fg}^*(i \to k) W_{\xi f}(i \to k)/\theta_f^i, \qquad (A11.24)$$

where $z_{fg}^*(i \to k)$ is the number of possible rotations between the states $i \to k$ for cell f, and the rate of this rotation $W_{\xi f}(i \to k)$ is defined by the formulas in Appendix 7.

In the non-equilibrium conditions, we introduce the following 'local' in space (site f) and rotation angles (for the transition from i to k) coefficients rotational diffusion $D_{\xi f}^{i\lambda}(i \to k)$, where $1 \le \lambda \le \Phi + 2$, and $1 \le \lambda \le \Phi$ corresponds to the mole fractions x_f^λ, $\lambda = \Phi + 1$ corresponds to the pressure and $\lambda = \Phi + 2$ to temperature. They are obtained by considering the flow $J_{ik}(f)$ of the reorientation of the particles i to the particles k as the difference in the velocity of rotations (A11.24) around the centre of mass of the molecule m in forward and reverse directions between the appropriate orientations λ_1 (for i) and λ_2 (for k). We expand the flow with respect to the independent parameters of state of the system A_λ, where A_λ are mole fractions $(1 \leq \lambda \leq \Phi)$, pressure and temperature:

$$J_{ik}(f) = z_{fg}^*(i \to k) \frac{W_{\mathcal{E}f}(k \to i) - W_{\mathcal{E}f}(i \to k)}{v_0} =$$

$$= z_{fg}^*(i \to k) \sum_{j=1}^{\Phi} \frac{W_{\mathcal{E}f}^{(j)kv}(k \to i) - W_{\mathcal{E}f}^{(j)iv}(i \to k)}{v_0} =$$

$$= \sum_{\lambda=1}^{\Phi} D_{\mathcal{E}f}^{i\lambda}(i \to k)\,\mathrm{grad}(A_\lambda),$$

$$(A11.25)$$

where we introduced the 'partial' local diffusion coefficients of rotational diffusion

$$D_{\mathcal{E}f}^{i\lambda}(i \to k) = \sum_{j=1}^{\Phi} D_{\mathcal{E}f}^{(j)i\lambda}(i \to k),$$

$$D_{\mathcal{E}f}^{(j)i\lambda}(i \to k) = z_{fg}^*(i \to k) W_{\mathcal{E}f}^{(j)kv} \frac{1}{\theta_f v_0} \frac{d}{dA_\lambda} \ln\left[Y_{f\xi}^{ij}(\rho) a_f^{i*} p_i\right]$$

$$(A11.26)$$

and all functions under the logarithm were defined above in (A11.22).

The mutual diffusion coefficient of rotational movement is derived by analogy with formula (A11.23):

$$D_{\mathcal{E}f}^{i\lambda}(i \to k) = D_{\mathcal{E}f}^{i\lambda}(i \to k) - \mathbf{z}_f^i(\gamma\varphi) \sum_{j=1}^{\Phi} D_{\mathcal{E}f}^{ij}(i \to k), \quad (A11.27)$$

where components $(\gamma\varphi)$ are unambiguously associated with the type of rotation i in k. It should be noted that previously the rotational diffusion coefficients 'local' in the angles were not introduced. In macroscopic hydrodynamics they do not appear because of the assumption of the 'uniformity' changes of the rotational motions for different orientations. Instead we use the average rotational diffusion coefficient value, which characterizes changes in orientation of the molecule as a whole. (For example, most often this term refers to a change in the orientation of the 'director' in the nematic liquid crystals [22–24].) For the conventional rotational diffusion coefficient the values (A11.27) $D_{\mathcal{E}f}^i = \sum_{k,\lambda} D_{\mathcal{E}f}^{i,\lambda}(i \to k)$ must be averaged over all angles and the initial orientations. With this averaging the mean velocity of rotation $\Omega(f)$ of the local volume in the vicinity of the site f, with respect to which we introduced the mutual rotational diffusion coefficients (A11.27), is reduced.

In principle, the same procedure can be used to define (local and hydrodynamic) rotational diffusion coefficients in flows under the influence of pressure, external force and temperature.

When calculating the coefficients of viscosity and thermal conductivity, we assume (as for the single-site particles [1, 2, 29]) that the only process responsible for the relaxation of the medium at different external perturbations is the migration of molecules by the vacancy mechanism.

Then the contributions of the first and second channels for the shear viscosity coefficient are expressed as

$$\eta_{\xi fg}^{(1)} = \sum_{i=1}^{\Phi} m_i x_f^i D_{\xi fg}^{*i}(\chi), \quad \eta_{\xi fg}^{(2)} = \theta_f \sum_{ij} \frac{G_{fg}^{ij} \theta_{fg}^{ij}}{\lambda_c U_{\xi fg}(\chi)}, \quad \text{(A11.28)}$$

here $x_f^i = \theta_f^i / \theta_f$, $\theta_f = \sum^{\Phi} \theta_f^i$, $U_{\xi fg}(\chi) = \sum^{\Phi} U_{\xi fg}^{iv}(\chi)$, and G_{fg}^{ij} is the local component of the tensor of elasticity of the relationship $G_{fg}^{ij} = \dfrac{\partial^2 E_{fg}^{ij}}{\partial u_{\alpha\beta} \partial u_{\gamma\delta}}$, where $E = \langle H \rangle$ is the potential energy of the system with the Hamiltonian of type (A4.1), generalized for large molecules, as described in section 67 [1, 29]; $u_{\alpha\beta}$ is the component $(\alpha\beta)$ of the strain tensor in the vicinity of the pair of sites fg. If the effective interparticle potential is represented by the Lennard–Jones type potential $\varepsilon_{ij}(r) = 4\varepsilon_{ij}^0 \left\{ (\sigma_{ij}/r)^{12} - (\sigma_{ij}/r)^6 \right\}$, then by analogy with [1, 2, 29], we obtain an expression for the local coefficient of elasticity of the relationship $G_{fg}^{ij} = 48 E_{fg}^{*ij} / (r_{fg}^{min})^2$ of the neighbouring molecules ij, where r_{fg}^{min} is the equilibrium distance between the sites f and g, corresponding to the minimum of the potential

$$r_{fg}^{min} = \left[2 \left(\sum_r z_r \sum_j \left(\frac{\sigma_{ij}}{\eta_r} \right)^{12} \right) \middle/ \left(\sum_r z_r \sum_j \left(\frac{\sigma_{ij}}{\eta_r} \right)^6 \right) \right]^{1/6}, \quad \text{and the average}$$

value of the binding energy of the central molecule in the site f can be written as

$$E_{fg}^{*ij} = Q_j^i + \varepsilon_{fg}^{ij} \Delta_{fg} + \sum_{g \in \Pi_i - S_{rr} - \Delta_{fg}} \frac{1}{1 + \Delta_{fg}} \sum_k t_{fg}^{ik} \varepsilon_{fg}^{ik}. \quad \text{(A11.29)}$$

Similarly, we obtain a formula for the coefficient of bulk viscosity [2]:

$$\eta_{\xi fg}^{bulk} = b\theta_f \sum_{ij} \frac{G_{fg}^{ij}\theta_{fg}^{ij}}{\lambda_c U_{\xi fg}^{iv}(\chi)}, \qquad (A11.30)$$

where the numerical ratio b depends on the lattice structure, and also the size and shape of the particle. Formula (A11.30) shows that if we do not take into account the contributions of the internal degrees of freedom, the shear and bulk viscosities for a mixture of molecules have comparable values in the dense phase (this conclusion is in good agreement with experimental data for simple fluids and mixtures [31], for which $b = 4/3$).

Contributions of the two channels for the rotational viscosity $\eta_r(f) = \eta_{\xi f}^{(1)} + \eta_{\xi f}^{(2)}$ is written as follows:

$$\eta_{\xi f}^{(1)} = \sum_{i=1}^{\Phi} \mathbf{z}_f^i D_{\xi f}^{i*}, \qquad \eta_{\xi f}^{(2)} = \theta_f \sum_i \frac{H_f^i \theta_f^i}{W_{\xi f}}, \qquad (A11.31)$$

where $W_{\xi f} = \sum_{i=1}^{\Phi} \sum_k W_{\xi f}(i \to k)$, $k \in s_i(1)$, and H_f^i is the local component of the elasticity tensor of the relationship $H_f^i = \dfrac{\partial^2 E_f^i}{\partial \lambda_{\alpha\beta} \partial \lambda_{\gamma\delta}}$ ($\lambda_{\alpha\beta}$ is the component ($\alpha\beta$) of the bending tensor bending – rotation in the vicinity of the pair of sites fg). Here E_f^i is the potential energy of the molecule i, located in site f:

$$E_f^i = Q_f^i + \sum_{g \in \Pi_i - S_{f_1} - \Delta_{f_1}} \frac{1}{1 + \Delta_{fg}} \sum_k t_{fg}^{ij} \varepsilon_{fg}^{ij}, \qquad (A11.32)$$

It consists of the contribution of the adsorbate–adsorbent interaction and intermolecular interactions.

The thermal conductivity coefficient of the translational motion is expressed as

$$\kappa_{\xi fg}^{(1)} = \sum_{i=1}^{\Phi} x_f^i C_v^i(f) D_{\xi fg}^{*i}(\rho)/v_0, \quad \kappa_{\xi fg}^{(2)} = z_{fg}^* \lambda_c^2 \sum_{ij} C_v^i(f) v_{fg}^{ij} \theta_{fg}^{ij}/3, \,(A11.33)$$

similarly for rotational movement $\kappa_{\xi fg} = \kappa_{\xi fg}^{(1)} + \kappa_{\xi fg}^{(2)}$:

$$\kappa_{\xi fg}^{(1)} = \sum_{i=1}^{\Phi} \mathbf{z}_f^i C_v^i(f) D_{\xi f}^{i*}, \qquad \kappa_{\xi fg}^{(2)} = z_{fg}^* \sum_g \sum_{ij} C_v^i(f) \zeta_{fg}^{ij} \theta_{fg}^{ij}/3, (A11.34)$$

where the local specific heat $C_v^i(f)$ is composed of two contributions $C_v^i(f) = C_v^{i(1)}(f) + C_v^{i(2)}(f)$, corresponding to the contributions of the kinetic and potential energies. And the kinetic energy consists of the energy of translational and rotational motions:

$$C_v^{i(1)}(f) = 3k\left[1 + \left(S_{i\gamma}\Delta_{f1} + \sum_{g \in \Pi_i - S_{i\gamma} - \Delta_{f1}} \sum_j t_{fg}^{ij}\right) \middle/ \Pi_i\right], \quad C_v^{i(2)}(f) = \frac{dE_f^i}{dT},$$

$$(A11.35)$$

where $S_{i\gamma}$ is the area of the face γ of particle i, this value is excluded for particles near the surface ($f = 1$); Δ_{ij} is the Kronecker symbol: $\Delta_{ii} = 1$ and $\Delta_{ij} = 0$ for $i \neq j$. In the bulk phase at low densities the expression (A11.35) gives $C_v^{i(1)} = 3k/2$, corresponding to the translational motion of the particles in the gas volume, and for high densities $C_v^{i(1)} = 3k\sum_{j=1}^{\Phi} t_{ij}$ – to the oscillatory motion of the liquid phase [12, 14]. In formula (A11.36) the contribution of the kinetic energy is presented for a heterogeneous mixture of molecules in the form of interpolation, which reflects the local oscillations of the particles in the Einstein model (their frequencies are denoted by v_{fg}^{ij}) and ensures compliance with the limiting values of filling for all partial contributions of the mixture components. For rotary motion we adopted a similar interpolation dependence (diatomic molecules are excluded here) for 'pure' free rotation at low densities and high temperatures $C_v^{i(1)} = 3k/2$ for local 'swinging' of three-dimensional particles at low temperatures (their frequency is denoted by ζ_{fg}^{ij}) and high densities $C_v^{i(1)} = 3k$ [14]. Both interpolations exclude the contribution of the neighbouring vacancies to the vibrational motion of the centre of mass and the rotational oscillations of the central particle i: $1 \leq j \leq \Phi$.

Expression for $C_v^{i(2)}(f)$ is the configuration contribution to the specific heat in a heterogeneous system due to intermolecular interaction. In calculating the derivative with respect to temperature, the local density of the adsorbate is assumed constant. The internal energy of the fluid in the site f can be written as $E(f) = \sum_i E_i(f)\theta_f^i$ where the terms are defined in (A11.32).

To calculate the frequency of harmonic oscillations v_{fg}^{ij} and swinging ζ_{fg}^{ij} for adjacent pairs of the particles ij we use the averaged quasi-dimer model in which all the particles, surrounding central particle i, form the second part of the 'dimer' [2, 29]. This interpolation structure together with the formulas (A11.32) and (A11.35) makes it possible to evaluate the frequency of oscillation and swinging of molecule i at site f as follows:

$$v_{fg}^{ij} = \frac{\left\{ G_{fg}^{ij}/\mu_{fg}^{ij} \right\}^{1/2}}{2\pi r_{fg}^{\min}}, \quad \zeta_{fg}^{ij} = \frac{\left\{ H_f^i/I_i \right\}^{1/2}}{2\pi r_{fg}^{\min}}, \tag{A11.36}$$

where $\left(\mu_{fg}^{ij} \right)^{-1} = m_i^{-1} + 1 \Bigg/ \left[m_j + \sum_{h \in \Pi_i - S_{i\gamma} - \Delta_{f1} - \Delta gh} \sum_k t_{fh}^{ik} m_k + S_{i\gamma} \Delta_{f1} m_s \right]$, μ_{fg}^{ij} is the

reduced mass of the particle i in the layer f; m_s is the mass of the atom (or atoms) of the solid (adsorbent), if the adsorbate is in the surface layer, as one of its neighbors is necessarily an atom (or atoms) of the solid (depending on the type of bonds of the particles with the substrate), and the remaining nearest neighbours g contribute when averaging over all components of the mixture. The differences between E_f^i and E_{fg}^{*ij} follow from different functional relationships for the average potential energy and the oscillation frequency of the molecules as a function of the density of the fluid. In calculating the contribution of the energy transfer through collisions of particles in a dense fluid to the thermal conductivity it is assumed that the characteristic time of vibrational relaxation in the activation hopping of all components of the mixture is less than the characteristic time of their local redistribution (this condition is also necessary for use as a discrete analogue of the Smoluchowski equation (A11.6)) [1].

All derived dissipative coefficients are similarly rewritten (including specifics of the ρ-scale) for elementary rotations of the relative contact with the wall and with constraints on the angles of rotation determine by the pore wall when it is considered that one of the neighbouring particles is replaced by the corresponding contact, consisting of a group of the surface atoms.

References

1.   Tovbin Yu.K. // Khim. Fizika. 2004. V. 23, No. 12. P. 82.
2.   Tovbin Yu.K. // Khim. Fizika. 2005. V. 24, No. 9. P. 91.
3.   Chapman S., Cowling T. G., Mathematical Theory of Non-Uniform Gases. –
        Moscow: Izd. inostr. lit., 1960. – 541 p. [Cambridge University Press, Cambridge,
        1952].
4.   Ferziger J. H. H., Kaper H. G., Mathematical theory of transport processes in
        gases. – Moscow: Mir, 1976. – 554 p. [North_Holland, Amsterdam, The Nether-
        lands, 1972].
5.   Rott L.A., Statistical theory of molecular systems. Moscow: Nauka, 1979.
6.   Lifshitz E.M., LP Pitaevskii L.P., Theoretical Physics. X. Physical kinetics. Mos-
        cow: Nauka, 1979.
7.   Eisenschitz R., Statistical theory of irreversible processes. – Moscow: Izd. inostr.

lit., 1963. – 128 p. [Oxford University Press, London, 1948].

8.   Croxton C. A., Physics of the liquid state: A Statistical Mechanical Introduction. – Moscow: Mir, 1979. – 400 p. [Cambridge University Press, Cambridge, 1974].

9.   Bogolyubov N.N., Problems of dynamical theory in statistical physics. – Moscow: Gostekhizdat, 1946 [Wiley Interscience, New York, 1962].

10. Gurov K.P., Kartashkin B.A., Ugaste E.Yu., Interdiffusion in multiphase tion metal systems. Moscow: Nauka, 1981.

11. Landau L.D., Lifshitz E.M., Theoretical Physics. I. Mechanics. Moscow: Nauka, 1965.

12. Hirschfelder J. O., Curtiss C. F., Bird R. B., Molecular Theory of Gases and Liquids. – Moscow: Izd. inostr. lit., 1961. – 929 p. [Wiley, New York, 1954].

13. Bird R. B., Stewart W., Lightfoot E. N., Transport Phenomena. – Moscow: Khimiya, 1974. – 687 p. [John Wiley and Sons, New York, 1965].

14. Frenkel Ya. I., Kinetic Theory of Liquid. Leningrad: Nauka, 1975. – 592 p. [Oxford University Press, Oxford, 1946].

15. Kagan Yu.M., Afanasiev A.M. // Zh. eksp. teor. fiz. 1961. V. 41, No. 11. P. 1536.

16. Landau L.D., Lifshitz E.M., Theoretical Physics. VI. Hydrodynamics. Moscow: Nauka, 1986.

17. Aero E.L., Bulygin A.N., Kuvshinskii E.V. // Prikl. Matem. Mekh. 1965. V. 29. P. 297.

18. Aero E.L., Bessonov N.M. // Itogi nauki i tekhniki. Mekh. zhidk. gazov. 1989. V. 23. P. 237.

19. Tovbin Yu.K., Zhidkova L.K., Gvozdeva E.E. // Inzh.-fiz. zh. 2003. V. 76, No. 3. P. 124 [J. Engin. Physics and Thermophys., 2003, V. 76, No. 3, P. 619] .

20. Lamb G. Hydrodynamics. Leningrad: OGIZ 1947.

21. de Groot S. R., Mazur P., Non-Equilibrium Thermodynamics. Moscow: Mir, 1964. – 456 p. [(North-Holland, Amsterdam, Amsterdam, 1962].

22. Chandrasekhar S., Liquid Crystals. – Moscow: Mir, 1980. – 344 p. [Cambridge Univer. Press, Cambridge, 1977].

23. Kats E.I., Lebedev V.V., Dynamics of liquid crystals. Moscow: Nauka, 1988.

24. Pikin S.A., Structural transformations in liquid crystals. Moscow: Nauka, 1981.

25. Curtiss C.F. // J. Chem. Phys. 1956. V. 24. P. 225.

26. Curtiss C.F., Muckenfuss C. // J. Chem. Phys. 1957. V. 26. P. 1619.

27. Muckenfuss C., Curtiss C.F. // J. Chem. Phys. 1958. V. 29. P. 1257.

28. Tovbin Yu.K. // Zh. fiz. khimii. 1992. V. 66, No. 5. P. 1395 [Russ. J. Phys. Chem., 1992 V. 66, No. 5, P. 741] .

29. Tovbin Yu.K. // Khim. fizika. 2006. V. 26, No. 12. P. 43-68.

30. Tovbin Yu.K. // Teoreticheski osnovy khim. tekhnol. 2005. V. 6. P. 613 [Theor. Found. Chem. Engin. 2013. V. 47. No 6. P. 734] .

31. Mikhailov I.G., Solov'ev V.A., Syrnikov Yu.P., Fundamentals of molecular acoustics. Moscow: Nauka, 1964.

Appendix 12

The mutual diffusion coefficient in binary mixtures

The proposed microhydrodynamic approach to describe the flows of simple molecules in narrow pores on the basis of a simple molecular model – the lattice gas model (LGM) – allows us to consider a wide range of the density of substances. Accordingly, it is necessary that the expressions for the transport coefficients, in particular for the mutual diffusion coefficient, were consistent with both the kinetic theory of gases and the kinetic theory of liquids and solids. We consider the process of diffusive transport of components of a binary mixture of simple molecules of spherical shape, having approximately the same dimensions. This system contains only one mutual diffusion coefficient and is a convenient example to illustrate the essence of the question.

Determination of the mutual diffusion coefficient
Rarefied gas
In the first-order approximation of the Chapman–Enskog theory of the gas mixture [1–5] the diffusion velocity of i-th component $\mathbf{V}_i$ at constant pressure P and temperature T of the system and in the absence of external fields is defined as follows:

$$\mathbf{V}_i = -\sum_j D_{ij} \operatorname{grad}(x_j), \qquad (A12.1)$$

where D_{ij} is the diffusion coefficient of multicomponent mixtures, which are related by $\sum x_i D_{ij} = 0$, $D_{ii} > 0$ and $D_{ij} = D_{ji}$; where $x_i = n_i/n$ is the mole fraction of component i, n_i is the numerical density of component i per unit volume (here we use the characteristic particle system). For a binary mixture $x_1 + x_2 = 1$ and $n = n_1 + n_2$, therefore it has a unique mutual diffusion coefficient $D_{1,2}$ in a coordinate system,

moving at an average velocity $w = \sum_i n_i v_i / n = v + \sum_i x_i V_i$, where v_i is

the average velocity of component of $i, v = \sum_i x_i v_i$ is the average

molar flow rate.

Mutual diffusion coefficient $D_{1,2}$ is based on the presented equations, if we express all the coefficients D_{ij} through D_{12}

$$
\mathbf{V}_1 - \mathbf{V}_2 = -\left(D_{11}\, \mathrm{grad}\,(x_1) + D_{12}\, \mathrm{grad}\,(x_2) \right) + \left(D_{21}\, \mathrm{grad}\,(x_1) + D_{22}\, \mathrm{grad}\,(x_2) \right) =
$$

$$
= -\left(-\frac{D_{12}n_2}{n_1}\, \mathrm{grad}\,(x_1) + D_{12}\mathrm{grad}\,(x_2) \right) + \left(D_{12}\, \mathrm{grad}\,(x_1) - \frac{D_{12}n_2}{n_2}\, \mathrm{grad}\,(x_2) \right) =
$$

$$
= \frac{D_{12}}{(x_1 x_2)}\, \mathrm{grad}\,(x_1) = -\frac{D_{1,2}}{(x_1 x_2)}\, \mathrm{grad}\,(x_1).
$$

(A12.2)

where it is assumed that $-D_{12} = D_{1,2} > 0$, i.e., the mutual diffusion coefficient depends on composition gradients and pressure of the system as well as the contributions from external fields.

In the hard-sphere model the coefficient $D_{1,2}$ is expressed as

$$
D_{1,2} = \left(\frac{3\pi}{8n\sigma_{12}} \right) w_{12}, \quad w_{12} = \left(\frac{kT}{2\pi\mu_{12}} \right)^{1/2},
$$

(A12.3)

where $\sigma_{ij} = \pi d_{ij}^2$, $d_{12} = (d_{11} + d_{22})/2$ is the distance of the closest approach of the molecules, d_{ii} is the diameter of the molecule i, μ_{12} is the reduced mass of the colliding molecules 1 and 2: $\mu_{ij} = m_1 m_2 / (m_1 + m_2)$, m_i is the mass component i.

Solid phase

Experimental measurement of the mutual diffusion coefficient in alloys is carried out at constant pressure and temperature. Two binary alloy annealed samples having the initial (non-equilibrium) state different molar composition $x_i (x_1 + x_2 = 1)$ are brought into close contact and the penetration of atoms on either side of the plane of contact of the samples (plane '0') is examined. The minimum of the Gibbs potential corresponds to the equalization of the alloy composition on both sides of the plane 0, which leads to the process of mixing the components. The alloy components differ in the masses

m_i, so as a result of unequal diffusion mobility of the atoms of different sorts there is macroscopic displacement of the crystal planes (the Kirkendall effect) [6–10], which is monitored by the displacement over macroscopic distances of the inert labels. The movement of the labels in the binary alloy indicates the displacement of the crystallographic layers (planes) as a whole. This is due to the uncompensated flow of vacancies whose sinks form during dislocation creep (actually the Kirkendahl effect) or formation of pores in the crystals (Frenkel effect) [9]. The collapse of the pores due to mechanical instability of the solid or climb of dislocations provides macromovements of the planes.

At the atomic level, the process develops by the mechanism in which lighter atoms rush into an area with a high content of heavy atoms and the heavy atoms begin to move in the region of high concentration of the light atoms to compensate the local increase the total concentration of atoms. The total flow rate of the lattice compensates for the difference in diffusion flows of different atoms. Derivation of the equation for the mutual diffusion coefficient in solid alloys is based on the scheme of random walks of the atoms [6–10]. (This pattern is common to all phases. Its equations for gas are given below.) Movements of atoms take place by hopping to neighbouring vacancies. Using Darken's theory [6–8] the following equation was derived for the mutual diffusion coefficient $D_{1,2}$:

$$D_{1,2} = x_2 D_1 + x_1 D_2, \qquad (A12.4)$$

where D_i is the diffusion coefficient of the atom type i in the fixed coordinate system. In alloys the expressions for individual diffusion coefficients are represented as $D_i = D_i^* \left(1 + \partial \ln \gamma_i / \partial \ln x_i\right)$, where D_i^* and γ_i are respectively the self-diffusion coefficients and activity of atom i in the alloy, respectively. Lack of the cross-diffusion terms in equation (A12.4) generally allows for easier transport models. More rigorous models take into account their contributions [7, 8, 10].

Mayer equation for the gas phase

Let there be a mixture of two gases with mole fractions of molecules x_1 and x_2, different masses m_i, at constant pressure (P = const) and temperature (T = const). In a non-equilibrium state the molar gas composition x_i differs on both sides of selected plane '0'. The situation is completely identical to the case with the non-equilibrium state of the alloy. Faster light molecules rush into an area with a higher content of heavy particles, and the latter begin to move to the region of high concentration of the light molecules to compensate for the local increase in the total concentration of the molecules. For the discussed process of mixing the components of the binary mixture the difference of the counter flows of both components creates the conditions to move the entire mass of the gas as a whole [11], i.e., the convective flow with the average rate w_0 is generated, depending on the difference in the diffusion coefficients D_i of the components in the fixed reference frame. If N_i denotes the number of molecules of type $i = 1$ and 2, the kinetic equations for their transport can be written as

$$J_i^{\Pi} = \frac{dN_i}{dt} = w_0 n_i - 0.5\left\{\left[n_i w_i l_i\right]_{\mathrm{I}} - \left[n_i w_i l_i\right]_{\mathrm{II}}\right\}, \qquad (A12.5)$$

where w_i is the average thermal velocity of the molecules i and l_i is the mean free path of the component i. The square bracket refers to the half-space I, to the left of the plane 0, and the second square bracket refers to the half-space II, to the right of the plane 0. Expanding the right-hand side of the equation over grad (n_i): n_i (II) $= n_i$ (I) $- l_i$ grad (n_i), we obtain

$$J_i^{\Pi} = \frac{dN_i}{dt} = w_0 n_i - D_i \frac{dn_i}{dZ}, \quad i = 1, 2, \qquad (A12.6)$$

where, for simplicity, we assume that Z is the coordinate along which the molecules flow. Here we have used the expressions for the coefficients D_i in the elementary kinetic theory, which uses the equilibrium distribution function of the velocity of the molecules:

$$D_i = \frac{w_i l_i}{2}, \quad w_i = \left(\frac{8kT}{\pi m_i}\right)^{1/2}. \qquad (A12.7)$$

Since $d(N_1 + N_2)/dt = 0$, and $n = n_1 + n_2$, then

$$w_0 = \left[D_1 \frac{dn_1}{dZ} + D_2 \frac{dn_2}{dZ}\right]\bigg/ n. \qquad (A12.8)$$

Equation (A12.6) using (A12.8) can be rewritten as follows:

$$J_i^{\Pi} = \frac{dN_i}{dt} = -D_{1,2}\frac{dn_i}{dZ}, \quad D_{1,2} = x_2 D_1 + x_1 D_2, \qquad (A12.9)$$

wherein $D_{1,2}$ is the mutual diffusion coefficient of the binary mixture. The result is the so-called Mayer equation (A12.9), which does not meet the experimental data for a rarefied gas, if the value of l_i is estimated account on the basis of counting the number of collisions of molecules i with both components $j = 1$ and 2 [11]:

$$l_1 \left[2^{1/2} n_1 \sigma_{11} + n_2 \sigma_{12} \left(1 + \frac{m_1}{m_2} \right)^{1/2} \right]^{-1},$$

$$l_2 \left[2^{1/2} n_2 \sigma_{22} + n_1 \sigma_{12} \left(1 + \frac{m_2}{m_1} \right)^{1/2} \right]^{-1}. \qquad \text{(A12.10)}$$

To circumvent the problem of inconsistency of (A12.9) and (A12.10) with the experimental data, Stefan and Maxwell postulated that when evaluating l_i we should consider only the collisions between molecules of another grade $j \neq i$, then the formula (A12.10) can be rewritten as [30]

$$l_1^{SM} = \left[n_2 \sigma_{12} \left(1 + \frac{m_1}{m_2} \right)^{1/2} \right]^{-1}, \quad l_2^{SM} = \left[n_1 \sigma_{12} \left(1 + \frac{m_2}{m_1} \right)^{1/2} \right]^{-1}. \text{(A12.11)}$$

As a result, the formula (A12.9) leads to the well-known Stefan-Maxwell equation

$$D_{1,2}^{SM} = \frac{w_{12}}{n \sigma_{12}}, \quad w_{12} = \left(\frac{2kT}{\pi \mu_{12}} \right)^{1/2}, \qquad \text{(A12.12)}$$

which is in good agreement with experiment, and the rigorous kinetic theory [1–5] provides a basis for this selection of the types of collisions.

Thus, despite the difference in the mechanisms of transport of particles in an ideal gas and a solid alloy, the mutual diffusion coefficient in (A12.4) and (A12.9) for the moving reference frame is expressed in the same way through the diffusion coefficients of the individual components D_i in the fixed reference frame. It follows that the difference between (A12.3) and (A12.9) is due to the very procedure for constructing the expressions for $D_{1,2}$, and not to the difference in the mechanisms of migration of molecules in different phases. Also note that the concept of the Kirkendall effect was repeatedly involved in the theory of gas flows in wide channels to describe the near-wall effects [12].

Analysis of definitions
The diffusion equations are a generalization of experimental data in terms of Fick's law. Their molecular interpretation based on consideration of the scheme of random walks of the molecules at relatively large time intervals. Such 'walk' processes refer to the free movement of the molecules in a gas between the collisions or elementary hopping of the atoms in the liquid and solid phases. A description of these processes is based on the analysis of the counter flows of particles resulting in a diffusion flow. The thermodynamics of irreversible processes [13, 14] gives the usual structure of the equations for the flow of molecules under conditions of fixed P, T:

$$J_i = -D_{ii}\text{grad}(n_i) - D_{ij}\text{grad}(n_j), \quad i = 1,2. \quad (A12.1a)$$

If we express the diffusion velocities of the components through the diffusion flows as $V_i = J_i/n_i$, then their difference gives

$$V_1 - V_2 = -\frac{D_{11} - D_{12}}{n_1}\text{grad}(n_1) - \frac{D_{22} - D_{21}}{n_2}\text{grad}(n_2) =$$

$$= -\frac{x_2 D_{11} + x_1 D_{22} - D_{12}}{x_1 x_2}\text{grad}(x_1). \quad (A12.2a)$$

This equation differs from (A12.2) by the presence of the diagonal terms. Equations (A12.2) and (A12.2a) coincide at $D_{ii} = 0$. In this case the flow rate (A12.8) $w_0 = (D_{11} - D_{22})\text{grad}(x_1)$ is also equal to 0 but it means the absence of convective motion of the mixture.

On the other hand, if it is assumed that $D_{12} = D_{1,2}$ and Mayer–Darken expression (A12.4), (A12.9) is fulfilled, the difference in the diffusion rates of both components is zero, which means the movement of the entire mixture as a whole. It takes place in the convective flow relative to the fixed coordinate system. Then equation (A12.2) is introduced for the diffusion velocity of the components in the moving coordinate system moving with the convective flow. Thus, the fulfillment of the Mayer–Darken equality means no mutual diffusion in terms of the kinetic theory of gases. In other words, the strict definition of the kinetic theory of gases and the purely diffusion interpretation of the mutual diffusion coefficient within the scheme of random walks of the mixture components are *mutually* exclusive.

The Stefan–Maxwell hypothesis points to the need to consider in more detail in the elementary kinetic theory the type of colliding molecules. But the formula (A12.12) can not be automatically transferred to the dense phase. Therefore, we discuss the modification of the elementary kinetic theory, arising from the specifics of the theory of dense phases and requirements of the Stefan–Maxwell hypothesis.

Modification of the scheme of random walks

To solve the problem how to overcome the shortcomings of the Mayer formula, we discuss the scheme for constructing this equation in terms of the theory of condensed phases. The kinetic theory of gases can not be directly generalized to the dense phases, so we use the kinetic theory within the framework of the LGM, which is applicable for all phases [15] to generalize the hopping scheme in the LGM to the rarefied gas and derive the expression for the mutual diffusion coefficient through 'jumps' of the molecules.

There are two positions that distinguish the kinetic theory of gas and solids.

1. In the kinetic theory of dense systems, instead of the collisions between the molecules we operate with the transition probabilities of the molecules from one point to any other point. These transition probabilities are expressed in terms of the state of the thermostat and the dedicated groups of molecules. They depend on the lateral interactions between all system components.

To calculate the mutual diffusion coefficients, there is a need for explicit consideration of collisions between the molecules of different types. This account is easy to combine with the construction of the transition probabilities of the molecules if we define accurately the type of molecule j with which the molecule i collided last time inside the region I (or II) *prior to* this molecule intersecting the chosen plane 0. In the conventional elementary kinetic theory we implicitly considered the collisions of all molecules *after* intersecting the plane 0, i.e., situated in a different plane II (or I). (In the rigorous kinetic theory of gases this aspect is not discussed, although the collision integral includes the characteristics of the molecules both before and after the collision, but they all refer to some elementary volume of the gas without reference to the position of the plane 0.)

2. From the viewpoint of the equilibrium distribution the ideal gas molecules is a rarefied system, in which on average all the particles are at about the *same* distance ρ. Otherwise, any deviation from the

uniform distribution of the molecules should lead to the fluctuations in the density only because of the difference in the masses of the components of the mixture. However, the latter is impossible without the influence of intermolecular interactions which are excluded in the very concept of the rarefied gas.

In the kinetic theory we operate with two main molecular linear dimensions: molecular size σ (uniquely linked to the size of cells in the LGM $\lambda \approx 2^{1/6}\sigma$ – scale λ) and mean free path l_i (scale l). Apart from these, there is a third characteristic linear value – average distance between the molecules of the gas phase ρ (scale ρ). It is defined as $\rho = v^{1/3}$, where v is the volume per one molecule of the mixture. In the LGM $\rho = \lambda/(\theta)^{1/3}$.

The quantity ρ is used [16] in the kinetic theories to determine the dimensionless density parameter as the ratio $(\lambda/\rho)^3$ which can be used in expansions. But this quantity does not play a functional role. In our modification of the LGM the very existence of the mean size ρ between the molecules is used as an analogue of the 'lattice constant' that depends on the density of the mixture θ. In dense phases $l = \rho = \lambda$. This allows us to introduce a uniform way to describe the system in the transition from the dense phase to the rarefied gas for which the following relations are satisfied: $\lambda \ll \rho \ll l_i$. The first sign $\ll$ should be understood to be rather arbitrary because for the rarefied gas at $\theta = 10^{-3}$–10^{-4} we have differences of about one order of magnitude $\rho = (10$–$22)\lambda$. At the same time, the differences between ρ and l_i reach two orders of

magnitude: $l_i = \dfrac{1}{2^{1/2}\pi n d_{ii}^2} = \dfrac{\rho^3}{2^{1/2}\pi d_{ii}^2} \approx (30-140)\rho.$

Consequences of the LGM
In dense phases the number of realizable transition probabilities of the molecules is relatively small – it depends on the number of bonds between adjacent sites of the lattice structure into which the particle can jump in the presence of a vacancy. In this case, the lateral interactions are taken into account fully. Increasing the average distance between the molecules dramatically increases the number of the transition probabilities of the molecules between different points in space and complicates consideration of the lateral interactions. To simplify the description of these probabilities, we use the LGM, built on a lattice with a constant of the ρ-scale. Then the contributions of lateral interactions on the l-scale will be replaced by the contributions on the ρ-scale, which greatly simplifies the

calculation, as if these contributions are small at the ρ-scale, then they can be neglected at the *l*-scale.

In the LGM there is a strict proof of the self-consistency of description of the dynamics of molecular processes and the equilibrium state of the mixture at any density and temperature [15]. Using the ρ-scale allows to save this important property of the LGM, and at its level we can describe the dynamic processes that are realized on the *l*-scale. (The very quantity ρ is an equilibrium characteristic, as part of the equation of state.) Thus, we can keep all the advantages of the LGM to consider the influence of lateral interactions on the dynamics of elementary processes.

Combining the positions 1 and 2, we see that for the movement of the molecule *i* from site *f* in region I in the direction of the site *g* in region II the molecule must first experience a collision with molecule *j* in the site ξ within region I. To describe such a movement instead of the traditional use in the elementary kinetic theory of average velocity w_i (A12.7) of the molecule *i* we need to use the average relative velocity of the molecule *i* after its collision with molecule *j*. This velocity of the molecule *i* is written as $w_{i(j)} = \left(kT/2\pi\mu_{ij}\right)^{1/2}$.

Another consequence of the LGM is that the distance between the regions I and II in which we examine the density gradients for both components and is measured in units of the ρ-scale, is the same for all components (this condition was used in the derivation of (A12.4) [6–10]). Therefore, in the construction of the diffusion equation for an ideal gas it is necessary to satisfy the equation $l_i = l = $ const. From this it immediately follows that in the mixture, unlike the formula (A12.11), *l* is expressed by the equation for one-component system: $l \sim (n(\langle\sigma\rangle))^{-1}$, where $\langle\sigma\rangle$ is the average value of the section for colliding molecules of the mixture, although the value of *l*, like formula (A12.10), can depend through $\langle\sigma\rangle$ on the molar composition of the mixture

Taking collisions into account

We derive Mayer-type equations for the flow of molecules *i*, intersecting plane 0, taking into account these adjustments for the case of commensurate size of components of the gas mixture [17]. Movement of this molecule occurs at a relative velocity $w_{i(j)}$ after the collision with the molecules *j* = 1 and 2. Accordingly, it is necessary to take into account the proportion x_j of each component *j* with which previously molecule *i* collided. Therefore, instead of the average velocity w_i the formulas (A12.5) and (A12.6) must include the following sum $(w_{1(1)} x_1 + w_{1(2)} x_2)$. The expression for the flow of the first component (A12.5) can be rewritten as follows:

$$J_1^{\Pi} = \frac{dN_1}{dt} = w_0 n_1 - 0.5\left\{\left[n_1\left(w_{1(1)}x_1 + w_{1(2)}x_2\right)l_1\right]_{\mathrm{I}} - \left[n_1\left(w_{1(1)}x_1 + w_{1(2)}x_2\right)l_2\right]_{\mathrm{II}}\right\}.$$

$$(A12.5a)$$

As above, the first square bracket refers to the half-space I situated to the left of the plane 0, and the second square bracket refers to the half-space II to the right of the plane 0. Expanding the right-hand side of this equation by grad (n_1), we obtain (for clarity, the value l_i is retained)

$$J_1^{\Pi} = w_0 n_1 - 0.5 l_1 \left[\left(2w_{1(1)}x_1 + w_{1(2)}x_2\right)\frac{dn_1}{dz} + w_{1(2)}x_1 \frac{dn_2}{dz}\right]. \; (A12.15)$$

Or, given that grad $(x_2) = -\mathrm{grad}(x_1)$, in the general form for both components (where $j \neq i$), we have

$$J_i^{\Pi} = w_0 n_i - 0.5 l_i \left[2w_{i(i)}x_i + w_{i(j)}(1 - 2x_i)\right]\frac{dn_i}{dz}, \; i = 1 \text{ and } 2.$$

$$(A12.16)$$

Hence, as above, it follows that

$$w_0 = \left\{l_1\left[2w_{1(1)}x_1 + w_{1(2)}\left(1 - 2x_1\right)\right]\frac{dn_1}{dz} + \right.$$

$$\left. + l_2\left[2w_{2(2)}x_2 + w_{2(1)}\left(1 - 2x_2\right)\right]\frac{dn_2}{dz}\right\}\bigg/(2n). \; (A12.17)$$

Then, instead of (A12.9), we obtain a new expression for the mutual diffusion coefficient

$$D_{1,2} = \frac{l_1 w_{1(2)}x_2 + l_2 w_{2(1)}x_1 + 2x_1 x_2\left[l_1\left(w_{1(1)} - w_{1(2)}\right) + l_2\left(w_{2(2)} - w_{2(1)}\right)\right]}{2}$$

$$(A12.18)$$

In formula (A12.18) l_i is replaced by l and we introduce, by analogy with (A12.7), $D_{i(j)} = w_{i(j)}l/2$ – the diffusion coefficients of molecules i, involved in transport after the last collision with molecule j. As a result, we obtain the final expression

$$D_{1,2} = x_2 \left(x_2 - x_1 \right) D_{1(2)} + x_1 \left(x_1 - x_2 \right) D_{2(1)} + 2x_1 x_2 \left(D_{1(1)} + D_{2(2)} \right) =$$

$$= D_{1(2)} \left(x_1 - x_2 \right)^2 + 2x_1 x_2 \left(D_{1(1)} + D_{2(2)} \right).$$

$$(A12.19)$$

In the second equality it is taken into account that due to $w_{1(2)} = w_{2(1)}$ the condition $D_{1(2)} = D_{2(1)}$ is satisfied (However, in dense phases the last equality can be violated.)

Analysis of the new equation

Consider the difference of the diffusion velocities $\mathbf{V}_1 - \mathbf{V}_2$. Taking into account the collisions of molecules leads to the expression for the flow of the molecules analogous to formula (A12.1a). However, the coefficients D_{ij}, according to (A12.15), depend themselves on the molar composition $D_{ii} = 2D_{i(i)}x_i + D_{i(j)}x_j$ and $D_{ij} = D_{i(i)}x_i$. As a result, we have

$$\mathbf{V}_1 - \mathbf{V}_2 = -\frac{D_{11} - D_{12}}{n_1} \operatorname{grad}\left(n_1\right) - \frac{D_{22} - D_{21}}{n_2} \operatorname{grad}\left(n_2\right) =$$

$$= -\frac{D_{1(2)} \left(x_1 - x_2\right)^2 + 2x_1 x_2 \left(D_{1(1)} + D_{2(2)}\right)}{\left(x_1 x_2\right)} \operatorname{grad}\left(x_1\right) =$$

$$= -\frac{D_{1,2}}{\left(x_1 x_2\right)} \operatorname{grad}\left(x_1\right).$$

$$(A12.20)$$

Thus, the modified scheme of random walks results in complete agreement with the rigorous kinetic theory, defines the relationship between the difference in the diffusion velocities of the components and the mutual diffusion coefficient.

We also discuss the behaviour of the velocity of the convective flow in the new modification of the LGM: $w_0 = 2D \operatorname{grad}\left(x_1\right)$, where $D = (D_{1(1)} - D_{1(2)}) x_1 - (D_{2(2)} - D_{2(1)}) x_2$. From the form of the proportionality coefficient D it follows that convective flow w_0 is determined by all types of collisions of molecules in the entire range of the mixture compositions. When the fraction of a component, such as 2, is small then $D = D_{1(1)} - D_{1(2)}$, i.e. the flow depends on the difference of the contributions from collisions of the first component with the first and send components. For the equimolar composition $D = (D_{1(1)} - D_{2(2)})/2$, i.e., the convective flow is determined only by the collisions of molecules of one type that corresponds to the maximum deviation from the $D_{1,2}^{SM}$. However, this in itself is a small deviation (about 15%), as there is a considerable compensation of the contributions from the collisions between the molecules of one sort.

Formula (A12.19) shows that the correct expression for the mutual diffusion coefficient can be obtained without using the Boltzmann equation (only through the average velocity of thermal motion of the molecules), for small values of the mole fraction of one of the components of a binary mixture. $M_{21} \gg 1$ at $x_2 \sim 1$ corresponds to a Lorentz mixture (light impurity is among the heavy particles), which is well studied in the rigorous kinetic theory of gases [3-5]. For this mixture the new approach gives the correct temperature dependence and the effect of the mass. Numerical differences (1/2 and 1/3) between the formulas (A12.19) (or (A12.12)) and the exact theory [1–3] for the Lorentz mixture are associated with the dispersion of the velocity but this systematic difference plays no role in the analysis of the temperature and concentration experimental data (and is reduced to a shift of the reference point for these dependences). Similarly, the new approach gives the correct result for the case $m_{21} \to 0$ at $x_2 \sim 0$, which corresponds to the anti–Lorentz mixture (heavy impurity amongst light particles).

Comparison of the new expressions with the experiment is given in section 61. Generalization of the expression for the velocity of thermal motion of molecules, taking into account collisions with the neighbouring molecules, is given in Appendix 7 [18], and the expressions for the transport coefficients of components in narrow channels, taking into account this modification of the LGM, are given in Appendix 11 [19].

References

1.   Chapman S., Cowling T. G., Mathematical Theory of Non-Uniform Gases. – Moscow: Izd. inostr. lit., 1960. – 541 p. [Cambridge University Press, Cambridge, 1952].

2.   Ferziger J. H. H., Kaper H. G., Mathematical theory of transport processes in gases. – Moscow: Mir, 1976. – 554 p. [North_Holland, Amsterdam, The Netherlands, 1972].

3.   Waldman L., Transportercheinungen in gases von mittlerem druck // Handbuch der Physik. 1958. Bd. 12. S. 295.

4.   Kogan M.N., Dynamics of rarefied gases. Moscow: Nauka, 1967.

5.   Hirschfelder J. O., Curtiss C. F., Bird R. B., Molecular Theory of Gases and Liquids. – Moscow: Izd. inostr. lit., 1961. – 929 p. [Wiley, New York, 1954].

6.   Borovsky I.B., Gurov K.P., Marchukova I.D., Ugaste E.Yu., Interdiffusion processes in alloys. Moscow: Nauka, 1973.

7.   Gurov K.P., Kartashkin B.A., Ugaste E.Yu., Interdiffusion in multiphase metal systems. Moscow: Nauka, 1981

8.   Bokshtein B.S., Bokshtein S.Z., Zhukhovitskii A., Thermodynamics and kinetics of diffusion in solids. Moscow: Metallurgiya, 1976.

9.   Geguzin Ya.E., Diffusion zone. Moscow: Nauka, 1979. 344 p.

10.   Manning J. R, Diffusion kinetics for atoms in crystals. Moscow: Mir, 1971. – 278 p. [D. van Nostrand Comp., Inc., Princeton, Toronto, 1968].

11.   Moelwyn-Hughes E.A., Physical Chemistry. – Moscow: Izd. Lit., 1962. – V.1 [2nd ed. Pergamon Press, London, 1961].

12.   Mason E., Sperling T., Virial equation of state. – Moscow: Mir, 1972. – 280 p. [The Internationsl Encyclopedia of Physical Chemistry. Topic 10. ed. J.S. Rowlinson, Vol. 2].

13.   de Groot S. R., Mazur P., Non-Equilibrium Thermodynamics. Moscow: Mir, 1964. – 456 p. [(North-Holland, Amsterdam, Amsterdam, 1962].

14.   Haase R., Thermodynamics of irreversible processes. Moscow: Mir, 1967.

15.   Tovbin Yu. K., Theory of physical and chemical processes at the gas-solid interface. – Moscow: Nauka, 1990. – 288 p. [CRC, Boca Raton, Florida, 1991].

16.   Klimontovich Yu.L., Kinetic theory of non-ideal gas and non-ideal plasma. Moscow: Nauka, 1975. 352 p.

17.   Tovbin Yu.K. // Teor. osnovy khim. tekhnol. 2005. V. 39, No. 5. P. 523 [Theor. Found. Chem. Engin. 2005. V. 39. No 5. P. 493].

18.   Tovbin Yu.K. // ibid. 2005. V. 39, No. 6. P. 613 [Theor. Found. Chem. Engin. 2013. V. 47. No 6. P. 734].

19.   Tovbin Yu.K. // Izv. RAN. Ser. khim. 2005. No. 8. P. 1717 [Russ. Chem. Bull., 2005. T. 54. № 8. C. 1768].

Index

flow
  diffusion flow 28
  flow of polymolecular films 31
  Free molecular flow 26
  Knudsen flow 26, 27, 396
  steady-state steam flow 28
  viscous flow 27, 28, 32, 340, 552
free volume 1, 2, 6, 404, 425, 498, 548

G

Gibbs ensemble 197, 198, 199, 200

H

hard sphere 101, 121, 147, 187, 224, 475, 476, 512

I

interaction
  intermolecular interaction 33, 73
  van der Waals interactions 35
isotherm
  Henry isotherm 65
  Langmuir isotherm 16, 63, 136
  logarithmic adsorption isotherm 66
  quasi-logarithmic isotherm 65

L

law
  Darcy law 25
  Fick's law 24
  Henry's law 25, 65, 96, 520, 522
  Poiseuille' law 31
level
  supramolecular level 398, 399, 406, 446, 460, 495
LGM (lattice-gas model) 44, 47, 53, 69, 77, 105, 107, 108, 109, 110, 111, 112,
       113, 118, 119, 120, 121, 123, 132, 135, 140, 141, 148, 151, 152, 159,
       160, 172, 175, 176, 178, 187, 198, 199, 206, 207, 208, 224, 225, 226,
       227, 228, 240, 244, 246, 247, 250, 256, 260, 263, 264, 265, 267, 272,
       273, 277, 279, 290, 291, 292, 294, 298, 303, 312, 317, 318, 319, 320,
       321, 322, 323, 324, 325, 326, 328, 329, 334, 398, 409, 431, 449, 459, 463,
       464, 468, 469, 479, 480, 481, 482, 483, 485, 486, 502, 550, 555, 556, 559
Lorentz–Berthelot combinational rules 38

M

macropores 4, 9, 11, 451, 553
mechanism
  Knudsen transport mechanism 30
mesopores 9, 10, 18, 191, 195, 228, 365, 539, 540, 541, 543

Milton Keynes UK
Ingram Content Group UK Ltd.
UKHW021936071024
449327UK00022B/1828